Electricity

$$F = \frac{kq_1q_2}{r^2}$$

$$R = \frac{\rho L}{A}$$

$$I = \frac{V}{R}$$

Series Circuits

(a) $I = I_1 = I_2 = I_3 = \cdots$

(b) $R = R_1 + R_2 + R_3 + \cdots$

(c) $E = V_1 + V_2 + V_3 + \cdots$

Parallel Circuits

(a) $I = I_1 + I_2 + I_3 + \cdots$

(b) $\frac{1}{R} = \frac{1}{R_1} + \frac{1}{R_2} + \frac{1}{R_3} + \cdots$

(c) $E = V_1 = V_2 = V_3 = \cdots$

Cells in Series

(a) $I = I_1 = I_2 = I_3 = \cdots$

(b) $r = r_1 + r_2 + r_3 + \cdots$

(c) $E = E_1 + E_2 + E_3 + \cdots$

Cells in Parallel

(a) $I = I_1 + I_2 + I_3 + \cdots$

(b) $r = \frac{r \text{ of one cell}}{\text{number of like cells}}$

(c) $E = E_1 = E_2 = E_3 = \cdots$

$V = E - Ir$

$$P = VI = I^2R = \frac{V^2}{R}$$

Magnetism

$$B = \frac{\mu_0 I}{2\pi R}$$

$$B = \pi_0 In$$

Transformers

$$\frac{V_P}{V_S} = \frac{N_P}{N_S}$$

$$\frac{I_S}{I_P} = \frac{N_P}{N_S}$$

ac Circuits

$$X_L = 2\pi fL$$

$$I = \frac{E}{X_L}$$

$$I = \frac{E}{Z}$$

$$Z = \sqrt{R^2 + X_L^2}$$

$$\tan \phi = \frac{X_L}{R}$$

$$X_C = \frac{1}{2\pi fC}$$

$$Z = \sqrt{R^2 + X_C^2}$$

$$\tan \phi = \frac{X_C}{R}$$

$$Z = \sqrt{R^2 + (X_L - X_C)^2}$$

$$\tan \phi = \frac{X_L - X_C}{R}$$

$$f = \frac{1}{2\pi\sqrt{LC}}$$

Light

$$c = \lambda f$$

$$E = hf$$

$$E = \frac{I}{4\pi r^2}$$

$$\frac{1}{f} = \frac{1}{s_o} + \frac{1}{s_i}$$

$$M = \frac{h_i}{h_o} = \frac{-s_i}{s_o}$$

$$n = \frac{\sin i}{\sin r} = \frac{\text{speed of light in vacuum}}{\text{speed of light in substance}}$$

$$\sin i_c = \frac{1}{n}$$

Modern Physics

$$E = -\frac{kZ^2}{n^2}$$

$$E = \Delta mc^2$$

$$Q = (M_p - M_d - m_\alpha)c^2$$

$$N = N_0 e^{-\lambda t}$$

$$T_{1/2} = \frac{0.693}{\lambda}$$

$$A = \lambda N = \lambda N_0 e^{-\lambda t} = A_0 e^{-\lambda t}$$

APPLIED PHYSICS

NINTH EDITION

DALE EWEN

Parkland Community College
Champaign, Illinois

NEILL SCHURTER

P. ERIK GUNDERSEN

Pascack Valley Regional High School District
Montvale, New Jersey

PEARSON

Prentice
Hall

Upper Saddle River, New Jersey
Columbus, Ohio

Library of Congress Cataloging-in-Publication Data

Ewen, Dale,
   Applied physics.— 9th ed. / Dale Ewen, Neil Schurter, P. Erik Gundersen.
      p. cm.
   Includes index.
   ISBN-13: 978-0-13-515733-6
   ISBN-10: 0-13-515733-1
   1. Physics—Textbooks. I. Schurter, Neill. II. Gundersen, P. Erik. III. Title.
   QC23.2.E88 2009
   530—dc22

                                                                      2008000199

Editor in Chief: Vernon R. Anthony
Senior Acquisitions Editor: Gary Bauer
Assistant Editor: Linda Cupp
Editorial Assistant: Kathleen Rowland
Project Manager: Louise N. Sette
Production Supervision: Aptara, Inc.
Art Director: Diane Ernsberger
Cover Designer: Jason Moore
Senior Operations Supervisor: Patricia Tonneman
Director of Marketing: David Gesell
Marketing Manager: Leigh Ann Sims
Marketing Coordinator: Alicia Dysert

Photo Credits: A. T. Willett, Getty Images, Inc.–Image Bank, 2; Ed Eckstein, Phototake NYC, 12; Will Hart, PhotoEdit Inc., 50; Mark D. Phillips, Photo Researchers, Inc., 68; Aerial Focus, Photolibrary.com, 98; Corbis Digital Stock, 128; Jim Cummins, Getty Images, Inc.-Taxi, 148; Jurgen Vogt, Getty Images, Inc.–Image Bank, 168; Jonathan Nourok, PhotoEdit Inc., p. 204; Kevin Horan, Getty Images, Inc.–Stone Allstock, 232; Robert Brenner, PhotoEdit Inc.; Jonathan Nourok, PhotoEdit Inc.; Richard Megna, Fundamental Photographs, NYC; EyeWire Collection, Getty Images–Photodisc, p. 266; Jan Halaska, Photo Researchers, Inc., 308; Spencer Grant, PhotoEdit Inc., 340; St. Meyers/Okapia, Photo Researchers, Inc., 366; Jim Steinberg, Photo Researchers, Inc., 406; EyeWire Collection, Getty Images–Photodisc, 420; Felix Velasquez, Felix Velasquez Photography, 448; Railway Technical Research Institute, Tokyo, Japan, 498; Jonathan Nourok, PhotoEdit Inc., 520; Ron Thomas, Getty Images, 554; Hank Morgan, Photo Researchers, Inc., 572; Carol and Mike Werner, Phototake NYC, 600; Getty Images, Inc.–Image Bank, 618; PhotoDisc Imaging, Getty Images, Inc.–Photodisc, 646.

This book was set in Times Roman by Aptara, Inc. It was printed and bound by Quebecor World Color/Versailles. The cover was printed by Phoenix Color Corp.

Pearson Education Ltd.   Pearson Education Australia Pty. Limited
Pearson Education Singapore Pte. Ltd.   Pearson Education North Asia Ltd.
Pearson Education Canada, Ltd.   Pearson Educación de Mexico, S.A. de C.V.
Pearson Education—Japan   Pearson Education Malaysia Pte. Ltd.

10 9 8 7 6 5 4 3 2
ISBN 13: 978-0-13-515733-6
ISBN 10:   0-13-515733-1

CONTENTS

Applied Physics, ninth edition, formerly *Physics for Career Education,* provides comprehensive and practical coverage of physics for students needing an applied physics approach or considering a vocational–technical career. It emphasizes physical concepts as applied to industrial–technical fields and uses common applications to improve the physics and mathematics competence of the student. This ninth edition has been carefully reviewed and special efforts have been taken to emphasize clarity and accuracy of presentation.

This text is divided into five major areas: mechanics, matter and heat, wave motion and sound, electricity and magnetism, and light and modern physics.

Key Features

- Real-world applications are used to motivate students.
- Topic coverage is clear and to the point.
- A unique problem-solving format is consistently used throughout the text.
- Detailed, well-illustrated examples in the problem-solving format support student understanding of skills and concepts.
- Problems and questions assist student learning, with extensive problem sets at the end of most sections that provide students with ample opportunity for practice.
- A four-color format with numerous drawings, diagrams, and photographs is used to illustrate the application of physics in the real world and improve student interest and comprehension.
- Try This Activity features provide students with opportunities to experiment with physics concepts. Activities involve a demonstration or mini-activity that can be performed by students on their own to experience a physics concept, allowing for more active versus passive learning.
- Physics Connections features apply physics to familiar real-world situations and events. These brief readings help students bridge the gap between what is taught in the chapter and real-world technical applications.
- Applied Concepts features provide application-based questions at the end of chapters that develop problem-solving skills in real-life physics applications.
- There is comprehensive discussion and consistent use of the results of working with measurements and significant digits.
- Biographical sketches of important scientists appear in most chapters.
- Answers to odd-numbered problems within the chapters and all chapter review questions and problems are given in Appendix E.
- A comprehensive glossary is given as a one-stop reference in Appendix D.
- Basic scientific calculator instructions are presented in Appendix B.
- A basic math review provides students with a refresher of the math needed for the course in Appendix A.

Changes made in this ninth edition include:

- More than 120 mostly metric problems and some problems of interest to automotive/ diesel and construction students have been added at the request of reviewers.
- More than 60 new photos are included.
- Problem Solving is covered in a separate chapter again, at the request of reviewers.
- The Vectors chapter has been reorganized and revised at the request of reviewers.
- Fourteen Try This Activity features have been added.

PROBLEM SOLVING

SKETCH

$$12 \text{ cm}^2 \quad w$$
$$4.0 \text{ cm}$$

DATA

$A = 12 \text{ cm}^2$, $l = 4.0$ cm, $w = ?$

BASIC EQUATION

$A = lw$

WORKING EQUATION

$w = \frac{A}{l}$

SUBSTITUTION

$w = \frac{12 \text{ cm}^2}{4.0 \text{ cm}} = 3.0$ cm

◆ Eight Physics Connections features have been added.
◆ The topic of collisions in two dimensions has been added to Section 6.2.
◆ Numerous improvements have been made throughout the text.

Examples of Key Features
Unique Problem-Solving Method

This textbook teaches students to use a proven effective problem-solving methodology. The consistent use of this method trains students to make a sketch, identify the data elements, select the appropriate equation, solve for the unknown quantity, and substitute the data in the working equation. An icon that outlines the method is placed in the margin of most problem sets as a reminder to students.

Figure P.1 and Figure P.2 show examples illustrating how the problem-solving method is used in the text. See Section 2.3 for the detailed presentation of the problem-solving method.

Figure P.1

A ball rolls at a constant speed of 0.700 m/s as it reaches the end of a 1.30-m-high table (Fig. 4.22). How far from the edge of the table does the ball land?

EXAMPLE 1

Sketch: Figure 4.22

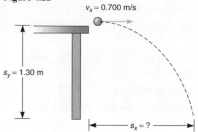

$v_x = 0.700$ m/s

$s_y = 1.30$ m

$s_x = ?$

Data:

$$v_{iy} = 0 \text{ m/s} \qquad\qquad v_x = 0.700 \text{ m/s}$$
$$s_y = 1.30 \text{ m} \qquad\qquad s_x = ?$$

Basic Equations:

$$s_y = v_{iy}\, t + \tfrac{1}{2} a_y t^2 \qquad\qquad s_x = v_x t$$

Working Equations (with $v_{iy} = 0$):

$$t = \sqrt{\frac{2s_y}{a}} \qquad\qquad s_x = v_x t$$

Substitution:

$$t = \sqrt{\frac{2(1.30 \text{ m})}{9.80 \text{ m/s}^2}}$$
$$= 0.515 \text{ s}$$

$$s_x = (0.700 \text{ m/s})(0.515 \text{ s})$$
$$= 0.361 \text{ m}$$

Figure P.2

Juan and Sonja use a push mower to mow a lawn. Juan, who is taller, pushes at a constant force of 33.1 N on the handle at an angle of 55.0° with the ground. Sonja, who is shorter, pushes at a constant force of 23.2 N on the handle at an angle of 35.0° with the ground. Assume they each push the mower 3000 m. Who does more work and by how much?

EXAMPLE 4

Sketch:

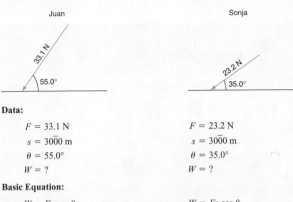

Juan   Sonja

Data:

$F = 33.1$ N	$F = 23.2$ N
$s = 30\overline{0}0$ m	$s = 30\overline{0}0$ m
$\theta = 55.0°$	$\theta = 35.0°$
$W = ?$	$W = ?$

Basic Equation:

$W = Fs \cos \theta$ $W - Fs \cos \theta$

Working Equation: Same   Same

Substitution:

$W = (33.1$ N$)(30\overline{0}0$ m$) \cos 55.0°$

$\qquad = 57,\overline{0}00$ N m

$\qquad = 57,\overline{0}00$ J $(1$ N m $= 1$ J$)$

$W = (23.2$ N$)(30\overline{0}0$ m$) \cos 35.0°$

$\qquad = 57,\overline{0}00$ N m

$\qquad = 57,\overline{0}00$ J

They do the same amount of work. However, Juan must exert more energy because he pushes into the ground more than Sonja, who pushes more in the direction of the motion.

....

Worked Examples

Worked examples are consistently displayed in the problem-solving format and used to illustrate and clarify basic concepts and problems. Since many students learn by example, a large number of examples are provided. The example in Figure P.3 shows how conversion factors are displayed and used.

Figure P.3

EXAMPLE 2

Find the depth in a lake at which the pressure is 105 lb/in^2.

Data:

$$P = 105 \text{ lb/in}^2$$
$$D_w = 62.4 \text{ lb/ft}^3$$
$$h = ?$$

Basic Equation:

$$P = hD_w$$

Working Equation:

$$h = \frac{P}{D_w}$$

Substitution:

$$h = \frac{105 \text{ lb/in}^2}{62.4 \text{ lb/ft}^3}$$

$$= 1.68 \frac{\text{ft}^3}{\text{in}^2}$$

$$\text{ft}$$

$$= 1.68 \frac{\text{ft}^3}{\text{in}^2} \times \left(\frac{12 \text{ in.}}{1 \text{ ft}}\right)^2$$

$$= 242 \text{ ft}$$

$$\boxed{\frac{\text{lb/in}^2}{\text{lb/ft}^3} = \frac{\text{lb}}{\text{in}^2} \div \frac{\text{lb}}{\text{ft}^3} = \frac{\text{lb}}{\text{in}^2} \times \frac{\text{ft}^3}{\text{lb}} = \frac{\text{ft}^3}{\text{in}^2}}$$

....

High-Interest Chapter Openers

Chapter opening photos feature topics of interest to students with hand-written formula notes relating the action in the photo to a physical principle discussed in the chapter.

Figure P.4

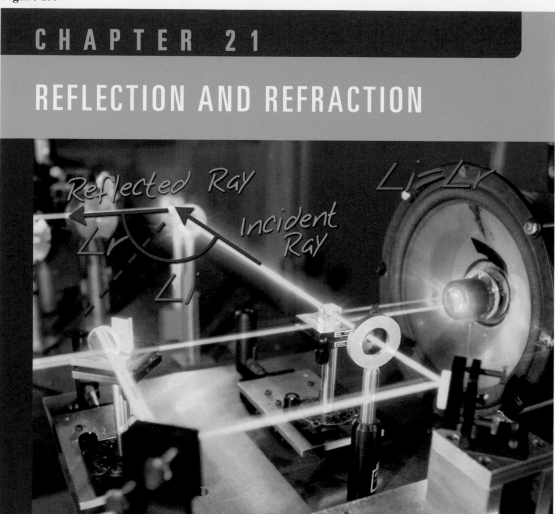

The nature of light may still be somewhat of a mystery. However, its characteristics have been the subject of intensive study for hundreds of years. Light may be transmitted, reflected, or absorbed by a medium.

Anyone wearing glasses can appreciate the refraction of light as it bends upon passing from one medium to another. The index of refraction is a tool that the scientist uses to describe the ability of certain substances to bend light as it passes through them.

Our examination of the behavior of light begins with the study of images and reflection.

Try This Activity

These activities provide students with opportunities to experiment with physics concepts. Activities involve a demonstration or mini-activity that can be performed by students on their own to experience a physics concept, allowing for active versus passive learning.

Figure P.5

TRY THIS ACTIVITY

Free Fall in a Vacuum

Drop a piece of paper and a book at the same time and note the relative time it takes for each to hit the floor. Now place that paper on top of the book as shown in Fig. 4.14. (**Note:** The top surface area of the book must be larger than that of the paper.) What happens to the time it takes the book and paper to fall? What does this show about objects falling in a vacuum? (A vacuum is a space in which there is no air resistance present.)

Figure 4.14   Place the paper on top of the book. The book must be larger than the paper.

Physics Connections

These features apply physics to familiar real-world situations and events. These brief readings help students bridge the gap between what is taught in the chapter and real-world technical applications.

Figure P.6

PHYSICS CONNECTIONS

Fiber Optic Cables

Most transmission of information travels as electric impulses through electric and telephone lines and fiber optic cables. Electric signals travel relatively slowly, cause wires to heat up, and need transformers to boost the voltage of signals traveling over long distances. Electric signals and wires are being replaced with light signals traveling through flexible, low-cost strands of glass. Because light travels through glass optical fibers, there is no electrical resistance to weaken the signal, and the signal travels at the speed of light, which is much faster than the speed of conventional electric signals. Such advances in fiber optics communications are revolutionizing the way we communicate.

Light traveling in the same medium travels in a straight line, whereas fiber optic cables can transmit a signal while twisting and turning, because of total internal reflection. The angle at which the light strikes the cladding of the fiber is always greater than the critical angle of the cladding and the core. The low critical angle allows the light to continually reflect and travel great distances without needing to be reamplified. In order for the cable to maintain a low critical angle, the glass must contain no imperfections or bubbles that would cause the light to be directed out or backward through the cable (see Fig. 21.26).

Fiber optic cables are used in telecommunications, computer networks, and medicine. A few strands of glass fiber can carry thousands of separate digital telephone conversations by slightly altering the frequency of the light for each phone conversation. A digital signal transmitted at one frequency cannot be confused by a signal carried at another frequency. Many computer networks and internal components in computers use fiber optic cables to carry data. By eliminating electric wiring, the fiber optic cable helps to reduce the temperature inside computers and servers. Finally, physicians use fiber optic bundles to perform minimally invasive procedures. A tool called an endoscope, composed of a bundle of fiber optic cables, transmits light into a patient's body while another bundle of fibers on the endoscope functions as a digital camera. The camera picks up the image and sends it back through the fiber optic cable to a monitor in the operating room.

Figure 21.26 (a) The red laser light entering the fiber optic cable is totally internally reflected, which results in the light emerging at the end of the cable. (b) An endoscope is a bundle of fiber optic cables used in many minimally invasive surgeries. Here an endoscope is used in the removal of nose adenoids with a laser therapy procedure.

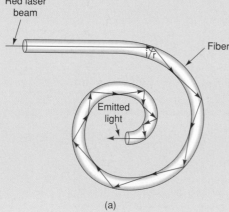

(a)

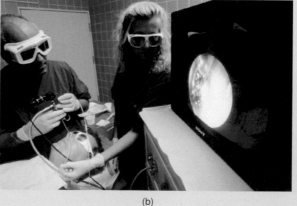

(b)

S. Elleringmann/Bilderberg/Aurora

Applied Concepts

Application-based questions at the end of each chapter develop problem-solving skills in real-life physics applications

Figure P.7

APPLIED CONCEPTS

1. Rosita needs to purchase a sump pump for her basement. (a) If the pump must carry 10.0 kg of water to a height of 2.75 m each minute, what minimum wattage pump is needed? (b) What three main factors determine power for a sump pump?

2. A roller coaster designer must carefully balance the desire for excitement and the need for safety. The most recent design is shown in Fig. 8.18. (a) If a 355-kg roller coaster car has zero velocity on the top of the first hill, determine its potential energy. (b) What is the velocity of the roller coaster car at the specified locations in the design? (c) Explain the relationship between velocity and the position on the track throughout the ride. (Consider the track to be frictionless.)

Figure 8.18

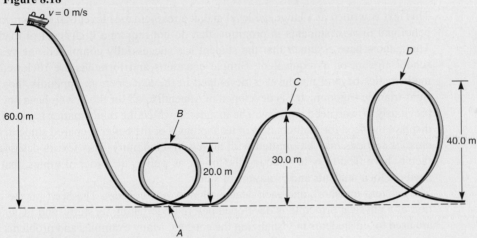

3. A 22,500-kg Navy fighter jet flying 235 km/h must catch an arresting cable to land safely on the runway strip of an aircraft carrier. (a) How much energy must the cable absorb to stop the fighter jet? (b) If the cable allows the jet to move 115 m before coming to rest, what is the average force that the cable exerts on the jet? (c) If the jet were given more than 115 m to stop, how would the force applied by the cable change?

4. The hydroelectric plant at the Itaipu Dam, located on the Parana River between Paraguay and Brazil, uses the transfer of potential to kinetic energy of water to generate electricity. (a) If 1.00×10^6 gallons of water (3.79×10^6 kg) flows down 142 m into the turbines each second, how much power does the hydroelectric power plant generate? (For comparison purposes, the Hoover Dam generates 1.57×10^6 W of power.) (b) How much power could the plant produce if the Itaipu Dam were twice its actual height? (c) Explain why the height of a dam is important for hydroelectric power plants.

5. A 1250-kg wrecking ball is lifted to a height of 12.7 m above its resting point. When the wrecking ball is released, it swings toward an abandoned building and makes an indentation of 43.7 cm in the wall. (a) What is the potential energy of the wrecking ball at a height of 12.7 m? (b) What is its kinetic energy as it strikes the wall? (c) If the wrecking ball transfers all of its kinetic energy to the wall, how much force does the wrecking ball apply to the wall? (d) Why should a wrecking ball strike a wall at the lowest point in its swing?

Ancillaries

◆ Companion Laboratory Manual (0-13-110353-9)
◆ Online Instructor's Resource Manual with Complete Solutions
◆ Online PowerPoint transparencies
◆ Test Item File
◆ The Prentice Hall TestGen, which provides the Test Item File on CD-ROM

Register today at www.prenhall.com to access instructor resources digitally.

Online Instructor's Manual

To access supplementary materials online, instructors need to request an instructor access code. Go to www.pearsonhighered.com/irc, where you can register for an instructor access code. Within 48 hours after registering, you will receive a confirming e-mail, including an instructor access code. Once you have received your code, go to the site and log on for full instructions on downloading the materials you wish to use.

To the Faculty

This text is written at a language level and at a mathematics level that is cognizant of and beneficial to *most* students in programs that do not require a high level of mathematics. The authors have assumed that the student has successfully completed one year of high school algebra or its equivalent. Simple equations and formulas are reviewed and any mathematics beyond this level is developed in the text or in an appendix. For example, right-triangle trigonometry is developed in Appendix A.5 for those who have not studied it previously or who need a review. The manner in which the mathematics is used in the text displays the need for mathematics in technology. For the better prepared student, the mathematics sections may be omitted with no loss in continuity. This text is designed so that faculty have flexibility in selecting the topics, as well as the order of topics, that meet the needs of their students and programs of study.

Sections are short, and each deals with only one concept. The need for the investigation of a physical principle is developed before undertaking its study, and many diagrams are used to aid students in visualizing the concept. Many examples and problems are given to help students develop and check their mastery of one concept before moving to another.

This text is designed to be used in a vocational–technical program in a community college, a technical institute, or a high school for students who plan to pursue a technical career or in a general physics course where an applied physics approach is preferred. The topics were chosen with the assistance of technicians and management in several industries and faculty consultants. Suggestions from users and reviewers of the previous edition were used extensively in this edition.

A general introduction to physics is presented in Chapter 0. Chapter 1 introduces students to basic units of measurement. For students who lack a metric background or who need a review, an extensive discussion of the metric system is given in Chapter 1, where it is shown how the results of measurements are approximate numbers, which are then used consistently throughout the text. Those who need to review some mathematical skills are referred to the appendices as necessary. Chapter 2 introduces students to a problem-solving method that is consistently used in the rest of the text. Vectors are developed in Chapter 3, followed by a comprehensive study of motion, force, work and energy, rotational energy, simple machines, and universal gravitation and satellite motion.

The treatment of matter includes a discussion of the three states of matter, density, fluids, pressure, and Pascal's principle. The treatment of heat includes temperature, specific heat, thermal expansion, change of state, and ideal gas laws.

The section on wave motion and sound deals with basic wave characteristics, the nature and speed of sound, the Doppler effect, and resonance.

The section on electricity and magnetism begins with a brief discussion of static electricity, followed by an extensive treatment of dc circuits and sources, Ohm's law, and series and parallel circuits. The chapter on magnetism, generators, and motors is largely descriptive, but it allows for a more in-depth study if desired. Then ac circuits and transformers are treated extensively.

The chapter on light briefly discusses the wave and particle nature of light, but deals primarily with illumination. The chapter on reflection and refraction treats the images formed by mirrors and lenses. A brief introduction to color includes diffraction, interference, and polarization of light.

The section on modern physics provides an introduction to the structure and properties of the atomic nucleus, radioactive decay, nuclear reactions, and radioactivity followed by a very brief introduction to relativity.

A companion laboratory manual is available from the publisher. An *Instructor's Resource Manual* that includes Complete Solutions, Transparency Masters, and a Test Item File is available at no charge to instructors using this text.

To the Student: Why Study Physics?

Physics is useful. Architects, mechanics, builders, carpenters, electricians, plumbers, and engineers are only some of the people who use physics every day in their jobs or professions. In fact, every person uses physics principles every hour of every day. The movement of an arm can be described using principles of the lever. All building trades, as well as the entire electronics industry, also use physics.

Physics is often defined as the study of matter, energy, and their transformations. The physicist uses scientific methods to observe, measure, and predict physical events and behaviors. However, gathered data left in someone's notebook in a laboratory are of little use to society.

Physics provides a universal means of describing and communicating about physical phenomena in the language of mathematics. Mechanics is the base on which almost all other areas of physics are built. Motion, force, work, electricity, and light are topics confronted daily in industry and technology. The basic laws of conservation of energy are needed to understand heat, sound, wave motion, electricity, and electromagnetic radiation.

Physics is always changing as new frontiers are being established in the study of the nature of matter. The topics studied in this course, however, will probably not greatly change with new research and will remain a classical foundation for work in many, many fields. We begin our study with the rules of the road—measurement, followed by a systematic problem-solving method. The end result should be a firm base on which to build a career in almost any field.

ACKNOWLEDGMENTS

The authors thank the many faculty and students who have used the previous editions, and especially those who have offered suggestions. If anyone wishes to correspond with us regarding suggestions, criticisms, questions, or errors, please contact Dale Ewen through Prentice Hall or through the web address http://www.prenhall.com.

We thank the following reviewers: Dr. A. M. Bloom, Hallmark Institute (TX); Robert H. Hadley, DeVry University (NJ); and Grace Wong, Heald College (CA).

We extend our sincere and special thanks to our Prentice Hall editor, Gary Bauer; senior production editor, Louise Sette; project manager, Sarvesh Mehrotra at Aptara, Inc.; and copyeditor, Philip Koplin.

Finally, we are especially grateful to Amy Gundersen for her assistance with the solutions manual and to our families for their encouragement.

Dale Ewen
Neill Schurter
P. Erik Gundersen

AN INTRODUCTION TO PHYSICS

velocity $v = s/t$

Force $F = ma$

Gear Ratios

Power $P = w/t$

Physics plays an important role in all aspects of our lives. For this mountain biker, the forces exerted by his legs, the air pressure in the tires, and the correct gearing combination will help determine the outcome of the race.

Before enrolling in a physics course, you may have taken physics for granted. In this chapter we will introduce physics to you and help you appreciate the impact that physics will have on your life and career.

Objectives

The major goals of this chapter are to enable you to:

1. Determine what physics governs and controls.
2. Conclude that physics is a building block of all the sciences.
3. Identify areas in your life that will be impacted by studying physics.
4. Differentiate between laws and theories.
5. Provide reasons why problem-solving techniques are vital in the study of physics.

0.1 Why Study Physics?

What do flying birds, automobiles, blue skies, and cellular phones have in common? They all involve physics. **Physics** is the branch of science that describes the motion and energy of all matter throughout the universe. Birds, for example, use the difference in air pressures above and below their wings to keep themselves aloft. Automobiles use the principles of mechanics and thermodynamics to transfer stored chemical energy in gasoline to moving energy in rotating tires. The sky appears blue when sunlight strikes and scatters off nitrogen and oxygen molecules in our atmosphere. Finally, cellular phones use electronic components and the principles of electromagnetic waves to transfer energy and information from one cellular phone to another (Fig. 0.1).

Physics is often considered to be the most fundamental of all the sciences. In order to study biology, chemistry, or any other natural science, one should have a firm understanding of the principles of physics. For example, **biology,** the branch of science that studies living organisms, uses the principles of fluid movement to understand how the blood moves through the heart, arteries, and veins. **Chemistry,** the branch of science that studies the composition, structure, properties, and reactions of matter, relies on the physics of subatomic particles to understand why chemical reactions take place. **Geology,** the branch of science that studies the origin, history, and structure of the earth, uses the physics of mechanical waves and energy transfer to determine the magnitude and location of earthquakes. Finally, **astronomy,** the field of science that studies everything that takes place outside the earth's atmosphere, relies on the laws of gravity and theory of relativity to describe the workings of the universe.

Students often wonder, "Why should I study physics? What is it going to do for me?" The answer is that physics plays an important role in everyday life and in the

Figure 0.1 Physics is involved in all aspects of cellular phone technology. It controls everything from the electrical circuits in the phone to the transmission of radio waves between phones.

TRY THIS ACTIVITY

Physics All Around Us

Look around and find something that may have to do with physics. Although you may not yet have studied many physics principles, you should know that physics governs things that move and transfer energy. Be as general as you need to be in your observations. The point is for you to see that physics plays a role in most everything.

Figure 0.2   A baseball player's understanding of physics can help improve all aspects of his game, including pitching, batting, and fielding.

Photo courtesy of AP/Worldwide Photos. Reprinted with permission

Figure 0.3   Albert Einstein (1879–1955) is often considered one of the most influential scientists of the twentieth century. His work on relativity, as made famous through the equation $E = mc^2$, and the photoelectric effect changed the way the world viewed physics.

Photo courtesy of the National Archives and Records Administration

Figure 0.4   Archimedes is best known for observing that water was displaced when he stepped into his bath. He proceeded to conduct experiments where he measured the amount of water that overflowed when objects were placed into a tub full of water. He established a principle that states that an object immersed in a fluid will experience a buoyant force equal to the weight of the displaced fluid. Archimedes is also recognized for his work with simple machines like the screw, lever, and pulley. Legend has it that Archimedes said, "Give me a firm spot on which to stand and I will move the earth" (referring to the use of a lever).

careers of many people. Choosing the right bat, golf club, or ski can be made easier with a bit of physics knowledge (Fig. 0.2). While on the job, architects, engineers, electricians, medical technicians, surveyors, and others use the principles of physics every day. When understood, physics can help us solve difficult physical problems and be better decision makers to determine the best design, tool, or process when working on a specific task.

A **physicist** is a person who is an expert in or who studies physics. It is a physicist's job to seek an understanding of how the physical universe behaves. Albert Einstein, perhaps one of the most famous physicists of all time, once said, "I am like a child, I always ask the simplest questions" (Fig. 0.3). Such **theoretical physicists** often spend their professional lives researching previous theories and mathematical models to form new theories in physics. **Experimental physicists,** however, focus on performing experiments to develop and confirm physical theories.

It is generally accepted that physics evolved from ancient Greek philosophers like Plato (c. 428–347 BC) and Aristotle (384–322 BC). Aristotle believed that there were two types of motion: Natural motion occurred because objects wanted to seek their "natural" resting place (smoke rising or rocks falling), whereas violent motion occurred when objects were unnaturally pulled or dragged from place to place (person dragging a crate). Although he was not correct in his analysis, it was the beginning of observing and documenting physical phenomena. Plato, Aristotle, and others like them can be considered theoretical physicists.

It was not until the days of Archimedes (287–212 BC) that experiments were conducted to document and prove physical theories (Fig. 0.4). Since then, thousands of physicists have built and improved upon the knowledge base developed by those before them. It is now your turn to use the physics that you will learn to help you understand, improve, and make advances in our technological world. You will see that physics has use!

0.2 Physics and Its Role in Technology

Although often discussed as though they are the same thing, science and technology are quite different. **Science** is a system of knowledge that is concerned with establishing accurate conclusions about the behavior of everything in the universe. It is a field in which **hypotheses** (scientifically based predictions) are made, information is gathered, and experiments are performed to determine how something in our natural world works or behaves. **Technology,** on the other hand, is a field that uses scientific knowledge to develop material products or processes that satisfy human needs and desires. Technology and science rely closely on one another to make further advances in their respective fields.

Thomas A. Edison (1847–1931) used scientific information and the discoveries of other scientists to create over 1000 inventions (Fig. 0.5). Edison's development of the first practical lighting system was made possible by applying the science of electricity and the science of materials and then putting that knowledge to use to satisfy his technological need.

The following illustrate how science has played a role in improving technology.

Robotics: Due to advances in electronics, materials, and machines, robots commonly perform a variety of tasks from assembling cars on a production line to exploring the surface of Mars. NASA's Sojourner was the first robotic device to explore the surface of another planet (Fig. 0.6).

Bridges: Work in materials science and structural engineering has paved the way for advances in bridge design and construction. The New Clark Bridge in Alton, Illinois, is just one example of a cable-stayed bridge that has used scientific breakthroughs in materials science and physics to increase the structural integrity of the bridge and cut costs (Fig. 0.7).

Superconductors: Superconductors allow electric current to travel with virtually no resistance through materials. Materials such as aluminum, lead, and niobium are

Figure 0.5   Thomas A. Edison

Photo courtesy of National Park Service, Edison National Historic Site

Figure 0.6   NASA's robotic Sojourner Rover on Mars in 1997

Photo courtesy of NASA Headquarters

Figure 0.7 The New Clark Bridge, a cable-stayed bridge in Alton, Illinois

Figure 0.8 A magnet levitates above a superconductor, demonstrating the Meissner effect.

Photo courtesy of Brookhaven National Laboratory

cooled by liquid helium to bring the temperature down to the low critical point. At the low critical point temperatures, the materials achieve zero electric resistance (Fig. 0.8). Scientific research is under way to develop superconducting materials that can operate closer to room temperature; this would bring about tremendous improvements in energy efficiency.

Active Noise Cancellation: Audiologists will tell you that noise increases stress levels. Acoustic and electrical engineers are now able to produce inverted noise patterns that cancel out disturbing noise (Fig. 0.9). Helicopter pilots, factory workers, and business travelers are using this technology to reduce stressful noise levels in their environment.

Liquid Crystal Displays: With advances in optics and electronics, physicists and chemists have created more advanced liquid crystal displays (LCDs), which are used as screens on laptop computers, personal digital assistants (PDAs), watches, and televisions (Fig. 0.10).

Figure 0.9 Active noise cancellation technology for the consumer can be found in small, noise-reduction headsets.

Figure 0.10 Advances in LCD panels allow small color PDAs to have color screens.

Figure 0.11 Maglev train technology has high-speed trains competing with airplane service.

Copyright of Apollo/PhotoEdit, Inc. Photo reprinted with permission

Figure 0.12 Gyroscope

Photo courtesy of Dorling Kindersley

Magnetic Levitation: The speed limitations of traditional trains have created a need for super-fast, magnetic levitation (maglev) trains (Fig. 0.11). Such a vehicle is levitated off a monorail by virtue of the magnetic repulsion between the train and the rail. Electromagnets are used to propel the train forward as it glides above the rail. Such improvements greatly reduce frictional resistance and allow trains to travel at twice the speed of conventional trains—up to 250 miles per hour.

Gyroscope: A gyroscope is a heavy wheel that uses rotational inertia to prevent tilting and is used to steady compasses, ships, airplanes, and rockets (Fig. 0.12). Advances in gyroscopes and electronic sensors have made it possible to create gyrostabilizers for ships. Such devices send signals to the ship's computer specifying how its fins should be positioned to prevent significant rolling motions.

0.3 Physics and Its Connection to Other Fields and Sciences

Ancient Greeks like Plato and Aristotle did not specialize in physics. In fact, it was not until the 1800s that physics was considered a science. Prior to the 1800s, Plato, Aristotle, Copernicus, and Galileo were considered natural philosophers, not physicists. Today, virtually every physicist specializes in a subdivision of physics. There is simply too much information to allow someone to study every type of physics.

The following is a listing of the 18 subdivisions of physics:

Mechanics: Study of forces, motion, and energy.
Thermodynamics: Study of heat energy transfer.
Cryogenics: Study of matter at extremely low temperatures.
Plasma Physics: Study of electrically charged, ionized gas.
Solid State Physics: Study of the physical properties of solid materials, also known as condensed matter physics.
Geophysics: Study of the interaction of forces and energy found within the earth; closely related to geology.
Astrophysics: Study of the interaction of forces and energy between interstellar objects; closely related to astronomy.

PHYSICS CONNECTIONS

Physics, Technology, and Sports

Physics plays a major role in sports. From the padding in a baseball glove to the stance of a wrestler, a good working knowledge of physics helps athletes and sports equipment companies achieve greater successes. Ski companies employ engineers who focus solely on the physics and engineering of improving a skier's time down the mountain. At the 2006 Winter Olympics in Turino, Italy, adjustments made to the length, shape, and composition of the skis played an important role in the success of the skiers. Such variables determine the amount of pressure the skier places on the snow and the friction of the ski. There are tradeoffs as well. Whereas a wider ski front increases its turning abilities, it also creates large vibrations that can slow down the skier. The use of titanium and various fibers and adhesives decreases those vibrations and results in lighter, stiffer skis. The application of physics has taken the once-simple wooden ski and has created a complex, high-performance device (Fig. 0.13).

Figure 0.13 Each ski design is created for use based on scientific knowledge.

Copyright of Cosmo Condina/Getty Images, Inc.

Acoustics: Study of the creation and transmission of sound under various conditions.
Optics: Study of the behavior of light in a variety of conditions.
Electromagnetism: Study of the relationship between electricity and magnetism.
Fluid Dynamics: Study of how liquids and gases move from one location to another.
Mathematical Physics: Study of the mathematics of physics and its related fields.
Statistical Mechanics: Study of the development of statistical models that simulate the effects of systems composed of many particles.
High-Energy Physics: Study of new fundamental, subatomic particles using high-energy machines that send known subatomic particles colliding into one another; simulation of what the universe was like close to the time of the "big bang."
Atomic Physics: Study of the structure of the atom based on the knowledge gained in the field of high-energy physics.
Molecular Physics: Study of the structure of molecules based on the knowledge gained in atomic physics.
Nuclear Physics: Study of nuclear interactions.
Quantum Physics: Study of small particles and their energy.

0.4 Theories, Laws, and Problem Solving

Physics is constantly being refined. Although the major principles of physics do not change drastically over time, newer theories requiring a tremendous amount of experimentation can modify our understanding of physics. A **theory** is a scientific conclusion that attempts to explain natural occurrences. Typically it has been tested in the laboratory but has not been proven with absolute certainty. A **principle** is a step closer to a law in physics. Principles have been experimentally proven in the laboratory, have stood the test of various conditions, and continue to hold true. Laws are the final degree of scientific certainty. **Laws** are often defined using formulas. For example, Newton's second law of motion, $F = ma$, has been proven to be true and is considered a law of physics.

The **scientific method** is an orderly procedure used by scientists in collecting, organizing, and analyzing new information which refutes or supports a scientific hypothesis.

The constant use of the scientific method and the development of theories, principles, and laws is similar to the problem-solving method discussed in detail in Chapter 2. The **problem-solving method** is an orderly procedure that aids in understanding questions and solving problems. Nonscientists use the problem-solving method more often than the scientific method. The problem-solving method is helpful when a problem arises in this text, in class, or on the job. An individual or a team must develop the skills needed to collect data, analyze a problem, and work toward finding its solution in a logical and orderly fashion. In order to find solutions to problems, tools are needed to make the job easier. In the next chapter, we will familiarize ourselves with two important tools of physics: measurement and mathematics.

Glossary

Astronomy The branch of science that studies everything that takes place outside of the earth's atmosphere. (p. 3)

Biology The branch of science that studies living organisms. (p. 3)

Chemistry The branch of science that studies the composition, structure, properties, and reactions of matter. (p. 3)

Experimental Physicist A physicist who performs experiments to develop and confirm physical theories. (p. 4)

Geology The branch of science that studies the origin, history, and structure of the earth. (p. 3)

Hypothesis A scientifically based prediction that needs testing to verify its validity. (p. 5)

Law The highest level of certainty for an explanation of physical occurrences. A law is often accompanied by a formula. (p. 8)

Physics The branch of science that describes the motion and energy of all matter throughout the universe. (p. 3)

Physicist A person who is an expert in or who studies physics. (p. 4)

Principle A rule or fundamental assumption that has been proven in the laboratory. (p. 8)

Problem Solving Method An orderly procedure that aids in understanding questions and solving problems. (p. 9)

Science A system of knowledge that is concerned with establishing accurate conclusions about the behavior of everything in the universe. (p. 5)

Scientific Method An orderly procedure used by scientists in collecting, organizing, and analyzing new information which refutes or supports a scientific hypothesis. (p. 8)

Technology The field that uses scientific knowledge to develop material products or processes that satisfy human needs and desires. (p. 5)

Theoretical Physicist A physicist who predominantly uses previous theories and mathematical models to form new theories in physics. (p. 4)

Theory A scientifically accepted principle that attempts to explain natural occurrences. (p. 8)

Review Questions

1. Physics is a field of study that governs
    (a) how the planets orbit the sun.
    (b) the rate at which blood flows through a person's veins.
    (c) how quickly a helium balloon will rise into the air.
    (d) all of the above.
2. Who among the following is an example of a theoretical physicist?
    (a) Archimedes, who measured the volume of water that was displaced after placing objects in a tub of water.
    (b) Albert Einstein, who performed various thought experiments in his mind to arrive at his theories of relativity.
    (c) Marie Curie, who, along with her husband, was credited with discovering radioactivity through a series of laboratory experiments.
    (d) Benjamin Franklin, who through various laboratory experiments determined that electricity is the flow of microscopic charged particles.
3. Why are Isaac Newton's conclusions on motion considered laws of physics?
    (a) Newton himself declared them laws.
    (b) Newton performed various thought experiments on motion.
    (c) The formulas accompanying Newton's laws have proved correct in experiments for years.
    (d) Newton's reputation alone made his scientific conclusions laws.

4. Which of the following is not considered a branch of physics?
   (a) Thermodynamics   (b) Astronomy
   (c) Geophysics   (d) Atomic physics
5. Analyzing the braking distance of a sports car would most likely utilize which field of physics?
   (a) Molecular physics   (b) Quantum physics
   (c) Fluid dynamics   (d) Mechanics
6. Who is considered to be the first true physicist and what did he do to deserve this recognition in scientific history?
7. Explain the difference between science and technology. Are the two fields related?
8. Provide two examples of scientific knowledge and a technological development that relies on that scientific knowledge.
9. What is the difference between the scientific method and the problem-solving method?
10. Why is it important to study physics? Provide a few examples of what an understanding of the physical world can do for you today and in your future.

CHAPTER 1

THE PHYSICS TOOL KIT

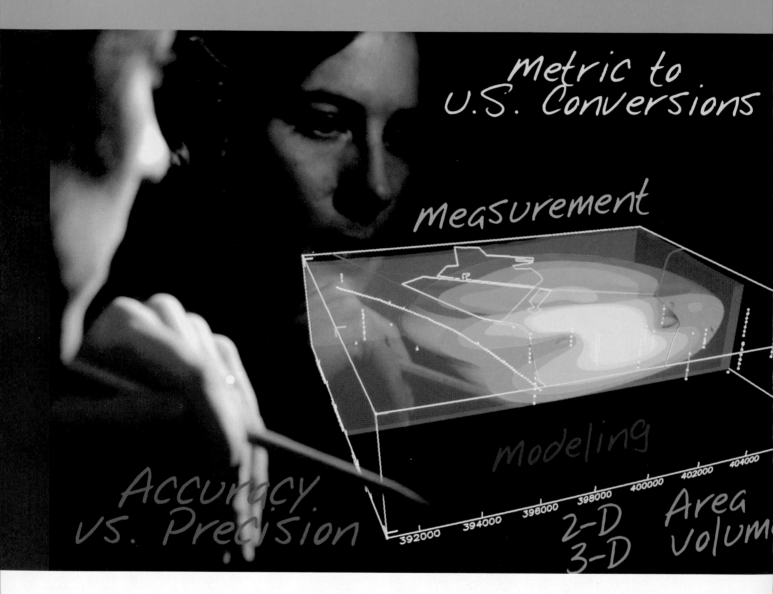

Metric to U.S. Conversions

Measurement

Modeling

Accuracy vs. Precision

2-D
3-D

Area
Volume

A good mechanic needs not only the right tools for the job but also to be proficient in using those tools. The same applies to physics. Analyzing the problem, choosing the correct formula, and manipulating the equation will help you become a good physics student. In this chapter, we discuss mathematical techniques, significant digits, accuracy, precision, and the problem-solving method. These will be your basic tools for physics.

Objectives

The major goals of this chapter are to enable you to:

1. Explain the need for standardization of measurement.
2. Use the metric system of measurement.
3. Convert measurements from one system to another.
4. Solve problems involving length, area, and volume.
5. Distinguish between mass and weight.
6. Use significant digits to determine the accuracy of measurements.
7. Differentiate between accuracy and precision.
8. Solve problems with measurements and consistently express the results with the correct significant digits.
9. Use a systematic approach to solving physics problems.
10. Analyze problems using the problem-solving method.

1.1 Standards of Measure

When two people work together on the same job, they should both use the same standards of measure. If not, the result can be a problem (Fig. 1.1).

Figure 1.1   The trouble with inconsistent systems of measurement

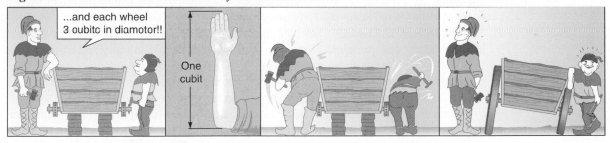

Standards of measure are sets of units of measurement for length, weight, and other quantities defined in a way that is useful to a large number of people. Throughout history, there have been many standards by which measurements have been made:

◆ *Chain:* A measuring instrument of 100 links used in surveying. One chain has a length of 66 feet.
◆ *Rod:* A length determined by having each of 16 men put one foot behind the foot of the man before him in a straight line [Fig. 1.2(a)]. The rod is now standardized as $16\frac{1}{2}$ feet.
◆ *Yard:* The distance from the tip of the king's nose to the fingertips of his outstretched hand [Fig. 1.2(b)].
◆ *Foot:* The rod divided by 16; it was also common to use the length of one's own foot as the unit foot.
◆ *Inch:* The length of three barley corns, round and dry, taken from the center of the ear, and laid end to end [Fig. 1.2(c)].

The U.S. system of measure, which is derived from and sometimes called the English system, is a combination of makeshift units of Anglo-Saxon, Roman, and French-Norman weights and measures.

Figure 1.2   Definitions of some old units

(a) A rod used to be 16 "people feet." This distance divided by 16 equals one foot.

(b) The "old" yard

(c) At one time, three barley corns were used to define one inch.

After the standards based on parts of the human body and on other gimmicks, basic standards were accepted by world governments. They also agreed to construct and distribute accurate standard copies of all the standard units. During the 1790s, a decimal system based on our number system, the metric system, was developed in France. Its acceptance was gained mostly because it was easy to use and easy to remember. Many nations began adopting it as their official system of measurement. By 1900, most of Europe and South America were metric. In 1866, metric measurements for official use were legalized in the United States. In 1893, the Secretary of the Treasury, by administrative order, declared the new metric standards to be the nation's "fundamental standards" of mass and length. Thus, indirectly, the United States *officially* became a metric nation. Even today, the U.S. units are officially defined in terms of the standard metric units.

Throughout U.S. history, several attempts have been made to convert the nation to the metric system. By the 1970s, the United States found itself to be the only nonmetric industrialized country left in the world. However, government inaction to implement the metric system resulted in the United States regularly using a greater variety of confusing units than any other country. Industry and business, however, found their international markets drying up because metric products were preferred. Now many segments of American industry and business have independently gone metric because world trade is geared toward the metric system of measurement. The inherent simplicity of the metric system of measurement and standardization of weights and measures has led to major cost savings in industries that have converted to it.

Most major U.S. industries, such as the automotive, aviation, and farm implement industries, as well as the Department of Defense and other federal agencies have effectively converted to the metric system. In this text approximately 70% of our examples and exercises are metric in the sections where both U.S. and metric systems are still commonly used. In some industries, you—the student and worker—will need to know and use both systems.

TRY THIS ACTIVITY

Stepping Off

When a short distance needs to be measured and a tape measure is not available, some people measure the approximate length by using the length of their own foot as a unit. Measure the distance between two points approximately 15 to 25 ft apart by placing one foot in front of the other and counting the steps. Then, measure the same distance with a tape measure. How close to the standard foot is the length of your own foot? How much error did you generate?

For longer distances, some measure the approximate length by pacing off the distance using one stride as approximately 1 yd or 3 ft. Measure the distance between two points approximately 50 to 75 ft apart by pacing off the distance and counting the strides. Then, measure the same distance with a tape measure. How close to the standard yard is the length of your own stride? How much error did you generate?

PHYSICS CONNECTIONS

Lost in Space

Even professional engineers and scientists sometimes forget to include units or make mistakes when converting from one system of measurement to another. NASA's Mars Climate Orbiter (Fig. 1.3) was lost in space on September 23, 1999, after engineers from Lockheed Martin used U.S. measurements when calculating rocket thrusts and did not convert those measurements to the metric units used by NASA engineers. Such a simple mistake, common to students on physics exams, cost Lockheed Martin and NASA a $125 million space probe and a great deal of embarrassment.

According to officials, NASA assumed that the twice-daily rocket thrust calculations were based on metric units, whereas Lockheed Martin had neither converted the numbers to metric nor labeled them in U.S. units. As a result, the Mars Climate Orbiter, which was to orbit Mars, was inadvertently sent off course and lost. The orbiter was intended to collect atmospheric and surface data while also serving as a communications link for the Mars Lander.

Figure 1.3 NASA's Mars Climate Orbiter—artist's conception

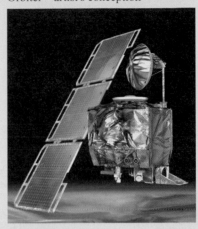

Photo courtesy of NASA Headquarters/Jet Propulsion Laboratory

1.2 Introduction to the Metric System

The modern metric system is identified in all languages by the abbreviation **SI** (for Système International d'Unités — the international system of units of measurement written in French). The SI metric system has seven *basic units* [Table 1.1(a)]. All other SI units are called *derived units*; that is, they can be defined in terms of these seven basic units (see Appendix C, Table 19). For example, the newton (N) is defined as 1 kg m/s^2 (kilogram metre per second per

Gabriel Mouton (1618–1694),

a French vicar who spent much of his time studying mathematics and astronomy, is credited by many for originating the metric system. French scientists in the late 18th century are credited with replacing the chaotic collection of systems then in use with the metric system.

Table 1.1   SI Units of Measure

(a) Basic Unit	SI Abbreviation	Used for Measuring
metre*	m	Length
kilogram	kg	Mass
second	s	Time
ampere	A	Electric current
kelvin	K	Temperature
candela	cd	Light intensity
mole	mol	Molecular substance

(b) Derived Unit	SI Abbreviation	Used for Measuring
litre*	L or ℓ	Volume
cubic metre	m^3	Volume
square metre	m^2	Area
newton	N	Force
metre per second	m/s	Speed
joule	J	Energy
watt	W	Power

*At present, there is some difference of opinion in the United States on the spelling of metre and litre. We have chosen the "re" spellings for two reasons. First, this is the internationally accepted spelling for all English-speaking countries. Second, the word "meter" already has many different meanings—parking meter, electric meter, odometer, and so on. Many feel that the metric units of length and volume should be distinctive and readily recognizable—thus the spellings "metre" and "litre."

Table 1.2 **Prefixes for SI Units**

Multiple or Submultiple[a] Decimal Form	Power of 10	Prefix[b]	Prefix Symbol	Pronunciation	Meaning
1,000,000,000,000	10^{12}	tera	T	tĕr'ă	One trillion times
1,000,000,000	10^{9}	giga	G	jĭg'ă	One billion times
1,000,000	10^{6}	mega	M	mĕg'ă	One million times
1,000	10^{3}	kilo	k	kĭl'ō	One thousand times
100	10^{2}	hecto	h	hĕk'tō	One hundred times
10	10^{1}	deka	da	dĕk'ă	Ten times
0.1	10^{-1}	deci	d	dĕs'ĭ	One tenth of
0.01	10^{-2}	centi	c	sĕnt'ĭ	One hundredth of
0.001	10^{-3}	milli	m	mĭl'ĭ	One thousandth of
0.000001	10^{-6}	micro	μ	mī'krō	One millionth of
0.000000001	10^{-9}	nano	n	năn'ō	One billionth of
0.000000000001	10^{-12}	pico	p	pē'kō	One trillionth of

[a]Factor by which the unit is multiplied.
[b]The same prefixes are used with all SI metric units.

second). Many derived units will be presented and discussed in this text. Some derived SI units are given in Table 1.1(b). **Gabriel Mouton** is often credited for originating the metric system.

Because the metric system is a decimal or base 10 system, it is very similar to our decimal number system and any decimal money system. It is an easy system to use because calculations are based on the number 10 and its multiples. Special prefixes are used to name these multiples and submultiples, which may be used with most all SI units. Because the same prefixes are used repeatedly, the memorization of many conversions has been significantly reduced. Table 1.2 shows these prefixes and the corresponding symbols.

EXAMPLE 1

Write the SI abbreviation for 36 centimetres.

The symbol for the prefix *centi* is c.
The symbol for the unit *metre* is m.

Thus, 36 cm is the SI abbreviation for 36 centimetres.

· · · · · · · · · · · · · · · · ·

EXAMPLE 2

Write the SI metric unit for the abbreviation 45 kg.

The prefix for k is *kilo*; the unit for g is *gram*.

Thus, 45 kilograms is the SI metric unit for 45 kg.

· · · · · · · · · · · · · · · · ·

PROBLEMS 1.2

Give the metric prefix for each value.

1.	1000	2.	0.01	3.	100	4.	0.1
5.	0.001	6.	10	7.	1,000,000	8.	0.000001

Give the metric symbol, or abbreviation, for each prefix.

9.	hecto	10.	kilo	11.	milli	12.	deci
13.	mega	14.	deka	15.	centi	16.	micro

Write the abbreviation for each quantity.

17.	135 millimetres	18.	83 dekagrams	19.	28 kilolitres
20.	52 centimetres	21.	49 centigrams	22.	85 milligrams
23.	75 hectometres	24.	15 decilitres		

Write the SI unit for each abbreviation.

25. 24 m   26. 185 L   27. 59 g   28. 125 kg
29. 27 mm   30. 25 dL   31. 45 dam   32. 27 mg
33. 26 Mm   34. 275 μg
35. The basic metric unit of length is _____.
36. The basic unit of mass is _____.
37. Two common metric units of volume are _____ and _____.
38. The basic unit for electric current is _____.
39. The basic metric unit for time is _____.
40. The common metric unit for power is _____.

1.3 Scientific Notation

Scientists and technicians often need to use very large or very small numbers. For example, the thickness of an oil film on water is about 0.0000001 m. **Scientific notation** is a useful method of expressing such very small (or very large) numbers. Expressed this way, the thickness of the film is 1×10^{-7} m or 10^{-7} m. For example:

$$0.1 = 1 \times 10^{-1} \quad \text{or} \quad 10^{-1}$$
$$10,000 = 1 \times 10^{4} \quad \text{or} \quad 10^{4}$$
$$0.001 = 1 \times 10^{-3} \quad \text{or} \quad 10^{-3}$$

> A number in scientific notation is written as a product of a number between 1 and 10 and a power of 10. General form: $M \times 10^{n}$, where
>
> $M =$ a number between 1 and 10
> $n =$ the exponent or power of 10

The following numbers are written in scientific notation:

EXAMPLE 1

(a) $325 = 3.25 \times 10^{2}$
(b) $100,000 = 1 \times 10^{5}$ or 10^{5}

· · · · · · · · · · · · · · · · ·

To write any decimal number in scientific notation:

1. Reading from left to right, place a decimal point after the first nonzero digit.
2. Place a caret ($\wedge$) at the position of the *original* decimal point.
3. If the newly added decimal point is to the *left* of the caret, the exponent of 10 is the number of places from the caret to the decimal point.
    Example: $83,662 = 8.3662_{\wedge} \times 10^{\underline{4}}$
4. If the newly added decimal point is to the *right* of the caret, the exponent of 10 is the negative of the number of places from the caret to the decimal point.
    Example: $0.00683 = {}_{\wedge}006.83 \times 10^{\underline{-3}}$
5. If the decimal point and the caret coincide, the exponent of 10 is zero.
    Example: $5.12 = 5.12 \times 10^{0}$

A number greater than 10 is expressed in scientific notation as a product of a decimal between 1 and 10 and a *positive* power of 10.

EXAMPLE 2

Write each number greater than 10 in scientific notation.

(a) $2580 = 2.58 \times 10^3$
(b) $54,600 = 5.46 \times 10^4$
(c) $42,000,000 = 4.2 \times 10^7$
(d) $715.8 = 7.158 \times 10^2$
(e) $34.775 = 3.4775 \times 10^1$

. . . .

A number between 0 and 1 is expressed in scientific notation as a product of a decimal between 1 and 10 and a *negative* power of 10.

EXAMPLE 3

Write each positive number less than 1 in scientific notation.

(a) $0.0815 = 8.15 \times 10^{-2}$
(b) $0.00065 = 6.5 \times 10^{-4}$
(c) $0.73 = 7.3 \times 10^{-1}$
(d) $0.0000008 = 8 \times 10^{-7}$

. . . .

A number between 1 and 10 is expressed in scientific notation as a product of a decimal between 1 and 10 and the *zero* power of 10.

EXAMPLE 4

Write each number between 1 and 10 in scientific notation.

(a) $7.33 = 7.33 \times 10^0$
(b) $1.06 = 1.06 \times 10^0$

. . . .

To change a number from scientific notation to decimal form:

1. Multiply the decimal part by the power of 10 by moving the decimal point *to the right* the same number of decimal places as indicated by the power of 10 if it is *positive*.
2. Multiply the decimal part by the power of 10 by moving the decimal point *to the left* the same number of decimal places as indicated by the power of 10 if it is *negative*.
3. Supply zeros as needed.

EXAMPLE 5

Write 7.62×10^2 in decimal form.

$$7.62 \times 10^2 = 762$$ Move the decimal point two places to the right.

. . . .

EXAMPLE 6

Write 6.15×10^{-4} in decimal form.

$$6.15 \times 10^{-4} = 0.000615$$ Move the decimal point four places to the left and insert three zeros.

. . . .

EXAMPLE 7

Write each number in decimal form.

(a) $3.75 \times 10^2 = 375$
(b) $1.09 \times 10^5 = 109,000$
(c) $2.88 \times 10^{-2} = 0.0288$
(d) $9.4 \times 10^{-6} = 0.0000094$
(e) $6.7 \times 10^0 = 6.7$

. . . .

Since calculators used in science and technology accept numbers entered in scientific notation and give some results in scientific notation, it is essential that you fully understand this topic before going to the next section. See Appendix B, Section B.1, for using a calculator with numbers in scientific notation.

PHYSICS CONNECTIONS

Powers of Ten

The importance of scientific notation is illustrated by a famous short documentary film called *Powers of Ten*. The film begins with a 1-metre-wide overhead view of a man relaxing on a blanket while on a picnic. The camera then zooms out to a shot that covers a width of 10 metres (10^1 m) with the picnicking man still in the center of the image but now surrounded by green grass. Throughout the course of another 10 seconds, the camera continues to zoom out to 100 metres (10^2 m) wide. As the camera continues to zoom out, the viewer begins to grasp how enormous the changes in distance are when a number is raised by a power of 10. Eventually, the trip outward stops at 10^{24} m, and then the picture zooms back to 10^0 m before proceeding to enter the skin of the man by moving in by negative powers of 10. Although the majority of the images are artist's renditions of what scientists imagine these objects to look like, *Powers of Ten* is a powerful film that conceptualizes how large our universe is and how small atoms and subatomic particles are within our bodies. The series of photographs demonstrating the powers of ten can be found at http://www.powersof10.com.

Some of the views used in *Powers of Ten* include:

10^0 m or 1 metre: a view of a man on a blanket while on a picnic
10^7 m or 10 thousand kilometres: a view of the entire earth
10^{21} m or 100 thousand light-years: a view of our galaxy, the Milky Way
10^{-1} m or 10 centimetres: a view of a small section of skin on the man's hand
10^{-5} m or 10 microns: a view of a white blood cell inside a capillary in the man's hand
10^{-14} m or 10 fermis: the nucleus of a typical carbon-12 atom

PROBLEMS 1.3

Write each number in scientific notation.

1. 326
2. 798
3. 2650
4. 14,500
5. 826.4
6. 24.97
7. 0.00413
8. 0.00053
9. 6.43
10. 482,300
11. 0.000065
12. 0.00224
13. 540,000
14. 1,400,000
15. 0.0000075
16. 0.0000009
17. 0.00000005
18. 3,500,000,000
19. 732,000,000,000,000,000
20. 0.00000000000000000618

Write each number in decimal form.

21. 8.62×10^4
22. 8.67×10^2
23. 6.31×10^{-4}
24. 5.41×10^3
25. 7.68×10^{-1}
26. 9.94×10^1
27. 7.77×10^8
28. 4.19×10^{-6}
29. 6.93×10^1
30. 3.78×10^{-2}
31. 9.61×10^4
32. 7.33×10^3
33. 1.4×10^0
34. 9.6×10^{-5}
35. 8.4×10^{-6}
36. 9×10^8
37. 7×10^{11}
38. 4.05×10^0
39. 7.2×10^{-7}
40. 8×10^{-9}
41. 4.5×10^{12}
42. 1.5×10^{11}
43. 5.5×10^{-11}
44. 8.72×10^{-10}

1.4 Length

In most sections that introduce units of measure, we present the units in subsections as follows: metric units, U.S. units, and conversions between metric and U.S. units.

Metric Length

The basic SI unit of length is the **metre** (m) (Fig. 1.4). The first standard metre was chosen in the 1790s to be one ten-millionth of the distance from the earth's equator to either pole. Modern measurements of the earth's circumference show that the first length is off by about 0.02% from this initial standard. The current definition adopted in 1983 is based on the speed of light in a vacuum and reads "The metre is the length of path traveled by light in a vacuum during a time interval of 1/299,792,458 of a second." Long distances are measured in kilometres (km) (Fig. 1.5). We use the centimetre (cm) to measure short distances, such as the length of this book or the width of a board [Fig. 1.6(a)]. The millimetre (mm) is used to measure very small lengths, such as the thickness of this book or the depth of a tire tread [Fig. 1.6(b)]. A metric ruler is shown in Fig. 1.6(c).

Figure 1.4 One metre

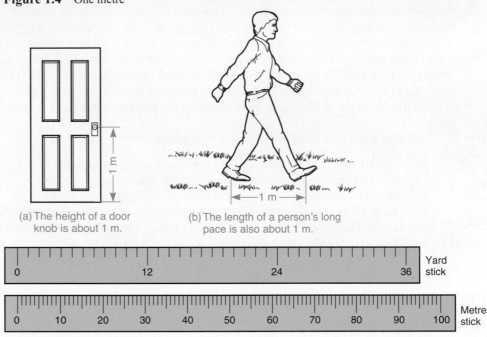

(a) The height of a door knob is about 1 m.

(b) The length of a person's long pace is also about 1 m.

(c) The length of 1 m is a little more than 1 yd.

Figure 1.5 The length of five city blocks is about 1 km.

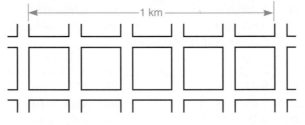

A **conversion factor** is an expression used to change from one unit or set of units to another. We know that we can multiply any number or quantity by 1 without changing the value of the original quantity. We also know that any fraction equals 1 when its numerator and denominator are equal. For example, $\frac{5}{5} = 1$, $\frac{12 \text{ m}}{12 \text{ m}} = 1$, and $\frac{6.5 \text{ kg}}{6.5 \text{ kg}} = 1$. In addition, since 1 m = 100 cm, $\frac{1 \text{ m}}{100 \text{ cm}} = 1$. Similarly, $\frac{100 \text{ cm}}{1 \text{ m}} = 1$, because the numerator equals the denominator. We call such names for 1 *conversion factors*. The information necessary for forming a conversion factor is usually found in tables. As in the case 1 m = 100 cm, there are two conversion factors for each set of data:

$$\frac{1 \text{ m}}{100 \text{ cm}} \quad \text{and} \quad \frac{100 \text{ cm}}{1 \text{ m}}$$

Figure 1.6   Small metric length units

(a) The width of your small fingernail is about 1 cm.   (b) The thickness of a dime is about 1 mm.

(c) The large numbered divisions are centimetres shown actual size.
Each centimetre is divided into 10 equal parts, called millimetres.

CONVERSION FACTORS

Choose a conversion factor in which the old units are in the numerator of the original expression and in the denominator of the conversion factor, or the old units are in the denominator of the original expression and in the numerator of the conversion factor. That is, we want the old units to cancel each other.

EXAMPLE 1

Change 215 cm to metres.

As we saw before, the two possible conversion factors are

$$\frac{1 \text{ m}}{100 \text{ cm}} \quad \text{and} \quad \frac{100 \text{ cm}}{1 \text{ m}}$$

We choose the conversion factor with centimetres in the *denominator* so that the cm units cancel each other.

$$215 \text{ cm} \times \frac{1 \text{ m}}{100 \text{ cm}} = 2.15 \text{ m}$$

Note: Conversions *within* the metric system involve only moving the decimal point.

· · · · · · · · · · · · · · · · · · ·

EXAMPLE 2

Change 4 m to centimetres.

$$4 \text{ m} \times \frac{100 \text{ cm}}{1 \text{ m}} = 400 \text{ cm}$$

· · · · · · · · · · · · · · · · · ·

EXAMPLE 3

Change 39.5 mm to centimetres.

Choose the conversion factor with millimetres in the denominator so that the mm units cancel each other.

$$39.5 \text{ mm} \times \frac{1 \text{ cm}}{10 \text{ mm}} = 3.95 \text{ cm}$$

· · · · · · · · · · · · · · · ·

EXAMPLE 4

Change 0.05 km to centimetres.

First, change to metres and then to centimetres.

$$0.05 \text{ km} \times \frac{1000 \text{ m}}{1 \text{ km}} = 50 \text{ m}$$

$$50 \text{ m} \times \frac{100 \text{ cm}}{1 \text{ m}} = 5000 \text{ cm}$$

Or,

$$0.05 \text{ km} \times \frac{1000 \text{ m}}{1 \text{ km}} \times \frac{100 \text{ cm}}{1 \text{ m}} = 5000 \text{ cm}$$

. . . .

U.S. Length

The basic units of the U.S. system are the foot, the pound, and the second. The foot is the basic unit of length and may be divided into 12 equal parts or inches. Common U.S. length conversions include

1 foot (ft) = 12 inches (in.)

1 yard (yd) = 3 ft

1 mile (mi) = 5280 ft

See Table 1 of Appendix C for U.S. weights and measures.

We also use a conversion factor to change from one U.S. length unit to another.

EXAMPLE 5

Change 84 in. to feet.

Choose the conversion factor with inches in the denominator and feet in the numerator.

$$84 \text{ in.} \times \frac{1 \text{ ft}}{12 \text{ in.}} = 7 \text{ ft}$$

. . . .

Metric–U.S. Conversions

To change from a U.S. unit to a metric unit or from a metric unit to a U.S. unit, again use a conversion factor, such as 1 in. = 2.54 cm.

EXAMPLE 6

Express 10 inches in centimetres.

$$1 \text{ in.} = 2.54 \text{ cm} \quad \text{so} \quad 10 \text{ in.} \times \frac{2.54 \text{ cm}}{1 \text{ in.}} = 25.4 \text{ cm}$$

. . . .

The conversion factors you will need are given in Appendix C. The following examples show you how to use these tables.

EXAMPLE 7

Change 15 miles to kilometres.

From Table 2 in Appendix C, we find 1 mile listed in the left-hand column. Moving over to the fourth column, under the heading "km," we see that 1 mile (mi) = 1.61 km. Then we have

$$15 \text{ mi} \times \frac{1.61 \text{ km}}{1 \text{ mi}} = 24.15 \text{ km}$$

. . . .

EXAMPLE 8

Change 220 centimetres to inches.

Find 1 centimetre in the left-hand column and move to the fifth column under the heading "in." We find that 1 centimetre = 0.394 in. Then

$$220 \text{ cm} \times \frac{0.394 \text{ in.}}{1 \text{ cm}} = 86.68 \text{ in.}$$

. . . .

EXAMPLE 9

Change 3 yards to centimetres.

Since there is no direct conversion from yards to centimetres in the tables, we must first change yards to inches and then inches to centimetres:

$$3 \text{ yd} \times \frac{36 \text{ in.}}{1 \text{ yd}} \times \frac{2.54 \text{ cm}}{1 \text{ in.}} = 274.32 \text{ cm}$$

. . . .

PROBLEMS 1.4

Which unit is longer?

1. 1 metre or 1 centimetre
2. 1 metre or 1 millimetre
3. 1 metre or 1 kilometre
4. 1 centimetre or 1 millimetre
5. 1 centimetre or 1 kilometre
6. 1 millimetre or 1 kilometre

Which metric unit (km, m, cm, or mm) would you use to measure the following?

7. Length of a wrench
8. Thickness of a saw blade
9. Height of a barn
10. Width of a table
11. Thickness of a hypodermic needle
12. Distance around an automobile racing track
13. Distance between New York and Miami
14. Length of a hurdle race
15. Thread size on a pipe
16. Width of a house lot

Fill in each blank with the most reasonable metric unit (km, m, cm, or mm).

17. Your car is about 6 _____ long.
18. Your pencil is about 20 _____ long.
19. The distance between New York and San Francisco is about 4200 _____.
20. Your pencil is about 7 _____ thick.
21. The ceiling in my bedroom is about 240 _____ high.
22. The length of a football field is about 90 _____.
23. A jet plane usually cruises at an altitude of 9 _____.
24. A standard film size for cameras is 35 _____.
25. The diameter of my car tire is about 60 _____.
26. The zipper on my jacket is about 70 _____ long.
27. Juan drives 9 _____ to school each day.
28. Jacob, our basketball center, is 203 _____ tall.
29. The width of your hand is about 80 _____.
30. A handsaw is about 70 _____ long.
31. A newborn baby is usually about 45 _____ long.
32. The standard metric piece of plywood is 120 _____ wide and 240 _____ long.

Fill in each blank.

33. 1 km = _____ m
34. 1 mm = _____ m
35. 1 m = _____ cm
36. 1 m = _____ hm
37. 1 dm = _____ m
38. 1 dam = _____ m
39. 1 m = _____ mm
40. 1 m = _____ dm

41. 1 hm = _____ m
42. 1 cm = _____ m
43. 1 m = _____ km
44. 1 m = _____ dam
45. 1 cm = _____ mm
46. Change 250 m to cm.
47. Change 250 m to km.
48. Change 546 mm to cm.
49. Change 178 km to m.
50. Change 35 dm to dam.
51. Change 830 cm to m.
52. Change 75 hm to km.
53. Change 375 cm to mm.
54. Change 7.5 mm to μm.
55. Change 4 m to μm.
56. State your height in centimetres and in metres.
57. The wheelbase of a certain automobile is 108 in. long. Find its length
    (a)   in feet.   (b)   in yards.
58. Change 43,296 ft
    (a)   to miles.   (b)   to yards.
59. Change 6.25 mi
    (a)   to yards.   (b)   to feet.
60. The length of a connecting rod is 7 in. What is its length in centimetres?
61. The distance between two cities is 256 mi. Find this distance in kilometres.
62. Change 5.94 m to feet.
63. Change 7.1 cm to inches.
64. Change 1.2 in. to centimetres.
65. The turning radius of an auto is 20 ft. What is this distance in metres?
66. Would a wrench with an opening of 25 mm be larger or smaller than a 1-in. wrench?
67. How many reamers, each 20 cm long, can be cut from a bar 6 ft long, allowing 3 mm for each saw cut?
68. If 214 pieces each 47 cm long are to be turned from $\frac{1}{4}$-in. round steel stock with $\frac{1}{8}$ in. of waste allowed on each piece, what length (in metres) of stock is required?

Figure 1.7   Actual sizes of 1 cm^2 and 1 in^2

One square centimetre (cm^2)

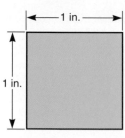

One square inch (in^2)

1.5 Area and Volume

Area

The **area** of a figure is the number of square units that it contains. To measure a surface area of an object, you must first decide on a standard unit of area. Standard units of area are based on the square and are called *square inches*, *square centimetres*, *square miles*, or some other square unit of measure. An area of 1 square centimetre (cm^2) is the amount of area found within a square 1 cm on each side. An area of 1 square inch (in^2) is the amount of area found within a square of 1 in. on each side (Fig. 1.7).

In general, when multiplying measurements of like units, multiply the numbers and then multiply the units as follows:

$$2 \text{ cm} \times 4 \text{ cm} = (2 \times 4)(\text{cm} \times \text{cm}) = 8 \text{ cm}^2$$
$$3 \text{ in.} \times 5 \text{ in.} = (3 \times 5)(\text{in.} \times \text{in.}) = 15 \text{ in}^2$$
$$1.4 \text{ m} \times 6.7 \text{ m} = (1.4 \times 6.7)(\text{m} \times \text{m}) = 9.38 \text{ m}^2$$

Metric Area

The basic unit of area in the metric system is the *square metre* (m^2), the area in a square whose sides are 1 m long (Fig. 1.8). The square centimetre (cm^2) and the square millimetre (mm^2) are smaller units of area. Larger units of area are the square kilometre (km^2) and the hectare (ha).

Figure 1.8

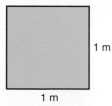

One square metre (m^2)

EXAMPLE 1

Find the area of a rectangle 5 m long and 3 m wide.

Each square in Fig. 1.9 represents 1 m^2. By simply counting the number of squares (square metres), we find that the area of the rectangle is 15 m^2. We can also find the area of the rectangle by using the formula

$$A = lw = (5 \text{ m})(3 \text{ m}) = 15 \text{ m}^2 \qquad (\textit{Note:} \ \text{m} \times \text{m} = \text{m}^2)$$

Figure 1.9

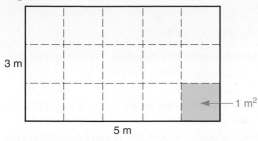

Find the area of the metal plate shown in Fig. 1.10.

EXAMPLE 2

Figure 1.10

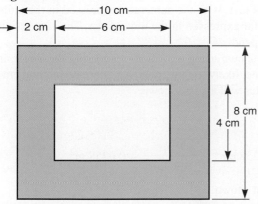

To find the area of the metal plate, find the area of each of the two rectangles and then find the difference of their areas. The large rectangle is 10 cm long and 8 cm wide. The small rectangle is 6 cm long and 4 cm wide. Thus,

area of large rectangle: $A = lw = (10 \text{ cm})(8 \text{ cm}) = 80 \text{ cm}^2$

area of small rectangle: $A = lw = (6 \text{ cm})(4 \text{ cm}) = \underline{24 \text{ cm}^2}$

area of metal plate: 56 cm^2

The surface that would be seen by cutting a geometric solid with a thin plate parallel to one side of the solid represents the *cross-sectional area* of the solid.

Find the smallest cross-sectional area of the box shown in Fig. 1.11(a).

EXAMPLE 3

Figure 1.11

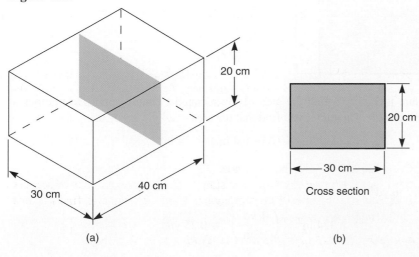

(a)   (b)

Cross section

The indicated cross section of this box is a rectangle 30 cm long and 20 cm wide [Fig. 1.11(b)]. Thus

$$A = lw = (30 \text{ cm})(20 \text{ cm}) = 600 \text{ cm}^2$$

The area of this rectangle is 600 cm², which represents the cross-sectional area of the box.

....

The formulas for finding the areas of other plane figures are found on the inside back cover.

The *hectare* is the fundamental SI unit for land area. An area of 1 hectare equals the area of a square 100 m on a side (Fig. 1.12). The hectare is used because it is more convenient to say and use than square hectometre. The metric prefixes are *not* used with the hectare unit. That is, instead of saying the prefix "kilo" with "hectare," we say "1000 hectares."

To convert area or square units, use a conversion factor. That is, the correct conversion factor will be in fractional form and equal to 1, with the numerator expressed in the units you wish to convert to and the denominator expressed in the units given. The conversion table for area is provided as Table 3 of Appendix C.

The conversion of area units will be shown using a method of squaring the linear or length conversion factor which you are most likely to remember. An alternate method emphasizing direct use of the conversion tables will also be shown.

Figure 1.12 One hectare

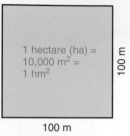

1 hectare (ha) = 10,000 m² = 1 hm²

100 m

100 m

EXAMPLE 4

Change 258 cm² to m².

$$258 \text{ cm}^2 \times \left(\frac{1 \text{ m}}{100 \text{ cm}} \right)^2 = 258 \text{ cm}^2 \times \frac{1^2 \text{ m}^2}{100^2 \text{ cm}^2} = 0.0258 \text{ m}^2$$

Note: The intermediate step is usually not shown.

Alternate Method:

$$258 \text{ cm}^2 \times \frac{1 \text{ m}^2}{10,000 \text{ cm}^2} = 0.0258 \text{ m}^2$$

....

U.S. Area

EXAMPLE 5

Find the area of a rectangle that is 6 in. long and 4 in. wide (Fig. 1.13).

Figure 1.13

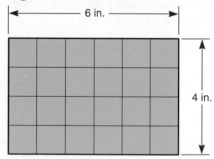

6 in.

4 in.

Each square is 1 in². To find the area of the rectangle, simply count the number of squares in the rectangle. Therefore, you find that the area = 24 in², or, by using the formula,

$$A = lw = (6 \text{ in.})(4 \text{ in.}) = 24 \text{ in}^2$$

....

EXAMPLE 6

Change 324 in² to yd².

$$324 \text{ in}^2 \times \left(\frac{1 \text{ yd}}{36 \text{ in.}} \right)^2 = 0.25 \text{ yd}^2$$

Alternate Method:

$$324 \text{ in}^2 \times \frac{1 \text{ yd}^2}{1296 \text{ in}^2} = \frac{324}{1296} \text{ yd}^2 = 0.25 \text{ yd}^2$$

....

Metric–U.S. Area Conversions

Change 25 cm² to in². ◀

EXAMPLE 7

$$25 \text{ cm}^2 \times \left(\frac{1 \text{ in.}}{2.54 \text{ cm}}\right)^2 = 3.875 \text{ in}^2$$

Alternate Method:

$$25 \text{ cm}^2 \times \frac{0.155 \text{ in}^2}{1 \text{ cm}^2} = 3.875 \text{ in}^2$$

....

Conversion factors found in tables are usually rounded. There are many rounding procedures in general use. We will use one of the simplest methods, stated as follows:

ROUNDING NUMBERS

To round a number to a particular place value:

1. If the digit in the next place to the right is less than 5, drop that digit and all other following digits. Replace any whole number places dropped with zeros.
2. If the digit in the next place to the right is 5 or greater, add 1 to the digit in the place to which you are rounding. Drop all other following digits. Replace any whole number places dropped with zeros.

Change 28.5 m² to in². ◀

EXAMPLE 8

$$28.5 \text{ m}^2 \times \left(\frac{39.4 \text{ in.}}{1 \text{ m}}\right)^2 = 44{,}242.26 \text{ in}^2$$

Alternate Method:

Figure 1.14

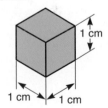

One cubic centimetre (cm³)

$$28.5 \text{ m}^2 \times \frac{1550 \text{ in}^2}{1 \text{ m}^2} = 44{,}175 \text{ in}^2$$

Note: The choice of rounded conversion factors will often lead to results that differ slightly. When checking your answers, you must allow for such rounding differences.

....

To convert between metric and U.S. land area units, use the relationship

$$1 \text{ hectare} = 2.47 \text{ acres}$$

Volume

The **volume** of a figure is the number of cubic units that it contains. Standard units of volume are based on the cube and are called *cubic centimetres, cubic inches, cubic yards,* or some other cubic unit of measure. A volume of 1 cubic centimetre (cm³) is the same as the amount of volume contained in a cube 1 cm on each side. One cubic inch (in³) is the volume contained in a cube 1 in. on each side (Fig. 1.14).

Note: When multiplying measurements of like units, multiply the numbers and then multiply the units as follows:

$$3 \text{ in.} \times 5 \text{ in.} \times 4 \text{ in.} = (3 \times 5 \times 4)(\text{in.} \times \text{in.} \times \text{in.}) = 60 \text{ in}^3$$
$$2 \text{ cm} \times 4 \text{ cm} \times 1 \text{ cm} = (2 \times 4 \times 1)(\text{cm} \times \text{cm} \times \text{cm}) = 8 \text{ cm}^3$$
$$1.5 \text{ ft} \times 8.7 \text{ ft} \times 6 \text{ ft} = (1.5 \times 8.7 \times 6)(\text{ft} \times \text{ft} \times \text{ft}) = 78.3 \text{ ft}^3$$

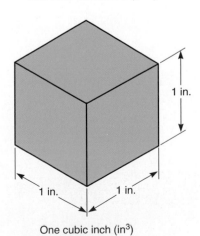

One cubic inch (in³)

EXAMPLE 9

Figure 1.15

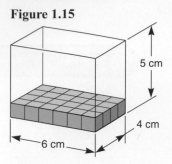

Metric Volume

Find the volume of a rectangular prism 6 cm long, 4 cm wide, and 5 cm high.

Each cube shown in Fig. 1.15 is 1 cm³. To find the volume of the rectangular solid, count the number of cubes in the bottom layer of the rectangular solid and then multiply that number by the number of layers that the solid can hold. Therefore, there are 5 layers of 24 cubes, which is 120 cubes or 120 cubic centimetres.

Or, by formula, $V = Bh$, where B is the area of the base and h is the height. However, the area of the base is found by lw, where l is the length and w is the width of the rectangle. Therefore, the volume of a rectangular solid can be found by the formula

$$V = lwh = (6 \text{ cm})(4 \text{ cm})(5 \text{ cm}) = 120 \text{ cm}^3$$

Note: cm × cm × cm = cm³.

· · · · · · · · · · · · · · · · · ·

A common unit of volume in the metric system is the *litre* (L) (Fig. 1.16). The litre is commonly used for liquid volumes.

Figure 1.16   One litre of milk is a little more than 1 quart of milk.

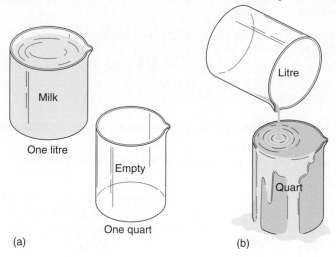

The cubic metre (m³) is used to measure large volumes. The cubic metre is the volume in a cube 1 m on an edge. For example, the usual teacher's desk could be boxed into 2 cubic metres side by side.

The relationship between the litre and the cubic centimetre deserves special mention. The litre is defined as the volume in 1 cubic decimetre (dm³). That is, 1 litre of liquid fills a cube 1 dm (10 cm) on an edge (Fig. 1.17). The volume of this cube can be found by using the formula

$$V = lwh = (10 \text{ cm})(10 \text{ cm})(10 \text{ cm}) = 1000 \text{ cm}^3$$

That is,

$$1 \text{ L} = 1000 \text{ cm}^3$$

Then

$$\frac{1}{1000} \text{ L} = 1 \text{ cm}^3$$

Figure 1.17   One litre contains 1000 cm³.

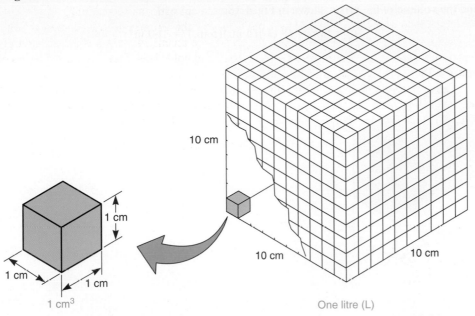

One litre (L)

But

$$\frac{1}{1000}\,L = 1\ mL$$

Therefore,

$$1\ mL = 1\ cm^3$$

Milk, soda, and gasoline are usually sold by the litre in countries using the metric system. Liquid medicine, vanilla extract, and lighter fluid are usually sold by the millilitre. Many metric cooking recipes are given in millilitres. Very large quantities of oil are sold by the kilolitre (1000 L).

Change 0.75 L to millilitres.   **◄ EXAMPLE 10**

$$0.75\ \cancel{L} \times \frac{1000\ mL}{1\ \cancel{L}} = 750\ mL$$

· · · · · · · · · · · · · · · · ·

Similarly, the conversion of volume cubic units will be shown using a method of cubing the linear, or length, conversion factor that you are most likely to remember. An alternate method emphasizing direct use of the conversion tables will also be shown.

Change 0.65 cm³ to cubic millimetres.   **◄ EXAMPLE 11**

$$0.65\ cm^3 \times \left(\frac{10\ mm}{1\ cm}\right)^3 = 0.65\ \cancel{cm^3} \times \frac{10^3\ mm^3}{1^3\ \cancel{cm^3}} = 650\ mm^3$$

Note: The intermediate step is usually not shown.

Alternate Method:

$$0.65\ \cancel{cm^3} \times \frac{1000\ mm^3}{1\ \cancel{cm^3}} = 650\ mm^3$$

· · · · · · · · · · · · · · · ·

EXAMPLE 12

U.S. Volume
Find the volume of the prism shown in Fig. 1.18.

$$V = lwh = (8 \text{ in.})(4 \text{ in.})(5 \text{ in.}) = 160 \text{ in}^3$$

Figure 1.18

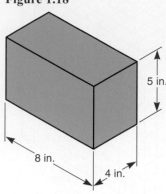

5 in.

8 in.

4 in.

EXAMPLE 13

Change 24 ft³ to in³.

$$24 \text{ ft}^3 \times \left(\frac{12 \text{ in.}}{1 \text{ ft}}\right)^3 = 41{,}472 \text{ in}^3$$

Alternate Method:

$$24 \text{ ft}^3 \times \frac{1728 \text{ in}^3}{1 \text{ ft}^3} = 41{,}472 \text{ in}^3$$

EXAMPLE 14

Metric–U.S. Volume Conversions
Change 56 in³ to cm³.

$$56 \text{ in}^3 \times \left(\frac{2.54 \text{ cm}}{1 \text{ in.}}\right)^3 = 917.68 \text{ cm}^3$$

Alternate Method:

$$56 \text{ in}^3 \times \frac{16.4 \text{ cm}^3}{1 \text{ in}^3} = 918.4 \text{ cm}^3$$

EXAMPLE 15

Change 28 m³ to ft³.

$$28 \text{ m}^3 \times \left(\frac{3.28 \text{ ft}}{1 \text{ m}}\right)^3 = 988.1 \text{ ft}^3$$

Alternate Method:

$$28 \text{ m}^3 \times \frac{35.3 \text{ ft}^3}{1 \text{ m}^3} = 988.4 \text{ ft}^3$$

Surface Area
The **lateral** (side) **surface area** of any geometric solid is the area of all the lateral faces. The **total surface area** of any geometric solid is the lateral surface area plus the area of the bases.

Find the lateral surface area of the prism shown in Fig. 1.19. ◄ **EXAMPLE 16**

$$\text{area of lateral face 1} = (6 \text{ in.})(5 \text{ in.}) = 30 \text{ in}^2$$
$$\text{area of lateral face 2} = (5 \text{ in.})(4 \text{ in.}) = 20 \text{ in}^2$$
$$\text{area of lateral face 3} = (6 \text{ in.})(5 \text{ in.}) = 30 \text{ in}^2$$
$$\text{area of lateral face 4} = (5 \text{ in.})(4 \text{ in.}) = \underline{20 \text{ in}^2}$$
$$\text{lateral surface area} = 100 \text{ in}^2$$

Figure 1.19

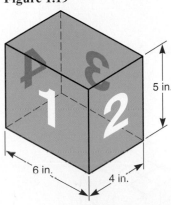

Find the total surface area of the prism shown in Fig. 1.19. ◄ **EXAMPLE 17**

$$\text{total surface area} = \text{lateral surface area} + \text{area of the bases}$$
$$\text{area of base} = (6 \text{ in.})(4 \text{ in.}) = 24 \text{ in}^2$$
$$\text{area of both bases} = 2(24 \text{ in}^2) = 48 \text{ in}^2$$
$$\text{total surface area} = 100 \text{ in}^2 + 48 \text{ in}^2 = 148 \text{ in}^2$$

Area formulas, volume formulas, and lateral surface area formulas are provided on the inside back cover.

PROBLEMS 1.5

Find the area of each figure.

1.

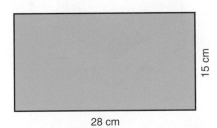

8 cm

2.

15 cm

28 cm

3.

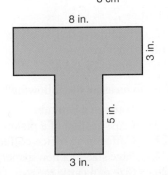

8 in.

3 in.

5 in.

3 in.

4.

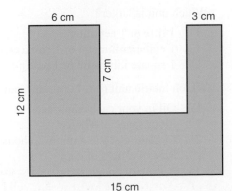

6 cm   3 cm

12 cm

7 cm

15 cm

5. Find the cross-sectional area of the I-beam.

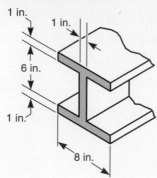

6. Find the largest cross-sectional area of the figure.

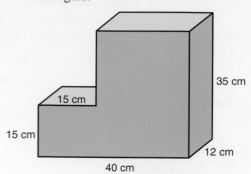

Find the volume in each figure.

7.

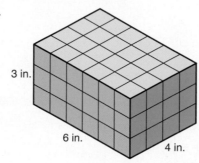

8.

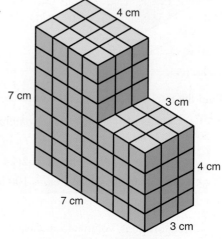

9.

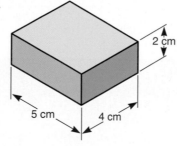

10.

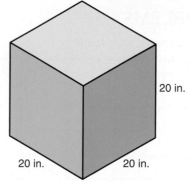

Which unit is larger?

11. 1 litre or 1 centilitre
12. 1 millilitre or 1 kilolitre
13. 1 cubic millimetre or 1 cubic centimetre
14. 1 cm^3 or 1 m^3
15. 1 square kilometre or 1 hectare
16. 1 mm^2 or 1 dm^2

Which metric unit (m^3, L, mL, m^2, cm^2, ha) would you use to measure the following?

17. Oil in your car's crankcase
18. Water in a bathtub
19. Floor space in a house
20. Cross section of a piston
21. Storage space in a miniwarehouse
22. Coffee in an office coffeepot
23. Size of a field of corn
24. Page size of a newspaper
25. A dose of cough syrup
26. Size of a cattle ranch

27. Cargo space in a truck
28. Gasoline in your car's gas tank
29. Piston displacement of an engine
30. Paint needed to paint a house
31. Dose of eye drops
32. Size of a plot of timber

Fill in the blank with the most reasonable metric unit (m^3, L, mL, m^2, cm^2, ha).

33. Go to the store and buy 4 _____ of root beer for the party.
34. I drank 200 _____ of orange juice for breakfast.
35. Craig bought a 30-_____ tarpaulin for his truck.
36. The cross section of a log is 3200 _____.
37. A farmer's gasoline storage tank holds 4000 _____.
38. Our city water tower holds 500 _____ of water.
39. Brian planted 60 _____ of soybeans this year.
40. David needs some copper tubing with a cross section of 3 _____.
41. Paula ordered 15 _____ of concrete for her new driveway.
42. Barbara heats 420 _____ of living space in her house.
43. Joyce's house has 210 _____ of floor space.
44. Kurt mows 5 _____ of grass each week.
45. Amy is told by her doctor to drink 2 _____ of water each day.
46. My favorite coffee cup holds 225 _____ of coffee.

Fill in each blank.

47. 1 L = _____ mL
48. 1 kL = _____ L
49. 1 L = _____ daL
50. 1 L = _____ kL
51. 1 L = _____ hL
52. 1 L = _____ dL
53. 1 daL = _____ L
54. 1 mL = _____ L
55. 1 mL = _____ cm^3
56. 1 L = _____ cm^3
57. 1 m^3 = _____ cm^3
58. 1 cm^3 = _____ mL
59. 1 cm^3 = _____ L
60. 1 dm^3 = _____ L
61. 1 m^2 = _____ cm^2
62. 1 km^2 = _____ m^2
63. 1 cm^2 = _____ mm^2
64. 1 mm^2 = _____ m^2
65. 1 dm^2 = _____ m^2
66. 1 ha = _____ m^2
67. 1 km^2 = _____ ha
68. 1 ha = _____ km^2
69. Change 7500 mL to L.
70. Change 0.85 L to mL.
71. Change 1.6 L to mL.
72. Change 9 mL to L.
73. Change 275 cm^3 to mm^3.
74. Change 5 m^3 to cm^3.
75. Change 4 m^3 to mm^3.
76. Change 520 mm^3 to cm^3.
77. Change 275 cm^3 to mL.
78. Change 125 cm^3 to L.
79. Change 1 m^3 to L.
80. Change 150 mm^3 to L.
81. Change 7.5 L to cm^3.
82. Change 450 L to m^3.
83. Change 5000 mm^2 to cm^2.
84. Change 1.75 km^2 to m^2.
85. Change 5 m^2 to cm^2.
86. Change 250 cm^2 to mm^2.
87. Change 4×10^8 m^2 to km^2.
88. Change 5×10^7 cm^2 to m^2.
89. Change 5 yd^2 to ft^2.
90. How many m^2 are in 225 ft^2?
91. Change 15 ft^2 to cm^2.
92. How many ft^2 are in a rectangle 15 m long and 12 m wide?
93. Change 108 in^2 to ft^2.
94. How many in^2 are in 51 cm^2?
95. How many in^2 are in a square 11 yd on a side?
96. How many m^2 are in a doorway whose area is 20 ft^2?
97. Change 19 yd^3 to ft^3.
98. How many in^3 are in 29 cm^3?
99. How many yd^3 are in 23 m^3?
100. How many cm^3 are in 88 in^3?
101. Change 8 ft^3 to in^3.
102. How many in^3 are in 12 m^3?
103. The volume of a casting is 38 in^3. What is its volume in cm^3?
104. How many castings of 14 cm^3 can be made from a 12-ft^3 block of steel?
105. Find the lateral surface area of the figure in Problem 9.
106. Find the lateral surface area of the figure in Problem 10.
107. Find the total surface area of the figure in Problem 9.
108. Find the total surface area of the figure in Problem 10.
109. How many mL of water would the figure in Problem 9 hold?
110. How many mL of water would the figure in Problem 8 hold?

1.6 Other Units

Mass and Weight

The **mass** of an object is the quantity of material making up the object. One unit of mass in the metric system is the *gram* (g) (Fig. 1.20). The gram is defined as the mass of 1 cubic centimetre (cm^3) of water at its maximum density. Since the gram is so small, the **kilogram** (kg) is the basic unit of mass in the metric system. One kilogram is defined as the mass of 1 cubic decimetre (dm^3) of water at its maximum density. The standard kilogram is a special platinum–iridium cylinder at the International Bureau of Weights and Measures near Paris, France. Since 1 dm^3 = 1 L, 1 litre of water has a mass of 1 kilogram.

Figure 1.20

(a) A common paper clip has a mass of about 1 g.

(b) Three aspirin have a mass of about 1 g.

For very, very small masses, such as medicine dosages, we use the *milligram* (mg). One grain of salt has a mass of about 1 mg. The *metric ton* (1000 kg) is used to measure the mass of very large quantities, such as the coal on a barge, a trainload of grain, or a shipload of ore.

EXAMPLE 1

Change 74 kg to grams.

Choose the conversion factor with kilograms in the denominator so that the kg units cancel each other.

$$74 \text{ kg} \times \frac{1000 \text{ g}}{1 \text{ kg}} = 74,000 \text{ g}$$

. . . .

EXAMPLE 2

Change 600 mg to grams.

$$600 \text{ mg} \times \frac{1 \text{ g}}{1000 \text{ mg}} = 0.6 \text{ g}$$

. . . .

Figure 1.21 Spring balance

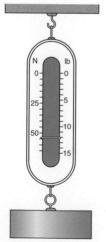

The **weight** of an object is a measure of the gravitational force or pull acting on it. The weight unit in the metric system is the *newton* (N). A small apple weighs about one newton.

The *pound* (lb), a unit of force, is one of the basic U.S. system units. It is defined as the pull of the earth on a cylinder of a platinum–iridium alloy that is stored in a vault at the U.S. Bureau of Standards. The *ounce* (oz) is another common unit of weight in the U.S. system. The relationship between pounds and ounces is

$$1 \text{ lb} = 16 \text{ oz}$$

The following relationships can be used for conversion between systems of units:

$$1 \text{ N} = 0.225 \text{ lb} \quad \text{or} \quad 1 \text{ lb} = 4.45 \text{ N}$$

The mass of an object remains constant, but its weight changes according to its distance from the earth or another planet. Mass and weight and their units of measure are discussed in more detail in Section 5.5.

A **spring balance** (Fig. 1.21) is an instrument containing a spring with a pointer attached to it. The spring stretches in proportion to the weight of the item being weighed. The weight is shown on a calibrated scale read directly in pounds or newtons. The common bathroom scale uses this principle to measure weight.

A **platform balance** (Fig. 1.22) consists of two platforms connected by a horizontal rod that balances on a knife edge. This device compares the pull of gravity on objects that are on the two platforms. The platforms are at the same height only when the unknown mass of the object on the left is equal to the known mass placed on the right. It is also possible to use one platform and a mass that slides along a calibrated scale. Variations of this basic design are found in some meat market and truck scales.

Figure 1.22   Platform balance

Photo courtesy of Dorling Kindersley

EXAMPLE 3

The weight of the intake valve of an auto engine is 0.18 lb. What is its weight in ounces and in newtons?

To find the weight in ounces, we simply use a conversion factor as follows:

$$0.18 \text{ lb} \times \frac{16 \text{ oz}}{1 \text{ lb}} = 2.88 \text{ oz}$$

To find the weight in newtons, we again use a conversion factor:

$$0.18 \text{ lb} \times \frac{4.45 \text{ N}}{1 \text{ lb}} = 0.801 \text{ N}$$

. . . .

Time

Airlines and other transportation systems run on time schedules that would be meaningless if we did not have a common unit for time measurement. All the common units for time measurement are the same in both systems. These units are based on the motion of the earth (Fig. 1.23). The year is the amount of time required for one complete revolution of the earth about the sun. The month is the amount of time for one complete revolution of the moon about the earth. The day is the amount of time for one rotation of the earth about its axis.

Figure 1.23   Time units are based on the motion of the earth.

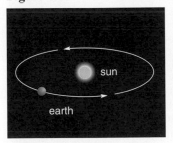 

(a) One year is the amount of time it takes for the earth to revolve about the sun.

(b) One day is the amount of time it takes for the earth to rotate about its axis.

The basic time unit is the **second** (s). For many years, the second was defined as $\frac{1}{86,400}$ of a mean solar day. The standard second adopted in 1967 is defined more precisely in terms of the frequency of radiation emitted by cesium atoms when they pass between two particular states; this is the time required for 9,192,631,770 periods of this radiation. The second is not always convenient to use, so other units are necessary. The *minute* (min) is 60 seconds, the *hour* (h) is 60 minutes, and the *day* is 24 hours. The *year* is 365 days in length except for every fourth year, when it is 366 days long. This difference is necessary to keep the seasons at the same time each year, since one revolution of the earth about the sun takes approximately $365\frac{1}{4}$ days.

The Julian calendar introduced by Julius Caesar in 46 BC provided for an ordinary year of 365 days and a leap year of 366 days every fourth year. Astronomers have found that the length of a year has varied from 365.24253 days in 5000 BC to 365.24219 days in 2000 AD. The Gregorian calendar now used in most countries of the world was introduced by Pope Gregory XIII in 1582 to correct the Julian calendar discrepancies. The Gregorian calendar provides for an ordinary year of 365 days and a leap year of 366 days in years divisible by four except in century years not divisible by 400. Thus, years 1600 and 2000 are leap years but years 1700, 1800, 1900, and 2100 are not. In 1582, 10 days were omitted from the calendar to adjust for the accumulated difference of the Julian calendar since 46 BC. By decree, the day following October 4, 1582, became October 15, 1582. Now, small fractions-of-a-second adjustments are made in the calendar annually by international agreement to compensate for the variation of the earth's orbit around the sun.

Common devices for time measurement are the electric clock, the mechanical watch, and the quartz crystal watch. The accuracy of an electric clock depends on how accurately the 60-Hz (hertz = cycles per second) line voltage is controlled. In the United States this is controlled very accurately. Most mechanical watches have a balance wheel that oscillates near a given frequency, usually 18,000 to 36,000 vibrations per hour, and drives the hands of the watch (Fig. 1.24). The quartz crystal in a watch is excited by a small power cell and vibrates 32,768 times per second. The accuracy of the watch depends on how well the frequency of oscillation is controlled.

Figure 1.24 Oscillation of balance wheel

EXAMPLE 4

Change 2 h 15 min to seconds.

First,

$$2 \, \text{h} \times \frac{60 \, \text{min}}{1 \, \text{h}} = 120 \, \text{min}$$

Then

$$2 \, \text{h} \, 15 \, \text{min} = 120 \, \text{min} + 15 \, \text{min} = 135 \, \text{min}$$

and

$$135 \, \text{min} \times \frac{60 \, \text{s}}{1 \, \text{min}} = 8100 \, \text{s}$$

Very short periods of time are measured in parts of a second, given with the appropriate metric prefix. Such units are commonly used in electronics.

EXAMPLE 5

What is the meaning of each unit?

(a) 1 ms = 1 millisecond = 10^{-3} s and means one one-thousandth of a second.
(b) 1 μs = 1 microsecond = 10^{-6} s and means one one-millionth of a second.
(c) 1 ns = 1 nanosecond = 10^{-9} s and means one one-billionth of a second.
(d) 1 ps = 1 picosecond = 10^{-12} s and means one one-trillionth of a second.

Note: The Greek letter μ is pronounced "my$\overline{oo}$." However, 1 μs is stated or read as "one microsecond."

EXAMPLE 6

Change 45 ms to seconds.

Since 1 ms = 10^{-3} s,

$$45 \, \text{ms} \times \frac{10^{-3} \, \text{s}}{1 \, \text{ms}} = 45 \times 10^{-3} \, \text{s} = 0.045 \, \text{s}$$

EXAMPLE 7

Change 0.000000025 s to nanoseconds.

Since 1 ns $= 10^{-9}$ s,

$$0.000000025 \cancel{s} \times \frac{1 \text{ ns}}{10^{-9} \cancel{s}} = 25 \text{ ns}$$

PROBLEMS 1.6

Which unit is larger?

1. 1 gram or 1 centigram
2. 1 gram or 1 milligram
3. 1 gram or 1 kilogram
4. 1 centigram or 1 milligram
5. 1 centigram or 1 kilogram
6. 1 milligram or 1 kilogram

Which metric unit (kg, g, mg, or metric ton) would you use to measure the following?

7. Your mass
8. An aspirin
9. A bag of lawn fertilizer
10. A bar of hand soap
11. A trainload of grain
12. A sewing needle
13. A small can of corn
14. A channel catfish
15. A vitamin capsule
16. A car

Fill in each blank with the most reasonable metric unit (kg, g, mg, or metric ton).

17. A newborn's mass is about 3 _____.
18. An elevator in a local department store has a load limit of 2000 _____.
19. Margie's diet calls for 250 _____ of meat.
20. A 200-car train carries 11,000 _____ of soybeans.
21. A truckload shipment of copper pipe has a mass of 900 _____.
22. A carrot has a mass of 75 _____.
23. A candy recipe calls for 175 _____ of chocolate.
24. My father has a mass of 70 _____.
25. A pencil has a mass of 10 _____.
26. Postage rates for letters would be based on the _____.
27. A heavyweight boxing champion has a mass of 93 _____.
28. A nickel has a mass of 5 _____.
29. My favorite spaghetti recipe calls for 1 _____ of ground beef.
30. My favorite spaghetti recipe calls for 150 _____ of tomato paste.
31. Our local grain elevator shipped 10,000 _____ of wheat last year.
32. A slice of bread has a mass of about 25 _____.
33. I bought a 5-_____ bag of potatoes at the store today.
34. My grandmother takes 250-_____ capsules for her arthritis.

Fill in each blank.

35. 1 kg = _____ g
36. 1 mg = _____ g
37. 1 g = _____ cg
38. 1 g = _____ hg
39. 1 dg = _____ g
40. 1 dag = _____ g
41. 1 g = _____ mg
42. 1 g = _____ dg
43. 1 hg = _____ g
44. 1 cg = _____ g
45. 1 g = _____ kg
46. 1 g = _____ dag
47. 1 g = _____ μg
48. 1 mg = _____ μg
49. Change 575 g to mg.
50. Change 575 g to kg.
51. Change 650 mg to g.
52. Change 375 kg to g.
53. Change 50 dg to g.
54. Change 485 dag to dg.
55. Change 30 kg to mg.
56. Change 4 metric tons to kg.
57. Change 25 hg to kg.
58. Change 58 μg to g.
59. Change 400 μg to mg.
60. Change 30,000 kg to metric tons.
61. What is the mass of 750 mL of water?
62. What is the mass of 1 m^3 of water?
63. The weight of a car is 3500 lb. Find its weight in newtons.

64. A certain bridge is designed to support 150,000 lb. Find the maximum weight that it will support in newtons.
65. Jose weighs 200 lb. What is his weight in newtons?
66. Change 80 lb to newtons.   67. Change 2000 N to pounds.
68. Change 2000 lb to newtons.   69. Change 120 oz to pounds.
70. Change 3.5 lb to ounces.   71. Change 10 N to ounces.
72. Change 25 oz to newtons.
73. Find the metric weight of a 94-lb bag of cement.
74. What is the weight in newtons of 500 blocks if each weighs 3 lb?

Fill in each blank.

75. The basic metric unit of time is _____. Its abbreviation is _____.
76. The basic metric unit of mass is _____. Its abbreviation is _____.
77. The common metric unit of weight is _____. Its abbreviation is _____.

Which is larger?

78. 1 second or 1 millisecond   79. 1 millisecond or 1 nanosecond
80. 1 ps or 1 μs   81. 1 ms or 1 μs

Write the abbreviation for each unit.

82. 8.6 microseconds   83. 45 nanoseconds   84. 75 picoseconds
85. Change 345 μs to s.   86. Change 1 h 25 min to min.
87. Change 4 h 25 min 15 s to s.   88. Change 7×10^6 s to h.
89. Change 4 s to ns.   90. Change 1 h to ps.

1.7 Measurement: Significant Digits and Accuracy

Up to this time in your studies, probably all numbers and all measurements have been treated as exact numbers. An **exact number** is a number that has been determined as a result of counting, such as 24 students are enrolled in this class, or by some definition, such as 1 h = 60 min or 1 in. = 2.54 cm, a conversion definition agreed to by the world governments' bureaus of standards. Generally, the treatment of the addition, subtraction, multiplication, and division of exact numbers is the emphasis or main content of elementary mathematics.

However, nearly all data of a technical nature involve **approximate numbers**; that is, numbers determined as a result of some measurement process—some direct, as with a ruler, and some indirect, as with a surveying transit or reading an electric meter. First, realize that no measurement can be found exactly. The length of the cover of this book can be found using many instruments. The better the measuring device used, the better is the measurement.

A measurement may be expressed in terms of its accuracy or its precision. The **accuracy** of a measurement refers to the number of digits, called **significant digits**, which indicates the number of units that we are reasonably sure of having counted when making a measurement. The greater the number of significant digits given in a measurement, the better is the accuracy and vice versa.

EXAMPLE 1

The average distance between the moon and the earth is 385,000 km. This measurement indicates measuring 385 thousands of kilometres; its accuracy is indicated by three significant digits.

· · · · · · · · · · · · · · · · · ·

EXAMPLE 2

A measurement of 0.025 cm indicates measuring 25 thousandths of a centimetre; its accuracy is indicated by two significant digits.

· · · · · · · · · · · · · · · · · ·

EXAMPLE 3

A measurement of 0.0500 s indicates measuring 50̄0 ten-thousandths of a second; its accuracy is indicated by three significant digits.

Notice that sometimes a zero is significant and sometimes it is not. To clarify this, we use the following rules for significant digits:

SIGNIFICANT DIGITS

1. All nonzero digits are significant: 156.4 m has four significant digits (this measurement indicates 1564 tenths of metres).
2. All zeros between significant digits are significant: 306.02 km has five significant digits (this measurement indicates 30,602 hundredths of kilometres).
3. In a number greater than 1, a zero that is specially tagged, such as by a bar above it, is significant: 23̄0,000 km has three significant digits (this measurement indicates 23̄0 thousands of kilometres).
4. All zeros to the right of a significant digit *and* a decimal point are significant: 86.10 cm has four significant digits (this measurement indicates 861̄0 hundredths of centimetres).
5. In whole-number measurements, zeros at the right that are not tagged are *not* significant: 2500 m has two significant digits (25 hundreds of metres).
6. In measurements of less than 1, zeros at the left are *not* significant: 0.00752 m has three significant digits (752 hundred-thousandths of a metre).

When a number is written in scientific notation, the decimal part indicates the number of significant digits. For example, 20̄0,000 m would be written in scientific notation as 2.00×10^5 m.

In summary:

To find the number of significant digits:

1. All nonzero digits are significant.
2. Zeros are significant when they
   (a) are between significant digits;
   (b) follow the decimal point and a significant digit; or
   (c) are in a whole number and a bar is placed over the zero.

EXAMPLE 4

Determine the accuracy (the number of significant digits) of each measurement.

Measurement	Accuracy (significant digits)
(a) 2642 ft	4
(b) 2005 m	4 (Both zeros are significant.)
(c) 2050 m	3 (Only the first zero is significant.)
(d) 2500 m	2 (No zero is significant.)
(e) 25̄00 m	3 (Only the first zero is significant.)
(f) 250̄0 m	4 (Both zeros are significant.)
(g) 34,000 mi	2 (No zeros are significant.)
(h) 15,670,000 lb	4 (No zeros are significant.)
(i) 203.05 km	5 (Both zeros are significant.)
(j) 0.000345 kg	3 (No zeros are significant.)

(continued)

Measurement	Accuracy (significant digits)
(k) 75 N	2
(l) 2.3 s	2
(m) 0.02700 g	4 (Only the right two zeros are significant.)
(n) 2.40 cm	3 (The zero is significant.)
(o) 4.050 μs	4 (All zeros are significant.)
(p) 100.050 km	6 (All zeros are significant.)
(q) 0.004 s	1 (No zeros are significant.)
(r) 2.03×10^4 m^2	3 (The zero is significant.)
(s) 1.0×10^{-3} N	2 (The zero is significant.)
(t) 5×10^6 kg	1
(u) 3.060×10^8 m^3	4 (Both zeros are significant.)

· · · · · · · · · · · · · · · · · · ·

PROBLEMS 1.7

Determine the accuracy (the number of significant digits) of each measurement.

1.	536 ft	2.	307.3 mi	3.	5007 m
4.	5.00 cm	5.	0.0070 in.	6.	6.010 cm
7.	84$\overline{0}$0 km	8.	30$\overline{0}$0 ft	9.	187.40 m
10.	5$\overline{0}$0 g	11.	0.00700 in.	12.	10.30 cm
13.	376.52 m	14.	3.05 mi	15.	4087 kg
16.	35.00 mm	17.	0.0160 in.	18.	37$\overline{0}$ lb
19.	4$\overline{0}$00 N	20.	5010 ft^3	21.	7 N
22.	32,000 tons	23.	70.00 m^2	24.	0.007 m
25.	2.4×10^3 kg	26.	1.20×10^{-5} ms	27.	3.00×10^{-4} kg
28.	4.0×10^6 ft	29.	5.106×10^7 kg	30.	1×10^{-9} m

1.8 Measurement: Precision

The **precision** of a measurement refers to the smallest unit with which a measurement is made, that is, the position of the last significant digit.

EXAMPLE 1

The precision of the measurement 385,000 km is 1000 km. (The position of the last significant digit is in the thousands place.)

· · · · · · · · · · · · · · · ·

EXAMPLE 2

The precision of the measurement 0.025 cm is 0.001 cm. (The position of the last significant digit is in the thousandths place.)

· · · · · · · · · · · · · · · ·

EXAMPLE 3

The precision of the measurement 0.0500 s is 0.0001 s. (The position of the last significant digit is in the ten-thousandths place.)

· · · · · · · · · · · · · · · ·

Unfortunately, the terms *accuracy* and *precision* have several different common meanings. Here we will use each term consistently as we have defined them. A measurement of

0.0004 cm has good precision and poor accuracy when compared with the measurement 378.0 cm.

Measurement	Precision	Accuracy
0.0004 cm	0.0001 cm	1 significant digit
378.0 cm	0.1 cm	4 significant digits

Determine the precision of each measurement given in Example 4 of Section 1.7. ◄   **EXAMPLE 4**

Measurement	Precision	Accuracy (significant digits)
(a) 2642 ft	1 ft	4
(b) 2005 m	1 m	4
(c) 2050 m	10 m	3
(d) 2500 m	100 m	2
(e) 2500 m	10 m	3
(f) 2500 m	1 m	4
(g) 34,000 mi	1000 mi	2
(h) 15,670,000 lb	10,000 lb	4
(i) 203.05 km	0.01 km	5
(j) 0.000345 kg	0.000001 kg	3
(k) 75 N	1 N	2
(l) 2.3 s	0.1 s	2
(m) 0.02700 g	0.00001 g	4
(n) 2.40 cm	0.01 cm	3
(o) 4.050 μs	0.001 μs	4
(p) 100.050 km	0.001 km	6
(q) 0.004 s	0.001 s	1
(r) 2.03×10^4 m^2	0.01×10^4 m^2 or 100 m^2	3
(s) 1.0×10^{-3} N	0.1×10^{-3} N or 0.0001 N	2
(t) 5×10^6 kg	1×10^6 kg or 1,000,000 kg	1
(u) 3.060×10^8 m^3	0.001×10^8 m^3 or 1×10^5 m^3 or 100,000 m^3	4

PHYSICS CONNECTIONS

Precision and the New Clark Bridge

Bridges are usually not built from one end to the other. Construction typically begins in the middle of the bridge or on each end and meets in the middle. In doing so, it becomes extremely important that every section is in precise alignment so that the bridge will meet at the critical connection points.

For example, the New Clark Bridge crossing the Mississippi at Alton, Illinois, spans over 4600 ft and was designed to high precision so that each member of the bridge would be no more than $\frac{1}{8}$ in. out of alignment. As frames and towers were built in the flowing river, teams of surveyors used fixed points of reference and laser beams to survey the placement of each tower and pier. Ignoring the importance of accuracy and precision would have caused serious problems.

TRY THIS ACTIVITY

Accuracy and Precision

Measure the time it takes someone to run the 100-yard dash. Use a digital stopwatch and a regular wrist-watch with a second hand. How accurate is each of the measurements? How precise is each of the measurements?

PROBLEMS 1.8

Determine the precision of each measurement.

1. 536 ft
2. 307.3 mi
3. 5007 m
4. 5.00 cm
5. 0.0070 in.
6. 6.010 cm
7. 84$\overline{0}$0 km
8. 30$\overline{0}$0 ft
9. 187.40 m
10. 5$\overline{0}$0 g
11. 0.00700 in.
12. 10.30 cm
13. 376.52 m
14. 3.05 mi
15. 4087 kg
16. 35.00 mm
17. 0.0160 in.
18. 37$\overline{0}$ lb
19. 40$\overline{0}$0 N
20. 5010 ft^3
21. 7 N
22. 32,000 tons
23. 70.00 m^2
24. 0.007 m
25. 2.4×10^3 kg
26. 1.20×10^{-5} ms
27. 3.00×10^{-4} kg
28. 4.0×10^6 ft
29. 5.106×10^7 kg
30. 1×10^{-9} m

In each set of the measurements, find the measurement that is (a) the most accurate and (b) the most precise.

31. 15.7 in.; 0.018 in.; 0.07 in.
32. 368 ft; 600 ft; 180 ft
33. 0.734 cm; 0.65 cm; 16.01 cm
34. 3.85 m; 8.90 m; 7.00 m
35. 0.035$\overline{0}$ s; 0.025 s; 0.00040 s; 0.051 s
36. 125.00 g; 8.50 g; 9.000 g; 0.05 g
37. 27,00$\overline{0}$ L; 350 L; 27.6 L; 4.75 L
38. 8.4 m; 15 m; 180 m; 0.40 m
39. 500 N; 10,000 N; 500,000 N; 50 N
40. 7.5 ms; 14.2 ms; 10.5 ms; 120.0 ms

In each set of measurements, find the measurement that is (a) the least accurate and (b) the least precise.

41. 16.4 in.; 0.075 in.; 0.05 in.
42. 475 ft; 300 ft; 360 ft
43. 27.5 m; 0.65 m; 12.02 m
44. 5.7 kg; 120 kg; 0.025 kg
45. 0.0250 g; 0.015 g; 0.00005 g; 0.75 g
46. 185.0 m; 6.75 m; 5.000 m; 0.09 m
47. 45,$\overline{0}$00 N; 250 N; 16.8 N; 0.25 N; 3 N
48. 2.50 kg; 42.0 kg; 15$\overline{0}$ kg; 0.500 kg
49. 200$\overline{0}$ kg; 10,$\overline{0}$00 kg; 400,$\overline{0}$00 kg; 20 kg
50. 80 ft; 250 ft; 12,550 ft; 260$\overline{0}$ ft

Figure 1.25 Micrometer with precision 0.01 mm

1.9 Calculations with Measurements

If one person measures the length of one of two parts of a shaft with a micrometer calibrated in 0.01 mm as 42.28 mm and another person measures the second part with a ruler calibrated in mm as 54 mm, would the total length be 96.28 mm? Note that the sum 96.28 mm indicates a precision of 0.01 mm. The precision of the ruler is 1 mm, which means that the measurement 54 mm with the ruler could actually be anywhere between 53.50 mm and 54.50 mm using the micrometer (which has a precision of 0.01 mm and is shown in Fig. 1.25). That is, using the ruler, any measurement between 53.50 mm and 54.50 mm can only be read as 54 mm. Of course, this means that the tenths and hundredths digits in the sum 96.28 mm are really meaningless. In other words, *the sum or difference of measurements can be no more precise than the least precise measurement.* That is,

To add or subtract measurements:

1. Make certain that all the measurements are expressed in the same unit. If they are not, convert them all to the same unit.
2. Add or subtract.
3. Round the results to the same precision as the least precise measurement.

Add the measurements 16.6 mi, 124 mi, 3.05 mi, and 0.837 mi.

EXAMPLE 1

All measurements are in the same unit, so add,

$$
\begin{array}{r}
16.6 \ \text{mi} \\
124 \ \ \ \text{mi} \\
3.05 \ \text{mi} \\
\underline{0.837 \ \text{mi}} \\
144.487 \ \text{mi} \rightarrow 144 \ \text{mi}
\end{array}
$$

Then, round this sum to the same precision as the least precise measurement, which is 124 mi. Thus, the sum is 144 mi.

· · · · · · · · · · · · · · · · · ·

Add the measurements 1370 cm, 1575 mm, 2.374 m, and 8.63 m.

EXAMPLE 2

First, convert all measurements to the same unit, say m.

$$
\begin{aligned}
1370 \ \text{cm} &= 13.7 \ \text{m} \\
1575 \ \text{mm} &= 1.575 \ \text{m}
\end{aligned}
$$

Then add,

$$
\begin{array}{r}
13.7 \ \ \ \text{m} \\
1.575 \ \text{m} \\
2.374 \ \text{m} \\
\underline{8.63 \ \ \ \text{m}} \\
26.279 \ \text{m} \rightarrow 26.3 \ \text{m}
\end{array}
$$

Then, round this sum to the same precision as the least precise measurement, which is 13.7 m. Thus, the sum is 26.3 m.

· · · · · · · · · · · · · · · · · ·

Subtract the measurements 3457.8 g − 2.80 kg.

EXAMPLE 3

First, convert both measurements to the same unit, say g.

$$
2.80 \ \text{kg} = 28\overline{0}0 \ \text{g}
$$

Then subtract.

$$
\begin{array}{r}
3457.8 \ \text{g} \\
\underline{28\overline{0}0 \ \ \ \ \text{g}} \\
657.8 \ \text{g} \rightarrow 660 \ \text{g}
\end{array}
$$

Then, round this difference to the same precision as the least precise measurement, which is $28\overline{0}0$ g. Thus, the difference is 660 g.

· · · · · · · · · · · · · · · · · ·

Now suppose that you wish to find the area of the base of a rectangular building. You measure its length as 54.7 m and its width as 21.5 m. Its area is then

$$
\begin{aligned}
A &= lw \\
A &= (54.7 \ \text{m})(21.5 \ \text{m}) \\
&= 1176.05 \ \text{m}^2
\end{aligned}
$$

Note that the result contains six significant digits, whereas each of the original measurements contains only three significant digits. To rectify this inconsistency, we say that the product or quotient of measurements can be no more accurate than the least accurate measurement. That is,

To multiply or divide measurements:

1. Multiply or divide the measurements as given.
2. Round the result to the same number of significant digits as the measurement with the least number of significant digits.

Using the preceding rules, we find that the area of the base of the rectangular building is 1180 m^2.

Note: We assume throughout that you are using a calculator to do all calculations.

EXAMPLE 4

Multiply the measurements (124 ft)(187 ft).

$$(124 \text{ ft})(187 \text{ ft}) = 23{,}188 \text{ ft}^2$$

Round this product to three significant digits, which is the accuracy of the least accurate measurement (and also the accuracy of each measurement in the example). That is,

$$(124 \text{ ft})(187 \text{ ft}) = 23{,}200 \text{ ft}^2$$

EXAMPLE 5

Multiply the measurements (2.75 m)(1.25 m)(0.75 m).

$$(2.75 \text{ m})(1.25 \text{ m})(0.75 \text{ m}) = 2.578125 \text{ m}^3$$

Round this product to two significant digits, which is the accuracy of the least accurate measurement (0.75 m). That is,

$$(2.75 \text{ m})(1.25 \text{ m})(0.75 \text{ m}) = 2.6 \text{ m}^3$$

EXAMPLE 6

Divide the measurements 144,000 ft^3 ÷ 108 ft.

$$144{,}000 \text{ ft}^3 \div 108 \text{ ft} = 1333.333\ldots \text{ ft}^2$$

Round this quotient to three significant digits, which is the accuracy of the least accurate measurement (the accuracy of both measurements in this example). That is,

$$144{,}000 \text{ ft}^3 \div 108 \text{ ft} = 1330 \text{ ft}^2$$

EXAMPLE 7

Find the value of $\dfrac{(68 \text{ ft})(10{,}\overline{0}00 \text{ lb})}{95.6 \text{ s}}$

$$\frac{(68 \text{ ft})(10{,}\overline{0}00 \text{ lb})}{95.6 \text{ s}} = 7112.9707\ldots \frac{\text{ft lb}}{\text{s}}$$

Round this result to two significant digits, which is the accuracy of the least accurate measurement (68 ft). That is,

$$\frac{(68 \text{ ft})(10{,}\overline{0}00 \text{ lb})}{95.6 \text{ s}} = 7100 \text{ ft lb/s}$$

Find the value of $\dfrac{(58.0\ \text{kg})(2.40\ \text{m/s})^2}{5.40\ \text{m}}$.

EXAMPLE 8

$$\frac{(58.0\ \text{kg})(2.40\ \text{m/s})^2}{5.40\ \text{m}} = 61.8666\ldots\frac{\text{kg m}}{\text{s}^2}$$

Carefully simplify the units:

$$\frac{(\text{kg})(\text{m/s})^2}{\text{m}} = \frac{(\text{kg})(\text{m}^2/\text{s}^2)}{\text{m}} = \frac{\text{kg m}}{\text{s}^2}$$

Round this result to three significant digits, which is the accuracy of the least accurate measurement (the accuracy of all measurements in this example). That is,

$$\frac{(58.0\ \text{kg})(2.40\ \text{m/s})^2}{5.40\ \text{m}} = 61.9\ \text{kg m/s}^2$$

· · · · · · · · · · · · · · · · · ·

Note: To multiply or divide measurements, the units do not need to be the same. The units must be the same to add or subtract measurements. Also, the units are multiplied and/or divided in the same manner as the corresponding numbers.

Any power or root of a measurement should be rounded to the same accuracy as the given measurement.

COMBINATIONS OF OPERATIONS WITH MEASUREMENTS

For combinations of additions, subtractions, multiplications, divisions, and powers involving measurements, follow the usual order of operations used in mathematics:

1. Perform all operations inside parentheses first.
2. Evaluate all powers.
3. Perform any multiplications or divisions, in order, from left to right; then express each product or quotient using its correct accuracy.
4. Perform any additions or subtractions, in order, from left to right; then express the final result using the correct precision.

Find the value of $(4.00\ \text{m})(12.65\ \text{m}) + (24.6\ \text{m})^2 + \dfrac{235.0\ \text{m}^3}{16.00\ \text{m}}$.

EXAMPLE 9

$$(4.00\ \text{m})(12.65\ \text{m}) + (24.6\ \text{m})^2 + \frac{235.0\ \text{m}^3}{16.00\ \text{m}} =$$

$$50.6\ \text{m}^2 + 605\ \text{m}^2 \quad + 14.69\ \text{m}^2 = 67\overline{0}\ \text{m}^2$$

· · · · · · · · · · · · · · · · · ·

Obviously, such calculations with measurements should be done with a calculator. When no calculator is available, you may round the original measurements or any intermediate results to one more digit than the required accuracy or precision as required in the final result.

If both exact numbers and approximate numbers (measurements) occur in the same calculation, only the approximate numbers are used to determine the accuracy or precision of the result.

The procedures for operations with measurements shown here are based on methods followed and presented by the American Society for Testing and Materials. There are even more sophisticated methods for dealing with the calculations of measurements. The method one uses, and indeed whether one should even follow any given procedure, depends on the number of measurements and the sophistication needed for a particular situation.

In this book, we generally follow the customary practice of expressing measurements in terms of three significant digits, which is the accuracy used in most engineering and design work.

PROBLEMS 1.9

Use the rules for addition of measurements to add each set of measurements.

1.	3847 ft	2.	8,560 m	3.	42.8	cm	4.	0.456 g
	5800 ft		84,000 m		16.48	cm		0.93 g
	4520 ft		18,476 m		1.497	cm		0.402 g
			12,500 m		12.8	cm		0.079 g
					9.69	cm		0.964 g

5. 39,000 N; 19,600 N; 8470 N; 2500 N
6. 6800 ft; 2760 ft; 4$\overline{0}$00 ft; 20$\overline{0}$0 ft
7. 467 m; 970 cm; 12$\overline{0}$0 cm; 1352 cm; 30$\overline{0}$ m
8. 36.8 m; 147.5 cm; 1.967 m; 125.0 m; 98.3 cm
9. 12 s; 1.004 s; 0.040 s; 3.9 s; 0.87 s
10. 160,000 N; 84,200 N; 4300 N; 239,000 N; 17,450 N

Use the rules for subtraction of measurements to subtract each second measurement from the first.

11.	2876 kg	12.	14.73 m	13.	45.585 g	14.	34,500 kg
	2400 kg		9.378 m		4.6 g		9,5$\overline{0}$0 kg

15. 4200 km − 975 km
16. 64.73 g − 9.4936 g
17. 1,600,000 kg − 685,000 kg
18. 170 mm − 10.2 cm
19. 3.00 m − 26$\overline{0}$ cm
20. 1.40 ms − 0.708 ms

Use the rules for multiplication of measurements to multiply each set of measurements.

21. (125 m)(39 m)
22. (470 ft)(1200 ft)
23. (1637 km)(857 km)
24. (9100 m)(6$\overline{0}$0 m)
25. (18.70 m)(39.45 m)
26. (565 cm)(180 cm)
27. (14.$\overline{5}$ cm)(18.7 cm)(20.5 cm)
28. (0.046 m)(0.0317 m)(0.0437 m)
29. (45$\overline{0}$ in.)(315 in.)(205 in.)
30. (18.7 kg)(217 m)

Use the rules for division of measurements to divide.

31. $360 \text{ ft}^3 \div 12 \text{ ft}^2$
32. $125 \text{ m}^2 \div 3.0 \text{ m}$
33. $275 \text{ cm}^2 \div 90.0 \text{ cm}$

34. $185 \text{ mi} \div 4.5 \text{ h}$
35. $\dfrac{347 \text{ km}}{4.6 \text{ h}}$
36. $\dfrac{2700 \text{ m}^3}{9\overline{0}0 \text{ m}^2}$

37. $\dfrac{8800 \text{ mi}}{8.5 \text{ h}}$
38. $\dfrac{4960 \text{ ft}}{2.95 \text{ s}}$

Use the rules for multiplication and division of measurements to find the value of each of the following.

39. $\dfrac{(18 \text{ ft})(290 \text{ lb})}{4.6 \text{ s}}$
40. $\dfrac{(18.5 \text{ kg})(4.65 \text{ m})}{19.5 \text{ s}}$

41. $\dfrac{4500 \text{ mi}}{12.3 \text{ h}}$
42. $\dfrac{48.9 \text{ kg}}{(1.5 \text{ m})(3.25 \text{ m})}$

43. $\dfrac{(48.7 \text{ m})(68.5 \text{ m})(18.4 \text{ m})}{(35.5 \text{ m})(40.0 \text{ m})}$
44. $\frac{1}{2}(270 \text{ kg})(16.4 \text{ m/s})^2$

45. $\dfrac{(85.7 \text{ kg})(25.7 \text{ m/s})^2}{12.5 \text{ m}}$
46. $\dfrac{(45.2 \text{ kg})(13.7 \text{ m})}{(2.65 \text{ s})^2}$

47. $\frac{4}{3}\pi(13.5 \text{ m})^3$
48. $\dfrac{140 \text{ g}}{(3.4 \text{ cm})(2.8 \text{ cm})(5.6 \text{ cm})}$

49. $(213 \text{ m})(65.3 \text{ m}) - (175 \text{ m})(44.5 \text{ m})$
50. $(4.5 \text{ ft})(7.2 \text{ ft})(12.4 \text{ ft}) + (5.42 \text{ ft})^3$

51. $\dfrac{(125 \text{ ft})(295 \text{ ft})}{44.7 \text{ ft}} + \dfrac{(215 \text{ ft})^3}{(68.8 \text{ ft})(12.4 \text{ ft})} + \dfrac{(454 \text{ ft})^3}{(75.5 \text{ ft})^2}$

52. $(12.5 \text{ m})(46.75 \text{ m}) + \dfrac{(6.76 \text{ m})^3}{4910 \text{ m}} - \dfrac{(41.5 \text{ m})(21 \text{ m})(28.8 \text{ m})}{31.7 \text{ m}}$

Glossary

Accuracy The number of digits, called significant digits, in a measurement, which indicates the number of units that we are reasonably sure of having counted. The greater the number of significant digits, the better is the accuracy. (p. 38)

Approximate Number A number that has been determined by some measurement or estimation process. (p. 38)

Area The number of square units contained in a figure. (p. 24)

Conversion Factor An expression used to convert from one set of units to another. Often expressed as a fraction whose numerator and denominator are equal to each other although in different units. (p. 20)

Exact Number A number that has been determined as a result of counting, such as 21 students enrolled in a class, or by some definition, such as 1 h = 60 min. (p. 38)

Kilogram The basic metric unit of mass. (p. 34)

Lateral Surface Area The area of all the lateral (side) faces of a geometric solid. (p. 30)

Mass A measure of the quantity of material making up an object. (p. 34)

Metre The basic metric unit of length. (p. 20)

Platform Balance An instrument consisting of two platforms connected by a horizontal rod that balances on a knife edge. The pull of gravity on objects placed on the two platforms is compared. (p. 35)

Precision Refers to the smallest unit with which a measurement is made, that is, the position of the last significant digit. (p. 40)

Scientific Notation A form in which a number can be written as a product of a number between 1 and 10 and a power of 10. The general form is $M \times 10^n$, where M is a number between 1 and 10 and n is the exponent or power of 10. (p. 17)

Second The basic unit of time. (p. 35)

SI (Système International d'Unités) The international modern metric system of units of measurement. (p. 15)

Significant Digits The number of digits in a measurement, which indicates the number of units we are reasonably sure of having counted. (p. 38)

Spring Balance An instrument containing a spring, which stretches in proportion to the force applied to it, and a pointer attached to the spring with a calibrated scale read directly in given units. (p. 34)

Standards of Measure A set of units of measurement for length, weight, and other quantities defined in such a way as to be useful to a large number of people. (p. 13)

Total Surface Area The total area of all the surfaces of a geometric solid; that is, the lateral surface area plus the area of the bases. (p. 30)

Volume The number of cubic units contained in a figure. (p. 27)

Weight A measure of the gravitational force or pull acting on an object. (p. 34)

Review Questions

1. What are the basic metric units for length, mass, and time?
   - (a) Foot, pound, hour
   - (b) Newton, litre, second
   - (c) Metre, kilogram, second
   - (d) Mile, ton, day
2. When a value is multiplied or divided by 1, the value is
   - (a) increased.
   - (b) unchanged.
   - (c) decreased.
   - (d) none of the above.
3. The lateral surface area of a solid is
   - (a) always equal to total surface area.
   - (b) never equal to total surface area.
   - (c) usually equal to total surface area.
   - (d) rarely equal to total surface area.
4. Accuracy is
   - (a) the same as precision.
   - (b) the smallest unit with which a measurement is made.

      (c)   the number of significant digits.

      (d)   all of the above.

5.   When multiplying or dividing two or more measurements, the units

      (a)   must be the same.   (b)   must be different.   (c)   can be different.

6.   Cite three examples of problems that would arise in the construction of a home by workers using different systems of measurement.

7.   Why is the metric system preferred worldwide to the U.S. system of measurement?

8.   List a very large and a very small measurement that could be usefully written in scientific notation.

9.   When using conversion factors, can units be treated like other algebraic quantities?

10.   What is the meaning of cross-sectional area?

11.   Can a brick have more than one cross-sectional area?

12.   What is the fundamental metric unit for land area?

13.   Which is larger, a litre or a quart?

14.   List three things that might conveniently be measured in millilitres.

15.   How do weight and mass differ?

16.   What is the basic metric unit of weight?

17.   A microsecond is one-_____ of a second.

18.   Why must we concern ourselves with significant digits?

19.   Can the sum or difference of two measurements ever be more precise than the least precise measurement?

20.   When rounding the product or quotient of two measurements, is it necessary to consider significant digits?

Review Problems

Give the metric prefix for each value:

1.   1000   2.   0.001

Give the metric symbol, or abbreviation, for each prefix:

3.   micro   4.   mega

Write the abbreviation for each quantity:

5.   45 milligrams   6.   138 centimetres

Which is larger?

7.   1 L or 1 mL   8.   1 kg or 1 mg   9.   1 L or 1 m^3

Fill in each blank (round to three significant digits when necessary):

10.   250 m = _____ km   11.   850 mL = _____ L

12.   5.4 kg = _____ g   13.   0.55 s = _____ μs

14.   25 kg = _____ g   15.   75 μs = _____ ns

16.   275 cm^2 = _____ mm^2 17.   350 cm^2 = _____ m^2

18.   0.15 m^3 = _____ cm^3 19.   500 cm^3 = _____ mL

20.   150 lb = _____ kg   21.   36 ft = _____ m

22.   250 cm = _____ in.   23.   150 in^2 = _____ cm^2

24.   24 yd^2 = _____ ft^2 25.   6 m^3 = _____ ft^3

26.   16 lb = _____ N   27.   15,600 s = _____ h _____ min

Determine the accuracy (the number of significant digits) in each measurement:

28.   5.08 kg   29.   20,570 lb   30.   0.060 cm   31. 2.00×10^{-4} s

Determine the precision of each measurement:

32.   30.6 ft   33.   0.0500 s   34.   18,000 mi   35. 4×10^5 N

For each set of measurements, find the measurement that is

(a) the most accurate. (b) the least accurate.
(c) the most precise. (d) the least precise.

36. 12.00 m; 0.150 m; 2600 m; 0.008 m
37. 208 L; 18,050 L; 21.5 L; 0.75 L

Use the rules of measurements to add the following measurements:

38. 0.0250 s; 0.075 s; 0.00080 s; 0.024 s
39. 2100 N; 36,800 N; 24,000 N; 14.5 N; 470 N

Use the rules for multiplication and division of measurements to find the value of each of the following:

40. (450 cm)(18.5 cm)(215 cm)   41. $\dfrac{1480 \text{ m}^3}{9.6 \text{ m}}$ 42. $\dfrac{(25.0 \text{ kg})(1.20 \text{ m/s})^2}{3.70 \text{ m}}$

43. Find the area of a rectangle 4.50 m long and 2.20 m wide.
44. Find the volume of a rectangular box 9.0 cm long, 6.0 cm wide, and 13 cm high.

CHAPTER 2

PROBLEM SOLVING

A formula is an equation, usually expressed in letters, called *variables,* and numbers. Much technical work includes the substitution of measured data into known formulas or relationships to find solutions to problems. A systematic approach to solving problems is a valuable tool.

The problem-solving method presented will assist you in processing data, analyzing the problems present, and finding the solution in an orderly manner.

Objectives

The major goals of this chapter are to enable you to:

1. Use formulas in problem solving.
2. Use a systematic approach to solving technical problems.
3. Analyze technical problems using a problem-solving method.

2.1 Formulas

A **formula** is an equation, usually expressed in letters (called *variables*) and numbers. A **variable** is a symbol, usually a letter, used to represent some unknown number or quantity.

The formula $s = vt$ states that the distance traveled, s, equals the product of the velocity, v, and the time, t.

EXAMPLE 1

· · · · · · · · · · · · · · · ·

The formula $I = \dfrac{Q}{t}$ states that the current, I, equals the quotient of the charge, Q, and the time, t.

EXAMPLE 2

· · · · · · · · · · · · · · · ·

To solve a formula for a given letter means to express the given letter or variable in terms of all the remaining letters. That is, by using the equation-solving principles, rewrite the formula so that the given letter appears on one side of the equation by itself and all the other letters appear on the other side.

Solve $s = vt$ for v.

EXAMPLE 3

$$s = vt$$

$$\frac{s}{t} = \frac{vt}{t} \qquad \text{Divide both sides by } t.$$

$$\frac{s}{t} = v$$

· · · · · · · · · · · · · · · ·

Solve $I = Q/t$

EXAMPLE 4

(a)　for Q.　　(b)　for t.

(a)

$$I = \frac{Q}{t}$$

$$(I)t = \left(\frac{Q}{t}\right)t \qquad \text{Multiply both sides by } t.$$

$$It = Q$$

(b)   Starting with $It = Q$, we obtain

$$\frac{It}{I} = \frac{Q}{I} \qquad \text{Divide both sides by } I.$$

$$t = \frac{Q}{I}$$

· · · · · · · · · · · · · · · · · ·

EXAMPLE 5

Solve $V = E - Ir$ for r.

Method 1:

$$V = E - Ir$$

$$V - E = E - Ir - E \qquad \text{Subtract } E \text{ from both sides.}$$

$$V - E = -Ir$$

$$\frac{V - E}{-I} = \frac{-Ir}{-I} \qquad \text{Dvide both sides by } -I.$$

$$\frac{V - E}{-I} = r$$

Method 2:

$$V = E - Ir$$

$$V + Ir = E - Ir + Ir \qquad \text{Add } Ir \text{ to both sides.}$$

$$V + Ir = E$$

$$V + Ir - V = E - V \qquad \text{Subtract } V \text{ from both sides.}$$

$$Ir = E - V$$

$$\frac{Ir}{I} = \frac{E - V}{I} \qquad \text{Divide both sides by } I.$$

$$r = \frac{E - V}{I}$$

Note that the two results are equivalent. Take the first result and multiply both numerator and denominator by -1. That is,

$$\frac{V - E}{-I} = \left(\frac{V - E}{-I}\right)\left(\frac{-1}{-1}\right) = \frac{-V + E}{I} = \frac{E - V}{I}$$

which is the same as the second result.

· · · · · · · · · · · · · · · · · ·

     We often use the same quantity in more than one way in a formula. For example, we may wish to use a certain measurement of a quantity, such as velocity, at a given time, say at $t = 0$ s, then use the velocity at a later time, say at $t = 6$ s. To write these desired values of the velocity is rather awkward. We simplify this written statement by using *subscripts* (small letters or numbers printed a half space below the printed line and to the right of the variable) to shorten what we must write.

     For the example given, v at time $t = 0$ s will be written as v_i (initial velocity); v at time $t = 6$ s will be written as v_f (final velocity). Mathematically, v_i and v_f are two different quantities, which in most cases are unequal. The sum of v_i and v_f is written as $v_i + v_f$. The product of v_i and v_f is written as $v_i v_f$. The subscript notation is used only to distinguish the general quantity, v, velocity, from the measure of that quantity at certain specified times.

EXAMPLE 6

Solve the formula $x = x_i + v_i t + \frac{1}{2}at^2$ for v_i.

Method 1:

$$x = x_i + v_i t + \tfrac{1}{2}at^2$$

$$x - v_i t = x_i + v_i t + \tfrac{1}{2}at^2 - v_i t \qquad \text{Subtract } v_i t \text{ from both sides.}$$

$$x - v_i t = x_i + \tfrac{1}{2}at^2$$

$$x - v_i t - x = x_i + \tfrac{1}{2}at^2 - x \qquad \text{Subtract } x \text{ from both sides.}$$

$$-v_i t = x_i + \tfrac{1}{2}at^2 - x$$

$$\frac{-v_i t}{-t} = \frac{x_i + \tfrac{1}{2}at^2 - x}{-t} \qquad \text{Divide both sides by } -t.$$

$$v_i = \frac{x_i + \tfrac{1}{2}at^2 - x}{-t}$$

Method 2:

$$x = x_i + v_i t + \tfrac{1}{2}at^2$$

$$x - x_i - \tfrac{1}{2}at^2 = x_i + v_i t + \tfrac{1}{2}at^2 - x_i - \tfrac{1}{2}at^2 \qquad \text{Subtract } x_i \text{ and } \tfrac{1}{2}at^2 \text{ from both sides.}$$

$$x - x_i - \tfrac{1}{2}at^2 = v_i t$$

$$\frac{x - x_i - \tfrac{1}{2}at^2}{t} = \frac{v_i t}{t} \qquad \text{Divide both sides by } t.$$

$$\frac{x - x_i - \tfrac{1}{2}at^2}{t} = v_i$$

· · · · · · · · · · · · · · · · ·

EXAMPLE 7

Solve the formula $v_{\text{avg}} = \frac{1}{2}(v_f + v_i)$ for v_f (avg is used here as a subscript meaning average).

$$v_{\text{avg}} = \tfrac{1}{2}(v_f + v_i)$$

$$2v_{\text{avg}} = v_f + v_i \qquad \text{Multiply both sides by 2.}$$

$$2v_{\text{avg}} - v_i = v_f \qquad \text{Subtract } v_i \text{ from both sides.}$$

· · · · · · · · · · · · · · · · ·

EXAMPLE 8

Solve $A = \dfrac{\pi d^2}{4}$ for d, where d is a diameter.

$$A = \frac{\pi d^2}{4}$$

$$4A = \left(\frac{\pi d^2}{4}\right)(4) \qquad \text{Multiply both sides by 4.}$$

$$4A = \pi d^2$$

$$\frac{4A}{\pi} = \frac{\pi d^2}{\pi} \qquad \text{Divide both sides by } \pi.$$

$$\frac{4A}{\pi} = d^2$$

$$\pm\sqrt{\frac{4A}{\pi}} = d \qquad \text{Take the square root of both sides.}$$

In this case, a negative diameter has no physical meaning, so the result is

$$d = \sqrt{\frac{4A}{\pi}}$$

· · · · · · · · · · · · · · · ·

PROBLEMS 2.1

Solve each formula for the quantity given.

1. $v = \dfrac{s}{t}$ for s

2. $a = \dfrac{v}{t}$ for v

3. $w = mg$ for m

4. $F = ma$ for a

5. $E = IR$ for R

6. $V = lwh$ for w

7. $PE = mgh$ for g

8. $PE = mgh$ for h

9. $v^2 = 2gh$ for h

10. $X_L = 2\pi fL$ for f

11. $P = \dfrac{W}{t}$ for W

12. $p = \dfrac{F}{A}$ for F

13. $P = \dfrac{W}{t}$ for t

14. $p = \dfrac{F}{A}$ for A

15. $KE = \frac{1}{2}mv^2$ for m

16. $KE = \frac{1}{2}mv^2$ for v^2

17. $W = Fs$ for s

18. $v_f = v_i + at$ for a

19. $V = E - Ir$ for I

20. $v_2 = v_1 + at$ for t

21. $R = \dfrac{\pi}{2P}$ for P

22. $R = \dfrac{kL}{d^2}$ for L

23. $F = \frac{9}{5}C + 32$ for C

24. $C = \frac{5}{9}(F - 32)$ for F

25. $X_C = \dfrac{1}{2\pi fC}$ for f

26. $R = \dfrac{\rho L}{A}$ for L

27. $R_T = R_1 + R_2 + R_3 + R_4$ for R_3

28. $Q_1 = P(Q_2 - Q_1)$ for Q_2

29. $\dfrac{I_S}{I_P} = \dfrac{N_P}{N_S}$ for I_P

30. $\dfrac{V_P}{V_S} = \dfrac{N_P}{N_S}$ for N_S

31. $v_{avg} = \frac{1}{2}(v_f + v_i)$ for v_i

32. $2a(s - s_i) = v^2 - v_i^2$ for a

33. $2a(s - s_i) = v^2 - v_i^2$ for s

34. $Ft = m(V_2 - V_1)$ for V_1

35. $Q = \dfrac{I^2 Rt}{J}$ for R

36. $x = x_i + v_i t + \frac{1}{2}at^2$ for x_i

37. $A = \pi r^2$ for r, where r is a radius

38. $V = \pi r^2 h$ for r, where r is a radius

39. $R = \dfrac{kL}{d^2}$ for d, where d is a diameter

40. $V = \frac{1}{3}\pi r^2 h$ for r, where r is a radius

41. $Q = \dfrac{I^2 Rt}{J}$ for I

42. $F = \dfrac{mv^2}{r}$ for v

2.2 Substituting Data into Formulas

An important part of problem solving is substituting the given data into the appropriate formula to find the value of the unknown quantity. Basically, there are two ways of substituting data into formulas to solve for the unknown quantity:

1.   Solve the formula for the unknown quantity and then make the substitution of the data.
2.   Substitute the data into the formula first and then solve for the unknown quantity.

When using a calculator, the first way is more useful. We will be using this way most of the time in this text.

Given the formula $A = bh$, $A = 120$ m², and $b = 15$ m, find h.

EXAMPLE 1

First, solve for h:

$$A = bh$$

$$\frac{A}{b} = \frac{bh}{b} \qquad \text{Divide both sides by } b.$$

$$\frac{A}{b} = h$$

Then substitute the data:

$$h = \frac{A}{b} = \frac{120 \text{ m}^2}{15 \text{ m}} = 8.0 \text{ m}$$

(Remember to follow the rules of measurement discussed in Chapter 1. We use them consistently throughout.)

· · · · · · · · · · · · · · · · ·

Given the formula $P = 2a + 2b$, $P = 824$ cm, and $a = 292$ cm, find b.

EXAMPLE 2

First, solve for b:

$$P = 2a + 2b$$

$$P - 2a = 2a + 2b - 2a \qquad \text{Subtract } 2a \text{ from both sides.}$$

$$P - 2a = 2b$$

$$\frac{P - 2a}{2} = \frac{2b}{2} \qquad \text{Divide both sides by 2.}$$

$$\frac{P - 2a}{2} = b \qquad \left(\text{or } b = \frac{P}{2} - a \right)$$

Then substitute the data:

$$b = \frac{P - 2a}{2} = \frac{824 \text{ cm} - 2(292 \text{ cm})}{2}$$

$$= \frac{824 \text{ cm} - 584 \text{ cm}}{2}$$

$$= \frac{24\overline{0} \text{ cm}}{2} = 12\overline{0} \text{ cm}$$

· · · · · · · · · · · · · · · · ·

Given the formula $A = \left(\dfrac{a + b}{2} \right) h$, $A = 15\overline{0}$ m², $b = 18.0$ m, and $h = 10.0$ m, find a.

EXAMPLE 3

First, solve for a:

$$A = \left(\frac{a + b}{2} \right) h$$

$$2A = \left[\left(\frac{a + b}{2} \right) h \right] (2) \qquad \text{Multiply both sides by 2.}$$

$$2A = (a + b)h$$

$$2A = ah + bh \qquad \text{Remove the parentheses.}$$

$$2A - bh = ah + bh - bh \qquad \text{Subtract } bh \text{ from both sides.}$$

$$2A - bh = ah$$

$$\frac{2A - bh}{h} = \frac{ah}{h} \qquad \text{Divide both sides by } h.$$

$$\frac{2A - bh}{h} = a \qquad \left(\text{or } a = \frac{2A}{h} - b \right)$$

Then substitute the data:

$$a = \frac{2A - bh}{h}$$

$$= \frac{2(15\overline{0} \text{ m}^2) - (18.0 \text{ m})(10.0 \text{ m})}{10.0 \text{ m}}$$

$$= \frac{30\overline{0} \text{ m}^2 - 18\overline{0} \text{ m}^2}{10.0 \text{ m}}$$

$$= \frac{12\overline{0} \text{ m}^2}{10.0 \text{ m}} = 12.0 \text{ m}$$

· · · · · · · · · · · · · · · · · ·

EXAMPLE 4

Given the formula $V = \frac{1}{3}\pi r^2 h$, $V = 64{,}400 \text{ mm}^3$, and $h = 48.0 \text{ mm}$, find r, where r is a radius.

First, solve for r:

$$V = \tfrac{1}{3}\pi r^2 h$$

$$3V = \left(\tfrac{1}{3}\pi r^2 h \right)(3) \qquad \text{Multiply both sides by 3.}$$

$$3V = \pi r^2 h$$

$$\frac{3V}{\pi h} = \frac{\pi r^2 h}{\pi h} \qquad \text{Divide both sides by } \pi h.$$

$$\frac{3V}{\pi h} = r^2$$

$$\pm \sqrt{\frac{3V}{\pi h}} = r \qquad \text{Take the square root of both sides.}$$

In this case, a negative radius has no physical meaning, so the result is

$$r = \sqrt{\frac{3V}{\pi h}}$$

Then substitute the data:

$$r = \sqrt{\frac{3(64{,}400 \text{ mm}^3)}{\pi (48.0 \text{ mm})}}$$

$$= 35.8 \text{ mm}$$

· · · · · · · · · · · · · · · · · ·

PROBLEMS 2.2

For each formula, (a) solve for the indicated letter and then (b) substitute the given data to find the value of the indicated letter. Follow the rules of calculations with measurements.

  Note: In Problems 14 and 16, r is a radius, and in Problem 15, b is the length of the side of a square.

Formula	Data	Find
1. $A = bh$	$b = 14.5$ cm, $h = 11.2$ cm	A
2. $V = lwh$	$l = 16.7$ m, $w = 10.5$ m, $h = 25.2$ m	V
3. $A = bh$	$A = 34.5$ cm^2, $h = 4.60$ cm	b
4. $P = 4b$	$P = 42\overline{0}$ in.	b
5. $P = a + b + c$	$P = 48.5$ cm, $a = 18.2$ cm, $b = 24.3$ cm	c
6. $C = \pi d$	$C = 495$ ft	d
7. $C = 2\pi r$	$C = 68.5$ yd	r
8. $A = \frac{1}{2}bh$	$A = 468$ m^2, $b = 36.0$ m	h
9. $P = 2(a + b)$	$P = 88.7$ km, $a = 11.2$ km	b
10. $V = \pi r^2 h$	$r = 61.0$ m, $h = 125.3$ m	V
11. $V = \pi r^2 h$	$V = 368$ m^3, $r = 4.38$ m	h
12. $A = 2\pi rh$	$A = 51\overline{0}$ cm^2, $r = 14.0$ cm	h
13. $V = Bh$	$V = 2185$ m^3, $h = 14.2$ m	B
14. $A = \pi r^2$	$A = 463.5$ m^2	r
15. $A = b^2$	$A = 465$ in^2	b
16. $V = \frac{1}{3}\pi r^2 h$	$V = 2680$ m^3, $h = 14.7$ m	r
17. $C = 2\pi r$	$r = 19.36$ m	C
18. $V = \frac{4}{3}\pi r^3$	$r = 25.65$ m	V
19. $V = \frac{1}{3}Bh$	$V = 19{,}850$ ft^3, $h = 486.5$ ft	B
20. $A = \left(\dfrac{a + b}{2}\right)h$	$A = 205.2$ m^2, $a = 16.50$ m, $b = 19.50$ m	h

2.3 Problem-Solving Method

Problem solving in technical fields is more than substituting numbers and units into formulas. You must develop skill in taking data, analyzing the problem, and finding the solution in an orderly manner. Understanding the principle involved in solving a problem is more important than blindly substituting into a formula. By following an orderly procedure for problem solving, we develop an approach to problem solving that you can use in your studies and on the job.

The following **problem-solving method** aids in understanding and solving problems and will be applied to all problems in this book where appropriate.

1. **Read the problem carefully.** This might appear obvious, but it is the most important step in solving a problem. As a matter of habit, you should read the problem at least twice.
   (a) The first time you should read the problem straight through from beginning to end. Do not stop to think about setting up an equation or formula. You are only trying to get a general overview of the problem during this first reading.
   (b) Read through a second time slowly and *completely*, beginning to think ahead to the following steps.
2. **Make a sketch.** Some problems may not lend themselves to a sketch. However, make a sketch whenever possible. Many times, seeing a sketch will show if you have forgotten important parts of the problem and may suggest the solution. This is a *very important* part of problem solving and is often overlooked.
3. **Write all given information including units.** This is necessary to have all essential facts in mind before looking for the solution. There are some common phrases that have understood physical meanings. For example, the term *from rest* means the initial velocity equals zero or $v_i = 0$; the term *smooth surface* means assume that no friction is present.
4. **Write the unknown or quantity asked for in the problem.** Many students have difficulty solving problems because they don't know what they are looking for and solve for the wrong quantity.

5. ***Write the basic equation or formula that relates the known and unknown quantities.*** Find the basic formula or equation to use by studying what is given and what you are asked to find. Then look for a formula or equation that relates these quantities. Sometimes, you may need to use more than one equation or formula in a problem.

6. ***Find a working equation by solving the basic equation or formula for the unknown quantity.***

7. ***Substitute the data in the working equation, including the appropriate units.*** It is important that you *carry the units all the way through the problem* as a check that you have solved the problem correctly. For example, if you are asked to find the weight of an object in newtons and the units of your answer work out to be metres, you need to review your solution for the error. (When the unit analysis is not obvious, we will go through it step by step in a box nearby.)

8. ***Perform the indicated operations and work out the solution.*** Although this will be your final written step, you should always ask yourself, "Is my answer reasonable?" Here and on the job you will be dealing with practical problems. A quick estimate will often reveal an error in your calculations.

9. ***Check your answer.*** Ask yourself, "Did I answer the questions?"

To help you recall this procedure, with almost every problem set that follows, you will find Fig. 2.1 as shown here. This figure is not meant to be complete, and is only an outline to assist you in remembering and following the procedure for solving problems. *You should follow this outline in solving all problems in this course.*

This problem-solving method will now be demonstrated in terms of relationships and formulas with which you are probably familiar. The formulas for finding area and volume can be found on the inside back cover.

Figure 2.1

SKETCH

12 cm² | w

4.0 cm

DATA

$A = 12$ cm², $l = 4.0$ cm, $w = ?$

BASIC EQUATION

$A = lw$

WORKING EQUATION

$w = \frac{A}{l}$

SUBSTITUTION

$w = \frac{12 \text{ cm}^2}{4.0 \text{ cm}} = 3.0$ cm

EXAMPLE 1

Find the volume of concrete required to fill a rectangular bridge abutment whose dimensions are 6.00 m $\times$ 3.00 m $\times$ 15.0 m.

Sketch:

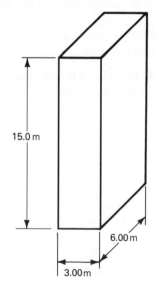

15.0 m

6.00 m

3.00 m

Data:

$\left. \begin{array}{l} l = 6.00 \text{ m} \\ w = 3.00 \text{ m} \\ h = 15.0 \text{ m} \end{array} \right\}$ This is a listing of the information that is known.

$V = ?$ This identifies the unknown.

Basic Equation:

$$V = lwh$$

Working Equation: Same

Substitution:

$$V = (6.00 \text{ m})(3.00 \text{ m})(15.0 \text{ m})$$
$$= 27\overline{0} \text{ m}^3$$

Note: m $\times$ m $\times$ m = m^3

....

A rectangular holding tank 24.0 m in length and 15.0 m in width is used to store water for short periods of time in an industrial plant. If 2880 m^3 of water is pumped into the tank, what is the depth of the water?

EXAMPLE 2

Sketch:

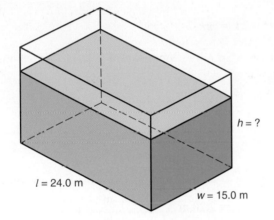

$h = ?$

$l = 24.0$ m

$w = 15.0$ m

Data:

$$V = 2880 \text{ m}^3$$
$$l = 24.0 \text{ m}$$
$$w = 15.0 \text{ m}$$
$$h = ?$$

Basic Equation:

$$V = lwh$$

Working Equation:

$$h = \frac{V}{lw}$$

Substitution:

$$h = \frac{2880 \text{ m}^3}{(24.0 \text{ m})(15.0 \text{ m})}$$

$$= 8.00 \text{ m}$$

$$\boxed{\frac{\text{m}^3}{\text{m} \times \text{m}} = \text{m}}$$

....

EXAMPLE 3

A storage bin in the shape of a cylinder contains 814 m³ of storage space. If its radius is 6.00 m, find its height.

Sketch:

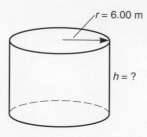

r = 6.00 m

h = ?

Data:

$$V = 814 \text{ m}^3$$
$$r = 6.00 \text{ m}$$
$$h = ?$$

Basic Equation:

$$V = \pi r^2 h$$

Working Equation:

$$h = \frac{V}{\pi r^2}$$

Substitution:

$$h = \frac{814 \text{ m}^3}{\pi (6.00 \text{ m})^2}$$

$$= 7.20 \text{ m} \qquad \boxed{\frac{\text{m}^3}{\text{m}^2} = \text{m}}$$

. . . .

EXAMPLE 4

A rectangular piece of sheet metal measures 45.0 cm by 75.0 cm. A 10.0-cm square is then cut from each corner. The metal is then folded to form a box-like container without a top. Find the volume of the container.

Sketch:

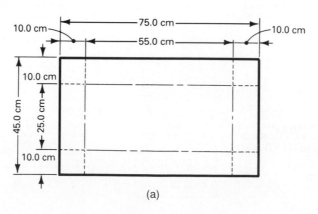

(a)

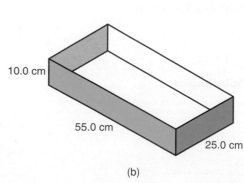

10.0 cm

55.0 cm

25.0 cm

(b)

Data:

$$l = 55.0 \text{ cm}$$
$$w = 25.0 \text{ cm}$$
$$h = 10.0 \text{ cm}$$
$$V = ?$$

Basic Equation:

$$V = lwh$$

Working Equation: Same

Substitution:

$$V = (55.0 \text{ cm})(25.0 \text{ cm})(10.0 \text{ cm})$$
$$= 13,800 \text{ cm}^3$$

$$\boxed{\text{cm} \times \text{cm} \times \text{cm} = \text{cm}^3}$$

· · · · · · · · · · · · · · · ·

The cross-sectional area of a hole is 725 cm². Find its radius. ◀

EXAMPLE 5

Sketch:

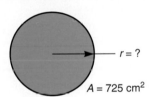

$r = ?$

$A = 725 \text{ cm}^2$

Data:

$$A = 725 \text{ cm}^2$$
$$r = ?$$

Basic Equation:

$$A = \pi r^2$$

Working Equation:

$$r = \sqrt{\frac{A}{\pi}}$$

Substitution:

$$r = \sqrt{\frac{725 \text{ cm}^2}{\pi}}$$
$$= 15.2 \text{ cm}$$

$$\boxed{\sqrt{\text{cm}^2} = \text{cm}}$$

· · · · · · · · · · · · · · · ·

PHYSICS CONNECTIONS

Eratosthenes, a third-century Egyptian, used a problem-solving method to determine that the earth was not flat, but round, a fact that Columbus has been credited with discovering more than 1000 years later. Eratosthenes wondered why it was that at noon of the summer solstice, towers in Syene, Egypt (modern Aswan on the Nile), made no shadows, whereas documentation showed that towers in Alexandria, Egypt, did make distinct shadows. Eratosthenes decided to determine why towers in one city would cast shadows while towers in another city would not.

Eratosthenes sketched the problem, gathered data, and collected geometrical equations to solve this complex problem. He hired a person to pace the distance between the two cities (800 km) and used geometry to solve the problem (Fig. 2.2). After calculating the difference in the positions of the two cities to be approximately 7° out of a 360° sphere, he concluded that the earth's circumference was 40,000 km— remarkably accurate when compared to today's calculations. Eratosthenes had a problem to solve, so he, like all good scientists and problem solvers, followed several steps that included analyzing the problem, collecting data, selecting appropriate equations, and making the calculations.

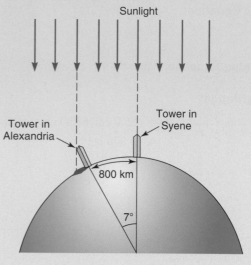

Figure 2.2 Towers in Alexandria and Syene, the shadow cast by the tower in Alexandria, and the curvature of the earth with the angle and the distance between the two cities.

SKETCH

4.0 cm

DATA

$A = 12$ cm^2, $l = 4.0$ cm, $w = ?$

BASIC EQUATION

$A = lw$

WORKING EQUATION

$w = \frac{A}{l}$

SUBSTITUTION

$w = \frac{12 \text{ cm}^2}{4.0 \text{ cm}} = 3.0$ cm

PROBLEMS 2.3

Use the problem-solving method to work each problem. (Here, as throughout the text, follow the rules for calculations with measurements.)

1. Find the volume of the box in Fig. 2.3.
2. Find the volume of a cylinder whose height is 7.50 in. and diameter is 4.20 in. (Fig. 2.4).
3. Find the volume of a cone whose height is 9.30 cm if the radius of the base is 5.40 cm (Fig. 2.5).

Figure 2.3

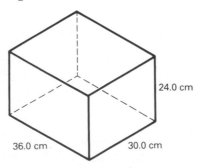

24.0 cm

36.0 cm   30.0 cm

Figure 2.4

4.20 in. diameter

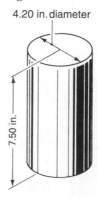

7.50 in.

Figure 2.5

9.30 cm

5.40 cm radius

The cylinder in an engine of a road grader as shown in Fig. 2.6 is 11.40 cm in diameter and 24.00 cm high. Use Fig. 2.6 for Problems 4 through 6.

4. Find the volume of the cylinder.
5. Find the cross-sectional area of the cylinder.
6. Find the lateral surface area of the cylinder.
7. Find the total volume of the building shown in Fig. 2.7.
8. Find the cross-sectional area of the concrete retaining wall shown in Fig. 2.8.

Figure 2.6

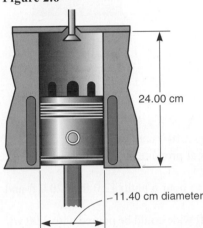

24.00 cm

11.40 cm diameter

Figure 2.7

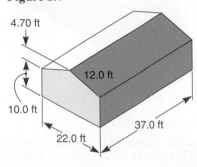

4.70 ft

12.0 ft

10.0 ft

22.0 ft

37.0 ft

Figure 2.8

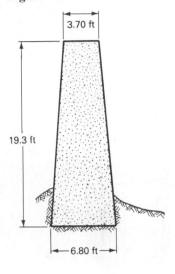

3.70 ft

19.3 ft

6.80 ft

9. Find the volume of a rectangular storage facility 9.00 ft by 12.0 ft by 8.00 ft.
10. Find the cross-sectional area of a piston head with a diameter of 3.25 cm.
11. Find the area of a right triangle that has legs of 4.00 cm and 6.00 cm.
12. Find the length of the hypotenuse of the right triangle in Problem 11.
13. Find the cross-sectional area of a pipe with outer diameter 3.50 cm and inner diameter 3.20 cm.
14. Find the volume of a spherical water tank with radius 8.00 m.
15. The area of a rectangular parking lot is 900 m². If the length is 25.0 m, what is the width?
16. The volume of a rectangular crate is 192 ft³. If the length is 8.00 ft and the width is 4.00 ft, what is the height?
17. Find the volume of a brake cylinder whose diameter is 4.00 cm and whose length is 4.20 cm.
18. Find the volume of a tractor engine cylinder whose radius is 3.90 cm and whose length is 8.00 cm.
19. A cylindrical silo has a circumference of 29.5 m. Find its diameter.
20. If the silo in Problem 19 has a capacity of 1000 m³, what is its height?
21. A wheel 30.0 cm in diameter moving along level ground made 145 complete rotations. How many metres did the wheel travel?
22. The side of the silo in Problems 19 and 20 needs to be painted. If each litre of paint covers 5.0 m², how many litres of paint will be needed? (Round up to the nearest litre.)
23. You are asked to design a cylindrical water tank that holds $500,000$ gal with radius 18.0 ft. Find its height. (1 ft³ = 7.50 gal)
24. If the height of the water tank in Problem 23 were 42.0 ft, what would be its radius?
25. A ceiling is 12.0 ft by 15.0 ft. How many suspension panels 1.00 ft by 3.00 ft are needed to cover the ceiling?
26. Find the cross-sectional area of the dovetail slide shown in Fig. 2.9.

Figure 2.9

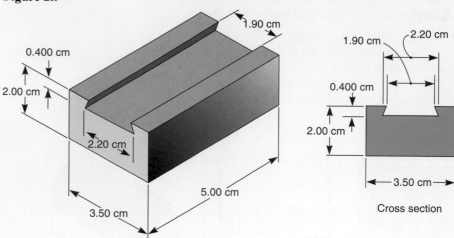

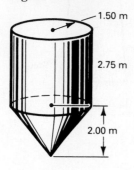

Figure 2.10

27. Find the volume of the storage bin shown in Fig. 2.10.
28. The maximum cross-sectional area of a spherical propane storage tank is 3.05 m². Will it fit into a 2.00-m-wide trailer?
29. How many cubic yards of concrete are needed to pour a patio 12.0 ft × 20.0 ft and 6.00 in. thick?
30. What length of sidewalk 4.00 in. thick and 4.00 ft wide could be poured with 2.00 yd³ of concrete?

Find the volume of each figure.

31.

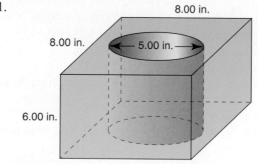

32.

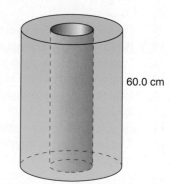

Inside diameter: 20.0 cm
Outside diameter: 50.0 cm

Glossary

Formula An equation, usually expressed in letters (called *variables*) and numbers. (p. 51)
Problem-Solving Method An orderly procedure that aids in understanding and solving problems. (p. 57)
Variable A symbol, usually a letter, used to represent some unknown number or quantity. (p. 51)

Review Questions

1. A formula is
   (a) the amount of each value needed.
   (b) a solution for problems.
   (c) an equation usually expressed in letters and numbers.
2. Subscripts are
   (a) the same as exponents.
   (b) used to shorten what must be written.
   (c) used to make a problem look hard.
3. A working equation
   (a) is derived from the basic equation.
   (b) is totally different from the basic equation.
   (c) comes before the basic equation in the problem.
   (d) none of the above.
4. Cite two examples in industry in which formulas are used.
5. How are subscripts used in measurement?
6. Why is reading the problem carefully the most important step in problem solving?
7. How can making a sketch help in problem solving?
8. What do we call the relationship between data that are given and what we are asked to find?
9. How is a working equation different from a basic equation?
10. How can analysis of the units in a problem assist in solving the problem?
11. How can making an estimate of your answer assist in the correct solution of problems?

Review Problems

1. Solve $F = ma$ for (a) m and (b) a.
2. Solve $v = \sqrt{2gh}$ for h.
3. Solve $s = \frac{1}{2}(v_f + v_i)t$ for v_f.
4. Solve $KE = \frac{1}{2}mv^2$ for v.
5. Given $P = a + b + c$, with $P = 36$ ft, $a = 12$ ft, and $c = 6$ ft, find b.
6. Given $A = \left(\dfrac{a + b}{2}\right)h$, with $A = 21\overline{0}$ m^2, b $= 16.0$ m, and $h = 15.0$ m, find a.
7. Given $A = \pi r^2$, if $A = 15.0$ m^2, find r.
8. Given $A = \frac{1}{2}bh$, if $b = 12.2$ cm and $h = 20.0$ cm, what is A?
9. A cone has a volume of 314 cm^3 and radius of 5.00 cm. What is its height?
10. A right triangle has a side of 41.2 mm and a side of 9.80 mm. Find the length of the hypotenuse.
11. Given a cylinder with a radius of 7.20 cm and a height of 13.4 cm, find the lateral surface area.
12. A rectangle has a perimeter of 40.0 cm. One side has a length of 14.0 cm. What is the length of an adjacent side?

SKETCH

12 cm^2 w

4.0 cm

DATA

$A = 12$ cm^2, $l = 4.0$ cm, w = ?

BASIC EQUATION

$A = lw$

WORKING EQUATION

$w = \frac{A}{l}$

SUBSTITUTION

$w = \frac{12 \text{ cm}^2}{4.0 \text{ cm}} = 3.0$ cm

13. The formula for the volume of a cylinder is $V = \pi r^2 h$. If $V = 21\overline{0}0 \text{ m}^3$ and $h = 17.0$ m, find r.

14. The formula for the area of a triangle is $A = \frac{1}{2}bh$. If $b = 12.3$ m and $A = 88.6 \text{ m}^2$, find h.

15. Find the volume of the lead sleeve with the cored hole in Fig. 2.11.

Figure 2.11

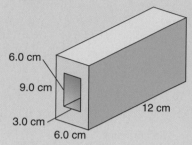

16. A rectangular plot of land measures 40.0 m by $12\overline{0}$ m with a parcel 10.0 m by 12.0 m out of one corner for an electrical transformer. What is the area of the remaining plot?

APPLIED CONCEPTS

Use the problem-solving method outlined in Section 2.3 to solve each problem.

1. You run a landscaping business and know that you want to charge $50.00 to mow a person's lawn whose property is 100 ft × 200 ft. If the house dimensions take up a 35.0 ft × 80.0 ft area, how much are you charging per square yard?

2. A room that measures 10.0 ft wide, 32.0 ft long, and 8.00 ft high needs a certain amount of air pumped into it per minute to keep the air quality up to regulations. If the room needs completely new air every 20.0 minutes, what is the volume of air per second that is being pumped into the room?

3. Instead of using a solid iron beam, structural engineers and contractors use I-beams to save materials and money. How many I-beams can be molded from the same amount of iron contained in the solid iron beam as shown in Fig. 2.12?

Figure 2.12

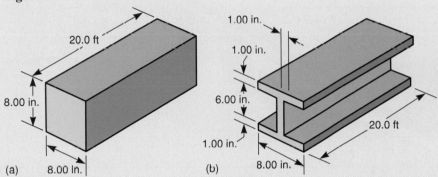

4. A shipping specialist at a craft store needs to pack Styrofoam balls of radius 4.00 in. into a 1.40 ft × 2.80 ft × 1.40 ft rectangular cardboard container. What is the maximum number of balls that can fit in the container? Hint: Spherical balls have spaces around them when packed in rectangular containers.

5. A crane needs to lift a spool of fine steel cable to the top of a bridge deck. The type of steel in the cable has a density of 7750 kg/m³. The maximum lifting mass of the crane is 43,400 kg. (a) Given the dimensions of the spool in Fig. 2.13, find the volume of the spool. (b) Can the crane safely lift the spool?

Figure 2.13

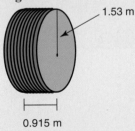

CHAPTER 3

VECTORS

$$\vec{U} = \vec{U}_x + \vec{U}_y$$

$$|\vec{U}_x| = |\vec{U}| \cos \theta$$
$$|\vec{U}_y| = |\vec{U}| \sin \theta$$

Some physical quantities, called *scalars,* may be described by and involve calculations with numerical quantities alone. Other physical quantities, called *vectors,* require both a numerical quantity and a direction to be completely described and often involve calculations using trigonometry. Vectors are developed in this chapter prior to their use in the following chapters.

Objectives

The major goals of this chapter are to enable you to:

1. Distinguish between a vector and a scalar quantity.
2. Add vectors graphically.
3. Find the components of a vector.
4. Work with vectors in standard position.
5. Apply the basic concepts of right-triangle trigonometry using displacement vectors.

3.1 Vectors and Scalars*

Every physical quantity can be classified as either a scalar or a vector quantity. A **scalar** is a quantity that can be completely described by a number (called its magnitude) and a unit. Examples of scalars are length, temperature, and volume. All these quantities can be expressed by a number with the appropriate units. For example, the length of a steel beam is expressed as 18 ft; the temperature at 11:00 A.M. is 15°C; and the volume of a room is 300 m³.

A **vector** is a quantity that requires both *magnitude* (size) and *direction* to be completely described. Examples of vectors are force, displacement, and velocity. To completely describe a force, you must give not only its magnitude (size or amount), but also its direction.

To describe the change of position of an object, such as an airplane flying from one city to another, we use the term *displacement*. **Displacement** is the net change in position of an object, or the direct distance and direction it moves. For example, to completely describe the flight of a plane between two cities requires both the *distance* between them and the *direction from* the first city *to* the second (Fig. 3.1). *The units of displacement are length units,* such as metres, kilometres, feet, or miles.

Suppose that a friend asks you how to reach your home from school. If you replied that he should walk four blocks, you would not have given him enough information [Fig. 3.2(a)].

Figure 3.1   Displacement

Figure 3.2   Displacement involves both a distance and a direction.

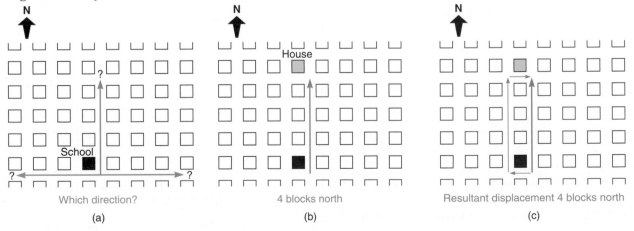

(a) Which direction?   (b) 4 blocks north   (c) Resultant displacement 4 blocks north

---

*Right-triangle trigonometry is developed in Appendix A.5, and instructions for using sin, cos, and tan keys on a scientific calculator are included in Appendix B.3 for those who have not studied this before or who need a review.

Figure 3.3

Obviously, you would need to tell him which direction to go. If you had replied, "Four blocks north," your friend could then find your home [Fig. 3.2(b)]. If your friend decides to walk one block west, four blocks north, and then one block east, he will still arrive at your house. This resultant displacement is the same as if he had walked four blocks north [Fig. 3.2(c)].

The magnitude of the displacement vector "15 miles NE" is 15 miles and its direction is northeast (Fig. 3.3).

To represent a vector in a diagram, we draw an arrow that points in the correct direction. The magnitude of the vector is indicated by the length of the arrow. We usually choose a scale, such as 1.0 cm = 25 mi, for this purpose (Fig. 3.4). Thus, a displacement of $10\bar{0}$ mi west is drawn as an arrow (pointing west) 4.0 cm long [Fig. 3.4(a)] since

$$100 \text{ mi} \times \frac{1.0 \text{ cm}}{25 \text{ mi}} = 4.0 \text{ cm}$$

Displacements of $5\bar{0}$ mi north [Fig. 3.4(b)] and $5\bar{0}$ mi east [Fig. 3.4(c)] using the same scale are also shown.

Figure 3.4 Use a scale to draw the proper length of a given vector.

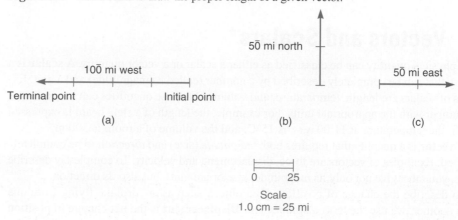

EXAMPLE 1

Using the scale 1.0 cm = $5\bar{0}$ km, draw the displacement vector 275 km at 45° north of west.
First, find the length of the vector.

$$275 \text{ km} \times \frac{1.0 \text{ cm}}{50 \text{ km}} = 5.5 \text{ cm}$$

Then draw the vector at an angle 45° north of west (Fig. 3.5).

Figure 3.5

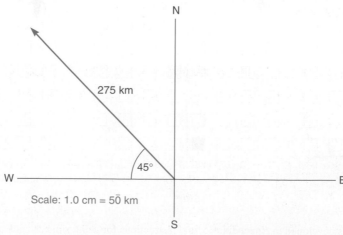

EXAMPLE 2

Using the scale $\frac{1}{4}$ in. $= 2\overline{0}$ mi, draw the displacement vector $15\overline{0}$ mi at 22° east of south.
   First, find the length of the vector.

$$15\overline{0} \text{ mi} \times \frac{\frac{1}{4} \text{ in.}}{2\overline{0} \text{ mi}} = 1\frac{7}{8} \text{ in.}$$

Then draw the vector at 22° east of south (Fig. 3.6).

Figure 3.6

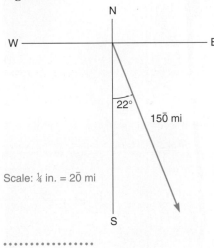

Scale: ¼ in. = 2̄0 mi

A vector may be denoted by a single letter with a small arrow above, such as $\vec{A}$, $\vec{v}$, or $\vec{R}$ [Fig. 3.7(a)]. This notation is especially useful when writing vectors on paper or on a chalkboard. In this book we use the traditional boldface type to denote vectors, such as **A**, **v**, or **R** [Fig. 3.7(b)]. The length of vector $\vec{A}$ is written $|\vec{A}|$; the length of vector **A** is written $|\mathbf{A}|$.

Figure 3.7

 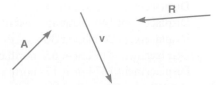

(a) Vector quantities $\vec{A}$, $\vec{v}$, and $\vec{R}$ usually have arrows when they are written on paper or on a chalkboard.

(b) Vector quantities **A**, **v**, and **R** usually are written in boldface type in textbooks.

TRY THIS ACTIVITY

New York Vectors

In 1811, a comprehensive plan was mapped out to create a rectangular grid of roadways on the island of Manhattan in New York City. As a result, giving directions in Manhattan can be done in a number of different ways while still achieving the same result. Assuming that the streets and avenues in the map in Fig. 3.8 are at right angles with one another and that the distance between streets is 0.05 mi and the distance between avenues in 0.20 mi, determine three different ways that someone could travel from Macy's at Herald Square to Times Square. What would be the distance traveled and the displacement for each of these paths?

Figure 3.8 A mid-town Manhattan map is a good way to demonstrate the usefulness of vectors.

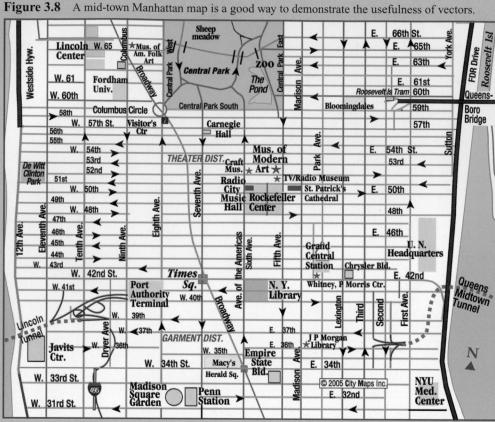

PROBLEMS 3.1

Using the scale 1.0 cm = $5\overline{0}$ km, find the length of the vector that represents each displacement.

1. Displacement $10\overline{0}$ km east — length = _____ cm
2. Displacement 125 km south — length = _____ cm
3. Displacement $14\overline{0}$ km at 45° east of south — length = _____ cm
4. Displacement $26\overline{0}$ km at $3\overline{0}$° south of west — length = _____ cm
5. Displacement 315 km at 65° north of east — length = _____ cm
6. Displacement 187 km at 17° north of west — length = _____ cm

7–12. Draw the vectors in Problems 1 through 6 using the scale indicated.

Using the scale $\frac{1}{4}$ in. = $2\overline{0}$ mi, find the length of the vector that represents each displacement.

13. Displacement $10\overline{0}$ mi west — length = _____ in.
14. Displacement $17\overline{0}$ mi north — length = _____ in.
15. Displacement $21\overline{0}$ mi at 45° south of west — length = _____ in.
16. Displacement 145 mi at $6\overline{0}$° north of east — length = _____ in.
17. Displacement 75 mi at 25° west of north — length = _____ in.
18. Displacement $16\overline{0}$ mi at 72° west of south — length = _____ in.

19–24. Draw the vectors in Problems 13 through 18 using the scale indicated.

3.2 Components of a Vector

Before we study vectors further, we need to discuss components of vectors and the number plane. The **number plane** (sometimes called the *Cartesian coordinate system,* after René Descartes) consists of a horizontal line called the *x*-axis and a vertical line called the *y*-axis intersecting at a right angle at a point called the *origin* as shown in Fig. 3.9. These two lines divide the number plane into four quadrants, which we label as quadrants I, II, III, and IV.

René Descartes (1596–1650),

mathematician and philosopher, was born in France. He founded analytic or coordinate geometry, often called Cartesian geometry, and made major contributions in optics.

Figure 3.9 Number plane

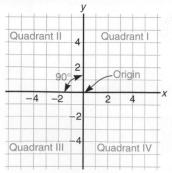

The *x*-axis contains positive numbers to the right of the origin and negative numbers to the left of the origin. The *y*-axis contains positive numbers above the origin and negative numbers below the origin.

Graphically, a vector is represented by a directed line segment. The length of the line segment indicates the magnitude of the quantity. An arrowhead indicates the direction. If *A* and *B* are the end points of a line segment as in Fig. 3.10, the symbol **AB** denotes the *vector from A to B*. Point *A* is called the *initial point*. Point *B* is called the *terminal point* or *end point* of the vector. Vector **BA** has the same length as vector **AB** but has the opposite direction. Vectors may also be denoted by a single letter, such as **u**, **v**, or **R**.

The sum of two or more vectors is called the **resultant vector**. When two or more vectors are added, each of these vectors is called a **component** of the resultant vector. The components of vector **R** in Fig. 3.11(a) are vectors **A**, **B**, and **C**. *Note:* A vector may have more than one set of component vectors. The components of vector **R** in Fig. 3.11(b) are vectors **E** and **F**.

Figure 3.10   Vector from *A* to *B*

Figure 3.11

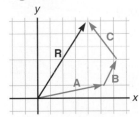 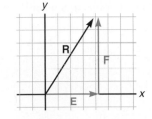

(a) Vectors **A**, **B**, and **C** are
   components of the
   resultant vector **R**.

(b) Vector **E** is a horizontal
   component and vector **F** is
   a vertical component of
   the resultant vector **R**.

We are often interested in the components of a vector that are perpendicular to each other and that are on or parallel to the *x*- and *y*-axes. In particular, we are interested in the type of component vectors shown in Fig. 3.11(b) (component vectors **E** and **F**). The horizontal component vector that lies on or is parallel to the *x*-axis is called the ***x*-component**. The vertical component vector that lies on or is parallel to the *y*-axis is called the ***y*-component**. Three examples are shown in Fig. 3.12.

Figure 3.12

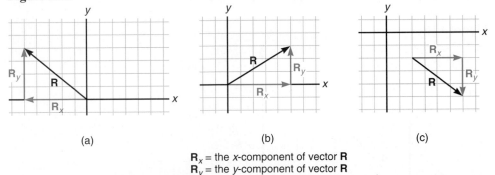

(a)   (b)   (c)

R_x = the *x*-component of vector **R**
R_y = the *y*-component of vector **R**

The *x*- and *y*-components of vectors can also be expressed as signed numbers. The absolute value of the signed number corresponds to the magnitude (length) of the component vector. The sign of the number corresponds to the direction of the component as follows:

x-component	*y*-component
+, if right	+, if up
−, if left	−, if down

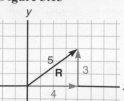

EXAMPLE 1

Figure 3.13

Find the *x*- and *y*-components of vector **R** in Fig. 3.13.

$$\mathbf{R}_x = x\text{-component of } \mathbf{R} = +4$$
$$\mathbf{R}_y = y\text{-component of } \mathbf{R} = +3$$

· · · · · · · · · · · · · · · · ·

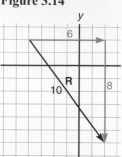

EXAMPLE 2

Figure 3.14

Find the *x*- and *y*-components of vector **R** in Fig. 3.14.

$$\mathbf{R}_x = x\text{-component of } \mathbf{R} = +6$$
$$\mathbf{R}_y = y\text{-component of } \mathbf{R} = -8$$

(The *y*-component points in a negative direction.)

· · · · · · · · · · · · · · · ·

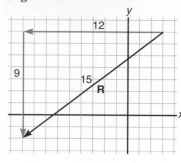

EXAMPLE 3

Figure 3.15

Find the *x*- and *y*-components of vector **R** in Fig. 3.15.

$$\mathbf{R}_x = -12$$
$$\mathbf{R}_y = -9$$

(Both *x*- and *y*-components point in a negative direction.)

· · · · · · · · · · · · · · · ·

A vector may be placed in any position in the number plane as long as its magnitude and direction are not changed. The vectors in each set in Fig. 3.16 are equal because they have the same magnitude (length) and the same direction.

Figure 3.16

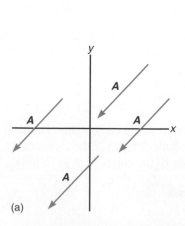

(a)

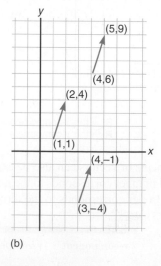

(b)

A vector is in **standard position** when its initial point is at the origin of the number plane. A vector in standard position is expressed in terms of its magnitude (length) and its angle θ, where θ *is measured counterclockwise from the positive x-axis to the vector.* The vectors shown in Fig. 3.17 are in standard position.

Figure 3.17   Vectors in standard position

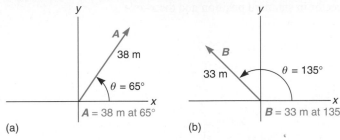

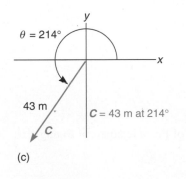

(a)

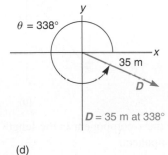

(b)

(c)

(d)

Finding the Components of a Vector

Find the x- and y-components of the vector **A** = 10.0 m at 60.0°.

First, draw the vector in standard position [Fig. 3.18(a)]. Then, draw a right triangle where the legs represent the x- and y-components [Fig. 3.18(b)]. The absolute value of the x-component of the vector is the length of the side adjacent to the 60.0° angle. Therefore, to find the x-component,

EXAMPLE 4

Figure 3.18

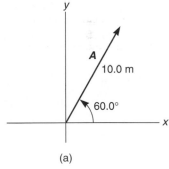

(a)

$$\cos 60.0° = \frac{\text{side adjacent to } 60.0°}{\text{hypotenuse}} = \frac{|\mathbf{A}_x|}{10.0 \text{ m}}$$

$$\cos 60.0° = \frac{|\mathbf{A}_x|}{10.0 \text{ m}}$$

$$(\cos 60.0°)(10.0 \text{ m}) = \left(\frac{|\mathbf{A}_x|}{10.0 \text{ m}} \right)(10.0 \text{ m}) \quad \text{Multiply both sides by 10.0 m.}$$

$$5.00 \text{ m} = |\mathbf{A}_x|$$

Since the x-component is pointing in the positive x-direction, $\mathbf{A}_x = +5.00$ m.

The absolute value of the y-component of the vector is the length of the side opposite the 60.0° angle. Therefore, to find the y-component,

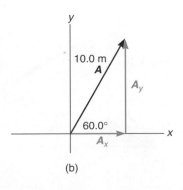

$$\sin 60.0° = \frac{\text{side opposite } 60.0°}{\text{hypotenuse}} = \frac{|\mathbf{A}_y|}{10.0 \text{ m}}$$

$$\sin 60.0° = \frac{|\mathbf{A}_y|}{10.0 \text{ m}}$$

$$(\sin 60.0°)(10.0 \text{ m}) = \left(\frac{|\mathbf{A}_y|}{10.0 \text{ m}} \right)(10.0 \text{ m}) \quad \text{Multiply both sides by 10.0 m.}$$

$$8.66 \text{ m} = |\mathbf{A}_y|$$

(b)

Since the y-component is pointing in the positive y-direction, $\mathbf{A}_y = +8.66$ m.

EXAMPLE 5

Find the *x*- and *y*-components of the vector **B** = 13.0 km at 220.0°.

First, draw the vector in standard position [Fig. 3.19(a)]. Then, complete a right triangle with the *x*- and *y*-components being the two legs [Fig. 3.19(b)].

We will let angle α (Greek letter alpha) be the acute angle (an angle whose measure is less than 90°) between the vector in standard position and the *x*-axis.

Find angle α as follows:

$$180° + \alpha = 220.0°$$
$$\alpha = 40.0°$$

Figure 3.19

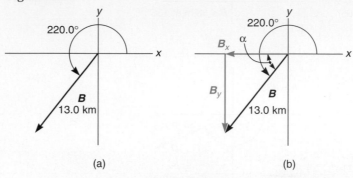

(a)   (b)

The absolute value of the *x*-component is the length of the side adjacent to angle α. Therefore, to find the *x*-component,

$$\cos \alpha = \frac{\text{side adjacent to } \alpha}{\text{hypotenuse}}$$

$$\cos 40.0° = \frac{|\mathbf{B}_x|}{13.0 \text{ km}}$$

$$(\cos 40.0°)(13.0 \text{ km}) = \left(\frac{|\mathbf{B}_x|}{13.0 \text{ km}}\right)(13.0 \text{ km}) \qquad \text{Multiply both sides by 13.0 km.}$$

$$9.96 \text{ km} = |\mathbf{B}_x|$$

Since the *x*-component is pointing in the negative *x*-direction, $\mathbf{B}_x = -9.96$ km.

The absolute value of the *y*-component of the vector is the length of the side opposite angle α. Therefore, to find the *y*-component,

$$\sin \alpha = \frac{\text{side opposite } \alpha}{\text{hypotenuse}}$$

$$\sin 40.0° = \frac{|\mathbf{B}_y|}{13.0 \text{ km}}$$

$$(\sin 40.0°)(13.0 \text{ km}) = \left(\frac{|\mathbf{B}_y|}{13.0 \text{ km}}\right)(13.0 \text{ km}) \qquad \text{Multiply both sides by 13.0 km.}$$

$$8.36 \text{ km} = |\mathbf{B}_y|$$

Figure 3.20

Since the *y*-component is pointing in the negative *y*-direction, $\mathbf{B}_y = -8.36$ km.

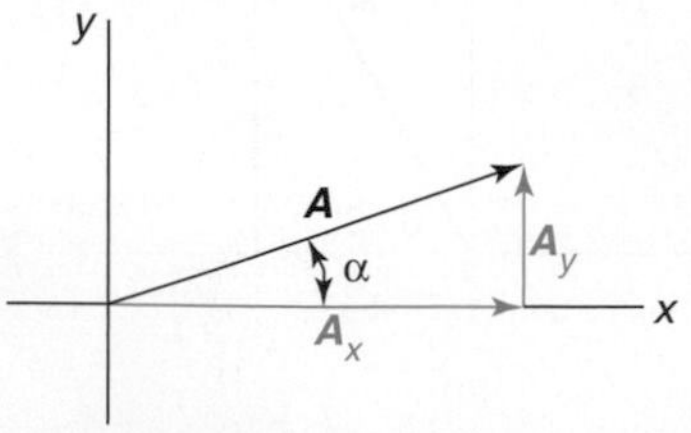

Vector **v** in standard position with its horizontal component **v**$_x$ and its vertical component **v**$_y$

In general, find the *x*- and *y*-components of a vector as follows. First, draw any vector **A** in standard position; then, draw its *x*- and *y*-components as shown in Fig. 3.20. Use the right triangle to find the *x*-component as follows:

$$\cos \alpha = \frac{\text{side adjacent to } \alpha}{\text{hypotenuse}}$$

$$\cos \alpha = \frac{|\mathbf{A}_x|}{|\mathbf{A}|}$$

$$|\mathbf{A}|(\cos \alpha) = \left(\frac{|\mathbf{A}_x|}{|\mathbf{A}|}\right)|\mathbf{A}| \qquad \text{Multiply both sides by } |\mathbf{A}|.$$

$$|\mathbf{A}|(\cos \alpha) = |\mathbf{A}_x|$$

Similarly, we use the right triangle to find the *y*-component as follows:

$$\sin \alpha = \frac{\text{side opposite } \alpha}{\text{hypotenuse}}$$

$$\sin \alpha = \frac{|\mathbf{A}_y|}{|\mathbf{A}|}$$

$$|\mathbf{A}|(\sin \alpha) = \left(\frac{|\mathbf{A}_y|}{|\mathbf{A}|}\right)|\mathbf{A}| \qquad \text{Multiply both sides by } |\mathbf{A}|.$$

$$|\mathbf{A}|(\sin \alpha) = |\mathbf{A}_y|$$

The signs of the *x*- and *y*-components are determined by the quadrants in which the vector in standard position lies.

In general:

To find the *x*- and *y*-components of a vector **A** given in standard position:

1. Complete the right triangle with the legs being the *x*- and *y*-components of the vector.
2. Find the lengths of the legs of the right triangle as follows:

$$|\mathbf{A}_x| = |\mathbf{A}|(\cos \alpha)$$
$$|\mathbf{A}_y| = |\mathbf{A}|(\sin \alpha)$$

where angle α is the acute angle between vector **A** in standard position and the *x*-axis.
3. Determine the signs of the *x*- and *y*-components.

EXAMPLE 6

Find the *x*- and *y*-components of the vector $\mathbf{C} = 27.0$ ft at $125.0°$.

First, draw the vector in standard position [Fig. 3.21(a)]. Then, complete a right triangle with the *x*- and *y*-components being the two legs [Fig. 3.21(b)]. Find angle α as follows:

$$\alpha + 125.0° = 180°$$
$$\alpha = 55.0°$$

Figure 3.21

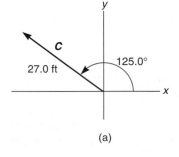

(a)

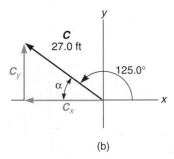

(b)

Next, find the *x*-component as follows:

$$|\mathbf{C}_x| = |\mathbf{C}|(\cos \alpha)$$
$$|\mathbf{C}_x| = (27.0 \text{ ft})(\cos 55.0°)$$
$$= 15.5 \text{ ft}$$

Since the *x*-component is pointing in the negative *x*-direction,

$$\mathbf{C}_x = -15.5 \text{ ft}$$

Then, find the *y*-component as follows:

$$|\mathbf{C}_y| = |\mathbf{C}|(\sin \alpha)$$
$$|\mathbf{C}_y| = (27.0 \text{ ft})(\sin 55.0°)$$
$$= 22.1 \text{ ft}$$

Since the *y*-component is pointing in the positive *y*-direction,

$$\mathbf{C}_y = +22.1 \text{ ft}$$

PROBLEMS 3.2

Find the *x*- and *y*-components of each vector in the following diagram. (Express them as signed numbers and then graph them as vectors.)

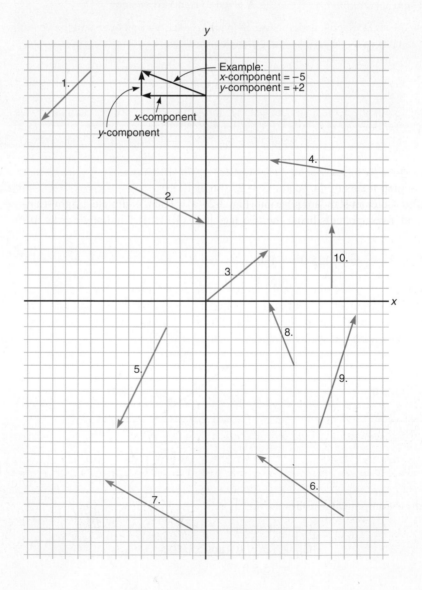

Make a sketch of each vector in standard position. Use the scale 1.0 cm = $1\overline{0}$ m.

11. **A** = $2\overline{0}$ m at 25°   12. **B** = 25 m at 125°
13. **C** = 25 m at 245°   14. **D** = $2\overline{0}$ m at 345°
15. **E** = 15 m at 105°   16. **F** = 35 m at 291°
17. **G** = $3\overline{0}$ m at 405°   18. **H** = 25 m at 525°

Find the *x*- and *y*-components of each vector.

19.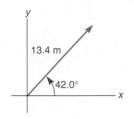

20.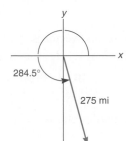

21.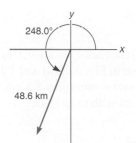

22.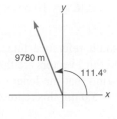

23.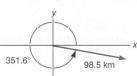

24.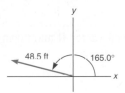

Find the *x* and *y* components of each vector given in standard position.

25. **A** = 38.9 m at 10.5°   26. **B** = 478 ft at 195.0°
27. **C** = 9.60 km at 310.0°   28. **D** = 5430 mi at 153.7°
29. **E** = 29.5 m at 101.5°   30. **F** = 154 mi at 273.2°

3.3 Addition of Vectors

Any given displacement can be the result of many different combinations of displacements. In Fig. 3.22, the displacement represented by the arrow labeled **R** for resultant is the result of either of the two paths shown. The resultant vector, **R**, is the sum of the vectors **A**, **B**, **C**, and **D**. It is also the sum of vectors **E** and **F**. That is,

$$\mathbf{A} + \mathbf{B} + \mathbf{C} + \mathbf{D} = \mathbf{R} \quad \text{and} \quad \mathbf{E} + \mathbf{F} = \mathbf{R}$$

Figure 3.22

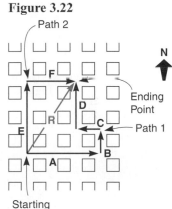

The resultant vector **R** is the graphic sum of the component sets of vectors **A**, **B**, **C**, and **D**, and **E** and **F**. That is, **A** + **B** + **C** + **D** = **R** and **E** + **F** = **R**.

To solve a vector addition problem graphically such as displacement:

1. Choose a suitable scale and calculate the length of each vector.
2. Draw the north–south reference line. Graph paper should be used.
3. Using a ruler and protractor, draw the first vector and then draw the other vectors so that the initial end of each vector is placed at the terminal end of the previous vector.
4. Draw the resultant vector from the initial end of the first vector to the terminal end of the last vector.
5. Measure the length of the resultant and use the scale to find the magnitude of the vector. Use a protractor to measure the angle of the resultant.

EXAMPLE 1

Find the resultant displacement of an airplane that flies $2\bar{0}$ mi due east, then $3\bar{0}$ mi due north, and then $1\bar{0}$ mi at 60° west of south.

We choose a scale of 1.0 cm = 5.0 mi so that the vectors are large enough to be accurate and small enough to fit on the paper. (Here each block represents 0.5 cm.) The length of the first vector is

$$|\mathbf{A}| = 2\bar{0} \text{ mi} \times \frac{1.0 \text{ cm}}{5.0 \text{ mi}} = 4.0 \text{ cm}$$

The length of the second vector is

$$|\mathbf{B}| = 3\bar{0} \text{ mi} \times \frac{1.0 \text{ cm}}{5.0 \text{ mi}} = 6.0 \text{ cm}$$

The length of the third vector is

$$|\mathbf{C}| = 1\bar{0} \text{ mi} \times \frac{1.0 \text{ cm}}{5.0 \text{ mi}} = 2.0 \text{ cm}$$

Draw the north–south reference line, and draw the first vector as shown in Fig. 3.23(a). The second and third vectors are then drawn as shown in Fig. 3.23(b) and 3.23(c).

Using a ruler, we find that the length of the resultant vector measures 5.5 cm [Fig. 3.23(d)]. Since 1.0 cm = 5.0 mi, this represents a displacement with magnitude

$$|\mathbf{R}| = 5.5 \text{ cm} \times \frac{5.0 \text{ mi}}{1.0 \text{ cm}} = 28 \text{ mi}$$

The angle between vector **R** and north measures 24°, so the resultant is 28 mi at 24° east of north.

Figure 3.23

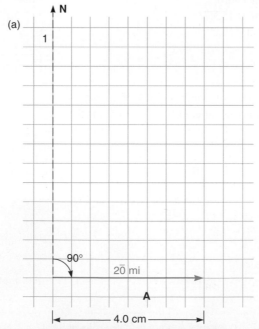

1. Draw the north–south reference line and the first vector: $2\bar{0}$ mi due east.

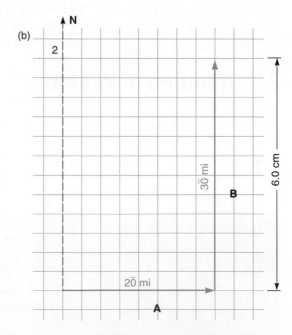

2. Draw the second vector: $3\bar{0}$ mi due north.

Figure 3.23 *(Continued)*

(c)

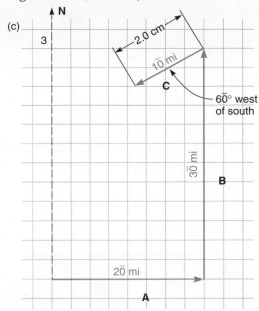

(d)

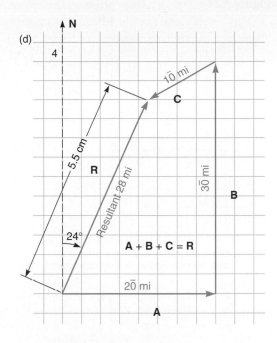

3. Draw the third vector: $1\overline{0}$ mi at $6\overline{0}°$ west of south.

4. Draw the resultant vector, which is 28 mi at 24° east of north.

Scale: 1.0 cm = 5.0 mi

· · · · · · · · · · · · · · · · ·

EXAMPLE 2

Find the resultant of the displacements $15\overline{0}$ km due west, then $20\overline{0}$ km due east, and then 125 km due south.

Choose a scale of 1.0 cm = $5\overline{0}$ km. The length of the first vector is

$$|\mathbf{A}| = 15\overline{0} \text{ km} \times \frac{1.0 \text{ cm}}{5\overline{0} \text{ km}} = 3.0 \text{ cm}$$

The length of the second vector is

$$|\mathbf{B}| = 20\overline{0} \text{ km} \times \frac{1.0 \text{ cm}}{5\overline{0} \text{ km}} = 4.0 \text{ cm}$$

The length of the third vector is

$$|\mathbf{C}| = 125 \text{ km} \times \frac{1.0 \text{ cm}}{5\overline{0} \text{ km}} = 2.5 \text{ cm}$$

Draw the north–south reference line, and draw the first vector as shown in Fig. 3.24(a). Then, draw the second and third vectors as shown in Fig. 3.24(b) and 3.24(c).

The length of the resultant vector measures 2.6 cm in Fig. 3.24(d). Since 1.0 cm = $5\overline{0}$ km,

$$|\mathbf{R}| = 2.6 \text{ cm} \times \frac{5\overline{0} \text{ km}}{1.0 \text{ cm}} = 130 \text{ km}$$

The angle between vector **R** and south measures 22°, so the resultant vector is 130 km at 22° east of south.

Figure 3.24

(a)

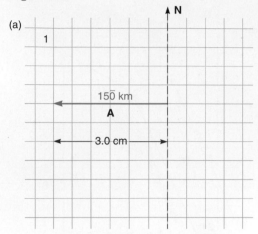

1. Draw the north–south reference line and the first vector: 15̄0 km due west.

(b)

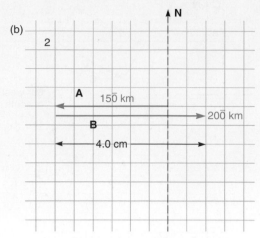

2. Draw the vector: 20̄0 km due east.

(c)

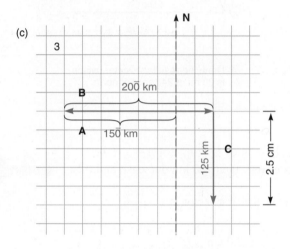

3. Draw the vector: 125 mi due south.

(d)

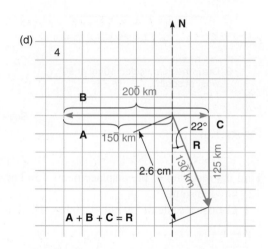

4. The length of the resultant is 2.6 cm, which represents 13̄0 km at 22° east of south.

Scale: 1.0 cm = 5̄0 km

· · · · · · · · · · · · · · · · · · ·

Expressing the x- and y-components as signed numbers, we find the resultant vector of several vectors as follows:

1. Find the x-component of each vector and then find the sum of these x-components. This sum is the x-component of the resultant vector.
2. Find the y-component of each vector and then find the sum of these y-components. This sum is the y-component of the resultant vector.

EXAMPLE 3

Given vectors **A** and **B** in Fig. 3.25, graph and find the x- and y-components of the resultant vector **R**.

Graph resultant vector **R** by connecting the initial point of vector **A** to the end point of vector **B** [Fig. 3.26(a)]. The resultant vector **R** is shown in Fig. 3.26(b).

Find the x-component of **R** by finding and adding the x-components of **A** and **B**.

$$A_x = +3$$
$$B_x = +2$$
$$R_x = +5$$

Figure 3.25

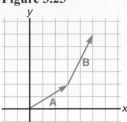

Figure 3.26

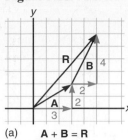

(a)   **A + B = R**

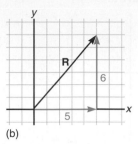

(b)

Find the y-component of **R** by finding and adding the y-components of **A** and **B**.

$$A_y = +2$$
$$B_y = +4$$
$$R_y = +6$$

· · · · · · · · · · · · · · · ·

Given vectors **A**, **B**, and **C** in Fig. 3.27, graph and find the x- and y-components of the resultant vector **R**.

Graph resultant vector **R** by connecting the initial point of vector **A** to the end point of vector **C** [Fig. 3.28(a)]. The resultant vector **R** is shown in Fig. 3.28(b).

EXAMPLE 4

Figure 3.27

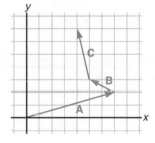

Figure 3.28

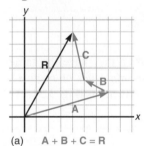

(a)   **A + B + C = R**

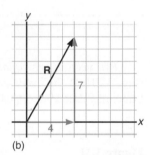

(b)

Find the x-component of **R** by finding and adding the x-components of **A**, **B**, and **C** as shown below. Find the y-component of **R** by finding and adding the y-components of **A**, **B**, and **C**.

Vector	x-component	y-component
A	+7	+2
B	−2	+1
C	−1	+4
R	+4	+7

· · · · · · · · · · · · · · · ·

Given vectors **A**, **B**, **C**, and **D** in Fig. 3.29, graph and find the x- and y-components of the resultant vector **R**.

Graph resultant vector **R** by connecting the initial point of vector **A** to the end point of vector **D** [Fig. 3.30(a)]. The resultant vector **R** is shown in Fig. 3.30(b).

EXAMPLE 5

Figure 3.29

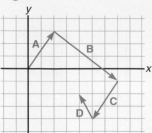

Figure 3.30

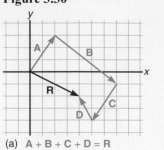

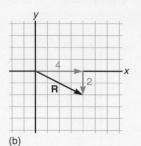

(a) A + B + C + D = R

(b)

Find the *x*-component of **R** by finding and adding the *x*-components of **A**, **B**, **C**, and **D** as shown below. Find the *y*-component of **R** by finding and adding the *y*-components of **A**, **B**, **C**, and **D**.

Vector	*x*-component	*y*-component
A	+2	+3
B	+5	−4
C	−2	−3
D	−1	+2
R	+4	−2

· · · · · · · · · · · · · · · · · · · ·

Two vectors are equal when they have the same magnitude and the same direction [Fig. 3.31(a)]. *Two vectors are opposites or negatives* of each other when they have the same magnitude but opposite directions [Fig. 3.31(b)].

Figure 3.31

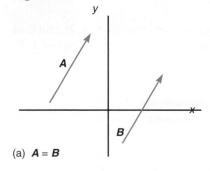

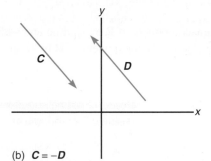

(a) **A** = **B**

(b) **C** = −**D**

To add two or more vectors in any position graphically, construct the first vector with its initial point at the origin and parallel to its given position. Then, construct the second vector with its initial point on the end point of the first vector and parallel to its given position. Then, construct the third vector with its initial point on the end point of the second vector and parallel to its given position. Continue this process until all vectors have been so constructed. The resultant vector is the vector joining the initial point of the first vector (origin) to the end point of the last vector. (The order of adding or constructing the given vectors does not matter.)

Given vectors **A**, **B**, and **C** in Fig. 3.32(a), graph and find the *x*- and *y*-components of the resultant vector **R**.

EXAMPLE 6

Figure 3.32

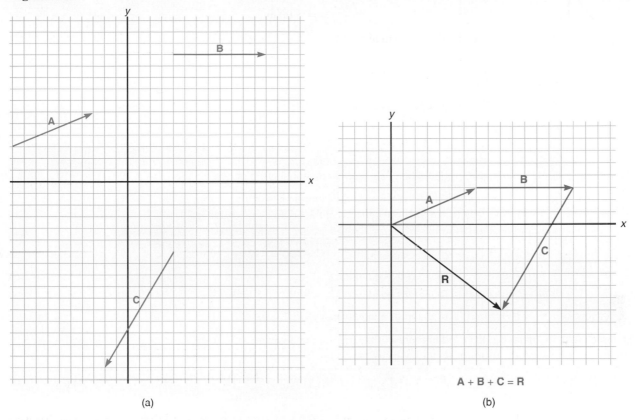

(a)   (b)

A + B + C = R

Construct vector **A** with its initial point at the origin and parallel to its given position as in Fig. 3.32(b). Next, construct vector **B** with its initial point on the end point of vector **A** and parallel to its given position. Then, construct vector **C** with its initial point on the end point of vector **B** and parallel to its given position. The resultant vector **R** is the vector with its initial point at the origin and its end point at the end point of vector **C**.

From the graph in Fig. 3.32(b), we read the *x*-component of **R** as +9 by counting the number of squares *to the right* between the *y*-axis and the end point of vector **R**. We read the *y*-component of **R** as −7 by counting the number of squares *below* between the *x*-axis and the end point of vector **R**.

Figure 3.33

$R = v − w = v + (−w)$

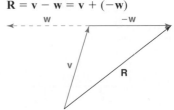

· · · · · · · · · · · · · · · ·

One vector may be subtracted from a second vector by adding its negative to the first. That is, $v − w = v + (−w)$. Construct **v** as usual. Then construct −**w** and find the resultant **R** as shown in Fig. 3.33.

Finding a Vector from Its Components

Find vector **R** in standard position with $R_x = +3.00$ m and $R_y = +4.00$ m.

EXAMPLE 7

First, graph the *x*- and *y*-components (Fig. 3.34) and complete the right triangle. The hypotenuse is the resultant vector. Find angle α as follows:

$$\tan \alpha = \frac{\text{side opposite } \alpha}{\text{side adjacent to } \alpha}$$

$$\tan \alpha = \frac{4.00 \text{ m}}{3.00 \text{ m}} = 1.333$$

$$\alpha = 53.1° \qquad \text{(see Appendix B, Section B.3)}$$

Figure 3.34

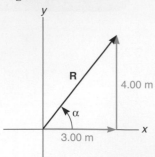

Find the magnitude of **R** using the Pythagorean theorem:

$$|\mathbf{R}| = \sqrt{|\mathbf{R}_x|^2 + |\mathbf{R}_y|^2}$$
$$|\mathbf{R}| = \sqrt{(3.00 \text{ m})^2 + (4.00 \text{ m})^2}$$
$$= 5.00 \text{ m}$$

That is, **R** = 5.00 m at 53.1°.

· · · · · · · · · · · · · · · · · ·

In general:

To find resultant vector **R** in standard position when its x- and y-components are given:

1. Complete the right triangle with the legs being the x- and y-components of the vector.
2. Find the acute angle α of the right triangle whose vertex is at the origin by using tan α.
3. Find angle θ in standard position as follows:

$$\theta = \alpha \qquad (\theta \text{ in first quadrant})$$
$$\theta = 180° - \alpha \qquad (\theta \text{ in second quadrant})$$
$$\theta = 180° + \alpha \qquad (\theta \text{ in third quadrant})$$
$$\theta = 360° - \alpha \qquad (\theta \text{ in fourth quadrant})$$

The Greek letter θ (theta) is often used to represent the measure of an angle.

4. Find the magnitude of the vector using the Pythagorean theorem:

$$\mathbf{R} = \sqrt{|\mathbf{R}_x|^2 + |\mathbf{R}_y|^2}$$

EXAMPLE 8

Find vector **R** in standard position whose x-component is +7.00 mi and y-component is −5.00 mi.

First, graph the x- and y-components (Fig. 3.35) and complete the right triangle. The hypotenuse is the resultant vector. Find angle α as follows:

Figure 3.35

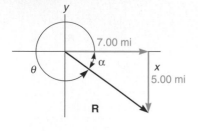

$$\tan \alpha = \frac{\text{side opposite } \alpha}{\text{side adjacent to } \alpha}$$

$$\tan \alpha = \frac{5.00 \text{ mi}}{7.00 \text{ mi}} = 0.7143$$

$$\alpha = 35.5°$$

Then

$$\theta = 360° - \alpha \qquad \mathbf{R} \text{ is in the fourth quadrant.}$$
$$= 360° - 35.5°$$
$$= 324.5°$$

Find the magnitude of **R** using the Pythagorean theorem:

$$|\mathbf{R}| = \sqrt{|\mathbf{R}_x|^2 + |\mathbf{R}_y|^2}$$
$$|\mathbf{R}| = \sqrt{|7.00 \text{ mi}|^2 + |-5.00 \text{ mi}|^2}$$
$$= 8.60 \text{ mi}$$

That is, **R** = 8.60 mi at 324.5°.

· · · · · · · · · · · · · · · · · ·

Find vector **R** in standard position with $R_x = -115$ km and $R_y = +175$ km.

EXAMPLE 9

First, graph the *x*- and *y*-components (Fig. 3.36) and complete the right triangle. The hypotenuse is the resultant vector. Find angle α as follows:

$$\tan \alpha = \frac{\text{side opposite } \alpha}{\text{side adjacent to } \alpha}$$

$$\tan \alpha = \frac{175 \text{ km}}{115 \text{ km}} = 1.522$$

$$\alpha = 56.7°$$

Then

$$\theta = 180° - \alpha \qquad \text{**R** is in the second quadrant.}$$

$$= 180° - 56.7°$$

$$= 123.3°$$

Figure 3.36

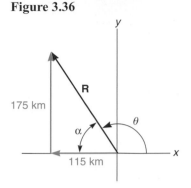

Find the magnitude of **R** using the Pythagorean theorem:

$$|\mathbf{R}| = \sqrt{|\mathbf{R}_x|^2 + |\mathbf{R}_y|^2}$$

$$|\mathbf{R}| = \sqrt{|-115 \text{ km}|^2 + |175 \text{ km}|^2}$$

$$= 209 \text{ km}$$

That is, **R** = 209 km at 123.3°.

· · · · · · · · · · · · · · · · · ·

To find the resultant vector **R** of any set of vectors, such as **R** = **A** + **B** + **C**, using right-triangle trigonometry:

1. Find the *x*-component of each vector and add: $\mathbf{R}_x = \mathbf{A}_x + \mathbf{B}_x + \mathbf{C}_x$.
2. Find the *y*-component of each vector and add: $\mathbf{R}_y = \mathbf{A}_y + \mathbf{B}_y + \mathbf{C}_y$.
3. Find the magnitude of the resultant vector **R** using the Pythagorean theorem $|\mathbf{R}| = \sqrt{|\mathbf{R}_x|^2 + |\mathbf{R}_y|^2}$.
4. Find the direction of the resultant vector **R** using right-triangle trigonometry by (a) first finding the acute α between the resultant vector and the *x*-axis and then finding angle θ in standard position or (b) expressing the direction of the resultant vector using some other reference.

A ship travels 105 km from port on a course of 55.0° west of north to an island. Then it travels 124 km due west to a second island. Then it travels 177 km on a course of 24.0° east of south to a third island. Find the displacement from the starting point to the ending point.

EXAMPLE 10

First, draw a vector diagram as in Fig. 3.37. Then, find the *x*- and *y*-components of each of the three vectors as follows:

Figure 3.37

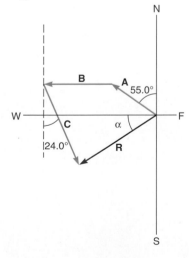

$|\mathbf{A}_x| = |\mathbf{A}| \cos 35.0°$ The acute angle between vector **A** and the *x*-axis is $90° - 55.0° = 35.0°$.

$= (105 \text{ km})(\cos 35.0°)$

$= -86.0 \text{ km}$ The *x*-component is in the negative *x*-direction.

$|\mathbf{A}_y| = |\mathbf{A}| \sin 35.0°$

$= (105 \text{ km})(\sin 35.0°)$

$= 60.2 \text{ km}$ The *y*-component is in the positive *y*-direction.

$|\mathbf{B}_x| = |\mathbf{B}| = -124 \text{ km}$ This *x*-component is in the negative *x*-direction.

$|\mathbf{B}_y| = |\mathbf{B}| = 0 \text{ km}$ This *y*-component of due west is 0.

$|\mathbf{C}_x| = |\mathbf{C}| \cos 66.0°$ The acute angle between vector **C** in standard position and the *x*-axis is $90° - 24.0° = 66.0°$.

$= (177 \text{ km})(\cos 66.0°)$

$= 72.0 \text{ km}$ The *x*-component is in the positive *x*-direction.

$$|C_y| = |C| \sin 66.0°$$
$$= (177 \text{ km})(\sin 66.0°)$$
$$= -162 \text{ km} \qquad \text{The } y\text{-component is in the negative } y\text{-direction.}$$

Thus

$$\mathbf{R}_x = \mathbf{A}_x + \mathbf{B}_x + \mathbf{C}_x = -86.0 \text{ km} + (-124 \text{ km}) + 72.0 \text{ km} = -138 \text{ km}$$
$$\mathbf{R}_y = \mathbf{A}_y + \mathbf{B}_y + \mathbf{C}_y = 60.2 \text{ km} + 0 + (-162 \text{ km}) = -102 \text{ km}$$
$$|\mathbf{R}| = \sqrt{|\mathbf{R}_x|^2 + |\mathbf{R}_y|^2}$$
$$= \sqrt{|-138 \text{ km}|^2 + |-102 \text{ km}|^2}$$
$$= 172 \text{ km}$$

$$\tan \alpha = \frac{102 \text{ km}}{138 \text{ km}} = 0.7391$$

$$\alpha = 36.5°$$

So, the displacement is 172 km at 36.5° south of west. That is, the ship stops at a port that is 172 km at 36.5° south of west from its starting point.

··················

PROBLEMS 3.3

Use graph paper to find the resultant of each displacement pair.

1.   35 km due east, then $5\overline{0}$ km due north
2. $6\overline{0}$ km due west, then $9\overline{0}$ km due south
3. $50\overline{0}$ mi at 75° east of north, then $150\overline{0}$ mi at $2\overline{0}°$ west of south
4. $2\overline{0}$ mi at 3° north of east, then 17 mi at 9° west of south
5.   67 km at 55° north of west, then 46 km at 25° south of east
6.   4.0 km at 25° west of south, then 2.0 km at 15° north of east

Use graph paper to find the resultant of each set of displacements.

7. $6\overline{0}$ km due south, then $9\overline{0}$ km at 15° north of west, and then 75 km at 45° north of east
8.   110 km at $5\overline{0}°$ north of east, then 170 km at $3\overline{0}°$ east of south, and then 145 km at $2\overline{0}°$ north of east
9.   1700 mi due north, then 2400 mi at $1\overline{0}°$ north of east, and then $20\overline{0}0$ mi at $2\overline{0}°$ south of west
10. $9\overline{0}$ mi at $1\overline{0}°$ west of north, then 75 mi at $3\overline{0}°$ west of south, and then 55 mi at $2\overline{0}°$ east of south
11. 75 km at 25° north of east, then 75 km at 65° south of west, and then 75 km due south
12. 17 km due north, then $1\overline{0}$ km at 7° south of east, and then 15 km at $1\overline{0}°$ west of south
13. 12 mi at 58° north of east, then 16 mi at 78° north of east, then $1\overline{0}$ mi at 45° north of east, and then 14 mi at $1\overline{0}°$ north of east
14. $1\overline{0}$ km at 15° west of south, then 27 km at 35° north of east, then 31 km at 5° north of east, and then 22 km at $2\overline{0}°$ west of north

Find the x- and y-components of each resultant vector **R** and graph the resultant vector **R**.

	Vector	x-component	y-component		Vector	x-component	y-component
15.	A	+2	+3	16.	A	+9	−5
	B	+7	+2		B	−4	−6
	R				R		
17.	A	−2	+13	18.	A	+10	−5
	B	−11	+1		B	−13	−9
	C	+3	−4		C	+4	+3
	R				R		

19.				20.			
	A	+17	+7		A	+1	+7
	B	−14	+11		B	+9	−4
	C	+7	+9		C	−4	+13
	D	−6	−15		D	−11	−4
	R				R		
21.				22.			
	A	+1.5	−1.5		A	+1	−1
	B	−3	−2		B	−4	−2
	C	+7.5	−3		C	+2	+4
	D	+2	+2.5		D	+5	−3
	R				E	+3	+5
					R		
23.				24.			
	A	+1.5	+2.5		A	−7	+15
	B	−2	−3		B	+13.5	−17.5
	C	+3.5	−7.5		C	−7.5	−20
	D	−4	+6		D	+6	+13.5
	E	−5.5	+2		E	+2.5	+2.5
	R				F	−11	+11.5
					R		

For each set of vectors, graph and find the *x*- and *y*-components of the resultant vector **R**.

25.

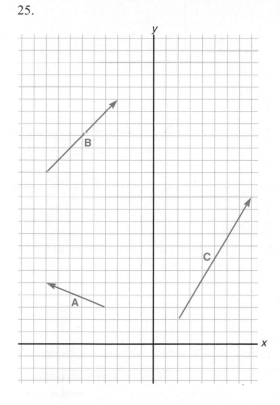

26.

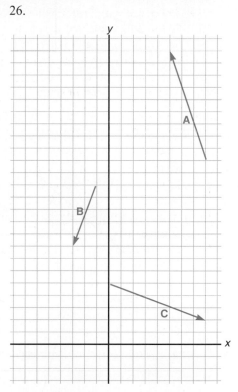

27.

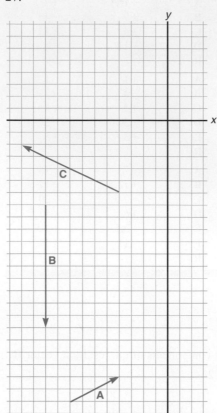

28.

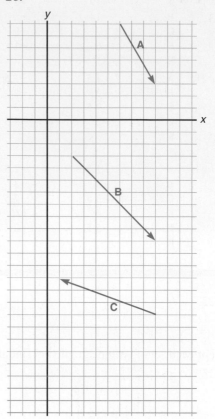

29.

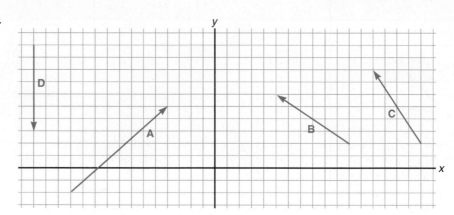

30.

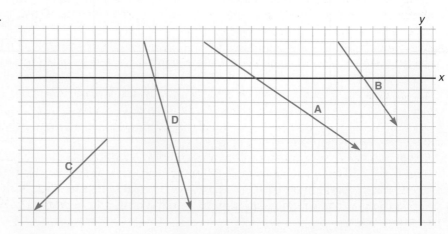

In Problems 31 through 42, find each resultant vector **R**. Give **R** in standard position.

31.

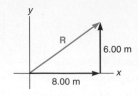
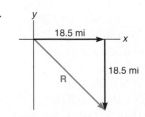

32.

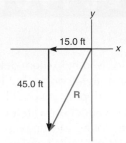

33.

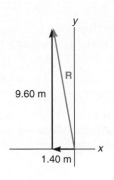

34.

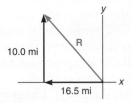

35.

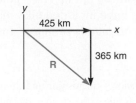

36.

	x-component	y-component		x-component	y-component
37.	+19.5 m	−49.6 m	38.	−158 km	+236 km
39.	+14.7 mi	+16.8 mi	40.	−3240 ft	−1890 ft
41.	−9.65 m	+4.36 m	42.	+375 km	−408 km

43. A road grader must go around a pond by traveling $10\overline{0}$ m south and then $15\overline{0}$ m east. If the road grader could go directly from the beginning point to the end point, how far would it travel?

44. An earthmover must go north $35\overline{0}$ m and then west 275 m to avoid a pipeline hazard. What distance would it travel if it could go directly to the endpoint?

45. An airplane takes off and flies 225 km on a course of 25.0° north of west and then changes direction and flies 135 km due north where it lands. Find the displacement from its starting point to its landing point.

46. A ship travels 50.0 mi on a course of 15.0° south of east and then travels 85.5 mi on a course of 60.0° west of south. Find the displacement from its starting point to its ending point.

47. A ship travels 135 km from port on a course of 25.0° south of east to an island. It then travels 122 km on a course of 35.5° west of south to a second island. Then it travels 135 km on a course of 10.4° north of west to a third island. Find the displacement from its starting point to its ending point.

48. A ship travels 145 km from port on a course of 65.0° north of east to an island. It then travels 112 km on a course of 30.5° west of north to a second island. Then it travels 182 km on a course of 10.4° west of south to a third island. Then it travels 42.5 km due south to a fourth island. Find the displacement from its starting point to its ending point.

PHYSICS CONNECTIONS

Global Positioning Satellites

Navigators continually struggle to find better tools to help them determine their location. The first explorers used the sun and stars to help them steer a straight course, but this method of navigation only worked under clear skies. Magnetic compasses were developed, yet could only be used to determine longitude, not latitude. Finally, the mechanical clock, in conjunction with the compass, provided navigators with the most accurate method of determining location. Today, most navigators use a hand-held device that functions in concert with a series of 24 orbiting satellites. This network, the Global Positioning System (GPS), can determine your position and altitude anywhere on earth.

The GPS pinpoints your location by sending out radio signals to locate any 4 of the 24 orbiting GPS satellites. Once the satellites are found, the GPS measures the length of time it takes for a radio signal to reach the hand-held receiver. When the time is determined for each of 4 satellites, the distance is calculated, and the longitude, latitude, and altitude are displayed on the screen [Fig. 3.38(a)].

GPS was first developed solely for military use. Eventually, the GPS system was made available for civilian businesses. Shipping, airline, farming, surveying, and geological companies made use of the technology. Today, GPS receivers are affordable and are used by the general public [Fig. 3.38(b)]. More sophisticated receivers not only locate a position, but can also guide the navigator to a predetermined location. Several automobile manufacturers have included GPS receivers as an option in their cars. Such receivers come complete with voice commands such as, "Turn left at the next traffic light," as part of their option packages.

Figure 3.38   (a) The screen on the GPS receiver shows the position and strength of the signal between the receiver and the various satellites. At the time this photograph was taken, the receiver picked up 7 of the 12 overhead satellites, bringing the precision to within 20 ft of the actual location. (Photo by William Brouhle.) (b) Global Positioning Systems have allowed for an enormous step forward in navigation. The GPS receiver shown has monitored and recorded precisely where the person has traveled and is now helping the user find his way back to camp.

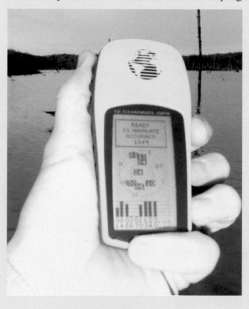

(a)

(b)

Glossary

Component Vector When two or more vectors are added, each of the vectors is called a component of the resultant, or sum, vector. (p. 73)

Displacement The net change in position of an object, or the direct distance and direction it moves; a vector. (p. 69)

Number Plane A plane determined by the horizontal line called the *x*-axis and a vertical line called the *y*-axis intersecting at a right angle at a point called the origin. These two lines divide the number plane into four quadrants. The *x*-axis contains positive numbers to the right of the origin and negative numbers to the left of the origin. The *y*-axis contains positive numbers above the origin and negative numbers below the origin. (p. 72)

Resultant Vector The sum of two or more vectors. (p. 73)

Scalar A physical quantity that can be completely described by a number (called its magnitude) and a unit. (p. 69)

Standard Position A vector is in standard position when its initial point is at the origin of the number plane. The vector is expressed in terms of its length and its angle θ, where θ is measured counterclockwise from the positive *x*-axis to the vector. (p. 74)

Vector A physical quantity that requires both magnitude (size) and direction to be completely described. (p. 69)

x-component The horizontal component of a vector that lies along the *x*-axis. (p. 73)

y-component The vertical component of a vector that lies along the *y*-axis. (p. 73)

Formulas

3.2 To find the *x*- and *y*-components of a vector **v** given in standard position (Fig. 3.39):

1. Complete the right triangle with the legs being the *x*- and *y*-components of the vector.
2. Find the lengths of the legs of the right triangle as follows:

$$|A_x| = |A|(\cos \alpha)$$
$$|A_y| = |A|(\sin \alpha)$$

3. Determine the signs of the *x*- and *y*- components.

Figure 3.39

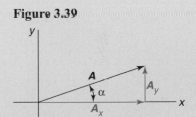

3.3 To find resultant vector **R** in standard position when its *x*- and *y*-components are given:

1. Complete the right triangle with the legs being the *x*- and *y*-components of the vector.
2. Find the acute angle α of the right triangle whose vertex is at the origin by using $\tan \alpha$.
3. Find angle θ in standard position as follows:

$$\theta = \alpha \qquad (\theta \text{ in first quadrant})$$
$$\theta = 180° - \alpha \qquad (\theta \text{ in second quadrant})$$
$$\theta = 180° + \alpha \qquad (\theta \text{ in third quadrant})$$
$$\theta = 360° - \alpha \qquad (\theta \text{ in fourth quadrant})$$

4. Find the magnitude of the vector using the Pythagorean theorem:

$$\mathbf{R} = \sqrt{|R_x|^2 + |R_y|^2}$$

To find the resultant vector **R** of any set of vectors, such as **R** = **A** + **B** + **C**, using right-triangle trigonometry:

1. Find the *x*-component of each vector and add: $R_x = A_x + B_x + C_x$.
2. Find the *y*-component of each vector and add: $R_y = A_y + B_y + C_y$.
3. Find the magnitude of the resultant vector **R** using the Pythagorean theorem
   $$|R| = \sqrt{|R_x|^2 + |R_y|^2}.$$

4. Find the direction of the resultant vector **R** using right-triangle trigonometry by
   (a) first finding the acute angle α between the resultant vector and the *x*-axis and then finding angle θ in standard position or (b) expressing the direction of the resultant vector using some other reference.

Review Questions

1. Displacement
    (a) can be interchanged with direction.
    (b) is a measurement of volume.
    (c) can be described only with a number.
    (d) is the net distance an object travels, showing direction and distance.
2. When adding vectors, the order in which they are added
    (a) is not important.
    (b) is important.
    (c) is important only in certain cases.
3. A vector is in standard position when its initial point is
    (a) at the origin.
    (b) along the *x*-axis.
    (c) along the *y*-axis.
4. Discuss number plane, origin, and axis in your own words.
5. Can every vector be described in terms of its components?
6. Can a vector have more than one set of component vectors?
7. Describe how to add two or more vectors graphically.
8. Describe how to find a resultant vector if given its *x*- and *y*-components.
9. Is a vector limited to a single position in the number plane?
10. Is the angle of a vector in standard position measured clockwise or counterclockwise?
11. What are the limits on the angle measure of a vector in standard position in the third quadrant?
12. Describe how to find the *x*- and *y*-components of a vector given in standard position.
13. Describe how to find a vector in standard position when the *x*- and *y*-components are given.

Review Problems

1. Find the *x*- and *y*-components of vector **R** which has a length of 13.0 cm at 30.0°.
2. Find the *x*- and *y*-components of vector **R**, which has a length of 10.0 cm at 60.0°.
3. Find the *x*- and *y*-components of vector **R**, which has a length of 20.0 cm at 30.0°.
4. Vector **R** has length 9.00 cm at 240.0°. Find its *x*- and *y*-components.
5. Vector **R** has length 9.00 cm at 40.0°. Find its *x*- and *y*-components.

6.   Vector **R** has length 18.0 cm at 305.0°. Find its *x*- and *y*-components.
7.   A hiker is plotting his course on a map with a scale of 1.00 cm = 3.00 km. If the hiker walks 2.50 cm north, then turns south and walks 1.50 cm, what is the actual displacement of the hiker in km?
8.   A hiker is plotting his course on a map with a scale of 1.00 cm = 3.00 km. If the hiker walks 1.50 cm north, then turns south and walks 2.50 cm, what is the actual displacement of the hiker in km?
9.   A co-pilot is charting her course on a map with a scale of 1.00 cm = 20.0 km. If the plane is charted to head 13.0 cm west, 9.00 cm north, and 2.00 cm east, what is the actual displacement of the plane in km?
10. A co-pilot is charting her course on a map with a scale of 1.00 cm = 20.0 km. If the plane is charted to head 25.0° north of east for 16.0 cm, north for 6.00 cm, and west for 5.00 cm, what is the actual displacement of the plane in km?
11. Vector **R** has *x*-component = +14.0 and *y*-component = +3.00. Find its length.
12. Vector **R** has *x*-component = −5.00 and *y*-component = +10.0. Find its length.
13. Vector **R** has *x*-component = +8.00 and *y*-component = −2.00. Find its length.
14. Vector **R** has *x*-component = −3.00 and *y*-component = −4.00. Find its length.
15. Vectors **A**, **B**, and **C** are given. Vector **A** has *x*-component = +3.00 and *y*-component = +4.00. Vector **B** has *x*-component = +5.00 and *y*-component = −7.00. Vector **C** has *x*-component = −2.00 and *y*-component = +1.00. Find the resultant vector **R**.
16. Vectors **A**, **B**, and **C** are given. Vector **A** has *x*-component = +5.00 and *y*-component = +7.00. Vector **B** has *x*-component = +9.00 and *y*-component = −3.00. Vector **C** has *x*-component = −5.00 and *y*-component = +5.00. Find the *x*- and *y*-components of the resultant vector **R**.
17. Vectors **A**, **B**, and **C** are given. Vector **A** has *x*-component = −3.00 and *y*-component = −4.00. Vector **B** has *x*-component = −5.00 and *y*-component = +7.00. Vector **C** has *x*-component = +2.00 and *y*-component = −1.00. Find the *x*- and *y*-components of the resultant vector **R**.
18. Vectors **A**, **B**, and **C** are given. Vector **A** has *x*-component = −5.00 and *y*-component = −7.00. Vector **B** has *x*-component = −9.00 and *y*-component = +3.00. Vector **C** has *x*-component = +5.00 and *y*-component = −5.00. Find the *x*- and *y*-components of the resultant vector **R**.

Graph and find the *x*- and *y*-components of each resultant vector **R**, where **R** = **A** + **B** + **C** + **D**.

19.

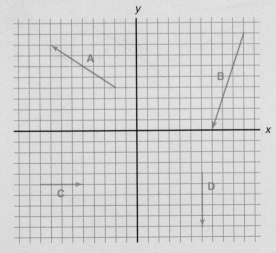

20.

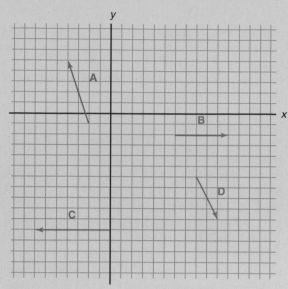

21.   An airplane takes off and flies 245 km on a course of 45.0° south of west and then changes direction and flies 175 km due south, where it lands. Find the displacement from its starting point to its landing point.

22.   A ship travels 155 km from port on a course of 35.0° south of west to an island. It then travels 142 km on a course of 55.5° east of south to a second island. Then it travels 138 km on a course of 9.4° north of east to a third island. Then it travels 185 km due east to a fourth island. Find the displacement from its starting point to its ending point.

APPLIED CONCEPTS

1. The New Clark Bridge is an elegant cable-stayed bridge. Its design requires cables to reach from the road deck up to the tower and back down to the road deck on the other side of the tower as shown in Fig. 3.40. In order to determine the best method for shipping the cables, the shipping company needs to know the lengths of the shortest and longest cables. Given the measurements in the diagram, determine the indicated total lengths BEC and AED, respectively.

Figure 3.40

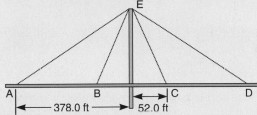

Tower's height above the road deck = 176 ft

2. Frank just learned that the $80\overline{0}$-m section of Broadway that he uses to get to work will be closed for several days. Given the information from a map of Manhattan (Fig. 3.41), what is the distance of Frank's next shortest route?

Figure 3.41

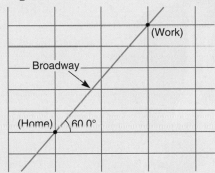

3. Power cables need to be suspended by the power company across a river to a new condominium development. Find the distance across the river in Fig. 3.42.

Figure 3.42

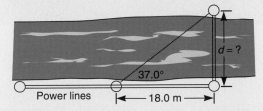

4. Bill has set his GPS to track his route. At the conclusion of his hike, the receiver indicates that he walked 3.50 mi north, 1.00 mi northeast, and 1.50 mi south. How far away is Bill from his original position?

5. With the airplane cruising at $30,\overline{0}00$ ft, the navigator indicates to the captain that the plane should continue traveling north for $50\overline{0}$ km and then turn to a heading of 45.0° east of north for $20\overline{0}$ km. What will be the resultant distance traveled?

CHAPTER 4

MOTION

Air Resistance

Acceleration
g = 9.80 m/s²

g

Motion is a change of position. Velocity and acceleration describe important kinds of motion. An analysis of motion helps introduce the real nature of physics—to understand the nature and behavior of the physical world.

Objectives

The major goals of this chapter are to enable you to:

1. Distinguish between speed and velocity.
2. Use vectors to illustrate and solve velocity problems.
3. Distinguish between velocity and acceleration.
4. Utilize vectors to illustrate and solve acceleration problems.
5. Analyze the motion of an object in free fall.
6. Solve two-dimensional motion problems.
7. Calculate the range of projectile motion.

4.1 Speed Versus Velocity

This chapter begins our study of mechanics, the study of motion. **Motion** can be defined as an object's change in position. How quickly the object changes its position is called its speed.

The ability to analyze and determine the speed of an object is important in many areas of science and technology. Automotive engineers are concerned not only with the motion of the entire vehicle, but also with the motion of the pistons, valves, driveshaft, and so on. The particular speeds of all the internal parts have a direct and very important effect on the motion of the vehicle.

Speed, as measured on a speedometer, is the distance traveled per unit of time. The speed of an automobile is represented in either miles per hour or kilometres per hour (Fig. 4.1). These units actually help define the formula for calculating speed:

$$\text{speed} = \frac{\text{distance traveled}}{\text{time to move that distance}}$$

Speed is a scalar value, for it shows only the magnitude of the position change per unit of time and does not indicate a direction. The unit for speed is a distance unit divided by a time unit, such as miles per hour (mi/h), kilometres per hour (km/h), metres per second (m/s), and feet per second (ft/s). For example, if you drive $35\overline{0}$ mi in 7.00 h, your average speed is

$$\frac{35\overline{0}\ \text{mi}}{7.00\ \text{h}} = 50.0\ \text{mi/h}$$

Speed represents how fast something is moving, yet it does not indicate the direction in which it is traveling. Suppose you started driving from Chicago at a speed of 50 mi/h for 6 h. Where did you end your trip? You may have driven 50 mi/h southwest toward St. Louis, 50 mi/h northeast toward Detroit, 50 mi/h southeast toward Louisville, or 50 mi/h in a loop that brought you back to Chicago. Although speed may indicate how fast you are moving, it may not give you all the information you need to solve a problem.

Distance traveled must be distinguished from displacement. Whereas *distance* traveled may follow a path that is not straight, *displacement* is the net change of position of an object. It is represented by a straight line from the initial position to the final position and is a vector because it has both magnitude and direction.

Figure 4.1 A speedometer measures speed, but not velocity.

The **velocity** of an object is the rate of motion in a particular direction. Velocity is a vector that not only represents the speed, but also indicates the direction of motion. The relationship may be expressed by the equation

$$v_{avg} = \frac{s}{t}$$

or

$$s = v_{avg}t$$

where s = displacement
v_{avg} = average velocity
t = time

This equation is used to find either average speed (a scalar quantity) or the magnitude of the velocity (a vector quantity). Remember that if indicating velocity, the direction must be included with the speed. Therefore, a speed of 50 mi/h would be written 50 mi/h northeast, 50 mi/h up, or 50 mi/h 30° east of south as a velocity.

Figure 4.2 shows an illustration of a car traveling at a constant velocity of 10 m/s to the right. Note that it travels 10 m to the right during each second.

Figure 4.2   The velocity, distance, and time for a car traveling at a constant velocity of 10 m/s to the right is shown in 1-s intervals.

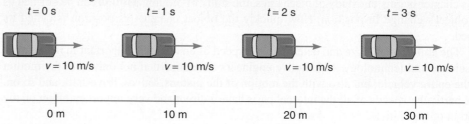

EXAMPLE 1

Find the average speed of an automobile that travels 160 km in 2.0 h.

Data:

$$s = 160 \text{ km}$$
$$t = 2.0 \text{ h}$$
$$v_{avg} = ?$$

Basic Equation:

$$s = v_{avg}t$$

Working Equation:

$$v_{avg} = \frac{s}{t}$$

Substitution:

$$v_{avg} = \frac{160 \text{ km}}{2.0 \text{ h}}$$
$$= 80 \text{ km/h}$$

An airplane flies $35\overline{0}0$ mi in 5.00 h. Find its average speed.

EXAMPLE 2

Data:

$$s = 35\overline{0}0 \text{ mi}$$
$$t = 5.00 \text{ h}$$
$$v_{avg} = ?$$

Basic Equation:

$$s = v_{avg}t$$

Working Equation:

$$v_{avg} = \frac{s}{t}$$

Substitution:

$$v_{avg} = \frac{35\overline{0}0 \text{ mi}}{5.00 \text{ h}}$$
$$= 70\overline{0} \text{ mi/h}$$

· · · · · · · · · · · · · · · · ·

Find the velocity of a plane that travels $60\overline{0}$ km due north in 3 h 15 min.

EXAMPLE 3

Data:

$$s = 60\overline{0} \text{ km}$$
$$t = 3 \text{ h } 15 \text{ min} = 3.25 \text{ h}$$
$$v_{avg} = ?$$

Basic Equation:

$$s = v_{avg}t$$

Working Equation:

$$v_{avg} = \frac{s}{t}$$

Substitution:

$$v_{avg} = \frac{60\overline{0} \text{ km}}{3.25 \text{ h}}$$
$$= 185 \text{ km/h}$$

The direction is north. Thus, the velocity is 185 km/h due north.

· · · · · · · · · · · · · · · · ·

Until now, our study of velocity has assumed a fixed observation point. The frame of reference can also be important for determining relative velocity. Paddling a canoe into a headwind may produce a net velocity of zero for the paddler. Another example is the flight of an airplane in which a crosswind, or in fact wind from any direction, will affect the airplane's velocity with respect to the ground. The airplane's final velocity is calculated by taking into account the velocity of the airplane in calm air and the velocity of any wind that the airplane encounters.

To find the sum (resultant vector) of velocity vectors, use the component method as outlined in Chapter 3.

EXAMPLE 4

A plane is flying due north (at 90°) at 265 km/h and encounters a wind from the east (at 180°) at 55.0 km/h. What is the plane's new velocity with respect to the ground in standard position? Assume that the plane's new velocity is the vector sum of the plane's original velocity and the wind velocity.

First, graph the plane's old velocity as the y-component and the wind velocity as the x-component (Fig. 4.3). The resultant vector is the plane's new velocity with respect to the ground. Find angle α as follows:

Figure 4.3

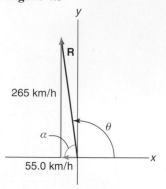

$$\tan \alpha = \frac{\text{side opposite } \alpha}{\text{side adjacent to } \alpha}$$

$$\tan \alpha = \frac{265 \text{ km/h}}{55.0 \text{ km/h}} = 4.818$$

$$\alpha = 78.3°$$

then

$$\theta = 180° - 78.3° = 101.7°$$

Find the magnitude of the new velocity (ground speed) using the Pythagorean theorem:

$$|\mathbf{R}| = \sqrt{|\mathbf{R}_x|^2 + |\mathbf{R}_y|^2}$$
$$|\mathbf{R}| = \sqrt{(55.0 \text{ km/h})^2 + (265 \text{ km/h})^2}$$
$$= 271 \text{ km/h}$$

That is, the new velocity of the plane is 271 km/h at 101.7°.

· · · · · · · · · · · · · · · · ·

EXAMPLE 5

A plane is flying northwest (at 135.0°) at 315 km/h and encounters a wind from 30.0° south of west (at 30.0°) at 65.0 km/h. What is the plane's new velocity with respect to the ground in standard position? Assume that the plane's new velocity is the vector sum of the plane's original velocity and the wind velocity.

First, graph the plane's old velocity and the wind velocity as vectors in standard position (Fig. 4.4). The resultant vector is the plane's new velocity with respect to the ground.

Figure 4.4

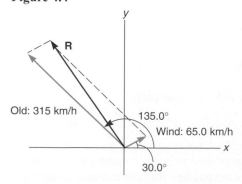

Then, find the x- and y-components of the plane's old velocity and the wind velocity using Fig. 4.5.

Figure 4.5

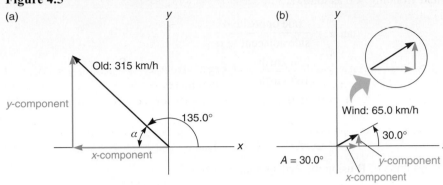

(a)

Old: 315 km/h

y-component

135.0°

α

x-component

(b)

Wind: 65.0 km/h

30.0°

A = 30.0° y-component

x-component

Plane: See Fig. 4.5(a). $\alpha = 180° - 135.0° = 45.0°$

x-component	y-component
$\cos \alpha = \dfrac{\text{side adjacent to } \alpha}{\text{hypotenuse}}$	$\sin \alpha = \dfrac{\text{side opposite } \alpha}{\text{hypotenuse}}$
$\cos 45.0° = \dfrac{x\text{-component}}{315 \text{ km/h}}$	$\sin 45.0° = \dfrac{y\text{-component}}{315 \text{ km/h}}$
$(315 \text{ km/h})(\cos 45.0°) = x\text{-component}$	$(315 \text{ km/h})(\sin 45.0°) = y\text{-component}$
$223 \text{ km/h} = x\text{-component}$	$223 \text{ km/h} = y\text{-component}$
Thus, $x\text{-component} = -223 \text{ km/h}$	$y\text{-component} = +223 \text{ km/h}$

Wind: See Fig. 4.5(b).

x-component	y-component
$\cos \alpha = \dfrac{\text{side adjacent to } \alpha}{\text{hypotenuse}}$	$\sin \alpha = \dfrac{\text{side opposite } \alpha}{\text{hypotenuse}}$
$\cos 30.0° = \dfrac{x\text{-component}}{65.0 \text{ km/h}}$	$\sin 30.0° = \dfrac{y\text{-component}}{65.0 \text{ km/h}}$
$(65.0 \text{ km/h})(\cos 30.0°) = x\text{-component}$	$(65.0 \text{ km/h})(\sin 30.0°) = y\text{-component}$
$56.3 \text{ km/h} = x\text{-component}$	$32.5 \text{ km/h} = y\text{-component}$
Thus, $x\text{-component} = +56.3 \text{ km/h}$	$y\text{-component} = +32.5 \text{ km/h}$

To find **R**:

x-component	y-component	
Plane:	-223 km/h	$+223$ km/h
Wind:	$+56.3$ km/h	$+32.5$ km/h
Sum:	-167 km/h	$+256$ km/h	(Round each component sum to its least precise component.)

Figure 4.6

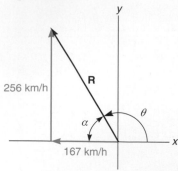

Find angle α from Fig. 4.6 as follows:

$$\tan \alpha = \frac{\text{side opposite } \alpha}{\text{side adjacent to } \alpha}$$

$$\tan \alpha = \frac{256 \text{ km/h}}{167 \text{ km/h}} = 1.533$$

$$\alpha = 56.9°$$

and

$$\theta = 180° - 56.9° = 123.1°$$

Find the magnitude of **R** using the Pythagorean theorem:

$$|\mathbf{R}| = \sqrt{|\mathbf{R}_x|^2 + |\mathbf{R}_y|^2}$$

$$|\mathbf{R}| = \sqrt{(167 \text{ km/h})^2 + (256 \text{ km/h})^2}$$

$$= 306 \text{ km/h}$$

That is, the new velocity of the plane is 306 km/h at 123.1°.

· · · · · · · · · · · · · · · · · ·

PHYSICS CONNECTIONS

Vectors Across Rivers

Crossing the Hudson River, which separates New York City from New Jersey, can be challenging. A working knowledge of velocity and vectors is absolutely essential, especially when attempting to cross the river in strong currents, brisk winds, driving rain, and dense fog. In addition, maneuvering between barges, cruise ships, recreational boaters, and driftwood can make the job even more difficult.

Figure 4.7 An example of how the velocity of a boat and the velocity of the current are combined so the resultant velocity is directed toward the desired location.

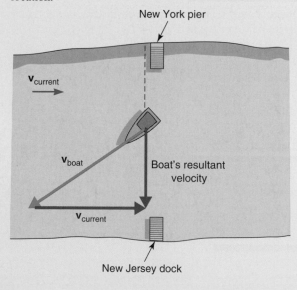

Figure 4.8 Although the boat is not pointed toward the dock, the combination of the boat's velocity (green vector) plus the current's velocity (blue vector) results in a perfect docking (red vector).

Ferry captains like Mike and John combine vectors every time they cross the river. Captain Mike said, "At times crossing the river can be quite tricky. Your ferry might be pointing directly across the river, but the current is pushing you farther down river. In order to combat this, you need to change the boat's heading so when your velocity and the current's velocity combine, you arrive at your planned destination" (Fig. 4.7).

Sometimes, when the current and wind are strong and headed in the same direction, ferryboats can appear to be heading up to 45° away from their destination, yet still travel directly across the river. In such situations, docking can be nerve-racking (Fig. 4.8).

The two experienced sea captains say vectors are even more important in the open seas. Captain Mike said, "Combining your boat's velocity vector with the current and wind vectors can mean the difference between arriving at your home port or at a port 100 miles away." These days the process is easier with the use of radar and computer navigation equipment and software. Looking at the monitor, Captain Mike said, "We can see the velocity of the current and our boat's intended velocity directly on the monitor. Instead of our combining the vectors, the computer can instantaneously combine them and provide the resultant velocity" (Fig. 4.9).

Figure 4.9 (a) Captain Mike combines velocity vectors every time he navigates across the Hudson River.

Figure 4.9 (b) Captain John's high-speed ferry is equipped with computers that automatically combine and display the boat's resultant velocity vector.

PROBLEMS 4.1

Find the average speed (in the given units) of an auto that travels each distance in the given time.

1. Distance of 150 mi in 3.0 h (in mi/h)
2. Distance of 190 m in 8.5 s (in m/s)
3. Distance of 8550 m in 6 min 35 s (in m/s)
4. Distance of 45 km in 0.50 h (in km/h)
5. Distance of 785 ft in 11.5 s (in ft/s)
6. Find the average speed (in mi/h) of a racing car that turns a lap on a 1.00-mi oval track in 30.0 s.
7. While driving at $9\overline{0}$ km/h, how far can you travel in 3.5 h?
8. While driving at $9\overline{0}$ km/h, how far (in metres) do you travel in 1.0 s?
9. An automobile is traveling at 55 mi/h. Find its speed
   (a) in ft/s.   (b) in m/s.   (c) in km/h.

SKETCH

12 cm² | w
4.0 cm

DATA
$A = 12 \text{ cm}^2$, $l = 4.0$ cm, w = ?

BASIC EQUATION
$A = lw$

WORKING EQUATION
$w = \frac{A}{l}$

SUBSTITUTION
$w = \frac{12 \text{ cm}^2}{4.0 \text{ cm}} = 3.0$ cm

10. An automobile is traveling at 22.0 m/s. Find its speed
    (a) in km/h. (b) in mi/h. (c) in ft/s.
11. A semi-trailer truck traveling 100 km/h continues for 2.75 h. How far does it go?
12. A flatbed truck travels for 3.85 h at 105 km/h. How far does it go?
13. The average speed of a garbage truck is 60.0 km/h. How long does it take for the truck to travel 265 km?
14. A highway maintenance truck has an average speed of 55.0 km/h. How far does it travel in 3.65 h?

Find the velocity for each displacement and time.

15. 160 km east in 2.0 h
16. 100 km north in 3.0 h
17. 1000 mi south in 8.00 h
18. 31.0 mi west in 0.500 h
19. 275 km at 30° south of east in 4.50 h
20. 426 km at 45° north of west in 2.75 h
21. Milwaukee is 121 mi (air miles) due west of Grand Rapids. Maria drives 255 mi in 4.75 h from Grand Rapids to Milwaukee around Lake Michigan. Find (a) her average driving speed and (b) her average travel velocity.
22. Telluride, Colorado, is 45 air miles at 11° east of north of Durango. On a winter day, Chuck drove 120 mi from Durango to Telluride around a mountain in $4\frac{1}{4}$ h including a traffic delay. Find (a) his average driving speed and (b) his average travel velocity.

In Problems 23–30, assume that the plane's new velocity is the vector sum of the plane's original velocity and the wind velocity.

23. A plane is flying due north at 325 km/h and encounters a wind from the south at 45 km/h. What is the plane's new velocity with respect to the ground in standard position?
24. A plane is flying due west at 275 km/h and encounters a wind from the west at 80 km/h. What is the plane's new velocity with respect to the ground in standard position?
25. A plane is flying due west at 235 km/h and encounters a wind from the north at 45.0 km/h. What is the plane's new velocity with respect to the ground in standard position?
26. A plane is flying due north at 185 mi/h and encounters a wind from the west at 35.0 mi/h. What is the plane's new velocity with respect to the ground in standard position?
27. A plane is flying southwest at 155 mi/h and encounters a wind from the west at 45.0 mi/h. What is the plane's new velocity with respect to the ground in standard position?
28. A plane is flying southeast at 215 km/h and encounters a wind from the north at 75.0 km/h. What is the plane's new velocity with respect to the ground in standard position?
29. A plane is flying at 25.0° north of west at 190 km/h and encounters a wind from 15.0° north of east at 45.0 km/h. What is the plane's new velocity with respect to the ground in standard position?
30. A plane is flying at 36.0° south of west at 150 mi/h and encounters a wind from 75.0° north of east at 55.0 mi/h. What is the plane's new velocity with respect to the ground in standard position?

4.2 Acceleration

When the dragster shown in Fig. 4.10 travels down a quarter-mile track, its velocity changes. Its velocity at the end of the race is much greater than its velocity near the start. The faster the velocity of the dragster changes, the less its travel time will be.

Figure 4.10 The velocity of the dragster changes in magnitude from zero at the start to its final velocity at the finish.

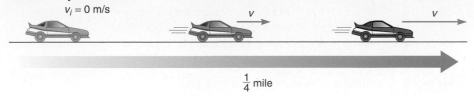

$v_i = 0$ m/s

v

v

$\frac{1}{4}$ mile

The faster its velocity changes, the larger its acceleration will be. **Acceleration** *is the change in velocity per unit time.* That is,

$$\text{average acceleration} = \frac{\text{change in velocity (or speed)}}{\text{elapsed time}}$$

$$= \frac{\text{final velocity} - \text{initial velocity}}{\text{time}}$$

This relationship can be expressed by the equation

$$a = \frac{\Delta v}{t} = \frac{v_f - v_i}{t}$$

or

$$\boxed{\Delta v = at}$$

where Δv = change in velocity (or speed)
$\quad\ a$ = acceleration
$\quad\ t$ = time

The Greek letter Δ (capital delta) is used to mean "change in."

A dragster starts from rest (velocity = 0 ft/s) and attains a speed of 150 ft/s in 10.0 s. Find its acceleration.

EXAMPLE 1

Data:

$$\Delta v = 15\overline{0} \text{ ft/s} - 0 \text{ ft/s} = 15\overline{0} \text{ ft/s}$$

$$t = 10.0 \text{ s}$$

$$a = ?$$

Basic Equation:

$$\Delta v = at$$

Working Equation:

$$a = \frac{\Delta v}{t}$$

Substitution:

$$a = \frac{15\overline{0} \text{ ft/s}}{10.0 \text{ s}}$$

$$= 15.0 \frac{\text{ft/s}}{\text{s}} \text{ or } 15.0 \text{ feet per second per second}$$

Recall from arithmetic that to simplify fractions in the form

$$\frac{\dfrac{a}{b}}{\dfrac{c}{d}}$$

we divide by the denominator; that is, invert and multiply:

$$\frac{\dfrac{a}{b}}{\dfrac{c}{d}} = \frac{a}{b} \div \frac{c}{d} = \frac{a}{b} \cdot \frac{d}{c} = \frac{ad}{bc}$$

Use this idea to simplify the units 15.0 feet per second per second:

$$\frac{\dfrac{15.0 \text{ ft}}{s}}{\dfrac{s}{1}} = \frac{15.0 \text{ ft}}{s} \div \frac{s}{1} = \frac{15.0 \text{ ft}}{s} \cdot \frac{1}{s} = \frac{15.0 \text{ ft}}{s^2} \text{ or } 15.0 \text{ ft/s}^2$$

The units of acceleration are usually ft/s^2 or m/s^2.

· · · · · · · · · · · · · · · · ·

When the speed of an automobile increases from rest to 5 mi/h in the first second, to 10 mi/h in the next second, and to 15 mi/h in the third second, its acceleration is $5 \dfrac{\text{mi/h}}{s}$. That is, its increase in speed is 5 mi/h during each second. In Fig. 4.11 an automobile increases in speed from 6 m/s to 9 m/s in the first second, to 12 m/s in the next second, and to 15 m/s in the third second, so, its acceleration is $3 \dfrac{\text{m/s}}{s}$, usually written 3 m/s^2. This means that the speed of the automobile increases 3 m/s during each second.

Figure 4.11   This car is speeding up with a constant acceleration. Note how the distance covered and the velocity change during each time interval.

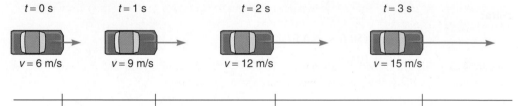

$t = 0 \text{ s}$ $t = 1 \text{ s}$ $t = 2 \text{ s}$ $t = 3 \text{ s}$

$v = 6 \text{ m/s}$ $v = 9 \text{ m/s}$ $v = 12 \text{ m/s}$ $v = 15 \text{ m/s}$

EXAMPLE 2

A car accelerates from 45 km/h to $8\overline{0}$ km/h in 3.00 s. Find its acceleration (in m/s^2).

Data:

$$\Delta v = 8\overline{0} \text{ km/h} - 45 \text{ km/h} = 35 \text{ km/h}$$
$$t = 3.00 \text{ s}$$
$$a = ?$$

Basic Equation:

$$\Delta v = at$$

Working Equation:

$$a = \frac{\Delta v}{t}$$

Substitution:

$$a = \frac{35 \ \cancel{km}/\cancel{h}}{3.00 \ s} \times \frac{1000 \ m}{1 \ \cancel{km}} \times \frac{1 \ \cancel{h}}{3600 \ s}$$

$$= 3.2 \ m/s^2$$

Note the use of the conversion factors to change the units km/h/s to m/s².

· · · · · · · · · · · · · · · · ·

A plane accelerates at 8.5 m/s² for 4.5 s. Find its increase in speed (in m/s). ◄

EXAMPLE 3

Data:

$$a = 8.5 \ m/s^2$$
$$t = 4.5 \ s$$
$$\Delta v = ?$$

Basic Equation:

$$\Delta v = at$$

Working Equation: Same

Substitution:

$$\Delta v = (8.5 \ m/s^2)(4.5 \ s)$$
$$= 38 \ m/s$$

$$\boxed{\frac{m}{s^2} \times s = \frac{m}{s}}$$

· · · · · · · · · · · · · · · · ·

Acceleration means more than just an increase in speed. In fact, since velocity is the speed of an object and its direction of motion, acceleration can mean speeding up, slowing down, or changing direction. The next example illustrates negative acceleration (sometimes called deceleration). **Deceleration** is an acceleration that usually indicates that an object is slowing down (Fig. 4.12). Acceleration when an object changes direction will be discussed in Chapter 9.

Figure 4.12 This car is slowing down with a constant acceleration of −10 m/s². Note how the distance covered and the velocity change during each unit of time interval.

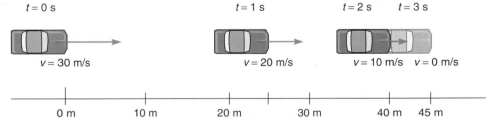

Sometimes the direction of the velocity or the acceleration of an object is understood and does not need to be stated explicitly. For example, if a car is accelerating on a straight road, both the direction of the velocity and the direction of the acceleration are along the road in the direction in which the car is moving. When the car begins to slow down, the direction of the velocity is understood to be along the road in the direction pointing ahead of the car, and the direction of its acceleration (deceleration) is understood to be along the road in the direction pointing behind the car.

PHYSICS CONNECTION

Shuttle Launch

Newton's Second and Third Laws of motion play a vital role during the launch of the Space Shuttle. (See Fig. 5.12.) Just as a massive gun recoils as a less-massive bullet is fired, the massive Shuttle recoils or "lifts off" as the less-massive, high-velocity gaseous exhaust is ejected from the rocket engines.

In order to overcome the force of gravity that the earth applies to all objects, the Space Shuttle must carry enough fuel in its external tank and solid rocket boosters to overcome the gravitational pull of the earth. To do so, the Shuttle carries more than 4.4 million pounds of rocket fuel for use during its launch. Compare this to the weight of the Space Shuttle itself, which is 170,000 pounds, and it is clear that the Shuttle uses a tremendous amount of fuel to lift off.

In order to take advantage of Newton's Third Law and send the Shuttle into space, the gaseous exhaust from the burned fuel is ejected from the Shuttle's rocket boosters at speeds of nearly 6000 mi/h. As the rocket fuel is burned and its gaseous exhaust is ejected out of the rocket, the overall weight of the shuttle's fuel decreases. According to Newton's Second Law, if the same force is applied but the mass is reduced, the acceleration increases.

Figure 5.12 The gas particles are expelled at a high velocity, causing the shuttle to recoil or launch into space.

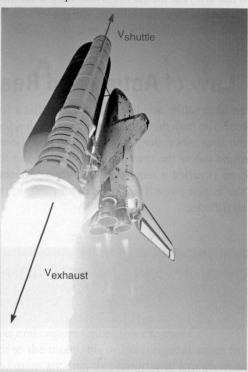

Glossary

Coefficient of Friction The ratio between the frictional force and the normal force of an object. The number represents how rough or smooth two surfaces are when moving across one another. (p. 135)

Force A push or a pull that tends to change the motion of an object or prevent an object from changing motion. Force is a vector quantity and thus has both magnitude and direction. (p. 129)

Friction A force that resists the relative motion of two objects in contact caused by the irregularities of two surfaces sliding or rolling across each other. (p. 134)

Inertia The property of a body that causes it to remain at rest if it is at rest or to continue moving with a constant velocity unless an unbalanced force acts upon it. (p. 130)

Law of Acceleration The total force acting on a body is equal to the mass of the body times its acceleration. (Newton's second law). (p. 131)

Law of Action and Reaction For every force applied by object A to object B (action), there is an equal but opposite force exerted by object B on object A (reaction). (Newton's third law). (p. 143)

Law of Inertia A body that is in motion continues in motion with the same velocity (at constant speed and in a straight line) and a body at rest continues at rest unless an unbalanced (outside) force acts upon it (Newton's first law). (p. 129)

Mass A measure of the inertia of a body. (p. 130)

Normal Force Force perpendicular to the contact surface. (p. 135)

Weight The amount of gravitational pull exerted on an object by the earth or by another large body. (p. 140)

Formulas

5.2 $F = ma$

5.3 $F_f = \mu F_N$

5.5 $F_w = mg$

where $g = 9.80 \text{ m/s}^2$ (metric)

$g = 32.2 \text{ ft/s}^2$ (U.S.)

Review Questions

1. Force
   (a) is a vector quantity.
   (b) may be different from weight.
   (c) does not always cause motion.
   (d) all of the above.
2. The metric weight of a 10-lb bag of sugar is approximately
   (a) 4.45 N.
   (b) 44.5 N.
   (c) 445 N.
   (d) none of the above.
3. Mass and weight
   (a) are the same.
   (b) are different.
   (c) do not change wherever you are.
4. According to Newton's second law, the law of acceleration,
   (a) acceleration is equal to mass times force.
   (b) mass is equal to mass times acceleration.
   (c) force is equal to mass times acceleration.
   (d) none of the above.

5.   Friction
      (a)   always acts parallel to the surface of contact and opposite to the direction of motion.
      (b)   acts in the direction of motion.
      (c)   is smaller when starting than moving.
      (d)   is an imaginary force.
6.   Cite three examples of forces acting without motion being produced.
7.   (a)   Does a pound of feathers have more inertia than a pound of lead?
      (b)   Does the pound of feathers have more mass than the pound of lead?
8.   How is inertia a factor in multicar pileups?
9.   Using your own words, state Newton's first law, the law of inertia.
10.   Distinguish between velocity and acceleration.
11.   When the same force is applied to two different masses, which will have a greater acceleration?
12.   Is 3 pounds heavier than 10 newtons?
13.   Explain how life would be easier or more difficult without friction.
14.   Explain how the weight of an astronaut is different on the moon than on the earth. Would the astronaut's mass be different?
15.   Explain the difference between action and reaction forces.
16.   State Newton's third law of motion, the law of action and reaction, in your own words.

Review Problems

PROBLEM SOLVING

SKETCH

| 12 cm² | w |

4.0 cm

DATA

$A = 12 \text{ cm}^2, l = 4.0 \text{ cm}, w = ?$

BASIC EQUATION

$A = lw$

WORKING EQUATION

$w = \frac{A}{l}$

SUBSTITUTION

$w = \frac{12 \text{ cm}^2}{4.0 \text{ cm}} = 3.0 \text{ cm}$

1.   A crate of mass 6.00 kg is moved by a force of 18.0 N. What is its acceleration?
2.   A 825-N force is required to pedal a bike with an acceleration of 11.0 m/s². What is the mass of the bike and person?
3.   A block of mass 0.89 slug moves with a force of 17.0 lb. Find the block's acceleration.
4.   What is the force necessary for a 2400-kg truck to accelerate at a rate of 8.0 m/s²?
5.   Two movers push a piano across a frictionless surface. One pushes with 29.0 N of force and the other mover exerts 35.0 N. What is the total force?
6.   A 340-N box has a frictional force of 57 N. Find the coefficient of kinetic friction.
7.   A truck pulls a trailer with a frictional force of 870 N and a coefficient of friction of 0.23. What is the trailer's normal force?
8.   A steel box is slid along a steel surface. It has a normal force of 57 N. What is the frictional force?
9.   A rock of a mass 13.0 kg is dropped from a cliff. Find its weight.
10.   A projectile has a mass of 0.37 slug. Find its weight.
11.   What force is required to produce an acceleration of 4.00 m/s² on a wrecking ball with a mass of 50.0 kg?
12.   Find the total force necessary to give a 2$\overline{8}$0-kg motorcycle an acceleration of 3.20 m/s².
13.   A force of 175 N is needed to keep a 6$\overline{4}$0-N stationary engine on wooden skids from sliding on a wooden floor. What is the coefficient of static friction?
14.   A crated garden tractor weighs 375 N. What force is needed to start the crate sliding on a wooden floor when the coefficient of static friction is 0.40?
15.   Find the acceleration of a forklift of mass 14$\overline{0}$0 kg pushed by a force of 21$\overline{0}$0 N that is opposed by a frictional force of 425 N.
16.   What is the weight of a 375-kg air compressor?
17.   What is the mass of a 405-N welder?
18.   What is the mass of a 12.0-N hammer?

APPLIED CONCEPTS

1. Engineers at Boeing developing specs for their "next-generation" 737 aircraft needed to know the acceleration of the 737-900 during a typical take-off. (a) What acceleration would they calculate given the plane's 78,200-kg mass and its maximum engine force of 121,000 N? (b) How fast would the plane be traveling after the first $50\overline{0}$ m of runway? (c) How fast would it travel after the first $150\overline{0}$ m of runway?

2. The Apollo spacecrafts were launched toward the moon using the Saturn rocket, the most powerful rocket available. Each rocket had five engines producing a total of 33.4×10^6 N of force to launch the 2.77×10^6-kg spacecraft toward the moon. (a) Find the average acceleration of the spacecraft. (b) Calculate the altitude of the rocket 2.50 min after launch—the point when the spacecraft loses its first stage.

3. Kirsten's mass is 3.73 slugs. Being the physics fan that she is, she decides to see what her apparent weight will be during an elevator ride. Beginning at rest, the elevator accelerates upward at 4.50 ft/s^2 for 3.00 s and then continues at a constant upward velocity. Finally, as the elevator comes to a stop at the top floor, the elevator slows down (accelerates downward but continues to move upward) at a rate of -5.5 ft/s^2 (the negative sign represents the downward direction). Find Kirsten's weight while the elevator is (a) at rest, (b) speeding up, (c) moving at a constant velocity, and (d) slowing down. The next time you ride in an elevator, concentrate on when you feel heavier and when you feel lighter.

4. A motorcycle racer traveling at 145 km/h loses control in a corner of the track and slides across the concrete surface. The combined mass of the rider and bike is 243 kg. The steel of the motorcycle rubs against the concrete road surface. (a) What is the frictional force between the road and the motorcycle and rider? (b) What would be the acceleration of the motorcycle and rider during the wipeout? (c) Assuming there were no barriers to stop the motorcycle and rider, how long would it take the bike and the rider to slow to a stop?

5. The motorcycle and rider are sliding with the same acceleration as found in Problem 4. If the motorcycle and rider have been sliding for 4.55 s, what will be the force applied to the motorcycle and the rider when they strike the side barrier and come to rest in another 0.530 s?

CHAPTER 6

MOMENTUM

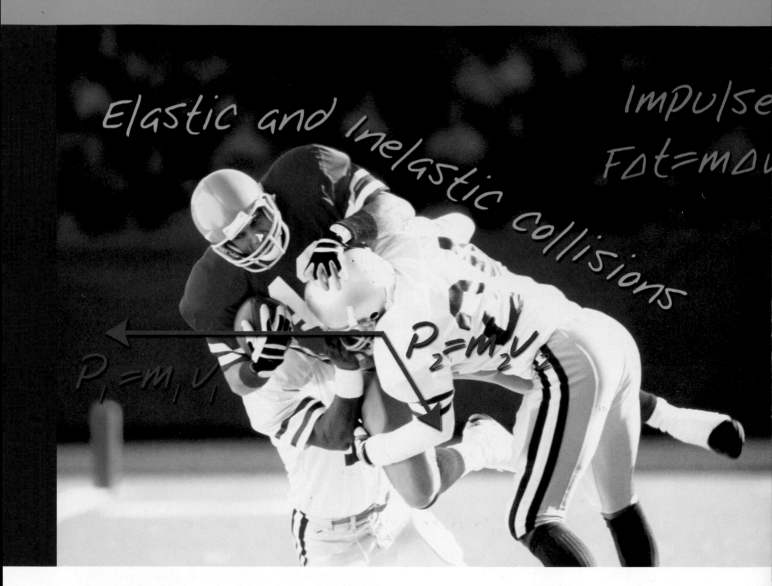

Momentum and impulse are important concepts in describing and understanding the motion of objects and the related effects on those objects. The law of conservation of momentum is an important law of physics, which helps us analyze how two objects interact with each other when they are in contact with each other and when they collide.

Objectives

The major goals of this chapter are to enable you to:

1. Use momentum and impulse in describing motion.
2. State the law of conservation of momentum and apply it to physical problems.
3. Analyze elastic and inelastic collisions of two objects.

6.1   Momentum and Impulse

We know that it is much more difficult to stop a large truck than a small car traveling at the same speed. The truck has more *inertia* and is more difficult to bring to a stop or to begin moving than the car. Momentum is a measure of the amount of inertia and motion an object has or of the difficulty in bringing a moving object to rest. **Momentum** *equals the product of the mass times the velocity of an object.*

$$p = mv$$

where p = momentum
  m = mass
  v = velocity

The momentum of a train makes it impossible to stop within a short distance, and this explains why it cannot stop at a railroad crossing when the engineer sees someone stopped or stalled at it.

The units of momentum are kg m/s in the metric system and slug ft/s in the U.S. system. Momentum is a vector quantity whose direction is the same as the velocity.

Find the momentum of an auto with mass 105 slugs traveling 60.0 mi/h.   ◀ **EXAMPLE 1**

Data:

$$m = 105 \text{ slugs}$$
$$v = 60.0 \text{ mi/h} = 88.0 \text{ ft/s}$$
$$p = ?$$

Basic Equation:

$$p = mv$$

Working Equation: Same

Substitution:

$$p = (105 \text{ slugs})(88.0 \text{ ft/s})$$
$$= 9240 \text{ slugs ft/s}$$

EXAMPLE 2

Find the momentum of an auto with mass 1350 kg traveling 75.0 km/h.

Data:

$$m = 1350 \text{ kg}$$

$$v = 75.0 \; \frac{\text{km}}{\text{h}} \times \frac{1000 \text{ m}}{1 \text{ km}} \times \frac{1 \text{ h}}{3600 \text{ s}} = 20.8 \text{ m/s}$$

$$p = ?$$

Basic Equation:

$$p = mv$$

Working Equation: Same

Substitution:

$$p = (1350 \text{ kg})(20.8 \text{ m/s})$$
$$= 28{,}100 \text{ kg m/s}$$

· · · · · · · · · · · · · · · ·

EXAMPLE 3

Find the velocity that a bullet of mass 1.00×10^{-2} kg would have to have so that it has the same momentum as a lighter bullet of mass 1.80×10^{-3} kg and velocity 325 m/s.

Sketch:

$m_1 = 1.00 \times 10^{-2}$ kg $m_2 = 1.80 \times 10^{-3}$ kg

$v_1 = ?$ $v_2 = 325$ m/s

Data:

Heavier Bullet	Lighter Bullet
$m_1 = 1.00 \times 10^{-2}$ kg	$m_2 = 1.80 \times 10^{-3}$ kg
$v_1 = ?$	$v_2 = 325$ m/s
$p_1 = ?$	$p_2 = ?$

Basic Equations:

$$p_1 = m_1 v_1$$
$$p_2 = m_2 v_2$$

We want

$$p_1 = p_2$$

or

$$m_1 v_1 = m_2 v_2$$

Working Equation:

$$v_1 = \frac{m_2 v_2}{m_1}$$

Substitution:

$$v_1 = \frac{(1.80 \times 10^{-3} \text{ kg})(325 \text{ m/s})}{1.00 \times 10^{-2} \text{ kg}}$$

$$= 58.5 \text{ m/s}$$

· · · · · · · · · · · · · · · ·

The **impulse** on an object is the product of the force applied and the time interval during which the force acts on the object. That is,

$$\text{impulse} = Ft$$

where F = force
        t = time interval during which the force acts

How are impulse and momentum related? Recall that

$$a = \frac{v_f - v_i}{t}$$

If we substitute this equation into Newton's second law of motion, we have

$$F = ma$$

$$F = m\left(\frac{v_f - v_i}{t}\right)$$

$$F = \frac{mv_f - mv_i}{t} \qquad \text{Remove parentheses.}$$

$$Ft = mv_f - mv_i \qquad \text{Multiply both sides by } t.$$

Note that mv_f is the final momentum and mv_i is the initial momentum. That is,

$$\text{impulse} = \Delta p \text{ (change in momentum)} = Ft = mv_f - mv_i$$

Note: The Greek letter Δ ("delta") is used to designate "change in."

The impulse is the measure of the change in momentum of an object in response to an exerted force. To change an object's momentum or motion, a force must be applied to the object for a given period of time. The amount of force and the length of time the force is applied will determine the change in momentum. A common example that illustrates this relationship is a golf club hitting a golf ball (Fig. 6.1). When a golf ball is on the tee, it has zero momentum because its velocity is zero. To give it or change its momentum (impulse), you apply a force for a given period of time. During the time that the club and ball are in contact, the force of the swinging club is transferring most of its momentum to the ball. The impulse given to the ball is the product of the *force* with which the ball is hit and the length of *time* that the club and ball are in direct contact. You can increase its momentum by increasing the *force* (by swinging the golf club faster) or increasing the *time* (by keeping the golf club in contact with the ball longer, which shows the importance of "followthrough").

Figure 6.1   When a person hits a golf ball with a golf club, the club applies a force F during the time t that the club is in contact with the ball. The impulse (change in momentum) is $Ft = mv_f - mv_i = mv_f$ because $v_i = 0$.

(a)   (b)

EXAMPLE 4

A 17.5-g bullet is fired at a muzzle velocity of 582 m/s from a gun with a mass of 8.00 kg and a barrel length of 75.0 cm.

(a) How long is the bullet in the barrel?
(b) What is the force on the bullet while it is in the barrel?
(c) Find the impulse exerted on the bullet while it is in the barrel.
(d) Find the bullet's momentum as it leaves the barrel.

Sketch:

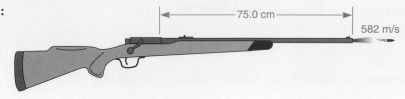

75.0 cm

582 m/s

(a) Data:

$$s = 75.0 \text{ cm} = 0.750 \text{ m}$$
$$v_f = 582 \text{ m/s}$$
$$v_i = 0 \text{ m/s}$$
$$v_{avg} = \frac{v_f + v_i}{2} = \frac{582 \text{ m/s} + 0 \text{ m/s}}{2} = 291 \text{ m/s}$$
$$t = ?$$

Basic Equation:

$$s = v_{avg}t$$

Working Equation:

$$t = \frac{s}{v_{avg}}$$

Substitution:

$$t = \frac{0.750 \text{ m}}{291 \text{ m/s}}$$
$$= 0.00258 \text{ s}$$

Note: This is the length of time that the force is applied to the bullet.

(b) Data:

$$t = 0.00258 \text{ s}$$
$$m = 17.5 \text{ g} = 0.0175 \text{ kg}$$
$$v_f = 582 \text{ m/s}$$
$$v_i = 0 \text{ m/s}$$
$$F = ?$$

Basic Equation:

$$Ft = mv_f - mv_i$$

Working Equation:

$$F = \frac{mv_f - mv_i}{t}$$

Substitution:

$$F = \frac{(0.0175 \text{ kg})(582 \text{ m/s}) - (0.0175 \text{ kg})(0 \text{ m/s})}{0.00258 \text{ s}}$$

$$= 3950 \text{ kg m/s}^2$$

$$= 3950 \text{ N} \qquad (1 \text{ N} = 1 \text{ kg m/s}^2)$$

(c) Data:

$$t = 0.00258 \text{ s}$$
$$F = 3950 \text{ N}$$
$$\text{impulse} = ?$$

Basic Equation:

$$\text{impulse} = Ft$$

Working Equation: Same

Substitution:

$$\text{impulse} = (3950 \text{ N})(0.00258 \text{ s})$$

$$= 10.2 \text{ N s}$$

$$= 10.2 \text{ (kg m/s}^2)(\text{s}) \qquad (1 \text{ N} = 1 \text{ kg m/s}^2)$$

$$= 10.2 \text{ kg m/s}$$

(d) Data:

$$m = 17.5 \text{ g} = 0.0175 \text{ kg}$$
$$v - 582 \text{ m/s}$$
$$p = ?$$

Basic Equation:

$$p = mv$$

Working Equation: Same

Substitution:

$$p = (0.0175 \text{ kg})(582 \text{ m/s})$$

$$= 10.2 \text{ kg m/s}$$

Note: The impulse equals the change in momentum.

· · · · · · · · · · · · · · · · · ·

TRY THIS ACTIVITY

Scrambled Eggs

Drop a raw egg from a height of a few feet onto a surface that can be cleaned. Observe the motion of the egg as it hits the surface and note the time the egg takes to come to rest. Drop another raw egg from the same height into a suspended bed sheet. Again observe the motion of the egg as it hits the bed sheet and note the time the egg takes to come to rest (Fig. 6.2). Explain the connection between what happened to these eggs and how airbags in automobiles work.

Figure 6.2

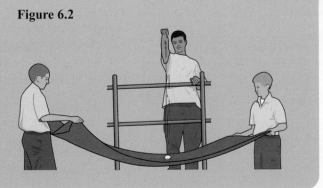

One of the most important laws of physics, the **law of conservation of momentum**, is the following:

LAW OF CONSERVATION OF MOMENTUM
When no outside forces are acting on a system of moving objects, the total momentum of the system remains constant.

For example, consider a 35-kg boy and a 75-kg man standing next to each other on ice skates on "frictionless" ice (Fig. 6.3). The man pushes on the boy, which gives the boy a velocity of 0.40 m/s. What happens to the man? Initially, the total momentum was zero because the initial velocity of each was zero. According to the law of conservation of momentum, the total momentum must still be zero. That is,

$$m_{boy}v_{boy} + m_{man}v_{man} = 0$$
$$(35 \text{ kg})(0.40 \text{ m/s}) + (75 \text{ kg})v_{man} = 0$$
$$v_{man} = -0.19 \text{ m/s}$$

Figure 6.3 Momentum is conserved by the lighter boy moving faster than the heavier man.

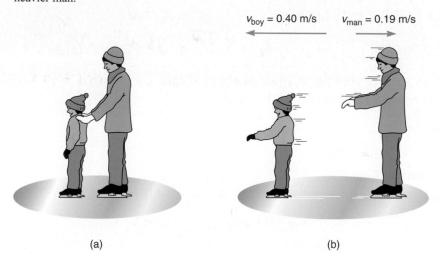

$v_{boy} = 0.40$ m/s $v_{man} = 0.19$ m/s

(a)   (b)

Note: The minus sign indicates that the man's velocity and the boy's velocity are in opposite directions.

Rocket propulsion is another illustration of conservation of momentum. **Wernher von Braun** was a pioneering rocket scientist. As in the example of the skaters, the total momentum of a rocket on the launch pad is zero. When the rocket engines are fired, hot exhaust gases (actually gas molecules) are expelled downward through the rocket nozzle at tremendous speeds. As the rocket takes off, the sum of the total momentums of the rocket and the gas particles must remain zero. The total momentum of the gas particles is the sum of the products of each mass and its corresponding velocity and is directed down. The momentum of the rocket is the product of its mass and its velocity and is directed up.

When the rocket is in space, its propulsion works in the same manner. The conservation of momentum is still valid except that when the rocket engines are fired, the total momentum is a nonzero constant. This is because the rocket has velocity.

Actually, repair work is more difficult in space than it is on the earth because of the conservation of momentum and the "weightlessness" of objects in orbit. On the earth, when

Wernher von Braun (1912–1977),

engineer and rocket expert, was born in Germany. He was chiefly responsible for the manufacture and launching of the first American artificial earth satellite, Explorer I, in 1958. As director of the Marshall Space Flight Center from 1960 to 1970, he developed the Saturn rocket for the Apollo 8 moon landing in 1969.

a hammer is swung, the person is coupled to the earth by frictional forces, so that the person's mass includes that of the earth. In space orbit, because the person is weightless, there is no friction to couple him or her to the spaceship. A person in space has roughly the same problem driving a nail as a person on the earth would have wearing a pair of "frictionless" roller skates.

A change in momentum takes force and time because

$$\text{change in momentum} = \text{impulse} = Ft$$

As we noted earlier in this section, it is more difficult to stop a large truck than a small car traveling at the same speed and impossible to stop a rapidly moving train within a short distance. These events can be explained in terms of the **impulse–momentum theorem** as follows.

If the mass of an object is constant, then a change in its velocity results in a change in its momentum. That is,

$$\Delta p = m\Delta v$$

The impulse of an object equals its change in momentum. That is,

$$F\Delta t = \Delta p$$

Then,

IMPULSE–MOMENTUM THEOREM

$$F\Delta t = \Delta p = m\Delta v = mv_f - mv_i$$

What force is required to slow a 1450-kg car traveling 115 km/h to 45.0 km/h within 3.00 s? How far does the car travel during its deceleration?

EXAMPLE 5

Data:

$$m = 1450 \text{ kg}$$

$$v_f = 45\frac{\text{km}}{\text{h}} \times \frac{1\text{ h}}{3600\text{ s}} \times \frac{1000\text{ m}}{1\text{ km}} = 12.5 \text{ m/s}$$

$$v_i = 115\frac{\text{km}}{\text{h}} \times \frac{1\text{ h}}{3600\text{ s}} \times \frac{1000\text{ m}}{1\text{ km}} = 31.9 \text{ m/s}$$

$$\Delta t = 3.00 \text{ s}$$

$$F = ?$$

Basic Equation:

$$F\Delta t = mv_f - mv_i$$

Working Equation:

$$F = \frac{mv_f - mv_i}{\Delta t}$$

Substitution:

$$F = \frac{(1450 \text{ kg})(12.5 \text{ m/s}) - (1450 \text{ kg})(31.9 \text{ m/s})}{3.00 \text{ s}}$$

$$= -9380 \text{ kg m/s}^2 = -9380 \text{ N}$$

Note: The negative sign indicates a deceleration force.

Basic Equation:

$$s = \tfrac{1}{2}(v_f + v_i)t$$

Working Equation: Same

Substitution:

$$s = \tfrac{1}{2}(12.5 \text{ m/s} + 31.9 \text{ m/s})(3.00 \text{ s})$$
$$= 66.6 \text{ m}$$

·················

PHYSICS CONNECTIONS

Airbags

During an automobile front-end collision, passengers will continue to travel forward until the dashboard, seat belt, or airbag applies a force on them to stop them. Airbags are designed to provide a cushion-like effect to gradually bring passengers to rest. Airbags increase the time it takes to bring passengers to a stop and reduce the force of the impact (Fig. 6.4). Airbags used in conjunction with seat belts help prevent death and serious injury.

Airbags expand from the steering wheel or dashboard when a sudden impulse or a change in momentum of the vehicle triggers a sensor that is connected to a heating element. The heating element causes a chemical reaction with a propellant that fills the airbag with nitrogen gas within $\frac{1}{20}$ s. This short inflation time gives the airbag enough time to inflate before the passenger strikes it. Within $\frac{1}{2}$ s, the collision is completed and the airbag deflates.

Airbags are designed to strike the average seat-belted man in the midsection of the body. An airbag, which expands at a rate of 150 mi/h, can be quite dangerous if the bag strikes short individuals in the face. Injuries to women and children caused by airbags are a serious problem. Efforts are being made to automatically adjust airbag deployment to make airbags safer for all passengers. Airbags are also used for side-impact collisions.

Figure 6.4   An airbag increases the time it takes to bring a passenger to a stop in a collision by reducing the force of the impact applied to the passenger.

Photo Courtesy of Insurance Institute for Highway Safety. Reprinted with Permission

PROBLEMS 6.1

Find the momentum of each object.

1. $m = 2.00$ kg, $v = 40.0$ m/s
2. $m = 5.00$ kg, $v = 90.0$ m/s
3. $m = 17.0$ slugs, $v = 45.0$ ft/s
4. $m = 38.0$ kg, $v = 97.0$ m/s
5. $m = 3.8 \times 10^5$ kg, $v = 2.5 \times 10^3$ m/s
6. $m = 3.84$ kg, $v = 1.6 \times 10^5$ m/s
7. $F_w = 1.50 \times 10^5$ N, $v = 4.50 \times 10^4$ m/s
8. $F_w = 3200$ lb, $v = 6\overline{0}$ mi/h (change to ft/s)
9. (a) Find the momentum of a heavy automobile of mass $18\overline{0}$ slugs traveling 70.0 ft/s.
   (b) Find the velocity of a light auto of mass 80.0 slugs so that it has the same momentum as the auto in part (a).
   (c) Find the weight (in lb) of each auto in parts (a) and (b).
10. (a) Find the momentum of a bullet of mass 1.00×10^{-3} slug traveling $70\overline{0}$ ft/s.
    (b) Find the velocity of a bullet of mass 5.00×10^{-4} slug so that it has the same momentum as the bullet in part (a).
11. (a) Find the momentum of an automobile of mass 2630 kg traveling 21.0 m/s.
    (b) Find the velocity (in km/h) of a light auto of mass 1170 kg so that it has the same momentum as the auto in part (a).
12. A ball of mass 0.50 kg is thrown straight up at 6.0 m/s.
    (a) What is the initial momentum of the ball?
    (b) What is the momentum of the ball at its peak?
    (c) What is the momentum of the ball as it hits the ground?
13. A bullet with mass 60.0 g is fired with an initial velocity of 575 m/s from a gun with mass 4.50 kg. What is the speed of the recoil of the gun?
14. A cannon is mounted on a railroad car. The cannon shoots a 1.75-kg ball with a muzzle velocity of $30\overline{0}$ m/s. The cannon and the railroad car together have a mass of $450\overline{0}$ kg. If the ball, cannon, and railroad car are initially at rest, what is the recoil velocity of the car and cannon?
15. A 125-kg pile driver falls from a height of 10.0 m to hit a piling.
    (a) What is its speed as it hits the piling?
    (b) With what momentum does it hit the piling?
16. A person is traveling 75.0 km/h in an automobile and throws a bottle of mass 0.500 kg out the window.
    (a) With what momentum does the bottle hit a roadway sign?
    (b) With what momentum does the bottle hit an oncoming automobile traveling 85.0 km/h in the opposite direction?
    (c) With what momentum does the bottle hit an automobile passing and traveling 85.0 km/h in the same direction?
17. A 75.0-g bullet is fired with a muzzle velocity of $46\overline{0}$ m/s from a gun with mass 3.75 kg and barrel length of 66.0 cm.
    (a) How long is the bullet in the barrel?
    (b) What is the force on the bullet while it is in the barrel?
    (c) Find the impulse exerted on the bullet while it is in the barrel.
    (d) Find the bullet's momentum as it leaves the barrel.
18. A 60.0-g bullet is fired at a muzzle velocity of 525 m/s from a gun with mass 4.50 kg and a barrel length of 55.0 cm.
    (a) How long is the bullet in the barrel?
    (b) What is the force on the bullet while it is in the barrel?
    (c) Find the impulse exerted on the bullet while it is in the barrel.
    (d) Find the bullet's momentum as it leaves the barrel.
19. (a) What force is required to stop a 1250-kg car traveling 95.0 km/h within 4.00 s?
    (b) How far does the car travel during its deceleration?
20. (a) What force is required to slow a 1350-kg car traveling 90.0 km/h to 25.0 km/h within 4.00 s?
    (b) How far does the car travel during its deceleration?

SKETCH

12 cm² | w

4.0 cm

DATA

$A = 12$ cm², $l = 4.0$ cm, $w = ?$

BASIC EQUATION

$A = lw$

WORKING EQUATION

$w = \frac{A}{l}$

SUBSTITUTION

$w = \frac{12 \text{ cm}^2}{4.0 \text{ cm}} = 3.0$ cm

course. The second vehicle's final path is 90° to the right of the final path of the first vehicle.

  (a) What is the momentum of the first vehicle after the collision?
  (b) What is the momentum of the second vehicle after the collision?
  (c) What is the velocity of the first vehicle after the collision?
  (d) What is the velocity of the second vehicle after the collision?

16. Two vehicles of equal mass collide at a 90° intersection. If the momentum of vehicle A is 1.20×10^5 kg km/h east and the momentum of vehicle B is 8.50×10^4 kg km/h north, what is the resulting momentum of the final mass?

17. A vehicle with mass of 950 kg is driving east with velocity 12.0 m/s. It crashes into a stationary vehicle of the same mass. Assume an elastic collision. The first vehicle is deflected at an angle of 40.0° north of its original path. The second vehicle's path is 90° to the right of the first vehicle's final path.

  (a) What is the momentum of the first vehicle after the crash?
  (b) What is the momentum of the second vehicle after the crash?
  (c) What is the velocity of the first vehicle after the crash?
  (d) What is the velocity of the second vehicle after the crash?

Glossary

Elastic Collision A collision in which two objects return to their original shape without being permanently deformed. (p. 158)

Impulse The product of the force exerted and the time interval during which the force acts on the object. Impulse equals the change in momentum of an object in response to the exerted force. (p. 151)

Impulse–Momentum Theorem If the mass of an object is constant, then a change in its velocity results in a change of its momentum. That is, $F\Delta t = \Delta p = m\Delta v = mv_f - mv_i$. (p. 155)

Inelastic Collision A collision in which two objects couple together. (p. 159)

Law of Conservation of Momentum When no outside forces are acting on a system of moving objects, the total momentum of the system remains constant. (p. 154)

Momentum A measure of the amount of inertia and motion an object has or the difficulty in bringing a moving object to rest. Momentum equals the mass times the velocity of an object. (p. 149)

Formulas

6.1 $p = mv$
    impulse $= Ft$
    impulse $=$ change in momentum
    impulse–momentum theorem

$$F\Delta t = \Delta p = m\Delta v = mv_f - mv_i$$

6.2   total momentum $_{before\ collision}$ = total momentum $_{after\ collision}$

Review Questions

1. Momentum is
   (a) equal to speed times weight.
   (b) equal to mass times velocity.
   (c) the same as force.
2. Impulse is
   (a) a force applied to an object.
   (b) the initial force applied to an object.
   (c) the initial momentum applied to an object.
   (d) the change in momentum due to a force being applied to an object during a given time.
3. Why do a slow-moving loaded truck and a speeding rifle bullet each have a large momentum?
4. How are impulse and change in momentum related?
5. Why is "followthrough" important in hitting a baseball or a golf ball?
6. Describe in your own words the law of conservation of momentum.
7. Describe conservation of momentum in terms of a rocket being fired.
8. One billiard ball striking another is an example of a(n) _____ collision.
9. One moving loaded railroad car striking and coupling with a parked empty railroad car and then both moving on down the track is an example of a(n) _____ collision.
10. A father and 8-year-old son are standing on ice skates in an ice arena. The father then pushes the son on the back to give him a quick start. What do we know about the momentum of each person?

Review Problems

1. A truck with mass 1475 slugs travels 57.0 mi/h. Find its momentum.
2. A projectile with mass 27.0 kg is fired with a momentum of 5.50 kg m/s. Find its velocity.
3. A box is pushed with a force of 125 N for 2.00 min. What is the impulse?
4. What is the momentum of a bullet of mass 0.034 kg traveling at 250 m/s?
5. A 4.00-g bullet is fired from a 4.50-kg gun with a muzzle velocity of 625 m/s. What is the speed of the recoil of the gun?
6. A 150-kg pile driver falls from a height of 7.5 m to hit a piling.
   (a) What is its speed as it hits the piling?
   (b) With what momentum does it hit the piling?
7. A 15.0-g bullet is fired at a muzzle velocity of 3250 m/s from a high-powered rifle with a mass of 4.75 kg and barrel of length 75.0 cm.
   (a) How long is the bullet in the barrel?
   (b) What is the force on the bullet while it is in the barrel?
   (c) Find the impulse exerted on the bullet while it is in the barrel.
   (d) Find the bullet's momentum as it leaves the barrel.
8. What force is required to slow a 1250-kg car traveling 115 km/h to 30.0 km/h within 3.50 s? (a) How far does the car travel during its deceleration? (b) How long does it take for the car to come to a complete stop at this same rate of deceleration?
9. One ball of mass 575 g traveling 3.50 m/s to the right collides with another ball of mass 425 g that is initially at rest. After the collision, the lighter ball is traveling 4.03 m/s. What is the velocity of the heavier ball after the collision?
10. A railroad car of mass 2.25×10^4 kg is traveling east 5.50 m/s and collides with a railroad car of mass 3.00×10^4 kg traveling west 1.50 m/s. Find the velocity of the railroad cars that become coupled after the collision.
11. A 195-g ball traveling 4.50 m/s to the right collides with a 125-g ball traveling 12.0 m/s to the left. After the collision, the heavier ball is traveling 8.40 m/s to the left. What is the velocity of the lighter ball after the collision?
12. Two trucks of equal mass collide at a 90° intersection. If the momentum of truck A is 9.50×10^4 kg km/h east and the momentum of truck B is 1.05×10^5 kg km/h north, what is the resulting momentum of the final mass, assuming the trucks remain joined together following the crash?
13. Ball A, of mass 0.35 kg, has a velocity 0.75 m/s east. It strikes a stationary ball, also of mass 0.35 kg. Ball A deflects off B at an angle of 37.0° north of A's original path. Ball B moves in a line 90° right of the final path of A.
   (a) Find ball A's momentum after the collision.
   (b) Find ball B's momentum after the collision.
   (c) Find the velocity of A after the collision.
   (d) Find the velocity of B after the collision.

APPLIED CONCEPTS

1. A coach knows it is vital that the volleyballs be fully inflated before a match. (a) Calculate the impulse on a spiked 0.123-slug volleyball when the incoming velocity of the ball is −11.5 ft/s and the outgoing velocity is 57.3 ft/s. (b) Using physics terms, explain what would happen to the outgoing velocity and the impulse on the ball if it were not fully inflated.

2. An automobile accident causes both the driver and passenger front airbags to deploy. (a) If the vehicle was traveling at a speed of 88.6 km/h and is now at rest, find the change in momentum for both the 68.4-kg adult driver and the 34.2-kg child passenger. (b) The adult took 0.564 s and the child took 0.260 s to come to rest. Find the force that the airbag exerted on each individual. Explain why airbags tend to be dangerous for children.

3. Several African tribes engage in a ritual much like bungee jumping, in which a tree vine is used instead of a bungee cord. (a) A 70.8-kg person falls from a cliff with such a vine attached to his ankles. Find the force applied to his ankles if it takes 0.355 s to change his velocity from −18.5 m/s to rest (the negative sign represents the downward direction). (b) Find the force applied to the person's ankles when he takes 1.98 s to change his velocity from −18.5 m/s to +9.75 m/s using a manufactured bungee cord (remember that a bungee cord causes the jumper to bounce back upward). (c) Using physics terms, describe why it is safer to use a bungee cord than a tree vine.

4. Sally, who weighs 125 lb, knows that getting out of a 65.5-lb canoe can be a difficult experience. (a) What happens to the canoe's velocity if she attempts to step out of the canoe and onto the dock with a velocity of 3.50 ft/s? (b) If the canoe were heavier, would it be easier or harder to step out of it?

5. An automobile accident investigator needs to determine the initial westerly velocity of a Jeep (m = 1720 kg) that may have been speeding before colliding head-on with a Volkswagen (m = 1510 kg) that was moving with a velocity of 75.7 km/h east. The speed limit on this road is 90 km/h. After the collision, the Jeep and the Volkswagen stuck together and continued to travel with a velocity of 15.5 km/h west. (a) Find the initial westerly velocity of the Jeep. (b) Was the Jeep speeding?

Figure 7.7

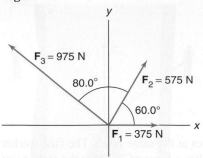

To find $\mathbf{F}_R$, find the x- and y-components of each vector and add the components as follows:

Vector	x-component		y-component	
$\mathbf{F}_1$		375 N		0 N
$\mathbf{F}_2$	$\|\mathbf{F}_2\| \cos \alpha =$		$\|\mathbf{F}_2\| \sin \alpha =$	
	$(575 \text{ N}) \cos 60.0° =$	288 N	$(575 \text{ N}) \sin 60.0° =$	498 N
$\mathbf{F}_3$	$\|\mathbf{F}_3\| \cos \alpha =$		$\|\mathbf{F}_3\| \sin \alpha =$	
	$-(975 \text{ N}) \cos 40.0° =$	-747 N	$(975 \text{ N}) \sin 40.0° =$	627 N
	Note for $\mathbf{F}_3$: $\alpha = 180° - 140.0° = 40.0°$			
$\mathbf{F}_R$		-84 N		1125 N

Find angle α of the resultant vector as follows:

$$\tan \alpha = \frac{|\mathbf{F}_{Ry}|}{|\mathbf{F}_{Rx}|} = \frac{1125 \text{ N}}{84 \text{ N}} = 13.39$$

$$\alpha = 85.7°$$

Note: The x-component of $\mathbf{F}_R$ is negative and its y-component is positive; this means that $\mathbf{F}_R$ is in the second quadrant. Its angle in standard position is $180° - 85.7° = 94.3°$.
Find the magnitude of $\mathbf{F}_R$ using the Pythagorean theorem:

$$|\mathbf{F}_R| = \sqrt{|\mathbf{F}_{Rx}|^2 + |\mathbf{F}_{Ry}|^2}$$
$$|\mathbf{F}_R| = \sqrt{(84 \text{ N})^2 + (1125 \text{ N})^2}$$
$$= 1130 \text{ N}$$

That is, $\mathbf{F}_R = 1130$ N at $94.3°$, or $94.3°$ from $\mathbf{F}_1$.
The resultant vector is shown in Fig. 7.8.

Figure 7.8

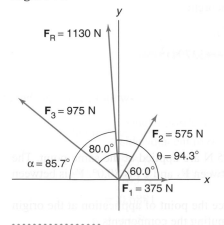

TRY THIS ACTIVITY

Tension Forces

An easy way to experiment with concurrent forces in cables is to suspend a 1.0-kg mass from two equal-length strings (the actual length of the strings is not important) (see Fig. 7.9). Attach a spring scale at the end of each string and observe the reading on the scales as they suspend the mass vertically. In this position, each spring scale should have the same reading, which is half the weight of the suspended mass.

Now, slowly separate the two strings from one another, gradually creating a larger and larger angle between them. Use the spring scales to observe the tension force in the strings and use a protractor to determine the angle between the two strings. What happens to the tension force as the angle between the suspending cables increases? What is the resultant upward force from the two strings?

Figure 7.9 Two spring scales suspending a mass by two cables that are separated from the vertical at equal angles

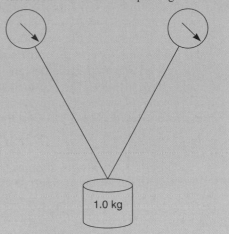

1.0 kg

PROBLEMS 7.1

Find the sum of each set of forces acting at the same point in a straight line.

1.   355 N (right); 475 N (right); 245 N (left); 555 N (left)
2.   703 N (right); 829 N (left); 125 N (left); 484 N (left)
3.   Forces of 225 N and 175 N act at the same point.
     (a)   What is the magnitude of the maximum net force the two forces can exert together?
     (b)   What is the magnitude of the minimum net force the two forces can exert together?
4.   Three forces with magnitudes of 225 N, 175 N, and 125 N act at the same point.
     (a)   What is the magnitude of the maximum net force the three forces can exert together?
     (b)   What is the magnitude of the minimum net force the three forces can exert together?

Find the sum of each set of vectors. Give angles in standard position.

5.

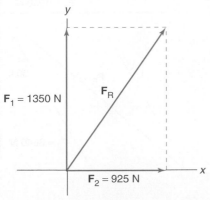

F_1 = 1350 N

F_R

F_2 = 925 N

6.

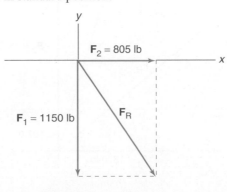

F_2 = 805 lb

F_1 = 1150 lb

F_R

SKETCH

12 cm² | w

4.0 cm

DATA

A = 12 cm², l = 4.0 cm, w = ?

BASIC EQUATION

A = lw

WORKING EQUATION

$w = \frac{A}{l}$

SUBSTITUTION

$w = \frac{12 \text{ cm}^2}{4.0 \text{ cm}} = 3.0$ cm

PHYSICS CONNECTIONS

The Cable-Stayed Bridge

All bridges are designed and constructed according to the needs of the community, the desired aesthetics, the costs, and the geographic and geological conditions around the bridge site. One of the most popular, attractive, and cost-effective designs is the cable-stayed bridge. The physical strength and relatively low cost of the design made the cable-stayed bridge ideal for the midlength span across the Mississippi River at Alton, Illinois (Fig. 7.10).

Cable-stayed bridges support the roadbed by attaching one end of multiple cables directly to the deck, passing them through a vertical tower, and attaching them to the deck on the opposite side of the tower. Through the use of lighter, stronger materials, engineers are able to avoid the need for the heavy and expensive steel and massive anchorages that are needed to support more traditional suspension bridges.

Figure 7.10 The new cable-stayed design of the New Clark Bridge at Alton, Illinois

The combination of compression, tension, shear, and bending forces keeps the cable-stayed New Clark Bridge static. This particular cable-stayed bridge was designed to replace a deteriorating truss bridge that the community had outgrown. The New Clark Bridge meets the needs of the growing community and a busy shipping channel, is aesthetically pleasing, and is economically viable. The bridge also met the geographic and geological conditions as dictated by the Mississippi River and surrounding landscape.

7. If forces of $10\overline{0}0$ N acting in a northerly direction and $15\overline{0}0$ N acting in an easterly direction both act on the same point, what is the resultant force?

8. If two forces of $10\overline{0}$ N and 50.0 N, respectively, act in a westerly direction on a point and a force of 175 N acts in a northerly direction on the same point, what is the resultant force?

Find the sum of each set of vectors. Give angles in standard position.

9.   10.

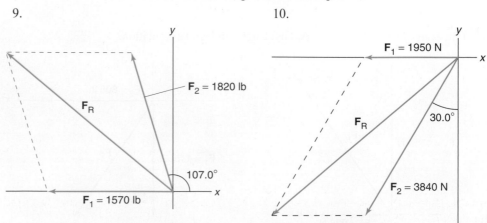

11.

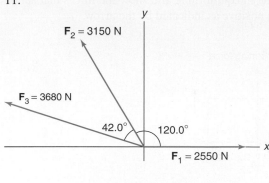

12.

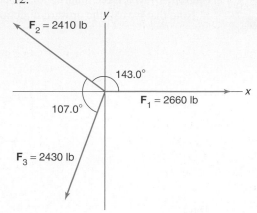

13. Forces of $F_1 = 1150$ N, $F_2 = 875$ N, and $F_3 = 1450$ N are applied at the same point. The angle between F_1 and F_2 is $90.0°$ and the angle between F_2 and F_3 is $120.0°$. F_2 is between F_1 and F_3. Find the resultant force.
14. Four forces, each of magnitude 2750 lb, act at the same point. The angle between adjacent forces is $30.0°$. Find the resultant force.

7.2 Concurrent Forces in Equilibrium

Equilibrium in One Dimension

Equilibrium *is the state of a body in which there is no change in its motion.* A body is in equilibrium when the net force acting on it is zero. That is, it is not accelerating; it is either at rest or moving at a constant velocity. The study of objects in equilibrium is called **statics**.

The forces applied to an object in one dimension act in the same direction or in opposite directions. For the net force to be zero, the forces in one direction must equal the forces in the opposite direction. We can write the equation for equilibrium in one dimension as

$$\boxed{F_+ = F_-}$$

where F_+ = the sum of all forces acting in one direction (call it the positive direction).
  F_- = the sum of all the forces acting in the opposite (negative) direction.

Note in Fig. 7.11 that the downward force (weight of the bridge) must equal the sum of the upward forces produced by the two bridge supports for the bridge to be in equilibrium.

Figure 7.11

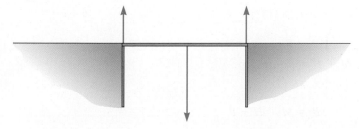

A cable supports a large crate of weight 1250 N (Fig. 7.12). What is the upward force on the crate if it is in equilibrium?

EXAMPLE 1

Figure 7.12

1250 N

Sketch: Draw a force diagram of the crate in equilibrium, and show the forces that act on it. Note that we call the upward direction positive as indicated by the arrow.

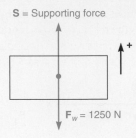

S = Supporting force

$\mathbf{F}_w$ = 1250 N

Data:

$$\mathbf{F}_w = 1250 \text{ N}$$
$$\mathbf{S} = ?$$

Basic Equation:

$$\mathbf{F}_+ = \mathbf{F}_-$$

Working Equation:

$$\mathbf{S} = \mathbf{F}_w$$

Substitution:

$$\mathbf{S} = 1250 \text{ N}$$

· · · · · · · · · · · · · · · · ·

EXAMPLE 2

Four persons are having a tug-of-war with a rope. Harry and Mary are on the left; Bill and Jill are on the right. Mary pulls with a force of 105 lb, Harry pulls with a force of 255 lb, and Jill pulls with a force of 165 lb. With what force must Bill pull to produce equilibrium?

Sketch:

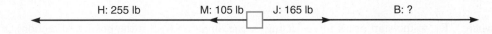

H: 255 lb   M: 105 lb   J: 165 lb   B: ?

Data:

$$M = 105 \text{ lb}$$
$$H = 255 \text{ lb}$$
$$J = 165 \text{ lb}$$
$$B = ?$$

Basic Equation:

$$\mathbf{F}_+ = \mathbf{F}_- \quad \text{or}$$
$$M + H = J + B$$

Working Equation:

$$B = M + H - J$$

Substitution:

$$B = 105 \text{ lb} + 255 \text{ lb} - 165 \text{ lb}$$
$$= 195 \text{ lb}$$

· · · · · · · · · · · · · · · ·

Equilibrium in Two Dimensions

A body is in equilibrium when it is either at rest or moving at a constant speed in a straight line. Figure 7.13(a) shows the resultant force of the sum of two forces from Example 1 in Section 7.1. When two or more forces act at a point, the **equilibrant force** is the force that, when applied at that same point as the resultant force, produces equilibrium. *The equilibrant force is equal in magnitude to that of the resultant force but it acts in the opposite direction* [see Fig. 7.13(b)]. In this case, the equilibrant force is 926 N at 214.5° (180° + 34.5°).

Figure 7.13

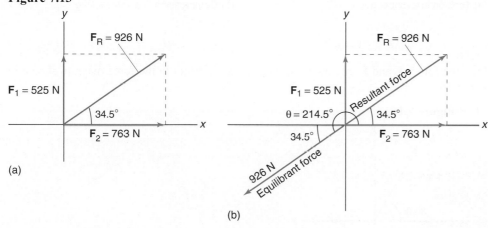

(a)

(b)

If an object is in equilibrium in two dimensions, the net force acting on it must be zero. For the net force to be zero, the sum of the x-components must be zero and the sum of the y-components must be zero. For forces **A**, **B**, and **C** with x-components A_x, B_x, and C_x, respectively, and with y-components A_y, B_y, and C_y, respectively, to be in equilibrium, both of the following conditions must hold:

CONDITIONS FOR EQUILIBRIUM

1. The sum of x-components = 0; that is, $A_x + B_x + C_x = 0$.
2. The sum of y-components = 0; that is, $A_y + B_y + C_y = 0$.

In general, to solve equilibrium problems:

1. Draw a force diagram from the point at which the unknown forces act.
2. Find the x- and y-component of each force.
3. Substitute the components in the equations

$$\text{sum of } x\text{-components} = 0$$
$$\text{sum of } y\text{-components} = 0$$

4. Solve for the unknowns. This may involve two simultaneous equations.

We may need to find the tension or compression in part of a structure, such as in a beam or a cable. **Tension** is a stretching force produced by forces pulling outward on the ends of an object [Fig. 7.14(a)]. **Compression** is a force produced by forces pushing inward on the ends of an object [Fig. 7.14(b)]. A rubber band being stretched is an example of tension [Fig. 7.15(a)]. A valve spring whose ends are pushed together is an example of compression [Fig. 7.15(b)].

Figure 7.14 Tension and compression forces

(a)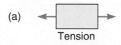

Tension

(b)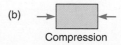

Compression

Figure 7.15

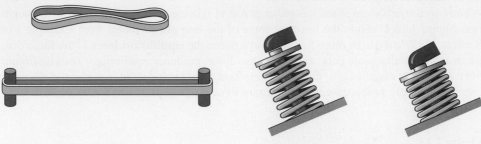

(a) Tension in a rubber band   (b) Compression in a valve spring

EXAMPLE 3

Find the forces **F** and **F'** necessary to produce equilibrium in the force diagram shown in Fig. 7.16.

1. **Figure 7.16**

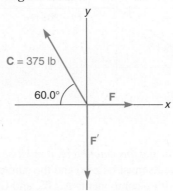

2. *x-components*

$\mathbf{F}_x = \mathbf{F}$

$\mathbf{F}'_x = 0$

$\mathbf{C}_x = -(375 \text{ lb})(\cos 60.0°)$

$\qquad = -188 \text{ lb}$

3. Sum of *x*-components = 0

$\mathbf{F} + 0 + (-188 \text{ lb}) = 0$

4. $\qquad\qquad\qquad \mathbf{F} = 188 \text{ lb}$

y-components

$\mathbf{F}_y = 0$

$\mathbf{F}'_y = -\mathbf{F}'$

$\mathbf{C}_y = (375 \text{ lb})(\sin 60.0°)$

$\qquad = 325 \text{ lb}$

Sum of *y*-components = 0

$0 + (-\mathbf{F}') + 325 \text{ lb} = 0$

$\qquad\qquad\qquad \mathbf{F}' = 325 \text{ lb}$

· · · · · · · · · · · · · · · · ·

EXAMPLE 4

Find the forces **F** and **F'** necessary to produce equilibrium in the force diagram shown in Fig. 7.17.

1. **Figure 7.17**

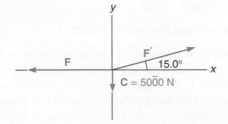

2. *x-components*   *y-components*

$$\mathbf{F}_x = -\mathbf{F}$$ $$\mathbf{F}_y = 0$$

$$\mathbf{F}'_x = \mathbf{F}' \cos 15.0°$$ $$\mathbf{F}'_y = \mathbf{F}' \sin 15.0°$$

$$\mathbf{C}_x = 0$$ $$\mathbf{C}_y = -50\overline{0}0 \text{ N}$$

3.   Sum of *x*-components $= 0$ Sum of *y*-components $= 0$

$$(-\mathbf{F}) + \mathbf{F}' \cos 15.0° + 0 = 0 \qquad 0 + \mathbf{F}' \sin 15.0° + (-50\overline{0}0 \text{ N}) = 0$$

4. *Note:* Solve for **F'** in the right-hand equation first. Then substitute this value in the left-hand equation to solve for **F**:

$$\mathbf{F}' = \frac{50\overline{0}0 \text{ N}}{\sin 15.0°}$$

$$= 19{,}300 \text{ N}$$

$$\mathbf{F} = \mathbf{F}' \cos 15.0°$$

$$= (19{,}300 \text{ N})(\cos 15.0°)$$

$$= 18{,}600 \text{ N}$$

· · · · · · · · · · · · · · · · ·

The crane shown in Fig. 7.18 is supporting a beam that weighs 60$\overline{0}$0 N. Find the tension in the horizontal supporting cable and the compression in the boom. The horizontal cable is attached to the boom at point *A*. The separate vertical cable holding the beam is attached through the pulley at point *A*.

EXAMPLE 5

Figure 7.18

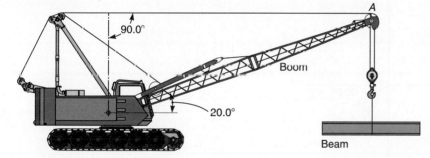

1. Draw the force diagram showing the forces acting at point *A*.

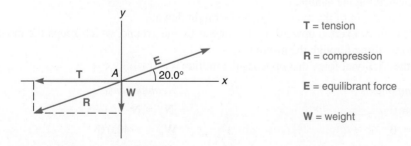

T = tension

R = compression

E = equilibrant force

W = weight

T is the force exerted at *A* by the horizontal supporting cable.
E is the force exerted by the boom at *A*.
W is the force (weight of the beam) pulling straight down at *A*.
R is the sum of forces **W** and **T**, which is equal in magnitude but opposite in direction to force **E** ($\mathbf{R} = -\mathbf{E}$).

2. *x-components*   *y-components*

$$\mathbf{E}_x = \mathbf{E} \cos 20.0° \qquad\qquad \mathbf{E}_y = \mathbf{E} \sin 20.0°$$
$$\mathbf{T}_x = -\mathbf{T} \qquad\qquad\qquad \mathbf{T}_y = 0$$
$$\mathbf{W}_x = 0 \qquad\qquad\qquad\quad \mathbf{W}_y = -60\overline{0}0 \text{ N}$$

3. Sum of *x*-components = 0   Sum of *y*-components = 0

$$\mathbf{E} \cos 20.0° + (-\mathbf{T}) = 0 \qquad \mathbf{E} \sin 20.0° + (-60\overline{0}0 \text{ N}) = 0$$

4. $\mathbf{T} = \mathbf{E} \cos 20.0°$ $\mathbf{E} = \dfrac{60\overline{0}0 \text{ N}}{\sin 20.0°}$

$$= 17{,}500 \text{ N}$$

$$\mathbf{T} = (17{,}500 \text{ N})(\cos 20.0°)$$
$$= 16{,}400 \text{ N}$$

· · · · · · · · · · · · · · · ·

EXAMPLE 6

A homeowner pushes a 40.0-lb lawn mower at a constant velocity (Fig. 7.19). The frictional force on the mower is 20.0 lb. What force must the person exert on the handle, which makes an angle of 30.0° with the ground? Also, find the normal (perpendicular to ground) force.
    This is an equilibrium problem because the mower is not accelerating and the net force is zero.

Figure 7.19

1. Draw the force diagram.

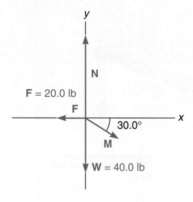

  M is the force exerted on the mower by the person; this compression force is directed down along the handle.
  W is the weight of the mower directed straight down.
  N is the force exerted upward on the mower by the ground, which keeps the mower from falling through the ground.
  F is the frictional force that opposes the motion.

2. *x-components*   *y-components*

$$\mathbf{N}_x = 0 \qquad\qquad\qquad \mathbf{N}_y = \mathbf{N}$$
$$\mathbf{W}_x = 0 \qquad\qquad\qquad \mathbf{W}_y = -40.0 \text{ lb}$$
$$\mathbf{F}_x = -20.0 \text{ lb} \qquad\qquad \mathbf{F}_y = 0$$
$$\mathbf{M}_x = \mathbf{M} \cos 30.0° \qquad\quad \mathbf{M}_y = -\mathbf{M} \sin 30.0°$$

3.   Sum of *x*-components = 0   Sum of *y*-components = 0

$$0 + 0 + (-20.0 \text{ lb}) + \mathbf{M} \cos 30.0° = 0 \qquad \mathbf{N} + (-40.0 \text{ lb}) + 0$$
$$+ (-\mathbf{M} \sin 30.0°) = 0$$
$$\mathbf{N} = \mathbf{M} \sin 30.0 + 40.0 \text{ lb}$$

4. $M = \dfrac{20.0\ \text{lb}}{\cos 30.0°}$

    $= 23.1\ \text{lb}$

$N = (23.1\ \text{lb})(\sin 30.0°) + 40.0\ \text{lb}$

    $= 51.6\ \text{lb}$

· · · · · · · · · · · · · · · · ·

The crane shown in Fig. 7.20 is supporting a beam that weighs $60\overline{0}0$ N. Find the tension in the supporting cable and the compression in the boom.

EXAMPLE 7

1. Draw the force diagram showing the forces acting at point A.

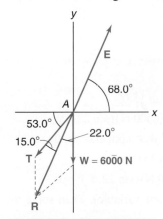

Figure 7.20

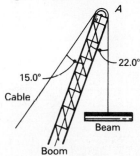

W is the weight of the beam, which pulls straight down.
T is the force exerted at *A* by the supporting cable.
E is the force exerted by the boom at *A*.
R is the sum of forces **W** and **T**, which is equal in magnitude but opposite in direction to force **E**,($\mathbf{R} = -\mathbf{E}$).

2. *x-components*   *y-components*

    $\mathbf{E}_x = \mathbf{E} \cos 68.0°$ $\mathbf{E}_y = \mathbf{E} \sin 68.0°$

    $\mathbf{T}_x = -\mathbf{T} \cos 53.0°$ $\mathbf{T}_y = -\mathbf{T} \sin 53.0°$

    $\mathbf{W}_x = 0$ $\mathbf{W}_y = -60\overline{0}0$ N

3. Sum of *x*-components $= 0$ Sum of *y*-components $= 0$

    $\mathbf{E} \cos 68.0° +$ $\mathbf{E} \sin 68.0° +$

      $(-\mathbf{T} \cos 53.0°) + 0 = 0$ $(-\mathbf{T} \sin 53.0°) + (-60\overline{0}0\ \text{N}) = 0$

4. *Note:* Solve the left equation for **E**. Then substitute this quantity in the right equation and solve for **T**:

$\mathbf{E} = \dfrac{\mathbf{T} \cos 53.0°}{\cos 68.0°}$ $\left(\dfrac{\mathbf{T} \cos 53.0°}{\cos 68.0°} \right)(\sin 68.0°) - \mathbf{T} \sin 53.0° = 60\overline{0}0\ \text{N}$

$1.490\mathbf{T} - 0.799\mathbf{T} = 60\overline{0}0\ \text{N}$

$0.691\mathbf{T} = 60\overline{0}0\ \text{N}$

$\mathbf{T} = \dfrac{60\overline{0}0\ \text{N}}{0.691}$

$= 8680\ \text{N}$

$\mathbf{E} = \dfrac{(8680\ \text{N})(\cos 53.0°)}{\cos 68.0°}$

    $= 13{,}900\ \text{N}$

Alternate Method: You can orient a force diagram any way you want on the x–y axes. You should orient it so that as many of the vectors as possible are on an x- or a y-axis. The result will be the same. Let's rework Example 7 as follows:

1. Draw the force diagram showing the forces acting at point A using the same notation as follows:

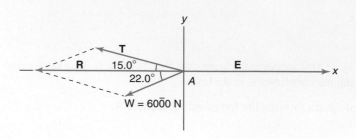

2. *x-components*   *y-components*

$$\mathbf{E}_x = \mathbf{E} \qquad\qquad\qquad\qquad \mathbf{E}_y = 0$$
$$\mathbf{T}_x = -\mathbf{T}\cos 15.0° \qquad\qquad \mathbf{T}_y = \mathbf{T}\sin 15.0°$$
$$\mathbf{W}_x = -(60\overline{0}0\ \text{N})(\cos 22.0°) \qquad \mathbf{W}_y = -(60\overline{0}0\ \text{N})(\sin 22.0°)$$

3. Sum of x-components $= 0$ Sum of y-components $= 0$

$$\mathbf{E} + (-\mathbf{T}\cos 15.0°) + \qquad\qquad 0 + \mathbf{T}\sin 15.0° +$$
$$(-60\overline{0}0\ \text{N})(\cos 22.0°) = 0 \qquad (-60\overline{0}0\ \text{N})(\sin 22.0°) = 0$$

4. Solve the right equation for $\mathbf{T}$ (since it has only one variable). Then solve the left equation for $\mathbf{E}$ and substitute this quantity:

$$\mathbf{T} = \frac{(60\overline{0}0\ \text{N})(\sin 22.0°)}{\sin 15.0°}$$
$$= 8680\ \text{N}$$

$$\mathbf{E} = \mathbf{T}\cos 15.0° + (60\overline{0}0\ \text{N})(\cos 22.0°)$$
$$\mathbf{E} = (8680\ \text{N})(\cos 15.0°) + (60\overline{0}0\ \text{N})(\cos 22.0°)$$
$$= 13{,}900\ \text{N}$$

·················

PROBLEMS 7.2

Find the force $\mathbf{F}$ that will produce equilibrium in each force diagram.

SKETCH

4.0 cm

DATA

$A = 12\ \text{cm}^2, l = 4.0\ \text{cm}, w = ?$

BASIC EQUATION

$A = lw$

WORKING EQUATION

$w = \frac{A}{l}$

SUBSTITUTION

$w = \frac{12\ \text{cm}^2}{4.0\ \text{cm}} = 3.0\ \text{cm}$

1.

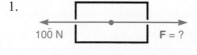

2. $F = ?$

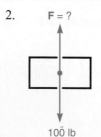

$10\overline{0}\ \text{lb}$

3.

4.

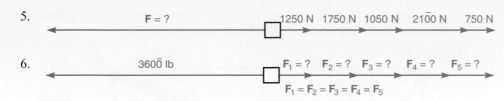

5.

6. $F_1 = F_2 = F_3 = F_4 = F_5$

7. Five persons are having a tug-of-war. Kurt and Brian are on the left; Amy, Barbara, and Joyce are on the right. Amy pulls with a force of 225 N, Barbara pulls with a force of 495 N, Joyce pulls with a force of 455 N, and Kurt pulls with a force of 605 N. With what force must Brian pull to produce equilibrium?

8. A certain wire can support 6450 lb before it breaks. Seven 820-lb weights are suspended from the wire. Can the wire support an eighth weight of 820 lb?

9. The frictional force of a loaded pallet in a warehouse is 385 lb. Can three workers, each exerting a force of 135 lb, push it to the side?

10. A bridge has a weight limit of 7.0 tons. How heavy a load can a 2.5-ton truck carry across?

11. A tractor transmission weighing $26\overline{0}$ N and a steering gear box weighing 62.0 N are on a workbench. What upward force must the bench exert to maintain equilibrium?

12. A skid loader lifts a compressor weighing 672 N and a hose weighing 26.0 N. What upward force must the loader exert to maintain equilibrium?

Find the forces F_1 and F_2 that produce equilibrium in each force diagram.

13.

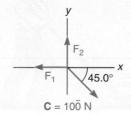

14.

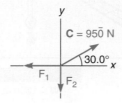

15.

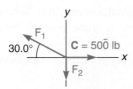

16.

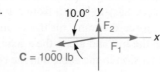

17.

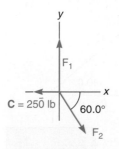

18.

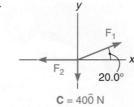

19. A rope is attached to two buildings and supports a $50\overline{0}$-lb sign (Fig. 7.21). Find the tensions in the two ropes T_1 and T_2. (*Hint:* Draw the force diagram of the forces acting at the point labeled A.)

20. If the angle between the horizontal and the ropes in Problem 19 is changed to 10.0°, what are the tensions in the two ropes T_1 and T_2?

21. If the angles between the horizontal and the ropes in Problem 19 are changed to 20.0° and 30.0°, find the tension in each rope.

Figure 7.21

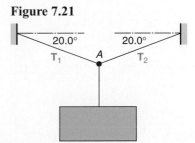

22. Find the tension in the horizontal supporting cable and the compression in the boom of the crane which supports an 8900-N beam shown in Fig. 7.22.

Figure 7.22

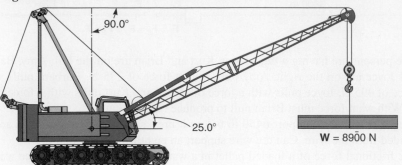

23. Find the tension in the horizontal supporting cable and the compression in the boom of the crane which supports a 1500-lb beam shown in Fig. 7.23.

24. The frictional force of the mower shown in Fig. 7.24 is 20 lb. What force must the man exert along the handle to push it at a constant velocity?

Figure 7.23   **Figure 7.24**

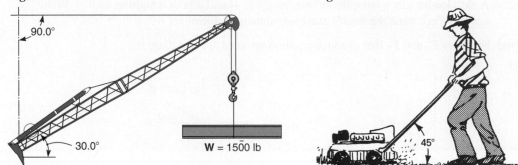

25. A vehicle that weighs 16,200 N is parked on a 20.0° hill (Fig. 7.25). What braking force is necessary to keep it from rolling? Neglect frictional forces. (*Hint:* When you draw the force diagram, tilt the *x*- and *y*-axes as shown. **B** is the braking force directed up the hill and along the *x*-axis.)

26. Find the tension in the cable and the compression in the support of the sign shown in Fig. 7.26.

Figure 7.25   **Figure 7.26**

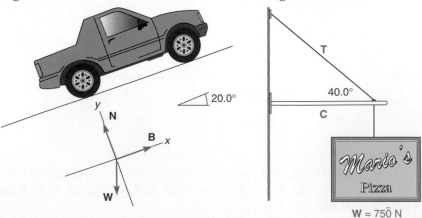

27. The crane shown in Fig. 7.27 is supporting a load of 1850 lb. Find the tension in the supporting cable and the compression in the boom.

28. The crane shown in Fig. 7.28 is supporting a load of 11,500 N. Find the tension in the supporting cable and the compression in the boom.

Figure 7.27   **Figure 7.28**

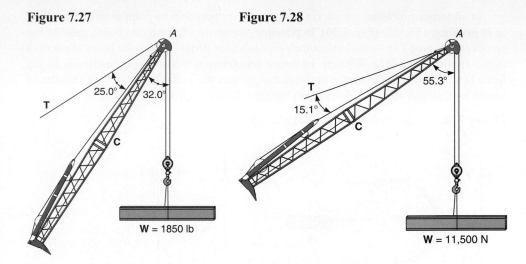

7.3 Torque

A *torque* is produced when a force is applied to produce a rotation, as, for example, when a wrench is used to turn a bolt or a claw hammer is used to pull a nail from wood. **Torque** is the tendency to produce change in rotational motion. The torque developed depends on two factors:

1.   The amount of force applied
2.   How far from the point of rotation the force is applied

Torque is expressed by the equation

$$\tau = F s_t$$

where τ = torque (N m or lb ft) (τ is the lowercase Greek letter "tau.")
    F = applied force (N or lb)
    s_t = length of torque arm (m or ft)

Note that s_t, the length of the torque arm, is different from s in the equation defining work ($W = Fs$). Recall that s in the work equation is the linear distance over which the force acts.

When you use a wrench to turn a bolt, less effort is used (greater torque is produced) as the distance you place your hand from the bolt increases (Fig. 7.29). Plumbers often use a wrench with a long torque arm to loosen or tighten large bolts and fittings.

Figure 7.29   Even though the same force is used, the torque applied to the bolt increases as the distance from your hand to the bolt increases. In (b), you produce more torque by placing your hand on the end of the wrench handle. In (c), you produce even more torque by using an extender sleeve.

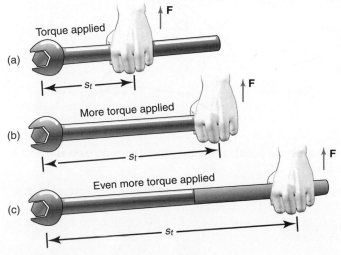

In all torque problems, we are concerned with motion about a point or axis of rotation as in pedaling a bicycle (Fig. 7.30). In pedaling, we apply a force to the pedal, causing the sprocket to rotate. The torque arm is the *perpendicular* distance from the point of rotation to the applied force [Fig. 7.30(a)]. In torque problems, s_t is always perpendicular to the force [Fig. 7.30(a)]. Note that s_t is the distance from the pedal to the axle. The units of torque look similar to those of work, but note the difference between s and s_t.

Figure 7.30   Torque produced in pedaling a bicycle

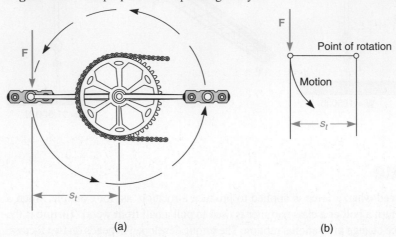

(a)   (b)

Figure 7.31   Maximum torque is only produced when the pedal reaches a position perpendicular to the applied force.

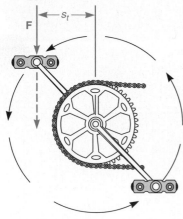

If the force is not exerted tangent to the circle made by the pedal (Fig. 7.31), the length of the torque arm is *not* the length of the pedal arm. The torque arm, s_t, is measured as the perpendicular distance to the force. Since s_t is therefore shorter, the product $F \cdot s_t$ is smaller, and the turning effect, the torque, is less in the pedal position shown in Fig. 7.30. Maximum torque is produced when the pedals are horizontal and the force applied is straight down.

Torque is a vector quantity that acts along the axis of rotation (not along the force) and points in the direction in which a right-handed screw would advance if turned by the torque as in Fig. 7.32(a). The *right-hand rule* is often used to determine the direction of the torque as follows: Grasp the axis of rotation with your right hand so that your fingers circle it in the direction that the torque tends to induce rotation. Your thumb will point in the direction of the torque vector [Fig. 7.32(b)]. Thus, the torque vector in Fig. 7.32(a) is perpendicular to and points out of the page.

Figure 7.32   Torque is a vector quantity that acts along the axis of rotation according to the right-hand rule.

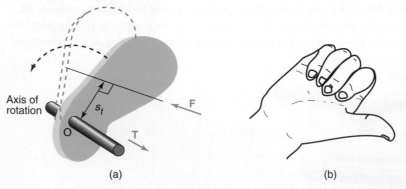

(a)   (b)

EXAMPLE

A force of 10.0 lb is applied to a bicycle pedal. If the length of the pedal arm is 0.850 ft, what torque is applied to the shaft?

Sketch:

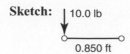

Data:

$$F = 10.0 \text{ lb}$$
$$s_t = 0.850 \text{ ft}$$
$$\tau = \text{?}$$

Basic Equation:

$$\tau = Fs_t$$

Working Equation: Same

Substitution:

$$\tau = (10.0 \text{ lb})(0.850 \text{ ft})$$
$$= 8.50 \text{ lb ft}$$

· · · · · · · · · · · · · ·

TRY THIS ACTIVITY

Hammers and Screwdrivers

Torque is an essential component of most hand tools. Drive a nail into a piece of wood by holding a hammer near its head. Count the number of hits it takes to drive the nail into the wood. Then, drive a like nail into the wood by holding the hammer near the end of the handle. Count the number of hits it takes. Using physics terminology, explain which handle grip is better for driving nails into wood.

  Use two screwdrivers with different diameter handles to screw two similar screws into a board. Which screwdriver is able to apply more torque to a screw? If one screwdriver applies more torque to a screw than the other, why would anyone want to use a screwdriver that cannot exert the maximum torque?

PROBLEMS 7.3

Assume that each force is applied perpendicular to the torque arm.

1. Given: $F = 16.0$ lb
    $s_t = 6.00$ ft
    $\tau = \text{?}$

2. Given: $F = 100 \text{ N}$
    $s_t = 0.420$ m
    $\tau = \text{?}$

3. Given: $\tau = 60.0$ N m
    $F = 30.0$ N
    $s_t = \text{?}$

4. Given: $\tau = 35.7$ lb ft
    $s_t = 0.0240$ ft
    $F = \text{?}$

5. Given: $\tau = 65.4$ N m
    $s_t = 35.0$ cm
    $F = \text{?}$

6. Given: $F = 630 \text{ N}$
    $s_t = 74.0$ cm
    $\tau = \text{?}$

7. If the torque on a shaft of radius 2.37 cm is 38.0 N m (Fig. 7.33), what force is applied to the shaft?

8. If a force of 56.2 lb is applied to a torque wrench 1.50 ft long (Fig. 7.34), what torque is indicated by the wrench?

Figure 7.33

F = ?

r = 2.37 cm

Figure 7.34

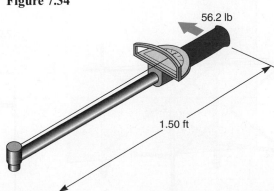

56.2 lb

1.50 ft

SKETCH

12 cm² | w

4.0 cm

DATA

A = 12 cm², l = 4.0 cm, w = ?

BASIC EQUATION

A = lw

WORKING EQUATION

$w = \frac{A}{l}$

SUBSTITUTION

$w = \frac{12 \text{ cm}^2}{4.0 \text{ cm}} = 3.0 \text{ cm}$

9. A motorcycle head bolt is torqued to 25.0 N m. What length shaft do we need on a wrench to exert a maximum force of 70.0 N?

10. A force of 112 N is applied to a shaft of radius 3.50 cm. What is the torque on the shaft?

11. A torque of 175 lb ft is needed to free a large rusted-on nut. The length of the wrench is 1.10 ft. What force must be applied to free it?

12. A torque wrench reads 14.5 N m. If its length is 25.0 cm, what force is being applied to the handle?

13. The torque on a shaft of radius 3.00 cm is 12.0 N m. What force is being applied to the shaft?

14. An engine bolt is torqued to 30.0 N m. If the length of the wrench is 29.0 cm, what force is applied to the wrench?

15. A mower bolt is torqued to 65.0 N m. If the length of the wrench is 30.0 cm, what force is applied to the wrench?

16. An automobile bolt is torqued to 27.0 N m. If the length of the wrench is 30.0 cm, what force is applied to the wrench?

17. A torque wrench reads 25 lb ft. (a) If its length is 1.0 ft, what force is being applied to the wrench? (b) What is the force if the length is doubled? Explain the results.

18. If 13 N m of torque is applied to a bolt with an applied force of 28 N, what is the length of the wrench?

19. If the torque required to loosen a nut on the wheel of a pickup truck is 40.0 N m, what minimum force must be applied to the end of a wrench 30.0 cm long to loosen the nut?

20. How is the required force to loosen the nut in Problem 19 affected if the length of the wrench is doubled?

21. A truck mechanic must loosen a rusted lug nut. If the torque required to loosen the nut is 60.0 N m, what force must be applied to a 35.0-cm wrench?

22. An ag mechanic tries to loosen a nut on a tractor wheel with a wrench that is 32.5 cm long. If the torque required to loosen the nut is 55.0 N m, what force must she apply to the wrench?

7.4 Parallel Forces

A painter stands 2.00 ft from one end of a 6.00-ft plank that is supported at each end by a scaffold [Fig. 7.35(a)]. How much of the painter's weight must each end of the scaffold support? Problems of this kind are often faced in the construction industry, particularly in the design of bridges and buildings. Using some things we learned about torques and equilibrium, we can now solve problems of this type.

Figure 7.35 Parallel forces shown by the example of a painter on a scaffold

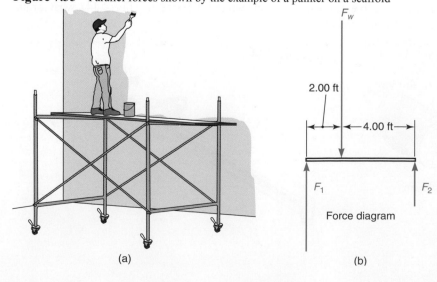

(a)

(b)

Let's look more closely at the painter problem. The force diagram [Fig. 7.35(b)] shows the forces and distances involved. The arrow pointing down represents the weight of the person, F_w. The arrows pointing up represent the forces exerted by each end of the scaffold in supporting the plank and painter. (For now, we will neglect the weight of the plank.) We have a condition of equilibrium. The plank and painter are not moving. The sum of the forces exerted by the ends of the scaffold is equal to the weight of the painter (Fig. 7.36). Since these forces are vectors and are parallel, we can show that their sum is zero. Using engineering notation, we write

$$\Sigma\mathbf{F} = 0$$

where Σ (Greek capital letter sigma) means summation or "the sum of" and $\mathbf{F}$ is force, a vector quantity. So $\Sigma\mathbf{F}$ means "the sum of forces," in this case the sum of parallel forces.

Figure 7.36 In equilibrium, the sum of the forces is zero.

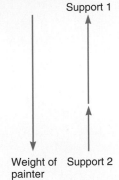

Support 1

Weight of painter   Support 2

FIRST CONDITION OF EQUILIBRIUM
The sum of all parallel forces on a body in equilibrium must be zero.

If the vector sum is not zero (forces up unequal to forces down), we have an unbalanced force tending to cause motion.

Now consider this situation: One end of the scaffold remains firmly in place, supporting the man, and the other is removed. What happens to the painter? The plank, supported only on one end, falls (Fig. 7.37), and the painter has a mess to clean up!

Figure 7.37 The position of the supporting force is important!

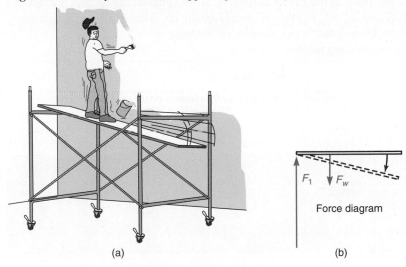

F_1 F_w

Force diagram

(a)   (b)

A sign of weight $150\overline{0}$ lb is supported by two cables (Fig. 7.38). If one cable has a tension of $60\overline{0}$ lb, what is the tension in the other cable?

Sketch: Draw the force diagram.

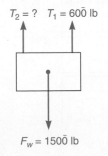

$T_2 = ?$ $T_1 = 60\overline{0}$ lb

$F_w = 150\overline{0}$ lb

EXAMPLE 1

Figure 7.38

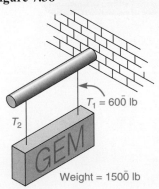

$T_1 = 60\overline{0}$ lb

T_2

Weight = $150\overline{0}$ lb

Data:

$$F_w = 150\overline{0} \text{ lb}$$
$$T_1 = 60\overline{0} \text{ lb}$$
$$T_2 = ?$$

Basic Equation:

$$F_+ = F_-$$

Working Equation:

$$T_1 + T_2 = F_w$$
$$T_2 = F_w - T_1$$

Substitution:

$$T_2 = 150\overline{0} \text{ lb} - 60\overline{0} \text{ lb}$$
$$= 90\overline{0} \text{ lb}$$

· · · · · · · · · · · · · · · · ·

Not only must the forces balance each other (vector sum = 0), but they must also be positioned so that there is no rotation in the system. To avoid rotation, we can have no unbalanced torques.

Sometimes there will be a natural point of rotation, as in our painter problem. We can, however, choose any point as our center of rotation as we consider the torques present. We will soon see that one of any number of points could be selected. What is necessary, though, is that there be no rotation (no unbalanced torques).

Again, using engineering notation, we write

$$\Sigma\tau_{\text{any point}} = 0$$

where $\Sigma\tau_{\text{any point}}$ is the sum of the torques about any chosen point or,

SECOND CONDITION OF EQUILIBRIUM

The sum of the clockwise torques on a body in equilibrium must equal the sum of the counterclockwise torques about any point.

$$\Sigma\tau_{\text{clockwise (cw)}} = \Sigma\tau_{\text{counterclockwise (ccw)}}$$

EXAMPLE 2

To illustrate these principles, we will find how much weight each end of the scaffold must support if our painter weighs $15\overline{0}$ lb.

Sketch:

Data:

$$F_w = 15\overline{0} \text{ lb}$$
$$\text{plank} = 6.00 \text{ ft}$$
$$F_w \text{ is } 2.00 \text{ ft from one end}$$

Basic Equations:

1.
$$\Sigma F = 0$$

sum of forces $= 0$

$$F_1 + F_2 - F_w = 0$$

(*Note:* F_w is negative because its direction is opposite F_1 and F_2.)

$$\text{or } F_1 + F_2 = F_w$$
$$F_1 + F_2 = 15\overline{0} \text{ lb}$$

2. $\Sigma\tau_{\text{clockwise}} = \Sigma\tau_{\text{counterclockwise}}$

First, select a point of rotation. Choosing an end is usually helpful in simplifying the calculations. Choose the left end (point A) where F_1 acts. What are the clockwise torques about this point?

The force due to the weight of the painter tends to cause clockwise motion. The torque arm is 2.00 ft (Fig. 7.39). Then $\tau = (15\overline{0} \text{ lb})(2.00 \text{ ft})$. This is the only clockwise torque.

The only counterclockwise torque is F_2 times its torque arm, 6.00 ft (Fig. 7.40). $\tau = (F_2)(6.00 \text{ ft})$. There is no torque involving F_1 because its torque arm is zero. Setting $\Sigma\tau_{\text{clockwise}} = \Sigma\tau_{\text{counterclockwise}}$ we have the equation:

$$(15\overline{0} \text{ lb})(2.00 \text{ ft}) = (F_2)(6.00 \text{ ft})$$

Note that by selecting an end as the point of rotation, we were able to have an equation with just one variable (F_2). Solving for F_2 gives the working equation:

$$F_2 = \frac{(15\overline{0} \text{ lb})(2.00 \text{ ft})}{6.00 \text{ ft}} = 50.0 \text{ lb}$$

Since $\Sigma F = F_1 + F_2 = F_w$, substitute for F_2 and F_w to find F_1:

$$F_1 + 50.0 \text{ lb} = 15\overline{0} \text{ lb}$$
$$F_1 = 15\overline{0} \text{ lb} - 50.0 \text{ lb}$$
$$= 10\overline{0} \text{ lb}$$

Figure 7.39   Torque arm of painter about point A

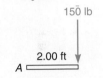

Figure 7.40   Torque arm of F_2 about point A

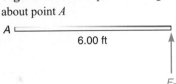

· · · · · · · · · · · · · · · · ·

To solve parallel force problems:

1.   Sketch the problem.
2.   Write an equation setting the sums of the opposite forces equal to each other.
3.   Choose a point of rotation. Eliminate a variable, if possible (by making its torque arm zero).
4.   Write the sum of all clockwise torques.
5.   Write the sum of all counterclockwise torques.
6.   Set $\Sigma\tau_{\text{clockwise}} = \Sigma\tau_{\text{counterclockwise}}$.
7.   Solve the equation $\Sigma\tau_{\text{clockwise}} = \Sigma\tau_{\text{counterclockwise}}$ for the unknown quantity.
8.   Substitute the value found in step 7 into the equation in step 2 to find the other unknown quantity.

EXAMPLE 3

A bricklayer weighing 175 lb stands on an 8.00-ft scaffold 3.00 ft from one end (Fig. 7.41). He has a pile of bricks, which weighs 40.0 lb, 3.00 ft from the other end. How much weight must each end support?

Figure 7.41

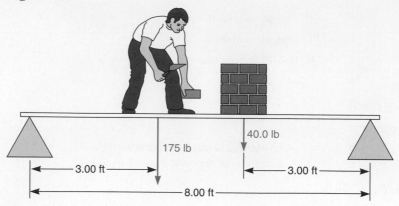

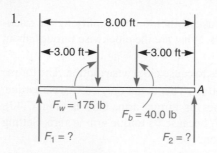

2. $\Sigma F = F_1 + F_2 = 175 \text{ lb} + 40.0 \text{ lb}$

3.   Choose a point of rotation. Choose either end to eliminate one of the variables F_1 or F_2. Let us choose the right end and label it A.

4. $\Sigma\tau_{\text{clockwise}} = (F_1)(8.00 \text{ ft})$

5. $\Sigma\tau_{\text{counterclockwise}} = (40.0 \text{ lb})(3.00 \text{ ft}) + (175 \text{ lb})(5.00 \text{ ft})$

Note that there are two counterclockwise torques.

6. $\Sigma\tau_{\text{clockwise}} = \Sigma\tau_{\text{counterclockwise}}$

$F_1(8.00 \text{ ft}) = (40.0 \text{ lb})(3.00 \text{ ft}) + (175 \text{ lb})(5.00 \text{ ft})$

7. $F_1 = \dfrac{(40.0 \text{ lb})(3.00 \text{ ft}) + (175 \text{ lb})(5.00 \text{ ft})}{8.00 \text{ ft}}$

$= \dfrac{12\overline{0} \text{ lb ft} + 875 \text{ lb ft}}{8.00 \text{ ft}} = \dfrac{995 \text{ lb ft}}{8.00 \text{ ft}} = 124 \text{ lb}$

8. $F_1 + F_2 = 175 \text{ lb} + 40.0 \text{ lb}$

$124 \text{ lb} + F_2 = 215 \text{ lb} \qquad\qquad (F_1 = 124 \text{ lb})$

$F_2 = 91 \text{ lb}$

· · · · · · · · · · · · · · · · · · ·

TRY THIS ACTIVITY

The Physics of Window Washing

Window washers on tall buildings typically stand on a platform that is suspended from the top of the building by ropes or cables. Designers of these devices must make sure that the cables can safely suspend the workers while they are all working on one side of the platform. To simulate this activity, take the two strings, the scales, and the mass from the previous Try This Activity and use a metre stick or a wooden dowel as the platform.

Set up the activity as indicated in Fig. 7.42, making sure both suspending strings are the same length and that the metre stick is horizontal. Place the weight directly in the middle of the metre stick and note the readings in the scales. Then gradually shift the mass over to one side and observe what happens to the readings in the spring scales. What happens to the sum of the tension forces as the mass is shifted to one of the sides?

Figure 7.42   After suspending the mass from the middle of the platform, shift the mass toward one side and observe what happens to the tension in the strings.

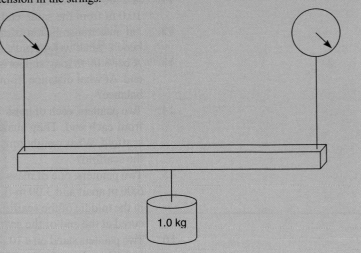

1.0 kg

PROBLEMS 7.4

Find the force F that will produce equilibrium for each force diagram. Use the same procedure as in Example 1.

1. $10\overline{0}$ lb $F = ?$

2. $F = ?$
    $20\overline{0}$ N

3. $F = ?$ $20\overline{0}$ N
    $70\overline{0}$ N

4. $F = ?$
    $20\overline{0}$ N $15\overline{0}$ N

5. $90\overline{0}$ N $45\overline{0}$ N
    $F = ?$

6. $65\overline{0}$ lb $10\overline{0}$ lb
    $25\overline{0}$ lb
    $F = ?$

7. $F = ?$ $75\overline{0}$ N
    $150\overline{0}$ N
    $210\overline{0}$ N $25\overline{0}$ N

8. 50.0 N 10.0 N
    35.0 N $F = ?$
    15.0 N 75.0 N

9. A 90.0-kg painter stands 3.00 m from one end of an 8.00-m scaffold. If the scaffold is supported at each end by a stepladder, how much of the weight of the painter must each ladder support?

16. A 125-kg horizontal beam is supported at each end. A 325-kg mass rests one-fourth of the way from one end. What weight must be supported at each end?

17. The sign shown in Fig. 7.48 is 4.00 m long, weighs $155\overline{0}$ N, and is made of uniform material. A weight of 245 N hangs 1.00 m from the end. Find the tension in each support cable.

18. The uniform bar in Fig. 7.49 is 5.00 m long and weighs 975 N. A weight of 255 N is attached to one end while a weight of 375 N is attached 1.50 m from the other end. (a) Find the tension in the cable. (b) Where should the cable be tied to lift the bar and its weights so that the bar hangs in a horizontal equilibrium position?

Figure 7.48

1.00 m   *A*   2.00 m   *B* 1.00 m

Main Street Cut and Style

1.00 m

Hours by Appointment

Figure 7.49

F_4

A

2.50 m ——— 1.50 m

$F_1 = 255$ N

$F_3 = 375$ N

$F_2 = 975$ N

Find the magnitude, direction, and placement (from point *A*) of a parallel vector F_6 that will produce equilibrium in each force diagram.

19. $F_1 = 125\overline{0}$ N

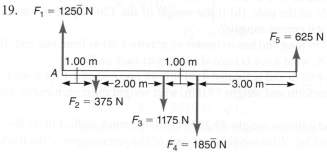

$F_5 = 625$ N

1.00 m   1.00 m

A

2.00 m ——— 3.00 m

$F_2 = 375$ N

$F_3 = 1175$ N

$F_4 = 185\overline{0}$ N

20.

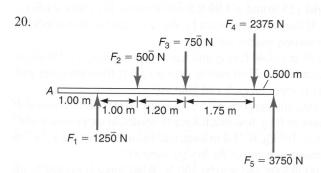

$F_4 = 2375$ N

$F_3 = 75\overline{0}$ N

$F_2 = 50\overline{0}$ N

0.500 m

A

1.00 m

1.00 m   1.20 m   1.75 m

$F_1 = 125\overline{0}$ N

$F_5 = 375\overline{0}$ N

Glossary

Center of Gravity The point of any body at which all of its weight can be considered to be concentrated. (p. 194)

Compression A force produced by forces pushing inward on the ends of an object (p. 177)

Concurrent Forces Two or more forces applied to, or acting at, the same point. (p. 169)

Equilibrant Force The force that, when applied at the same point as the resultant force, produces equilibrium. (p. 177)

Equilibrium An object is said to be in equilibrium when the net force acting on it is zero. A body that is in equilibrium is either at rest or moving at a constant velocity. (p. 175)

First Condition of Equilibrium The sum of all parallel forces on a body in equilibrium must be zero. (p. 189)

Resultant Force The sum of the forces applied at the same point. The single force that has the same effect as two or more forces acting together. (p. 169)

Second Condition of Equilibrium The sum of the clockwise torques on a body in equilibrium must be equal to the sum of the counterclockwise torques about any point. (p. 190)

Statics The study of objects that are in equilibrium. (p. 175)

Tension A stretching force produced by forces pulling outward on the ends of an object. (p. 177)

Torque The tendency to produce change in rotational motion. Equal to the applied force times the length of the torque arm. (p. 185)

Formulas

7.1 To find the resultant vector $\mathbf{F}_R$ of two or more vectors:

(a) find the x- and y-components of each vector and add the components.

(b) find angle A as follows:

$$\tan A = \frac{|\text{sum of } y\text{-components}|}{|\text{sum of } x\text{-components}|} = \frac{|\mathbf{F}_{Ry}|}{|\mathbf{F}_{Rx}|}$$

Determine the quadrant of the angle from the signs of the sum of the x- and y-components.

(c) find the magnitude of $\mathbf{F}_R$ using the Pythagorean theorem:

$$|\mathbf{F}_R| = \sqrt{|\mathbf{F}_{Rx}|^2 + |\mathbf{F}_{Ry}|^2}$$

7.2 Condition for equilibrium in one dimension:

$$\mathbf{F}_+ = \mathbf{F}_-$$

where $\mathbf{F}_+$ is the sum of the forces acting in one direction (call it the positive direction) and $\mathbf{F}_-$ is the sum of the forces acting in the opposite (negative) direction.

Conditions for equilibrium in two dimensions:

(a) The sum of x-components = 0; that is, $\mathbf{A}_x + \mathbf{B}_x + \mathbf{C}_x = 0$; and

(b) The sum of y-components = 0; that is, $\mathbf{A}_y + \mathbf{B}_y + \mathbf{C}_y = 0$.

To solve equilibrium problems:

1. Draw a force diagram from the point at which the unknown forces act.
2. Find the x- and y-components of each force.
3. Substitute the components in the equations

$$\text{sum of } x\text{-components} = 0$$
$$\text{sum of } y\text{-components} = 0$$

4. Solve for the unknowns. This may involve two simultaneous equations.

7.3 $\tau = F s_t$

7.4 *First condition of equilibrium:* The sum of all parallel forces on an object must be zero.

$$\Sigma F = 0$$

Second condition of equilibrium: The sum of the clockwise torques on an object must equal the sum of the counterclockwise torques.

$$\Sigma \tau_{clockwise} = \Sigma \tau_{counterclockwise}$$

Review Questions

1. Concurrent forces act at
   (a) two or more different points.   (b) the same point.   (c) the origin.
2. The resultant force is
   (a) the last force applied.
   (b) the single force that has the same effect as two or more forces acting together.
   (c) equal to either diagonal when using the parallelogram method to add vectors.
3. A moving object
   (a) can be in equilibrium.   (b) is never in equilibrium.
   (c) has no force being applied.
4. The study of an object in equilibrium is called
   (a) dynamics.   (b) astronomy.   (c) statics.   (d) biology.
5. Torque is
   (a) applied force in rotational motion.
   (b) the length of the torque arm.
   (c) applied force times the length of the torque arm.
   (d) none of the above.
6. The first condition of equilibrium states that
   (a) all parallel forces must be zero.
   (b) all perpendicular forces must be zero.
   (c) all frictional forces must be zero.
7. In the second condition of equilibrium,
   (a) clockwise and counterclockwise torques are unequal.
   (b) clockwise and counterclockwise torques are equal.
   (c) there are no torques.
8. The center of gravity of an object
   (a) is always at its geometric center.
   (b) does not have to be at the geometric center.
   (c) exists only in symmetrical objects.
9. Is motion produced every time a force is applied to an object?
10. What is the relationship between opposing forces on a body that is in equilibrium?
11. Define *equilibrium*.
12. In what direction does the force due to gravity always act?
13. What may be said about concurrent forces whose sum of x-components equals zero and whose sum of y-components equals zero?
14. What is a force diagram?
15. Is the length of the pedal necessarily the true length of the torque arm in pedaling a bicycle?
16. In your own words, explain the second condition of equilibrium.
17. What is the primary consideration in the selection of a point of rotation in an equilibrium problem?
18. List three examples from daily life in which you use the concept of center of gravity.
19. Is the center of gravity of an object always at its geometric center?
20. On a 3.00-m scaffold of uniform mass, with supports at each end, there is a pile of bricks 0.500 m from one end. Which support will exert a greater force: the one closer to the bricks or the one farther away?

Review Problems

1. Find the sum of the following forces acting at the same point in a straight line: 345 N (right); 108 N (right); 481 N (left); 238 N (left); 303 N (left).

2. Forces of 275 lb and 225 lb act at the same point.
   (a) What is the magnitude of the maximum net force the two forces can exert together?
   (b) What is the magnitude of the minimum net force the two forces can exert together?

Find the sum of each set of vectors. Give angles in standard position.

3.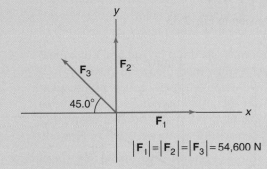

4.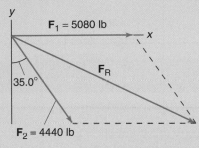

SKETCH

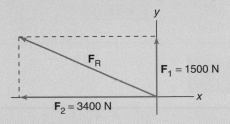

DATA

$A = 12$ cm^2, $l = 4.0$ cm, $w = ?$

BASIC EQUATION

$A = lw$

WORKING EQUATION

$w = \frac{A}{l}$

SUBSTITUTION

$w = \frac{12 \text{ cm}^2}{4.0 \text{ cm}} = 3.0$ cm

5.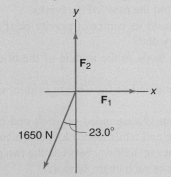

$|\mathbf{F}_1| = |\mathbf{F}_2| = |\mathbf{F}_3| = 54{,}600$ N

6. Forces of $\mathbf{F}_1 = 1250$ N, $\mathbf{F}_2 = 625$ N, and $\mathbf{F}_3 = 1850$ N are applied at the same point. The angle between $\mathbf{F}_1$ and $\mathbf{F}_2$ is 120.0° and the angle between $\mathbf{F}_2$ and $\mathbf{F}_3$ is 30.0°. $\mathbf{F}_2$ is between $\mathbf{F}_1$ and $\mathbf{F}_3$. Find the resultant force.

7. Eight people are involved in a tug-of-war. The blue team members pull with forces of 220 N, 340 N, 180 N, and 560 N. Three members of the red team pull with forces of 250 N, 160 N, and 420 N. With what force must the fourth person pull to maintain equilibrium?

8. A bridge has a weight limit of 14.0 tons. What is the maximum weight an 8.0-ton truck can carry across and still maintain equilibrium?

9. The x-components of three vectors are $\mathbf{F}_x$, 375 units, and 150 units. If their sum is equal to zero, what is $\mathbf{F}_x$?

10. If $\mathbf{W}_y = 600$ N and $\mathbf{W}_x = 900$ N, what are the magnitude and direction of the resultant $\mathbf{W}$?

Find forces $\mathbf{F}_1$ and $\mathbf{F}_2$ that produce equilibrium in each force diagram.

11.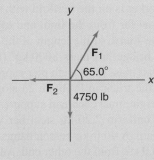

12.

PROBLEM SOLVING

TRY THIS ACTIVITY

Human Horsepower

How much equivalent horsepower do you possess? Using a tape measure, a stopwatch, and a flight of stairs, determine the horsepower in your legs when climbing a flight of stairs. First, measure the vertical height of the stairs. Use this distance and your weight to determine the work required to move your body's weight up the stairs. Clock the time it takes to walk and to run up the stairs. Then, find and compare the power. Convert your power to horsepower. How does your running horsepower compare to the horsepower of a lawnmower or a car?

EXAMPLE 1

A freight elevator with operator weighs $50\overline{0}0$ N. If it is raised to a height of 15.0 m in 10.0 s, how much power is developed?

Data:

$$F = 50\overline{0}0 \text{ N}$$
$$s = 15.0 \text{ m}$$
$$t = 10.0 \text{ s}$$
$$P = ?$$

Basic Equations:

$$P = \frac{W}{t} \quad \text{and} \quad W = Fs$$

Working Equation:

$$P = \frac{Fs}{t}$$

Substitution:

$$P = \frac{(50\overline{0}0 \text{ N})(15.0 \text{ m})}{10.0 \text{ s}}$$
$$= 75\overline{0}0 \text{ N m/s}$$

. . . .

EXAMPLE 2

The power expended in lifting an 825-lb girder to the top of a building $10\overline{0}$ ft high is 10.0 hp. How much time is required to raise the girder?

Data:

$$F = 825 \text{ lb}$$
$$s = 10\overline{0} \text{ ft}$$
$$P = 10.0 \text{ hp}$$
$$t = ?$$

Basic Equations:

$$P = \frac{W}{t} \quad \text{and} \quad W = Fs$$

Working Equation:

$$t = \frac{W}{P} = \frac{Fs}{P}$$

Substitution:

$$t = \frac{(825 \text{ lb})(10\overline{0} \text{ ft})}{10.0 \text{ hp}}$$

$$= \frac{(825 \text{ lb})(10\overline{0} \text{ ft})}{10.0 \text{ hp}} \times \frac{1 \text{ hp}}{550 \dfrac{\text{ft lb}}{\text{s}}}$$

$$= 15.0 \text{ s}$$

$$\boxed{\frac{\text{lb ft}}{\text{hp}} \times \frac{\text{hp}}{\dfrac{\text{ft lb}}{\text{s}}} = \frac{\text{lb ft}}{\text{hp}} \times \left(\text{hp} \div \frac{\text{ft lb}}{\text{s}} \right) = \frac{\cancel{\text{lb ft}}}{\cancel{\text{hp}}} \times \left(\cancel{\text{hp}} \times \frac{\text{s}}{\cancel{\text{ft lb}}} \right) = \text{s}}$$

Note: We use a conversion factor to obtain time units.

. . . .

The mass of a large steel wrecking ball is $20\overline{0}0$ kg. What power is used to raise it to a height of 40.0 m if the work is done in 20.0 s?

EXAMPLE 3

Data:

$$m = 20\overline{0}0 \text{ kg}$$
$$s = 40.0 \text{ m}$$
$$t = 20.0 \text{ s}$$
$$F = ?$$

Basic Equations:

$$P = \frac{W}{t} \quad \text{and} \quad W = Fs$$

Working Equation:

$$P = \frac{Fs}{t}$$

Substitution: Note that we cannot directly substitute into the working equation because our data are given in terms of *mass* and we must find *force* to substitute in $P = Fs/t$. The force is the weight of the ball:

$$F = mg = (20\overline{0}0 \text{ kg})(9.80 \text{ m/s}^2) = 19{,}600 \text{ kg m/s}^2 = 19{,}600 \text{ N}$$

Then

$$P = \frac{Fs}{t} = \frac{(19{,}600 \text{ N})(40.0 \text{ m})}{20.0 \text{ s}}$$

$$= 39{,}200 \text{ N m/s}$$

$$= 39{,}200 \text{ W} \quad \text{or} \quad 39.2 \text{ kW}$$

. . . .

EXAMPLE 4

A machine is designed to perform a given amount of work in a given amount of time. A second machine does the same amount of work in half the time. Find the power of the second machine compared with the first.

Data (for the second machine given in terms of the first):

$$W = W$$

$$t = \tfrac{1}{2}t = \frac{t}{2}$$

$$P = ?$$

Basic Equation:

$$P = \frac{W}{t}$$

Working Equation: Same

Substitution:

$$P = \frac{W}{\dfrac{t}{2}} = W \div \frac{t}{2} = W \times \frac{2}{t} = 2\left(\frac{W}{t}\right) = 2P$$

Thus, the power is doubled when the time is halved.

· · · · · · · · · · · · · · · · · ·

EXAMPLE 5

A motor is capable of developing 10.0 kW of power. How large a mass can it lift 75.0 m in 20.0 s?

Data:

$$P = 10.0 \text{ kW} = 10{,}\overline{0}00 \text{ W}$$

$$s = 75.0 \text{ m}$$

$$t = 20.0 \text{ s}$$

$$F = ?$$

Basic Equations:

$$P = \frac{W}{t} \quad \text{and} \quad W = Fs \quad \text{or} \quad P = \frac{Fs}{t}$$

Working Equation:

$$F = \frac{Pt}{s}$$

Substitution:

$$F = \frac{(10{,}\overline{0}00 \text{ W})(20.0 \text{ s})}{75.0 \text{ m}}$$

$$= 2670 \frac{W\,s}{m} \times \frac{1 \text{ N m/s}}{1 \text{ W}} \qquad (1 \text{ W} = 1 \text{ J/s} = 1 \text{ N m/s})$$

$$= 2670 \text{ N}$$

Next, change the weight to mass as follows:

Data:

$$F = 2670 \text{ N}$$
$$g = 9.80 \text{ m/s}^2$$
$$m = ?$$

Basic Equation:

$$F = mg$$

Working Equation:

$$m = \frac{F}{g}$$

Substitution:

$$m = \frac{2670 \text{ N}}{9.80 \text{ m/s}^2} \times \frac{1 \text{ kg m/s}^2}{1 \text{ N}} \qquad (1 \text{ N} = 1 \text{ kg m/s}^2)$$
$$= 272 \text{ kg}$$

· · · · · · · · · · · · · · · · ·

EXAMPLE 6

A pump is needed to lift $150\overline{0}$ L of water per minute a distance of 45.0 m. What power, in kW, must the pump be able to deliver? (1 L of water has a mass of 1 kg.)

Data:

$$m = 150\overline{0} \text{ L} \times \frac{1 \text{ kg}}{1 \text{ L}} = 150\overline{0} \text{ kg}$$

$$s = 45.0 \text{ m}$$
$$t = 1 \text{ min} = 60.0 \text{ s}$$
$$g = 9.80 \text{ m/s}^2$$
$$P = ?$$

Basic Equations:

$$P = \frac{W}{t}, \quad W = Fs, \quad \text{and} \quad F = mg, \quad \text{or} \quad P = \frac{mgs}{t}$$

Working Equation:

$$P = \frac{mgs}{t}$$

Substitution:

$$P = \frac{(150\overline{0} \text{ kg})(9.80 \text{ m/s}^2)(45.0 \text{ m})}{60.0 \text{ s}}$$

$$= 1.10 \times 10^4 \text{ kg m}^2/\text{s} \qquad \left(1 \text{ W} = \frac{1 \text{ J}}{\text{s}} = \frac{1 \text{ N m}}{\text{s}} = \frac{1 \text{ (kg m/s}^2)(\text{m})}{\text{s}} = 1 \text{ kg m}^2/\text{s}\right)$$

$$= 1.10 \times 10^4 \text{ W} \times \frac{1 \text{ kW}}{10^3 \text{ W}}$$

$$= 11.0 \text{ kW}$$

· · · · · · · · · · · · · · · · ·

SKETCH

12 cm² | w

4.0 cm

DATA

A = 12 cm², l = 4.0 cm, w = ?

BASIC EQUATION

A = lw

WORKING EQUATION

w = $\frac{A}{l}$

SUBSTITUTION

w = $\frac{12\ cm^2}{4.0\ cm}$ = 3.0 cm

PROBLEMS 8.2

1. Given: W = 132 J
   t = 7.00 s
   P = ?

2. Given: P = 231 ft lb/s
   t = 14.3 s
   W = ?

3. Given: P = 75.0 W
   W = 40.0 J
   t = ?

4. Given: W = 55.0 J
   t = 11.0 s
   P = ?

5. The work required to lift a crate is $31\overline{0}$ J. If the crate is lifted in 25.0 s, what power is developed?

6. When a 3600-lb automobile runs out of gas, it is pushed by its unhappy driver and a friend a quarter of a mile (0.250 mi). To keep the car rolling, they must exert a constant force of 175 lb.
   (a) How much work do they do?
   (b) If it takes them 15.0 min, how much power do they develop?
   (c) Expressed in horsepower, how much power do they develop?

7. An electric golf cart develops 1.25 kW of power while moving at a constant speed.
   (a) Express its power in horsepower.
   (b) If the cart travels $20\overline{0}$ m in 35.0 s, what force is exerted by the cart?

8. How many seconds would it take a 7.00-hp motor to raise a 475-lb boiler to a platform 38.0 ft high?

9. How long would it take a $95\overline{0}$-W motor to raise a $36\overline{0}$-kg mass to a height of 16.0 m?

10. A 1500-lb casting is raised 22.0 ft in 2.50 min. Find the required horsepower.

11. What is the rating in kW of a 2.00-hp motor?

12. A wattmeter shows that a motor is drawing $220\overline{0}$ W. What horsepower is being delivered?

13. A 525-kg steel beam is raised 30.0 m in 25.0 s. How many kilowatts of power are needed?

14. How long would it take a 4.50-kW motor to raise a 175-kg boiler to a platform 15.0 m above the floor?

15. A 475-kg prestressed concrete beam is to be raised 10.0 m in 24.0 s. How many kilowatts of power are needed for the job?

16. A 50.0-kg welder is to be raised 15.0 m in 12.0 s. How many kilowatts of power are needed for the job?

17. An escalator is needed to carry 75 passengers per minute a vertical distance of 8.0 m. Assume that the mass of each passenger is $7\overline{0}$ kg.
    (a) What is the power (in kW) of the motor needed?
    (b) Express this power in horsepower.
    (c) What is the power (in kW) of the motor needed if 35% of the power is lost to friction and other losses?

18. A pump is needed to lift $75\overline{0}$ L of water per minute a distance of 25.0 m. What power (in kW) must the pump be able to deliver? (1 L of water has a mass of 1 kg.)

19. A machine is designed to perform a given amount of work in a given amount of time. A second machine does twice the same amount of work in half the time. Find the power of the second machine compared with the first.

20. A certain machine is designed to perform a given amount of work in a given amount of time. A second machine does 2.5 times the same amount of work in one-third the time. Find the power of the second machine compared with the first.

21. A motor on an escalator is capable of developing 12 kW of power.
    (a) How many passengers of mass 75 kg each can it lift a vertical distance of 9.0 m per min, assuming no power loss?
    (b) What power, in kW, motor is needed to move the same number of passengers at the same rate if 45% of the actual power developed by the motor is lost to friction and heat loss?

22. A pump is capable of developing 4.00 kW of power. How many litres of water per minute can be lifted a distance of 35.0 m? (1 L of water has a mass of 1 kg.)

23. A pallet weighing 575 N is lifted a distance of 20.0 m vertically in 10.0 s. What power is developed in kilowatts?
24. A pallet is loaded with bags of cement; the total weight of 875 N is lifted 21.0 m vertically in 11.0 s. What power in kilowatts is required to lift the cement?
25. A bundle of steel reinforcing rods weighing 175 N is lifted 32.0 m in 16.0 s. What power in kilowatts is required to lift the steel?
26. An ironworker carries a 7.50-kg toolbag up a vertical ladder on a high-rise building under construction.
    (a) After 30.0 s, he is 8.20 m above his starting point. How much work does the worker do on the toolbag?
    (b) If the worker weighs 645 N, how much work does he do in lifting himself and the toolbag?
    (c) What is the average power developed by the worker?

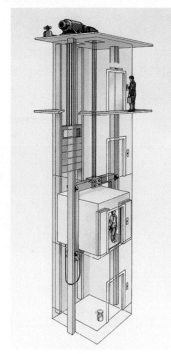

Figure 8.10 Here the counterweight working with the electric motor balances the load in the elevator.

8.3 Energy

Energy is defined as the ability to do work. There are many forms of energy, such as mechanical, electrical, thermal, fluid, chemical, atomic, and sound.

The mechanical energy of a body or a system is due to its position, its motion, or its internal structure. There are two kinds of mechanical energy: potential energy and kinetic energy. **Potential energy** is the stored energy of a body due to its internal characteristics or its position. **Kinetic energy** is the energy due to the mass and the velocity of a moving object.

Internal potential energy is determined by the nature or condition of the substance; for example, gasoline, a compressed spring, or a stretched rubber band has internal potential energy due to its internal characteristics. **Gravitational potential energy** is determined by the position of an object relative to a particular reference level; for example, a rock lying on the edge of a cliff, the raised counterweight on an elevator (Fig. 8.10), or a raised pile driver has potential energy due to its position. Each weight has the ability to do work because of the pull of gravity on it. The unit of energy is the joule (J) in the metric system and the foot-pound (ft lb) in the U.S. system.

The formula for gravitational potential energy is

$$PE = mgh$$

where   PE = potential energy
        m = mass
        g = 9.80 m/s^2 or 32.2 ft/s^2
        h = height above reference level

In position 1 in Fig. 8.11, the crate is at rest on the floor. It has no ability to do work because it is in its lowest position. To raise the crate to position 2, work must be done to lift it. In the raised position, however, it now has stored ability to do work (by falling to the floor). Its PE (potential energy) can be calculated by multiplying the mass of the crate times acceleration of gravity (g) times height above reference level (h). Note that we can calculate the potential energy of the crate with respect to any level we choose. Here we have chosen the floor as the zero or lowest reference level.

Figure 8.11 Work done in raising the crate gives it potential energy.

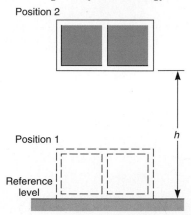

A wrecking ball of mass $20\overline{0}$ kg is poised 4.00 m above a concrete platform whose top is 2.00 m above the ground.

(a) With respect to the platform, what is the potential energy of the ball?
(b) With respect to the ground, what is the potential energy of the ball?

EXAMPLE 1

Sketch:

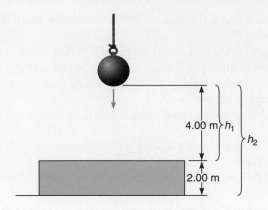

Data:

$$m = 20\overline{0} \text{ kg}$$
$$h_1 = 4.00 \text{ m}$$
$$h_2 = 6.00 \text{ m}$$
$$PE = ?$$

Basic Equation:

$$PE = mgh$$

Working Equation: Same

(a) Substitution:

$$PE = (20\overline{0} \text{ kg})(9.80 \text{ m/s}^2)(4.00 \text{ m})$$
$$= 7840 \frac{\text{kg m}^2}{\text{s}^2} \times \frac{1 \text{ J}}{\text{kg m}^2/\text{s}^2} \quad [1 \text{ J} = 1 \text{ N m} = 1 \text{ (kg m/s}^2)(\text{m}) = 1 \text{ kg m}^2/\text{s}^2]$$
$$= 7840 \text{ J}$$

(b) Substitution:

$$PE = (20\overline{0} \text{ kg})(9.80 \text{ m/s}^2)(6.00 \text{ m})$$
$$= 11{,}800 \frac{\text{kg m}^2}{\text{s}^2} \times \frac{1 \text{ J}}{\text{kg m}^2/\text{s}^2}$$
$$= 11{,}800 \text{ J}$$

· · · · · · · · · · · · · · · · · · ·

Kinetic energy is due to the mass and the velocity of a moving object and is given by the formula

$$KE = \tfrac{1}{2}mv^2$$

where   KE = kinetic energy
        m = mass of moving object
        v = velocity of moving object

A pile driver (Fig. 8.12) shows the relation of energy of motion to useful work. The energy of the driver is its kinetic energy as it hits. When the driver strikes the pile, work is done on the pile, and it is forced into the ground. The depth it goes into the ground is determined by the force applied to it. The force applied is determined by the energy of the driver. If all the kinetic energy of the driver is converted to useful work, then

$$\tfrac{1}{2}mv^2 = Fs$$

Figure 8.12   Energy of motion becomes useful work in the pile driver.

A pile driver with mass $10{,}\overline{0}00$ kg strikes a pile with velocity 10.0 m/s.

EXAMPLE 2

(a)   What is the kinetic energy of the driver as it strikes the pile?

(b)   If the pile is driven 20.0 cm into the ground, what force is applied to the pile by the driver as it strikes the pile? Assume that all the kinetic energy of the driver is converted to work.

Sketch:

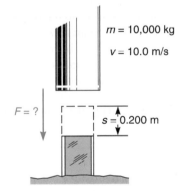

$m = 10{,}000$ kg

$v = 10.0$ m/s

$F = ?$

$s = 0.200$ m

Data:

$m = 1.00 \times 10^4$ kg

$v = 10.0$ m/s

$s = 20.0$ cm $= 0.200$ m

$F = ?$

(a) Basic Equation:

$KE = \frac{1}{2}mv^2$

Working Equation: Same

Substitution:

$KE = \frac{1}{2}(1.00 \times 10^4 \text{ kg})(10.0 \text{ m/s})^2$

$= 5.00 \times 10^5 \dfrac{\text{kg m}^2}{\text{s}^2} \times \dfrac{1 \text{ J}}{\text{kg m}^2/\text{s}^2}$ $[1 \text{ J} = 1 \text{ N m} = 1 \text{ (kg m/s}^2)(\text{m}) = 1 \text{ kg m}^2/\text{s}^2]$

$= 5.00 \times 10^5 \text{ J}$ or $5\overline{0}0$ kJ

13. A 4.000-kg mass is dropped from a hot air balloon 300.0 m above the ground. Find its kinetic energy, its potential energy, and the sum of the kinetic energy and the potential energy in 1-s intervals until the mass hits the ground. (Assume no air resistance.)

14. A 2.00-kg projectile is fired vertically upward with an initial velocity of 98.0 m/s. Find its kinetic energy, its potential energy, and the sum of its kinetic and potential energies at each of the following times:
    (a) the instant of its being fired
    (b) $t = 1.00$ s
    (c) $t = 2.00$ s
    (d) $t = 5.00$ s
    (e) $t = 10.00$ s
    (f) $t = 12.00$ s
    (g) $t = 15.00$ s
    (h) $t = 20.00$ s

Glossary

Energy The ability to do work. There are many forms of energy, such as mechanical, electrical, thermal, fluid, chemical, atomic, and sound. (p. 217)

Gravitational Potential Energy The energy determined by the position of an object relative to a particular reference level. (p. 217)

Internal Potential Energy The energy determined by the nature or condition of a substance. (p. 217)

Kinetic Energy The energy due to the mass and the velocity of a moving object. (p. 217)

Law of Conservation of Mechanical Energy The sum of the kinetic energy and the potential energy in a system is constant if no resistant forces do work. (p. 222)

Potential Energy The stored energy of a body due to its internal characteristics or its position. (p. 217)

Power The rate of doing work (work divided by time). (p. 211)

Work The product of the force in the direction of motion and the displacement. (p. 206)

Formulas

8.1 $W = Fs$

$W = Fs \cos \theta$

8.2 $P = \dfrac{W}{t}$

8.3 $PE = mgh$

$KE = \frac{1}{2}mv^2$

8.4 $v = \sqrt{2gh}$

Review Questions

1. Work is done when
    (a) a force is applied.
    (b) a person tries unsuccessfully to move a crate.
    (c) force is applied and an object is moved.
2. Power
    (a) is work divided by time.
    (b) is measured in newtons.
    (c) is time divided by work.
    (d) none of the above.
3. A large boulder at rest possesses
    (a) potential energy.
    (b) kinetic energy.
    (c) no energy.
4. A large boulder rolling down a hill possesses
    (a) potential energy.
    (b) kinetic energy.
    (c) no energy.
    (d) both kinetic and potential energy.
5. With no air resistance and no friction, a pendulum would
    (a) not swing.
    (b) swing for a short time.
    (c) swing forever.

6. Can work be done by a moving object on itself?
7. Has a man swinging a sledgehammer done work if he misses the stake at which he is swinging?
8. Develop the units associated with work from the components of the definition: work = force × displacement.
9. Is work a vector quantity?
10. Is work being done on a boulder by gravity?
11. Is work being done by the weight of a grandfather clock?
12. How could the power developed by a man pushing a stalled car be measured?
13. How does water above a waterfall possess potential energy?
14. What are two devices possessing gravitational potential energy?
15. Is kinetic energy dependent on time?
16. At what point is the kinetic energy of a swinging pendulum bob at a maximum?
17. At what point is the potential energy of a swinging pendulum bob at a maximum?
18. Is either kinetic or potential energy a vector quantity?
19. Can an object possess both kinetic and potential energy at the same time?
20. Why is a person more likely to be severely injured by a bolt falling from the fourth floor of a job site than by one falling from the second floor?

Review Problems

PROBLEM SOLVING

SKETCH

12 cm²	w

4.0 cm

DATA

$A = 12 \text{ cm}^2, l = 4.0 \text{ cm}, w = ?$

BASIC EQUATION

$A = lw$

WORKING EQUATION

$w = \frac{A}{l}$

SUBSTITUTION

$w = \frac{12 \text{ cm}^2}{4.0 \text{ cm}} = 3.0 \text{ cm}$

1. How many joules are in one kilowatt-hour?
2. An endloader holds $15\overline{0}0$ kg of sand 2.00 m off the ground for 3.00 min. How much work does it do?
3. How high can a 10.0-kg mass be lifted by $10\overline{0}0$ J of work?
4. A 40.0-kg pack is carried up a $25\overline{0}0$-m-high mountain in 10.0 h. How much work is done?
5. Find the average power output in Problem 4 in (a) watts; (b) horsepower.
6. A 10.0-kg mass has a potential energy of 10.0 J when it is at what height?
7. A 10.0-lb weight has a potential energy of 20.0 ft lb at what height?
8. At what speed does a 1.00-kg mass have a kinetic energy of 1.00 J?
9. At what speed does a 10.0-N weight have a kinetic energy of 1.00 J?
10. What is the kinetic energy of a 3000-lb automobile moving at 55.0 mi/h?
11. What is the potential energy of an 80.0-kg diver standing 3.00 m above the water?
12. What is the kinetic energy of a 0.020-kg bullet having a velocity of 550 m/s?
13. What is the potential energy of an 85.0-kg high jumper clearing a 2.00-m bar?
14. A worker pulls a crate 10.0 m by exerting a force of $30\overline{0}$ N.
    (a) How much work does the worker do?
    (b) How much work does the worker do pulling the crate a distance of 10.0 m by exerting the same force at an angle of 20.0° with the horizontal?
15. A hammer falls from a scaffold on a building 50.0 m above the ground. Find its speed as it hits the ground.

APPLIED CONCEPTS

1. Rosita needs to purchase a sump pump for her basement. (a) If the pump must carry 10.0 kg of water to a height of 2.75 m each minute, what minimum wattage pump is needed? (b) What three main factors determine power for a sump pump?

2. A roller coaster designer must carefully balance the desire for excitement and the need for safety. The most recent design is shown in Fig. 8.18. (a) If a 355-kg roller coaster car has zero velocity on the top of the first hill, determine its potential energy. (b) What is the velocity of the roller coaster car at the specified locations in the design? (c) Explain the relationship between velocity and the position on the track throughout the ride. (Consider the track to be frictionless.)

Figure 8.18

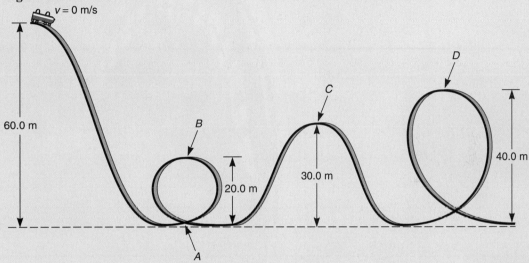

3. A 22,500-kg Navy fighter jet flying 235 km/h must catch an arresting cable to land safely on the runway strip of an aircraft carrier. (a) How much energy must the cable absorb to stop the fighter jet? (b) If the cable allows the jet to move 115 m before coming to rest, what is the average force that the cable exerts on the jet? (c) If the jet were given more than 115 m to stop, how would the force applied by the cable change?

4. The hydroelectric plant at the Itaipu Dam, located on the Parana River between Paraguay and Brazil, uses the transfer of potential to kinetic energy of water to generate electricity. (a) If 1.00×10^6 gallons of water (3.79×10^6 kg) flows down 142 m into the turbines each second, how much power does the hydroelectric power plant generate? (For comparison purposes, the Hoover Dam generates 1.57×10^6 W of power.) (b) How much power could the plant produce if the Itaipu Dam were twice its actual height? (c) Explain why the height of a dam is important for hydroelectric power plants.

5. A 1250-kg wrecking ball is lifted to a height of 12.7 m above its resting point. When the wrecking ball is released, it swings toward an abandoned building and makes an indentation of 43.7 cm in the wall. (a) What is the potential energy of the wrecking ball at a height of 12.7 m? (b) What is its kinetic energy as it strikes the wall? (c) If the wrecking ball transfers all of its kinetic energy to the wall, how much force does the wrecking ball apply to the wall? (d) Why should a wrecking ball strike a wall at the lowest point in its swing?

CHAPTER 9

ROTATIONAL MOTION

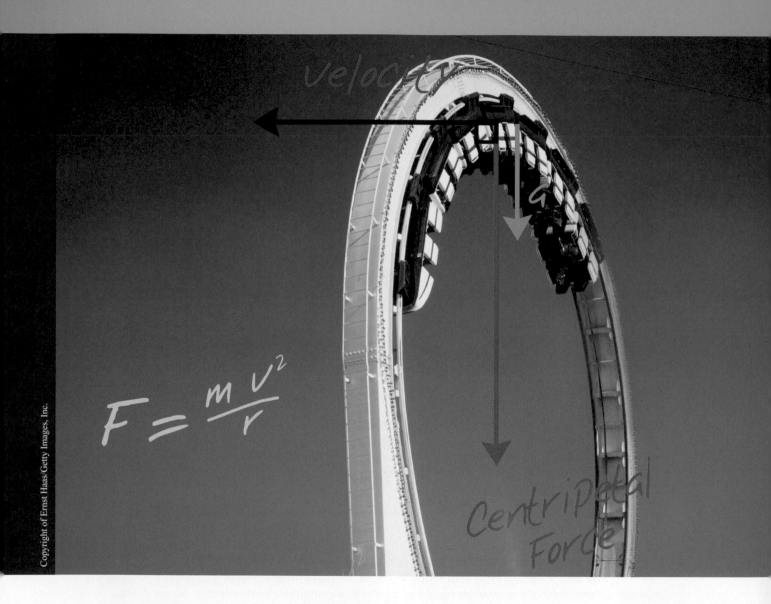

velocity

$$F = \frac{m\,v^2}{r}$$

Centripetal Force

Τhe concepts of displacement, velocity, acceleration, vectors, and forces in a straight line also apply to motion in a curved path and rotational motion.

Objectives

The major goals of this chapter are to enable you to:

1. Distinguish between rectilinear, curvilinear, and rotational motion.
2. Find angular displacement, velocity, and acceleration.
3. Use conservation of angular momentum to describe rotational motion.
4. Find centripetal force.
5. Find power in rotational systems.
6. Analyze how gears, gear trains, and pulleys are used to transfer rotational motion.

9.1 Measurement of Rotational Motion

Until now we have considered only motion in a straight line, called **rectilinear motion**. Technicians are often faced with many problems with motion along a curved path or with objects rotating about an axis. Although these kinds of motion are similar, we must distinguish between them.

Motion along a curved path is called **curvilinear motion**. A satellite in orbit around the earth is an example of curvilinear motion [Fig. 9.1(a)].

Figure 9.1

 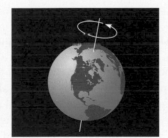 

(a) Curvilinear motion of an orbiting satellite

(b) Rotational motion of earth spinning on its axis

(c) Rotational motion of a wheel spinning on its axle

Rotational motion occurs when the body itself is spinning. Examples of rotational motion are the earth spinning on its axis, a turning wheel, a turning driveshaft, and the turning shaft of an electric motor [Fig. 9.1(b) and (c)].

We can see a wheel turn, but to gather useful information about its motion, we need a system of measurement. There are three basic systems of defining angle measurement. The **revolution** is one complete rotation of a body. This unit of measurement of rotational motion is the number of rotations—how many times the object goes around. The unit of rotation (most often used in industry) is the revolution (rev). A second system of angular measurement divides the circle of rotation into 360 degrees (360° = 1 rev). One **degree** is 1/360 of a complete revolution.

The **radian** (rad), which is approximately 57.3° or exactly $\left(\dfrac{360}{2\pi}\right)^{\circ}$, is a third angular unit of measurement. A radian is defined as that angle with its vertex at the center of a circle whose sides cut off an arc on the circle equal to its radius (Fig. 9.2), where $s = r$ and $\theta = 1$ rad.

Figure 9.2

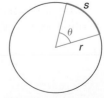

When $s = r$, $\theta = 1$ rad

Stated as a formula,

$$\theta = \frac{s}{r}$$

where
θ = angle determined by s and r
s = length of the arc of the circle
r = radius of the circle

Technically, angle θ measured in radians is defined as the ratio of two lengths: the lengths of the arc and the radius of a circle. Since the length units in the ratio cancel, the radian is a dimensionless unit. As a matter of convenience, "rad" is often used to show radian measurement. A useful relationship is 2π rad equals one revolution. Therefore,

$$1 \text{ rev} = 360° = 2\pi \text{ rad}$$

You may need to use this conversion between systems of measurement.

EXAMPLE 1

Convert the angle 10π rad (a) to rev and (b) to degrees.

Using 1 rev = 360° = 2π rad, form conversion factors so that the old units are in the denominator and the new units are in the numerator.

(a) $\theta = (10\pi \text{ rad})\left(\dfrac{1 \text{ rev}}{2\pi \text{ rad}}\right)$

   $= 5 \text{ rev}$

(b) $\theta = (10\pi \text{ rad})\left(\dfrac{360°}{2\pi \text{ rad}}\right)$

   $= 1800°$

· · · · · · · · · · · · · · · · ·

Figure 9.3 The angular displacements of points A and B on the flywheel are always the same.

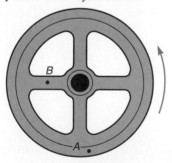

Angular displacement is the angle through which any point on a rotating body moves. Note that on any rotating body, all points on that body move through the same angle in any given amount of time—even though each may travel different linear distances. Point A on the flywheel shown in Fig. 9.3 travels much farther than point B (along a curved line), but during one revolution both travel through the same angle (have equal angular displacements).

In the automobile industry, technicians are concerned with the *rate* of rotational motion. Recall that in the linear system, velocity is the rate of motion (displacement/time). Similarly, **angular velocity** in the rotational system is the rate of angular displacement. Angular velocity (designated ω, the Greek lowercase letter omega) is usually measured in rev/min (rpm) for relatively slow rotations (e.g., automobile engines) and rev/s or rad/s for high-speed instruments. We use the term *angular velocity* when referring to a vector that includes the direction of rotation. We use the term *angular speed* in referring to a magnitude when the direction of rotation is either not known or not important.

$$\omega = \text{angular velocity} = \frac{\text{number of revolutions}}{\text{time}} = \frac{\text{angular displacement}}{\text{time}}$$

Written as a formula,

$$\omega = \frac{\theta}{t}$$

where
ω = angular velocity or speed (rad/s)   or ω = angular velocity or speed (rev/min)
θ = angle (in radians) θ = angle (in revolutions)
t = time (in seconds) t = time (in minutes)

EXAMPLE 2

A motorcycle wheel turns $36\overline{0}0$ times while being ridden for 6.40 min. What is the angular speed in rev/min?

Data:

$$t = 6.40 \text{ min}$$
$$\text{number of revolutions} = 36\overline{0}0 \text{ rev}$$
$$\omega = ?$$

Basic Equation:

$$\omega = \frac{\theta}{t}$$

Working Equation: Same

Substitution:

$$\omega = \frac{36\overline{0}0 \text{ rev}}{6.40 \text{ min}}$$
$$= 563 \text{ rev/min or } 563 \text{ rpm}$$

· · · · · · · · · · · · · · · · ·

Formulas for linear speed of a rotating point on a circle and angular speed are related as follows. We know

$$(1) \quad \theta = \frac{s}{r} \qquad (2) \quad v = \frac{s}{t} \qquad (3) \quad \omega = \frac{\theta}{t}$$

Therefore, combining and substituting s/r for θ in (3), we obtain

$$\omega = \frac{s/r}{t}$$

$$\omega(r) = \frac{(s/r)(r)}{t} \qquad \text{Multiply both sides by } r.$$

$$\omega r = \frac{s}{t}$$

$$\omega r = v \qquad \text{Recall that } v = s/t.$$

Thus,

$$\boxed{v = \omega r}$$

where v = linear velocity of a point on the circle
   ω = angular speed
   r = radius

EXAMPLE 3

A wheel of 1.00 m radius turns at $10\overline{0}0$ rpm.

(a)   Express the angular speed in rad/s.
(b)   Find the angular displacement in 2.00 s.
(c)   Find the linear speed of a point on the rim of the wheel.

Sketch:

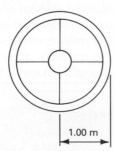

1.00 m

(a) Data:

$$\omega = 10\overline{0}0 \text{ rpm} \qquad \text{(change to rad/s)}$$

$$\omega = 10\overline{0}0 \frac{\text{rev}}{\text{min}} \times \frac{2\pi \text{ rad}}{1 \text{ rev}} \times \frac{1 \text{ min}}{60 \text{ s}} = 105 \text{ rad/s}$$

(b) Data:

$$\omega = 105 \text{ rad/s}$$
$$t = 2.00 \text{ s}$$
$$\theta = ?$$

Basic Equation:

$$\omega = \frac{\theta}{t}$$

Working Equation:

$$\theta = \omega t$$

Substitution:

$$\theta = (105 \text{ rad/s})(2.00 \text{ s})$$
$$= 21\overline{0} \text{ rad}$$

(c) Data:

$$\omega = 105 \text{ rad/s}$$
$$r = 1.00 \text{ m}$$
$$v = ?$$

Basic Equation:

$$v = \omega r$$

Working Equation: Same

Substitution:

$$v = (105 \text{ rad/s})(1.00 \text{ m})$$
$$= 105 \text{ m/s}$$

> (rad/s)(m) = m/s because the rad is a dimensionless unit.

· · · · · · · · · · · · · · · · ·

A device called a *stroboscope* or strobe light may be used to measure or check the speed of rotation of a shaft or other machinery part. Repeating motion is "slowed down" so it can be observed more easily. The light flashes rapidly, and the rate of flash can be adjusted to coincide with the rotation of a point or points on the rotating object. Knowing the rate of flashing will also then reveal the rate of rotation. A slight variation in the rate of rotation and flash will cause the observed point to appear to move either forward or backward as the stagecoach wheels in old western movies sometimes appear to do. Figure 9.4 shows a stroboscopic linear motion time-lapse photo of a woman running.

In linear motion, we found a change in velocity results in an acceleration. Similarly, in rotational motion, changing the rate of rotation involves a change in angular velocity and results in an angular acceleration. For uniformly accelerated rotational motion, **angular acceleration** is the rate of change of angular velocity. That is,

$$\alpha = \frac{\Delta \omega}{t}$$

Figure 9.4

Photo courtesy of Corbis Corporation. Reprinted with permission

where α = angular acceleration
     $\Delta\omega$ = change in angular velocity
       t = time

The equations for uniformly accelerated linear motion in Section 4.3 may easily be transformed into the corresponding equations for uniformly accelerated rotational motion by substituting θ for s, ω for v, and α for a.

Linear Motion	Rotational Motion
$s = v_{avg}t$	$\theta = \omega_{avg}t$
$s = v_i t + \frac{1}{2}at^2$	$\theta = \omega_i t + \frac{1}{2}\alpha t^2$
$v_{avg} = \dfrac{v_f + v_i}{2}$	$\omega_{avg} = \dfrac{\omega_f + \omega_i}{2}$
$v_f = v_i + at$	$\omega_f = \omega_i + \alpha t$
$a = \dfrac{v_f - v_i}{t}$	$\alpha = \dfrac{\omega_f - \omega_i}{t}$
$s = \frac{1}{2}(v_f + v_i)t$	$\theta = \frac{1}{2}(\omega_f + \omega_i)t$
$2as = v_f^2 - v_i^2$	$2\alpha\theta = \omega_f^2 - \omega_i^2$

where s = linear displacement
     v_f = final linear velocity
     v_i = initial linear velocity
   v_{avg} = average linear velocity
     a = linear acceleration
     t = time

where θ = angular displacement
     ω_f = final angular velocity
     ω_i = initial angular velocity
   ω_{avg} = average angular velocity
     α = angular acceleration
     t = time

While these rotational motion equations are somewhat intuitive, their derivations are beyond the scope of this text.

A rotating pulley 24.0 cm in diameter is rotating at an initial angular speed of 30.5 rad/s. The speed is steadily increased to 41.5 rad/s within 6.30 s. (a) Find the pulley's angular acceleration. (b) Find the final linear speed of a point on its rim.

EXAMPLE 4

(a) Data:

$$\omega_i = 30.5 \text{ rad/s}$$
$$\omega_f = 41.5 \text{ rad/s}$$
$$t = 6.30 \text{ s}$$
$$\alpha = ?$$

Basic Equation:

$$\omega_f = \omega_i + \alpha t$$

Working Equation:

$$\alpha = \frac{\omega_f - \omega_i}{t}$$

Substitution:

$$\alpha = \frac{41.5 \text{ rad/s} - 30.5 \text{ rad/s}}{6.30 \text{ s}}$$
$$= 1.75 \text{ rad/s}^2$$

(b) Data:

$$\omega = 41.5 \text{ rad/s}$$
$$r = 12.0 \text{ cm}$$
$$v = ?$$

Basic Equation:

$$v = \omega r$$

Working Equation: Same

Substitution:

$$v = (41.5 \text{ rad/s})(12.0 \text{ cm})$$
$$= 498 \text{ cm/s}$$

···················

TRY THIS ACTIVITY

Global Rotational Physics

On a globe depicting the earth, place a push pin or a small mark on northern Canada, another on Florida, and another on Ecuador. Rotate the globe about its central axis. Of the places that you marked, where would a person experience the greatest linear velocity as the earth is rotating? Where would a person experience the greatest angular velocity?

Finally, what would happen to a person's angular velocity if the rotational speed of the earth changed from one rotation per day to two rotations per day?

PROBLEMS 9.1

1. Convert $6\frac{1}{2}$ revolutions
   (a) to radians.
   (b) to degrees.

2. Convert 2880°
   (a) to revolutions.
   (b) to radians.

3. Convert 25π rad
   (a) to revolutions.
   (b) to degrees.

4. Convert 12.0 revolutions
   (a) to radians.
   (b) to degrees.

Find the angular speed in Problems 5–10.

5. Number of revolutions = 525
    $t = 3.42$ min
    $\omega =$ _____ rpm

6. Number of revolutions = 7360
    $t = 37.0$ s
    $\omega =$ _____ rev/s

7. Number of revolutions = 4.00
    $t = 3.00$ s
    $\omega =$ _____ rad/s

8. Number of revolutions = 325
    $t = 5.00$ min
    $\omega =$ _____ rpm

9. Number of revolutions = 6370
    $t = 18.0$ s
    $\omega =$ _____ rev/s

10. Number of revolutions = 6.25
    $t = 5.05$ s
    $\omega =$ _____ rad/s

11. Convert 675 rad/s to rpm.
12. Convert 285 rpm to rad/s.
13. Convert 136 rpm to rad/s.
14. Convert 88.4 rad/s to rpm.
15. A motor turns at a rate of 11.0 rev/s. Find its angular speed in rpm.
16. A rotor turns at a rate of 180 rpm. Find its angular speed in rev/s.
17. A rotating wheel completes one revolution in 0.150 s. Find its angular speed
    (a) in rev/s.   (b) in rpm.   (c) in rad/s.
18. A rotor completes 50.0 revolutions in 3.25 s. Find its angular speed
    (a) in rev/s.   (b) in rpm.   (c) in rad/s.
19. A flywheel rotates at 1050 rpm.
    (a) How long (in s) does it take to complete one revolution?
    (b) How many revolutions does it complete in 5.00 s?
20. A wheel rotates at 36.0 rad/s.
    (a) How long (in s) does it take to complete one revolution?
    (b) How many revolutions does it complete in 8.00 s?
21. A shaft of radius 8.50 cm rotates 7.00 rad/s. Find its angular displacement (in rad) in 1.20 s.
22. A wheel of radius 0.240 m turns at 4.00 rev/s. Find its angular displacement (in rev) in 13.0 s.
23. A pendulum of length 1.50 m swings through an arc of 5.0°. Find the length of the arc through which the pendulum swings.
24. An airplane circles an airport twice while 5.00 mi from the control tower. Find the length of the arc through which the plane travels.
25. A wheel of radius 27.0 cm has an angular speed of 47.0 rpm. Find the linear speed (in m/s) of a point on its rim.
26. A belt is placed around a pulley that is 30.0 cm in diameter and rotating at 275 rpm. Find the linear speed (in m/s) of the belt. (Assume no belt slippage on the pulley.)
27. A flywheel of radius 25.0 cm is rotating at 655 rpm.
    (a) Express its angular speed in rad/s.
    (b) Find its angular displacement (in rad) in 3.00 min.
    (c) Find the linear distance traveled (in cm) by a point on the rim in one complete revolution.
    (d) Find the linear distance traveled (in m) by a point on the rim in 3.00 min.
    (e) Find the linear speed (in m/s) of a point on the rim.
28. An airplane propeller with blades 2.00 m long is rotating at 1150 rpm.
    (a) Express its angular speed in rad/s.
    (b) Find its angular displacement in 4.00 s.
    (c) Find the linear speed (in m/s) of a point on the end of the blade.
    (d) Find the linear speed (in m/s) of a point 1.00 m from the end of the blade.
29. An automobile is traveling at 60.0 km/h. Its tires have a radius of 33.0 cm.
    (a) Find the angular speed of the tires (in rad/s).
    (b) Find the angular displacement of the tires in 30.0 s.
    (c) Find the linear distance traveled by a point on the tread in 30.0 s.
    (d) Find the linear distance traveled by the automobile in 30.0 s.
30. Find the angular speed (in rad/s) of the following hands on a clock.
    (a) Second hand   (b) Minute hand   (c) Hour hand

SKETCH

$\boxed{12\text{ cm}^2}$ w

4.0 cm

DATA

A = 12 cm², l = 4.0 cm, w = ?

BASIC EQUATION

$A = lw$

WORKING EQUATION

$w = \frac{A}{l}$

SUBSTITUTION

$w = \frac{12\text{ cm}^2}{4.0\text{ cm}} = 3.0$ cm

PROBLEM SOLVING

31. A bicycle wheel of diameter 30.0 in. rotates twice each second. Find the linear velocity of a point on the wheel.

32. A point on the rim of a flywheel with radius 1.50 ft has a linear velocity of 30.0 ft/s. Find the time for it to complete 4π rad.

33. The earth rotates on its axis at an angular speed of 1 rev/24 h. Find the linear speed (in km/h)
    (a) of Singapore, which is nearly on the equator.
    (b) of Houston, which is approximately 30.0° north latitude.
    (c) of Minneapolis, which is approximately 45.0° north latitude.
    (d) of Anchorage, which is approximately 60.0° north latitude.

34. A truck tire rotates at an initial angular speed of 21.5 rad/s. The driver steadily accelerates, and after 3.50 s the tire's angular speed is 28.0 rad/s. What is the tire's angular acceleration during its linear acceleration?

35. Find the angular acceleration of a radiator fan blade as its angular speed increases from 8.50 rad/s to 15.4 rad/s in 5.20 s.

36. A wheel of radius 20.0 cm starts from rest and makes 6.00 revolutions in 2.50 s. (a) Find its angular velocity in rad/s. (b) Find its angular acceleration. (c) Find the final linear speed of a point on the rim of the wheel. (d) Find the linear acceleration of a point on the rim of the wheel.

37. A circular disk 30.0 cm in diameter is rotating at 275 rpm and then uniformly stopped within 8.00 s. (a) Find its angular acceleration. (b) Find the initial linear speed of a point on its rim. (c) How many revolutions does the disk make before it stops?

38. A rotating flywheel of diameter 40.0 cm uniformly accelerates from rest to 250 rad/s in 15.0 s. (a) Find its angular acceleration. (b) Find the linear velocity of a point on the rim of the wheel after 15.0 s. (c) How many revolutions does the wheel make during the 15.0 s?

9.2 Angular Momentum

Recall that in Section 5.1 we saw that *inertia* is the property of a body that causes it to remain at rest if it is at rest or to continue moving with a constant velocity unless an unbalanced force acts upon it. Similarly, *rotational inertia,* called the **moment of inertia**, is the property of a rotating body that causes it to continue to turn until a torque causes it to change its rotational motion. A freely spinning wheel on an upside-down bicycle continues to spin after you stop hand cranking the pedals because of its rotational inertia.

In Section 7.3, we saw that a *torque* is produced when a force is applied to produce a rotation ($\tau = Fs_t$). To increase the speed of the spinning bicycle wheel above, additional torque must be applied by increasing the applied force, F. The angular acceleration of a rotating body is found to be directly proportional to the torque applied to it. This applied torque can be expressed as follows:

$$\tau = I\alpha$$

where τ = applied torque
I = moment of inertia (rotational inertia)
α = angular acceleration

This is the rotational equivalent of Newton's second law of motion and applies to a rigid body rotating about a fixed axis.

The moment of inertia, I, is a measure of the rotational inertia of a body. The rotational inertia is determined by the mass of the rotating object and how far away that mass is from its axis of rotation. Figure 9.5 shows two cylinders of equal mass, one solid and one hollow. The hollow cylinder has more rotational inertia because its mass is concentrated farther from its axis of rotation. A flywheel is a mechanical device, which is usually a heavy metal rotating wheel attached to a drive shaft with most of its weight concentrated at its circumference. A small motor can slowly increase the speed of a flywheel to store up kinetic en-

Figure 9.5 Even though the two cylinders are of equal mass, the hollow cylinder has more rotational inertia because its mass is concentrated farther from its axis of rotation.

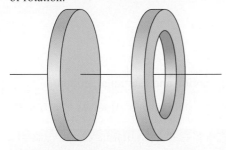

ergy; then, the small motor uses this inertia to perform a task for which it is ordinarily too small. A flywheel is also used to minimize rotational variations due to fluctuations in load and applied torque. Figure 9.6 shows a flywheel that is used to produce a steady rotation where the applied force of the piston is intermittent in this two-cylinder engine.

Figure 9.6 The large flywheel on this 1913 Case 30–60 tractor produces steady rotation of the crankshaft between fuel ignition cycles in its two-cylinder engine.

In Section 6.1 we saw that (linear) *momentum* is a measure of the amount of inertia and motion an object has or of the difficulty in bringing a moving body to rest. The formula for linear momentum is $p = mv$. There we studied applications involving linear motion and found the impulse is the change in linear momentum, $Ft = mv_f - mv_i$. Similarly, **angular momentum** for a rotating body about a fixed axis is defined as

$$L = I\omega$$

where L = angular momentum
I = moment of inertia (rotational inertia)
ω = angular velocity

Note the comparison with linear dynamics:

angular momentum = (moment of inertia) × (angular velocity)

linear momentum = (mass, a measure of inertia) × (linear velocity)

Furthermore, the *angular impulse* is the change in angular momentum.

$$\tau t = I\omega_f - I\omega_i$$

where τ = torque
  t = time
  I = moment of inertia
  ω_f = final angular velocity
  ω_i = initial angular velocity

Compare the following pairs of equations for linear motion and rotational motion:

Linear Motion	Rotational Motion
$F = ma$	$\tau = I\alpha$
$p = mv$	$L = I\omega$
$Ft = mv_f - mv_i$	$\tau t = I\omega_f - I\omega_i$

where F = applied force
  m = mass (inertia)
  a = linear acceleration
  p = linear momentum
  v = linear velocity
  t = time
  v_f = final linear velocity
  v_i = initial linear velocity

where τ = applied torque
  I = moment of inertia
  α = angular acceleration
  L = angular momentum
  ω = angular velocity
  t = time
  ω_f = final angular velocity
  ω_i = initial angular velocity

Conservation of Angular Momentum

In Section 6.1 we learned from the law of conservation of momentum that the total linear momentum ($p = mv$) of a system remains unchanged unless an external force acts on it. Similarly, the **law of conservation of angular momentum** states that the total angular momentum ($L = I\omega$) of a system remains unchanged unless an external torque acts on it.

LAW OF CONSERVATION OF ANGULAR MOMENTUM

The angular momentum of a system remains unchanged unless an external torque acts on it.

A spinning ice skater is an interesting example. When the skater's arms are extended, the rotational inertia, I, is relatively large and the angular velocity, ω, is relatively small. Often at the end of a spin, the skater pulls his or her arms tight to the body resulting in a much faster spin (larger angular velocity) because of a much smaller rotational inertia, I. When a rotating body contracts, its angular velocity, ω, increases; and when a rotating body expands, its angular velocity decreases. This phenomenon is the result of the conservation of angular momentum.

Similarly, gymnasts and divers generate their spins (torque) from a solid base or a diving board after which the angular momentum remains unchanged. The usual somersaults and twists result from making variations in their rotational inertia. Astronauts must also learn to control their spins as they maneuver their bodies to work in space.

Spinning Non-Ice Skaters

Ice skaters are able to control their rotational speed by changing their rotational inertia. To do this without a pair of skates, sit in an office chair that rotates freely and hold a couple of weights or heavy books close to your chest. Have someone spin the chair and gradually increase your rotational speed. After reaching your optimum rotational speed, stretch out your arms holding the weights in your hands. What happens to your rotational speed as you move your arms from close to your chest to an out-stretched position? Try this a few times. Explain how you can change your rotational speed like ice skaters do.

9.3 Centripetal Force

Newton's laws of motion apply to motion along a curved path as well as in a straight line. Recall that a moving body tends to continue in a straight line because of inertia. If we are to cause the body to move in a circle, we must constantly apply a force perpendicular to the line of motion of the body. A simple example is a rock on the end of a string being swung in a circle (Fig. 9.7). By Newton's first law, the rock tends to go in a straight line but the string exerts a constant force on the rock perpendicular to this line of travel. The resulting path of the rock is a circle [Fig. 9.8(a)]. The force of the string on the rock is the *centripetal* (toward the center) force. The **centripetal force** acting on a body in circular motion causes it to move in a circular path. This force is exerted toward the center of the circle. If the string should break, however, there would no longer be a centripetal force acting on the rock, which would fly off tangent to the circle [Fig. 9.8(b)].

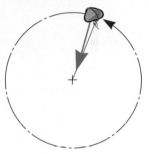

Figure 9.7 Rock on a string being swung in a circle. The centripetal force is directed toward the center.

Figure 9.8 Centripetal force on a rock being swung in a circle

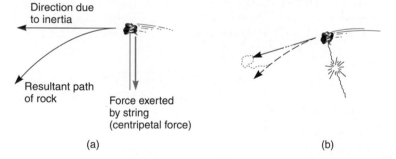

Direction due to inertia

Resultant path of rock

Force exerted by string (centripetal force)

(a)

(b)

The equation for finding the centripetal force on any body moving along a curved path is

$$F = \frac{mv^2}{r}$$

where
F = centripetal force
m = mass of the body
v = velocity of the body
r = radius of curvature of the path of the body

TRY THIS ACTIVITY

Whirling a Bucket of Water

Hold a bucket with some water in your hand and quickly swing it vertically in a circular motion. Why does the water not fall out of the bucket when the bucket is upside down? Gradually decrease the speed of the whirling bucket while measuring the approximate rotational speed of the bucket. Measure the radius of the circular arc from your shoulder to the bottom of the bucket, the mass of the bucket and water, and the rotational speed of the bucket to find the minimum centripetal force needed to keep the water in the bucket. Next, double the amount of water in the bucket and repeat the experiment. How does the amount of water influence the rotational speed needed to keep the water in the bucket?

EXAMPLE

An automobile of mass 1640 kg rounds a curve of radius 25.0 m with a velocity of 15.0 m/s (54.0 km/h). What centripetal force is exerted on the automobile while rounding the curve?

Sketch:

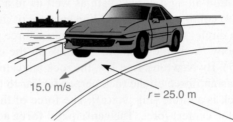

15.0 m/s

$r = 25.0$ m

Data:

$$m = 1640 \text{ kg}$$
$$v = 15.0 \text{ m/s}$$
$$r = 25.0 \text{ m}$$
$$F = ?$$

Basic Equation:

$$F = \frac{mv^2}{r}$$

Working Equation: Same

Substitution:

$$F = \frac{(1640 \text{ kg})(15.0 \text{ m/s})^2}{25.0 \text{ m}}$$
$$= 14{,}800 \text{ kg m/s}^2 \qquad (\text{Recall: } 1 \text{ N} = 1 \text{ kg m/s}^2)$$
$$= 14{,}800 \text{ N}$$

PROBLEMS 9.3

1. Given: $m = 64.0$ kg
   $v = 34.0$ m/s
   $r = 17.0$ m
   $F = \underline{\hspace{1cm}}$ N

2. Given: $m = 11.3$ slugs
   $v = 3.00$ ft/s
   $r = 3.24$ ft
   $F = \underline{\hspace{1cm}}$ lb

3. Given: $F = 2500$ lb
   $v = 47.6$ ft/s
   $r = 72.0$ ft
   $m = \underline{\hspace{1cm}}$ slugs

4. Given: $F = 587$ N
   $v = 0.780$ m/s
   $m = 67.0$ kg
   $r = \underline{\hspace{1cm}}$ m

5. Given: $F = 602$ N
$m = 63.0$ kg
$r = 3.20$ m
$v =$ _____ m/s

6. Given: $m = 37.5$ kg
$v = 17.0$ m/s
$r = 3.75$ m
$F =$ _____ N

7. Given: $F = 75.0$ N
$v = 1.20$ m/s
$m = 10\overline{0}$ kg
$r =$ _____ m

8. Given: $F = 80.0$ N
$m = 43.0$ kg
$r = 17.5$ m
$v =$ _____ m/s

9. An automobile of mass 117 slugs follows a curve of radius 79.0 ft with a speed of 49.3 ft/s. What centripetal force is exerted on the automobile while it is rounding the curve?

10. Find the centripetal force exerted on a 7.12-kg mass moving at a speed of 2.98 m/s in a circle of radius 2.72 m.

11. The centripetal force on a car of mass $80\overline{0}$ kg rounding a curve is 6250 N. If its speed is 15.0 m/s, what is the radius of the curve?

12. The centripetal force on a runner is 17.0 lb. If the runner weighs 175 lb and his speed is 14.0 mi/h, find the radius of the curve.

13. An automobile with mass 1650 kg is driven around a circular curve of radius $15\overline{0}$ m at 80.0 km/h. Find the centripetal force of the road on the automobile.

14. A cycle of mass 510 kg rounds a curve of radius $4\overline{0}$ m at 95 km/h. What is the centripetal force on the cycle?

15. What is the centripetal force exerted on a rock with mass 3.2 kg moving at 3.5 m/s in a circle of radius 2.1 m?

16. What is the centripetal force on a $15\overline{0}0$-kg vehicle driven around a curve of radius 35.0 m at 60.0 km/h?

17. What is the centripetal force on a $75\overline{0}$-kg vehicle rounding a curve of radius 40.0 m at 30.0 km/h?

18. A truck with mass 215 slugs rounds a curve of radius 53.0 ft with a speed of 62.5 ft/s. (a) What centripetal force is exerted on the truck while rounding the curve? (b) How does the centripetal force change when the velocity is doubled? (c) What is the new force?

19. A 225-kg dirt bike is rounding a curve with linear velocity of 35 m/s and an angular speed of 0.25 rad/s. Find the centripetal force exerted on the bike.

20. A $55,\overline{0}00$-kg truck rounds a curve at 62.0 km/h. If the radius of the curve is 38.0 m, what is the centripetal force on the truck?

21. The radius of a curve is 27.5 m. What is the centripetal force on a 10,000-kg truck going around it at 35.0 km/h?

SKETCH

12 cm^2 | w

4.0 cm

DATA

$A = 12$ cm^2, $l = 4.0$ cm, $w = ?$

BASIC EQUATION

$A = lw$

WORKING EQUATION

$w = \frac{A}{l}$

SUBSTITUTION

$w = \frac{12 \text{ cm}^2}{4.0 \text{ cm}} = 3.0$ cm

9.4 Power in Rotational Systems

One of the most important aspects of rotational motion is the power developed. Recall that torque was discussed in Section 7.3. Power, however, must be considered whenever an engine or motor is used to turn a shaft. Some common examples are winches and drive trains (Fig. 9.9).

Figure 9.9 This driveshaft connects the engine transmission with the axle to supply power to the wheels and other components of this tractor. (Courtesy of Deere & Company)

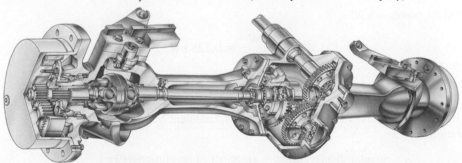

Earlier we learned that

$$\text{power} = \frac{\text{force} \times \text{displacement}}{\text{time}} = \frac{\text{work}}{\text{time}}$$

in the linear system. In the rotational system

$$P = \frac{(\text{torque})(\text{angular displacement})}{\text{time}}$$
$$= (\text{torque})(\text{angular velocity})$$
$$= \tau\omega$$

Recall that angular displacement is the angle through which a shaft is turned. In the metric system, angular displacement must be expressed in radians (1 rev = 2π radians). Substituting symbols and units, we have, in watts (W),

$$P = \tau\omega$$
$$= (\text{N m})\left(\frac{1}{\text{s}}\right) = \frac{\text{N m}}{\text{s}} = \frac{\text{J}}{\text{s}} = \text{W}$$

To find the power in kilowatts (kW), multiply the number of watts by the conversion factor

$$\frac{1 \text{ kW}}{1000 \text{ W}}$$

Note: In problem solving, the radian unit is a dimensionless unit; ω is expressed with the unit /s.

EXAMPLE 1

How many watts of power are developed by a mechanic tightening bolts using 50.0 N m of torque at a rate of 2.50 rad/s? How many kW?

Data:
$$\tau = 50.0 \text{ N m}$$
$$\omega = 2.50/\text{s}$$
$$P = ?$$

Basic Equation:
$$P = \tau\omega$$

Working Equation: Same

Substitution:
$$P = (50.0 \text{ N m})(2.50/\text{s})$$
$$= 125 \text{ N m/s}$$
$$= 125 \text{ W} \qquad (1 \text{ W} = 1 \text{ N m/s})$$

To find the power in kW:

$$125 \text{ W} \times \frac{1 \text{ kW}}{1000 \text{ W}} = 0.125 \text{ kW}$$

· · · · · · · · · · · · · · · · · ·

In the U.S. system, we measure angular displacement by multiplying the number of revolutions by 2π:

$$\text{angular displacement} = (\text{number of revolutions})(2\pi)$$

For the rotational system

$$\text{power} = \frac{(\text{torque})(2\pi \text{ revolutions})}{\text{time}}$$

When time is in minutes

$$\text{power} = \text{torque} \times 2\pi \times \frac{\text{rev}}{\text{min}} \times \frac{1 \text{ min}}{60 \text{ s}}$$

$$\boxed{\text{power in } \frac{\text{ft lb}}{\text{s}} = \text{torque in lb ft} \times \frac{\text{number of revolutions}}{\text{min}} \times 0.105 \frac{\text{min}}{\text{rev s}}}$$

$$\underbrace{2\pi \times \frac{1 \text{ min}}{60 \text{ s}}}$$

Another common unit of power is the horsepower (hp). The conversion factor between $\frac{\text{ft lb}}{\text{s}}$ and hp is

$$\boxed{\text{power in hp} = \text{power in } \frac{\text{ft lb}}{\text{s}} \times \frac{\text{hp}}{550 \text{ ft lb/s}}}$$

What power (in ft lb/s) is developed by an electric motor with torque 5.70 lb ft and speed 425 rpm?

> **EXAMPLE 2**

Data:

$$\tau = 5.70 \text{ lb ft}$$
$$\omega = 425 \text{ rpm}$$
$$P = ?$$

Basic Equation:

$$P = \text{torque} \times \frac{\text{rev}}{\text{min}} \times 0.105 \frac{\text{min}}{\text{rev s}}$$

Working Equation: Same

Substitution:

$$P = (5.70 \text{ lb ft})\left(425 \frac{\text{rev}}{\text{min}}\right)\left(0.105 \frac{\text{min}}{\text{rev s}}\right)$$
$$= 254 \text{ ft lb/s}$$

· · · · · · · · · · · · · · · · ·

What horsepower is developed by a racing engine with torque 545 lb ft at $6\overline{5}00$ rpm?
   First, find power in ft lb/s and then convert to hp.

> **EXAMPLE 3**

Data:

$$\tau = 545 \text{ lb ft}$$
$$\omega = 6\overline{5}00 \text{ rpm}$$
$$P = ?$$

Basic Equation:

$$P = \text{torque} \times \frac{\text{rev}}{\text{min}} \times 0.105 \frac{\text{min}}{\text{rev s}}$$

Working Equation: Same

Substitution:

$$P = (545 \text{ lb ft})\left(6500 \frac{\text{rev}}{\text{min}}\right)\left(0.105 \frac{\text{min}}{\text{rev s}}\right)$$

$$= 372{,}000 \frac{\text{ft lb}}{s} \times \frac{1 \text{ hp}}{550 \frac{\text{ft lb}}{s}}$$

$$= 676 \text{ hp}$$

. . . .

PROBLEMS 9.4

SKETCH

12 cm² | w

4.0 cm

DATA

$A = 12 \text{ cm}^2$, $l = 4.0 \text{ cm}$, $w = ?$

BASIC EQUATION

$A = lw$

WORKING EQUATION

$w = \frac{A}{l}$

SUBSTITUTION

$w = \frac{12 \text{ cm}^2}{4.0 \text{ cm}} = 3.0 \text{ cm}$

1. Given: $\tau = 125$ lb ft
   $\omega = 555$ rpm
   $P = \underline{\quad}$ ft lb/s

2. Given: $\tau = 39.4$ N m
   $\omega = 6.70/s$
   $P = \underline{\quad}$ W

3. Given: $\tau = 372$ lb ft
   $\omega = 264$ rpm
   $P = \underline{\quad}$ hp

4. Given: $\tau = 650$ N m
   $\omega = 45.0/s$
   $P = \underline{\quad}$ kW

5. Given: $P = 8950$ W
   $\omega = 4.80/s$
   $\tau = \underline{\quad}$

6. Given: $P = 650$ W
   $\tau = 540$ N m
   $\omega = \underline{\quad}$

7. What horsepower is developed by an engine with torque $40\overline{0}$ lb ft at $45\overline{0}0$ rpm?

8. What torque must be applied to develop 175 ft lb/s of power in a motor if $\omega = 394$ rpm?

9. Find the angular velocity of a motor developing 649 W of power with torque 131 N m.

10. A high-speed industrial drill develops 0.500 hp at $16\overline{0}0$ rpm. What torque is applied to the drill bit?

11. An engine has torque of 550 N m at 8.3 rad/s. What power in watts does it develop?

12. Find the angular velocity of a motor developing 33.0 N m/s of power with a torque of 6.0 N m.

13. What power (in hp) is developed by an engine with torque 524 lb ft
    (a) at $30\overline{0}0$ rpm?   (b) at $60\overline{0}0$ rpm?

14. Find the angular velocity of a motor developing 650 W of power with a torque of 130 N m.

15. A drill develops 0.500 kW of power at $18\overline{0}0$ rpm. What torque is applied to the drill bit?

16. What power is developed by an engine with torque $75\overline{0}$ N m applied at $45\overline{0}0$ rpm?

17. A tangential force of 150 N is applied to a flywheel of diameter 45 cm to maintain a constant angular velocity of 175 rpm. How much work is done per minute?

18. Find the power developed by an engine with a torque of $12\overline{0}0$ N m applied at $20\overline{0}0$ rpm.

19. Find the power developed by an engine with a torque of $16\overline{0}0$ N m applied at $15\overline{0}0$ rpm.

20. Find the power developed by an engine with torque 1250 N m applied at $50\overline{0}0$ rpm.

21. Find the angular velocity of a motor developing $100\overline{0}$ W of power with a torque of $15\overline{0}$ N m.

22. A motor develops 0.75 kW of power at $20\overline{0}0$ revolutions per $1\overline{0}$ min. What torque is applied to the motor shaft?

23. What power is developed when a tangential force of 175 N is applied to a flywheel of diameter 86 cm, causing it to have an angular velocity of 36 revolutions per 6.0 s?

24. What power is developed when a tangential force of $25\overline{0}$ N is applied to a wheel 57.0 cm in diameter with an angular velocity of 25.0 revolutions in 13.0 s?

25. An engine develops 1.50 kW of power at 10,000 revolutions per 5.00 min. What torque is applied to the engine's crankshaft?
26. A mechanic tightens engine bolts using 45.5 N m of torque at a rate of 2.75 rad/s. How many watts of power is used to tighten the bolts?
27. An ag mechanic tightens implement bolts using 52.5 N m of torque at a rate of 2.25 rad/s. What power does the mechanic develop in tightening the bolts?

9.5 Transferring Rotational Motion

Suppose that two disks are touching each other as in Fig. 9.10. Disk A is driven by a motor and turns disk B (wheel) by making use of the friction between them. The relationship between the diameters of the two disks and their number of revolutions is

$$D \cdot N = d \cdot n$$

where D = diameter of the driver disk
d = diameter of the driven disk
N = number of revolutions of the driver disk
n = number of revolutions of the driven disk

Figure 9.10 In the self-propelled lawn mower, disk A, driven by the motor, turns disk B, the wheel, which results in the mower moving along the ground.

However, using two disks to transfer rotational motion is not very efficient due to slippage that may occur between them. The most common ways to prevent disk slippage are placing teeth on the edge of the disk and connecting the disks with a belt. Therefore, instead of using disks, we use gears or belt-driven pulleys to transfer this motion. The teeth on the gears eliminate the slippage; the belt connecting the pulleys helps reduce the slippage and provides for distance between rotating centers (Fig. 9.11).

We can change the equation $D \cdot N = d \cdot n$ to the form $D/d = n/N$ by dividing both sides by dN. The left side indicates the ratio of the diameters of the disks. If the ratio is 2, this means that the larger disk must have a diameter two times the diameter of the smaller disk. The same ratio would apply to gears and pulleys. The ratio of the diameters of the gears must be 2 to 1, and the ratio of the diameters of the pulleys must be 2 to 1. In fact, the ratio of the number of teeth on the gears must be 2 to 1.

The right side of the equation indicates the ratio of the number of revolutions of the two disks. If the ratio is 2, this means that the smaller disk makes two revolutions while the larger disk makes one revolution. The same would be true for gears and for pulleys connected by a belt.

Figure 9.11 Gears and pulleys are used to reduce slippage in transferring rotational motion.

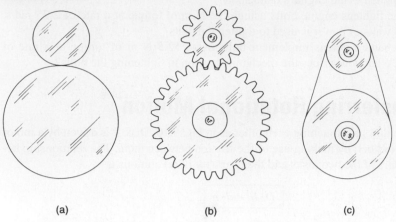

(a)   (b)   (c)

Gears and pulleys are used to increase or reduce the angular velocity of a rotating shaft or wheel. When two gears or pulleys are connected, the speed at which each turns compared to the other is inversely proportional to the diameter of that gear or pulley. The larger the diameter of a pulley or gear, the slower it turns. The smaller the diameter of a pulley or gear, the faster it will turn when connected to a larger one.

9.6 Gears

Gears are used to transfer rotational motion from one gear to another. The gear that causes the motion is called the *driver gear*. The gear to which the motion is transferred is called the *driven gear*.

There are many different sizes, shapes, and types of gears. Some examples are shown in Fig. 9.12. For any type of gear, we use one basic formula:

$$T \cdot N = t \cdot n$$

where T = number of teeth on the driver gear
 N = number of revolutions of the driver gear
 t = number of teeth on the driven gear
 n = number of revolutions of the driven gear

Figure 9.12 Examples of different types of gears. (Courtesy of Foote-Jones/Illinois Gear, Chicago, IL)

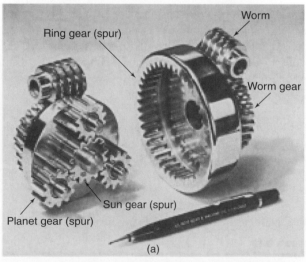

(a)

Figure 9.12 (*Continued*)

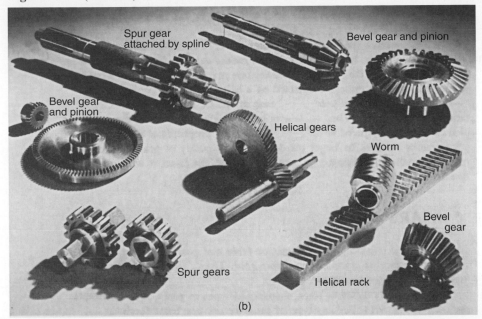

Spur gear attached by spline

Bevel gear and pinion

Bevel gear and pinion

Helical gears

Worm

Bevel gear

Spur gears

Helical rack

(b)

A driver gear has 30 teeth. How many revolutions does the driven gear with 20 teeth make while the driver makes one revolution?

EXAMPLE 1

Data:

$$T = 30 \text{ teeth} \qquad t = 20 \text{ teeth}$$
$$N = 1 \text{ revolution} \qquad n = ?$$

Basic Equation:

$$T \cdot N = t \cdot n$$

Working Equation:

$$n = \frac{T \cdot N}{t}$$

Substitution:

$$n = \frac{(30 \text{ teeth})(1 \text{ rev})}{20 \text{ teeth}}$$
$$= 1.5 \text{ rev}$$

· · · · · · · · · · · · · · · · ·

A driven gear of 70 teeth makes 63.0 revolutions per minute (rpm). The driver gear makes 90.0 rpm. What is the number of teeth required for the driver gear?

EXAMPLE 2

Data:

$$N = 90.0 \text{ rpm}$$
$$t = 70 \text{ teeth}$$
$$n = 63.0 \text{ rpm}$$
$$T = ?$$

Basic Equation:

$$T \cdot N = t \cdot n$$

Working Equation:

$$T = \frac{t \cdot n}{N}$$

Figure 9.13 Meshed gears shown as cylinders for simplicity

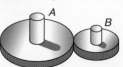

Substitution:

$$T = \frac{(70 \text{ teeth})(63.0 \text{ rpm})}{90.0 \text{ rpm}}$$

$$= 49 \text{ teeth}$$

Figure 9.14 Gear train of three gears

Gear Trains

When two gears mesh (Fig. 9.13),* they turn in opposite directions. If gear *A* turns clockwise, gear *B* turns counterclockwise. If gear *A* turns counterclockwise, gear *B* turns clockwise. If a third gear is inserted between the two (Fig. 9.14), then gears *A* and *B* are rotating in the same direction. This third gear is called an *idler*. A **gear train** is a series of gears that transfers rotational motion from one gear to another.

When the number of shafts in a gear train is odd (such as 1, 3, 5, . . .), the first gear and the last gear rotate in the same direction. When the number of shafts is even, the gears rotate in opposite directions.

When a complex gear train is considered, the relationship between revolutions and number of teeth is still present. This relationship is: The number of revolutions of the first driver times the product of the number of teeth of all the driver gears equals the number of revolutions of the final driven gear times the product of the number of teeth on all the driven gears. That is,

$$NT_1T_2T_3T_4 \cdots = nt_1t_2t_3t_4 \cdots$$

where N = number of revolutions of first driver gear
   T_1 = teeth on first driver gear
   T_2 = teeth on second driver gear
   T_3 = teeth on third driver gear
   T_4 = teeth on fourth driver gear
   n = number of revolutions of last driven gear
   t_1 = teeth on first driven gear
   t_2 = teeth on second driven gear
   t_3 = teeth on third driven gear
   t_4 = teeth on fourth driven gear

EXAMPLE 3

Determine the relative motion of gears *A* and *B* in Fig. 9.15.

Figure 9.15
(a)   (b)

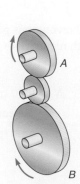

*Although gears have teeth, in technical work they are often shown as cylinders.

Figure 9.15 (*Continued*)

(c)

(d)

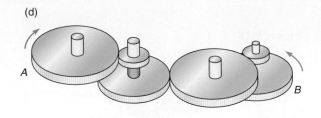

Find the number of revolutions per minute of gear D in Fig. 9.16 if gear A rotates at 20.0 rpm. Gears A and C are drivers and gears B and D are driven.

EXAMPLE 4

Figure 9.16

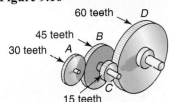

Data:

$$N = 20.0 \text{ rpm} \qquad t_1 = 45 \text{ teeth}$$
$$T_1 = 30 \text{ teeth} \qquad t_2 = 60 \text{ teeth}$$
$$T_2 = 15 \text{ teeth} \qquad n = ?$$

Basic Equation:

$$NT_1T_2 = nt_1t_2$$

Working Equation:

$$n = \frac{NT_1T_2}{t_1t_2}$$

Substitution:

$$n = \frac{(20.0 \text{ rpm})(30 \text{ teeth})(15 \text{ teeth})}{(45 \text{ teeth})(60 \text{ teeth})}$$
$$= 3.33 \text{ rpm}$$

Find the rpm of gear D in the train shown in Fig. 9.17. Gears A and C are drivers and gears B and D are driven.

EXAMPLE 5

Data:

$$N = 16\overline{0}0 \text{ rpm} \qquad t_1 = 30 \text{ teeth}$$
$$T_1 = 60 \text{ teeth} \qquad t_2 = 48 \text{ teeth}$$
$$T_2 = 15 \text{ teeth} \qquad n = ?$$

Figure 9.17

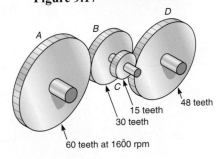

Basic Equation:

$$NT_1T_2 = nt_1t_2$$

Working Equation:

$$n = \frac{NT_1T_2}{t_1t_2}$$

Substitution:

$$n = \frac{(16\overline{0}0 \text{ rpm})(60 \text{ teeth})(15 \text{ teeth})}{(30 \text{ teeth})(48 \text{ teeth})}$$
$$= 10\overline{0}0 \text{ rpm}$$

EXAMPLE 6

In the gear train shown in Fig. 9.18, find the speed in rpm of gear A.

Figure 9.18

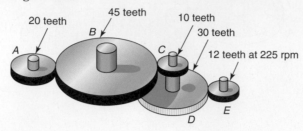

Data:

$$t_1 = 45 \text{ teeth} \qquad T_1 = 20 \text{ teeth}$$
$$t_2 = 10 \text{ teeth} \qquad T_2 = 45 \text{ teeth}$$
$$t_3 = 12 \text{ teeth} \qquad T_3 = 30 \text{ teeth}$$
$$n = 225 \text{ rpm} \qquad N = ?$$

Gear B is both a driver and a driven gear.

Basic Equation:

$$NT_1T_2T_3 = nt_1t_2t_3$$

Working Equation:

$$N = \frac{nt_1t_2t_3}{T_1T_2T_3}$$

Substitution:

$$N = \frac{(225 \text{ rpm})(45 \text{ teeth})(10 \text{ teeth})(12 \text{ teeth})}{(20 \text{ teeth})(45 \text{ teeth})(30 \text{ teeth})}$$
$$= 45.0 \text{ rpm}$$

· · · · · · · · · · · · · · · · ·

In a gear train, when a gear is both a driver gear and a driven gear, it may be omitted from the computation.

EXAMPLE 7

The problem in Example 6 could have been worked as follows because gear B is both a driver and a driven.

Basic Equation:

$$NT_1T_3 = nt_2t_3$$

Working Equation:

$$N = \frac{nt_2t_3}{T_1T_3}$$

Substitution:

$$N = \frac{(225 \text{ rpm})(10 \text{ teeth})(12 \text{ teeth})}{(20 \text{ teeth})(30 \text{ teeth})}$$
$$= 45.0 \text{ rpm}$$

· · · · · · · · · · · · · · · · ·

Bicycle Gears

The gearing system on a bicycle allows a cyclist to choose how much force he or she would like to exert when riding a bicycle. What gear ratio should be used when riding uphill, downhill, or on level ground? Prior to the use of gears on bicycles, a rider needed to sit directly above the front wheel in order to pedal. Gears, chains, and other advances have made the modern bicycle much more comfortable and efficient.

To find the gear ratio of a bicycle's gearing system, divide the number of teeth on the rear, driven gear by the number of teeth on the front, driver gear. When the number of teeth on the front gear is larger, the gear ratio is less than one and the rear wheel turns faster than the pedals. This results in high speeds for going down slight inclines or for level ground and allows the cyclist to pedal fewer revolutions while traveling a greater distance. When the number of teeth in the rear gear is larger, the gear ratio is greater than one. Here, the pedals turn faster than the rear wheel. This results in low speeds for going up large hills and allows the cyclist to pedal more revolutions while traveling a shorter distance but exerting a more manageable leg force. The following table illustrates the differences:

Number of Teeth on Front Gear (Driver)	Number of Teeth on Rear Gear (Driven)	Gear Ratio (Driven/Driver)	Number of Teeth on Front Gear (Driver)	Number of Teeth on Rear Gear (Driven)	Gear Ratio (Driven/Driver)
44	11	1/4	15	30	2/1
When the front gear turns once, the back gear turns four times (good for traveling at high speeds).			When the front gear turns once, the back gear turns only $\frac{1}{2}$ a rotation (good for reducing the force needed to pedal while going uphill).		

Most beginner bicycles simply connect the chain on the front gear to the rear gear with no option for changing the gear ratios. Children must stand up to pedal with more force. Geared bicycles, including road and mountain bikes, are engineered with a variety of gear ratios to allow a cyclist to travel more easily over many terrains (Fig. 9.19).

Figure 9.19 (a) Beginner bicycle   (b) Mountain bike

Courtesy of Pearson Education/PH College

Copyright of Getty Images. Photo reprinted with permission

Find the number of teeth for gear D in each gear train.

34.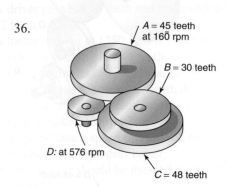

D: at 1500 rpm

B = 30 teeth
C = 15 teeth
A = 60 teeth
at 1850 rpm

35.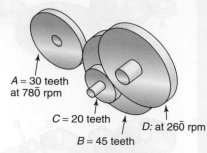

A = 30 teeth
at 780 rpm
C = 20 teeth
D: at 260 rpm
B = 45 teeth

36.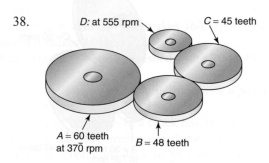

A = 45 teeth
at 160 rpm
B = 30 teeth
D: at 576 rpm
C = 48 teeth

37.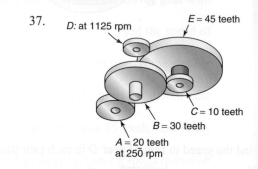

D: at 1125 rpm
E = 45 teeth
C = 10 teeth
B = 30 teeth
A = 20 teeth
at 250 rpm

38.

D: at 555 rpm
C = 45 teeth

A = 60 teeth
at 370 rpm
B = 48 teeth

Figure 9.20 A single belt drives several components from this engine. (Courtesy of Deere & Company)

39. Find the direction of rotation of gear B if gear A is turned counterclockwise in Problems 22 through 28.

40. Find the effect of doubling the number of teeth on gear A in Problem 38.

9.7 Pulleys Connected with a Belt

Pulleys connected with a belt are used to transfer rotational motion from one shaft to another (Fig. 9.20). Assuming no slippage, the linear speed of any point on the belt equals the linear speed of any point on the rim of each pulley as the belt travels around each pulley. The larger the pulley, the larger is its circumference: $C = \pi d$. The larger the circumference of the pulley, the longer a point on the belt stays in contact with the pulley. The smaller the circumference of the pulley, the shorter a point on the belt stays in contact with the pulley, which causes the smaller pulley to rotate faster than the larger pulley. Two pulleys connected with a belt have a relationship similar to gears. Assuming no slippage, when two pulleys are connected

$$D \cdot N = d \cdot n$$

where D = diameter of the driver pulley
N = number of revolutions per minute of the driver pulley
d = diameter of the driven pulley
n = number of revolutions per minute of the driven pulley

The preceding equation may be generalized in the same manner as for gear trains as follows:

$$ND_1D_2D_3 \cdots = nd_1d_2d_3 \cdots$$

Find the speed in rpm of pulley A shown in Fig. 9.21.

EXAMPLE

Data:

$$D = 6.00 \text{ in.}$$
$$d = 30.0 \text{ in.}$$
$$n = 35\overline{0} \text{ rpm}$$
$$N = ?$$

Figure 9.21
Driver diameter = 6.00 in.

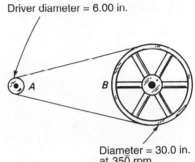

Diameter = 30.0 in.
at 35$\overline{0}$ rpm

Basic Equation:

$$D \cdot N = d \cdot n$$

Working Equation:

$$N = \frac{dn}{D}$$

Substitution:

$$N = \frac{(30.0 \text{ in.})(35\overline{0} \text{ rpm})}{6.00 \text{ in.}}$$
$$= 1750 \text{ rpm}$$

· · · · · · · · · · · · · · · · ·

When two pulleys are connected with an open-type belt, the pulleys turn in the same direction. When two pulleys are connected with a cross-type belt, the pulleys turn in opposite directions. See Fig. 9.22.

Figure 9.22   Crossing a belt reverses direction.

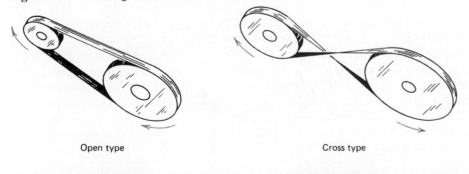

Open type

Cross type

(a) Pulleys rotate in
same direction.

(b) Pulleys rotate in
opposite directions.

SKETCH

12 cm²	w

4.0 cm

DATA

$A = 12$ cm², $l = 4.0$ cm, $w = ?$

BASIC EQUATION

$A = lw$

WORKING EQUATION

$w = \frac{A}{l}$

SUBSTITUTION

$w = \frac{12 \text{ cm}^2}{4.0 \text{ cm}} = 3.0$ cm

PROBLEMS 9.7

Find each missing quantity using $D \cdot N = d \cdot n$.

	D	N	d	n
1.	18.0	1500̄	12.0	___
2.	36.0	___	9.00	972
3.	12.0	1800̄	6.00	___
4.	___	2250	9.00	1125
5.	49.0	1860	___	620̄

6. A driver pulley of diameter 6.50 in. revolves at 1650 rpm. Find the speed of the driven pulley if its diameter is 26.0 in.

7. A driver pulley of diameter 25.0 cm revolves at 120̄ rpm. At what speed will the driven pulley turn if its diameter is 48.0 cm?

8. One pulley of diameter 36.0 cm revolves at 600̄ rpm. Find the diameter of the second pulley if it rotates at 360̄ rpm.

9. One pulley rotates at 450̄ rpm. The diameter of the second pulley is 15.0 in. and rotates at 675 rpm. Find the diameter of the first pulley.

10. A pulley with a radius of 10.0 cm rotates at 120̄ rpm. The radius of the second pulley is 15.0 cm; find its rpm.

Determine the direction of pulley B in each pulley system.

11.

12.
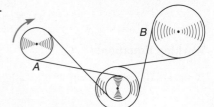

13.
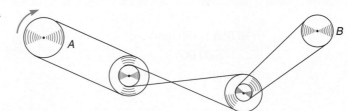

14.
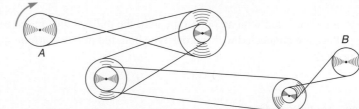

15.
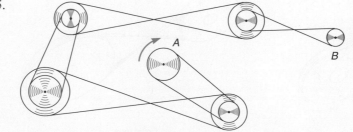

16. What size pulley should be placed on a countershaft turning 150̄ rpm to drive a grinder with a 12.0-cm pulley that is to turn at 1200̄ rpm?

Glossary

Angular Acceleration The rate of change of angular velocity (change in angular velocity/time). (p. 236)

Angular Displacement The angle through which any point on a rotating body moves. (p. 234)

Angular Momentum For a rotating body about a fixed axis, the angular momentum is the product of the moment of inertia and the angular velocity of the body. (p. 241)

Angular Velocity The rate of angular displacement (angular displacement/time). (p. 234)

Centripetal Force The force acting on a body in circular motion that causes it to move in a circular path. This force is exerted toward the center of the circle. (p. 243)

Curvilinear Motion Motion along a curved path. (p. 233)

Degree An angular unit of measure. Defined as 1/360 of one complete revolution. (p. 233)

Gear Train A series of gears that transfers rotational motion from one gear to another. (p. 232)

Law of Conservation of Angular Momentum The angular momentum of a system remains unchanged unless an outside torque acts on it. (p. 242)

Moment of Inertia Rotational inertia; the property of a rotating body that causes it to continue to turn until a torque causes it to change its rotational motion. (p. 240)

Radian An angular unit of measurement. Defined as that angle with its vertex at the center of a circle whose sides cut off an arc on the circle equal to its radius. Equal to $(360°/2\pi)$ or approximately 57.3°. (p. 233)

Rectilinear Motion Motion in a straight line. (p. 233)

Revolution A unit of measurement in rotational motion. One complete rotation of a body. (p. 233)

Rotational Motion Spinning motion of a body. (p. 233)

Formulas

9.1 $\quad \theta = \dfrac{s}{r}$

$\quad$ 1 rev $= 360° = 2\pi$ rad

$\quad \omega = \dfrac{\theta}{t}$

$\quad v = \omega r$

$\quad \alpha = \dfrac{\Delta\omega}{t}$

$\quad \theta = \omega_{avg}t$

$\quad \theta = \omega_i t + \frac{1}{2}\alpha t^2$

$\quad \omega_{avg} = \dfrac{\omega_f + \omega_i}{2}$

$\quad \omega_f = \omega_i + \alpha t$

$\quad \alpha = \dfrac{\omega_f - \omega_i}{t}$

$\quad \theta = \frac{1}{2}(\omega_f + \omega_i)t$

$\quad 2\alpha\theta = \omega_f^2 - \omega_i^2$

9.2 $\quad \tau = I\alpha$

$\quad L = I\omega$

$\quad \tau t = I\omega_f - I\omega_i$

ply speed, as with the gears on a bicycle [Fig. 10.2(b)]. Machines are used to change direction. When we use a single fixed pulley on a flag pole to raise a flag [Fig. 10.2(c)], the only advantage we get is the change in direction. (We pull the rope down, and the flag goes up.)

A **simple machine** is any one of six mechanical devices in which an applied force results in useful work (Fig. 10.3). All other machines—no matter how complex—are combinations of two or more of these simple machines.

Figure 10.3   Six simple machines

1. Lever

2. Wheel and axle

4. Inclined plane

3. Pulley

5. Screw

6. Wedge

In every machine we are concerned with two forces—effort and resistance. The **effort** is the force applied *to* the machine. The **resistance** is the force overcome *by* the machine. A person applies $3\overline{0}$ lb on the jack handle in Fig. 10.4 to produce a lifting force of $60\overline{0}$ lb on the car. The effort force is $3\overline{0}$ lb. The resistance force is $60\overline{0}$ lb.

Figure 10.4   This lever multiplies force.

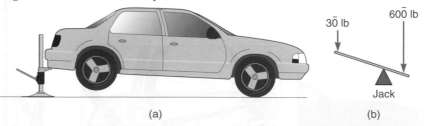

$3\overline{0}$ lb $60\overline{0}$ lb

Jack

(a)   (b)

LAW OF SIMPLE MACHINES

resistance force × resistance distance = effort force × effort distance

Mechanical Advantage and Efficiency

The **mechanical advantage** (MA) is the ratio of the resistance force to the effort force. By formula,

$$\text{MA} = \frac{\text{resistance force}}{\text{effort force}}$$

The MA of the jack in Fig. 10.4 is found as follows:

$$\text{MA} = \frac{\text{resistance force}}{\text{effort force}} = \frac{600\ \text{lb}}{30\ \text{lb}} = \frac{20}{1}$$

This MA means that, for each pound applied by the person, he or she lifts 20 pounds. Note that MA has no units. Why?

Each time a machine is used, part of the energy or effort applied to the machine is lost due to friction (Fig. 10.5). The **efficiency** of a machine is the ratio of the work output to the work input. By formula,

$$\text{efficiency} = \frac{\text{work output}}{\text{work input}} \times 100\% = \frac{F_{\text{output}} \times s_{\text{output}}}{F_{\text{input}} \times s_{\text{input}}} \times 100\%$$

Figure 10.5   Some work is always lost to friction.

10.2 The Lever

A **lever** consists of a rigid bar free to turn on a pivot called a **fulcrum** (Fig. 10.6). The mechanical advantage (MA) is the ratio of the effort arm (s_E) to the resistance arm (s_R):

$$\text{MA}_{\text{lever}} = \frac{\text{effort arm}}{\text{resistance arm}} = \frac{s_E}{s_R}$$

Figure 10.6   Mechanical advantage of the lever: $MA = \dfrac{s_E}{s_R}$.

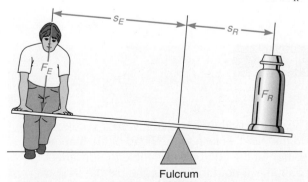

Fulcrum

The **effort arm** is the distance from the effort force to the fulcrum. The **resistance arm** is the distance from the fulcrum to the resistance force. The three types or classes of levers are shown in Fig. 10.7. The law of simple machines as applied to levers (basic equation) is

$$F_R \cdot s_R = F_E \cdot s_E$$

where F_R = resistance force
 s_R = length of resistance arm
 F_E = effort force
 s_E = length of effort arm

Figure 10.7 Three classes of levers

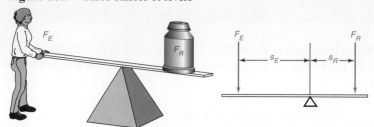

First class: The fulcrum is between the resistance force
(F_R) and the effort force (F_E).

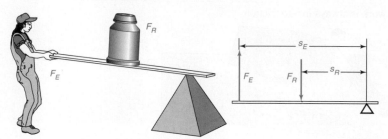

Second class: The resistance force (F_R) is between the
fulcrum and the effort force (F_E).

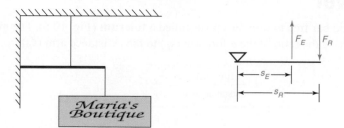

Third class: The effort force (F_E) is between the fulcrum
and the resistance force (F_R).

TRY THIS ACTIVITY

Pulling Nails

Hammer two identical nails into a piece of wood so that the heads are slightly above the wood. Wedge the jaws of the hammer beneath the head of the first nail and hold the handle close to the head of the hammer. Before removing the nail, note the resistance distance, the position of the fulcrum, and the effort distance. Note the amount of effort force necessary to remove the first nail. Next, hold the end of the handle as you remove the second nail. Explain why simple machines, such as hammers pulling nails, are able to reduce the effort force.

A bar is used to raise a 12$\overline{0}$0-N stone. The pivot is placed 30.0 cm from the stone. The worker pushes 2.50 m from the pivot. What is the mechanical advantage? What force is exerted?

EXAMPLE 1

Sketch:

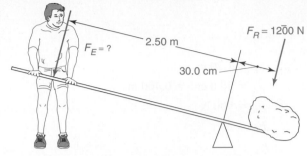

First, find MA:

$$MA_{lever} = \frac{s_E}{s_R} = \frac{2.50 \text{ m}}{0.300 \text{ m}} = \frac{8.33}{1}$$

To find the force:

Data:

$$s_E = 2.50 \text{ m}$$
$$s_R = 30.0 \text{ cm} = 0.300 \text{ m}$$
$$F_R = 12\overline{0}0 \text{ N}$$
$$F_E = ?$$

Basic Equation:

$$F_R \cdot s_R = F_E \cdot s_E$$

Working Equation:

$$F_E = \frac{F_R \cdot s_R}{s_E}$$

Substitution:

$$F_E = \frac{(12\overline{0}0 \text{ N})(0.300 \text{ m})}{2.50 \text{ m}}$$
$$= 144 \text{ N}$$

· · · · · · · · · · · · · · · ·

A wheelbarrow 1.20 m long has a 90$\overline{0}$-N load 40.0 cm from the axle. What is the MA? What force is needed to lift the wheelbarrow?

EXAMPLE 2

Sketch:

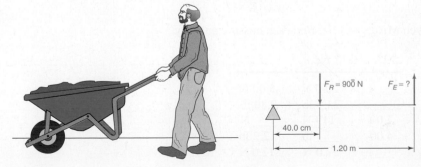

First, find MA:

$$MA = \frac{s_E}{s_R} = \frac{1.20 \ \cancel{m}}{0.400 \ \cancel{m}} = \frac{3.00}{1}$$

To find the force:

Data:

$$s_E = 1.20 \ m$$
$$s_R = 40.0 \ cm = 0.400 \ m$$
$$F_R = 90\overline{0} \ N$$
$$F_E = ?$$

Basic Equation:

$$F_R \cdot s_R = F_E \cdot s_E$$

Working Equation:

$$F_E = \frac{F_R \cdot s_R}{s_E}$$

Substitution:

$$F_E = \frac{(90\overline{0} \ N)(0.400 \ \cancel{m})}{1.20 \ \cancel{m}}$$

$$= 30\overline{0} \ N$$

················

EXAMPLE 3

The MA of a pair of pliers is 6.0/1. A force of 8.0 lb is exerted on the handle. What force is exerted on a wire in the pliers?

MA = 6.0/1 means that for each pound of force applied on the handle, 6.0 lb is exerted on the wire. Therefore, if a force of 8.0 lb is applied on the handle, a force of (6.0)(8.0 lb) or 48 lb is exerted on the wire.

················

PROBLEMS 10.2

Given $F_R \cdot s_R = F_E \cdot s_E$, find each missing quantity.

	F_R	F_E	s_R	s_E
1.	20.0 N	5.00 N	3.70 cm	____ cm
2.	____ N	176 N	49.2 cm	76.3 cm
3.	37.0 N	12.0 N	____ cm	112 cm
4.	23.4 lb	9.80 lb	____ in.	53.9 in.
5.	119 N	____ N	29.7 cm	67.4 cm

Given $MA_{lever} = \dfrac{F_R}{F_E}$, find each missing quantity.

	MA	F_R	F_E
6.	____	20.0 N	5.00 N
7.	____	23.4 lb	9.80 lb
8.	7.00	119 N	____ N
9.	4.00	____ lb	12.2 lb
10.	____	37.0 N	12.0 N

SKETCH

12 cm² | w

4.0 cm

DATA

$A = 12 \ cm^2, l = 4.0 \ cm, w = ?$

BASIC EQUATION

$A = lw$

WORKING EQUATION

$w = \frac{A}{l}$

SUBSTITUTION

$w = \frac{12 \ cm^2}{4.0 \ cm} = 3.0 \ cm$

PHYSICS CONNECTIONS

The Human Body—A Complex Machine

The human arm is a classic example of a human simple machine. Figure 10.8(a) illustrates the forces and lever arms that are in place while lifting a weight. The elbow's hinge joint acts as a relatively low-friction fulcrum. The bicep is the muscle that exerts the effort force, and the barbell is the resistance force. According to the equation for mechanical advantage, when the resistance force is farther from the fulcrum than the effort force, the mechanical advantage is less than one. Instead of reducing the force needed to lift the object, the bicep actually exerts more force than the weight of the barbell.

Machines with mechanical advantages less than one, such as the human arm, are useful because the effort force does not have to move far in order for the weight to move large distances. The bicep muscle contracts a relatively small distance, whereas the barbell moves a larger distance. When the arm is used to throw a ball, the muscles exert a large force over a small distance, whereas the end of the forearm moves a relatively large distance with a high velocity. As seen in Fig. 10.8(b), the arm acts like several levers when throwing a ball. Although the force needed to throw the ball is greater than the weight of the ball, the great distance covered by the ball in a short period of time translates into a high velocity of the ball.

Figure 10.8   (a) The bicep moves very little while exerting a large force to lift the heavy barbell.
(b) The parts of the arm act as several simple machines to throw a ball with a high velocity.

(a)   (b)

Given $MA_{lever} = \dfrac{s_E}{s_R}$, find each missing quantity.

	MA	s_R	s_E
11.	____	49.2 cm	76.3 cm
12.	7.00	29.7 in.	____ in.
13.	____	29.7 cm	67.4 cm
14.	4.00	____ cm	67.4 cm

15.   A pole is used to lift a car that fell off a jack (Fig. 10.9). The pivot is 2.00 ft from the car. Two people together exert 275 lb of force 8.00 ft from the pivot. (a) What force is applied to the car? (Ignore the weight of the pole.) (b) Find the MA.

Figure 10.9

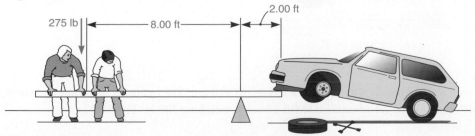

Figure 10.10

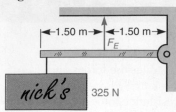

16. A bar is used to lift a $10\overline{0}$-kg block of concrete. The pivot is 1.00 m from the block. (a) If the worker pushes down on the other end of the bar a distance of 2.50 m from the pivot, what force (in N) must the worker apply? (b) Find the MA.

17. A wheelbarrow 6.00 ft long is used to haul a 180-lb load. (a) How far from the wheel is the load placed so that a person can lift the load with a force of 45.0 lb? (b) Find the MA.

18. (a) Find the force, F_E, pulling up on the beam holding the sign shown in Fig. 10.10. (b) Find the MA.

10.3 The Wheel-and-Axle

The **wheel-and-axle** consists of a large wheel attached to an axle so that both turn together (Fig. 10.11). Other examples include a doorknob and a screwdriver with a thick handle.

The law of simple machines as applied to the wheel-and-axle (basic equation) is

$$F_R \cdot r_R = F_E \cdot r_E$$

where F_R = resistance force
r_R = radius of resistance force
F_E = effort force
r_E = radius of effort force

Figure 10.11 Examples of the wheel-and-axle

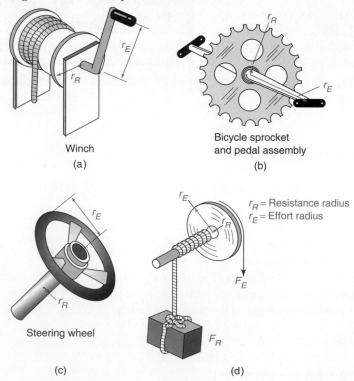

Winch
(a)

Bicycle sprocket and pedal assembly
(b)

Steering wheel
(c)

r_R = Resistance radius
r_E = Effort radius

(d)

EXAMPLE 1

A winch has a handle that turns in a radius of 30.0 cm. The radius of the drum or axle is 10.0 cm. Find the force required to lift a bucket weighing $50\overline{0}$ N (Fig. 10.12).

Data:

$$F_R = 50\bar{0} \text{ N}$$
$$r_E = 30.0 \text{ cm}$$
$$r_R = 10.0 \text{ cm}$$
$$F_E = ?$$

Basic Equation:

$$F_R \cdot r_R = F_E \cdot r_E$$

Working Equation:

$$F_E = \frac{F_R \cdot r_R}{r_E}$$

Substitution:

$$F_E = \frac{(50\bar{0} \text{ N})(10.0 \text{ cm})}{30.0 \text{ cm}}$$
$$= 167 \text{ N}$$

·················

Figure 10.12

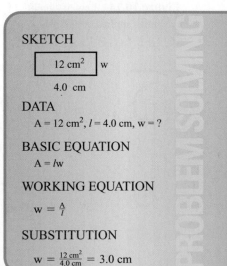

$r_E = 30.0$ cm
$r_R = 10.0$ cm

$50\bar{0}$ N

The mechanical advantage (MA) of the wheel-and-axle is the ratio of the radius of the effort force to the radius of the resistance force.

$$\text{MA}_{\text{wheel-and-axle}} = \frac{\text{radius of effort force}}{\text{radius of resistance force}} = \frac{r_E}{r_R}$$

Calculate the MA of the winch in Example 1. ◄

EXAMPLE 2

$$\text{MA}_{\text{wheel-and-axle}} = \frac{r_E}{r_R} = \frac{30.0 \text{ cm}}{10.0 \text{ cm}} = \frac{3.00}{1}$$

·················

PROBLEMS 10.3

Given $F_R \cdot r_R = F_E \cdot r_E$, find each missing quantity.

	F_R	F_E	r_R	r_E
1.	20.0 N	5.30 N	3.70 cm	_____ cm
2.	37$\bar{0}$ N	12$\bar{0}$ N	_____ m	1.12 m
3.	_____ N	175 N	49.2 cm	76.3 cm
4.	23.4 lb	9.80 lb	_____ in.	53.9 in.
5.	1190 N	_____ N	29.7 cm	67.4 cm

Given $\text{MA}_{\text{wheel-and-axle}} = \dfrac{r_E}{r_R}$, find each missing quantity.

	MA	r_E	r_R
6.	7.00	119 mm	_____ mm
7.	4.00	_____ in.	12.2 in.
8.	_____	49.2 cm	31.7 cm
9.	3.00	61.3 cm	_____ cm
10.	_____	67.4 mm	29.7 mm

SKETCH

12 cm² w

4.0 cm

DATA

A = 12 cm², *l* = 4.0 cm, w = ?

BASIC EQUATION

A = *l*w

WORKING EQUATION

$w = \frac{A}{l}$

SUBSTITUTION

$w = \frac{12 \text{ cm}^2}{4.0 \text{ cm}} = 3.0 \text{ cm}$

11. A wheel with radius 75.0 cm is attached to an axle of radius 13.6 cm. What force must be applied to the rim of the wheel to raise a $100\overline{0}$-N weight?

12. An axle of radius 12.0 cm is used with a wheel of radius 62.0 cm. What force must be applied to the rim of the wheel to lift a weight of 975 N?

13. The radius of the axle of a winch is 3.00 in. The length of the handle (radius of wheel) is 1.50 ft. (a) What weight will be lifted by an effort of 73.0 lb? (b) Find the MA.

14. A wheel of radius of 70.0 cm is attached to an axle of radius 20.0 cm. (a) What force must be applied to the rim of the wheel to raise a weight of $150\overline{0}$ N? (b) Find the MA. (c) What weight can be lifted if a force of 575 N is applied?

15. The diameter of the wheel of a wheel-and-axle is 10.0 cm. (a) If a force of 475 N is raised by applying a force of 142 N, find the diameter of the axle. (b) Find the MA.

16. Two persons use a large winch to raise a mass of 470 kg. The radius of the wheel is 48 cm and the radius of the axle is 4.0 cm. (a) What force is required to lift the load? (b) Find the MA of the windlass. (c) If the efficiency of the windlass is 60% and each person exerts the same force, how much force must each apply?

10.4 The Pulley

A **pulley** is a grooved wheel that turns readily on an axle and is supported in a frame. It can be fastened to a fixed object or to the resistance that is to be moved. A **fixed pulley** is a pulley fastened to a fixed object [Fig. 10.13(a)]. A **movable pulley** is fastened to the object to be moved [Fig. 10.13(b)]. A pulley system consists of combinations of fixed and movable pulleys [Fig. 10.13(c)–(e)].

Figure 10.13 Pulleys and pulley systems

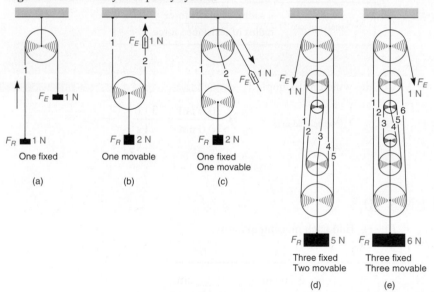

| (a) One fixed | (b) One movable | (c) One fixed One movable | (d) Three fixed Two movable | (e) Three fixed Three movable |

Figure 10.14 Law of simple machines applied to pulleys: $F_R \cdot s_R = F_E \cdot s_E$

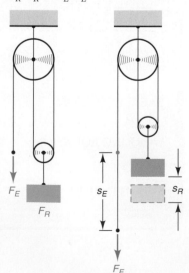

The law of simple machines as applied to pulleys (Fig. 10.14) is

$$F_R \cdot s_R = F_E \cdot s_E$$

Here, s refers to the distance moved. From the preceding equation,

$$\frac{F_R}{F_E} = \frac{s_E}{s_R} = \text{MA}_{\text{pulley}}$$

However, when one continuous cord is used, this ratio reduces to the number of strands holding the resistance in the pulley system. Therefore,

$$\text{MA}_{\text{pulley}} = \text{number of strands holding the resistance}$$

This result may be explained as follows: When a weight is supported by two strands, each individual strand supports one-half of the total weight. Thus, the MA = 2. If a weight is supported by three strands, each individual strand supports one-third of the total weight. Thus, the MA = 3. If a weight is supported by four strands, each individual strand supports one-fourth of the total weight. In general, when a weight is supported by n strands, each individual strand supports $\frac{1}{n}$ of the total weight. Thus, the MA = n.

Stated another way, the resistance force, F_R, is spread equally among the supporting strands. Thus, $F_R = nT$, where n is the number of strands holding the resistance and T is the tension in each supporting strand.

The effort force, F_E, equals the tension, T, in each supporting strand. The equation may then be written

$$MA_{\text{pulley}} = \frac{F_R}{F_E} = \frac{nT}{T} = n$$

Note: The mechanical advantage of the pulley does not depend on the diameter of the pulley.

A number of examples are shown in Fig. 10.15.

Figure 10.15 Mechanical advantage of pulleys and pulley systems

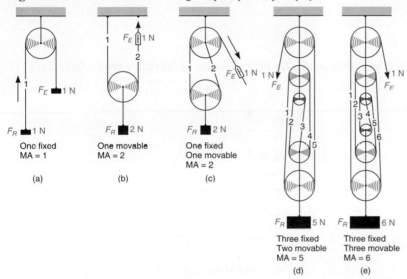

Draw two different sets of pulleys, each with an MA of 4.

EXAMPLE 1

Sketch:

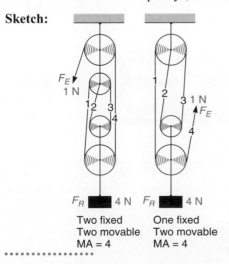

EXAMPLE 2

What effort will lift a resistance of 480 N in the pulley systems in Example 1?

Data:

$$MA_{pulley} = 4$$
$$F_R = 480 \text{ N}$$
$$F_E = ?$$

Basic Equation:

$$MA_{pulley} = \frac{F_R}{F_E}$$

Working Equation:

$$F_E = \frac{F_R}{MA_{pulley}}$$

Substitution:

$$F_E = \frac{480 \text{ N}}{4}$$
$$= 120 \text{ N}$$

· · · · · · · · · · · · · · · · ·

EXAMPLE 3

If the resistance moves 7.00 ft, what is the effort distance of the pulley system in Example 1?

Data:

$$MA_{pulley} = 4$$
$$s_R = 7.00 \text{ ft}$$
$$s_E = ?$$

Basic Equation:

$$MA_{pulley} = \frac{s_E}{s_R}$$

Working Equation:

$$s_E = s_R(MA_{pulley})$$

Substitution:

$$s_E = (7.00 \text{ ft})(4)$$
$$= 28.0 \text{ ft}$$

· · · · · · · · · · · · · · · · ·

EXAMPLE 4

The pulley system in Fig. 10.16 is used to raise a 650-lb object 25 ft. What is the mechanical advantage? What force is exerted?

$$MA_{pulley} = \text{number of strands holding the resistance}$$
$$= 5$$

To find the force exerted:

Data:

$$MA_{pulley} = 5$$
$$F_R = 650 \text{ lb}$$
$$F_E = ?$$

Basic Equation:

$$MA_{pulley} = \frac{F_R}{F_E}$$

Working Equation:

$$F_E = \frac{F_R}{MA_{pulley}}$$

Substitution:

$$F_E = \frac{650 \text{ lb}}{5}$$

$$= 130 \text{ lb}$$

·················

PROBLEMS 10.4

Find the mechanical advantage of each pulley system.

1.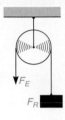

2.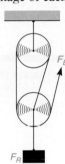

3.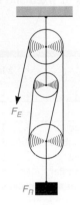

4.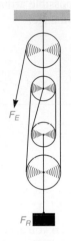

5.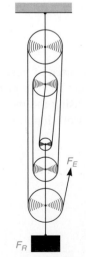

6.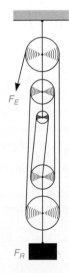

7.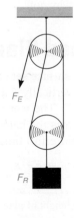

8.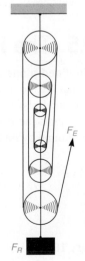

Draw each pulley system for Problems 9–14.

9. One fixed and two movable. Find the system's MA.
10. Two fixed and two movable with an MA of 5.
11. Three fixed and three movable with an MA of 6.
12. Four fixed and three movable. Find the system's MA.

Figure 10.16

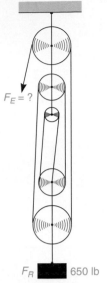

$F_E = ?$

F_R 650 lb

Three fixed
Two movable
MA = ?

SKETCH

12 cm² w

4.0 cm

DATA

A = 12 cm², *l* = 4.0 cm, w = ?

BASIC EQUATION

A = *l*w

WORKING EQUATION

$w = \frac{A}{l}$

SUBSTITUTION

$w = \frac{12 \text{ cm}^2}{4.0 \text{ cm}} = 3.0$ cm

PROBLEM SOLVING

13. Four fixed and four movable with an MA of 8.
14. Three fixed and four movable with an MA of 8.
15. What is the MA of a single movable pulley?
16. (a) What effort will lift a 250-lb weight by using a single movable pulley? (b) If the weight is moved 15.0 ft, how many feet of rope are pulled by the person exerting the effort?
17. A system consisting of two fixed pulleys and two movable pulleys has a mechanical advantage of 4. (a) If a force of 97.0 N is exerted, what weight is raised? (b) If the weight is raised 20.5 m, what length of rope is pulled?
18. A 400-lb weight is lifted 30.0 ft. (a) Using a system of one fixed and two movable pulleys, find the effort force and effort distance. (b) If an effort force of 65.0 N is applied through an effort distance of 13.0 m, find the weight of the resistance and the distance it is moved.
19. Can an effort force of 75.0 N lift a 275-N weight using the pulley system in Problem 3?
20. (a) What effort will lift a 1950-N weight using the pulley system in Problem 5? (b) If the weight is moved 3.00 m, how much rope must be pulled through the pulley system by the person exerting the force?
21. Complete the following pulley system mechanical advantage chart, which lists two possible arrangements of fixed and movable pulleys for each given mechanical advantage.

	Mechanical Advantage (MA)							
Pulleys	**1**	**2**	**3**	**4**	**5**	**6**	**7**	**8**
Fixed	1	1	2					
Movable	0	1	1					
Fixed		0	1					
Movable		1	1					

22. Can you arrange a pulley system containing 10 pulleys and obtain a mechanical advantage of 12? Why or why not?

10.5 The Inclined Plane

An **inclined plane** is a plane surface set at an angle from the horizontal used to raise objects that are too heavy to lift vertically. Gangplanks, chutes, and ramps are all examples of the inclined plane (Fig. 10.17). The work done in raising a resistance using the inclined plane equals the resistance times the height. This must also equal the work input, which can be found by multiplying the effort times the length of the plane.

$$F_R \cdot s_R = F_E \cdot s_E \qquad \text{(law of machines)}$$

$$\boxed{F_R \cdot \text{height of plane} = F_E \cdot \text{length of plane}}$$

Figure 10.17 Inclined plane

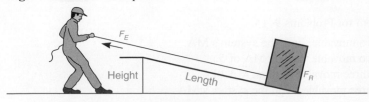

From the preceding equation,

$$\frac{F_R}{F_E} = \frac{\text{length of plane}}{\text{height of plane}} = \text{MA}_{\text{inclined plane}}$$

EXAMPLE 1

A worker is pushing a box weighing $15\overline{0}0$ N up a ramp 6.00 m long onto a platform 1.50 m above the ground. What is the mechanical advantage? What effort is applied?

Sketch:

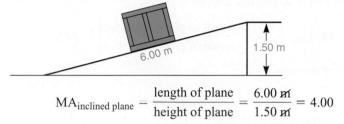

$$\text{MA}_{\text{inclined plane}} = \frac{\text{length of plane}}{\text{height of plane}} = \frac{6.00 \text{ m}}{1.50 \text{ m}} = 4.00$$

To find the effort force:

Data:

$$F_R = 15\overline{0}0 \text{ N}$$
$$\text{MA}_{\text{inclined plane}} = 4.00$$
$$F_E = \,?$$

Basic Equation:

$$\text{MA}_{\text{inclined plane}} = \frac{F_R}{F_E}$$

Working Equation:

$$F_E = \frac{F_R}{\text{MA}_{\text{inclined plane}}}$$

Substitution:

$$F_E = \frac{15\overline{0}0 \text{ N}}{4.00}$$
$$= 375 \text{ N}$$

· · · · · · · · · · · · · · · ·

Find the length of the shortest ramp that can be used to push a $60\overline{0}$-lb resistance onto a platform 3.50 ft high by exerting a force of 72.0 lb.

EXAMPLE 2

Data:

$$F_R = 60\overline{0} \text{ lb}$$
$$F_E = 72.0 \text{ lb}$$
$$\text{height} = 3.50 \text{ ft}$$
$$\text{length} = \,?$$

Basic Equation:

$$F_R \cdot \text{height} = F_E \cdot \text{length}$$

Working Equation:

$$\text{length} = \frac{F_R \cdot \text{height}}{F_E}$$

Substitution:

$$\text{length} = \frac{(60\overline{0}\text{ lb})(3.50\text{ ft})}{72.0\text{ lb}}$$

$$= 29.2\text{ ft}$$

· · · · · · · · · · · · · · · ·

EXAMPLE 3

An inclined plane is 13.0 m long and 5.00 m high. What is its mechanical advantage and what weight can be raised by exerting a force of 375 N?

$$\text{MA}_{\text{inclined plane}} = \frac{\text{length of plane}}{\text{height of plane}} = \frac{13.0\text{ m}}{5.00\text{ m}} = 2.60$$

To find the weight of the resistance:

Data:

$$\text{MA}_{\text{inclined plane}} = 2.60$$
$$F_E = 375\text{ N}$$
$$F_R = ?$$

Basic Equation:

$$\text{MA}_{\text{inclined plane}} = \frac{F_R}{F_E}$$

Working Equation:

$$F_R = (F_E)(\text{MA}_{\text{inclined plane}})$$

Substitution:

$$F_R = (375\text{ N})(2.60)$$
$$= 975\text{ N}$$

· · · · · · · · · · · · · · · ·

PHYSICS CONNECTION

The Physics of Handicap Ramps

Handicap ramps are inclined planes. They make locations that are at different elevations accessible to people with physical disabilities. There are many factors related to physics that play a part in the design and construction of a wheelchair ramp.

The most important factor in designing a wheelchair ramp is to make sure the ideal mechanical advantage of the incline is at least 12. The general guideline is that for each 1 in. of height, there should be a 12-in. "run" or horizontal component to the incline. However, this does not mean that if someone needs to go up 5 feet to a door, there should be a 60-ft-long incline. Imagine someone unable to continue upward along the incline and rolling back down a 60-ft inclined plane. That would be a very scary and dangerous situation for a person in a wheelchair. As a result, if the length is more than 30 ft, then a 5-ft-long landing must be placed somewhere in the incline.

A feature that is often overlooked is the landing at the top of the incline. A landing must be placed at the top of any wheelchair ramp to allow the wheelchair rider room to open a door without being on the incline.

Finally, frictional forces must be taken into consideration. Most manufacturers of wheelchair ramps use concrete with a brushed finish or aluminum with adhesive pads or ridges in the metal so the coefficient of friction between the ramp and the wheels prevents the wheels from losing traction as the wheelchair makes its way up the incline.

Figure 10.18 A school with appropriately sloped inclines and landings

PROBLEMS 10.5

Given $F_R \cdot$ height $= F_E \cdot$ length, find each missing quantity.

	F_R	F_E	Height of Plane	Length of Plane
1.	20.0 N	5.30 N	3.40 cm	____ cm
2.	9800 N	2340 N	____ m	3.79 m
3.	119 lb	____ lb	13.2 in.	74.0 in.
4.	____ N	1760 N	82.1 cm	3.79 m
5.	3700 N	1200 N	____ cm	112 cm

Given $MA_{inclined\ plane} = \dfrac{\text{length of plane}}{\text{height of plane}}$, find each missing quantity.

	MA	Length of Plane	Height of Plane
6.	9.00	3.40 ft	____ ft
7.	____	3.79 m	0.821 m
8.	1.30	____ ft	9.72 ft
9.	____	74.0 cm	13.2 cm
10.	17.4	____ in.	13.4 in.

11. An inclined plane is 10.0 m long and 2.50 m high. (a) Find its mechanical advantage. (b) A resistance of 727 N is pushed up the plane. What effort is needed? (c) An effort of 200 N is applied to push an 815-N resistance up the inclined plane. Is the effort enough?

12. A safe is loaded onto a truck whose bed is 5.50 ft above the ground. The safe weighs 538 lb. (a) If the effort applied is 140 lb, what length of ramp is needed? (b) What is the MA of the inclined plane? (c) Another safe weighing 257 lb is loaded onto the same truck. If the ramp is 21.1 ft long, what effort is needed?

13. A 3.00-m-long plank is used to raise a cooling unit 1.00 m. What is the MA of the ramp made by the plank?

14. A 2.75-m-long board is used to slide a compressor a vertical distance of 0.750 m. What is the MA of the ramp made by the board?

15. A resistance of 325 N is raised by using a ramp 5.76 m long and by applying a force of 75.0 N. (a) How high can it be raised? (b) Find the MA of the ramp.

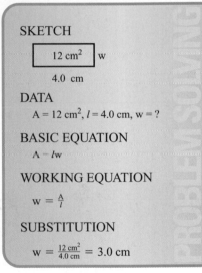

SKETCH

12 cm² | w

4.0 cm

DATA

$A = 12$ cm², $l = 4.0$ cm, $w = ?$

BASIC EQUATION

$A = lw$

WORKING EQUATION

$w = \dfrac{A}{l}$

SUBSTITUTION

$w = \dfrac{12\ cm^2}{4.0\ cm} = 3.0$ cm

PROBLEM SOLVING

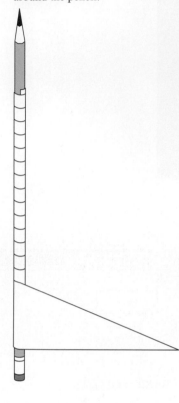

Figure 10.19 The screw is an inclined plane wound around a cylinder. The hypotenuse of the triangular section of paper corresponds to the inclined plane (threads) of a screw as it is wound around the pencil.

16. A plank 12 ft long is used as an inclined plane to a platform 3.0 ft high. (a) What force must be used to push a load weighing 480 lb up the plank? (b) Find the MA of the inclined plane.

17. A pallet stacked with bags of cement weighing a total of 5500 N must be pushed up a 2.30-m incline to a 75.0-cm-high platform. What force must be applied to get the job done?

18. A nursery loading dock is 1.20 m above the ground. What force must be used to push a 255-N pallet of fertilizer up a 3.80-m incline to the platform?

10.6 The Screw

A **screw** is an inclined plane wrapped around a cylinder. To illustrate, cut a sheet of paper in the shape of a right triangle and wind it around a pencil as shown in Fig. 10.19. The jackscrew, wood screw, and auger are examples of this simple machine (Fig. 10.20). The distance a beam rises or the distance the wood screw advances into a piece of wood in one revolution is called the **pitch** of the screw. Therefore, the pitch of a screw is also the distance between two successive threads.

From the law of machines,

$$F_R \cdot s_R = F_E \cdot s_E$$

However, for advancing a screw with a screwdriver

$$s_R = \text{pitch of screw}$$
$$s_E = \text{circumference of the handle of the screwdriver}$$

or

$$s_E = 2\pi r$$

where r is the radius of the handle of the screwdriver. Therefore,

$$\boxed{F_R \cdot \text{pitch} = F_E \cdot 2\pi r}$$

so

$$\boxed{\frac{F_R}{F_E} = \frac{2\pi r}{\text{pitch}} = \text{MA}_{\text{screw}}}$$

In the case of a jackscrew, *r is the length of the handle turning the screw and not the radius of the screw.*

Figure 10.20

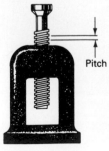

Jackscrew

(a)

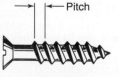

Wood screw

(b)

Drill bit for wood

(c)

Snow blower/auger system

(d)

Find the mechanical advantage of a jackscrew having a pitch of 25.0 mm and a handle radius of 35.0 cm.

EXAMPLE 1

Data:

$$\text{pitch} = 25.0 \text{ mm} = 2.50 \text{ cm}$$
$$r = 35.0 \text{ cm}$$
$$\text{MA}_{\text{screw}} = ?$$

Basic Equation:

$$\text{MA}_{\text{screw}} = \frac{2\pi r}{\text{pitch}}$$

Working Equation: Same

Substitution:

$$\text{MA}_{\text{screw}} = \frac{2\pi(35.0 \text{ cm})}{2.50 \text{ cm}}$$
$$= 88.0$$

· · · · · · · · · · · · · · · ·

What resistance can be lifted using the jackscrew in Example 1 if an effort of 203 N is exerted?

EXAMPLE 2

Data:

$$\text{MA}_{\text{screw}} = 88.0$$
$$F_E = 203 \text{ N}$$
$$F_R = ?$$

Basic Equation:

$$\text{MA}_{\text{screw}} = \frac{F_R}{F_E}$$

Working Equation:

$$F_R = (F_E)(\text{MA}_{\text{screw}})$$

Substitution:

$$F_R = (203 \text{ N})(88.0)$$
$$= 17,900 \text{ N}$$

· · · · · · · · · · · · · · · ·

A 19,400-N weight is raised using a jackscrew having a pitch of 5.00 mm and a handle length of 255 mm. What force must be applied?

EXAMPLE 3

Data:

$$\text{pitch} = 5.00 \text{ mm}$$
$$r = 255 \text{ mm}$$
$$F_R = 19,400 \text{ N}$$
$$F_E = ?$$

Basic Equation:

$$F_R \cdot \text{pitch} = F_E \cdot 2\pi r$$

Working Equation:

$$F_E = \frac{F_R\,(\text{pitch})}{2\pi r}$$

Substitution:

$$F_E = \frac{(19{,}400\ \text{N})(5.00\ \text{mm})}{2\pi\,(255\ \text{mm})}$$

$$= 60.5\ \text{N}$$

. . . .

SKETCH

| 12 cm² | w |

4.0 cm

DATA

$A = 12\ \text{cm}^2, l = 4.0\ \text{cm}, w = ?$

BASIC EQUATION

$A = lw$

WORKING EQUATION

$w = \frac{A}{l}$

SUBSTITUTION

$w = \frac{12\ \text{cm}^2}{4.0\ \text{cm}} = 3.0\ \text{cm}$

PROBLEMS 10.6

Given $F_R \cdot \text{pitch} = F_E \cdot 2\pi r$, find each missing quantity.

	F_R	F_E	Pitch	r
1.	20.7 N	5.30 N	3.70 mm	____ mm
2.	____ lb	17.6 lb	0.130 in.	24.5 in.
3.	234 N	9.80 N	____ mm	53.9 mm
4.	1190 N	____ N	2.97 mm	67.4 mm
5.	370 lb	12.0 lb	____ in.	11.2 in.

Given $\text{MA}_{\text{screw}} = \dfrac{2\pi r}{\text{pitch}}$, find each missing quantity.

	MA	r	Pitch
6.	7.00	34.0 mm	____ mm
7.	____	3.79 in.	0.812 in.
8.	9.00	____ in.	0.970 in.
9.	____	7.40 mm	1.32 mm
10.	13.0	____ mm	2.10 mm

11. A 3650-lb car is raised using a jackscrew having eight threads to the inch and a handle 15.0 in. long. (a) What effort must be applied? (b) What is the MA?
12. The mechanical advantage of a jackscrew is 97.0. (a) If the handle is 34.5 cm long, what is the pitch? (b) How much weight can be raised by applying an effort of 405 N to the jackscrew?
13. A wood screw with pitch 0.125 in. is advanced into wood using a screwdriver whose handle is 1.50 in. in diameter. (a) What is the mechanical advantage of the screw? (b) What is the resistance of the wood if 15.0 lb of effort is applied on the wood screw? (c) What is the resistance of the wood if 15.0 lb of effort is applied to the wood screw using a screwdriver whose handle is 0.500 in. in diameter?
14. The handle of a jackscrew is 60.0 cm long. (a) If the mechanical advantage is 78.0, what is the pitch? (b) How much weight can be raised by applying a force of $43\overline{0}$ N to the jackscrew handle?

10.7 The Wedge

A **wedge** is an inclined plane in which the plane is moved instead of the resistance. Examples are shown in Fig. 10.21.

Finding the mechanical advantage of a wedge is not practical because of the large amount of friction. A narrow wedge is easier to drive than a thick wedge. Therefore, the mechanical advantage depends on the ratio of its length to its thickness.

Figure 10.21 Inclined planes where the plane moves instead of the resistance

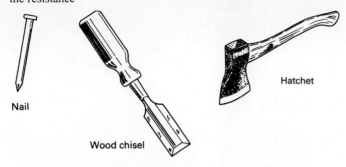

Nail

Wood chisel

Hatchet

10.8 Compound Machines

A **compound machine** is a combination of simple machines. Examples are shown in Fig. 10.22. In most compound machines, *the total mechanical advantage is the product of the mechanical advantage of each simple machine.*

$$MA_{\text{compound machine}} = (MA_1)(MA_2)(MA_3) \cdots$$

Figure 10.22 Compound machines multiply mechanical advantage. (Reprinted courtesy of Caterpillar Inc.)

(a)

(b)

A crate weighing $95\overline{0}0$ N is pulled up the inclined plane using the pulley system shown in Fig. 10.23.

EXAMPLE

(a) Find the mechanical advantage of the total system.
(b) What effort force (F_E) is needed?

Figure 10.23

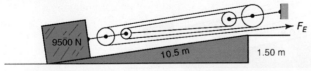

9500 N

10.5 m

1.50 m

F_E

(a) First, find the MA of the inclined plane.

$$MA_{\text{inclined plane}} = \frac{\text{length of plane}}{\text{height of plane}} = \frac{10.5 \text{ m}}{1.50 \text{ m}} = 7.00$$

The MA of the pulley system = 5 (the number of supporting strands).
The MA of the total system (compound machine) is

$$(MA_{\text{inclined plane}})(MA_{\text{pulley system}}) = (7.00)(5) = 35.0$$

(b) Data:

$$MA_{\text{compound machine}} = 35.0$$
$$F_R = 95\overline{0}0 \text{ N}$$
$$F_E = ?$$

Basic Equation:

$$MA_{\text{compound machine}} = \frac{F_R}{F_E}$$

Working Equation:

$$F_E = \frac{F_R}{MA_{\text{compound machine}}}$$

Substitution:

$$F_E = \frac{95\overline{0}0 \text{ N}}{35.0}$$
$$= 271 \text{ N}$$

· · · · · · · · · · · · · · · ·

PROBLEMS 10.8

1. The box shown in Fig. 10.24 being pulled up an inclined plane using the indicated pulley system (called a block and tackle) weighs 9790 N. If the inclined plane is 6.00 m long and the height of the platform is 2.00 m, find the mechanical advantage of this compound machine.

Figure 10.24

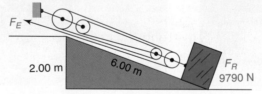

F_E

2.00 m   6.00 m F_R
                             9790 N

2. What effort force must be exerted to move the box to the platform in Problem 1?
3. Find the mechanical advantage of the compound machine shown in Fig. 10.25. The radius of the crank is 1.00 ft and the radius of the axle is 0.500 ft.

Figure 10.25

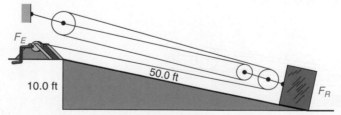

F_E

10.0 ft   50.0 ft F_R

4. If an effort of 30$\overline{0}$ lb is exerted, what weight can be moved up the inclined plane using the compound machine in Problem 3?
5. What effort is required to move a load of 1.50 tons up the inclined plane using the compound machine in Problem 3? (1 ton = 200$\overline{0}$ lb)
6. Find the mechanical advantage of the compound machine in Problem 1 if the inclined plane is 8.00 m long and 2.00 m high.
7. What effort force is needed to move a box of weight 25$\overline{0}$0 N to the platform in Problem 6?

8. Find the mechanical advantage of the compound machine in Problem 3 if the radius of the crank is 40.0 cm, the radius of the axle is 12.0 cm, the length of the inclined plane is 12.0 m, and the height of the inclined plane is 50.0 cm.
9. If an effort of $45\overline{0}$ N is exerted in Problem 8, what weight can be moved up the inclined plane?
10. What effort force (in N) is needed to move 2.50 metric tons up the inclined plane in Problem 8?

10.9   The Effect of Friction on Simple Machines

Figure 10.26   Friction has been ignored in our study of pulleys.

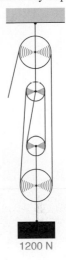

1200 N

The *law of simple machines* has been stated for a particular machine and in the general case in the previous sections. Each has been stated in terms of what is called *ideal mechanical advantage* (IMA), in which we have 100% efficiency. Actually, in every machine energy is lost through heat to overcome friction. This lost energy decreases the efficiency of the machine; that is, more work must be put into a machine than is gotten out of the machine. This lost energy is heat energy, which results in machine wear or even burning out of certain parts of the machine.

Throughout this chapter we have been discussing simple machines, mechanical advantage, resistance force, effort force, resistance distance, and effort distance in the ideal case while ignoring friction. For example, in the pulley system in Fig. 10.26, we find the IMA is 4 to 1. Ideally, it takes 300 N of effort force to lift the resistant force of 1200 N; that is, it ideally takes 1 N of force to raise 4 N of weight. However, it actually takes 400 N of effort to lift the 1200-N weight; the *actual mechanical advantage* (AMA) is then 3 to 1; that is, it actually takes 1 N of force to raise 3 N of weight.

In general, the actual mechanical advantage is found by the following formula:

$$\text{AMA} = \frac{F_R}{F_E} = \frac{\text{resistance force}}{\text{effort force}}$$

In Section 5.3, we studied the effects of sliding friction. The actual effects of sliding friction are substantial in inclined plane problems. In Example 1 of Section 10.5, the inclined plane (repeated in Fig. 10.27) has an IMA of 4 to 1. Actually, it takes 545 N of effort to move the 150̄0-N box up the ramp. Therefore, the AMA is

$$\text{AMA} = \frac{F_R}{F_E} = \frac{15\overline{0}0 \text{ N}}{545 \text{ N}} = 2.75$$

That is, it actually takes 1.00 N of force to push 2.75 N up the ramp.

Figure 10.27

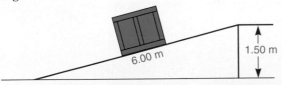

6.00 m

1.50 m

Glossary

Compound Machine A combination of simple machines. Its total mechanical advantage is the product of the mechanical advantage of each simple machine. (p. 287)

Efficiency The ratio of the work output to the work input of a machine. (p. 269)

Effort The force applied to a machine. (p. 268)

Effort Arm The distance from the effort force to the fulcrum of a lever. (p. 269)

Fixed Pulley A pulley that is fastened to a fixed object. (p. 276)

Fulcrum A pivot about which a lever is free to turn. (p. 269)

Inclined Plane A plane surface set at an angle from the horizontal used to raise objects that are too heavy to lift vertically. (p. 280)

Law of Simple Machines Resistance force × resistance distance = effort force × effort distance. (p. 268)

Lever A rigid bar free to turn on a pivot called a fulcrum. (p. 269)

Machine An object or system that is used to transfer energy from one place to another and allows work to be done that could not otherwise be done or could not be done as easily. (p. 267)

Mechanical Advantage The ratio of the resistance force to the effort force. (p. 269)

Movable Pulley A pulley that is fastened to the object to be moved. (p. 276)

Pitch The distance a screw advances in one revolution of the screw. Also the distance between two successive threads. (p. 284)

Pulley A grooved wheel that turns readily on an axle and is supported in a frame. (p. 276)

Resistance The force overcome by a machine. (p. 268)

Resistance Arm The distance from the resistance force to the fulcrum of a lever. (p. 269)

Screw An inclined plane wrapped around a cylinder. (p. 284)

Simple Machine Any one of six mechanical devices in which an applied force results in useful work. The six simple machines are the lever, the wheel and axle, the pulley, the inclined plane, the screw, and the wedge. (p. 268)

Wedge An inclined plane in which the plane is moved instead of the resistance. (p. 286)

Wheel-and-Axle A large wheel attached to an axle so that both turn together. (p. 274)

Formulas

10.1   resistance force × resistance distance = effort force × effort distance

$$MA = \frac{\text{resistance force}}{\text{effort force}}$$

$$\text{efficiency} = \frac{\text{work output}}{\text{work input}} \times 100\% = \frac{F_{\text{output}} \times s_{\text{output}}}{F_{\text{input}} \times s_{\text{input}}} \times 100\%$$

10.2 $MA_{\text{lever}} = \dfrac{\text{effort arm}}{\text{resistance arm}} = \dfrac{s_E}{s_R}$

$$F_R \cdot s_R = F_E \cdot s_E$$

10.3 $MA_{\text{wheel-and-axle}} = \dfrac{\text{radius of effort force}}{\text{radius of resistance force}} = \dfrac{r_E}{r_R}$

$$F_R \cdot r_R = F_E \cdot r_E$$

10.4 $MA_{\text{pulley}} = \text{number of strands holding the resistance}$

$$MA_{\text{pulley}} = \frac{s_E}{s_R}$$

10.5 $\text{MA}_{\text{inclined plane}} = \dfrac{\text{length of plane}}{\text{height of plane}}$

$F_R \cdot \text{height} = F_E \cdot \text{length}$

10.6 $\text{MA}_{\text{screw}} = \dfrac{2\pi r}{\text{pitch}}$

$F_R \cdot \text{pitch} = F_E \cdot 2\pi r$

10.8 $\text{MA}_{\text{compound machine}} = (\text{MA}_1)(\text{MA}_2)(\text{MA}_3) \cdots$

10.9 $\text{AMA} = \dfrac{F_R}{F_E} = \dfrac{\text{resistance force}}{\text{effort force}}$

Review Questions

1. Which of the following is not a simple machine?
   (a) Pulley
   (b) Lever
   (c) Wedge
   (d) Automobile
2. The force applied to the machine is the
   (a) effort.
   (b) frictional.
   (c) horizontal.
   (d) resistance.
3. Efficiency is
   (a) the same as mechanical advantage.
   (b) a percentage.
   (c) impossible to determine.
4. A second-class lever has
   (a) two fulcrums.
   (b) two effort arms.
   (c) two resistance arms.
   (d) a resistance arm shorter than the effort arm.
5. A pulley has eight strands holding the resistance. The mechanical advantage is
   (a) 4.   (b) 8.   (c) 16.   (d) 64.
6. The mechanical advantage of a compound machine
   (a) is the sum of the MA of each simple machine.
   (b) is the product of the MA of each simple machine.
   (c) cannot be found.
   (d) is none of the above.
7. Cite three examples of machines used to multiply speed.
8. What name is given to the force overcome by the machine?
9. State the law of simple machines in your own words.
10. What is the term used for the ratio of the resistance force to the effort force?
11. What is the term used for the ratio of the amount of work obtained from a machine to the amount of work put into the machine?
12. Does a friction-free machine exist?
13. What is the pivot point of a lever called?
14. In your own words, state how to find the MA of a lever.
15. Which type of lever do you think would be most efficient?
16. State the law of simple machines as it is applied to levers.

SKETCH

12 cm² | w

4.0 cm

DATA

$A = 12 \text{ cm}^2$, $l = 4.0 \text{ cm}$, $w = ?$

BASIC EQUATION

$A = lw$

WORKING EQUATION

$w = \dfrac{A}{l}$

SUBSTITUTION

$w = \dfrac{12 \text{ cm}^2}{4.0 \text{ cm}} = 3.0 \text{ cm}$

17. Where is the fulcrum located in a third-class lever?
18. In your own words, explain the law of simple machines as applied to the wheel-and-axle.
19. Does the MA of a wheel-and-axle depend on the force applied?
20. Describe the difference between a fixed pulley and a movable pulley.
21. Does the MA of a pulley depend on the radius of the pulley?
22. How can you find the MA of an inclined plane?
23. In your own words, describe the pitch of a screw.
24. How does the MA of a jackscrew differ from the MA of a screwdriver?

Review Problems

SKETCH

12 cm² | w

4.0 cm

DATA

$A = 12 \text{ cm}^2, l = 4.0 \text{ cm}, w = ?$

BASIC EQUATION

$A = lw$

WORKING EQUATION

$w = \frac{A}{l}$

SUBSTITUTION

$w = \frac{12 \text{ cm}^2}{4.0 \text{ cm}} = 3.0 \text{ cm}$

1. A girl uses a lever to lift a box. The box has a resistance force of $25\overline{0}$ N while she exerts an effort force of 125 N. What is the mechanical advantage of the lever?
2. A bicycle requires 1575 N m of input but only puts out 1150 N m of work. What is the bicycle's efficiency?
3. A lever uses an effort arm of 2.75 m and has a resistance arm of 72.0 cm. What is the lever's mechanical advantage?
4. Two people are on a teeter-totter. One person exerts a force of 540 N and is 2.00 m from the fulcrum. If they are to remain balanced, how much force does the other person exert if she is (a) also 2.00 m from the fulcrum? (b) 3.00 m from the fulcrum?
5. A wheel-and-axle has an effort force of 125 N and an effort radius of 17.0 cm. (a) If the resistance force is 325 N, what is the resistance radius? (b) Find the mechanical advantage.
6. What is the mechanical advantage of a pulley system having 12 strands holding the resistance?
7. A pulley system has a mechanical advantage of 5. What is the resistance force if an effort of 135 N is exerted?
8. An inclined plane has a height of 1.50 m and a length of 4.50 m. (a) What effort must be exerted to pull up an 875-N box? (b) What is the mechanical advantage?
9. What height must a 10.0-ft-long inclined plane be to lift a $100\overline{0}$-lb crate with 230 lb of effort?
10. A screw has a pitch of 0.0200 cm. An effort force of 29.0 N is used to turn a screwdriver whose handle diameter is 36.0 mm. What is the maximum resistance force?
11. A 945-N resistance force is overcome with a 13.5-N effort using a screwdriver whose handle is 24.0 mm in diameter. What is the pitch of the screw?
12. Find the mechanical advantage of a jackscrew with a 1.50-cm pitch and a handle 36.0 cm long.
13. A courier uses a bicycle with rear wheel radius 35.6 cm and gear radius 4.00 cm. If a force of 155 N is applied to the chain, the wheel rim moves 14.0 cm. (a) If the efficiency is 95.0%, what is the ideal mechanical advantage of the wheel and gear? (b) What is the actual mechanical advantage of the wheel and gear? (c) What is the force on the pavement applied by the wheel?
14. (a) If the gear radius is doubled on the courier's bicycle in Problem 13, how does the mechanical advantage change? (b) How far did the courier move the chain to produce the 14.0-cm linear movement of the rim?
15. A farmer uses a pulley system to raise a 225-N bale 16.5 m. A 129-N force is applied by pulling a rope 33.0 m. What is the mechanical advantage of the pulley system?
16. A laborer uses a lever to raise a 1250-N rock a distance of 13.0 cm by applying a force of 225 N. If the efficiency of the lever is 88.7%, how far does the laborer have to move his end of the lever?

Find the mechanical advantage of each pulley system.

17.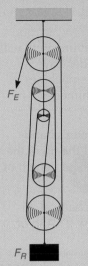

18.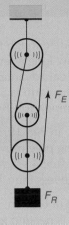

19. (a) Find the mechanical advantage of the compound machine in Fig. 10.28. The radius of the crank is 32.0 cm and the radius of the axle is 8.00 cm. (b) If an effort force of 75 N is applied to the handle of the crank, what force can be moved up the inclined plane?

Figure 10.28

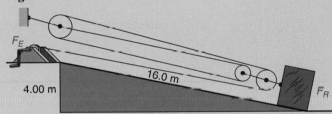

20. If an effort force of 45 N is applied to a simple machine and moves a resistance of 270 N, what is the actual mechanical advantage?

CHAPTER 11

UNIVERSAL GRAVITATION AND SATELLITE MOTION

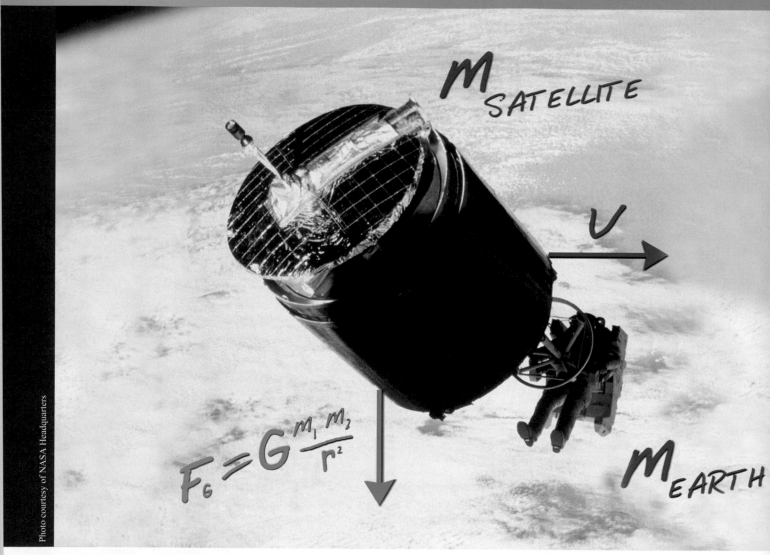

$M_{SATELLITE}$

v

$$F_G = G\frac{m_1 m_2}{r^2}$$

M_{EARTH}

What allows thousands of satellites to orbit the earth? Why does vertical motion on the moon appear to take place in slow motion? In this chapter we will discover how we can send massive objects into orbit and why a person weighs less on the moon than on the earth. In this chapter we appreciate the importance of understanding the difference between mass and weight.

Objectives

The major goals of this chapter are to enable you to:

1. Describe how gravitation acts between all objects with mass.
2. Calculate what your weight would be on various planets.
3. Explain the concept of a gravitational field.
4. Express how objects are able to orbit the earth.
5. Determine the connection between orbiting the earth and being in free fall.

11.1 Universal Gravitation

What makes an astronaut weigh less on the moon than on the earth? **Newton's law of universal gravitation** states that all objects that have mass are attracted to one another by a gravitational force. Newton determined that the greater the mass of two objects, the stronger is the attractive gravitational force between them. He also discovered that as objects move away from each other, the attraction between them diminishes dramatically. Newton's law of universal gravitation is

$$F_G = G\frac{m_1 m_2}{r^2}$$

where F_G = gravitational force between the two objects

$G = 6.67 \times 10^{-11}$ N m^2/kg$^2 = 3.44 \times 10^{-8}$ lb ft^2/slug^2 (universal gravitational constant)

m_1 = mass of the first object

m_2 = mass of the second object

r = distance between the centers of mass of the two objects.

The law of universal gravitation makes it possible to calculate an object's weight on any planet, the gravitational force exerted between the person sitting next to you in class and you, and the gravitational forces between planets, moons, and stars. Table 11.1 gives

Table 11.1   Table of Planetary Data

Object	Average Radius (m)	Mass (kg)	Mean Sun to Planet Distance (m)
Sun	6.96×10^8	1.99×10^{30}	—
Mercury	2.44×10^6	3.30×10^{23}	5.79×10^{10}
Venus	6.05×10^6	4.87×10^{24}	1.08×10^{11}
Earth	6.38×10^6	5.97×10^{24}	1.50×10^{11}
Mars	3.40×10^6	6.42×10^{23}	2.28×10^{11}
Jupiter	7.15×10^7	1.90×10^{27}	7.78×10^{11}
Saturn	6.03×10^7	5.69×10^{26}	1.43×10^{12}
Uranus	2.56×10^7	8.66×10^{25}	2.87×10^{12}
Neptune	2.48×10^7	1.03×10^{26}	4.50×10^{12}
Pluto	1.15×10^6	1.5×10^{22}	5.91×10^{12}

the average radius, the mass, and the mean distance from the sun of various objects in our solar system.

Other information:

$$earth\ to\ moon\ distance = 3.84 \times 10^8\ m$$
$$moon's\ radius\qquad\quad = 1.74 \times 10^6\ m$$
$$moon's\ mass\qquad\qquad = 7.35 \times 10^{22}\ kg$$

EXAMPLE 1

Compare (a) the gravitational force between Vince and the earth to (b) the gravitational force between Vince and Matt, who is sitting 2.34 m away from Vince. Both Matt and Vince have a mass of 85.5 kg.

Data:

$$m_{earth} = 5.97 \times 10^{24}\ kg$$
$$m_{Matt} = 85.5\ kg$$
$$m_{Vince} = 85.5\ kg$$
$$r_{Vince-earth} = 6.38 \times 10^6\ m$$
$$r_{Vince-Matt} = 2.34\ m$$

Basic Equation:

$$F_G = G\frac{m_1 m_2}{r^2}$$

Working Equation: Same

Substitution:

(a)
$$F_G = \left(6.67 \times 10^{-11}\ \frac{N\ m^2}{kg^2}\right)\frac{(85.5\ kg)(5.97 \times 10^{24}\ kg)}{(6.38 \times 10^6\ m)^2}$$
$$= 836\ N$$

(b)
$$F_G = \left(6.67 \times 10^{-11}\ \frac{N\ m^2}{kg^2}\right)\frac{(85.5\ kg)(85.5\ kg)}{(2.34\ m)^2}$$
$$= 8.90 \times 10^{-8}\ N$$

The gravitational force between Vince and Matt is virtually undetectable compared to the gravitational force between Vince and the earth.

· · · · · · · · · · · · · · · · · ·

Henry Cavendish (1731–1810),

English physicist and chemist, was born in France. He was the first person to experimentally determine the universal gravitational constant used in Isaac Newton's law of universal gravitation (Fig. 11.1). He also found that water was composed of hydrogen and oxygen and estimated the density of the earth.

Figure 11.1   The Cavendish experiment

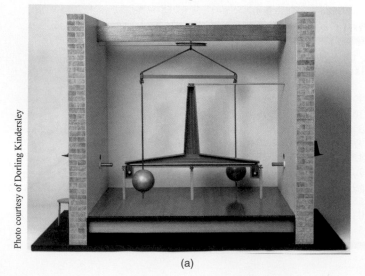

Photo courtesy of Dorling Kindersley

(a)

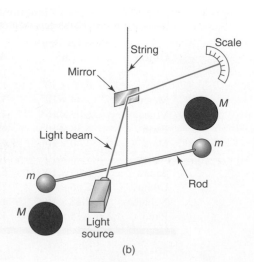

(b)

In 1798, Henry Cavendish attached a pair of 2-in. lead balls to the ends of a light horizontal rod, which was suspended by a string onto which a mirror was mounted as shown in Fig. 11.1. When this pair of lead balls was brought near two more massive 12-in. balls, the gravitational force caused the smaller balls to move toward the more massive balls, which caused the string and mirror to twist. The tiny movement was measured by reflecting a narrow light beam onto a scale. By carefully measuring the very small twisting motion of the suspended balls, Cavendish was able to determine G within 1% of today's accepted value.

One of Newton's preliminary studies of universal gravitation involved the moon and a falling apple. (a) Find the force between an apple ($m = 0.153$ kg) and the earth. (b) Compare this to the force between the moon and the earth, two objects that are quite massive yet far away from each other. See Table 11.1.

EXAMPLE 2

Data:

$$m_{apple} = 0.153 \text{ kg}$$
$$m_{moon} = 7.35 \times 10^{22} \text{ kg}$$
$$m_{earth} = 5.97 \times 10^{24} \text{ kg}$$
$$r_{apple-earth} = 6.38 \times 10^6 \text{ m}$$
$$r_{moon-earth} = 3.84 \times 10^8 \text{ m}$$

Basic Equation:

$$F_G = G\frac{m_1 m_2}{r^2}$$

Working Equation: Same

Substitutions:

(a)
$$F_G = \left(6.67 \times 10^{-11} \frac{\text{N m}^2}{\text{kg}^2}\right) \frac{(0.153 \text{ kg})(5.97 \times 10^{24} \text{ kg})}{(6.38 \times 10^6 \text{ m})^2}$$
$$= 1.50 \text{ N}$$

(b)
$$F_G = \left(6.67 \times 10^{-11} \frac{\text{N m}^2}{\text{kg}^2}\right) \frac{(7.35 \times 10^{22} \text{ kg})(5.97 \times 10^{24} \text{ kg})}{(3.84 \times 10^8 \text{ m})^2}$$
$$= 1.98 \times 10^{20} \text{ N}$$

The force between the earth and the moon is much greater than the force between the earth and the apple. Although the moon is much farther away from the earth than the apple is, the moon's much larger mass more than makes up for the difference in the distances.

· · · · · · · · · · · · · · · · ·

What would be the gravitational force exerted on a 65.0-kg person on Jupiter?

EXAMPLE 3

Data:

$$m_{person} = 65.0 \text{ kg}$$
$$m_{Jupiter} = 1.90 \times 10^{27} \text{ kg}$$
$$r_{Jupiter} = 7.15 \times 10^7 \text{ m}$$

Basic Equation:

$$F_G = G\frac{m_1 m_2}{r^2}$$

Working Equation: Same

Substitution:

$$F_G = \left(6.67 \times 10^{-11}\,\frac{\text{N m}^2}{\text{kg}^2}\right)\frac{(65.0\,\text{kg})(1.90 \times 10^{27}\,\text{kg})}{(7.15 \times 10^7\,\text{m})^2}$$

$$= 1610\,\text{N}$$

· · · · · · · · · · · · · · · · ·

As these examples show, the gravitational force becomes stronger as the mass of the objects increases and as the distance between them decreases.

PROBLEMS 11.1

1. Compare the gravitational force that (a) the earth exerts on an 84.3-kg person and (b) the force that the sun exerts on the same person.
2. Find the gravitational force between the sun and the earth.
3. Find the gravitational force between the sun and Mercury.
4. Find the gravitational force between the sun and Jupiter.
5. Find the gravitational force between the sun and Pluto.
6. Explain why the gravitational force between the sun and Jupiter is greater than the gravitational force between the sun and the earth even though the sun and the earth are much closer to one another than are the sun and Jupiter.
7. A satellite is orbiting 3.22×10^5 m above the surface of the earth. If the mass of the satellite is 3.80×10^4 kg, what is the weight or gravitational force exerted on the satellite by the earth?
8. If the satellite in Problem 7 is orbiting at twice its original distance from the earth, what would be the weight or gravitational force exerted on the satellite by the earth?
9. What is the gravitational force exerted between an electron ($m = 9.11 \times 10^{-31}$ kg) and a proton ($m = 1.67 \times 10^{-27}$ kg) in a hydrogen atom where the distance between the electron and proton is 5.3×10^{-9} m?
10. The Apollo 16 lunar module had a mass of 4240 kg. Using Newton's law of universal gravitation, find its weight (a) on the earth and (b) on the moon.

11.2 Gravitational Fields

Isaac Newton's contemporaries did not welcome Newton's laws. The thought that the earth could exert a force on the moon almost 240,000 mi away without touching it seemed outrageous to many at the time. Until then, it was thought that most objects needed to be physically touching in order to have an effect on one another.

The area around a massive body where an object experiences a gravitational force is called a **gravitational field.** By introducing the concept of a gravitational field, physicists have been able to determine that an object does not need to be in physical contact with another object to exert a force on it (Fig. 11.2). The change in velocity caused by the gravitational field is known as acceleration due to gravity.

The gravitational field around the surface of the earth is quite strong. However, if one travels out into space a bit, the distance alone affects the intensity of the field. That is, the acceleration due to gravity decreases significantly as shown in Fig. 11.3.

Figure 11.2   Gravitational field lines represent the strength of gravity around the earth. The closer the lines, the stronger is the gravitational attraction as shown by the force vectors.

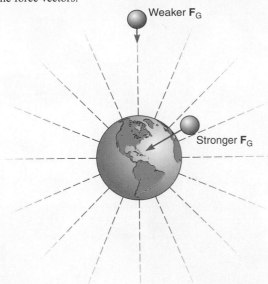

Weaker **F**G

Stronger **F**G

Figure 11.3   The gravitational force depends on the object's distance from the center of the earth. For example, if an object weighs 100 lb on the earth's surface, its weight becomes dramatically less the farther away the object is located.

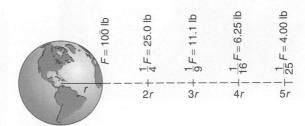

$F = 10\overline{0}$ lb

$\frac{1}{4}F = 25.0$ lb

$\frac{1}{9}F = 11.1$ lb

$\frac{1}{16}F = 6.25$ lb

$\frac{1}{25}F = 4.00$ lb

r $2r$ $3r$ $4r$ $5r$

The concept of a gravitational field inspired Albert Einstein to establish a theory that would describe a warped space-time field in our universe. This fascinating, yet conceptually challenging, theory will be covered in Chapter 24.

11.3 Satellite Motion

An **orbit** is the path taken by an object during its revolution around another object, such as the path of the moon or a satellite about the earth or of a planet about the sun. We have seen pictures of astronauts weightless and appearing to "float" in space as they orbit the earth (Fig. 11.4). The reason is because of gravity, not the lack of gravity, at that point in space.

In Chapter 4, we learned that when an object is dropped or falls from an elevated point on earth, nothing is pushing back up on the object, so it appears weightless (Fig. 11.5). Gravity may be pulling the object down, but there is no normal force to support the object. In essence, when astronauts orbit the earth, they are in a constant state of free fall.

A person in free fall feels weightless, just like the astronauts. However, as the person continues to fall, he or she will crash to the earth. Newton realized that the moon and other orbiting objects must sustain a large enough horizontal velocity to remain in orbit and avoid crashing to the earth.

Figure 11.4   Although an astronaut in orbit feels weightless, there is still a gravitational force acting on the astronaut. Here astronaut Joseph R. Tanner (right) stands fixed on the end of Discovery's remote manipulator system arm and aims a camera at the solar array panels of the Hubble Space Telescope as astronaut Gregory J. Harbaugh assists.

Photo courtesy of NASA/Johnson Space Center

TRY THIS ACTIVITY

Weightless Water

Place two small holes on opposite sides near the bottom of a Styrofoam cup. Cover the holes with your thumb and finger and fill the cup with water. While standing on a stepladder or a table, grasp the cup near the top with your other hand. Uncover the holes just long enough to see the water streaming out. Then drop the cup. What happens to the streams of water as the cup and water fall? How is this activity similar to what an astronaut orbiting the earth experiences? How is it different?

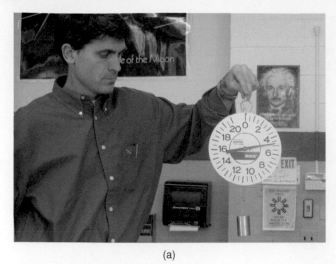

(a)

Figure 11.5 (a) When supported, the 0.50-kg mass weighs 4.90 N. (b) When no longer supported, the object appears weightless and the scale reads 0 N.

(b)

Figure 11.6

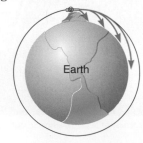

An object launched beyond the earth's atmosphere and given a horizontal velocity moves sideways while falling at the same time. Moving horizontally at just the right velocity while falling due to gravity allows the space shuttle, astronauts, and other objects to continually miss the surface of the earth and achieve orbit (Fig. 11.6).

Isaac Newton conceived this concept several centuries before the development of rockets and artificial satellites and predicted that orbiting the earth could theoretically be done. Newton made the connection between satellite motion and the orbit of the moon around the earth. The velocity of satellites orbiting the earth or any other planet can be calculated by combining centripetal motion and the law of universal gravitation:

$$F_c = \frac{mv^2}{r}, \qquad F_G = G\frac{m_1 m_2}{r^2}$$

Solve the centripetal force equation for velocity,

$$v = \sqrt{\frac{F_c r}{m}}$$

Since the centripetal force holding a satellite in orbit is actually the gravitational force between the earth and the satellite, F_G can be substituted for F_c:

$$v = \sqrt{\frac{Gm_1 m_2 r}{r^2 m_1}} \qquad \text{(Cancel like terms.)}$$

The mass of the satellite (m_1) is the term that is canceled in the last equation. As a result, the mass of the object being orbited (m_2) is the only mass that determines the orbital velocity of the satellite and is included in the following simplified formula:

$$v = \sqrt{\frac{Gm}{r}}$$

where v = velocity of the satellite
$G = 6.67 \times 10^{-11}$ N m^2/kg^2 = 3.44×10^{-8} lb ft^2/slug^2
m = mass of the object being orbited
r = distance from the center of the earth to the satellite

The time or period, T, to orbit the earth or other celestial body is

$$T = 2\pi\sqrt{\frac{r^3}{Gm}}$$ (The derivation of this equation is beyond the scope of this text.)

Note that just as acceleration due to gravity is independent of the mass of falling objects, the velocity and the period of the object are independent of the mass of the satellite as well.

If a satellite orbits the earth at $20\overline{0}0$ km above sea level, (a) how fast will the orbiting satellite travel and (b) how long will it take to orbit the earth once?

EXAMPLE 1

Data:

$r_{earth-satellite} = 6.38 \times 10^6$ m $+ 2.00 \times 10^6$ m $= 8.38 \times 10^6$ m
$m_{earth} = 5.97 \times 10^{24}$ kg

Basic Equations:

$$v = \sqrt{\frac{Gm}{r}}$$

$$T = 2\pi\sqrt{\frac{r^3}{Gm}}$$

Working Equations: Same

Substitutions:

(a) $v = \sqrt{\dfrac{(6.67 \times 10^{-11} \text{ N m}^2/\text{kg}^2)(5.97 \times 10^{24} \text{ kg})}{8.38 \times 10^6 \text{m}}}$

$= 6890$ m/s $\sqrt{\dfrac{\text{N m}^2/\text{kg}^2 \text{ kg}}{\text{m}}} = \sqrt{\dfrac{(\text{kg m/s}^2)\text{m}^2/\text{kg}^2 \text{ kg}}{\text{m}}} = \sqrt{\text{m}^2/\text{s}^2} = \text{m/s}$

(b) $T = 2\pi\sqrt{\dfrac{(8.38 \times 10^6 \text{ m})^3}{(6.67 \times 10^{-11} \text{ N m}^2/\text{kg}^2)(5.97 \times 10^{24} \text{ kg})}}$

$= 7640$ s $= 2.12$ h $\sqrt{\dfrac{\text{m}^3}{\text{N m}^2/\text{kg}^2 \text{ kg}}} = \sqrt{\dfrac{\text{m}^3}{(\text{kg m/s}^2)\text{m}^2/\text{kg}^2 \text{ kg}}} = \sqrt{\text{s}^2} = \text{s}$

EXAMPLE 2

An asteroid orbits the sun 8.35×10^{11} m from the sun. (a) How fast must the asteroid travel to maintain its orbit around the sun? (b) How long will it take the asteroid to orbit the sun?

Data:

$$r_{\text{sun–asteroid}} = 8.35 \times 10^{11} \text{ m}$$
$$m_{\text{sun}} = 1.99 \times 10^{30} \text{ kg}$$

Basic Equations:

$$v = \sqrt{\frac{Gm}{r}}$$

$$T = 2\pi\sqrt{\frac{r^3}{Gm}}$$

Working Equations: Same

Substitutions:

(a)

$$v = \sqrt{\frac{(6.67 \times 10^{-11} \text{ N m}^2/\text{kg}^2)(1.99 \times 10^{30} \text{ kg})}{8.35 \times 10^{11} \text{ m}}}$$

$$= 1.26 \times 10^4 \text{ m/s}$$

(b)

$$T = 2\pi\sqrt{\frac{(8.35 \times 10^{11} \text{ m})^3}{(6.67 \times 10^{-11} \text{ N m}^2/\text{kg}^2)(1.99 \times 10^{30} \text{ kg})}}$$

$$= 4.16 \times 10^8 \text{ s} = 13.2 \text{ yr}$$

PHYSICS CONNECTIONS

Satellite Orbits

Thousands of functioning artificial satellites currently orbit the earth. These satellites have many functions such as tracking weather patterns, collecting scientific data about the earth and space, and transmitting telephone, television, and Internet communications. Our society has become dependent upon satellites over the last 50 years; much of our use of technology today would not be possible without satellites orbiting the earth.

Several types of orbits allow artificial satellites to maintain a stable orbit (Fig. 11.7). Communication and weather satellites need to remain over a fixed point on the earth. A television satellite dish on a house always points at a particular satellite in a stationary orbit and common weather images are also taken by satellites in a stationary orbit. A stationary orbit is known as a geosynchronous orbit, and satellites in such an orbit are positioned 22,400 mi (36,000 km) above the equator.

Imaging and intelligence satellites need to take clear and detailed pictures of the entire earth. Such satellites are placed in low-altitude, north-south polar orbits. The earth rotates underneath the satellites, allowing pictures and images to be collected for most of the earth.

Finally, the most versatile orbit for satellites is the asynchronous orbit. The space shuttle, scientific research satellites, and global positioning satellites need to be in an orbit that allows them to pass over various locations on earth at different times. These satellites typically orbit at varying altitudes ranging from 200 to 12,000 mi above the surface of the earth.

Figure 11.7 Three types of orbits for artificial satellites

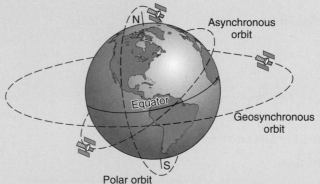

PROBLEMS 11.3

1. The moon orbits 3.84×10^8 m from the earth. How fast does the moon travel?
2. The moon orbits 3.84×10^8 m from the earth. How long does it take to orbit the earth?
3. Find the orbital velocity for Mercury as it orbits the sun.
4. Find the orbital velocity for the earth as it orbits the sun.
5. Find the orbital velocity for Saturn as it orbits the sun.
6. Find the orbital velocity for Uranus as it orbits the sun.
7. Find the time Mercury takes to orbit the sun.
8. Find the time the earth takes to orbit the sun.
9. Find the time Saturn takes to orbit the sun.
10. Find the time Uranus takes to orbit the sun.

SKETCH

12 cm² | w

4.0 cm

DATA

$A = 12$ cm², $l = 4.0$ cm, w = ?

BASIC EQUATION

$A = lw$

WORKING EQUATION

$w = \frac{A}{l}$

SUBSTITUTION

$w = \frac{12 \text{ cm}^2}{4.0 \text{ cm}} = 3.0$ cm

Glossary

Gravitational Field The area around a massive body in which an object experiences a gravitational force. The more massive and closer an object is to that body, the stronger is the gravitational field. (p. 300)

Newton's Law of Universal Gravitation All objects that have mass are attracted to one another by a gravitational force. (p. 297)

Orbit The path taken by an object during its revolution around another object, such as the path of the moon or a satellite about the earth or of a planet about the sun. (p. 301)

Formulas

11.1 $F_G = G\dfrac{m_1 m_2}{r^2}$

11.3 $v = \sqrt{\dfrac{Gm}{r}}$

$T = 2\pi\sqrt{\dfrac{r^3}{Gm}}$

Review Questions

1. What type of force is related to the mass of objects?
   (a) Electric force   (b) Strong force
   (c) Magnetic force   (d) Gravitational force

2. As the distance increases between two objects, the gravitational force between the objects
   (a) increases.   (b) decreases.   (c) remains constant.

3. As the mass of two objects increases, the gravitational force between the objects
   (a) increases.   (b) decreases.   (c) remains constant.

4. The mass of a satellite is increased. In order to maintain the same orbital period, its distance from the earth must
   (a) increase.   (b) decrease.   (c) remain constant.

5. As the distance increases between a satellite and the earth, what happens to the time it takes to complete an orbit?
   (a) Increases   (b) Decreases   (c) Remains constant

6. Explain why the gravitational force that exists between the person sitting next to you in class and you is much less than the gravitational force that exists between the earth and you.

7. What would happen to your weight on earth if the radius of the earth doubled, but its mass stayed the same?

8. What would happen to your weight on earth if the mass of the earth doubled, but its radius stayed the same?

9. Explain how a satellite in orbit is in a constant state of free fall, yet does not crash to the earth.

10. According to Isaac Newton, how is the motion of a falling apple different from the motion of the moon orbiting the earth?

11. Most planets actually have slightly elliptical orbits around the sun. What is the force exerted on a planet at its perigee (the point closest to the sun) compared to the force at its apogee (the point farthest from the sun)?

12. Does the mass of a satellite influence the time it takes to orbit the earth?

Review Problems

1. Two 0.300-kg apples are 25.0 cm apart from one center to the other. Find the gravitational attraction between the two apples.
2. Two 65.0-kg people are standing 1.00 m apart. What is the attractive gravitational force between them?
3. Find the weight of a 65.0-kg person on the earth (in newtons and pounds).
4. Find the weight of a 65.0-kg person on Jupiter (in newtons and pounds).
5. Find the weight of a 65.0-kg person on Pluto (in newtons and pounds).
6. If the moon orbited at one-half the present distance to the earth, what would be the orbiting time for the moon?
7. If the moon orbited at twice the present distance to the earth, what would be the orbiting time for the moon?
8. If the moon orbited at four times the present distance to the earth, what would be the orbiting time for the moon?
9. Using Newton's law of universal gravitation, find the amount of gravitational force on an 85.0-kg astronaut on the launch pad.
10. If an 85.0-kg astronaut in a space shuttle orbits the earth 362 km above sea level, what is the amount of gravitational force between the astronaut and the earth?

APPLIED CONCEPTS

1. The gravitational differences between the earth and Mars is a factor that engineers and scientists must consider before sending an astronaut to the "Red Planet." (a) What is the acceleration due to gravity on Mars? (b) If an 85.0-kg astronaut landed on Mars, how much would the astronaut weigh there compared to his weight on the earth? (c) Based on the acceleration due to gravity on Mars, what might happen to the strength of the astronaut's muscles while the astronaut is away from earth?

2. (a) How far from the center of the earth must a person be located in space so that his or her weight would be the same as when he or she is standing on the moon? (b) How many earth radii would this location be from the center of the earth?

3. A geosynchronous communication satellite orbits at a fixed point above the earth's surface. The earth takes 24.0 h to rotate on its axis. (a) Find the communication satellite's orbiting altitude from the center of the earth and from the surface of the earth. (b) Find its linear speed.

4. Flight engineers for the Apollo Lunar Orbiter placed the orbiter $15\overline{0}$ km above the surface of the moon. What was the Orbiter's (a) linear velocity and (b) period as it orbited the moon? (c) How would increasing the altitude affect the Orbiter's velocity and period?

5. (a) What is the gravitational force on a 65.7-kg space shuttle astronaut orbiting the earth 427 km above the surface of the earth? (b) Explain why astronauts orbiting the earth experience weightlessness even though the earth continually applies a force on their bodies.

CHAPTER 12

MATTER

$$\text{Weight Density} = \frac{F_w}{V}$$

$$\text{Specific Gravity} = \frac{D_{material}}{D_{water}}$$

Technology is used to take raw materials and shape, refine, mold, and transform them into products useful to our society. To do this, it is necessary to have some understanding of the nature and properties of matter and basic characteristics of its various forms of solids, liquids, and gases. Once these materials take shape and become useful products, they are subjected to various stresses and strains that can ultimately result in structural failure. We will show how to analyze these stresses and strains so that materials can be evaluated before being put to use.

Objectives

The major goals of this chapter are to enable you to:

1. Describe the properties of matter.
2. Apply Hooke's law.
3. Describe the properties of solids, liquids, and gases.
4. Solve density and specific gravity problems.
5. Calculate the amount of stress on objects.

12.1 Properties of Matter

What are the building blocks of matter? First, **matter** is anything that occupies space and has mass. Suppose that we take a cube of sugar and divide it into two pieces. Then we divide a resulting piece into another two pieces. Can we continue this process indefinitely and get smaller and smaller particles of sugar each time? No, at some point we will arrive at the building blocks of sugar.

An **element** is a substance that cannot be separated into simpler substances. A **compound** is a substance containing two or more elements.

A **molecule** is the smallest particle of an element that can exist in a free state and still retain the characteristics of that element or compound. Most simple molecules are about 3×10^{-10} m in diameter. An **atom** is the smallest particle of an element that can exist in a stable or independent state. The molecules of elements consist of one atom or two or more similar atoms; the molecules of compounds consist of two or more different atoms.

What do we get if we divide the sugar molecule? The resulting particles are carbon, hydrogen, and oxygen atoms. Models of water and sugar molecules are shown in Fig. 12.1.

Figure 12.1

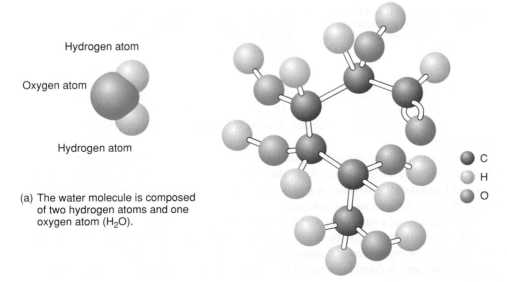

Hydrogen atom

Oxygen atom

Hydrogen atom

- C
- H
- O

(a) The water molecule is composed of two hydrogen atoms and one oxygen atom (H_2O).

(b) The sugar (glucose) molecule is composed of six carbon atoms, twelve hydrogen atoms, and six oxygen atoms ($C_6H_{12}O_6$).

4. A 17.0-N force stretches a wire 0.650 cm.
   (a)   What force will stretch a similar piece of wire 1.87 cm?
   (b)   A force of 21.3 N is applied to a similar piece of wire. How far will it stretch?
5. A force of 36.0 N stretches a spring 18.0 cm. Find the spring constant (in N/m).

PHYSICS CONNECTIONS

Development of Materials

Engineers choose materials based on two main factors, cost and strength. The strength of a material is chosen according to its proposed usage. For example, will the material be used underneath a roadway to support cars and trucks, or will it be used to suspend a roadway? Advances in the science and engineering of materials have helped in the development of more economical bridges that span greater and greater distances.

Ancient Roman, Mesopotamian, and Chinese bridge builders used stone to create their arch bridges. Stone was abundant and very strong under compression forces. However, the need for lighter and stronger materials sparked technological improvements in building materials during the Industrial Revolution of the 18th and 19th centuries. During that time, an English engineer, Abraham Darby, built the first cast-iron arch bridge (Fig. 12.19). The low cost and high compression strength of iron made it a much better alternative to the expensive and heavy stone arch. In later years, however, problems with cast-iron bridges became evident, as the brittle nature of iron under tension and shear stresses caused fractures in cast-iron structures.

In the late 19th century, an American engineer, James Eads, constructed the first steel arch bridge over the Mississippi River. The steel used in this bridge was not much stronger under compression than iron but was more than twice as strong as iron under tension and shearing stresses. Steel is used for beams and cables in almost all bridges today.

Although concrete is extremely weak under shearing and tension stresses, it is used to combat compression stresses. To improve concrete's ability to withstand tension and shearing stresses, stretched steel wire mesh or cables are embedded in the concrete. After the concrete dries, the tension in the wire mesh or cables is released, placing the concrete in a permanent state of compression. Therefore, when a tension force is placed on the prestressed concrete, the concrete itself remains under compression from the steel mesh or cables.

Figure 12.19

Photo courtesy of Dorling Kindersley

(a) Darby's cast iron bridge, constructed at Coalbrookdale, England, in 1779

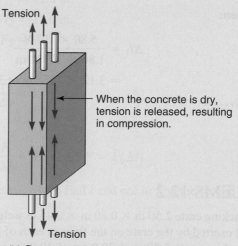

When the concrete is dry, tension is released, resulting in compression.

(b) Pre-stressed concrete requires tension on the cables while the concrete dries. After the concrete dries, the tension is released, placing the concrete in compression.

6. A force of 5.00 N is applied to a spring whose spring constant is 0.250 N/cm. Find its change in length (in cm).

7. Each vertical steel column of an office building supports 1.30×10^5 N and is compressed 5.90×10^{-3} cm.
    (a) Find the compression in each column if a weight of 5.50×10^5 N is supported.
    (b) If the compression of each steel column is 0.0710 cm, what weight is supported by each column?

8. Each vertical steel column of an office building supports $30,\overline{0}00$ lb and is compressed 0.00234 in. Find the compression of each column if a weight of 125,000 lb is supported.

9. If the compression of each steel column in Problem 8 is 0.0279 in., what weight is supported by each column?

10. A coiled spring is stretched 40.0 cm by a 5.00-N weight.
    (a) How far is it stretched by a 15.0-N weight?
    (b) What weight will stretch the spring 60.0 cm?

11. A 12,$\overline{0}$00-N load is hanging from a steel cable that is 10.0 m long and 16.0 mm in diameter. Find the stress.

12. A rectangular cast-iron column 25.0 m $\times$ 25.0 cm $\times$ 5.00 m supports a weight of 6.80×10^6 N. Find the stress in pascals on the column.

13. In a Hooke's law experiment, the following weights were attached to a spring, resulting in the following elongations:

Weight (N)	Elongation (cm)
$5\overline{0}$	2.0
75	3.3
105	4.2
125	5.0
$15\overline{0}$	6.0
175	7.4
225	9.5
275	11.1

    (a) Plot the graph of weight versus elongation and draw the best straight line through the data.
    (b) From the graph, what weight corresponds to an elongation of 7.5 cm?
    (c) From the graph, what elongation corresponds to a weight of 220 N?
    (d) From the graph, determine the spring constant.

14. What was the original length of a spring with spring constant 96.0 N/m that is stretched to 28.0 cm by a 15.0-N weight?

15. Two hanging springs, each 15.0 cm long, with spring constants 0.970 N/cm and 1.45 N/cm, respectively, are stretched by a bar weighing 26.0 N that connects them. The bar is 6.00 m long and has a center of mass 2.00 m from the spring with constant 0.970 N/cm. How far does each spring stretch?

16. A firefighter weighs 725 N. She wears shoes that each cover an area of 206 cm^2.
    (a) What is the average stress she applies to the ground on which she is standing?
    (b) How does the stress change if she stands on only one foot?

17. Two identical wires are 125 cm and 375 cm long, respectively. The first wire is broken by a force of 489 N. What force is needed to break the other?

18. The cross-sectional area of a wire is 2.50×10^{-3} cm^2 and its tensile strength is 1.00×10^5 N/cm^2. What force will break the wire?

19. A spring having a force constant of 1.25 N/cm is stretched through a distance of 11.5 cm. How much work is required to stretch the spring?

12.3 Properties of Liquids

As noted previously, a liquid is a substance that has a definite volume and takes the shape of its container. The molecules move in a flowing motion, yet are so close together that it is very difficult to compress a liquid. Most liquids share the following common properties.

Cohesion and Adhesion

Cohesion, the force of attraction between like molecules, causes a liquid like molasses to be sticky. Adhesion, the force of attraction between unlike molecules, causes the molasses to also stick to your finger. In the case of water, its adhesive forces are greater than its cohesive forces. Put a glass in water and pull it out. Some water remains on the glass.

In the case of mercury, the opposite is true. Mercury's cohesive forces are greater than its adhesive forces. If glass is submerged in mercury and then pulled out, virtually no mercury remains on the glass.

A liquid whose adhesive forces are greater than its cohesive forces tends to wet any surface that comes in contact with it. A liquid whose cohesive forces are greater than its adhesive forces tends to leave objects dry or not wet when any surface comes in contact with it.

Figure 12.20   (a) The surface tension of water will support a needle. (b) Adding soap reduces surface tension.

(a) Water   (b) Soap added

Surface Tension

Surface tension is the ability of the surface of a liquid to act like a thin, flexible film. The ability of the surface of water to support a needle is an example. The water's surface acts like a thin, flexible surface film. The surface tension of water can be reduced by adding soap to the water (Fig. 12.20). Soaps are added to laundry water to decrease the surface tension of water so that the water more easily penetrates the fibers of the clothes being washed.

Surface tension causes a raindrop to hold together and a small drop of mercury to keep an almost spherical shape. A liquid drop suspended in space is spherical. A falling raindrop's shape is due to the friction with the air (Fig. 12.21).

Figure 12.21   (a) Surface tension causes a drop of mercury to be more spherical. (b) Surface tension causes a liquid drop to hold together. (c) The shape of a falling raindrop is due to the friction with air.

Figure 12.22   Cold oil is more viscous than hot oil.

(a) Mercury drop on   (b) Liquid drop in space   (c) Falling raindrop
    a surface

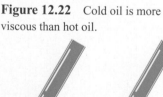

Cold oil   Hot oil

Viscosity

Viscosity is the internal friction of a fluid caused by molecular attraction, which makes it resist a tendency to flow. The greater the molecular attraction, the greater is the liquid's internal friction and viscosity. For example, it takes more force to move a block of wood through oil than through water. This is because oil is more viscous than water.

If a liquid's temperature is increased, its viscosity decreases. For example, the viscosity of oil in a car engine before it is started on a winter morning at $-10.0°C$ is greater than after the engine has been running for an hour (Fig. 12.22).

Higher viscosity does not mean higher density. For example, oil is more viscous, but water is denser. Therefore, oil floats on water.

Figure 12.23 A liquid keeps the same level in tubes of large enough diameter that are connected. Neither the shape nor the size of the containers makes a difference.

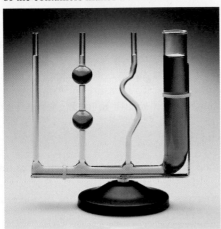

Figure 12.24 Liquids rise in a thin tube because of surface tension and adhesion. The height of the rise depends on the surface tension, the diameter of the tube, and the density of the liquid. Here you can see the varying heights of colored water in thin tubes of various diameters.

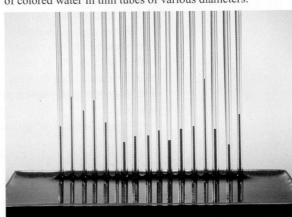

Capillary Action

Capillary action is the behavior of liquids that causes the liquid level in very small-diameter tubes to be different than that in larger-diameter tubes. A liquid keeps the same level in connected tubes filled with it if the tubes have a large enough diameter (Fig. 12.23). In tubes of different very small diameters, water does not stand at the same level. The smaller the diameter, the higher the water rises (Fig. 12.24). Mercury does not stand at the same level either, but instead of rising up the narrower tube, the mercury level falls or is depressed in the smaller-diameter tube. The smaller the diameter, the lower mercury's level is depressed (Fig. 12.25).

Figure 12.25 Comparison of the capillary action of water and mercury: In very thin tubes, water stands at higher levels, and the water surface is concave. In very thin tubes, mercury stands at lower levels, and the mercury surface is convex. [The tube diameters and differences in liquid levels are exaggerated in (a) and (b).]

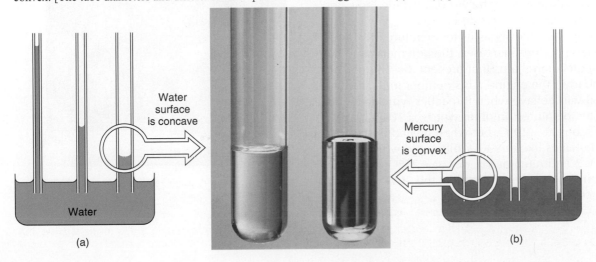

Water surface is concave

Water

(a)

Mercury surface is convex

(b)

Capillary action is due to both the adhesion of the liquid molecules with the tube and the surface tension of the liquid. For water, the adhesive forces are greater than the cohesive molecular forces. Thus, water creeps up the sides of the tube and produces a concave water surface. The surface tension of the water tends to flatten the concave surface. Together, these two forces raise the water up the tube until it is counterbalanced by the weight of the water column itself.

In the case of mercury, the cohesive molecular forces are greater than the adhesive forces and produce a convex mercury surface. The surface tension of the mercury tends to further hold down its level. This crescent-shaped surface of a liquid column in a tube, whether concave or convex, is called a **meniscus**. To measure the height of a liquid in a tube, measure to the lowest point of a concave meniscus or to the highest point of a convex meniscus.

Experimentally, scientists have found that:

1. Liquids rise in capillary tubes they tend to wet and are depressed in tubes they tend not to wet.
2. Elevation or depression in the tube is inversely proportional to the diameter of the tube.
3. The elevation or depression decreases as the temperature increases.

Capillary action causes the rise of oil (or kerosene) in the wick of an oil lamp. Towels also absorb water because of capillary action.

PHYSICS CONNECTIONS

Oil Viscosity

One of the most important factors in choosing the correct motor oil for an automobile is its viscosity grade (Fig. 12.26). The viscosity grade is indicated by a numbering system established by the Society of Automotive Engineers (S.A.E.), such as 10W-30. The first number followed by the "W" indicates that the oil remains viscous and does not lose its lubricating abilities at low temperatures. A 5W weight oil is guaranteed to remain viscous at −25°C, a 10W weight oil is guaranteed to remain viscous at −20°C, a 15W weight oil is guaranteed to remain viscous at −15°C, and so on.

The second number in the viscosity grade indicates that the oil has had polymers added to it to prevent the oil from losing its viscosity as its temperature increases. When the temperature increases, the polymers stretch into long molecular chains that prevent the oil from shearing apart in the engine. This number indicates how the oil will behave when it reaches typical engine operating temperatures. In other words, a 10W-30 oil remains viscous at −20°C, but when heated in the engine, it behaves like the 30W oil would when it is cold. A 5W-20 oil remains viscous at −25°C, but when heated in the engine, it behaves like the 20W oil would when it is cold.

Figure 12.26 Filling the engine with the motor oil of the correct viscosity is essential for maintaining a well-lubricated engine.

Although the addition of polymers to the engine oil is beneficial in achieving viscosity at large temperature ranges, those polymers also form deposits in the engine. A large range between the two numbers, such as 10W-40, indicates that more polymers are present in the oil as compared to a smaller range, such as 10W-20. If there are more polymers, then there is also an increased risk of thermal breakdown and the formation of unwanted deposits in the engine. As a result, a narrower range between the two numbers is typically desired, but consumers should refer to the recommendations from the car's manufacturer when they select the proper engine oil.

12.4 Properties of Gases

Expansion is a property of a gas in which the rapid random movement of its molecules causes the gas to completely occupy the volume of its container.

Diffusion is the process by which molecules of a gas mix with the molecules of a solid, a liquid, or another gas. If you remove the cap from a can of gasoline, you soon smell the fumes. The air molecules and the gasoline molecules mix throughout the room because of diffusion.

A balloon inflates due to the pressure of the air molecules on its inside surface. This pressure is caused by the bombardment on the walls by the moving molecules. The pressure may be increased by increasing the number of molecules by blowing more air into the balloon. Pressure may also be increased by heating the air molecules already in the balloon. Heat increases the velocity of the molecules.

The behavior of liquids and gases is often very similar. A **fluid** is a substance that takes the shape of its container. The term is used when discussing principles and behaviors common to both liquids and gases.

12.5 Density

Density is a property of all three states of matter. **Mass density**, D_m, is defined as mass per unit volume. **Weight density**, D_w, is defined as weight per unit volume, or,

$$D_m = \frac{m}{V} \qquad D_w = \frac{F_w}{V}$$

where D_m = mass density D_w = weight density
m = mass F_w = weight
V = volume V = volume

Although mass density and weight density can be expressed in both the metric system and the U.S. system, mass density is usually given in the metric units kg/m^3 and weight density is usually given in the U.S. units lb/ft^3 (Table 12.2).

The mass density of water is $10\overline{0}0 \ kg/m^3$; that is, 1 cubic metre of water has a mass of $10\overline{0}0$ kg. The weight density of water is $62.4 \ lb/ft^3$; that is, 1 cubic foot of water weighs 62.4 lb.

In nearly all forms of matter, the density usually decreases as the temperature increases and increases as the temperature decreases. Water does not follow the usual pattern of increasing density at lower temperatures; ice is actually less dense than liquid water. This phenomenon is discussed more fully in Section 14.7.

Note: Conversion factors must often be used to obtain the desired units.

Table 12.2   Densities for Various Substances

Substance	Mass Density (kg/m^3)	Weight Density (lb/ft^3)
Solids		
Copper	8,890	555
Iron	7,800	490
Lead	11,300	708
Aluminum	2,700	169
Brass	8,700	540
Ice	917	57
Wood, white pine	420	26
Concrete	2,300	140
Cork	240	15
Liquids		
Water	1,000	62.4
Seawater	1,025	64.0
Oil	870	54.2
Mercury	13,600	846
Alcohol	790	49.4
Gasoline	680	42.0
Gases[*]	At 0°C and 1 atm pressure	At 32°F and 1 atm pressure
Air	1.29	0.081
Carbon dioxide	1.96	0.123
Carbon monoxide	1.25	0.078
Helium	0.178	0.011
Hydrogen	0.0899	0.0056
Oxygen	1.43	0.089
Nitrogen	1.25	0.078
Ammonia	0.760	0.047
Propane	2.02	0.126

[*]The density of a gas is found by pumping the gas into a container, measuring its volume and mass or weight, and then using the appropriate density formula.

EXAMPLE 1

Find the weight density of a block of wood 3.00 in. × 4.00 in. × 5.00 in. with weight 0.700 lb.

Sketch:

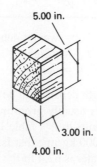

5.00 in.

3.00 in.

4.00 in.

Data:

$$l = 4.00 \text{ in.}$$
$$w = 3.00 \text{ in.}$$
$$h = 5.00 \text{ in.}$$
$$F_w = 0.700 \text{ lb}$$
$$D_w = ?$$

Basic Equations:

$$V = lwh \quad \text{and} \quad D_w = \frac{F_w}{V}$$

Working Equations: Same

Substitutions:

$$V = (4.00 \text{ in.})(3.00 \text{ in.})(5.00 \text{ in.})$$
$$= 60.0 \text{ in}^3$$

$$D_w = \frac{0.700 \text{ lb}}{60.0 \text{ in}^3}$$
$$= 0.0117 \frac{\text{lb}}{\text{in}^3} \times \left(\frac{12 \text{ in.}}{1 \text{ ft}}\right)^3$$
$$= 20.2 \text{ lb/ft}^3$$

· · · · · · · · · · · · · · · ·

Find the mass density of a ball bearing with mass 22.0 g and radius 0.875 cm.　◄ **EXAMPLE 2**

Data:

$$r = 0.875 \text{ cm}$$
$$m = 22.0 \text{ g}$$
$$D_m = ?$$

Basic Equations:

$$V = \tfrac{4}{3}\pi r^3 \quad \text{and} \quad D_m = \frac{m}{V}$$

Working Equations: Same

Substitutions:

$$V = \tfrac{4}{3}\pi (0.875 \text{ cm})^3$$
$$= 2.81 \text{ cm}^3$$

$$D_m = \frac{22.0 \text{ g}}{2.81 \text{ cm}^3}$$
$$= 7.83 \text{ g/cm}^3$$
$$= 7.83 \frac{\text{g}}{\text{cm}^3} \times \left(\frac{100 \text{ cm}}{1 \text{ m}}\right)^3 \times \frac{1 \text{ kg}}{10^3 \text{ g}} = 7830 \text{ kg/m}^3$$

· · · · · · · · · · · · · · · ·

Find the weight density of a gallon of water weighing 8.34 lb.　◄ **EXAMPLE 3**

Data:

$$F_w = 8.34 \text{ lb}$$
$$V = 1 \text{ gal} = 231 \text{ in}^3$$
$$D_w = ?$$

Basic Equation:

$$D_w = \frac{F_w}{V}$$

Working Equation: Same

Substitution:

$$D_w = \frac{8.34 \text{ lb}}{231 \text{ in}^3}$$

$$= 0.0361 \frac{\text{lb}}{\text{in}^3} \times \left(\frac{12 \text{ in.}}{1 \text{ ft}}\right)^3$$

$$= 62.4 \text{ lb/ft}^3$$

....

EXAMPLE 4

Find the weight density of a can of oil (1 quart) weighing 1.90 lb.

Data:

$$V = 1 \text{ qt} = \tfrac{1}{4} \text{ gal} = \tfrac{1}{4}(231 \text{ in}^3) = 57.8 \text{ in}^3$$
$$F_w = 1.90 \text{ lb}$$
$$D_w = ?$$

Basic Equation:

$$D_w = \frac{F_w}{V}$$

Working Equation: Same

Substitution:

$$D_w = \frac{1.90 \text{ lb}}{57.8 \text{ in}^3}$$

$$= 0.0329 \frac{\text{lb}}{\text{in}^3} \times \left(\frac{12 \text{ in.}}{1 \text{ ft}}\right)^3$$

$$= 56.9 \text{ lb/ft}^3$$

....

EXAMPLE 5

A quantity of gasoline weighs 5.50 lb with weight density 42.0 lb/ft^3. Find its volume.

Data:

$$D_w = 42.0 \text{ lb/ft}^3$$
$$F_w = 5.50 \text{ lb}$$
$$V = ?$$

Basic Equation:

$$D_w = \frac{F_w}{V}$$

Working Equation:

$$V = \frac{F_w}{D_w}$$

Substitution:

$$V = \frac{5.50 \text{ lb}}{42.0 \text{ lb/ft}^3}$$

$$= 0.131 \text{ ft}^3$$

....

The density of an irregular solid (rock) cannot be found directly because of the difficulty of finding its volume. However, we could find the amount of water the solid displaces, which is the same as the volume of the irregular solid. In Fig. 12.27, the volume of water in the small beaker equals the volume of the rock.

Figure 12.27   The volume of the rock can be found by measuring the volume of the liquid it displaces.

(a)

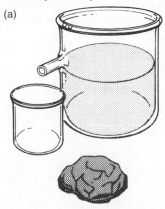

(b)

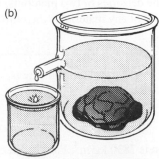

The volume of the rock = the volume of the water displaced into the overflow beaker.

A rock of mass 10.8 kg displaces $32\overline{0}0$ cm^3 of water. What is the mass density of the rock?

EXAMPLE 6

Data:

$$m = 10.8 \text{ kg}$$
$$V = 32\overline{0}0 \text{ cm}^3$$
$$D_m = ?$$

Basic Equation:

$$D_m - \frac{m}{V}$$

Working Equation: Same

Substitution:

$$D_m = \frac{10.8 \text{ kg}}{32\overline{0}0 \text{ cm}^3} \times \left(\frac{100 \text{ cm}}{1 \text{ m}}\right)^3$$
$$= 3380 \text{ kg/m}^3$$

· · · · · · · · · · · · · · · · ·

A rock displaces 3.00 gal of water and has a weight density of 156 lb/ft^3. What is its weight?

EXAMPLE 7

Data:

$$D_w = 156 \text{ lb/ft}^3$$
$$V = 3.00 \text{ gal}$$
$$F_w = ?$$

Basic Equation:

$$D_w = \frac{F_w}{V}$$

Working Equation:

$$F_w = D_w V$$

Substitution:

$$F_w = 156 \frac{\text{lb}}{\text{ft}^3} \times 3.00 \text{ gal} \times \frac{231 \text{ in}^3}{1 \text{ gal}} \times \left(\frac{1 \text{ ft}}{12 \text{ in.}}\right)^3$$

$$= 62.6 \text{ lb}$$

To compare the densities of two materials, we compare each with the density of water. The **specific gravity** of any material is the ratio of the density of the material to the density of water. That is,

$$\text{specific gravity (sp gr)} = \frac{D_{\text{material}}}{D_{\text{water}}}$$

Note that specific gravity is a unitless quantity.

EXAMPLE 8

The density of iron is 7830 kg/m³. Find its specific gravity.

Data:

$$D_{\text{material}} = 7830 \text{ kg/m}^3$$
$$D_{\text{water}} = 10\overline{0}0 \text{ kg/m}^3$$
$$\text{sp gr} = ?$$

Basic Equation:

$$\text{sp gr} = \frac{D_{\text{material}}}{D_{\text{water}}}$$

Working Equation: Same

Substitution:

$$\text{sp gr} = \frac{7830 \text{ kg/m}^3}{10\overline{0}0 \text{ kg/m}^3}$$

$$= 7.83$$

This means that iron is 7.83 times as dense as water, and thus it sinks in water.

EXAMPLE 9

The density of oil is 54.2 lb/ft³. Find its specific gravity.

$$D_{\text{material}} = 54.2 \text{ lb/ft}^3$$
$$D_{\text{water}} = 62.4 \text{ lb/ft}^3$$
$$\text{sp gr} = ?$$

Basic Equation:

$$\text{sp gr} = \frac{D_{\text{material}}}{D_{\text{water}}}$$

Working Equation: Same

Substitution:

$$\text{sp gr} = \frac{54.2 \text{ lb/ft}^3}{62.4 \text{ lb/ft}^3}$$

$$= 0.869$$

This means oil is 0.869 times as dense as water and thus it floats on water.

In general, the specific gravity of

$$\text{water} = 1$$

$$\text{a material denser than water} > 1$$

$$\text{a material less dense than water} < 1$$

When we check the antifreeze in a radiator in winter, we are really finding the specific gravity of the liquid. Specific gravity is a ratio comparison of the density of a substance to that of water. Because the density of antifreeze is different from the density of water, we find the concentration of antifreeze (and thus the amount of protection from freezing) by measuring the specific gravity of the solution in the radiator.

A **hydrometer** is a sealed glass tube weighted at one end so that it floats vertically in a liquid (Fig. 12.28). It sinks in the liquid until it displaces an amount of liquid equal to its own weight. The densities of the displaced liquids are inversely proportional to the depths to which the tube sinks. That is, the greater the density of the liquid, the less the tube sinks; the less the density of the liquid, the more the tube sinks. A hydrometer usually has a scale inside the tube and is calibrated so that it floats in water at the 1.000 mark. Anything with a specific gravity greater than 1 sinks in water. A substance with a specific gravity less than 1 floats in water; its specific gravity indicates the fractional volume that is under water.

Hydrometers are commonly used to measure the specific gravities of battery acid and antifreeze in radiators (Fig. 12.29). In a lead storage battery, the electrolyte is a solution of sulfuric acid and water, and the specific gravity of the solution varies with the amount of charge of the battery. Table 12.3 gives common specific gravities of conditions of a lead storage battery. Table 12.4 gives various specific gravities and the corresponding temperatures below which the antifreeze and water solution will freeze.

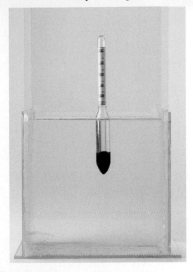

Figure 12.28 A hydrometer measures density of a liquid.

Table 12.3   Specific Gravities for a Lead Storage Battery

Condition	Specific Gravity
New (fully charged)	1.30
Old (discharged)	1.15

Table 12.4   Specific Gravities for Antifreeze and Water Solution

Temperature (°C)	Specific Gravity
−1.24	1.00
−2.99	1.01
−6.89	1.02
−19.82	1.05
−44.83	1.07
−51.23	1.08

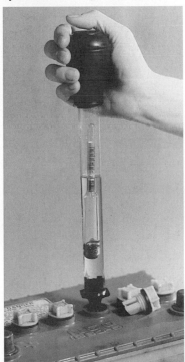

Figure 12.29   A common hydrometer.

One other factor must be considered in the use of the hydrometer—that of temperature. Significant differences in readings will occur over a range of temperatures. Specific gravities of some common liquids at room temperature are given in Table 12.5.

Table 12.5   Specific Gravities of Common Liquids at Room Temperature (20°C or 68°F)

Liquid	Specific Gravity
Benzene	0.9
Ethyl alcohol	0.79
Gasoline	0.68
Kerosene	0.82
Mercury	13.6
Seawater	1.025
Sulfuric acid	1.84
Turpentine	0.87
Water	1.000

TRY THIS ACTIVITY

Pour some water into a clear glass. Then pour in some cooking oil. Which liquid is less dense? (See Fig. 12.30.)

Figure 12.30

Oil and other fluids of different densities float on top of each other (Fig. 12.31, contrasting colors have been added for illustration purposes).

Figure 12.31

PROBLEMS 12.5

Express mass density in kg/m^3 and weight density in lb/ft^3.

1. Find the mass density of a chunk of rock of mass 215 g that displaces a volume of 75.0 cm^3 of water.
2. A block of wood is 55.9 in. × 71.1 in. × 25.4 in. and weighs 1810 lb. Find its weight density.

3. If a block of wood of the size in Problem 2 has a weight density of 30.0 lb/ft^3, what does it weigh?
4. Find the volume (in cm^3) of 1350 g of mercury.
5. Find the volume (in cm^3) of 1350 g of cork.
6. Find the volume (in m^3) of 1350 g of nitrogen at 0°C and 1 atm pressure.
7. A block of gold 9.00 in. × 8.00 in. × 6.00 in. weighs 302 lb. Find its weight density.
8. A cylindrical piece of copper is 9.00 in. tall and 1.40 in. in radius. How much does it weigh?
9. A piece of aluminum of mass 6.24 kg displaces water that fills a container 12.0 cm × 12.0 cm × 16.0 cm. Find its mass density.
10. If 1.00 pint of turpentine weighs 0.907 lb, what is its weight density?
11. Find the mass density of gasoline if 106 g occupies 155 cm^3.
12. How much does 1.00 gal of gasoline weigh?
13. Determine the volume (in m^3) of 3045 kg of oil.
14. How many ft^3 will 573 lb of water occupy?
15. If 20.4 in^3 of linseed oil weighs 0.694 lb, what is its weight density?
16. If 108 in^3 of ammonia gas weighs 0.00301 lb, what is its weight density?
17. Find the volume of 3.00 kg of propane at 0°C and 1 atm pressure.
18. Granite has a mass density of 2650 kg/m^3. Find its weight density in lb/ft^3.
19. Find the mass density of a metal block 18.0 cm × 24.0 cm × 8.00 cm with mass 9.76 kg.
20. Find the mass (in kg) of 1.00 m^3 of
    (a) water.   (b) gasoline.   (c) copper.
    (d) mercury.   (e) air at 0°C and 1 atm pressure.
21. What size tank (in litres) is needed for 10$\overline{0}$0 kg of
    (a) water?   (b) gasoline?   (c) mercury?
22. Copper has a mass density of 8890 kg/m^3. Find its mass density in g/cm^3.

Use Table 12.2 to find the specific gravity of each material.

23. What is the mass of gasoline in a 1250-litre gas tank?
24. If racing alcohol has a mass density of 790 kg/m^3, what mass will a 1250-litre tank hold?
25. Ice   26. Concrete   27. Iron
28. Air   29. Gasoline   30. Cork
31. The specific gravity of material X is 0.82. Does it sink in or float on water?
32. The specific gravity of material Y is 1.7. Does it sink in or float on water?
33. The specific gravity of material Z is 0.52. Does it sink in or float on gasoline?
34. The specific gravity of material W is 11.5. Does it sink in or float on mercury?
35. A proton has mass 1.67×10^{-27} kg and diameter 8.2×10^{-16} m. Find its specific gravity.
36. Find the mass density of a 315-g object that displaces 0.275 m^3 of water.
37. What is the mass density of a 50$\overline{0}$-g block that displaces 215 cm^3 of water?

SKETCH

| 12 cm² | w |

4.0 cm

DATA

$A = 12$ cm^2, $l = 4.0$ cm, $w = ?$

BASIC EQUATION

$A = lw$

WORKING EQUATION

$w = \frac{A}{l}$

SUBSTITUTION

$w = \frac{12 \text{ cm}^2}{4.0 \text{ cm}} = 3.0$ cm

PROBLEM SOLVING

Glossary

Adhesion The force of attraction between different or unlike molecules. (p. 311)

Atom The smallest particle of an element that can exist in a stable or independent state. (p. 309)

Bending Consists of both tension and compression stresses. It occurs when a force is placed on a beam causing it to sag. (p. 317)

Brinell Method Common industrial method used to measure the hardness of a metal. (p. 311)

Capillary Action The behavior of liquids that causes the liquid level in very small-diameter tubes to be different than in larger-diameter tubes. This behavior is due both to adhesion of the liquid molecules to the tube and to the surface tension of the liquid. (p. 325)

Cohesion The force of attraction between like molecules. Holds the closely packed molecules of a solid together. (p. 311)

Compound A substance containing two or more elements. (p. 309)

Compression A stress caused by two forces acting directly toward each other. This stress tends to cause objects to become shorter and thicker. (p. 316)

Diffusion The process by which molecules of a gas mix with the molecules of a solid, a liquid, or another gas. (p. 327)

Ductility A property of a metal that enables it to be drawn through a die to produce a wire. (p. 313)

Elastic Limit The point beyond which a deformed object cannot return to its original shape. (p. 314)

Elasticity A measure of a deformed object's ability to return to its original size and shape once the deforming force is removed. (p. 313)

Electron One of the particles that makes up atoms. Has a negative charge. (p. 310)

Element A substance that cannot be separated into simpler substances. (p. 309)

Expansion Property of a gas in which the rapid random movement of its molecules causes the gas to completely occupy the volume of its container. (p. 327)

Fluid A substance that takes the shape of its container. Either a liquid or a gas. (p. 327)

Gas A substance that takes the shape of its container and has the same volume as its container. (p. 310)

Hardness A measure of the internal resistance of the molecules of a solid being forced farther apart or closer together. (p. 311)

Hooke's Law A principle of elasticity in solids: The ratio of the force applied to an object to its change in length (resulting in its being stretched or compressed by the applied force) is constant as long as the elastic limit has not been exceeded. (p. 319)

Hydrometer A sealed glass tube weighted at one end so that it floats vertically in a liquid; an instrument used to determine specific gravity. (p. 333)

Liquid A substance that takes the shape of its container and has a definite volume. (p. 310)

Malleability A property of a metal that enables it to be hammered and rolled into a sheet. (p. 313)

Mass Density The mass per unit volume of a substance. (p. 327)

Matter Anything that occupies space and has mass. (p. 309)

Meniscus The crescent-shaped surface of a liquid column in a tube. (p. 326)

Molecule The smallest particle of a substance that exists in a stable and independent state. (p. 309)

Neutron One of the particles that makes up atoms. Does not carry an electric charge. (p. 310)

Nucleus The center part of an atom made up of protons and neutrons. (p. 310)

Proton One of the particles that makes up atoms. Has a positive charge. (p. 310)

Shearing A stress caused by two forces applied in parallel, opposite directions. (p. 316)

Solid A substance that has a definite shape and a definite volume. (p. 310)

Specific Gravity The ratio of the density of any material to the density of water. (p. 332)

Strain The deformation of an object due to an applied force. (p. 318)

Stress The ratio of an outside applied distorting force to the area over which the force acts. (p. 314)

Surface Tension The ability of the surface of a liquid to act like a thin, flexible film. (p. 324)

Tensile Strength A measure of a solid's resistance to being pulled apart. (p. 311)

Tension A stress caused by two forces acting directly opposite each other. This stress tends to cause objects to become longer and thinner. (p. 316)

Torsion A stress related to a twisting motion. This type of stress severely compromises the strength of most materials. (p. 316)

Viscosity The internal friction of a fluid caused by molecular attraction, which makes it resist a tendency to flow. (p. 324)

Weight Density The weight per unit volume of a substance. (p. 327)

Formulas

12.2 $S = \dfrac{F}{A}$

$\dfrac{F}{\Delta l} = k$

12.5 $D_m = \dfrac{m}{V}$

$D_w = \dfrac{F_w}{V}$

specific gravity (sp gr) $= \dfrac{D_{material}}{D_{water}}$

Review Questions

1. The most important particles that make up atoms include which of the following?
   - (a) Neutron
   - (b) Molecule
   - (c) Electron
   - (d) Hydrogen
   - (e) Proton

2. Matter exists in which of the following?
   - (a) Gas
   - (b) Neutrons
   - (c) Electrons
   - (d) Solid
   - (e) Liquid

3. The common industrial method used to measure the hardness of a metal is
   - (a) the Bernoulli method.
   - (b) Hooke's method.
   - (c) the capillary method.
   - (d) the Brinell method.
   - (e) none of the above.

4. Density is a property of
   - (a) gases.
   - (b) liquids.
   - (c) solids.
   - (d) all of the above.
   - (e) none of the above.

5. The process by which molecules of a gas mix with the molecules of a solid, a liquid, or a gas is called
   - (a) expansion.
   - (b) contraction.
   - (c) capillary action.
   - (d) diffusion.
   - (e) none of the above.

6. Capillary action refers to
   - (a) the mixing of molecules of different types.
   - (b) the behavior of liquids in small tubes.
   - (c) the attractive force between molecules.
   - (d) stretching beyond the elastic limit.

7. The relationship of the change in length of a stretched or compressed object to the force causing the change is given by
   - (a) Pascal's law.
   - (b) Brinell's law.
   - (c) the elastic limit.
   - (d) Hooke's law.
   - (e) none of the above.

8. The ability of the surface of water to support a needle is an example of
   (a) mass density.   (b) Hooke's law.   (c) diffusion.
   (d) stress.   (e) surface tension.
9. In your own words, describe the difference between mass density and weight density.
10. Would the mass density of an object be the same if the object were on the moon rather than on the earth? Would the weight density be the same?
11. In your own words, describe capillary action.
12. What is the difference between adhesion and cohesion?
13. Give one example of the effect of surface tension that is not described in this book.
14. The mass of a proton is approximately _____ times heavier than the mass of an electron.
15. The applied force divided by the area over which the force acts is called _____.
16. In your own words, state Hooke's law.
17. The commonly used unit of stress in the metric system is the _____.
18. Describe how to find the specific gravity of an object.
19. What is the ratio of mass to volume called?
20. What is friction in liquids called?
21. A spring that has been permanently deformed is said to have been deformed past its _____ _____.
22. List the three states of matter.
23. Distinguish between a molecule and an atom.
24. Distinguish between a neutron and a proton.
25. List the five basic stresses.
26. Explain how a hydrometer measures the charge in a lead storage battery. Does the temperature affect the measurement?

Review Problems

PROBLEM SOLVING

SKETCH

4.0 cm

DATA

$A = 12$ cm^2, $l = 4.0$ cm, $w = ?$

BASIC EQUATION

$A = lw$

WORKING EQUATION

$w = \frac{A}{l}$

SUBSTITUTION

1. A force of 32.5 N stretches a wire 0.470 cm. What force will stretch a similar piece of wire 2.39 cm?
2. A force of 7.33 N is applied to a spring whose spring constant is 0.298 N/cm. Find its change in length.
3. Each vertical steel column of an office building supports 42,100 lb and is compressed 0.0258 in. What is the spring constant of the steel? Find the compression if the weight were 51,700 lb.
4. A rectangular cast-iron column 16.0 cm × 16.0 cm × 4.50 m supports a weight of 7.95×10^6 N. Find the stress on the top of the column.
5. Find the weight density of a block of metal 7.00 in. × 6.50 in. × 8.00 in. that weighs 425 lb.
6. A cylindrical piece of aluminum is 4.25 cm tall and 1.95 cm in radius. How much does it weigh?
7. A piece of metal has a mass of 8.36 kg. If it displaces water that fills a container 9.34 cm × 10.0 cm × 10.0 cm, what is the mass density of the metal?
8. A block of wood is 27.7 in. × 36.3 in. × 12.4 in. and weighs 602 lb. Find its weight density.
9. Find the volume (in cm^3) of 759 g of mercury.
10. Find the volume (in m^3) of 1970 g of hydrogen at 0°C and 1 atm.
11. Find the mass of 1510 m^3 of oxygen at 0°C and 1 atm.
12. Find the weight of 951 ft^3 of water.
13. Find the weight density of a block of material 4.27 in. × 3.87 in. × 5.44 in. that weighs 0.982 lb.
14. Find the weight density of 2.00 quarts of liquid weighing 3.67 lb.
15. A quantity of liquid weighs 4.65 lb with a weight density of 39.8 lb/ft^3. What is its volume?

16. The density of a metal is 694 kg/m^3. Find its specific gravity.

17. A solid displaces 4.30 gal of water and has a weight density of 135 lb/ft^3. What is its weight?

18. Find the mass of a rectangular gold bar 4.00 cm × 6.00 cm × 20.00 cm. The mass density of gold is 19,300 kg/m^3.

19. Find the mass density of a chunk of rock using only a scale knowing the following information: mass of rock is 225 g; mass of water the rock displaces is 75.9 g.

20. The specific gravity of an unknown substance is 0.80. Will it float on or sink in gasoline?

APPLIED CONCEPTS

1. Instead of carrying a full-size spare tire, many automobiles are equipped with a significantly smaller spare tire with a warning that the car should not exceed a speed of 50 mi/h. (a) How much pressure does the smaller tire withstand if it must support one-fourth of a 1650-kg minivan and its area of contact with the road measures 8.26 cm × 15.9 cm? (b) Find the pressure on a full-size spare tire if its area of contact with the road measures 15.2 cm × 21.0 cm. (c) Why is it important to limit the speed of the smaller spare tire?

2. Observe the warped lines on asphalt pavement in front of a stoplight (Fig. 12.32). (a) What type of stress causes such curved lines? (b) Why are the warped lines, as seen in the photograph, more noticeable at stop signs and stoplights than on other sections of road?

Figure 12.32

3. Raul weighs 235 lb and is able to float in seawater but not in fresh water. Find his volume assuming that his density is the average density of seawater and fresh water.

4. A tanker truck with a cylindrical container 11.3 m long and 2.85 m in diameter transports several types of liquids. How much mass does it carry if it is transporting a full load of (a) water, (b) oil, and (c) gasoline?

5. Every morning Shakira weighs herself on a bathroom spring scale. Today, the scale reads 134 lb. (a) If the spring in the bathroom scale compresses 0.752 in., find its spring constant. (b) If her husband steps on the scale and the spring compresses 1.13 in., how much does he weigh? (c) Why is there a limit to how much weight a spring scale can measure?

CHAPTER 13

FLUIDS

Flow Rate

$Q = vA$

$A = \pi r^2$

Substances that flow are called fluids. Because liquids and gases behave in much the same manner, they will be presented together. From ships to airplanes to automobile brake systems to balloons, the utilization of fluids and their properties is important to technology and everyday life. We will now study hydraulics, hydrostatic pressure, buoyancy, and flow.

Objectives

The major goals of this chapter are to enable you to:

1. Describe the behavior of fluids.
2. Determine pressure using the hydraulic principle.
3. Distinguish between gauge pressure and absolute pressure.
4. Calculate buoyancy using Archimedes' principle.
5. Analyze fluid flow and Bernoulli's principle.

13.1 Hydrostatic Pressure

Since liquids and gases behave in much the same manner, they are often studied together as fluids. The gas and water piped to your home are fluids having several common characteristics.

Pressure is the force applied per unit area. **Hydrostatic pressure** is the pressure a liquid at rest exerts on a submerged object (Fig. 13.1). As you probably know, the pressure on an object increases as the water depth increases. Liquids differ from solids in that they exert force in all directions, whereas solids exert only a downward force due to gravity.

Figure 13.1 (a) The pressure on a submerged object is exerted by the liquid in all directions. The pressure at ocean-floor level is so great that very strong containers must be built to withstand the enormous pressures. (b) When a force is exerted on a liquid in a container, the force is exerted uniformly in all directions. Note that the liquid squirts out of the holes uniformly in all directions.

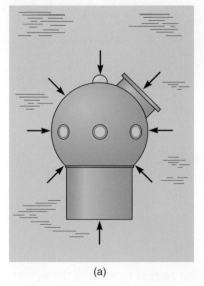
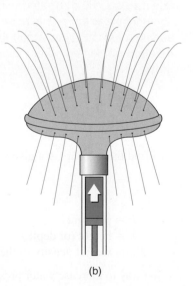

(a)	(b)

The pressure in a liquid depends *only* on the depth and weight density of the liquid and *not* on the surface area. Because the pressure exerted by water increases with depth, dams are built much thicker at the base than at the top (Fig. 13.2).

Figure 13.2 Dams must be built much thicker at the base because the pressure exerted by the water increases with depth.

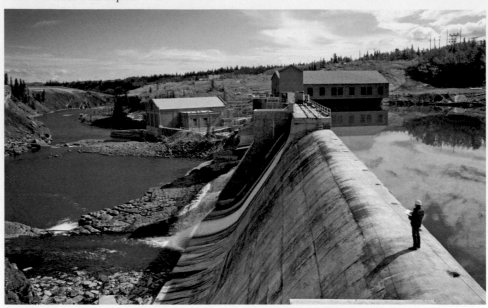

TRY THIS ACTIVITY

Water Pressure

The next time you go swimming in a pool, swim toward the bottom. As you swim deeper, notice the gradual change in pressure on your eardrums. Explain how the change in water pressure on your eardrums can be used to describe hydrostatic pressure. How would the pressure change if the pool were full of salt water?

This general pressure principle may also be illustrated with a pile of bricks. The weight of the pile and the pressure on the ground clearly increase as the pile is made taller. The force (weight) applied to the same area at the bottom of the pile increases the pressure at the bottom of the pile with the addition of each brick. Likewise, the pressure on a submerged object increases the farther it is pushed down due to the weight of the additional water above it.

To find the pressure at a given depth in a liquid, use the formula

$$P = hD_w$$

where P = pressure
h = height (or depth)
D_w = weight density of the liquid

One way to increase water pressure is to raise a water storage tank above the ground. This same formula applies except h is the sum of the depth of the water in the tank and the distance the bottom of the water tank is above the ground or some other reference such as a building. This is why community water storage tanks are placed on towers or on the tops of large hills.

TRY THIS ACTIVITY

Water Pressure and Depth

Figure 13.3

Punch three small holes near the top, the middle, and the bottom on the side of a large juice can with its lid removed. Next, fill the can with water and compare the streams of water spurting out each hole (Fig. 13.3). Pressure in a liquid also depends on the weight density of the liquid. Pressure on an object 50 m below the surface of fresh water is less than the pressure at the same depth in salt water because fresh water is less dense than salt water.

Find the pressure at the bottom of a water-filled drum 4.00 ft high.

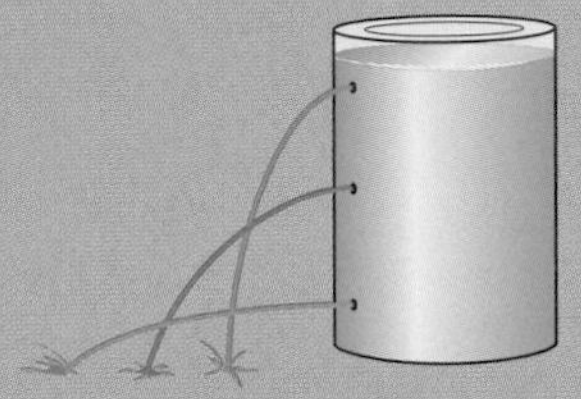

EXAMPLE 1

Sketch:

4.00 ft

Data:

$$h = 4.00 \text{ ft}$$
$$D_w = 62.4 \text{ lb/ft}^3$$
$$P = ?$$

Basic Equation:

$$P = hD_w$$

Working Equation: Same

Substitution:

$$P = (4.00 \text{ ft}) \left(62.4 \frac{\text{lb}}{\text{ft}^3}\right)$$
$$= 25\overline{0} \frac{\text{lb}}{\text{ft}^2} \times \left(\frac{1 \text{ ft}}{12 \text{ in.}}\right)^2$$
$$= 1.74 \text{ lb/in}^2$$

Note: The pressure depends only on the height, not the width or area of the container.

· · · · · · · · · · · · · · · · ·

Find the depth in a lake at which the pressure is 105 lb/in².

EXAMPLE 2

Data:

$$P = 105 \text{ lb/in}^2$$
$$D_w = 62.4 \text{ lb/ft}^3$$
$$h = ?$$

Basic Equation:

$$P = hD_w$$

Working Equation:

$$h = \frac{P}{D_w}$$

Substitution:

$$h = \frac{105 \text{ lb/in}^2}{62.4 \text{ lb/ft}^3}$$

$$= 1.68 \frac{\text{ft}^3}{\text{in}^2}$$

$$= 1.68 \frac{\cancel{\text{ft}^3}}{\text{in}^2} \times \left(\frac{12 \text{ in.}}{1 \cancel{\text{ft}}}\right)^2 \qquad \frac{\text{ft}}{\;}$$

$$= 242 \text{ ft}$$

$$\boxed{\frac{\text{lb/in}^2}{\text{lb/ft}^3} = \frac{\text{lb}}{\text{in}^2} \div \frac{\text{lb}}{\text{ft}^3} = \frac{\cancel{\text{lb}}}{\text{in}^2} \times \frac{\text{ft}^3}{\cancel{\text{lb}}} = \frac{\text{ft}^3}{\text{in}^2}}$$

· · · · · · · · · · · · · · · ·

EXAMPLE 3

Find the height of a water column where the pressure at the bottom of the column is $40\overline{0}$ kPa and the weight density of water is $980\overline{0}$ N/m³.

Data:

$$P = 40\overline{0} \text{ kPa}$$
$$D_w = 980\overline{0} \text{ N/m}^3$$
$$h = ?$$

Basic Equation:

$$P = hD_w$$

Working Equation:

$$h = \frac{P}{D_w}$$

Substitution:

$$h = \frac{40\overline{0} \text{ kPa}}{980\overline{0} \text{ N/m}^3}$$

$$= \frac{40\overline{0} \text{ kPa}}{980\overline{0} \text{ N/m}^3} \times \frac{10^3 \text{ N/m}^2}{1 \text{ kPa}} \qquad (\textit{Recall:} \ 1 \text{ kPa} = 10^3 \text{ N/m}^2)$$

$$= 40.8 \text{ m}$$

$$\boxed{\frac{\cancel{\text{kPa}}}{\text{N/m}^3} \times \frac{\text{N/m}^2}{\cancel{\text{kPa}}} = \text{N/m}^2 \div \text{N/m}^3 = \frac{\cancel{\text{N}}}{\text{m}^2} \times \frac{\text{m}^3}{\cancel{\text{N}}} = \text{m}}$$

· · · · · · · · · · · · · · · ·

Total Force Exerted by Liquids

The *total force* exerted by a liquid on a horizontal surface (such as the bottom of a barrel) depends on the area of the surface, the depth of the liquid, and the weight density of the liquid. By formula,

$$\boxed{F_t = AhD_w}$$

where F_t = total force
  A = area of bottom or horizontal surface
  h = height or depth of the liquid
  D_w = weight density

Find the total force on the bottom of a rectangular tank 10.0 ft by 5.00 ft by 4.00 ft deep filled with water.

EXAMPLE 4

Sketch:

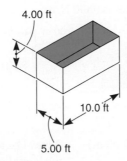

4.00 ft

10.0 ft

5.00 ft

Data:

$$A = lw = (10.0 \text{ ft})(5.00 \text{ ft}) = 50.0 \text{ ft}^2$$
$$h = 4.00 \text{ ft}$$
$$D_w = 62.4 \text{ lb/ft}^3$$
$$F_t = ?$$

Basic Equation:

$$F_t = AhD_w$$

Working Equation: Same

Substitution:

$$F_t = (50.0 \text{ ft}^2)(4.00 \text{ ft})\left(62.4\frac{\text{lb}}{\text{ft}^3}\right)$$
$$= 12,500 \text{ lb}$$

. . . .

The total force on a vertical surface F_s (such as the *side* of a tank) is found by using half the vertical height (average height):

$$\boxed{F_s = \tfrac{1}{2}AhD_w}$$

where A is the area of the side or vertical surface.

Find the total force on the small side of the rectangular tank in Example 4.

EXAMPLE 5

Data:

$$A = lw = (5.00 \text{ ft})(4.00 \text{ ft}) = 20.0 \text{ ft}^2$$
$$h = 4.00 \text{ ft}$$
$$D_w = 62.4 \text{ lb/ft}^3$$
$$F_s = ?$$

Basic Equation:

$$F_s = \tfrac{1}{2}AhD_w$$

Working Equation: Same

Substitution:

$$F_s = \tfrac{1}{2}(20.0 \text{ ft}^2)(4.00 \text{ ft})(62.4 \text{ lb/ft}^3)$$
$$= 25\overline{0}0 \text{ lb}$$

....

EXAMPLE 6

Find the total force on the side of a water-filled cylindrical tank 3.00 m high with radius 5.00 m. The weight density of water is $98\overline{0}0 \text{ N/m}^3$.

Sketch:

5.00 m

3.00 m

Data:

$$A = 2\pi rh = 2\pi(5.00 \text{ m})(3.00 \text{ m})$$

Note: The area of the vertical surface is the lateral surface of the cylinder.

$$h = 3.00 \text{ m}$$
$$D_w = 98\overline{0}0 \text{ N/m}^3$$
$$F_s = ?$$

Basic Equation:

$$F_s = \tfrac{1}{2}AhD_w$$

Working Equation: Same

Substitution:

$$F_s = \tfrac{1}{2}[2\pi(5.00 \text{ m})(3.00 \text{ m})](3.00 \text{ m})(98\overline{0}0 \text{ N/m}^3)$$
$$= 1.39 \times 10^6 \text{ N}$$

....

PROBLEMS 13.1

1. Find the pressure (in lb/in²) at the bottom of a tower with water 50.0 ft deep.
2. Find the height of a column of water where the pressure at the bottom of the column is 20.0 lb/in².
3. Find the density of a liquid that exerts a pressure of 0.400 lb/in² at a depth of 42.0 in.
4. (a) Find the total force on the bottom of a water-filled circular cattle tank 0.750 m high with radius 1.30 m where the weight density of water is $98\overline{0}0 \text{ N/m}^3$. (b) Find the total force on the side of the tank.
5. What must the water pressure be to supply water to the third floor of a building (35.0 ft up) with a pressure of 40.0 lb/in² at that level?
6. A small rectangular tank 5.00 in. by 9.00 in. is filled with mercury. (a) If the total force on the bottom of the tank is 165 lb, how deep is the mercury? (Weight density of mercury = 0.490 lb/in³.) (b) Find the total force on the larger side of the tank.
7. Find the water pressure (in kPa) at the 25.0-m level of a water tower containing water 50.0 m deep.
8. Find the height of a column of water where the pressure at the bottom is 115 kPa.
9. What is the height of a column of water if the pressure at the bottom of the column is 95.0 kPa?
10. What is the mass density of a liquid that exerts a pressure of $18\overline{0}$ kPa at a depth of 30.0 m?
11. What is the mass density of a liquid that exerts a pressure of 178 kPa at a depth of 24.0 m?

12. (a) Find the total force on the bottom of a cylindrical gasoline storage tank 15.0 m high with radius 23.0 m. (b) Find the total force on the side of the tank.

13. What must the water pressure be to supply the second floor (18.0 ft up) with a pressure of 50.0 lb/in^2 at that level?

14. Find the water pressure at ground level to supply water to the third floor of a building 8.00 m high with a pressure of 325 kPa at the third-floor level.

15. What pressure must a pump supply to pump water up to the thirtieth floor of a skyscraper with a pressure of 25 lb/in^2? Assume that the pump is located on the first floor and that there are 16.0 ft between floors.

16. A submarine is submerged to a depth of 3550 m in the Pacific Ocean. What air pressure (in kPa) is needed to blow water out of the ballast tanks?

A filled water tower sits on the top of the highest hill in a town (use Fig. 13.4 for Problems 17–21). The cylindrical tower has a radius of 12.0 m and a height of 50.0 m.

Figure 13.4

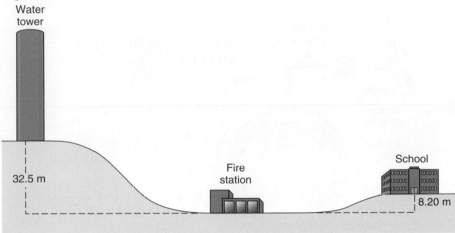

17. Find the total force on the bottom of the water tower.
18. Find the total force on the sides of the water tower.
19. Find the pressure (in kPa) on the bottom of the water tower.
20. What is the water pressure (in kPa) at the fire station?
21. What is the water pressure (in kPa) at the school?
22. A cylindrical grain bin 24.0 ft in diameter is filled with corn whose weight density is 45.1 lb/ft^3. How tall can the bin be for the floor to support 94.0 lb/in^2 of pressure?

13.2 Hydraulic Principle

Put a stopper in one end of a metal pipe. Fill it with a fluid such as water. Then put a second stopper into the open end. Put the pipe in a horizontal position as in Fig. 13.5 and push on stopper A. What happens? Stopper B is pushed out. This illustrates a basic principle of hydraulics: The liquid in the pipe transmits the pressure from one stopper or piston to another without measurable loss.

Figure 13.5   Pressure on a confined liquid is transmitted in all directions without measurable loss.

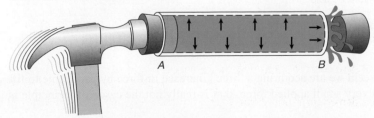

HYDRAULIC PRINCIPLE (PASCAL'S PRINCIPLE)

The pressure applied to a confined liquid is transmitted without measurable loss throughout the entire liquid to all inner surfaces of the container.

The rear-wheel hydraulic brake system of a front-wheel-drive automobile (Fig. 13.6) is an application of Pascal's principle. When the driver pushes the brake pedal, the pressure on the piston in the master cylinder is transmitted through the brake fluid to the two pistons in the brake cylinder. This transmitted pressure then forces the brake-cylinder pistons to push the brake shoes against the brake drum and stop the automobile. Releasing the brake pedal releases the pressure on the pistons in the brake cylinder. The spring pulls the brake shoes away from the brake drum, which allows the wheels to turn freely again.

Figure 13.6 Hydraulic brake system of an automobile.

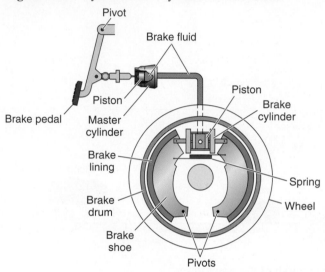

The hydraulic jack, lift, and press are applications of hydraulics being used as a simple machine to multiply force. If we apply a force to the small piston of the hydraulic lift in Fig. 13.7, the pressure is transmitted without measurable loss in all directions. The reason for this is the virtual noncompressibility of liquids. The *pressure* on the large piston is the same as the pressure on the small piston; however, the *total force* on the large piston is greater because of its larger surface area.

Figure 13.7 Hydraulic lift.

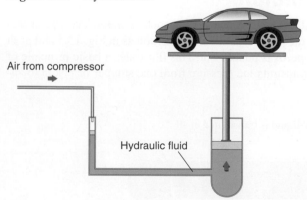

Although it may seem we are acquiring a "free" increase in force by being able to lift an automobile using a very small applied force, that is really not the case. The principle is similar to the mechanical advantage of simple machines; we can multiply force but at the expense of distance. The lift, although able to move a large weight, can only move it a relatively short distance.

From the diagram of the hydraulic jack in Fig. 13.8, find

EXAMPLE 1

(a)   the pressure on the small piston.
(b)   the pressure on the large piston.
(c)   the total force on the large piston.
(d)   the mechanical advantage of the jack.

Figure 13.8   Hydraulic jack.

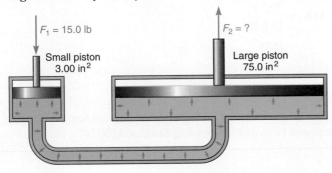

Data:

$$F_1 = 15.0 \text{ lb}$$
$$A_{\text{small piston}} = A_1 = 3.00 \text{ in}^2$$
$$A_{\text{large piston}} = A_2 = 75.0 \text{ in}^2$$
$$P_1 = ?$$
$$P_2 = ?$$
$$F_2 = ?$$
$$\text{MA} = ?$$

(a) Basic Equation:

$$P_1 = \frac{F_1}{A_1}$$

Working Equation: Same

Substitution:

$$P_1 = \frac{15.0 \text{ lb}}{3.00 \text{ in}^2}$$
$$= 5.00 \text{ lb/in}^2$$

(b) Applying Pascal's Principle:

$$P_2 = P_1 = 5.00 \text{ lb/in}^2$$

(c) Basic Equation:

$$P_2 = \frac{F_2}{A_2}$$

Working Equation:

$$F_2 = P_2 A_2$$

Substitution:

$$F_2 = \left(5.00 \, \frac{\text{lb}}{\text{in}^2}\right)(75.0 \, \text{in}^2)$$

$$= 375 \, \text{lb}$$

(d) Basic Equation:

$$\text{MA} = \frac{F_R}{F_E}$$

Working Equation: Same

Substitution:

$$\text{MA} = \frac{375 \, \text{lb}}{15.0 \, \text{lb}}$$

$$= 25.0$$

·················

EXAMPLE 2

The small piston of a hydraulic press has an area of 10.0 cm². If the applied force is 50.0 N, what must the area of the large piston be to exert a pressing force of 4800 N?

Data:

$$A_1 = 10.0 \, \text{cm}^2$$
$$F_1 = 50.0 \, \text{N}$$
$$F_2 = 4800 \, \text{N}$$
$$A_2 = \, ?$$

Basic Equations:

$$P_1 = \frac{F_1}{A_1}, \quad P_2 = \frac{F_2}{A_2}, \quad \text{and since } P_1 = P_2, \quad \frac{F_1}{A_1} = \frac{F_2}{A_2}$$

Working Equation:

$$A_2 = \frac{A_1 F_2}{F_1}$$

Substitution:

$$A_2 = \frac{(10.0 \, \text{cm}^2)(4800 \, \text{N})}{50.0 \, \text{N}}$$

$$= 960 \, \text{cm}^2$$

·················

SKETCH

12 cm² | w

4.0 cm

DATA

A = 12 cm², l = 4.0 cm, w = ?

BASIC EQUATION

A = lw

WORKING EQUATION

$w = \frac{A}{l}$

SUBSTITUTION

$w = \frac{12 \, \text{cm}^2}{4.0 \, \text{cm}} = 3.0 \, \text{cm}$

PROBLEMS 13.2

1. The area of the small piston in a hydraulic jack is 0.750 in². The area of the large piston is 3.00 in². If a force of 15.0 lb is applied to the small piston, what weight can be lifted by the large one?

2. The mechanical advantage of a hydraulic press is 25. What applied force is necessary to produce a pressing force of 2400 N?

3. Find the mechanical advantage of a hydraulic press that produces a pressing force of 8250 N when the applied force is 375 N.

4. The mechanical advantage of a hydraulic press is 18. What applied force is necessary to produce a pressing force of 990 lb?

5. Find the mechanical advantage of a hydraulic press that produces a pressing force of 1320 N when the applied force is 55.0 N.

6. The small piston of a hydraulic press has an area of 8.00 cm^2. If the applied force is 25.0 N, find the area of the large piston to exert a pressing force of 3600 N.

7. The MA of a hydraulic jack is 250. What force must be applied to lift an automobile weighing 12,000 N?

8. The small piston of a hydraulic press has an area of 4.00 in^2. If the applied force is 10.0 lb, what must the area of the large piston be to exert a pressing force of 865 lb?

9. The MA of a hydraulic jack is 420. Find the weight of the heaviest automobile that can be lifted by an applied force of 55 N.

10. The mechanical advantage of a hydraulic jack is 450. Find the weight of the heaviest automobile that can be lifted by an applied force of 60.0 N.

11. The pistons of a hydraulic press have radii of 2.00 cm and 12.0 cm, respectively.
    (a) What force must be applied to the smaller piston to exert a force of 5250 N on the larger?
    (b) What is the pressure (in N/cm^2) on each piston?
    (c) What is the mechanical advantage of the press?

12. The small circular piston of a hydraulic press has an area of 8.00 cm^2. If the applied force is 25.0 N, what must the area of the large piston be to exert a pressing force of 3650 N?

13. The large piston on a hydraulic lift has radius 40.0 cm. The small piston has radius 5.00 cm to which a force of 75.0 N is applied.
    (a) Find the force exerted by the large piston.
    (b) Find the pressure on the large piston.
    (c) Find the pressure on the small piston.
    (d) Find the mechanical advantage of the lift.
    (e) What happens when the area of the small piston is half as large?
    (f) What happens when the radius of the small piston is half as large?

14. In a hydraulic system a 20.0-N force is applied to the small piston with cross-sectional area 25.0 cm^2. What weight can be lifted by the large piston with cross-sectional area 50.0 cm^2?

15. If the diameter of the larger piston in Problem 14 is doubled, how is the weight able to be lifted changed?

16. If a dentist's chair weighs 1600 N and is raised by a large piston with cross-sectional area of 75.0 cm^2, what force must be exerted on a small piston of cross-sectional area 3.75 cm^2 to lift the chair?

17. A hydraulic jack whose piston has a cross-sectional area of 115 cm^2 supports a pickup truck weighing 1.20×10^4 N. Compressed air is used to apply a force on the second piston with cross-sectional area 25.0 cm^2. How large must this force be to support the truck?

18. Compressed air in a car lift applies a force to a piston with radius 5.00 cm. This pressure is transmitted through a hydraulic system to a second piston with radius 15.0 cm. (a) How much force must the compressed air exert to lift a vehicle weighing 1.33×10^4 N? (b) What pressure produces the lift?

19. The small piston of an automobile lift has an area of 12.0 cm^2. If the applied force on the lift is 75.0 N, find the area of the large piston needed to exert a lifting force of 9800 N.

20. If the lifting force of a hydraulic truck jack is 19,600 N and the ratio of the area of the large piston to the small piston is 150 to 1, what is the applied force on the jack?

13.3 Air Pressure

Since air has weight, as does any fluid, it exerts pressure. The atmosphere exerts pressure on objects on the surface of the earth. This atmospheric pressure can be illustrated by using a bell jar with a hole in the top over which a thin rubber membrane is stretched as in Fig. 13.9(a). As

Figure 13.9   Effects of air pressure.

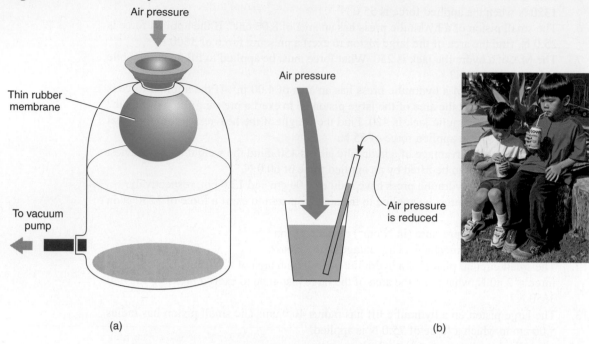

(a)   (b)

Figure 13.10   The barometer measures air pressure.

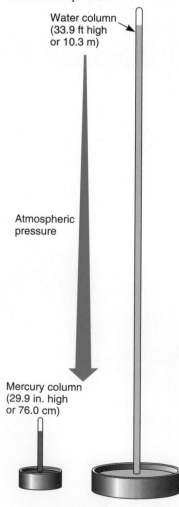

air is pumped out of the bell jar, the inside air pressure is reduced by removing a number of air molecules. Thus, there are fewer molecular bombardments on the inside surface of the rubber membrane than there are on its outside surface. The outside air pressure, now greater than the inside air pressure, pushes the rubber membrane down into the bell jar. When a straw is used to drink, the air pressure inside the straw is reduced [Fig. 13.9(b)]. As a result, the outside air pressure is higher than the pressure in the straw, which forces the fluid up the straw.

In Section 13.1 we saw that the pressure on a submerged body increases as the body goes deeper into the liquid. Some creatures live near the bottom of the ocean, where the pressure of the water is so great that it would collapse any human body and most submarines, but through the process of evolution, such creatures have adapted to this tremendous pressure. Similarly, we on the earth live at the bottom of a fluid, air, that is several miles deep. The pressure from this fluid is normally 14.7 lb/in² or 101.32 kPa at sea level. We do not feel this pressure because it normally is almost the same from all directions and also because living bodies maintain an internal pressure that balances the external pressure. **Atmospheric pressure** is the pressure caused by the weight of the air in the atmosphere. Air pressure and the amount of air decrease with altitude. Mountain climbers often must use oxygen tanks to help them breathe at high altitudes. Aircraft must have pressurized cabins for passengers to be able to breathe and function. When the air pressure becomes unequal, its force becomes quite evident in the form of wind. This wind may be a cool summer breeze or the tremendous concentrated force of a tornado.

What is the *pressure of our atmosphere* equivalent to? Experiments have shown that the atmosphere supports a column of water 33.9 ft or 10.3 m high in a tube from which the air has been removed. The atmosphere supports 29.9 in. or 76.0 cm of mercury in a similar tube (Fig. 13.10). This is not surprising; mercury is 13.6 times as dense as water and

$$\frac{1}{13.6} \times 33.9 \text{ ft} \times \frac{12 \text{ in.}}{1 \text{ ft}} = 29.9 \text{ in.}$$

The height of the mercury column in a barometer is independent of the width (or diameter or cross-sectional area) of the barometer tube. This "inches of mercury" measurement has been standard for many years on TV weather programs but is increasingly being replaced by the metric standard measurement in kilopascals (kPa).

The pressure of the atmosphere can be expressed in terms of the pressure of an equivalent column of mercury. Air pressure at sea level is normally 29.9 in. or 76.0 cm or 760 mm of mercury. How do we arrive at the 14.7 lb/in² measurement? In the case of mercury, the height of the column is 29.9 in. or 2.49 ft. Its density is 13.6×62.4 lb/ft³ or 849 lb/ft³. Therefore,

$$P = hD_w$$

$$P = 2.49 \text{ ft} \times 849 \frac{\text{lb}}{\text{ft}^3} \times \left(\frac{1 \text{ ft}}{12 \text{ in.}}\right)^2$$

$$= 14.7 \text{ lb/in}^2 \text{ at standard temperature}$$

A pressure of 2 atm is equivalent to 29.4 lb/in² or 202.64 kPa. If the pressure is $\frac{1}{2}$ atm at a given point in the atmosphere, it is 7.35 lb/in² or 50.66 kPa.

An interesting demonstration illustrates atmospheric pressure by using it to collapse a metal can. A can that can be tightly capped is placed on a burner with its cap off. A small amount of water is placed in the can and heated until it boils. The cap is then secured, and the can is removed from the heat source. The steam has pushed air from the can, and any remaining air has expanded from being heated. When the sealed can cools and the steam inside condenses, the pressure in the can is reduced. The greater pressure of the atmosphere outside the can crushes the can (Fig. 13.11). The same concept has been used for generations to seal jars of food in home canning, but the jars are strong enough to not collapse.

Different types of gauges read either atmospheric or gauge pressure. When we purchase bottled gas, the amount of gas and its density vary with the pressure. If the pressure is low, the amount of gas in the bottle is low. If the pressure is high, the bottle is "nearly full." The *gauge* that is usually used for checking the pressure in bottles and tires shows a reading of zero at normal atmospheric pressure. The pressure of the atmosphere is not included in this reading. Thus, **gauge pressure** is the amount of air pressure excluding the normal atmospheric pressure, as when one uses a tire gauge. The actual pressure, called **absolute pressure**, is the gauge pressure reading plus the normal atmospheric pressure, 101.32 kPa or 14.7 lb/in², as when one uses an atmospheric barometer. That is,

$$\text{absolute pressure} = \text{gauge pressure} + \text{atmospheric pressure}$$

or

$$\boxed{P_{abs} = P_{ga} + P_{atm}}$$

where $P_{atm} = 101.32$ kPa or 14.7 lb/in² at standard temperature.

Note: Atmospheric pressure has not been included in our previous pressure calculations.

Figure 13.11 This can was crushed by atmospheric pressure.

Courtesy of Peter Arnold, Inc. Reprinted with permission

EXAMPLE

What is the absolute pressure in a tire inflated to 32.0 lb/in²

(a) in lb/in²?
(b) in kPa?

(a) Data:

$$P_{ga} = 32.0 \text{ lb/in}^2$$
$$P_{atm} = 14.7 \text{ lb/in}^2$$
$$P_{abs} = ?$$

Basic Equation:

$$P_{abs} = P_{ga} + P_{atm}$$

Working Equation: Same

Substitution:

$$P_{abs} = 32.0 \text{ lb/in}^2 + 14.7 \text{ lb/in}^2$$
$$= 46.7 \text{ lb/in}^2$$

(b) We use the conversion factor:

$$101.32 \text{ kPa} = 14.7 \text{ lb/in}^2$$

Therefore,

$$P_{abs} = 46.7 \text{ lb/in}^2 \times \frac{101.32 \text{ kPa}}{14.7 \text{ lb/in}^2}$$
$$= 322 \text{ kPa}$$

· · · · · · · · · · · · · · · · ·

PROBLEMS 13.3

PROBLEM SOLVING

SKETCH

12 cm² | w

4.0 cm

DATA

A = 12 cm², l = 4.0 cm, w = ?

BASIC EQUATION

A = lw

WORKING EQUATION

w = A/l

SUBSTITUTION

w = 12 cm²/4.0 cm = 3.0 cm

1. Change 815 kPa to lb/in².
2. Change 64.3 lb/in² to kPa.
3. Change 42.5 lb/in² to kPa.
4. Change 215 kPa to lb/in².
5. Find the pressure of
   (a) 3 atm (in kPa).
   (b) 2 atm (in kPa).
   (c) 6 atm (in lb/in²).
   (d) 5 atm (in kPa).
   (e) $\frac{1}{3}$ atm (in kPa).
   (f) $\frac{1}{4}$ atm (in kPa).
6. A barometer in the Rocky Mountains reads 516 mm of mercury. Find this pressure (a) in kPa and (b) in lb/in².
7. Find the absolute pressure in a bicycle tire with a gauge pressure of 485 kPa.
8. Find the absolute pressure of a motorcycle tire with a gauge pressure of 255 kPa.
9. Find the gauge pressure of a tire with an absolute pressure of 45.0 lb/in².
10. Find the gauge pressure of a tire with an absolute pressure of 425 kPa.
11. Find the absolute pressure of a tire gauge that reads 205 kPa.
12. Find the absolute pressure of a tank whose gauge pressure reads 362 lb/in².
13. Find the gauge pressure of a tank whose absolute pressure is 1275 kPa.
14. Find the gauge pressure of a tank whose absolute pressure is 218 lb/in².
15. Find the absolute pressure of a cycle tire with gauge pressure 3.00×10^5 Pa.
16. Find the absolute pressure in a hydraulic jack with a small piston of area 23.0 cm² when a force of 125 N is applied. The area of the large piston is 46.0 cm².

13.4 Buoyancy

Archimedes was one of the first to study fluids and formulated what is now called **Archimedes' principle**.

ARCHIMEDES' PRINCIPLE
Any object placed in a fluid apparently loses weight equal to the weight of the displaced fluid.

In Fig. 13.12, the toy boat floats on water. Note that the weight of the boat equals the weight of the water it displaces. Three people in a boat displace more water than only one person in the boat. The boat rides lower due to the increased weight. Similarly, a loaded ship displaces more water than an empty ship and rides lower due to its increased weight (Fig. 13.13).

What happens to the weight of a brick or some other object that sinks in water? First, weigh the brick in air. Then, lower the brick under the water and weigh it again. It weighs less. The difference between the two weights is the buoyant (upward) force of the water. That is, the **buoyant force** is the upward force exerted on a submerged or partially submerged object. This is illustrated in the following example.

Figure 13.12 Weight of object = weight of displaced water.

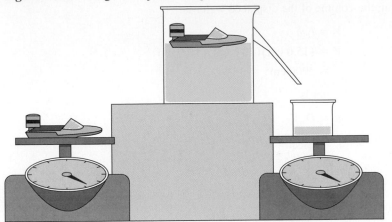

Archimedes (287 B.C.–212 B.C.),

Greek mathematician, was born in Syracuse. He is remembered for the construction of siege-engines against the Romans, Archimedes' screw (still used for raising water), discovering the principle for the buoyant force of a floating body, and founding the science of hydrostatics. In mathematics, he discovered the formulas for the areas and volumes of spheres, cylinders, parabolas, and other plane and solid figures.

Figure 13.13

(a) This loaded barge rides low in the water because it displaces an amount of water equal to its weight.

(b) This empty oil tanker rides high in the water. Note the mark on its side of its normal water level when it is full.

EXAMPLE 1

A solid concrete block 15.0 cm × 20.0 cm × 10.0 cm weighs 67.6 N in air (Fig. 13.14). When lowered into water, it weighs 38.2 N. Find the buoyant force. Using the scale readings,

Figure 13.14 Weight in water = weight in air − buoyant force (weight of displaced water).

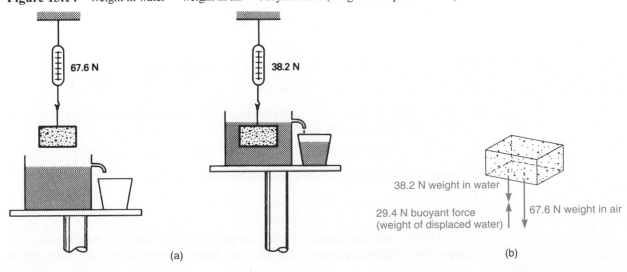

67.6 N

38.2 N

(a)

38.2 N weight in water

29.4 N buoyant force (weight of displaced water)

67.6 N weight in air

(b)

we see that the buoyant force is 67.6 N − 38.2 N = 29.4 N.
By calculation, the volume of the displaced water is

$$V = lwh$$
$$V = (15.0 \text{ cm})(20.0 \text{ cm})(10.0 \text{ cm})$$
$$= 30\overline{0}0 \text{ cm}^3 = 3.00 \times 10^{-3} \text{ m}^3$$

The mass of the displaced water is

$$m = D_m V$$
$$m = (100\overline{0} \text{ kg/m}^3)(3.00 \times 10^{-3} \text{ m}^3)$$
$$= 3.00 \text{ kg}$$

The weight of the displaced water is then

$$F_w = mg$$
$$F_w = (3.00 \text{ kg})(9.80 \text{ m/s}^2)$$
$$= 29.4 \text{ N} \qquad (1 \text{ N} = 1 \text{ kg m/s}^2)$$

which equals the buoyant force.

· · · · · · · · · · · · · · · · ·

A submarine uses these same principles to dive and rise in the ocean. A submarine has ballast tanks that are filled with seawater, making it heavier than the volume of water it displaces, and so it descends. For the submarine to rise, the water in the ballast tanks is blown out by compressed air. The submarine then becomes lighter than the volume of water it displaces and so it rises.

Primitive peoples made their boats of wood, which has densities less than water. How can a barge made of steel, which is approximately eight times as dense as water, float? The barge must be built so that it displaces at least its own weight plus its cargo load equivalent of water. The wider the barge and the deeper it is immersed, the more water is displaced and the greater is the buoyant force exerted on the barge by the water.

Imagine a loaded fishing boat coming in with the day's catch and just barely clearing a bridge on its way to the dock. After it is emptied of its cargo and heads back to sea, will it have more or less clearance under the bridge? Use Fig. 13.13 to explain your answer.

EXAMPLE 2

A rectangular boat is 4.00 m wide, 8.00 m long, and 3.00 m deep.

(a) How many m³ of water will it displace if the top stays 1.00 m above the water?
(b) What load (in newtons) will the boat contain under these conditions if the empty boat weighs 8.60 × 10⁴ N in dry dock?

(a) The volume of water displaced by the boat is

$$V = lwh$$
$$V = (8.00 \text{ m})(4.00 \text{ m})(2.00 \text{ m})$$
$$= 64.0 \text{ m}^3$$

(b) The load of the boat is the buoyant force of the displaced water ($D_w V$) minus the weight of the boat in dry dock.

$$(98\overline{0}0 \text{ N/m}^3)(64.0 \text{ m}^3) − (8.60 \times 10^4 \text{ N}) = 5.41 \times 10^5 \text{ N}$$

Note: The weight density of water is $98\overline{0}0 \text{ N/m}^3$.

· · · · · · · · · · · · · · · · ·

Archimedes' principle applies to gases as well as liquids. Lighter-than-air craft (such as the Goodyear blimps) operate on this principle. Since they are filled with helium, which is lighter than air, the buoyant force on them causes them to be supported by the air. Being "submerged" in the air, a blimp is buoyed up by the weight of the air it displaces, which equals the buoyant force of the air on the balloon.

PROBLEMS 13.4

1. A metal alloy weighs 81.0 lb in air and 68.0 lb when under water. Find the buoyant force of the water.
2. A piece of metal weighs 67.0 N in air and 62.0 N in water. Find the buoyant force of the water.
3. A rock weighs 25.7 N in air and 21.8 N in water. What is the buoyant force of the water?
4. A metal bar weighs 455 N in air and 437 N in water. What is the buoyant force of the water?
5. A rock displaces 1.21 ft^3 of water. What is the buoyant force of the water?
6. A metal displaces 16.8 m^3 of water. Find the buoyant force of the water.
7. A metal casting displaces 327 cm^3 of water. Find the buoyant force of the water.
8. A piece of metal displaces 657 cm^3 of water. Find the buoyant force of the water.
9. A metal casting displaces 2.12 ft^3 of alcohol. Find the buoyant force of the alcohol.
10. A metal cylinder displaces 515 cm^3 of gasoline. Find the buoyant force of the gasoline.
11. A 75.0-kg rock lies at the bottom of a pond. Its volume is 3.10×10^4 cm^3. How much force is needed to lift the rock?
12. A 125-lb rock lies at the bottom of a pond. Its volume is 0.800 ft^3. How much force is needed to lift the rock?
13. A flat-bottom river barge is 30.0 ft wide, 85.0 ft long, and 15.0 ft deep.
    (a) How many ft^3 of water will it displace while the top stays 3.00 ft above the water?
    (b) What load in tons will the barge contain under these conditions if the empty barge weighs $16\overline{0}$ tons in dry dock?
14. A flat-bottom river barge is 12.0 m wide, 30.0 m long, and 6.00 m deep.
    (a) How many m^3 of water will it displace while the top stays 1.00 m above the water?
    (b) What load (in newtons) will the barge contain under these conditions if the empty barge weighs 3.55×10^6 N in dry dock?
15. What is the volume (in m^3) of the water displaced by a submerged air tank that is acted on by a buoyant force of 7.50×10^4 N?
16. A lifeguard swims with her head just above the water. What is the volume of the submerged part of her body if she weighs $60\overline{0}$ N?
17. An underwater camera weighing 1250 N in air is submerged and supported by a tether line. If the volume of the camera is 8.30×10^{-2} m^3, what is the tension in the line?

SKETCH

12 cm^2 w

4.0 cm

DATA
A = 12 cm^2, l = 4.0 cm, w = ?

BASIC EQUATION
A = lw

WORKING EQUATION
$w = \frac{A}{l}$

SUBSTITUTION
$w = \frac{12 \text{ cm}^2}{4.0 \text{ cm}} = 3.0$ cm

13.5 Fluid Flow

Think about the motion of water flowing down a fast-moving mountain stream that contains boulders and rapids and about the motion of the air during a thunderstorm or during a tornado. These types of motion are complex, indeed. Our discussion will focus on the simpler examples of fluid flow.

Streamline flow, also known as *laminar flow*, is the smooth flow of a fluid through a tube (Fig. 13.15). By smooth flow we mean that all particles of the fluid follow the same uniform path. **Turbulent flow**, also known as *nonlaminar flow*, is the erratic, unpredictable flow of a fluid resulting from excessive speed of the flow or sudden changes in direction or size of the tube or pipe.

The **flow rate** of a fluid is the volume of fluid flowing past a given point in a pipe per unit time. Assume that we have a streamline flow through a straight section of pipe at speed v. During a time interval of t, each particle of fluid travels a distance vt. If A is the cross-sectional area of the pipe, the volume of fluid passing a given point during the (a)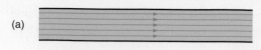

(b)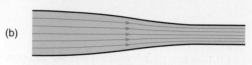

(c)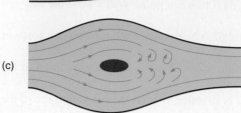

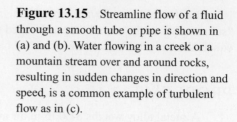

Figure 13.15 Streamline flow of a fluid through a smooth tube or pipe is shown in (a) and (b). Water flowing in a creek or a mountain stream over and around rocks, resulting in sudden changes in direction and speed, is a common example of turbulent flow as in (c).

time interval t is vtA. Thus, the flow rate, Q, is given by

$$Q = \frac{vtA}{t}$$

or

$$\boxed{Q = vA}$$

where Q = flow rate
v = speed of the fluid through the tube or pipe
A = cross-sectional area of the tube or pipe

EXAMPLE

Water flows through a fire hose of diameter 6.40 cm at a speed of 5.90 m/s. Find the flow rate of the fire hose in L/min.

Data:

$$v = 5.90 \text{ m/s}$$
$$r = 3.20 \text{ cm} = 0.0320 \text{ m}$$
$$A = \pi r^2 = \pi(0.0320 \text{ m})^2$$
$$Q = ?$$

Basic Equation:

$$Q = vA$$

Working Equation: Same

Substitution:

$$Q = \left(5.90 \frac{\text{m}}{\text{s}}\right) \pi(0.0320 \text{ m})^2 \times \frac{10^3 \text{ L}}{1 \text{ m}^3} \times \frac{60 \text{ s}}{1 \text{ min}}$$

$$= 1140 \text{ L/min}$$

· · · · · · · · · · · · · · · · ·

The overall volume rate of flow of water in a stream or creek is relatively constant; that is, the volume of water passing through the various sections is constant. Thus, the speed of the water increases within sections where the stream is narrow and decreases within sections where the stream is broad.

For an incompressible fluid, the flow rate is constant throughout the pipe. If the cross-sectional area of the pipe changes and streamline flow is maintained, the flow rate is the same all along the pipe. That is, as the cross-sectional area increases, the velocity decreases, and vice versa (Fig. 13.16).

Figure 13.16 For an incompressible fluid, the flow rate is constant throughout.

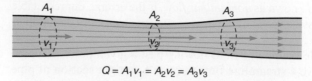

$$Q = A_1 v_1 = A_2 v_2 = A_3 v_3$$

TRY THIS ACTIVITY

The Venturi Hose

Turn on the water to a garden hose that does not have a nozzle. Note the speed of the water as it flows out of the hose. Without turning the faucet, pinch the hose just behind the metal coupling allowing only a small space for the water to flow. Describe what happens to the speed of the water as it now comes out of the hose. Explain why the speed changes although the faucet remained the same.

What happens to the pressure as the cross-sectional area of the pipe changes? This concept can be illustrated by use of a Venturi meter, named after **Giovanni Battista Venturi** (Fig. 13.17). Here the vertical tubes act like pressure gauges; the higher the column, the higher the pressure. As you can see, the higher the speed, the lower the pressure, and vice versa. This change in pressure of a fluid in streamline flow was first explained by **Daniel Bernoulli**.

Figure 13.17 A Venturi meter shows that the higher the speed of a fluid through a tube, the lower the pressure; the lower the speed of a fluid, the higher the pressure.

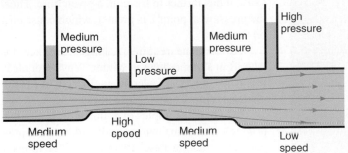

Giovanni Battista Venturi (1746–1822),

physicist, was born in Italy. He worked on the flow of fluids and is remembered for his discovery of the Venturi effect, the decrease of the pressure in a fluid in a pipe as the diameter is gradually reduced. The effect has many applications, such as in the carburetor and fluid-flow measuring instruments.

Daniel Bernoulli (1700–1782),

mathematician, was born in The Netherlands. He worked on trigonometry, mechanics, vibrating systems, and hydrodynamics (leading to the kinetic theory of gases), and developed Bernoulli's principle.

Do you recall walking into the wind in winter on a city street lined with several tall buildings when the wind seemed stronger than usual? The wind actually was stronger because it was acting as a fluid as illustrated using the Venturi meter principle in Fig. 13.17. That is, the speed of the wind increased as it was forced to flow between the tall buildings along the street.

BERNOULLI'S PRINCIPLE

For the horizontal flow of a fluid through a tube, the sum of the pressure and energy of motion (kinetic energy) per unit volume of the fluid is constant.

One application of **Bernoulli's principle** involves a small engine carburetor (Fig. 13.18). The volume of airflow is determined by the position of the butterfly valve. As the air flows through the throat, the air gains speed and loses pressure. The pressure in the fuel bowl equals the pressure above the throat. Due to the difference in pressure between the fuel bowl and the throat, gasoline is drawn into and is mixed with the airstream. The reduced pressure in the throat also helps the gasoline to vaporize.

Figure 13.18

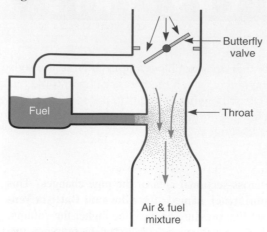

(a) Schematic of a small engine carburetor

(b) Top view of a butterfly valve on a carburetor

Figure 13.19 Air flowing past an airplane wing with its more-rounded curved upper surface creates lift.

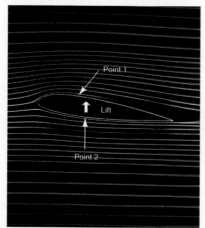

Another application of Bernoulli's principle involves airplane travel. Fig. 13.19 shows the flow of air rushing past the wing of an airplane. The velocity, v_1, of the air above the wing is greater than the velocity of the air below, v_2, because it has farther to travel in a given time. Thus, the pressure at point 2 is greater, which causes lift on the wing.

An airplane requires a longer distance for takeoff in summer than in winter. Why? Hot air is less dense and has fewer air molecules to lift the plane as it moves down the runway.

Two other examples that illustrate Bernoulli's principle are a curving baseball and a paint spray gun. The rotating baseball in Fig. 13.21 drags a layer of air immediately next to its surface. This reduces the speed of air at the bottom of the ball

TRY THIS ACTIVITY

Low Pressure

Tear a sheet of paper in half and position each half as shown in Fig. 13.20. What do you think will happen to the positions of the papers when a continuous stream of air is blown between them? Will the sheets separate or move closer to one another? Test your prediction by blowing a continuous stream of air between the two sheets of paper. Was your prediction correct? Use Bernoulli's principle to explain the behavior of the paper.

Figure 13.20

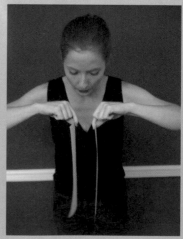

and increases the speed at the top of the ball causing the pressure to drop more at the top than at the bottom so that the ball curves as shown.

When the air in a spray gun is accelerated through a narrowing in the line, the pressure is reduced, and paint is drawn into the airstream and is then forced from the gun.

Figure 13.21 Bernoulli's principle explains why a baseball curves.

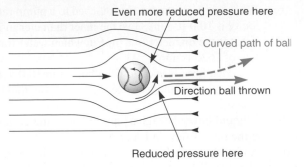

Even more reduced pressure here

Curved path of ball

Direction ball thrown

Reduced pressure here

PHYSICS CONNECTIONS

Blood Pressure

A fluid exerts pressure on the walls of its closed container. As a fluid, blood exerts pressure on the walls of the heart, arteries, vessels, and capillaries that make up the circulatory system. When the diameter of the arteries and vessels is narrowed, an increased pressure is exerted on the walls of the blood vessels.

A sphygmomanometer (Fig. 13.22) is a device used to measure a person's blood pressure. The sphygmomanometer cuff is placed around the upper arm, inflated, and then deflated while the meter measures the pressure of blood passing through that section of the arm. As seen in this chapter, liquid pressure is dependent on the depth of the fluid. Since a sphygmomanometer cannot be placed around the heart and the depth of the fluid must be at the same level as the heart, the upper arm becomes a convenient location to measure blood pressure. When a person is lying down, the depth of the fluid throughout the body is roughly the same so that the pressure could be measured on any limb of the body.

When standing on one's head, one notices that the blood vessels in the head experience a great deal of pressure. As a result, the blood vessels on the sides of the temple protrude. The vessels in the feet and legs typically withstand much higher pressures than the vessels in the head.

Figure 13.22 A person having his blood pressure measured. Notice the sphygmomanometer is at the same level as the heart.

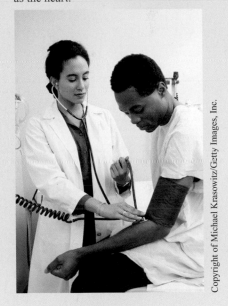

Copyright of Michael Krasowitz/Getty Images, Inc.

PROBLEMS 13.5

1. Water flows through a hose of diameter 3.90 cm at a velocity of 5.00 m/s. Find the flow rate of the hose in L/min.
2. Water flows through a 15.0-cm fire hose at a rate of 5.00 m/s.
   (a) Find the rate of flow through the hose in L/min.
   (b) How many litres pass through the hose in 30.0 min?
3. Water flows from a pipe at 650 L/min.
   (a) What is the diameter (in cm) of the pipe if the velocity of the water is 1.5 m/s?
   (b) Find the velocity (in m/s) of the water if the diameter of the pipe is 20.0 cm.
4. Water flows through a pipe of diameter 8.00 cm at 45.0 m/min. Find the flow rate
   (a) in m^3/min and (b) in L/s.

SKETCH

$$12 \text{ cm}^2 \quad | \quad w$$

4.0 cm

DATA

$A = 12 \text{ cm}^2$, $l = 4.0$ cm, $w = ?$

BASIC EQUATION

$A = lw$

WORKING EQUATION

$w = \frac{A}{l}$

SUBSTITUTION

$w = \frac{12 \text{ cm}^2}{4.0 \text{ cm}} = 3.0$ cm

5. A pump is rated to deliver 50.0 gal/min. Find the velocity of water in (a) a 6.00-in.-diameter pipe and (b) a 3.00-in.-diameter pipe.

6. What size pipe needs to be attached to a pump rated at 36.0 gal/min if the desired velocity is 10.0 ft/min? Give the inner diameter in inches.

7. What is the diameter of a pipe in which water travels 32.0 m in 16.0 s and has a flow rate of 2620 L/min?

8. A garden hose is used to fill a bucket in 30.0 s. If you cover part of the hose nozzle so the speed of the water leaving the hose doubles, how long does it take to fill the bucket?

9. A liquid flows through a pipe with a diameter of 0.50 m at a speed of 4.20 m/s. What is the rate of flow in L/min?

10. A pipe system with a radius of 0.060 m has a liquid flowing through it at a speed of 3.96 m/s. What is the rate of flow in L/min?

Glossary

Absolute Pressure The actual air pressure given by the gauge reading plus the normal atmospheric pressure. (p. 353)

Archimedes' Principle Any object placed in a fluid apparently loses weight equal to the weight of the displaced fluid. (p. 354)

Atmospheric Pressure The pressure caused by the weight of the air in the atmosphere. (p. 352)

Bernoulli's Principle For the horizontal flow of a fluid through a tube, the sum of the pressure and energy of motion (kinetic energy) per unit volume of the fluid is constant. (p. 359)

Buoyant Force The upward force exerted on a submerged or partially submerged object. (p. 354)

Flow Rate The volume of fluid flowing past a given point in a pipe per unit time. (p. 357)

Gauge Pressure The amount of air pressure excluding the normal atmospheric pressure. (p. 353)

Hydraulic Principle (Pascal's Principle) The pressure applied to a confined liquid is transmitted without measurable loss throughout the entire liquid to all inner surfaces of the container. (p. 348)

Hydrostatic Pressure The pressure a liquid at rest exerts on a submerged object. (p. 341)

Pressure The force applied per unit area. (p. 341)

Streamline Flow The smooth flow of a fluid through a tube. (p. 357)

Turbulent Flow The erratic, unpredictable flow of a fluid resulting from excessive speed of the flow or sudden changes in direction or size of the tube or pipe. (p. 357)

Formulas

13.1 $P = hD_w$

$F_t = AhD_w$

$F_s = \frac{1}{2}AhD_w$

13.3 $P_{abs} = P_{ga} + P_{atm}$

13.5 $Q = vA$

Review Questions

1. The force applied to a unit area is called
   - (a) strain.
   - (b) total force.
   - (c) pressure.
   - (d) none of the above.

2. The statement that the pressure applied to a confined liquid is transmitted without measurable loss throughout the entire liquid to all inner surfaces of the container is called
   - (a) Hooke's law.
   - (b) Pascal's principle.
   - (c) Archimedes' principle.
   - (d) none of the above.

3. For an incompressible fluid, the flow rate is
   - (a) equal for all surfaces.
   - (b) constant throughout the pipe.
   - (c) greater for the larger parts of the pipe.
   - (d) none of the above.

4. Bernoulli's principle states that for horizontal flow of a fluid through a tube, the sum of the pressure and energy of motion per unit volume is
   (a) increasing with time.   (b) decreasing with time.
   (c) constant.   (d) none of the above.

5. Bernoulli's principle explains
   (a) curving baseballs.   (b) the hydraulic principle.
   (c) absolute pressure.   (d) buoyant forces.
   (e) none of the above.

6. What is the metric unit for pressure?
7. In your own words, define *pressure*.
8. In your own words, state how to find the force exerted on the vertical side of a rectangular water tank.
9. In your own words, state the hydraulic principle.
10. Describe why a ship floats.
11. Describe how a rotating baseball follows a curved path.
12. How does an airplane wing provide lift?
13. What is the difference between streamline and turbulent flow?
14. Give an example of how Archimedes' principle applies to gases.
15. Describe the difference between absolute and gauge pressure. Which do you use when you measure the pressure in your automobile tires?
16. Is the pressure on a small piston different from the pressure on a large piston in the same hydraulic system? Are the forces on the two pistons the same?
17. On what does the total force exerted by a liquid on a horizontal surface depend?
18. Why must the thickness of a dam be greater at the bottom than at the top?
19. Is the hydraulic piston in the master brake cylinder in an automobile larger or smaller than the piston in the brake cylinder at the wheels? Why?
20. Would a drinking straw work in space where there is no gravity? Explain.

Review Problems

SKETCH

12 cm² | w

4.0 cm

DATA

$A = 12$ cm², $l = 4.0$ cm, $w = ?$

BASIC EQUATION

$A = lw$

WORKING EQUATION

$w = \frac{A}{l}$

SUBSTITUTION

$w = \frac{12 \text{ cm}^2}{4.0 \text{ cm}} = 3.0$ cm

1. Find the pressure (in kPa) at the bottom of a water-filled drum 3.24 m high.
2. Find the depth in a lake at which the pressure is 197 lb/in².
3. Find the height of a water column when the pressure at the bottom of the column is 297 kPa.
4. What is the total force exerted on the bottom of a rectangular tank 8.67 ft by 4.83 ft by 3.56 ft deep?
5. Find the water pressure (in kPa) at a point 35.0 m from the bottom of a 55.0-m-tall full water tower.
6. Find the total force on the bottom of a cylindrical water tower 55.0 m high and 7.53 m in radius.
7. Find the total force on the side of a cylindrical water tower 55.0 m high and 7.53 m in radius.
8. Find the total force on the side of a rectangular water trough 1.25 m high by 1.55 m by 2.95 m.
9. What must the water pressure (in kPa) be on the ground to supply a water pressure of 252 N/cm² on the third floor, which is 9.00 m above the ground?
10. What water pressure must a pump that is located on the first floor supply to have water on the twenty-fifth floor of a building with a pressure of 26 lb/in²? Assume that the distance between floors is 16.0 ft.
11. A submarine is submerged to a depth of 3150 ft in the Atlantic Ocean. What air pressure (in kPa) is needed to blow water out of the ballast tanks?
12. The area of the large piston in a hydraulic jack is 4.75 in². The area of the small piston is 0.564 in². (a) What force must be applied to the small piston if a weight of 650 lb is to be lifted? (b) What is the mechanical advantage of the hydraulic jack?

13. The MA of a hydraulic jack is 324. What force must be applied to lift an automobile weighing 11,500 N?

14. The pistons of a hydraulic press have radii of 0.543 cm and 3.53 cm, respectively. (a) What force must be applied to the smaller piston to exert a force of 4350 N on the larger? (b) What is the pressure (in N/cm^2) on each piston? (c) What is the mechanical advantage of the press?

15. Find the absolute pressure in a bicycle tire with a gauge pressure of 202 kPa.

16. Find the gauge pressure of a tire with an absolute pressure of 655 kPa.

17. Find the gauge pressure of a tank whose absolute pressure is 314 lb/in^2.

18. A rock weighs 55.4 N in air and 52.1 N in water. Find the buoyant force on the rock.

19. A metal displaces 643 cm^3 of water. Find the buoyant force of the water.

20. A rock displaces 314 cm^3 of alcohol. Find the buoyant force on the rock.

21. A flat-bottom barge is 22.3 ft wide, 87.5 ft long, and 16.5 ft deep. (a) How many ft^3 of water will it displace while the top stays 3.20 ft above the water? (b) What load in tons will the barge contain if the barge weighs 157 tons in dry dock?

22. Water flows through a hose of diameter 3.00 cm at a velocity of 4.43 m/s. Find the flow rate of the hose in L/min.

23. Water flows through a 13.0-cm-diameter fire hose at a rate of 4.53 m/s. (a) What is the rate of flow through the hose in L/min? (b) How many litres pass through the hose in 25.0 min?

24. (a) What is the weight density of a liquid that exerts a total force of 433 N on the sides of a 3.00-m-tall cylindrical tank? The radius of the tank is 0.913 m, and it is filled to 1.75 m. (b) What liquid might this be?

APPLIED CONCEPTS

1. An aquarium's main tank holds $20\overline{0},000$ gal or 758 m^3 of salt water. (a) What is the lateral surface area of the glass wall if the cylindrical saltwater tank is 14.5 m tall? (b) What is the force applied to the vertical glass surface? (c) Steel bands are often placed around aquariums to reinforce the glass walls. Explain how the steel bands should be spaced toward the bottom of the tank.

2. The piston in a master cylinder has a radius of 0.570 in. and the radius of each of the two brake cylinder pistons is 1.75 in. (a) How much pressure is created in the master cylinder if a driver quickly applies 45.5 lb of force to the automobile's brake pedal? (b) Given the area of the two brake cylinder pistons, what is the force applied to each of the brake drums? (See Fig. 13.6.)

3. A crane that can lift a maximum of $90\overline{0}0$ N is preparing to lift and move an underwater concrete mooring. The 1.25 m × 1.25 m × 0.450 m concrete block is located in seawater. Verify that the crane is strong enough to lift the mooring by finding (a) its volume, (b) its dry-dock weight, (c) the water's buoyant force, and (d) the force needed to lift the mooring while it is under water. (Refer to Table 11.2 for the mass density of concrete and of water.)

4. Wind tunnels are used to measure the aerodynamic properties of prototype models. (a) If a fan generates a wind speed of 25.0 mi/h inside an 8.75-ft^2 section of a wind tunnel, what is the wind speed as the air enters the narrower, 4.35-ft^2 section of the wind tunnel? (b) Explain why it is often windier on city streets surrounded by tall buildings than in more open areas.

5. A flexible hose with inside radius 0.250 in. leads to a shower fixture with 15 holes, each with radius 3.13×10^{-2} in. (a) If a faucet is opened to allow the water to move through the pipes at 3.94 ft/s, what is the speed of the water as it comes out of the holes? (b) What are two ways to increase the speed of the water from a shower fixture without opening the faucet farther? Explain.

CHAPTER 14

TEMPERATURE AND HEAT TRANSFER

Heat of vaporization

$$L_v = Q/m$$

Almost all forms of technology have concerns about temperature and heat transfer. The concern may be direct, as in refrigeration, or indirect, as in the thermal expansion of highways. Being able to measure heat transfer can mean the difference between success and failure of many things, from steam heat to space travel.

Objectives

The major goals of this chapter are to enable you to:

1. Distinguish between temperature and heat.
2. Express temperature using different scales.
3. Analyze heat transfer applications.
4. Determine final temperature using the method of mixtures.
5. Relate heat transfer to the expansion of solids and liquids.
6. Find the heat required for change of phase of solids, liquids, and gases.

14.1 Temperature

An understanding of temperature and heat and their differences is very important in many applications in technology. The automotive, truck, and heavy equipment industries are very concerned with the heat energy released by the fuel mixture in an engine cylinder's combustion chamber (Fig. 14.1). While the contained explosion drives the piston, the excess heat produced must be transferred to the atmosphere. This is accomplished in air-cooled small engines by cooling fins and in larger engines by liquid cooling systems. Allowing the engine temperature to become too hot causes the metal parts to excessively expand and freeze up or warp, permanently damaging the engine.

Basically, **temperature** is a measure of the hotness or coldness of an object. Temperature could be measured in a simple way by using your hand to sense the hotness or coldness of an object. However, the range of temperatures that your hand can withstand is too small, and your hand is not precise enough to measure temperature adequately. Therefore, other methods are used for measuring temperature.

Certain properties of matter vary with their temperature. For example, when objects are heated, they give off light of different colors. When an object is heated, in the absence of chemical reactions, it first gives off red light. As it is heated more, it appears white.

Chemical reactions sometimes cause different colors. When carbon steel is heated and exposed to air, several colors are observed before the rod appears red (see Fig. 14.2). This is due to a chemical reaction involving the carbon. If we could measure the color of the light, we could then determine the temperature. Although this works only for high temperatures, it is used in the production of metal alloys. The temperature of hot molten metals is determined this way.

Another property of matter that we use to find temperature is the change in volume of a liquid or a solid as its temperature changes. The liquid in glass thermometers is an example. This type of thermometer (Fig. 14.3) consists of a hollow glass bulb and a hollow glass tube joined together. A small amount of liquid such as alcohol is placed in the bulb. The air is removed from the tube. When the liquid is heated, it expands and rises up the glass tube. The height to which the liquid rises indicates the temperature.

The thermometer is standardized by marking two points on the glass that indicate the liquid level at two known temperatures. The temperatures used are the *freezing point* of water and the *boiling point* of water at sea level. The distance between these marks is then divided up into equal segments called *degrees*.

Figure 14.1 Force on a piston produced by hot expanding gas

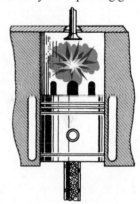

Anders Celsius (1701–1744),

astronomer, was born in Sweden. He devised the centigrade scale of temperature in 1742. The Celsius scale (formerly the centigrade scale) is named after him.

Figure 14.2 Metallurgy and heat treatment: temperatures, steel colors, and related processes (Courtesy of Allegheny Ludlum Steel Corp. Reprinted by permission)

	Colors	°C	°F	Processes
Heat colors	White	1371	2500	Welding
		1315	2400	High-speed steel hardening (2150–2450°F)
	Yellow white	1259	2300	
		1204	2200	
		1149	2100	
	Yellow	1093	2000	
		1036	1900	Alloy tool steel hardening (1500–1950°F)
	Orange red	981	1800	
		926	1700	
		871	1600	
	Light cherry red	815	1500	Carbon tool steel hardening (1350–1550°F)
	Cherry red	760	1400	
		704	1300	
	Dark red	648	1200	
		593	1100	High-speed steel tempering (1000–1100°F)
	Very dark red	538	1000	Carbon tool steel tempering (300–1050°F)
		482	900	
	Black red in dull light or blackness	426	800	
		371	700	
Temper colors	Pale blue (590°F)	315	600	
	Violet (545°F)			
	Purple (525°F)	260	500	
	Yellowish brown (490°F)			
	Dark straw (465°F)	204	400	
	Light straw (425°F)			
		149	300	
		93	200	
		38	100	
		18	0	

Figure 14.3 Common thermometer

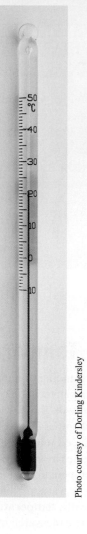

Photo courtesy of Dorling Kindersley

We will study the four temperature scales shown in Fig. 14.4. The common metric temperature scale is the **Celsius scale** with freezing point 0°C and boiling point 100°C. To write a temperature, we write the number followed by the degree symbol (°) followed by the capital letter of the scale used. Temperatures below zero on a scale are written as negative numbers. Thus, 20° below zero on the Celsius scale is written as −20°C.

The U.S. temperature scale is the **Fahrenheit scale** with freezing point 32°F and boiling point 212°F. The relationship between Fahrenheit temperatures (T_F) and Celsius temperatures (T_C) is given by

$$T_C = \frac{5}{9}(T_F - 32°)$$

$$T_F = \frac{9}{5}T_C + 32°$$

where T_C = Celsius temperature
T_F = Fahrenheit temperature

Gabriel Daniel Fahrenheit (1686–1736),

physicist, was born in Poland. He invented the alcohol thermometer in 1709 and the mercury thermometer in 1714.

Figure 14.4 Four basic temperature scales

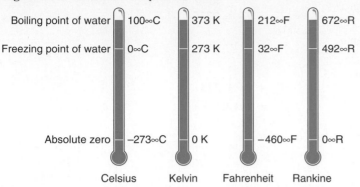

The human body average temperature is 98.6°F. What is it in degrees Celsius?

EXAMPLE 1

Data:

$$T_F = 98.6°F$$
$$T_C = ?$$

Basic Equation:

$$T_C = \frac{5}{9}(T_F - 32°)$$

Working Equation: Same

Substitution:

$$T_C = \frac{5}{9}(98.6° - 32°)$$
$$= \frac{5}{9}(66.6°)$$
$$= 37.0°C$$

. . . .

Lord Kelvin (Sir William Thomson) (1824–1907),

mathematician and physicist, was born in Belfast. He helped develop the law of conservation of energy and the absolute temperature scale (now named the Kelvin scale), did fundamental research in thermodynamics, presented the dynamic theory of heat, developed theorems for the mathematical analysis of electricity and magnetism, and designed several kinds of electrometers.

Sometimes it is necessary to use the *absolute temperature scales*, which are the Kelvin scale and the Rankine scale. These are called absolute scales because 0 on either scale refers to the lowest limit of temperature, called *absolute zero*.

The **Kelvin scale** is the metric absolute temperature scale on which absolute zero is 0 K and is closely related to the Celsius scale. The relationship is*

$$T_K = T_C + 273$$

The **Rankine scale** is the U.S. absolute temperature scale on which absolute zero is 0°R and is closely related to the Fahrenheit scale. The relationship is

$$T_R = T_F + 46\overline{0}°$$

William Rankine (1820–1872),

engineer and scientist, was born in Scotland. He is noted for his work on the steam engine, machinery, shipbuilding, applied mechanics, the new science of thermodynamics, and the theories of elasticity and of waves.

Change 18°C to Kelvin.

EXAMPLE 2

Data:

$$T_C = 18°C$$
$$T_K = ?$$

---

*The degree symbol (°) is not used when writing a temperature on the Kelvin scale.

Basic Equation:

$$T_K = T_C + 273$$

Working Equation: Same

Substitution:

$$T_K = 18 + 273$$
$$= 291 \text{ K}$$

· · · · · · · · · · · · · · · · ·

EXAMPLE 3

Change 535°R to degrees Fahrenheit.

Data:

$$T_R = 535°R$$
$$T_F = ?$$

Basic Equation:

$$T_R = T_F + 46\overline{0}°$$

Working Equation:

$$T_F = T_R - 46\overline{0}°$$

Substitution:

$$T_F = 535° - 46\overline{0}°$$
$$= 75°F$$

· · · · · · · · · · · · · · · · ·

PROBLEMS 14.1

Find each temperature as indicated.

SKETCH

12 cm²	w

4.0 cm

DATA

$A = 12$ cm², $l = 4.0$ cm, $w = ?$

BASIC EQUATION

$A = lw$

WORKING EQUATION

$w = \frac{A}{l}$

SUBSTITUTION

$w = \frac{12 \text{ cm}^2}{4.0 \text{ cm}} = 3.0$ cm

1. $T_F = 77°F$, $T_C = $ _____
2. $T_F = 113°F$, $T_C = $ _____
3. $T_F = 257°F$, $T_C = $ _____
4. $T_C = 15°C$, $T_F = $ _____
5. $T_C = 145°C$, $T_F = $ _____
6. $T_C = 35°C$, $T_F = $ _____
7. $T_F = 1\overline{0}°F$, $T_C = $ _____
8. $T_F = 2\overline{0}°F$, $T_C = $ _____
9. $T_C = 95°C$, $T_F = $ _____
10. $T_F = -50°F$, $T_C = $ _____
11. $T_C = 25°C$, $T_K = $ _____
12. $T_F = -45°F$, $T_R = $ _____
13. $T_K = 406$ K, $T_C = $ _____
14. $T_C = 75°C$, $T_K = $ _____
15. $T_C = -5\overline{0}°C$, $T_K = $ _____
16. $T_K = 175$ K, $T_C = $ _____
17. $T_K = 600\overline{0}$ K, $T_C = $ _____
18. The melting point of pure iron is 1505°C. What Fahrenheit temperature is this?
19. The melting point of mercury is −38.0°F. What Celsius temperature is this?
20. A welding white heat is approximately 140̄0°C. Find this temperature expressed in degrees Fahrenheit.
21. The temperature in a crowded room is 85°F. What is the Celsius reading?
22. The temperature of an iced tea drink is 5°C. What is the Fahrenheit reading?
23. The boiling point of liquid nitrogen is −196°C. What is the Fahrenheit reading?
24. The melting point of ethyl alcohol is −179°F. What is the Celsius reading?

During the forging and heat-treating of steel, the color of heated steel is used to determine its temperature. Complete the following table, which shows the color of heat-treated steel and the corresponding approximate temperatures in degrees Celsius and Fahrenheit. (Round to three significant digits.)

	Color	°C	°F
25.	White	——	220̄0
26.	Yellow	110̄0	——
27.	Orange	——	1725
28.	Cherry red	718	——
29.	Dark red	635	——
30.	Faint red	——	90̄0
31.	Pale blue	31̄0	——

14.2 Heat

When a hole is drilled in a metal block (Fig. 14.5), it becomes very hot. As the drill does mechanical work on the metal, the temperature of the metal increases. How can we explain this? Note the difference between the metal at low temperatures and at high temperatures. At high temperatures, the atoms in the metal vibrate more rapidly than at low temperatures. Their velocity is higher at high temperatures, and thus their kinetic energy ($KE = \frac{1}{2}mv^2$) is greater. To raise the temperature of a material, we must speed up the atoms; that is, we must add energy to them. **Heat** is a form of internal kinetic and potential energy contained in an object associated with the motion of its atoms or molecules and may be transferred from an object at a higher temperature to one at a lower temperature.

Figure 14.5 Friction causes a rise in temperature of the drill and plate.

Drilling a hole in a metal block causes a temperature increase. As the drill turns, it collides with atoms of the metal, causing them to speed up. This mechanical work done on the metal has caused an increase in the energy of the atoms. For this reason, any friction between two surfaces results in a temperature rise of the materials.

Since heat is a form of energy, we could measure it in joules or ft lb, which are energy units. However, before it was known that heat is a form of energy, special units for heat were developed, which are still in use. These units are the caloric and the kilocalorie in the metric system and the Btu (British thermal unit) in the U.S. system. The **kilocalorie** (kcal) is the amount of heat necessary to raise the temperature of 1 kg of water 1°C. *Note:* The precise definition is based on the amount of heat needed to raise the temperature of 1 kg of water from 14.5°C to 15.5°C; however, the variation for each 1°C change in temperature is so minimal that it can be ignored for all practical purposes. The **Btu** is the amount of heat (energy) necessary to raise the temperature of 1 lb of water 1°F. The **calorie** (cal) is the amount of heat (energy) necessary to raise the temperature of 1 g of water 1°C. *Note:* One food calorie is the same as 1 kcal.

To lower the temperature of a substance, we need to remove some of the heat, the energy of motion of the molecules. When we have removed all the heat possible (when the molecules are moving as slowly as possible), we have reached **absolute zero**, the lowest possible temperature. Lower temperatures cannot be reached because all the heat has been removed. However, there is no upper limit on temperature because we can always add more heat (energy) to a substance to increase its temperature.

As mentioned before, heat and work are somehow related. James Prescott Joule determined by experiments the relationship between heat and work, called the **mechanical equivalent of heat**. He found that

1. 1 cal of heat is produced by 4.19 J of work.
2. 1 kcal of heat is produced by 4190 J of work.
3. 1 Btu of heat is produced by 778 ft lb of work.

The following are some examples in which heat is converted into useful work:

1. *In our bodies.* When food is oxidized, heat energy is produced, which can be converted into muscular energy, which in turn can be turned into work. Experiments have

shown that only about 25% of the heat energy from our food is converted into muscular energy. That is, our bodies are about 25% efficient.

2. **By burning gases.** When a gas is burned, the gas expands and builds up a tremendous pressure that may convert heat to work by exerting a force to move a piston in an engine or turn the blades of a turbine. Since the burning of the fuel occurs within the cylinder or turbine, such engines are called *internal combustion engines*.

3. **By steam.** Heat from burning oil, coal, or wood may be used to generate steam. When water changes to steam under normal atmospheric pressure, it expands about 1700 times. When confined to a boiler, the pressure exerts a force against the piston in a steam engine or against the blades of a steam turbine. Since the fuel burns outside the engine, most steam engines or steam turbines are *external combustion engines*.

Technically, what is the difference between temperature and heat? *Temperature* is a measure of the hotness or coldness of an object. *Heat* is the total thermal energy (kinetic and potential) that can be transferred from an object at a higher temperature to one at a lower temperature. There are two basic ways of changing the temperature of an object:

1. By doing work *on* the object, such as the work done by the drill on the metal block in Fig. 14.5.
2. By supplying energy *to* the object, such as mechanical, chemical, or electrical energy.

EXAMPLE 1

Find the amount of work (in J) that is equivalent to 4850 cal of heat.

$$4850 \text{ cal} \times \frac{4.19 \text{ J}}{1 \text{ cal}} = 20{,}300 \text{ J} \qquad \text{or} \qquad 20.3 \text{ kJ}$$

EXAMPLE 2

How much work must a person do to offset eating a 775-calorie breakfast?
First, note that one food calorie equals one kilocalorie.

$$775 \text{ kcal} \times \frac{4190 \text{ J}}{1 \text{ kcal}} = 3.25 \times 10^6 \text{ J} \qquad \text{or} \qquad 3.25 \text{ MJ}$$

EXAMPLE 3

A given coal gives off 7150 kcal/kg of heat when burned. How many joules of work result from burning one metric ton, assuming that 35.0% of the heat is lost?
First, note that one metric ton equals 1000 kg.

$$7150 \frac{\text{kcal}}{\text{kg}} \times \frac{4190 \text{ J}}{\text{kcal}} \times 1000 \text{ kg} \times 0.350 = 1.05 \times 10^{10} \text{ J}$$

PROBLEMS 14.2

1. Find the amount of heat in cal generated by 95 J of work.
2. Find the amount of heat in kcal generated by 7510 J of work.
3. Find the amount of work that is equivalent to 1550 Btu.
4. Find the amount of work that is equivalent to 3850 kcal.
5. Find the mechanical work equivalent (in J) of 765 kcal of heat.
6. Find the mechanical work equivalent (in J) of 8550 cal of heat.
7. Find the heat equivalent (in Btu) of 3.46×10^6 ft lb of work.
8. Find the heat equivalent (in kcal) of 7.63×10^5 J of work.
9. How much work must a person do to offset eating a piece of cake containing 625 cal?

10. How much work must a person do to offset eating a $20\overline{0}$-g bag of potato chips if 28 g of chips contain 150 cal?

11. A fuel yields 1.15×10^4 cal/g when burned. How many joules of work are obtained by burning $10\overline{0}0$ g of the fuel?

12. A racing fuel produces 1.60×10^4 cal/g when burned. If $50\overline{0}$ g of the fuel is burned, how many joules of work are produced?

13. A given gasoline yields 1.15×10^4 cal/g when burned. How many joules of work are obtained by burning 875 g of gasoline?

14. A coal sample yields 1.25×10^4 Btu/lb. How many foot-pounds of work result from burning 1.00 ton of this coal?

15. Natural gas burned in a gas turbine has a heating value of 1.10×10^5 cal/g. If the turbine is 24.0% efficient and 2.50 g of gas is burned each second, find (a) how many joules of work are obtained and (b) the power output in kilowatts.

16. Find the amount of heat energy that must be produced by the body to be converted into muscular energy and then into $10\overline{0}0$ ft lb of work. Assume that the body is 25% efficient.

17. What is the mechanical work equivalent in $50,\overline{0}00$ kcal of heat produced by a stationary diesel engine?

18. An industrial engine produces $38,\overline{0}00$ kcal of heat. What is the mechanical work equivalent of the heat produced?

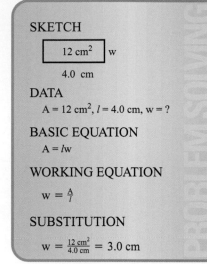

SKETCH

12 cm² | w

4.0 cm

DATA

$A = 12$ cm², $l = 4.0$ cm, $w = ?$

BASIC EQUATION

$A = lw$

WORKING EQUATION

$w = \frac{A}{l}$

SUBSTITUTION

$w = \frac{12 \text{ cm}^2}{4.0 \text{ cm}} = 3.0$ cm

14.3 Heat Transfer

The movement of heat from a hot engine to the air is necessary to keep the engine from overheating. The heat produced by a furnace must be transferred to the various rooms in a house. The movement of heat is a major technical application.

The transfer of heat from one object to another is always from the warmer object to the colder one or from the warmer part of an object to a colder part (Fig. 14.6). There are three methods of heat transfer: *conduction, convection*, and *radiation*. **Conduction** is the heat transfer from a warmer part of a substance to a cooler part as a result of molecular collisions, which cause the slower-moving molecules to move faster. Conduction is the usual method of heat transfer in solids. When one end of a metal rod is heated, the molecules in that end move faster than before. These molecules collide with other molecules and cause them to move faster also. In this way, the heat is transferred from one end of the metal to the other (Fig. 14.7). Another example of conduction is the transfer of the excess heat produced in the combustion chamber of an engine through the engine block into the coolant (Fig. 14.8).

Figure 14.6 Transfer of heat from a warmer to a colder area.

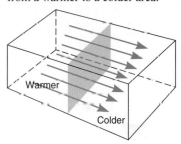

Figure 14.7 Heat flows by conduction in the metal rod.

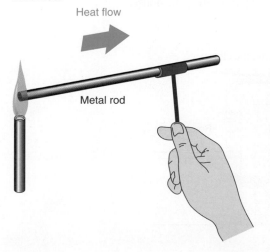

Heat flow

Metal rod

Figure 14.8 Heat conduction in an auto engine cooling system.

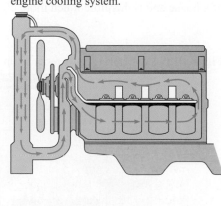

The conduction of heat through some materials is better than through others. A poor conductor of heat is called an *insulator*. A list of good conductors and poor conductors is given in Table 14.1.

Table 14.1   Good and Poor Heat Conductors

Good Heat Conductors	Poor Heat Conductors
Copper	Asbestos
Aluminum	Glass
Steel	Wood
	Air
	Snow

Sitting on aluminum bleachers on a cold day feels colder than sitting on wood bleachers at the same temperature because aluminum is a much better heat conductor than wood. Wood is a good insulator and is often used as handles for pots and pans. In winter, animals find shelter in snow banks by making snow holes because snow is a poor heat conductor. Snow causes not only the ground heat, but also the animals' body heat to be retained. Before health concerns were realized, asbestos was once widely used in insulating buildings because it is such a poor heat conductor.

Convection is the heat transfer by the movement of warm molecules from one region of a gas or a liquid to another. The wind carries heat along with it. The coolant in an engine carries hot antifreeze from the engine block to the radiator by a convection process. Heat transfer by the wind is a natural convection process. Heat transfer by the engine coolant is a forced convection process because it depends on a pump.

A dramatic illustration of the difference between convection and conduction can be shown by restraining some ice at the bottom of a test tube with some steel wool and filling the tube with water. When the water in the top of the tube is heated with a flame, it will boil without melting the ice. The poor conductivity of water keeps the less dense boiling water at the top and any heat transfer to the ice must be by conduction.

Convection currents are caused by the expansion of liquids or gases as they are heated. This expansion makes the hot gas or liquid less dense than the surrounding fluid. The lighter fluid is then forced upward by the heavier, surrounding fluid, which then flows in to replace it (Fig. 14.9). This type of behavior occurs in a fireplace as hot air goes up the chimney and is replaced by cool air from the adjacent room. The cool air draft, as this is called, is eventually supplied from outside air. This is why a fireplace is not very effective in heating a house. An airtight woodburning stove, however, draws little air from the inside of a house and is therefore much more efficient at heating the house.

Figure 14.9   A fireplace draws room air up into the chimney.

Room air   Room air

Conduction and convection require the presence of matter. All life on earth depends on the transfer of energy from the sun; this energy is transferred through nearly empty space. **Radiation** is heat transfer through energy being transmitted in the forms of rays, waves, or particles. Put your hand several inches from a hot iron (Fig. 14.10). The heat you feel is not transferred by conduction, because air is a poor conductor. It is not transferred by convection, because the hot air rises. This heat transfer is through radiation. This radiant heat is similar to light and passes through air, glass, and the vacuum of space. The energy that comes to us from the sun is in the form of radiant energy. At night, heat in the ground is radiated into the air. Dark objects absorb radiant heat and light objects reflect radiant heat. This is why we feel cooler on a hot day in light-colored clothing than in dark clothing (Fig. 14.11).

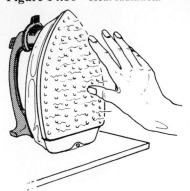

Figure 14.10   Heat radiation.

Heat flow calculations are important because of concern about energy conservation. All three of the heat-transfer mechanisms discussed here must be considered in any estimation of heat loss from a building. In addition, infiltration losses that arise from leakage through cracks and openings near doors, windows, and other such areas must also be considered. Infiltration losses are a form of convective transfer. Heat loss by conduction must also be considered. We will discuss methods for calculating heat loss by conduction.

The equations describing the flow of heat through an object are very similar to those for the flow of electricity, which are developed in later chapters. The driving potential for heat flow is the temperature difference between the hot and cold sides of the object. Heat flow is similar to the flow of electrical charge.

Figure 14.11   Dark objects absorb more radiant heat than light objects.

The ability of a material to transfer heat by conduction is called its **thermal conductivity**. Metals are good conductors of heat. Glass and air are poor conductors. The rate at which heat is transferred through an object depends on the following factors:

1. The thermal conductivity
2. The cross-sectional area through which the heat flows
3. The thickness of the material
4. The temperature difference between the two sides of the material

The total amount of heat transferred is given by the equation

$$Q = \frac{KAt(T_2 - T_1)}{L}$$

where Q = heat transferred in J or Btu
      K = thermal conductivity (from Table 14.2)
      A = cross-sectional area
      t = total time
      T_2 = temperature of the hot side
      T_1 = temperature of the cool side
      L = thickness of the material

Table 14.2 gives the thermal conductivities of some common materials.

Table 14.2 Thermal Conductivities

Substance	J/(s m °C)	Btu/(ft °F h)
Air	0.025	0.015
Aluminum	230	140
Brass	120	68
Brick/concrete	0.84	0.48
Cellulose fiber (loose fill)	0.039	0.023
Copper	380	220
Corkboard	0.042	0.024
Glass	0.75	0.50
Gypsum board (sheetrock)	0.16	0.092
Mineral wool	0.045	0.026
Plaster	0.14	0.083
Polystyrene foam	0.035	0.020
Polyurethane (expanded)	0.024	0.014
Steel	45	26
Water	0.56	0.32

EXAMPLE 1

Find the heat flow in an 8.0-h period through a 36 in. × 36 in. pane of glass (0.125 in. thick) if the temperature of the inner surface of the glass is 65°F and the temperature of the outer surface is 15°F.

Data:

$$K = 0.50 \text{ Btu/(ft °F h)}$$
$$A = 36 \text{ in.} \times 36 \text{ in.} = 3.0 \text{ ft} \times 3.0 \text{ ft} = 9.0 \text{ ft}^2$$
$$t = 8.0 \text{ h}$$
$$T_2 = 65°F$$
$$T_1 = 15°F$$
$$L = 0.125 \text{ in.} \times \left(\frac{1 \text{ ft}}{12 \text{ in.}}\right) = 0.0104 \text{ ft}$$
$$Q = ?$$

Basic Equation:

$$Q = \frac{KAt(T_2 - T_1)}{L}$$

Working Equation: Same

Substitution:

$$Q = \frac{[0.50 \text{ Btu/(ft °F h)}](9.0 \text{ ft}^2)(8.0 \text{ h})(65°F - 15°F)}{0.0104 \text{ ft}}$$
$$= 1.7 \times 10^5 \text{ Btu}$$

········

The insulation value of construction material is often expressed in terms of the *R value*, which indicates the ability of the material to resist the flow of heat and uses U.S. units. The *R* value is inversely proportional to the thermal conductivity and directly proportional to the thickness. Low thermal conductivity is characteristic of good insulators. This is described by the equation

$$R = \frac{L}{K}$$

where R = R value (in ft^2 °F/Btu/h)
　　　　K = thermal conductivity
　　　　L = thickness of the material (in ft)

Calculate the R value of 6.0 in. of mineral wool insulation. ◄

EXAMPLE 2

Data:

　L = 6.0 in. = 0.50 ft
　K = 0.026 Btu/(ft °F h)
　R = ?

Basic Equation:

$$R = \frac{L}{K}$$

Working Equation: Same

Substitution:

$$R = \frac{0.50 \text{ ft}}{0.026 \text{ Btu/(ft °F h)}}$$
$$= 19 \text{ ft}^2 \text{ °F/Btu/h}$$

$$\frac{\text{ft}}{\dfrac{\text{Btu}}{\text{ft °F h}}} = \text{ft} \div \frac{\text{Btu}}{\text{ft °F h}} = \text{ft} \cdot \frac{\text{ft °F h}}{\text{Btu}} = \frac{\text{ft}^2 \text{ °F h}}{\text{Btu}} = \frac{\text{ft}^2 \text{ °F}}{\text{Btu/h}}$$

This result could also have been written R-19. There is no equivalent in the metric system.

PROBLEMS 14.3

1. Find the R value of a pane of 0.125-in.-thick glass.
2. Find the R value of a brick wall 4.0 in. thick.
3. Find the R value of 0.50-in.-thick sheetrock.
4. Find the thermal conductivity of a piece of building material 0.25 in. thick that has an R value of 1.6 ft^2°F/Btu/h.
5. Find the R value of 0.50-in.-thick corkboard.
6. The dimensions of a rectangular building are $2\overline{0}$ ft × $10\overline{0}$ ft. The average outer wall temperature is $2\overline{0}$°F and the average inner wall temperature is 55°F. Find the amount of heat conducted through the walls of the building in 24 h if the R value of the walls is 11 ft^2°F/Btu/h.
7. Find the heat flow during 30.0 days through a glass window of thickness 0.20 in. with area 15 ft^2 if the average outer surface temperature is 25°F and the average inner glass surface temperature is $5\overline{0}$°F.
8. Find the heat flow in 30.0 days through a 0.25-cm-thick steel plate with cross section 45 cm × 75 cm. Assume a temperature differential of 95°C.
9. Find the heat flow in 75 s through a steel rod of length 85 cm and diameter 0.50 cm if the temperature of the hot end of the rod is $11\overline{0}$°C and the temperature of the cool end is −25°C.
10. Find the heat flow in 15 min through a 0.10-cm-thick copper plate with cross-sectional area 150 cm^2 if the temperature of the hot side is $99\overline{0}$°C and the temperature of the cool side is 5°C.
11. Find the heat flow in 24 h through a refrigerator door 30.0 in. × 58.0 in. insulated with cellulose fiber 2.0 in. thick. The temperature inside the refrigerator is 38°F. Room temperature is 72°F.

SKETCH

| 12 cm^2 | w |

4.0 cm

DATA
　A = 12 cm^2, l = 4.0 cm, w = ?

BASIC EQUATION
　$A = lw$

WORKING EQUATION
　$w = \frac{A}{l}$

SUBSTITUTION
　$w = \frac{12 \text{ cm}^2}{4.0 \text{ cm}} = 3.0$ cm

PHYSICS CONNECTIONS

Insulation

The purpose of insulating a home is to limit the amount of heat transfer through the walls and ceiling. Insulation limits heat transfer via conduction, convection, and radiation. Different insulating materials and methods are used. Less dense materials are often more effective in preventing heat transfer. Attic insulation must be fluffy and not compressed to be less dense. Molecules farther apart make fewer collisions, which results in less transfer of heat energy to a cooler environment. Other examples of less dense materials limiting heat transfer via conduction are argon gas placed between double-pane windows and evacuated linings in vacuum bottles. Argon has very low density; evacuated linings contain almost no air and therefore few molecules to transfer heat.

Fiberglass is one of the most popular types of insulation. The glass fiber is a poor conductor of heat. In addition, the air pockets in the fluffy fiberglass insulation prevent convection currents from transferring energy between the molecules and transferring heat (Fig. 14.12).

Figure 14.12 (a) Insulation is made of fiberglass, a poor conductor of heat. The fluffiness of the insulation creates air pockets to also eliminate convection currents. (b) The evacuated chamber inside a vacuum bottle prevents heat from transferring through the sides of the bottle.

Phil Degginger/Color-Pic, Inc.
(a)

(b)

12. Find the heat flow in 30.0 days through a freezer door 30.0 in. × 58.0 in. insulated with cellulose fiber 2.0 in. thick. The temperature inside the freezer is $-10°F$. Room temperature is 72°F.

13. Find the heat flow in 24 h through a refrigerator door 76.0 cm × 155.0 cm insulated with cellulose fiber 5.0 cm thick. The temperature inside the refrigerator is 3°C. Room temperature is 21°C.

14. Find the heat flow in 30.0 days through a freezer door 76.0 cm × 155.0 cm insulated with cellulose fiber 5.0 cm thick. The temperature inside the freezer is $-18°C$. Room temperature is 21°C.

15. Find the heat flow through the sides of an 18-cm-tall glass of ice water in 45 s. The glass is 6.00 mm thick; the temperature inside is 28.0°C. The temperature outside is 43.3°C and the radius is 7.0 cm.

14.4 Specific Heat

If we placed a piece of steel and a pan of water in the direct summer sunlight, we would find that the water becomes only slightly warmer whereas the steel gets quite hot. Why should one get so much hotter than the other? If equal masses of steel and water were placed over the same flame for 1 min, the temperature of the steel would increase almost 10 times more than that of the water. The water has a greater capacity to absorb heat.

Because water has a much higher capacity for storing energy than most common materials, it is very useful in cooling systems in engines and power plants. This property of water affects the climate of many places. Cities on large lakes, like Chicago, are warmer in the winter and cooler in the summer near the lake because of the high heat capacity of water. High temperatures in summer and low temperatures in winter in the middle of large continents are largely due to the absence of large bodies of water. Europe is warmer in the winter than mid-Canada because warm air from over the Atlantic Ocean is blown by prevailing westerly winds over the land.

The specific heat of a substance is a measure of its capacity to absorb or give off heat per degree change in temperature. This property of water to absorb or give off large amounts of heat makes it an effective substance for transferring heat in industrial processes.

The **specific heat** of a substance is the amount of heat necessary to change the temperature of 1 kg of it 1°C (1 lb of it 1°F in the U.S. system). By formula,

$$c = \frac{Q}{m\Delta T} \quad \text{(metric)} \qquad c = \frac{Q}{w\Delta T} \quad \text{(U.S.)}$$

To find the amount of heat added or taken away from a substance to produce a certain temperature change, we use

$$Q = cm\Delta T \quad \text{(metric)} \qquad Q = cw\Delta T \quad \text{(U.S.)}$$

where
c = specific heat
Q = heat
m = mass
w = weight
ΔT = change in temperature

A list of specific heats is given in Table 15 of Appendix C.

TRY THIS ACTIVITY

Cool Floors

A dramatic example of heat conduction is often experienced on cold winter mornings. While standing with your bare feet on a cold tile floor, note how quickly heat is transferred from your feet to the tile. Then, stand in a doorway with one bare foot on tile and one bare foot on wood and note the difference in the rate at which heat is transferred. What are the general characteristics that determine the heat capacity for your floors? Why are mats commonly placed on bathroom floors?

How many kilocalories of heat must be added to 10.0 kg of steel to raise its temperature 150°C?

EXAMPLE 1

Data:

m = 10.0 kg
ΔT = 150°C
c = 0.115 kcal/kg °C   (from Table 15 of Appendix C)
Q = ?

Basic Equation:

$$Q = cm\Delta T$$

Working Equation: Same

Substitution:

$$Q = \left(0.115 \frac{\text{kcal}}{\text{kg °C}}\right)(10.0 \text{ kg})(15\bar{0}°C)$$

$$= 173 \text{ kcal}$$

. . . .

EXAMPLE 2

How many joules of heat must be absorbed to cool 5.00 kg of water from 75.0°C to 10.0°C?

Data:

$$m = 5.00 \text{ kg}$$
$$\Delta T = 75.0°C - 10.0°C = 65.0°C$$
$$c = 4190 \text{ J/kg °C} \qquad \text{(from Table 15 of Appendix C)}$$
$$Q = ?$$

Basic Equation:

$$Q = cm\Delta T$$

Working Equation: Same

Substitution:

$$Q = \left(4190 \frac{\text{J}}{\text{kg °C}}\right)(5.00 \text{ kg})(65.0°C)$$

$$= 1.36 \times 10^6 \text{ J} \qquad \text{or} \qquad 1.36 \text{ MJ}$$

. . . .

PROBLEMS 14.4

Find Q for each material.

1. Steel, $w = 3.00$ lb, $\Delta T = 50\bar{0}°F$, $Q = $ _____ Btu
2. Copper, $m = 155$ kg, $\Delta T = 170°C$, $Q = $ _____ kcal
3. Water, $w = 19.0$ lb, $\Delta T = 200°F$, $Q = $ _____ Btu
4. Water, $m = 25\bar{0}$ g, $\Delta T = 17.0°C$, $Q = $ _____ cal
5. Ice, $m = 5.00$ kg, $\Delta T = 2\bar{0}°C$, $Q = $ _____ J
6. Steam, $w = 5.00$ lb, $\Delta T = 4\bar{0}°F$, $Q = $ _____ Btu
7. Aluminum, $m = 79.0$ g, $\Delta T = 16°C$, $Q = $ _____ cal
8. Brass, $m = 750$ kg, $\Delta T = 125°C$, $Q = $ _____ J
9. Steel, $m = 1250$ g, $\Delta T = 50.0°C$, $Q = $ _____ J
10. Aluminum, $m = 85\bar{0}$ g, $\Delta T = 115°C$, $Q = $ _____ kcal
11. Water, $m = 80\bar{0}$ g, $\Delta T = 80.0°C$, $Q = $ _____ kcal
12. Lead, $m = 475$ kg, $\Delta T = 245°C$, $Q = $ _____ J
13. How many Btu of heat must be added to 1200 lb of copper to raise its temperature from $10\bar{0}°F$ to $45\bar{0}°F$?
14. How many Btu of heat are given off by $50\bar{0}$ lb of aluminum when it cools from $65\bar{0}°F$ to 75°F?
15. How many kcal of heat must be added to 1250 kg of copper to raise its temperature from 25°C to 275°C?
16. How many joules of heat are absorbed by an electric freezer in lowering the temperature of 1850 g of water from 80.0°C to 10.0°C?
17. How many joules of heat are required to raise the temperature of $75\bar{0}$ kg of water from 15.0°C to 75.0°C?

18. How many kilocalories of heat must be added to $75\overline{0}$ kg of steel to raise its temperature from 75°C to $30\overline{0}$°C?

19. How many joules of heat are given off when 125 kg of steel cools from 1425°C to 82°C?

20. A 525-kg steam boiler is made of steel and contains 315 kg of water at 40.0°C. Assuming that 75% of the heat is delivered to the boiler and water, how many kilocalories are required to raise the temperature of both the boiler and water to 100.0°C?

21. Find the initial temperature of a 49.0-N cube of zinc, 16.0 cm on a side, that gives off 3.36×10^5 J of heat while cooling to 80.0°C.

22. A coolant lowers the temperature 13°C in a steel engine weighing 16,250 N and running at a temperature of $11\overline{0}$°C. What is the heat reduction (in joules) in the steel engine?

23. A block of iron with mass 0.400 kg is heated to 325°C from 295°C. How much heat is absorbed by the iron?

24. A block of copper is heated from 20.0°C to 80.0°C. How much heat is absorbed by the copper if its mass is 60.0 g?

25. The cooling system of a truck engine contains 20.0 L of water. (1 L of water has a mass of 1 kg.) (a) If the engine is run until 845 kJ of heat is added, what is the change in temperature of the water? (b) In the winter, the system was filled with 20.0 L of ethyl alcohol with density 0.800 g/cm³. If the ethyl alcohol absorbed the same 845 kJ of heat, what would be the increase in temperature of the alcohol? (c) Would ethyl alcohol or water be a better coolant? Why?

14.5 Method of Mixtures

When two substances at different temperatures are mixed together, heat flows from the warmer body to the cooler body until they reach the same temperature (Fig. 14.13). This is known as thermal equilibrium or the **method of mixtures**. Part of the heat lost by the warmer body is transferred to the cooler body and part is lost to the surrounding objects or the air. In most cases almost all the heat is transferred to the cooler body. We assume here that all the heat lost by the warmer body equals the heat gained by the cooler body. The amount of heat lost or gained by a body is

$$Q = cm\Delta T \quad \text{or} \quad Q = cw\Delta T$$

Figure 14.13 Heat flows from the warmer substance to the cooler.

By formula,

$$Q_{\text{lost}} = Q_{\text{gained}}$$
$$c_l m_l (T_l - T_f) = c_g m_g (T_f - T_g)$$

where the subscript l refers to the warmer body, which *loses* heat, the subscript g refers to the cooler body, which *gains* heat, and T_f is the final temperature of the mixture.

A 10.0-lb piece of hot copper is dropped into 30.0 lb of water at $5\overline{0}$°F. If the final temperature of the mixture is 65°F, what was the initial temperature of the copper?

EXAMPLE 1

Data:

$$w_l = 10.0 \text{ lb} \qquad\qquad w_g = 30.0 \text{ lb}$$
$$c_l = 0.093 \text{ Btu/lb °F} \qquad c_g = 1.00 \text{ Btu/lb °F}$$
$$T_l = ? \qquad\qquad\qquad T_g = 5\overline{0}°F$$
$$T_f = 65°F$$

Basic Equation:

$$c_l w_l (T_l - T_f) = c_g w_g (T_f - T_g)$$

Working Equation:

$$T_l = \frac{c_g w_g}{c_l w_l}(T_f - T_g) + T_f$$

Substitution:

$$T_l = \frac{(1.00 \text{ Btu/lb } °F)(30.0 \text{ lb})}{(0.093 \text{ Btu/lb } °F)(10.0 \text{ lb})}(65°F - 5\overline{0}°F) + 65°F$$

$$= 550°F$$

Some find it easier to find T_l using a second method. Substitute the data directly into the basic equation. Then solve for T_l as follows:

$$\left(0.093\frac{\text{Btu}}{\text{lb } °F}\right)(10.0 \text{ lb})(T_l - 65°F) = \left(1.00\frac{\text{Btu}}{\text{lb } °F}\right)(30.0 \text{ lb})(65°F - 5\overline{0}°F)$$

$$0.93T_l \text{ Btu/}°F - 6\overline{0} \text{ Btu} = 450 \text{ Btu}$$

$$0.93T_l \text{ Btu/}°F = 510 \text{ Btu}$$

$$T_l = \frac{510 \text{ Btu}}{0.93 \text{ Btu/}°F}$$

$$T_l = 550°F$$

· · · · · · · · · · · · · · · ·

EXAMPLE 2

If $20\overline{0}$ g of steel at $22\overline{0}°C$ is added to $50\overline{0}$ g of water at $10.0°C$, find the final temperature of this mixture.

Data:

$$c_l = 0.115 \text{ cal/g } °C \qquad c_g = 1.00 \text{ cal/g } °C$$
$$m_l = 20\overline{0} \text{ g} \qquad m_g = 50\overline{0} \text{ g}$$
$$T_l = 22\overline{0}°C \qquad T_g = 10.0°C$$
$$T_f = ?$$

Basic Equation:

$$c_l m_l(T_l - T_f) = c_g m_g(T_f - T_g)$$

Working Equation:

$$T_f = \frac{c_l m_l T_l + c_g m_g T_g}{c_l m_l + c_g m_g}$$

Substitution:

$$T_f = \frac{(0.115 \text{ cal/g } °C)(20\overline{0} \text{ g})(22\overline{0}°C) + (1.00 \text{ cal/g } °C)(50\overline{0} \text{ g})(10.0°C)}{(0.115 \text{ cal/g } °C)(20\overline{0} \text{ g}) + (1.00 \text{ cal/g } °C)(50\overline{0} \text{ g})}$$

$$= 19.2°C$$

To find T_f by the second method, substitute the data directly into the basic equation. Then, solve for T_f as follows:

$$\left(0.115\frac{\text{cal}}{\text{g } °C}\right)(20\overline{0} \text{ g})(22\overline{0}°C - T_f) = \left(1.00\frac{\text{cal}}{\text{g } °C}\right)(50\overline{0} \text{ g})(T_f - 10.0°C)$$

$$5060 \text{ cal} - 23.0\frac{\text{cal}}{°C}T_f = 50\overline{0}\frac{\text{cal}}{°C}T_f - 50\overline{0}0 \text{ cal}$$

$$10,060 \text{ cal} = 523\frac{\text{cal}}{°C}T_f$$

$$\frac{10,060 \text{ cal}}{523 \text{ cal/}°C} = T_f$$

$$19.2°C = T_f$$

· · · · · · · · · · · · · ·

PROBLEMS 14.5

Refer to Table 15 of Appendix C.

1. A 2.50-lb piece of steel is dropped into 11.0 lb of water at 75.0°F. The final temperature is 84.0°F. What was the initial temperature of the steel?

2. Mary mixes 5.00 lb of water at 200°F with 7.00 lb of water at 65.0°F. Find the final temperature of the mixture.

3. A 250-g piece of tin at 99°C is dropped in 100 g of water at 10°C. If the final temperature of the mixture is 20°C, what is the specific heat of the tin?

4. How many grams of water at 20°C are necessary to change 800 g of water at 90°C to 50°C?

5. A 159-lb piece of aluminum at 500°F is dropped into 400 lb of water at 60°F. What is the final temperature?

6. A 42.0-lb piece of steel at 670°F is dropped into 100 lb of water at 75.0°F. What is the final temperature of the mixture?

7. If 1250 g of copper at 20.0°C is mixed with 500 g of water at 95.0°C, find the final temperature of the mixture.

8. If 500 g of brass at 200°C and 300 g of steel at 150°C are added to 900 g of water in an aluminum pan of mass 150 g both at 20.0°C, find the final temperature, assuming no loss of heat to the surroundings.

9. The following data were collected in the laboratory to determine the specific heat of an unknown metal:

Mass of copper calorimeter	153 g
Specific heat of calorimeter	0.092 cal/g °C
Mass of water	275 g
Specific heat of water	1.00 cal/g °C
Mass of metal	236 g
Initial temperature of water and calorimeter	16.2°C
Initial temperature of metal	99.6°C
Final temperature of calorimeter, water, and metal	22.7°C

Find the specific heat of the unknown metal. *Note:* A calorimeter is usually a metal cup inside another metal cup that is insulated by the air between them (Fig. 14.14).

Figure 14.14 Apparatus for measuring the specific heat of a metal by the method of mixtures.

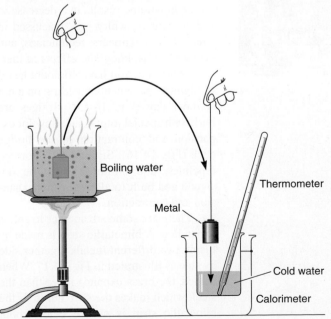

Boiling water

Thermometer

Metal

Cold water

Calorimeter

PROBLEM SOLVING

SKETCH

$$12 \text{ cm}^2 \quad | \quad w$$

4.0 cm

DATA

$A = 12 \text{ cm}^2, l = 4.0 \text{ cm}, w = ?$

BASIC EQUATION

$A = lw$

WORKING EQUATION

$w = \frac{A}{l}$

SUBSTITUTION

$w = \frac{12 \text{ cm}^2}{4.0 \text{ cm}} = 3.0 \text{ cm}$

10. The following data were collected in the laboratory to determine the specific heat of an unknown metal:

Mass of aluminum calorimeter	132 g
Specific heat of calorimeter	$92\overline{0}$ J/kg °C
Mass of water	285 g
Specific heat of water	4190 J/kg °C
Mass of metal	215 g
Initial temperature of water and calorimeter	12.6°C
Initial temperature of metal	99.1°C
Final temperature of calorimeter, water, and metal	18.6°C

Find the specific heat of the unknown metal.

11. Determine the original temperature of a $56\overline{0}$-g piece of lead placed in a 165-g brass calorimeter that contains 325 g of water. The initial temperature of the water and calorimeter was 18.0°C. The final temperature of the lead, calorimeter, and water is 31.0°C.

12. How much heat must be absorbed by its surroundings to cool a 565-g cube of iron from 100.0°C to 20.0°C?

13. How much water at 0°C would be needed to cool the iron in Problem 12 to 20.0°C?

14. The specific heat of water is 4190 J/kg °C. The specific heat of steel is 481 J/kg °C. Why may you burn your tongue on hot coffee but not on the spoon when both are at the same temperature?

Figure 14.15 Thermal expansion causes pavement to sometimes break up (buckle) in the summer. Air temperatures in the mid-90s (°F) or low-30s (°C) can easily translate into pavement temperatures in the 125°F or 52°C range. As the pavement absorbs heat, it expands and can reach the point where the concrete buckles across the traffic lanes.

14.6 Expansion of Solids

Most solids expand when heated and contract when cooled. They expand or contract in all three dimensions—length, width, and thickness. When a solid is heated, the expansion is due to the increased length of the vibrations of the atoms and molecules. This results in the solid expanding in all directions. This increase in volume results in a decrease in weight density, which was discussed in Chapter 12. Engineers, technicians, and designers must know the effects of thermal expansion. You have no doubt heard of highway pavements buckling on a hot summer day (Fig. 14.15). Bridges are built with special joints that allow for expansion and contraction of the bridge deck (Fig. 14.16). Similarly, TV towers, pipelines, and buildings must be designed and built to allow for this expansion and contraction.

There are some advantages to solids expanding. A bimetallic strip is made by fusing two different metals together side by side as illustrated in Fig. 14.17. When heated, the brass expands more than the steel, which makes the strip curve. If the bimetallic strip is cooled below room

Figure 14.16 Thermal expansion joint on a bridge.

Figure 14.17 When the thin bimetal strip is heated, it bends because of unequal expansion of the two metals. Here brass expands more than steel when heated, making the strip bend toward the steel side.

Courtesy of Peter Arnold, Inc. Reprinted with permission

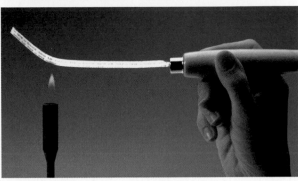

temperature, the brass will contract more than the steel, forcing the strip to curve in the opposite direction. The thermostat operates on this principle. As shown in Fig. 14.18, the basic parts of a thermostat are a bimetallic strip on the right and a regular metal strip on the left. The bimetallic strip of brass and steel bends with the temperature. The regular metal strip is moved by hand to set the temperature desired. This particular bimetallic strip is made and placed so that it bends to the left when cooled. As a result, when it comes in contact with the strip on the left, it completes a circuit, which turns on the furnace. When the room warms to the desired temperature, the bimetallic strip moves back to the right, which opens the contacts and shuts off the heat. Bimetallic strips are in spiral form in some thermostats (Fig. 14.19).

Figure 14.18 Simple thermostat.

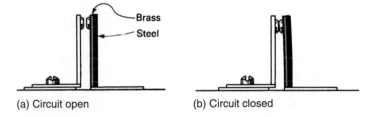

(a) Circuit open   (b) Circuit closed

Figure 14.19 These inner workings of a thermostat show the bimetal coil that expands and contracts as the temperature changes to activate the mercury bulb switch.

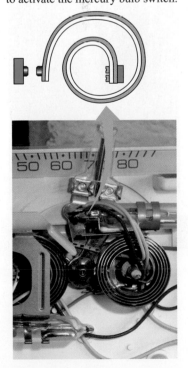

Linear Expansion

The amount that a solid expands depends on the following:

1. **Material.** Different materials expand at different rates. Steel expands at a rate less than that of brass.
2. **Length of the solid.** The longer the solid, the larger is the expansion. A 20.0-cm steel rod will expand twice as much as a 10.0-cm steel rod.
3. **Amount of change in temperature.** The greater the change in temperature, the greater is the expansion.

This can be written as a formula:

$$\Delta l = \alpha l \Delta T$$

where Δl = change in length
α = a constant called the **coefficient of linear expansion***
l = original length
ΔT = change in temperature

---

*Defined as change in unit length of a solid when its temperature is changed 1 degree.

Table 14.3 lists the coefficients of linear expansion for some common solids.

Table 14.3   Coefficients of Linear Expansion

Material	α (metric)	α (U.S.)
Aluminum	$2.3 \times 10^{-5}/C°$	$1.3 \times 10^{-5}/F°$
Brass	$1.9 \times 10^{-5}/C°$	$1.0 \times 10^{-5}/F°$
Concrete	$1.1 \times 10^{-5}/C°$	$6.0 \times 10^{-6}/F°$
Copper	$1.7 \times 10^{-5}/C°$	$9.5 \times 10^{-6}/F°$
Glass	$9.0 \times 10^{-6}/C°$	$5.1 \times 10^{-6}/F°$
Pyrex	$3.0 \times 10^{-6}/C°$	$1.7 \times 10^{-6}/F°$
Steel	$1.3 \times 10^{-5}/C°$	$6.5 \times 10^{-6}/F°$
Zinc	$2.6 \times 10^{-5}/C°$	$1.5 \times 10^{-5}/F°$

Comparing the coefficients of linear expansion of common glass and Pyrex, we can see that Pyrex expands and contracts approximately one-third as much as glass. This is why it is used in cooking and chemical laboratories.

EXAMPLE 1

A steel railroad rail is 40.0 ft long at 0°F. How much will it expand when heated to $10\overline{0}$°F?

Data:

$$l = 40.0 \text{ ft}$$
$$\Delta T = 10\overline{0}°F - 0°F = 10\overline{0}°F$$
$$\alpha = 6.5 \times 10^{-6}/F°$$
$$\Delta l = ?$$

Basic Equation:

$$\Delta l = \alpha l \Delta T$$

Working Equation: Same

Substitution:

$$\Delta l = (6.5 \times 10^{-6}/F°)(40.0 \text{ ft})(10\overline{0}°F)$$
$$= 0.026 \text{ ft} \quad \text{or} \quad 0.31 \text{ in.}$$

. . . .

Figure 14.20 Expansion loop in a steam pipe allows for temperature variations.

Pipes that undergo large temperature changes are installed to allow for expansion and contraction (Fig. 14.20).

EXAMPLE 2

What allowance for expansion must be made for a steel pipe $12\overline{0}$ m long that handles coolants and must undergo temperature changes of $20\overline{0}$°C?

Data:

$$\alpha = 1.3 \times 10^{-5}/C°$$
$$l = 12\overline{0} \text{ m}$$
$$\Delta T = 20\overline{0}°C$$
$$\Delta l = ?$$

Basic Equation:

$$\Delta l = \alpha l \Delta T$$

Working Equation: Same

Substitution:

$$\Delta l = (1.3 \times 10^{-5}/\mathrm{C}°)(12\overline{0}\text{ m})(20\overline{0}\text{ °C})$$
$$= 0.31\text{ m} \quad \text{or} \quad 31\text{ cm}$$

. . . .

Area and Volume Expansion of Solids

Solids expand in width and thickness as well as in length when heated. The area of a hole cut out of a metal sheet will expand in the same way as the surrounding material. To allow for this expansion the following formulas are used:

Area expansion: $\quad \Delta A = 2\alpha A\ \Delta T$
Volume expansion: $\quad \Delta V = 3\alpha V\ \Delta T$

where A = original area
       V = original volume

The top of a circular copper disk has an area of 64.2 in^2 at $2\overline{0}$°F. What is the change in area when the temperature is increased to $15\overline{0}$°F?

EXAMPLE 3

Data:

$$\alpha = 9.5 \times 10^{-6}/\text{F}°$$
$$A = 64.2\text{ in}^2$$
$$\Delta T = 15\overline{0}°\text{F} - 2\overline{0}°\text{F} = 13\overline{0}°\text{F}$$
$$\Delta A = ?$$

Basic Equation:

$$\Delta A = 2\alpha A\ \Delta T$$

Working Equation: Same

Substitution:

$$\Delta A = 2(9.5 \times 10^{-6}/\text{F}°)(64.2\text{ in}^2)(13\overline{0}°\text{F})$$
$$= 0.16\text{ in}^2$$

. . . .

A section of concrete in a bridge support measures 6.00 m × 12.0 m × 30.0 m at 38°C. What allowance for change in volume is necessary for a temperature drop to −15°C?

EXAMPLE 4

Data:

$$V = (6.00\text{ m})(12.0\text{ m})(30.0\text{ m}) = 2160\text{ m}^3$$
$$\alpha = 1.1 \times 10^{-5}/\text{C}°$$
$$\Delta T = 38°\text{C} - (-15°\text{C}) = 53°\text{C}$$

Basic Equation:

$$\Delta V = 3\alpha V\ \Delta T$$

Working Equation: Same

Substitution:

$$\Delta V = 3(1.1 \times 10^{-5}/\mathscr{C}°)(2160 \text{ m}^3)(53°\mathscr{C})$$
$$= 3.8 \text{ m}^3$$

· · · · · · · · · · · · · · · · ·

SKETCH

4.0 cm

DATA

$A = 12 \text{ cm}^2$, $l = 4.0 \text{ cm}$, $w = ?$

BASIC EQUATION

$A = lw$

WORKING EQUATION

$w = \frac{A}{l}$

SUBSTITUTION

$w = \frac{12 \text{ cm}^2}{4.0 \text{ cm}} = 3.0 \text{ cm}$

PROBLEMS 14.6

1. Find the increase in length of copper tubing 200.0 ft long at 40.0°F when it is heated to 200.0°F.

2. Find the increase in length of a zinc rod 50.0 m long at 15.0°C when it is heated to 130.0°C.

3. Find the increase in length of 300.00 m of copper wire when its temperature changes from 14°C to 34°C.

4. A steel pipe 8.25 m long is installed at 45°C. Find the decrease in length when coolants at −60°C pass through the pipe.

5. A steel tape measures 200.00 m at 15°C. What is its length at 55°C?

6. A brass rod 1.020 m long expands 3.0 mm when it is heated. Find the temperature change.

7. The road bed on a bridge 500.0 ft long is made of concrete. What allowance is needed for temperatures of −40°F in winter and 140°F in summer?

8. An aluminum plug has a diameter of 10.003 cm at 40.0°C. At what temperature will it fit precisely into a hole of constant diameter 10.000 cm?

9. The diameter of a steel drill at 45°F is 0.750 in. Find its diameter at 375°F.

10. A brass ball with diameter 12.000 cm is 0.011 cm too large to pass through a hole in a copper plate when the ball and plate are at a temperature of 20.0°C. What is the temperature of the ball when it will just pass through the plate, assuming that the temperature of the plate does not change? What is the temperature of the plate when the ball will just pass through, assuming that the temperature of the ball does not change?

11. A brass cylinder has a cross-sectional area of 482 cm² at −5°C. Find its change in area when heated to 95°C.

12. The volume of the cylinder in Problem 11 is 4820 cm³ at 240.0°C. Find its change in volume when cooled to −75.0°C.

13. An aluminum pipe has a cross-sectional area of 88.40 cm² at 15°C. What is its cross-sectional area when the pipe is heated to 155°C?

14. A steel pipe has a cross-sectional area of 127.20 in² at 25°F. What is its cross-sectional area when the pipe is heated to 175°F?

15. A glass plug has a volume of 60.00 cm³ at 12°C. What is its volume at 76°C?

16. The diameter of a hole drilled through brass at 21°C measures 6.500 cm. Find the diameter and area of the hole when the brass is heated to 175°C.

17. Steel rails 15.000 m long are laid at 10.0°C. How much space should be left between them if they are to just touch at 35.0°C?

18. Steel beams 60.000 ft long are placed in a highway overpass to allow for expansion and contraction. The temperature range allowance is −30°F to 130°F.
    (a) Find the space allowance (in inches) between the beams at −30°F if the beams touch at 130°F.
    (b) Find the space allowance between the beams if placed at 75°F and touch at 130°F.

19. The spaces between 13.00-m steel rails are 0.711 cm at −15°C. If the rails touch at 35.5°C, what is the coefficient of linear expansion?

20. A section of concrete dam is a rectangular solid 20.0 ft by 50.0 ft by 80.0 ft at 115°F. What allowance for change in volume is necessary for a temperature of −15°F?

21. A glass ball has a radius of 12.000 cm at 6.0°C. Find its change in volume when the temperature is increased to 81.0°C.

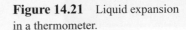

22. Find the final height of a concrete column that is 1.250 m × 1.250 m × 4.250 m at 0.0°C when the column is heated to 45.0°C.
23. What is the final volume of a glass right circular cylinder with original height 1.200 m and radius 30.00 cm that is heated from 13.0°C to 56.0°C?
24. A metal bar at 21.0°C is 2.6000 m long. If the bar is heated to 93.0°C, its change in length is 3.40 mm. What is the coefficient of linear expansion of the bar?

14.7 Expansion of Liquids

Liquids also generally expand when heated and contract when cooled. The thermometer is made using this principle. When a thermometer is placed under your tongue, the heat from your mouth causes the liquid in the bottom of the thermometer to expand. The liquid is then forced to rise up the thin calibrated tube (Fig. 14.21). Similarly, when the gasoline tank on a car is filled to capacity on a hot summer day and then parked in a hot parking lot, it overflows. The cold gas from the underground storage tanks expands as it warms up in the car's tank. An automobile radiator filled to the brim with cold water would likewise overflow as the engine heats the water. The formula for *volume expansion of liquids* is

$$\Delta V = \beta V \Delta T$$

where β = coefficient of volume expansion for liquids
         V = original volume

Table 14.4 lists the coefficients of volume expansion for some common liquids.

Figure 14.21 Liquid expansion in a thermometer.

Table 14.4   Coefficients of Volume Expansion

Liquid	β (metric)	β (U.S.)
Acetone	$1.49 \times 10^{-3}/\text{C}°$	$8.28 \times 10^{-4}/\text{F}°$
Alcohol, ethyl	$1.12 \times 10^{-3}/\text{C}°$	$6.62 \times 10^{-4}/\text{F}°$
Carbon tetrachloride	$1.24 \times 10^{-3}/\text{C}°$	$6.89 \times 10^{-4}/\text{F}°$
Mercury	$1.8 \times 10^{-4}/\text{C}°$	$1.0 \times 10^{-4}/\text{F}°$
Petroleum	$9.6 \times 10^{-4}/\text{C}°$	$5.33 \times 10^{-4}/\text{F}°$
Turpentine	$9.7 \times 10^{-4}/\text{C}°$	$5.39 \times 10^{-4}/\text{F}°$
Water	$2.1 \times 10^{-4}/\text{C}°$	$1.17 \times 10^{-4}/\text{F}°$

EXAMPLE 1

If petroleum at 0°C occupies $25\overline{0}$ L, what is its volume at $5\overline{0}$°C?

Data:

$$\beta = 9.6 \times 10^{-4}/\text{C}°$$
$$V = 25\overline{0}\ \text{L}$$
$$\Delta T = 5\overline{0}°\text{C}$$
$$\Delta V = ?$$

Basic Equation:

$$\Delta V = \beta V \Delta T$$

Working Equation: Same

Substitution:

$$\Delta V = (9.6 \times 10^{-4}/\cancel{\text{C}}°)(25\overline{0}\ \text{L})(5\overline{0}°\cancel{\text{C}})$$
$$= 12\ \text{L}$$
$$\text{volume at } 5\overline{0}°\text{C} = V + \Delta V$$
$$= 25\overline{0}\ \text{L} + 12\ \text{L} = 262\ \text{L}$$

EXAMPLE 2

Find the increase in volume of 18.2 in³ of water when the water is heated from $40^-°F$ to $180^-°F$.

Data:

$$\beta = 1.17 \times 10^{-4}/F°$$
$$V = 18.2 \text{ in}^3$$
$$\Delta T = 180^-°F - 40^-°F = 140^-°F$$
$$\Delta V = ?$$

Basic Equation:

$$\Delta V = \beta V \Delta T$$

Working Equation: Same

Substitution:

$$\Delta V = (1.17 \times 10^{-4}/F°)(18.2 \text{ in}^3)(140^-°F)$$
$$= 0.298 \text{ in}^3$$

Figure 14.22 Expansion of water in change from liquid to solid.

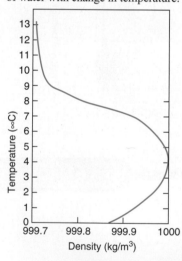

Figure 14.23 Change in density of water with change in temperature.

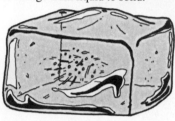

Expansion of Water

Water is unusual in its expansion characteristics. Recall the mound in the middle of each ice cube in an ice cube tray (Fig. 14.22). This is evidence of the expansion of water during its change of state from liquid to solid form.

Nearly all liquids are the most dense at their lowest temperature before a change of phase to become solids. As the temperature drops, the molecular motion slows and the substance becomes denser. Water does not follow this general rule. Because of its unusual structural characteristics, water is most dense at 4°C or 39.2°F instead of 0°C or 32°F. A graph of its change in density with increase in temperature is shown in Fig. 14.23. As ice melts and the water temperature is slightly increased, there are still groups of molecules that have the open crystallographic structure of ice, which is less dense than water. As the water is heated to 4°C, these groupings disappear and the water becomes denser. Above 4°C, water then expands normally as the temperature is raised.

This unique behavior of water is critically important in lakes that freeze in winter. If, as in most liquids, water were most dense at its freezing point, the coldest water would settle to the bottom and the lake would freeze from the bottom up. Any living creatures would be killed. Fortunately, the most dense water at 4°C is at the bottom and the less dense water at 0°C is above it, so ice forms at the surface, the water below remains liquid, and the lake freezes from the top down.

When ice melts at 0°C or 32°F, the water formed *contracts* as the temperature is raised to 4°C or 39.2°F. Then it begins to *expand*, as do most other liquids.

PHYSICS CONNECTIONS

Hot Fuel

The U.S. government has been investigating the possibility of regulations on the sale of "hot fuel." Petroleum, like all liquids, expands when its temperature increases; however, the amount of energy in a given amount of fuel does not increase as its volume expands. Fuel pricing is based on a price per gallon at a temperature of 60°F. Fuel at the distribution center is typically pumped from 60°F underground storage tanks into tanker trucks that transport the fuel to local gas stations, where the fuel tends to heat up and expand.

During the summer, the temperature of fuel often reaches 90°F as it is stored at local gas stations and as it enters an automobile's gas tank. (Fig. 14.24). As a result, a consumer receives 1 gallon of expanded gasoline, which does not have as much energy as contained in the gasoline stored in the 60°F underground storage tanks.

Figure 14.24 On a summer day, the consumer receives less energy per gallon of gas than on a cool day.

Let's put the formula for the volume expansion of liquids to work. One gallon is equal to 231 in³. When gas is pumped into a tank at 90°F, the gas becomes less dense and occupies a volume of 235 in³. However, the pump still only pumps 231 in³ into the tank, shortchanging the consumer of the energy contained in the missing 4 in³ of petroleum.

The impact on consumers of the sale of hot fuel is greater in warm climates than in cooler, northern climates. In the United States it is estimated that consumers are short-changed by as much as $1.5 billion over the course of a summer as a result of the sale of hot fuel.

PROBLEMS 14.7

1. A quantity of carbon tetrachloride occupies 625 L at 12°C. Find its volume at 48°C.
2. Some mercury occupies 157 in³ at −30°F. What is its change in volume when heated to 90°F?
3. Some petroleum occupies 11.7 m³ at −17°C. Find its volume at 28°C.
4. Find the increase in volume of 35 L of acetone heated from 28°C to 38°C.
5. Some water at 180°F occupies 3780 ft³. What is its volume at 122°F?
6. A 1200-L tank of petroleum is completely filled at 9°C. How much spills over if the temperature rises to 45°C?
7. Find the increase in volume of 215 cm³ of mercury when its temperature increases from 10°C to 25°C.
8. Find the decrease in volume of 2000 ft³ of alcohol in a railroad tank car if the temperature drops from 75°F to 54°F.
9. A gasoline service station owner receives a truckload of 34,000 L of gasoline at 32°C. It cools to 15°C in the underground tank. At 75 cents/L, how much money is lost as a result of the contraction of the gasoline?
10. A Pyrex container is completely filled with 275 cm³ of mercury at 10.0°C. How much mercury spills over when heated to 75.0°C?
11. What was the temperature of 180 mL of acetone before it was heated to 98°C and increased to a volume of 200 mL?
12. What is the increase in volume of 1200 L of petroleum as it warms from 3.0°C to 25.0°C?
13. Five hundred litres of petroleum at 4.0°C is heated to 30.0°C. What is its increase in volume?

SKETCH

12 cm² | w

4.0 cm

DATA

$A = 12$ cm², $l = 4.0$ cm, $w = ?$

BASIC EQUATION

$A = lw$

WORKING EQUATION

$w = \frac{A}{l}$

SUBSTITUTION

$w = \frac{12 \text{ cm}^2}{4.0 \text{ cm}} = 3.0$ cm

14.8 Change of Phase

Many industries are concerned with a change of phase in the materials they use. In foundries the principal activity is to change solid metals to liquid, pour the liquid metal into molds, and allow it to become solid again (Fig. 14.25). **Change of phase** (sometimes called *change of state*) is a change in a substance from one form of matter (solid, liquid, or gas) to another.

Figure 14.25 Molten pig iron from the blast furnace is poured into an open hearth furnace, refined, and purified into steel at temperatures about 2900°F.

Copyright of Michael Rosenfeld/Getty Images, Inc.

Fusion

The change of phase from solid to liquid is called **melting** or **fusion**. The change from liquid to solid is called **freezing** or **solidification**. Most solids have a crystalline structure and a definite melting point at any given pressure. Melting and solidification of these substances occur at the same temperature. For example, water at 0°C (32°F) changes to ice and ice changes to water at the same temperature. There is no temperature change during change of phase. Ice at 0°C changes to water at 0°C. Only a few substances, such as butter and glass, have no particular melting temperature but change phase gradually.

Although there is no temperature change during a change of phase, *there is a transfer of heat*. A melting solid *absorbs* heat and a solidifying liquid *gives off* heat. When 1 g of ice at 0°C melts, it absorbs $8\overline{0}$ cal of heat. Similarly, when 1 g of water freezes at 0°C, ice at 0°C is produced, and $8\overline{0}$ cal of heat is released.

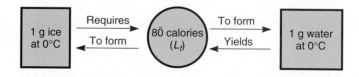

When 1 kg of ice at 0°C melts, it absorbs $8\overline{0}$ kcal of heat. Similarly, when 1 kg of water freezes at 0°C, ice at 0°C is produced and $8\overline{0}$ kcal of heat is released.

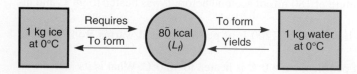

Or, when 1 kg of ice at 0°C melts, it absorbs 335 kilojoules (kJ) of heat. Then, when 1 kg of water freezes at 0°C, ice at 0°C is produced and 335 kJ of heat is released.

When 1 lb of ice at 32°F melts, it absorbs 144 Btu of heat. Similarly, when 1 lb of water freezes at 32°F, ice at 32°F is produced and 144 Btu of heat is released.

The amount of heat required to melt 1 g or 1 kg or 1 lb of a liquid is called its **heat of fusion**, designated L_f.

$$L_f = \frac{Q}{m} \quad \text{(metric)} \qquad L_f = \frac{Q}{w} \quad \text{(U.S.)}$$

where $L_f =$ heat of fusion (see Table 15 in Appendix C)
$Q =$ quantity of heat
$m =$ mass of substance (metric system)
$w =$ weight of substance (U.S. system)

If 1340 kJ of heat is required to melt 4.00 kg of ice at 0°C into water at 0°C, what is the heat of fusion of water?

EXAMPLE 1

Data:

$$Q = 1340 \text{ kJ}$$
$$m = 4.00 \text{ kg}$$
$$L_f = ?$$

Basic Equation:

$$L_f = \frac{Q}{m}$$

Working Equation: Same

Substitution:

$$L_f = \frac{1340 \text{ kJ}}{4.00 \text{ kg}}$$
$$= 335 \text{ kJ/kg}$$

. . . .

heat of fusion (water) $= 80\overline{}$ cal/g, or $80\overline{}$ kcal/kg, or 335 kJ/kg, or 144 Btu/lb

A very interesting (and delicious) change-of-phase activity is to make homemade ice cream. A sealed container with a mixture of milk, egg, vanilla, and sugar is submerged in a mixture of rock salt and crushed ice. The salt causes the ice to rapidly melt, which requires heat while the ice changes phase from solid to liquid. Most of this heat is transferred from the ice cream mixture, which hardens into ice cream.

Vaporization

The change of phase from a liquid to a gas or vapor is called **vaporization**. A pot of boiling water (Fig. 14.26) vividly shows this change of phase as the steam evaporates and leaves the liquid. Note that vaporization requires that heat be supplied; in this case heat is required to boil the water. The reverse process (change from a gas to a liquid) is called **condensation**. As steam condenses in radiators (Fig. 14.27), large amounts of heat are released.

Figure 14.26 Heat supplied to boiling water changes liquid water into steam—the gas form of water.

Figure 14.27 A large amount of heat is released by condensation of steam in a radiator.

At the point of condensation, the vapor becomes *saturated*; that is, the vapor cannot hold any more moisture. For example, water vapor is always present in some amount in the earth's atmosphere. The weather term **relative humidity** is the ratio of the actual amount of vapor in the atmosphere to the amount of vapor required to reach 100% of saturation at the existing temperature. As the air temperature decreases without change in pressure or vapor content, the relative humidity increases until it reaches 100% at saturation. The temperature at which saturation is reached is called the **dew point**. Once saturation is reached and the temperature continues to decrease, condensation occurs in the form of dew, fog, mist, clouds, and rain or other forms of precipitation.

While a liquid is boiling, the temperature of the liquid does not change. However, there is a transfer of heat. A liquid being vaporized (boiled) *absorbs* heat. As a vapor condenses, heat is given off.

The amount of heat required to vaporize 1 g or 1 kg or 1 lb of a liquid is called its **heat of vaporization**, designated L_v. So, when 1 g of water at $10\overline{0}°C$ changes to steam at $10\overline{0}°C$, it absorbs $54\overline{0}$ cal; when 1 g of steam at $100°C$ condenses to water at $100°C$, $54\overline{0}$ cal of heat is given off. The tremendous amount of heat released accounts for the potential for far more serious burns from steam than from hot water.

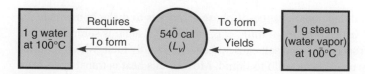

When 1 kg of water at $10\overline{0}°C$ changes to steam at $10\overline{0}°C$, it absorbs $54\overline{0}$ kcal of heat. Similarly, when 1 kg of steam at $10\overline{0}°C$ condenses to water at $100°C$, $54\overline{0}$ kcal of heat is given off.

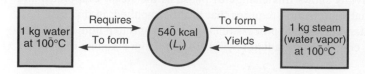

Or, when 1 kg of water at $10\overline{0}°C$ changes to steam at $10\overline{0}°C$, it absorbs 2.26 MJ (2.26×10^6 J) of heat. Then, when 1 kg of steam at $10\overline{0}°C$ condenses to water at $10\overline{0}°C$, 2.26 MJ of heat is given off.

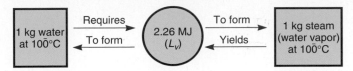

When 1 lb of water at 212°F changes to steam at 212°F, $97\overline{0}$ Btu of heat is absorbed; when 1 lb of steam at 212°F condenses to water at 212°F, $97\overline{0}$ Btu of heat is given off.

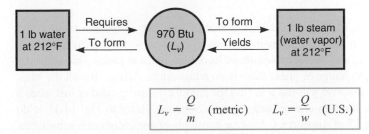

$$L_v = \frac{Q}{m} \quad \text{(metric)} \qquad L_v = \frac{Q}{w} \quad \text{(U.S.)}$$

where L_v = heat of vaporization (see Table 15 in Appendix C)
        Q = quantity of heat
        m = mass of substance (metric system)
        w = weight of substance (U.S. system)

If 135,000 cal of heat is required to vaporize $25\overline{0}$ g of water at $10\overline{0}°C$, what is the heat of vaporization of water?

EXAMPLE 2

Data:

$$Q = 135,000 \text{ cal}$$
$$m = 25\overline{0} \text{ g}$$
$$L_v = \text{?}$$

Basic Equation:

$$L_v = \frac{Q}{m}$$

Working Equation: Same

Substitution:

$$L_v = \frac{135,000 \text{ cal}}{25\overline{0} \text{ g}}$$
$$= 54\overline{0} \text{ cal/g}$$

· · · · · · · · · · · · · · · · ·

heat of vaporization (water) = $54\overline{0}$ cal/g, or $54\overline{0}$ kcal/kg, or 2.26 MJ/kg, or $97\overline{0}$ Btu/lb

If 15.8 MJ of heat is required to vaporize 18.5 kg of ethyl alcohol at 78.5°C (its boiling point), what is the heat of vaporization of ethyl alcohol?

EXAMPLE 3

Data:

$$Q = 15.8 \text{ MJ}$$
$$m = 18.5 \text{ kg}$$
$$L_v = \text{?}$$

Basic Equation:

$$L_v = \frac{Q}{m}$$

Working Equation: Same

Substitution:

$$L_v = \frac{15.8 \text{ MJ}}{18.5 \text{ kg}}$$

$$= 0.854 \text{ MJ/kg} \quad \text{or} \quad 854 \text{ kJ/kg} \quad \text{or} \quad 8.54 \times 10^5 \text{ J/kg}$$

·················

Figures 14.28 through 14.30 show the heat gained by one unit of ice at a temperature below its melting point as it warms to its melting point, changes to water, warms to its boiling point, changes to steam, and then is heated above its boiling point in joules, Btu, and calories. Note that during each change of phase there is no temperature change. Recall the basic shape of these graphs because we will use it to find the amount of heat gained or lost when a quantity of material goes through one or both changes of phase. Refer to Fig. 14.31 to do such problems. See Table 15 of Appendix C for heat constants of some common substances.

Figure 14.28

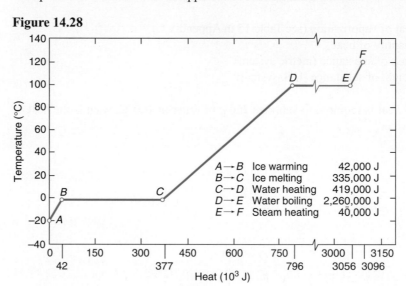

Heat gained by one kilogram of ice at −20°C as it is converted to steam at 120°C

Figure 14.29

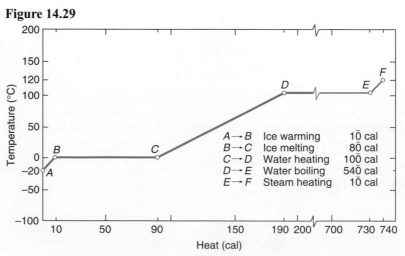

Heat gained by one gram of ice at −20°C as it is converted to steam at 120°C

Figure 14.30

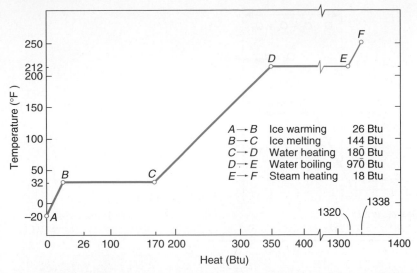

A→B   Ice warming   26 Btu
B→C   Ice melting   144 Btu
C→D   Water heating 180 Btu
D→E   Water boiling 970 Btu
E→F   Steam heating   18 Btu

Heat gained by one pound of ice at −20°F as it is converted to steam at 250°F

Figure 14.31   Graph of heat transfer during change of phase.

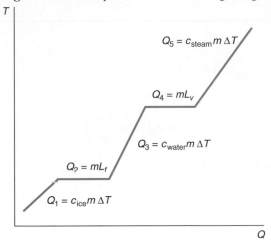

$Q_5 = c_{steam} m \Delta T$

$Q_4 = mL_v$

$Q_3 = c_{water} m \Delta T$

$Q_2 = mL_f$

$Q_1 = c_{ice} m \Delta T$

How many Btu of heat are released when 4.00 lb of steam at 222°F is cooled to water at 82°F?

EXAMPLE 4

To find the amount of heat released when steam at a temperature above its vaporization point is cooled to water below its boiling point, we need to consider three amounts (see Fig. 14.32):

$Q_5 = c_{steam} w \Delta T$ (amount of heat released as the steam changes temperature from 222°F to 212°F)

$Q_4 = wL_v$ (amount of heat released as the steam changes to water)

$Q_3 = c_{water} w \Delta T$ (amount of heat released as the water changes temperature from 212°F to 82°F)

So the total amount of heat released is

$$Q = Q_5 + Q_4 + Q_3$$

Data:

$$w = 4.00 \text{ lb}$$
$$T_i \text{ of steam} = 222°F$$
$$T_f \text{ of water} = 82°F$$
$$Q = ?$$

Figure 14.32

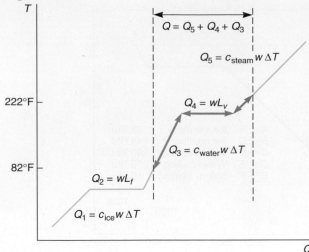

Basic Equation:

$$Q = Q_5 + Q_4 + Q_3$$

Working Equation:

$$Q = c_{steam}w\Delta T + wL_v + c_{water}w\Delta T$$

Substitution:

$$Q = \left(0.48\frac{Btu}{lb\ °F}\right)(4.00\ lb)(1\overline{0}°F) + (4.00\ lb)\left(97\overline{0}\ \frac{Btu}{lb}\right)$$

$$+ \left(1.00\frac{Btu}{lb\ °F}\right)(4.00\ lb)(13\overline{0}°F)$$

$$= 4420\ Btu$$

· · · · · · · · · · · · · · · · · ·

EXAMPLE 5

How many joules of heat are needed to change 3.50 kg of ice at $-15.0°C$ to steam at $120.0°C$?

Sketch:

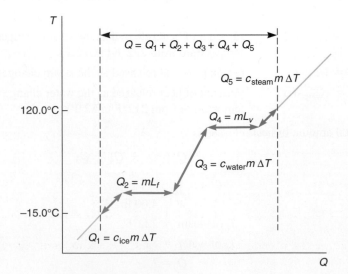

Data:

$$m = 3.50 \text{ kg}$$
$$T_i \text{ of ice} = -15.0°C$$
$$T_f \text{ of steam} = 120.0°C$$
$$Q = ?$$

Basic Equation:

$$Q = Q_1 + Q_2 + Q_3 + Q_4 + Q_5$$

Working Equation:

$$Q = c_{ice}m\Delta T + mL_f + c_{water}m\Delta T + mL_v + c_{steam}m\Delta T$$

Substitution:

$$Q = \left(2100 \frac{J}{kg \, °C}\right)(3.50 \text{ kg})(15.0°C) + (3.50 \text{ kg})\left(335 \frac{kJ}{kg}\right)$$
$$+ \left(4190 \frac{J}{kg \, °C}\right)(3.50 \text{ kg})(100.0°C) + (3.50 \text{ kg})\left(2.26 \frac{MJ}{kg}\right)$$
$$+ \left(2000 \frac{J}{kg \, °C}\right)(3.50 \text{ kg})(20.0°C)$$
$$= 1.080 \times 10^7 \text{ J} \quad \text{or} \quad 10.80 \text{ MJ}$$

Evaporation as a Cooling Process

Evaporation is the process by which high-energy molecules of a liquid continually leave its surface. This change of a liquid to a gas helps keep your body cool. When you become too warm, your sweat glands produce water, which evaporates from your skin. As the water evaporates, your body loses heat at the rate of $\frac{Q}{m} = L_v$. In a cool summer breeze, the perspiration evaporates more rapidly, which cools you faster. On a hot, humid day, you tend to remain hot because the perspiration does not evaporate as quickly.

Years ago when a person, especially a child, had a very high body temperature, the common medical practice was to rub the body with rubbing alcohol because it quickly evaporates from the skin. As it evaporates, it removes heat from the body. This practice is not recommended now because we know the alcohol is also absorbed in the body.

During evaporation the molecules of a liquid continually leave the surface of the liquid. Some molecules have enough energy to leave, freeing themselves from the liquid's surface. The rest, not having enough energy to leave, fall back and remain as part of the liquid. The rate of evaporation of a liquid depends on the following:

1. *Amount of surface area.* The larger the surface area, the greater is the number of molecules that have a chance to escape from the surface.
2. *Temperature.* The higher the temperature, the higher is the molecular energy of the molecules, which allows more molecules to escape.
3. *Surface currents.* Air currents blowing over the liquid's surface remove many of the molecules that have evaporated before they fall back into the liquid, which is why a cool summer breeze "feels so good."
4. *Volatility.* The **volatility** of a liquid is a measure of its ability to vaporize. Examples of highly volatile liquids are rubbing alcohol and gasoline. The more volatile the liquid, the greater is its rate of evaporation.
5. *Pressure on or above the liquid.* The lower the pressure, the greater is the rate of evaporation. Under a partial vacuum, there are fewer molecules available with which the liquid molecules may collide, allowing for a higher rate of escape and a higher rate of evaporation.
6. *Humidity.* If the liquid is exposed to the atmosphere, lower humidity values will provide for greater evaporation.

Heat Pump

A **heat pump** is a device that warms or cools by transferring heat from a lower-temperature source to a higher-temperature source or vice versa. It is often used for heating in the winter and cooling in the summer.

A heat pump contains a vapor, usually called a *refrigerant*, that is easily condensed to a liquid when under pressure. The liquid refrigerant gives up the heat gained during compression to the higher-temperature source. The liquid is then released to a low-pressure part of the heat pump, where it quickly evaporates and takes its heat of vaporization from the lower-temperature heat source. The vapor is then compressed again, and the cycle repeats. Work is done on the vapor for the heat pump to transfer the heat from a lower-temperature source to a higher-temperature one.

In winter, a house may be heated by a heat pump that extracts heat from the outside air and transfers it into the house. In summer, the heat pump is reversed and extracts heat from the inside air and transfers it to the outdoors (Fig. 14.33).

Refrigerators and freezers are forms of heat pumps. Heat is transferred from the appliance to the room.

Figure 14.33 The heat pump is used to heat a home in winter and cool it in summer.

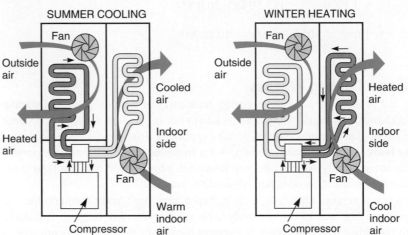

Effects of Pressure and Impurities on Change of Phase

Automobile cooling systems present important problems concerning change of phase. Most substances contract on solidifying. However, water and a few other substances expand. The tremendous force exerted by this expansion is shown by the number of cracked automobile blocks and burst radiators suffered by careless motorists every winter.

Impurities in water tend to *lower* the freezing point. Alcohol has a lower freezing point than water and is used in some types of antifreeze. By mixing water with antifreeze in the cooling system, an unknowing person can raise the freezing point of the coolant, allowing it to freeze and crack the engine block. Automobile engines may also be ruined in winter by overheating if the water in the radiator is frozen, preventing the engine from being cooled by circulation in the system.

An increase in the pressure on a liquid *raises* the boiling point. Automobile manufacturers utilize this fact by pressurizing their cooling systems and thereby raising the boiling point of the coolant used.

A decrease in the pressure on a liquid *lowers* the boiling point. Frozen concentrated orange juice is produced by subjecting the pure juice to very low pressures at which the water in the juice is evaporated. Then the consumer must restore the lost water before serving the juice.

PROBLEMS 14.8

1. How many calories of heat are required to melt 14.0 g of ice at 0°C?
2. How many pounds of ice at 32°F can be melted by the addition of 635 Btu of heat?
3. How many Btu of heat are required to vaporize 11.0 lb of water at 212°F?
4. How many grams of steam in a boiler at $10\overline{0}$°C can be condensed to water at $10\overline{0}$°C by the removal of 1520 cal of heat?
5. How many calories of heat are required to melt 320 g of ice at 0°C?
6. How many calories of heat are given off when 3250 g of steam is condensed to water at $10\overline{0}$°C?
7. How many joules of heat are required to melt 20.0 kg of ice at 0°C?
8. How many kilocalories of heat are required to melt 20.0 kg of ice at 0°C?
9. How many joules of heat need to be removed to condense 1.50 kg of steam at $10\overline{0}$°C?
10. How many litres of water at $10\overline{0}$°C are vaporized by the addition of 5.00 MJ of heat?
11. How many Btu of heat are required to melt 33.0 lb of ice at 32°F and to raise the temperature of the melted ice to 72°F?
12. How many Btu of heat are released when 20.0 lb of water at $8\overline{0}$°F is cooled to 32°F and then frozen in an ice plant?
13. How many Btu of heat are required to change 9.00 lb of ice at 10°F to steam at 232°F?
14. How many calories of heat are released when $20\overline{0}$ g of steam at $12\overline{0}$°C is changed to ice at −12°C?
15. How many kilocalories of heat are required to melt 50.0 kg of ice at 0°C and to raise the temperature of the melted ice to $2\overline{0}$°C?
16. How many joules of heat are required to melt 15.0 kg of ice at 0°C and to raise the temperature of the melted ice to 75°C?
17. How many joules of heat need to be removed from 1.25 kg of steam at 115°C to condense it to water and cool the water to $5\overline{0}$°C?
18. How many kcal of heat are needed to vaporize 5.00 kg of water at $10\overline{0}$°C and raise the temperature of the steam to 145°C?
19. How many calories of heat are needed to change 625 g of ice at −24.0°C to steam at 132.0°C?
20. How many Btu of heat must be withdrawn from 5.65 lb of steam at 236.0°F to change it to ice at 12.0°F?
21. How many kilocalories of heat are needed to change 143 N of ethyl alcohol at 65.0°C to vapor?
22. How many joules of heat does 620 g of mercury require to go from a solid at −38.9°C to vapor?

SKETCH

12 cm² | w

4.0 cm

DATA
$A = 12$ cm², $l = 4.0$ cm, $w = ?$

BASIC EQUATION
$A = lw$

WORKING EQUATION
$w = \frac{A}{l}$

SUBSTITUTION
$w = \frac{12 \text{ cm}^2}{4.0 \text{ cm}} = 3.0$ cm

Glossary

Absolute Zero The lowest possible temperature. (p. 371)

Btu (British thermal unit) The amount of heat (energy) necessary to raise the temperature of 1 lb of water 1°F. (p. 371)

Calorie The amount of heat necessary to raise the temperature of 1 g of water 1°C. (p. 371)

Celsius Scale The metric temperature scale on which ice melts at 0° and water boils at 100°. (p. 368)

Change of Phase (sometimes called *change of state*) A change in a substance from one form of matter (solid, liquid, or gas) to another. (p. 391)

Coefficient of Linear Expansion A constant that indicates the amount by which a solid expands or contracts when its temperature is changed 1 degree. (p. 385)

Condensation The change of phase from gas or vapor to a liquid. (p. 393)

Conduction A form of heat transfer from a warmer part of a substance to a cooler part as a result of molecular collisions, which cause the slower-moving molecules to move faster. (p. 373)

Convection A form of heat transfer by the movement of warm molecules from one region of a gas or a liquid to another. (p. 374)

Dew Point The temperature at which air becomes saturated with water vapor and condensation occurs. (p. 394)

Evaporation The process by which high-energy molecules of a liquid continually leave its surface. (p. 399)

Fahrenheit Scale The U.S. temperature scale on which ice melts at 32° and water boils at 212°. (p. 368)

Freezing The change of phase from liquid to solid. Also called *solidification*. (p. 392)

Fusion The change of phase from solid to liquid. Also called *melting*. (p. 392)

Heat A form of internal kinetic and potential energy contained in an object associated with the motion of its atoms or molecules and which may be transferred from an object at a higher temperature to one at a lower temperature. (p. 371)

Heat of Fusion The heat required to melt 1 g or 1 kg or 1 lb of a liquid. (p. 393)

Heat of Vaporization The amount of heat required to vaporize 1 g or 1 kg or 1 lb of a liquid. (p. 394)

Heat Pump A device that warms or cools by transferring heat. It contains a vapor (refrigerant) that is easily condensed to a liquid when under pressure. Produces heat during compression and cooling during vaporization. (p. 400)

Kelvin Scale The metric absolute temperature scale on which absolute zero is 0 K and the units are the same as on the Celsius scale. (p. 369)

Kilocalorie The amount of heat necessary to raise the temperature of 1 kg of water 1°C. (p. 371)

Mechanical Equivalent of Heat The relationship between heat and mechanical work. (p. 371)

Melting The change of phase from solid to liquid. Also called *fusion*. (p. 392)

Method of Mixtures When two substances at different temperatures are mixed together, heat flows from the warmer body to the cooler body until they reach the same temperature. Part of the heat lost by the warmer body is transferred to the cooler body and to surrounding objects. If the two substances are well insulated from surrounding objects, the heat lost by the warmer body is equal to the heat gained by the cooler body. (p. 381)

Radiation A form of heat transfer through energy being radiated or transmitted in the forms of rays, waves, or particles. (p. 375)

Rankine Scale The U.S. absolute temperature scale on which absolute zero is 0°R and the degree units are the same as on the Fahrenheit scale. (p. 369)

Relative Humidity Ratio of the actual amount of vapor in the atmosphere to the amount of vapor required to reach 100% of saturation at the existing temperature. (p. 394)

Solidification The change of phase from liquid to solid. Also called *freezing*. (p. 392)

Specific Heat The amount of heat necessary to change the temperature of 1 kg of a substance by 1°C in the metric system or 1 lb of a substance by 1°F in the U.S. system. (p. 379)

Temperature A measure of the hotness or coldness of an object. (p. 367)

Thermal Conductivity The ability of a material to transfer heat by conduction. (p. 375)

Vaporization The change of phase from liquid to a gas or vapor. (p. 393)

Volatility A measure of a liquid's ability to vaporize. The more volatile the liquid, the greater is its rate of evaporation. (p. 399)

Formulas

14.1 $T_C = \frac{5}{9}(T_F - 32°)$

$T_F = \frac{9}{5}T_C + 32°$

$T_K = T_C + 273$

$T_R = T_F + 460°$

14.3 $Q = \dfrac{KAt(T_2 - T_1)}{L}$

$R = \dfrac{L}{K}$

14.4 $Q = cm\Delta T$

$Q = cw\Delta T$

14.5 $Q_{lost} = Q_{gained}$

$c_l m_l (T_l - T_f) = c_g m_g (T_f - T_g)$

14.6 $\Delta l = \alpha l \Delta T$

$\Delta A = 2\alpha A \Delta T$

$\Delta V = 3\alpha V \Delta T$

14.7 $\Delta V = \beta V \Delta T$

14.8 $L_f = \dfrac{Q}{m}$ $L_f = \dfrac{Q}{w}$

$L_v = \dfrac{Q}{m}$ $L_v = \dfrac{Q}{w}$

Review Questions

1. Which of the following are methods of heat transfer?
    (a) Convection   (b) Conduction
    (c) Temperature   (d) Radiation
    (e) Potential energy
2. Which of the following are good conductors of heat?
    (a) Air   (b) Copper
    (c) Steel   (d) Aluminum
    (e) Brick   (f) Mineral wool
3. The amount that a solid expands when heated depends on
    (a) the type of material.   (b) the length of the solid.
    (c) the density of the solid.   (d) the amount of temperature change.
    (e) all of the above.
4. The rate of evaporation from the surface of a liquid depends on
    (a) the temperature.   (b) the volatility of the liquid.
    (c) the mass density of the liquid.   (d) the air pressure above the liquid.

5. The amount of heat required to melt 1 kg of a solid is called its
   (a) heat of vaporization.   (b) mass density.
   (c) weight density.   (d) heat of fusion.
   (e) volume expansion.
6. The operation of a simple thermostat depends on
   (a) the mechanical equivalent of heat.   (b) specific heat.
   (c) thermal expansion.   (d) the *R* value.
7. In your own words, describe the method of mixtures.
8. What is the mechanical equivalent of heat in the U.S. system?
9. Which other temperature scale is closely related to the Fahrenheit scale?
10. Which other temperature scale is closely related to the Celsius scale?
11. Distinguish between the Celsius and Fahrenheit temperature scales.
12. Distinguish between heat and temperature.
13. Give three examples of the conversion of heat into useful work.
14. Give three examples of the conversion of work into heat.
15. Should you wear light- or dark-colored clothing on a hot sunny summer day? Explain.
16. Does the area of a hole cut out of a metal block increase or decrease as the metal is heated? Explain.
17. (a) At what temperature does water have its highest density? (b) How does water differ from other liquids in this regard? (c) What is the impact of this unique characteristic of water?
18. Which would cool a hot object better: 10 kg of water at 0°C or 10 kg of ice at 0°C? Explain.
19. Steam can cause much more severe burns than hot water. Explain.
20. Why are ice cubes often observed to have a slight mound on the top of the cube?
21. In your own words, describe each method of heat transfer.
22. Describe why automotive cooling systems are designed to operate at elevated pressures.
23. Explain how a heat pump works to heat in the winter and cool in the summer.

Review Problems

SKETCH

4.0 cm

DATA

$A = 12 \text{ cm}^2, l = 4.0 \text{ cm}, w = ?$

BASIC EQUATION

$A = lw$

WORKING EQUATION

$w = \frac{A}{l}$

SUBSTITUTION

$w = \frac{12 \text{ cm}^2}{4.0 \text{ cm}} = 3.0 \text{ cm}$

1. Change 344 K to degrees Celsius.
2. Change 24°C to Kelvin.
3. Change 5110°C to degrees Fahrenheit.
4. Change 635°F to degrees Celsius.
5. Find the amount of heat in cal generated by 43.0 J of work.
6. Find the amount of heat in kcal generated by 6530 J of work.
7. Find the amount of work equivalent to 435 Btu.
8. Find the heat flow during 4.10 h through a glass window of thickness 0.15 in. with an area of 33 ft² if the average outer surface temperature is 22°F and the average inner glass surface is 48°F.
9. Find the heat flow in 25.0 days through a freezer door 80.0 cm × 144 cm insulated with cellulose fiber 4.0 cm thick. The temperature inside the freezer is −14°C. Room temperature is 22°C.
10. How many Btu of heat must be added to 835 lb of steel to raise its temperature from 20.0°F to 455°F?
11. How many kcal of heat must be added to 148 kg of aluminum to raise its temperature from 21.5°C to 485°C?
12. A 161-kg steam boiler is made of steel and contains 1127 N of water at 8.9°C. How much heat is required to raise the temperature of both the boiler and the water to 100‾°C?

13. A 3.80-lb piece of copper is dropped into 8.35 lb of water at 48.0°F. The final temperature is 98.2°F. What was the initial temperature of the copper?

14. A 355-g piece of metal at 48.0°C is dropped into 111 g of water at 15.0°C. If the final temperature of the mixture is 32.5°C, what is the specific heat of the metal?

15. A brass rod 45.2 cm long expands 0.734 mm when heated. Find the temperature change.

16. The length of a steel rod at 5°C is 12.500 m. What is its length when heated to 154°C?

17. The diameter of a hole drilled through aluminum at 22°C is 7.50 mm. Find the diameter and the area of the hole at 89°C.

18. A steel ball has a radius of 1.54 cm at 35°C. Find its change in volume when the temperature is increased to 84.5°C.

19. Find the increase in volume of 44.8 L of acetone when it is heated from 37.0°C to 75.5°C.

20. What is the decrease in volume of 3450 ft³ of alcohol in a railroad tank car if the temperature drops from 87.0°F to 33.0°F?

21. How many kcal of heat are required to vaporize 21.5 kg of water at $10\overline{0}$°C?

22. How many Btu of heat are required to melt 8.35 lb of ice at 32°F?

23. How many kcal of heat must be withdrawn from 4.56 kg of steam at 125°C to change it to ice at $\overline{4}4.5$°C?

24. How many joules of heat are required to change 336 g of ethyl alcohol from a solid at −117°C to a vapor at 78.5°C?

APPLIED CONCEPTS

1. A polystyrene foam cover prevents an ice–water mixture from absorbing heat from 25.3°C air outside a cooler. If the cover is 54.5 cm × 37.8 cm in cross section and 5.25 cm thick, how much heat will transfer through the foam in 60.0 min?

2. Every winter a local recreation department fills a parking lot with water so that the water will freeze and people can ice skate on it. Although the weather does not get cold enough for the water to freeze, the water temperature drops from 4.30°C to 0.00°C. If the water is 12.7 cm deep and the parking lot is 24.5 m × 33.8 m, how much heat energy does the water release into the air?

3. (a) What is the heat of fusion for the water in Problem 2 to freeze into a solid block of ice? (b) What is the total energy needed for the water at 4.30°C to change to ice at 0.00°C? (c) How much additional heat is removed from the ice to lower its temperature to −3.43°C?

4. Pedro, a contractor, is trying to choose between purchasing a steel or an aluminum 50-ft tape measure. (a) If he wants the tape measure that expands and contracts less, which tape measure should he purchase? Consider that the tape measures were calibrated at 68.0°F in the factory. (b) How much can he expect his new tape measure to contract if the temperature outside is 18.5°F?

5. In anticipation of winter snowstorms, Jamal fills his 2.50-gal gas can at the local gas station. (a) If the temperature is 65.3°F on the day he fills the gas can, what volume of gas will Jamal have when the temperature drops to 10.5°F? (b) If the gas cost $1.97/gal, how much money does Jamal lose?

CHAPTER 15

PROPERTIES OF GASES

Charles' Law:
$$\frac{V}{T} = \frac{V'}{T'}$$

High T
High V

Low T
Low V

Properties of gases are related to the temperature and the pressure under which the gas is contained. Gas laws concern the behavior of "ideal" gases at standard conditions of temperature and pressure. The focus of this chapter is on the behavior of gases as temperature and pressure are varied.

Objectives

The major goals of this chapter are to enable you to:

1. Use Charles' law to determine thermal expansion.
2. Apply Boyle's law to calculate volume changes of gases.
3. Relate gas density to pressure and temperature.

15.1 Charles' Law

Before making a long summer trip, you notice that the tires are low. You stop at a gas station around the corner and add air to 28 lb/in^2 gauge pressure. Later in the afternoon you stop for gas. Since your tires were low that morning, you decide to check them again. Now you notice that they look a bit larger, and the gauge pressure is 40 lb/in^2. What happened? When a gas is heated, the increased kinetic energy causes the volume to increase, the pressure to increase, or both to increase.

To study these concepts we will use the idea of an "ideal gas." When the density of a gas is sufficiently low, the pressure, volume, and temperature of the gas tend to be related in a rather simple way. These relationships are fairly accurate over a broad range of temperatures and pressures in real gases. They also apply over a broad range of different gases. We will therefore consider the wide variety of gases and their behaviors by using the ideal gas model. Charles' and Boyle's laws mathematically describe these relationships.

> **CHARLES' LAW**
>
> If the pressure on a gas is constant, the volume is directly proportional* to its absolute (Kelvin or Rankine) temperature.

By formula,

$$\frac{V}{T} = \frac{V'}{T'} \quad \text{or} \quad VT' = V'T$$

where V = original volume
   T = original temperature
   V' = final volume
   T' = final temperature

> **Jacques Charles (1746–1823),**
>
> *physicist, was born in France. He became famous for making the first manned ascent by hydrogen balloon in 1783. His interest in gases led him to formulate Charles' law.*

A gas occupies $45\overline{0}$ cm^3 at $3\overline{0}°$C. At what temperature will the gas occupy $48\overline{0}$ cm^3?

◄ EXAMPLE 1

Data:

   $V = 45\overline{0}$ cm^3
   $T = 3\overline{0}° + 273 = 303$ K   (*Note:* We must use Kelvin temperature.)
   $V' = 48\overline{0}$ cm^3
   $T' = ?$

---

*Directly proportional means that as temperature increases, volume increases, and as temperature decreases, volume decreases.

Basic Equation:

$$\frac{V}{T} = \frac{V'}{T'}$$

Working Equation:

$$T' = \frac{TV'}{V}$$

Substitution:

$$T' = \frac{(303 \text{ K})(48\overline{0} \text{ cm}^3)}{45\overline{0} \text{ cm}^3}$$

$$= 323 \text{ K} \quad \text{or} \quad 5\overline{0}°\text{C}$$

· · · · · · · · · · · · · · · · ·

EXAMPLE 2

At $4\overline{0}°$F, some helium occupies 15.0 ft³. What will be its volume at $9\overline{0}°$F?

Data:

$$V = 15.0 \text{ ft}^3$$
$$T = 4\overline{0}° + 46\overline{0}° = 50\overline{0}°\text{R} \quad (\textit{Note:} \text{ We must use Rankine temperature.})$$
$$T' = 9\overline{0}° + 46\overline{0}° = 55\overline{0}°\text{R}$$
$$V' = ?$$

Basic Equation:

$$\frac{V}{T} = \frac{V'}{T'}$$

Working Equation:

$$V' = \frac{VT'}{T}$$

Substitution:

$$V' = \frac{(15.0 \text{ ft}^3)(55\overline{0}°\text{R})}{50\overline{0}°\text{R}}$$

$$= 16.5 \text{ ft}^3$$

· · · · · · · · · · · · · · · · ·

TRY THIS ACTIVITY

Mylar® Balloons

Take a room-temperature, helium-filled Mylar® balloon and place it in a significantly colder environment, such as outside in the cold winter or in a freezer. As the temperature of the helium decreases, observe what happens to the volume of the balloon. After it has been outside several minutes, bring the balloon back into the room-temperature environment. Observe the change that occurs. Then, place the balloon in a warmer area such as near a hot radiator. Why does the volume of the balloon change in different temperature environments?

PROBLEMS 15.1

1. Change 15°C to K.
2. Change −14°C to K.
3. Change 317 K to °C
4. Change 235 K to °C.

5. Change 72°F to °R.
6. Change −55°F to °R.
7. Change 55$\overline{0}$°R to °F.
8. Change 375°R to °F.

Use $\dfrac{V}{T} = \dfrac{V'}{T'}$ to find each quantity:

9. $T = 315$ K, $V' = 225$ cm^3, $T' = 275$ K, find V.
10. $T = 615$°R, $V = 60.3$ in^3, $T' = 455$°R, find V'.
11. $V = 20\overline{0}$ ft^3, $T' = 95$°F, $V' = 25\overline{0}$ ft^3, find T.
12. $V = 19.7$ L, $T = 51$°C, $V' = 25.2$ L, find T'.
13. Some gas occupies a volume of 325 m^3 at 41°C. What is its volume at 94°C?
14. Some oxygen occupies 275 in^3 at 35°F. Find its volume at 95°F.
15. Some methane occupies 1575 L at 45°C. Find its volume at 15°C.
16. Some helium occupies 120$\overline{0}$ ft^3 at 70°F. At what temperature will its volume be 60$\overline{0}$ ft^3?
17. Some nitrogen occupies 14,300 cm^3 at 25.6°C. What is the temperature when its volume is 10,250 cm^3?
18. Some propane occupies 1270 cm^3 at 18.0°C. What is the temperature when its volume is 1530 cm^3?
19. Some carbon dioxide occupies 34.5 L at 49.0°C. Find its volume at 12.0°C.
20. Some oxygen occupies 28.7 ft^3 at 11.0°F. Find its temperature when its volume is 18.5 ft^3.
21. A balloon contains 26.0 L of hydrogen at 40.0°F. What is the Kelvin temperature change needed to make the balloon expand to 36.0 L?
22. Using Charles' law, determine the effect
    (a) on the temperature of a gas when the volume is doubled.
    (b) on the temperature of a gas when the volume is tripled.
    (c) on the volume when temperature is doubled.
    (d) Explain the relationship between the volume and the temperature.
23. If 38.0 L of hydrogen is heated to 11$\overline{0}$°C and expands to 90.0 L, what was the original temperature?
24. A tank contains 3.00 L of acetylene at 4.00°C. Find the volume at 12.0°C.
25. A hot air balloon contains 147 m^3 of air at 19.0°C. What is the volume of air at 32.0°C?
26. A tank with 139 L of propane is cooled from 91.0°C to 37.0°C. Find the original volume.
27. A 200$\overline{0}$-litre fuel tank filled with propane at 21°C is cooled to 9°C. What is the new volume of propane?
28. A propane nurse tank is left on a job site overnight. If the tank was filled in the heat of the day at 19°C and cools overnight to 3°C, what is the new propane volume if the tank was filled with 140$\overline{0}$ litres?
29. A propane tank now containing 25$\overline{0}$ L of propane was cooled from 35.0°C to 5.0°C. What was the original volume of the propane?
30. A tank with 50$\overline{0}$ L of gasoline is heated from 17.0°C to 31.0°C. What is the new volume of the gasoline in the tank?

15.2 Boyle's Law

BOYLE'S LAW

If the temperature of a gas is constant, the volume is inversely proportional* to the absolute pressure.

---

*Inversely proportional means that as volume increases, pressure decreases, and as volume decreases, pressure increases.

SKETCH

| 12 cm^2 | w |

4.0 cm

DATA

$A = 12$ cm^2, $l = 4.0$ cm, $w = ?$

BASIC EQUATION

$A = lw$

WORKING EQUATION

$w = \dfrac{A}{l}$

SUBSTITUTION

$w = \dfrac{12 \text{ cm}^2}{4.0 \text{ cm}} = 3.0$ cm

PROBLEM SOLVING

Robert Boyle (1627–1691),

chemist, was born in Ireland. Working with Robert Hooke as his assistant, he performed experiments on air, vacuums, combustion, and respiration. He developed Boyle's law in 1662. He also did research on properties of acids and alkalis, specific gravity, crystallography, and refraction.

By formula,

$$\frac{V}{V'} = \frac{P'}{P} \quad \text{or} \quad VP = V'P'$$

where
- V = original volume
- V' = final volume
- P = original pressure
- P' = final pressure

Note: The pressure must be expressed in terms of *absolute pressure*.

EXAMPLE 1

Some oxygen occupies $50\overline{0}$ in^3 at an absolute pressure of 40.0 lb/in^2 (psi). What is its volume at an absolute pressure of $10\overline{0}$ psi?

Data:

$$V = 50\overline{0} \text{ in}^3$$
$$P = 40.0 \text{ psi}$$
$$P' = 10\overline{0} \text{ psi}$$
$$V' = ?$$

Basic Equation:

$$\frac{V}{V'} = \frac{P'}{P}$$

Working Equation:

$$V' = \frac{VP}{P'}$$

Substitution:

$$V' = \frac{(50\overline{0} \text{ in}^3)(40.0 \text{ psi})}{10\overline{0} \text{ psi}}$$
$$= 20\overline{0} \text{ in}^3$$

• • • • • • • • • • • • • • • • •

EXAMPLE 2

Some nitrogen occupies 20.0 m^3 at a gauge pressure of 274 kPa. Find the absolute pressure when its volume is 30.0 m^3.

Data:

$$V = 20.0 \text{ m}^3$$
$$P_{\text{abs}} = 274 \text{ kPa} + 101 \text{ kPa} = 375 \text{ kPa} \quad \text{(See Section 13.3.)}$$
$$V' = 30.0 \text{ m}^3$$
$$P' = ?$$

Basic Equation:

$$\frac{V}{V'} = \frac{P'}{P}$$

Working Equation:

$$P' = \frac{VP}{V'}$$

Substitution:

$$P' = \frac{(20.0 \ \cancel{m^3})(375 \ \text{kPa})}{30.0 \ \cancel{m^3}}$$

$$= 25\overline{0} \ \text{kPa}$$

. . . .

Density and Pressure

If the pressure of a given amount (constant volume) of gas is increased, its density increases as the gas molecules are forced closer together. (Recall that density is discussed in Section 12.5.) Also, if the pressure is decreased, the density decreases. That is, the *density of a gas is directly proportional to its pressure* as long as there is no change in state. In equation form,

$$\boxed{\frac{D}{D'} = \frac{P}{P'} \quad \text{or} \quad DP' = D'P}$$

where D = original density
   D' = final density
   P = original pressure (absolute)
   P' = final pressure (absolute)

Some amount of carbon dioxide has a density of 1.60 kg/m³ at an absolute pressure of 95.0 kPa. What is the density when the pressure is decreased to 80.0 kPa?

◀ **EXAMPLE 3**

Data:

$$D = 1.60 \ \text{kg/m}^3$$
$$P = 95.0 \ \text{kPa}$$
$$P' = 80.0 \ \text{kPa}$$
$$D' = ?$$

Basic Equation:

$$\frac{D}{D'} = \frac{P}{P'}$$

Working Equation:

$$D' = \frac{DP'}{P}$$

Substitution:

$$D' = \frac{(1.60 \ \text{kg/m}^3)(80.0 \ \cancel{\text{kPa}})}{95.0 \ \cancel{\text{kPa}}}$$

$$= 1.35 \ \text{kg/m}^3$$

. . . .

A gas has a density of 2.00 kg/m³ at a gauge pressure of 16\overline{0} kPa. What is the density at a gauge pressure of 30\overline{0} kPa?

◀ **EXAMPLE 4**

Data:

$$D = 2.00 \ \text{kg/m}^3$$
$$P = 16\overline{0} \ \text{kPa} + 101 \ \text{kPa} = 261 \ \text{kPa}$$
$$P' = 30\overline{0} \ \text{kPa} + 101 \ \text{kPa} = 401 \ \text{kPa}$$
$$D' = ?$$

Basic Equation:

$$\frac{D}{D'} = \frac{P}{P'}$$

Working Equation:

$$D' = \frac{DP'}{P}$$

Substitution:

$$D' = \frac{(2.00 \text{ kg/m}^3)(401 \text{ kPa})}{261 \text{ kPa}}$$

$$= 3.07 \text{ kg/m}^3$$

· · · · · · · · · · · · · · · · ·

SKETCH

12 cm² w

4.0 cm

DATA

A = 12 cm², *l* = 4.0 cm, w = ?

BASIC EQUATION

A = *l*w

WORKING EQUATION

w = $\frac{A}{l}$

SUBSTITUTION

w = $\frac{12 \text{ cm}^2}{4.0 \text{ cm}}$ = 3.0 cm

PROBLEMS 15.2

Use $\dfrac{V}{V'} = \dfrac{P'}{P}$ or $\dfrac{D}{D'} = \dfrac{P}{P'}$ to find each quantity. (All pressures are absolute unless otherwise stated.)

1. $V' = 315$ cm³, $P = 101$ kPa, $P' = 85.0$ kPa; find V.
2. $V = 45\overline{0}$ L, $V' = 70\overline{0}$ L, $P = 75\overline{0}$ kPa; find P'.
3. $V = 76.0$ m³, $V' = 139$ m³, $P' = 41.0$ kPa; find P.
4. $V = 439$ in³, $P' = 38.7$ psi, $P = 47.1$ psi; find V'.
5. $D = 1.80$ kg/m³, $P = 108$ kPa, $P' = 125$ kPa; find D'.
6. $D = 1.65$ kg/m³, $P = 87.0$ kPa, $D' = 1.85$ kg/m³; find P'.
7. $P = 51.0$ psi, $P' = 65.3$ psi, $D' = 0.231$ lb/ft³; find D.
8. Some air at 22.5 psi occupies $140\overline{0}$ in³. What is its volume at 18.0 psi?
9. Some nitrogen at a pressure of 110.0 kPa occupies 185 m³. Find its pressure if its volume is changed to 225 m³.
10. Some methane at 185.0 kPa occupies 65.0 L. What is its volume at a pressure of 95.0 kPa?
11. Some carbon dioxide has a density of 3.75 kg/m³ at 815 kPa. What is its density if the pressure is decreased to 725 kPa?
12. Some oxygen has a density of 1.75 kg/m³ at normal atmospheric pressure. What is its pressure (in kPa) when the density is changed to 1.45 kg/m³?
13. Some methane at $50\overline{0}$ kPa gauge pressure occupies $75\overline{0}$ m³. What is its gauge pressure if its volume is $50\overline{0}$ m³?
14. Some helium at 15.0 psi gauge pressure occupies 20.0 ft³. Find its volume at 20.0 psi gauge pressure.
15. Some nitrogen at 80.0 psi gauge pressure occupies 13.0 ft³. Find its volume at 50.0 psi gauge pressure.
16. Some carbon dioxide has a density of 6.35 kg/m³ at 685 kPa gauge pressure. What is the density when the gauge pressure is 455 kPa?
17. Some propane has a density of 48.5 oz/ft³ at 265 psi gauge pressure. What is the gauge pressure when the density is 30.6 oz/ft³?
18. Some propane occupies 2.30 m³ at a gauge pressure of 36.0 kPa. What is the new volume at a gauge pressure of 40.0 kPa?
19. A quantity of oxygen at a gauge pressure of 20.0 kPa occupies 2.60 m³. What is the new volume at 26.0 kPa?
20. Some air occupies 4.5 m³ at a gauge pressure of 46 kPa. What is the volume at a gauge pressure of 13 kPa?
21. Some oxygen at 87.6 psi (absolute) occupies 75.0 in³. Find its volume if its absolute pressure is (a) doubled, (b) tripled, (c) halved.

22. A gas at $30\overline{0}$ kPa (absolute) occupies 40.0 m³. Find its absolute pressure if its volume is (a) doubled, (b) tripled, (c) halved.

23. A volume of 58.0 L of hydrogen is heated from 33°C to 68°C. If its original density is 4.85 kg/m³ and its original absolute pressure is $12\overline{0}$ kPa, what is the resulting density?

24. Some argon gas is in a 42.0-L container at a pressure of $32\overline{0}$ kPa. What is the pressure if it is transferred to a container with a volume of 51.0 L?

25. A 2.00-L plastic bottle contains air at a pressure of 33.0 kPa. A person squeezes the bottle resulting in a pressure of 57.0 kPa. What is the new volume of the container?

26. The 3.25-cm³ volume of air remaining in a water balloon is reduced to 2.75 cm³ upon impact when the air pressure is 48.3 kPa. (a) Is the original pressure more or less than that upon impact? (b) Find the original pressure.

27. A mass of 1.31 kg of neon is in a 3.00-m³ container at a pressure of 121 kPa. When the pressure is reduced to 97.4 kPa, what is the final density?

28. The air density in a tractor tire is 1.40 kg/m³ at a pressure of 314 kPa. (a) As the air pressure increases to $70\overline{0}$ kPa, does the final density of the air increase or decrease? (b) What is the resulting density of the air?

29. An unknown gas is in a tank at 13.3 kPa. (a) If the density of the gas changes from 1.45 kg/m³ to 1.35 kg/m³, will the resulting pressure be higher or lower than the original pressure? (b) Find the resulting pressure.

15.3 Charles' and Boyle's Laws Combined

Most of the time it is very difficult to keep the pressure constant or the temperature constant. In this case we combine Charles' law and Boyle's law as follows:

$$\frac{VP}{T} = \frac{V'P'}{T'} \quad \text{or} \quad VPT' = V'P'T$$

Note: Both pressure and temperature must be absolute.

We have 5.00 m³ of acetylene at 4.00°C at $16\overline{0}$ kPa (absolute). What is the pressure if its volume is changed to 4.00 m³ at 30.0°C?

EXAMPLE

Data:

$$V = 5.00 \text{ m}^3$$
$$P = 16\overline{0} \text{ kPa}$$
$$T = 4.00° + 273° = 277 \text{ K}$$
$$V' = 4.00 \text{ m}^3$$
$$T' = 30.0° + 273° = 303 \text{ K}$$
$$P' = ?$$

Basic Equation:

$$\frac{VP}{T} = \frac{V'P'}{T'}$$

Working Equation:

$$P' = \frac{VPT'}{TV'}$$

Substitution:

$$P' = \frac{(5.00 \text{ m}^3)(16\overline{0} \text{ kPa})(303 \text{ K})}{(277 \text{ K})(4.00 \text{ m}^3)}$$

$$= 219 \text{ kPa}$$

• • • • • • • • • • • • • • • • • •

The gas laws are reasonably accurate except at very low temperatures and under extreme pressures. A commonly used reference in gas laws is called **standard temperature and pressure** (STP). Standard temperature is the freezing point of water, 0°C or 32°F. Standard pressure is equivalent to atmospheric pressure, 101.32 kPa or 14.7 lb/in².

Vapor Pressure and Humidity

When a liquid evaporates, molecules of the liquid pass from its surface into the air above the liquid. This increase in the number of molecules in the air causes an increase in the pressure above the liquid. This increase in pressure is called *vapor pressure*.

Water boiling is an example of the change of phase of a liquid to a gas. Bubbles of gas form below the surface of the water, rise to the surface, and escape into the air. Recall that boiling water remains at a constant 100°C temperature at standard pressure. Water can, however, have its temperature raised above 100°C and be kept from boiling in a pressure cooker. This pot has a heavy, tight-fitting lid with a steam escape valve. As the water in the pot begins to boil, the pressure above the water builds up and interferes with further boiling. This way, the boiling temperature of the water can be raised above 100°C, allowing for fast cooking of a roast or other foods. Likewise, at a higher altitude, water boils at a lower temperature. In the mile-high city of Denver, water boils at 95°C, making for longer cooking times. Many premixed baking foods provide alternate baking times for high altitudes.

Evaporation of water molecules into the air from a lake followed by cooling as the air passes over cooler land may create a fog. Fog is basically a cloud that forms near the ground. The returning of some water vapor molecules in the air to a liquid is an example of a process called *condensation*. Condensation occurs at a point in temperature called the *dew point*. Dew forms when moist air is cooled by the earth's surface. It can be closely reproduced in the laboratory by taking a glass container partly filled with water and adding pieces of ice. When the water and glass reach the dew point in temperature, water will condense from the air on the outside of the glass. The condensation on a can of cold soda is a familiar example, in which water vapor in the air is chilled upon contact with the cold can. The water molecules slow when cooled and change phase to liquid.

Condensation of steam can be very dangerous. Since steam gives up considerable energy when it changes from vapor to liquid (540 cal/g °C), getting burned by steam can be much worse than getting burned by water at the same temperature. This same concept explains why a steam heating system produces much more heat than a hot water system.

The maximum amount of water vapor that air will hold at a given temperature is *absolute humidity*. Any increase becomes rainfall. *Relative humidity*, in contrast, is the ratio of the amount of water vapor a sample of air holds to the maximum amount it can hold if it is saturated.

Another example of the effect of condensation on people can be seen by comparing the "dry heat" in Arizona with the humid air at the same temperature on the coast of the Gulf of Mexico. In Phoenix, where the humidity is very low, evaporation is substantially greater than condensation. In New Orleans, however, where the humidity is much higher, condensation is much greater and offsets evaporation so people are more uncomfortable as condensation limits the cooling effect of the evaporation of a person's perspiration. For most people, a relative humidity of 50% to 60% at about 20°C is comfortable. When the relative humidity is higher, the air feels heavy and uncomfortable.

PHYSICS CONNECTIONS

Decompression Sickness: "The Bends"

When a scuba diver dives in deep water, he or she experiences a greater water pressure than when closer to the surface. According to Boyle's law, when the pressure is increased on an object, its volume decreases. In the case of a scuba diver, the volume of air inside the diver's lungs decreases as well. In order for the diver to breathe correctly, the air in the diver's tank must be highly pressurized and comprise 10% to 20% oxygen and 80% to 90% nitrogen.

Under normal atmospheric pressure, nitrogen does not dissolve in blood. Under pressure, nitrogen's volume decreases, thereby increasing its solubility in blood. As the diver swims deeper, the pressure in the lungs increases, and the blood accepts the smaller nitrogen bubbles. If the scuba diver returns too quickly to the surface, the diver experiences a quickly reduced pressure on his or her body, which results in enlarged nitrogen bubbles in the blood stream. The presence of nitrogen bubbles in the blood can cause generalized barotrauma, otherwise known as decompression sickness or "the bends."

Figure 15.1   A hyperbaric chamber.

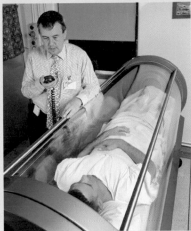

Copyright of James King-Holmes/Science Photo Library/Photo Researchers, Inc. Photo reprinted with permission

Common symptoms of "the bends" include joint pain, dizziness, and numbness. The only way to prevent "the bends" is to slowly rise to the surface. This gradually reduces the pressure on the diver and the nitrogen and gives the nitrogen bubbles time to dissolve and prevents them from forming large bubbles in the blood.

In the mid-1800s, workers on the Brooklyn Bridge were the first individuals to be documented with "the bends." They worked under the Hudson River in high-pressure compartments where they would enter and exit several times per day without slowly decompressing. Symptoms were documented and the cause of this extremely painful condition was eventually discovered. Today, someone suffering from "the bends" is taken to decompression chambers, called hyperbaric chambers, where the patient breathes highly pressurized oxygen that is gradually decompressed to normal atmospheric pressure (Fig. 15.1).

PROBLEMS 15.3

Use $\dfrac{VP}{T} = \dfrac{V'P'}{T'}$ to find each quantity. (All pressures are absolute unless otherwise stated.)

1. $P = 825$ psi, $T = 575°$R, $V' = 1550$ in^3, $P' = 615$ psi, $T' = 525°$R; find V.
2. $V = 50\overline{0}$ in^3, $T = 50\overline{0}°$R, $V' = 80\overline{0}$ in^3, $P' = 80\overline{0}$ psi, $T' = 45\overline{0}°$R; find P.
3. $V = 90\overline{0}$ m^3, $P = 105$ kPa, $T = 30\overline{0}$ K, $P' = 165$ kPa, $T' = 265$ K; find V'.
4. $V = 18.0$ m^3, $P = 112$ kPa, $V' = 15.0$ m^3, $P' = 135$ kPa, $T' = 235$ K; find T.
5. $V = 532$ m^3, $P = 135$ kPa, $T = 87°$C, $V' = 379$ m^3, $P' = 123$ kPa; find T'.
6. We have $60\overline{0}$ in^3 of oxygen at $150\overline{0}$ psi at 65°F. What is the volume at $120\overline{0}$ psi at $9\overline{0}°$F?
7. We have $80\overline{0}$ m^3 of natural gas at 235 kPa at $3\overline{0}°$C. What is the temperature if the volume is changed to $120\overline{0}$ m^3 at 215 kPa?
8. We have $140\overline{0}$ L of nitrogen at 135 kPa at 54°C. What is the temperature if the volume is changed to $80\overline{0}$ L at 275 kPa?
9. An acetylene welding tank has a pressure of $200\overline{0}$ psi at $4\overline{0}°$F. If the temperature rises to $9\overline{0}°$F, what is the new pressure?
10. What is the new pressure in Problem 9 if the temperature falls to $-3\overline{0}°$F?
11. An ideal gas occupies a volume of 5.00 L at STP. What is its gauge pressure (in kPa) if the volume is halved and its temperature increases to 400°C?

SKETCH

| 12 cm² | w |

4.0 cm

DATA

$A = 12$ cm^2, $l = 4.0$ cm, $w = ?$

BASIC EQUATION

$A = lw$

WORKING EQUATION

$w = \dfrac{A}{l}$

SUBSTITUTION

$w = \dfrac{12 \text{ cm}^2}{4.0 \text{ cm}} = 3.0 \text{ cm}$

12. An ideal gas occupies a volume of 5.00 L at STP.
    (a) What is its temperature if its volume is halved and its absolute pressure is doubled?
    (b) What is its temperature if its volume is doubled and its absolute pressure is tripled?

13. Some propane occupies 2.00 m^3 at 18.0°C at an absolute pressure of 3.50 × 10^5 N/m^2. (a) Find the absolute pressure (in kPa) at the same temperature when the volume is halved. (b) Find the new temperature when the absolute pressure is doubled and the volume is doubled. (c) Find the new volume when the absolute pressure is halved and the temperature is decreased to −12.0°C. (d) Find the new volume if the absolute pressure is 1.30 × 10^6 N/m^2 and the temperature is 31.0°C.

14. A balloon with volume 3200 mL of xenon gas is at a gauge pressure of 122 kPa and a temperature of 27°C. What is the volume when the balloon is heated to 65°C and the gauge pressure is decreased to 112 kPa?

15. A 7.85-L helium-filled balloon experiences a change in both pressure from normal atmospheric pressure to 60.5 kPa and temperature from 24.0°C to 6.00°C. What is the resulting volume?

Glossary

Boyle's Law If the temperature of a gas is constant, the volume is inversely proportional to the absolute pressure, $V/V' = P'/P$. (p. 409)

Charles' Law If the pressure on a gas is constant, the volume is directly proportional to its Kelvin or Rankine temperature, $V/T = V'/T'$. (p. 407)

Standard Temperature and Pressure (STP) A commonly used reference in gas laws. Standard temperature is the freezing point of water. Standard pressure is equivalent to atmospheric pressure. (p. 414)

Formulas

15.1 Charles' law: $\dfrac{V}{T} = \dfrac{V'}{T'}$

15.2 Boyle's law: $\dfrac{V}{V'} = \dfrac{P'}{P}$

$\dfrac{D}{D'} = \dfrac{P}{P'}$

15.3 $\dfrac{VP}{T} = \dfrac{V'P'}{T'}$

Review Questions

1. The gas law that relates volume and temperature is called
   (a) Boyle's law.　　(b) Hooke's law.
   (c) Charles' law.　　(d) none of the above.
2. The gas law that relates volume and pressure is called
   (a) Boyle's law.　　(b) Hooke's law.
   (c) Charles' law.　　(d) none of the above.
3. If the temperature of a gas is constant and the volume is decreased, the pressure will
   (a) stay the same.　　(b) decrease.
   (c) increase.　　(d) increase or decrease, depending on the gas.
4. If the temperature of a gas is constant and the pressure is decreased, the volume will
   (a) stay the same.　　(b) decrease.
   (c) increase.　　(d) increase or decrease, depending on the gas.
5. If the pressure on a gas is constant and the temperature is decreased, the volume will
   (a) stay the same.　　(b) decrease.
   (c) increase.　　(d) increase or decrease, depending on the gas.
6. If the pressure on a gas is constant and the volume is decreased, the temperature will
   (a) stay the same.　　(b) decrease.
   (c) increase.　　(d) increase or decrease, depending on the amount of gas.
7. Describe the conditions of standard temperature and pressure.
8. Describe what happens to the volume of a gas if its temperature and pressure increase.
9. Describe what happens to the temperature of a gas if its volume and pressure increase.
10. What causes the tendency of the volume and pressure of a gas to increase when it is heated?
11. What causes the tendency of the temperature of a gas to increase when it is compressed?
12. What causes the tendency of the pressure of a gas to decrease when the volume is increased?

Review Problems

SKETCH

$$12 \text{ cm}^2 \quad | \quad w$$

4.0 cm

DATA

$A = 12 \text{ cm}^2, l = 4.0 \text{ cm}, w = ?$

BASIC EQUATION

$A = lw$

WORKING EQUATION

$w = \frac{A}{l}$

SUBSTITUTION

$w = \frac{12 \text{ cm}^2}{4.0 \text{ cm}} = 3.0 \text{ cm}$

All pressures are absolute unless otherwise stated.

1. A gas occupies 13.5 ft^3 at 35.8°F. What will the volume of this gas be at 88.6°F if the pressure is constant?

2. A gas occupies 3.45 m^3 at 18.5°C. What will the volume of this gas be at 98.5°C if the pressure is constant?

3. Some hydrogen occupies 115 ft^3 at 54.5°F. What is the temperature when the volume is 132 ft^3 if the pressure is constant?

4. Some carbon dioxide occupies 45.3 L at 38.5°C. What is the temperature when the volume is 44.2 L if the pressure is constant?

5. Some propane occupies 145 cm^3 at 12.4°C. What is the temperature when the volume is 156 cm^3 if the pressure is constant?

6. Some air at 276 kPa occupies 32.4 m^3. What is its absolute pressure if its volume is doubled at constant temperature?

7. Some helium at 17.5 psi gauge pressure occupies 35.0 ft^3. What is the volume at 32.4 psi if the temperature is constant?

8. Some carbon dioxide has a density of 6.35 kg/m^3 at 685 kPa gauge pressure. What is the density when the gauge pressure is 355 kPa if the temperature is constant?

9. We have 435 in^3 of nitrogen at 1340 psi gauge pressure at 75°F. What is the volume at 1150 psi gauge pressure at 45°F?

10. We have 755 m^3 of carbon dioxide at 344 kPa at 25°C. Find the temperature if the volume is changed to 1330 m^3 at 197 kPa.

11. A welding tank has a gas pressure of 1950 psi at 38°F. (a) What is the new pressure if the temperature rises to 98°F? (b) What is the temperature if the gas pressure falls to 1870 psi?

12. An ideal gas occupies a volume of 4.50 L at STP. What is its gauge pressure (in kPa) if the volume is halved and the absolute temperature is doubled?

13. An ideal gas occupies a volume of 5.35 L at STP. What is its gauge pressure (in kPa) if the volume is halved and the temperature is increased by 45.5°C?

14. A volume of 1120 L of helium at 40$\overline{0}$0 Pa is heated from 45°C to 77°C, increasing the pressure to 60$\overline{0}$0 Pa. What is the resulting volume of the gas?

15. In a 47-cm-tall cylinder of radius 7.0 cm, hydrogen of density 2.50 kg/m^3 is at a gauge pressure of 327 kPa. What is the density when the absolute pressure is changed to 525 kPa?

APPLIED CONCEPTS

1. Fran purchases a 1.85-ft^3, helium-filled Mylar® balloon in a store with an inside temperature of 65.5°F. When she walks outside, the air temperature is only 11.0°F. (a) What happens to the balloon as Fran walks toward her car? (b) What is the volume of the balloon when the temperature of the helium reaches the air temperature? (c) What would have been the volume of the balloon if she had purchased it in the summer with an outside temperature of 101°F? (d) Depending on the season, what can the store clerk do to compensate for the outside temperature and its impact on the volume of the balloon?

2. An automobile tire is filled to an air pressure of 32 lb/in^2 at an air temperature of $40\overline{0}$°F. (a) After the car is driven for a period of time, the temperature of the air inside the tire increases to 145°F. If the volume of the tire remains constant, what is the pressure inside the tire? (b) If a person wants to maintain an air pressure of 32 lb/in^2 when the tire is warm, what initial air pressure should be put in the tire?

3. A 15.0-cm-long cylinder has a movable piston with radius 2.00 cm. (a) What is the volume of the cylinder? (b) If the initial air pressure is 101 kPa at 0.0°C, what is the air pressure inside the cylinder when the piston is pushed down 4.00 cm? (c) If the air temperature inside the compressed cylinder is heated to 20.0°C, what is the new air pressure?

4. A 0.0300-m^3 steel tank containing helium is stored at 20.0°C under a pressure of 22.5 × 10^3 kPa. (a) If the helium is released into a standard 20.0°C, 1010-kPa atmosphere, how much volume will the helium occupy? (b) How much volume will the released helium occupy if released into a −28.5°C, 285-Pa stratosphere?

5. A lightweight weather-collecting sensor is attached to a 2.50-L balloon. The balloon is released on a day when the temperature is 23.4°C and the atmospheric pressure is 1.35 × 10^5 Pa. The balloon will collect data until it reaches the stratosphere. (a) The weather sensor measures a temperature of −28.2°C and an atmospheric pressure of 285 Pa. What is the volume of the balloon? (b) Given the change in the volume of the balloon as it reaches the stratosphere, should the balloon be fully inflated when released from the ground?

Another type of wave motion is shown in Fig. 16.2. In this case the elastic medium is a long spring. If the left end of the spring is rapidly lifted up and then returned to its starting position, a crest is formed that travels to the right [Fig. 16.2(a)]. If the left end is displaced downward and rapidly returned to its original position, a trough is formed that travels to the right [Fig. 16.2(b)].

Figure 16.2 Pulses in a long spring.

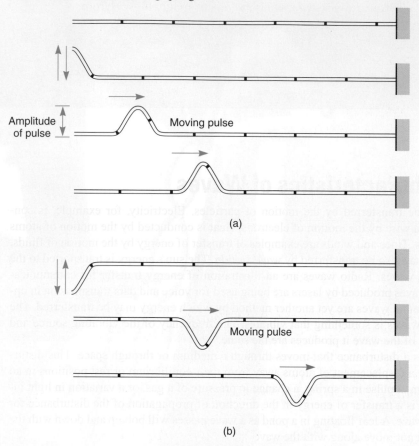

(a)

(b)

A **pulse** is a nonrepeated disturbance that carries energy through a medium or through space. If the pulse is repeated periodically, then a series of crests and troughs will travel through the medium, creating a traveling wave. A **transverse wave** is a disturbance in a medium in which the motion of the particles is perpendicular to the direction of the wave motion (Fig. 16.3). Water waves are another example of transverse waves. The **amplitude** of a wave is the maximum displacement of any part of the wave from its equilibrium, or rest, position.

Figure 16.3 Transverse waves.

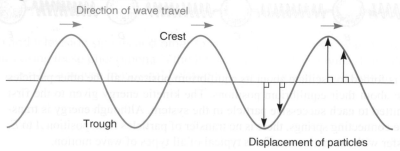

If a spring is compressed at the left end as shown in Fig. 16.4 and then released, the compression will travel to the right. Similarly, if the spring is stretched, a rarefaction (the stretched portion of the spring) is formed that will propagate or travel to the right. In this case the particle motion is along the direction of the wave travel. A **longitudinal wave** is a disturbance in a medium in which the motion of the particles is along the direction of the wave travel. Sound is another example.

Figure 16.4   Longitudinal wave in a spring.

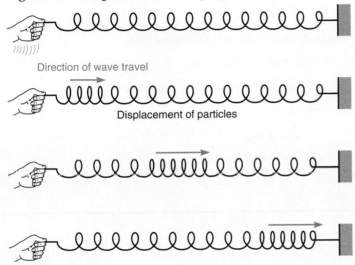

Direction of wave travel

Displacement of particles

The **wavelength** λ is the minimum distance between particles that have the same displacement and are moving in the same direction (Fig. 16.5).

Figure 16.5   The wavelength of a wave is the distance between successive corresponding points on a uniformly repeated wave.

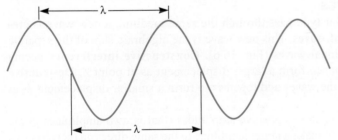

The **period** is the time required for a single wave to pass a given point. The **frequency** is the number of complete waves passing a given point per unit time. The common unit for frequency is the hertz (Hz) (named after **Heinrich Hertz**), where one oscillation per second is equal to 1 hertz (1 Hz = 1/s). Higher frequencies are measured in kilohertz (kHz), megahertz (MHz), and gigahertz (GHz). Radar and microwaves are measured in GHz, frequency-modulated (FM) radio waves are measured in MHz, and amplitude-modulated (AM) radio waves are measured in kHz. The period and the frequency are related by

$$f = \frac{1}{T}$$

where f = frequency
         T = period

Heinrich Hertz (1857–1894),

physicist, was born in Germany. His main work was on electromagnetic waves; he was the first to broadcast and receive radio waves. The unit of frequency, the hertz, is named after him.

The **propagation velocity** v of a wave is the velocity of the energy transfer and is given by the distance traveled by the wave in one period divided by the period, or

$$v = \frac{\lambda}{T} = \lambda f$$

where v = velocity
λ = wavelength
T = period
f = frequency

These relationships apply to sound, water, light, and all other waves.

EXAMPLE

Find the velocity of a wave with wavelength 2.5 m and frequency 44 Hz.

Data:

$$\lambda = 2.5 \text{ m}$$
$$f = 44 \text{ Hz} = 44/\text{s}$$
$$v = ?$$

Basic Equation:

$$v = \lambda f$$

Working Equation: Same

Substitution:

$$v = (2.5 \text{ m})(44/\text{s})$$
$$= 110 \text{ m/s}$$

Superposition of Waves

When two waves of a similar type pass through the same medium, a new wave is created by the **superposition of waves**. This new wave is the algebraic sum of the separate displacements of the individual waves (Fig. 16.6). **Constructive interference** occurs where the waves add together to form a larger displacement as at point A. **Destructive interference** occurs where the waves add together to form a smaller displacement as at point B.

The addition is algebraic, so two positive amplitudes (that is, two amplitudes in the same direction) add together to make a larger amplitude in the same direction (constructive interference). Where the amplitudes are in opposite directions, the smaller counteracts the larger with a net amplitude in the direction of the larger (destructive interference). See location (1) of Fig. 16.6(c) for an illustration of constructive interference and location (2) for an illustration of destructive interference.

Standing Waves

When a transverse pulse reaches the end of a spring that is fastened to a rigid support (Fig. 16.7), the pulse is reflected along the spring with the displacement of the reflected wave opposite in direction to that of the incident wave. A traveling wave is also reflected at the rigid end of a spring, producing two waves moving in opposite directions. Reflections from a free end are not inverted.

In one special case two waves combine so that there is no propagation of energy along the wave. The wave displacements are constant and remain fixed in location. This is called a **standing wave** (Fig. 16.8) because the two waves of equal amplitude and wavelength do not appear to be traveling. The points of destructive interference and

Figure 16.6 Superposition of two waves to form a new wave by adding their displacements.

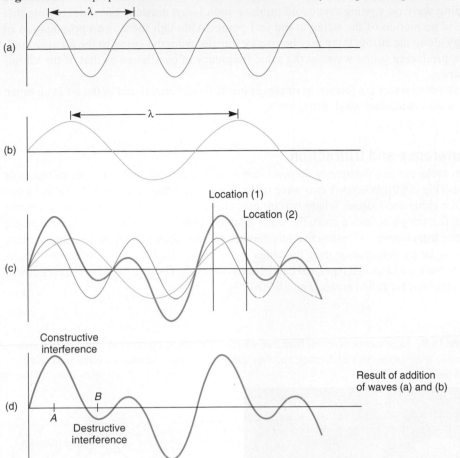

(a)

(b)

(c) Location (1) Location (2)

(d) Constructive interference

B

A

Destructive interference

Result of addition of waves (a) and (b)

Figure 16.7 The reflected wave in the spring has the same amplitude as the incident wave.

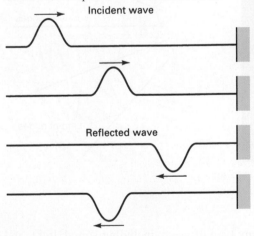

Incident wave

Reflected wave

Figure 16.8 Standing waves in a string generated by increasing the frequency of vibration.

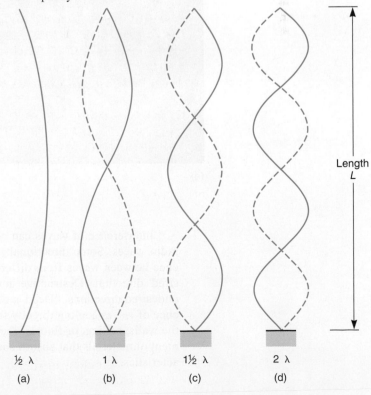

Length L

½ λ   1 λ   1½ λ   2 λ

(a)   (b)   (c)   (d)

constructive interference remain in fixed positions. Figure 16.8 shows an example of a standing wave on a string that could produce sound on a musical instrument. Note that there is no motion of the string at the end points. Although there is no propagation of energy along the string, there may be energy transfer from the string to the air surrounding it, producing sound waves at the same frequency of oscillation as that of the vibrating string.

Standing waves are formed in strings of musical instruments and in the air in an organ pipe, a flute, and other wind instruments.

Interference and Diffraction

If two rocks are simultaneously dropped into a pool, each will produce a set of waves or ripples (Fig. 16.9). Wherever two wave crests cross each other, the water height is higher than for either crest alone. Where two troughs cross, the water level is lower than for one alone. If a trough crosses a crest, the water level is nearly undisturbed. This is an example of wave **interference**. Constructive interference occurs when two crests or troughs meet, giving a larger disturbance than for either wave alone. Destructive interference occurs when a wave and a trough meet and cancel each other out. Those areas where waves cancel each other out are called *nodes* [Fig. 16.9(b)].

Figure 16.9 Interference of waves from two sources. The waves combine to form larger waves where two wave crests cross each other; and they cancel each other out where a wave crest and a trough meet.

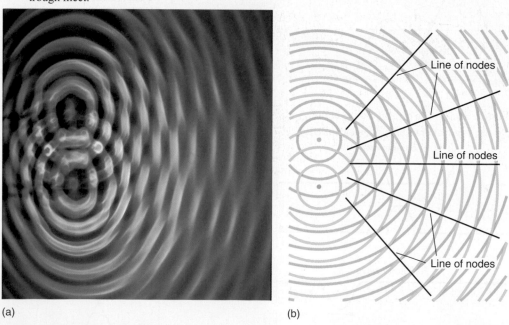

(a)

(b)

Interference of waves can occur for any type of wave, including sound, light, and radio waves. Some directional radio broadcast antennas rely on constructive interference between waves from different parts of the antenna to direct the signal in the desired direction. Destructive interference prevents the signal from propagating in undesired directions. "Dead spots" in an auditorium are caused by destructive interference of waves coming directly from the sound source with sound waves reflected from the walls, ceiling, or floor of the room. Proper choice of room shape and proper placement of materials that absorb sound well can lead to a room with good acoustical characteristics.

TRY THIS ACTIVITY

Interfering Waves

Place either a long Slinky® toy or a length of rope on a smooth, low-friction floor. With one person at each end of the rope, create a "wave pulse" toward each other. Observe what happens when the two waves interfere with each other. Explain what happens when two positive pulses interfere, two negative pulses interfere, and a positive and a negative pulse interfere with one another (Fig. 16.10).

Figure 16.10

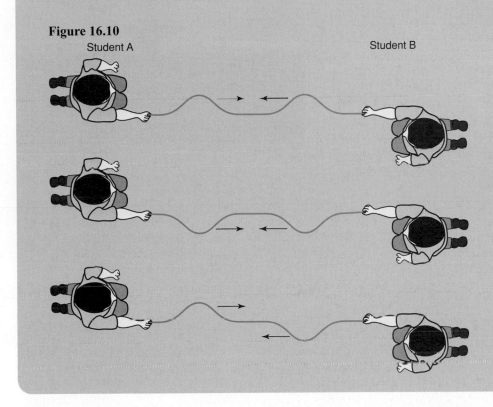

Student A   Student B

Diffraction is a property of a wave that describes its ability to bend around obstacles in its path. Water waves bend around the supports of a pier or a large rock (Fig. 16.11). Sound waves pass from one room through a door and spread into a second room (Fig. 16.12).

Figure 16.11 Water waves bend around obstacles and pass into the region behind (a) the pier supports or (b) a large rock. This property is called diffraction.

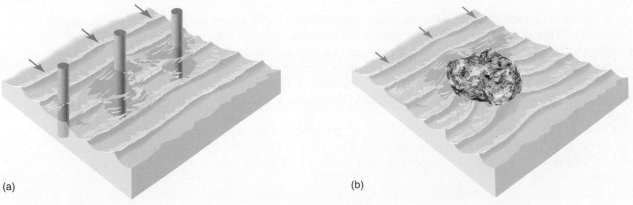

(a)   (b)

Figure 16.12 Diffraction of (a) sound waves entering a room through a door and (b) water waves passing through a small opening.

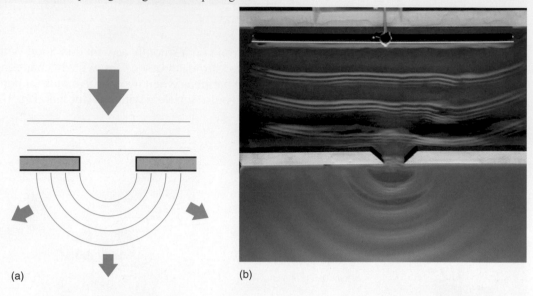

(a)

(b)

Wave diffraction is commonly observed only when the obstacle or opening is nearly the same size as the wavelength. Water waves and sound waves often have wavelengths in an easily observed range. Light, however, has a wavelength approximately 5×10^{-7} m. For this reason diffraction of light is not as easily recognized as diffraction of some other waves. The light waves from the sun when they encounter obstacles (air molecules in the atmosphere) are diffracted around them. The diffracted wave fronts are more or less spherical and are spread out or *scattered*.

Blue sky and a red setting sun are results of scattering. When we look up at the sky, we see scattered light. Those colors with the shortest wavelengths (blue) are scattered the most; the longer wavelengths (red and yellow) are transmitted with very little scattering. (These topics are covered in Chapter 22.)

PHYSICS CONNECTIONS

Wave Pools

Land-locked beach enthusiasts enjoy visiting water parks that generate large waves and allow them to body surf as if they are at the beach. Huge pumps, reservoirs, and pools are necessary to create such waves.

At Typhoon Lagoon in Disney World, a large pool shaped like a funnel holds over 2 million gallons of water (Fig. 16.13). Giant pumps and a large reservoir hold and then release 40,000 to 90,000 gal of water into the pool. The massive amount of water creates a surge in the pool that results in a high-amplitude wave. Water is lifted into the reservoir by a number of large pumps. Once the reservoir is full, giant valves are moved away from an opening in the bottom of the reservoir. Since water seeks to achieve its own level, the water rushes out of the bottom of the reservoir and into a narrow but deep section of the pool.

Once the giant surge of water enters the pool, the slope of the pool and the gradual increase in width of the pool promote the surf-like wave (Fig. 16.14). Once the wave passes and the water level flattens out, water is drawn out of the pool and back to the pumps so the process can be repeated.

Figure 16.13   Overhead view of a typical man-made wave pool.

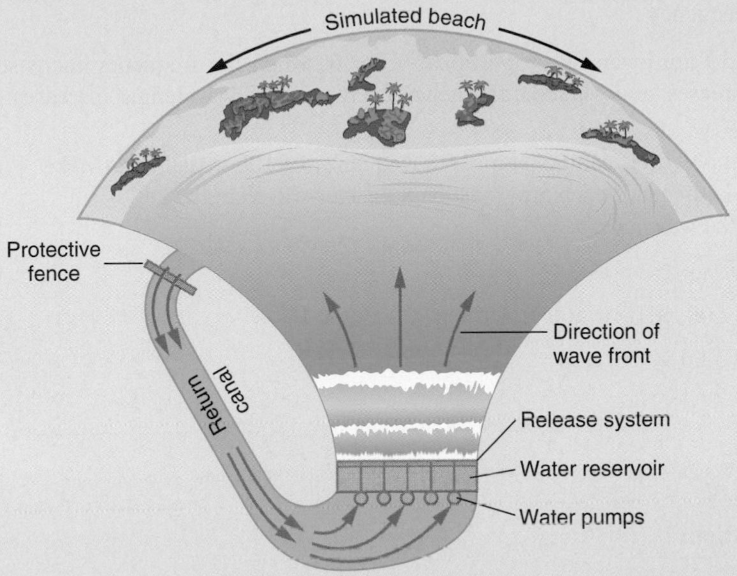

Simulated beach

Protective
fence

Return
canal

Direction of
wave front

Release system

Water reservoir

Water pumps

Figure 16.14   Swimmers enjoying a man-made wave at Typhoon Lagoon in Disney World.

16.2 Electromagnetic Waves

An **electromagnetic wave** is a transverse wave resulting from a periodic disturbance in an electromagnetic field, which has an electric component and a magnetic component, each being perpendicular to the other and both perpendicular to the direction of travel. All electromagnetic waves travel with velocity $v = c =$ **speed of light** $= 3.00 \times 10^8$ m/s. So, for electromagnetic waves

$$c = \lambda f$$

where c = speed of light (3.00×10^8 m/s)
 λ = wavelength
 f = frequency

Note that λ and f are inversely proportional. That is, when the frequency increases, the wavelength decreases; and when the frequency decreases, the wavelength increases.

EXAMPLE

The FM band of a radio is centered around a frequency of $10\overline{0}$ megahertz (MHz). Find the length of an FM antenna if each arm must be a quarter-wavelength.
 First, find the wavelength, λ.

Data:

$$f = 10\overline{0} \text{ MHz} = 10\overline{0} \times 10^6 \text{ Hz} = 1.00 \times 10^8/\text{s} \qquad (1 \text{ Hz} = 1/\text{s})$$
$$c = 3.00 \times 10^8 \text{ m/s}$$
$$\lambda = ?$$

Basic Equation:

$$c = \lambda f$$

Working Equation:

$$\lambda = \frac{c}{f}$$

Substitution:

$$\lambda = \frac{3.00 \times 10^8 \text{ m/s}}{1.00 \times 10^8/\text{s}}$$
$$= 3.00 \text{ m}$$

Therefore,

$$\frac{\lambda}{4} = \frac{3.00 \text{ m}}{4} = 0.750 \text{ m}$$

The **electromagnetic spectrum** is the entire range of electromagnetic waves classified according to frequency (Fig. 16.15). All electromagnetic waves travel at the same speed. The radio broadcast band is in the region of 1 MHz. The very high frequency (VHF) television band starts in the region of 50 MHz; the ultra high frequency (UHF) band is even higher. The highest-frequency waves generated by electronic oscillators are microwaves. Microwaves may be used to relay TV signals from a remote location to the studio transmitter.

Figure 16.15   The electromagnetic spectrum.

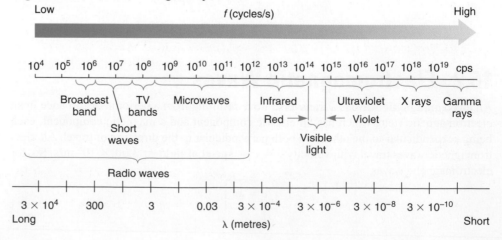

Molecular and atomic oscillations create waves of even higher frequencies, including infrared, visible light, ultraviolet, and X-ray waves. Visible light is electromagnetic radiation in the range 4.3×10^{14} to 7×10^{14} Hz.

We are surrounded by and bombarded by a sea of electromagnetic radiation. The light we see is only a very tiny part of this radiation.

PHYSICS CONNECTIONS

Evading Radar

Radar (RAdio Detecting And Ranging) is used as a navigation, early-warning, and detection device. Radar transmitters send electromagnetic radio waves into the atmosphere to measure the time and the frequency of radio waves after they have reflected off objects. (See Section 16.4 for more information on the "Doppler effect.") The reflected radio waves provide the observer with the object's position, size, and velocity.

A futuristic French fighter plane known as Rafale uses a device to help the plane evade radar. The plane uses a technology called active cancellation. As radio waves from radar strike the Rafale, an inverted version of the reflected radio wave is emitted. When two identical but opposite waves meet, they cancel each other, which results in destructive interference. Through destructive interference, the producer of a radar signal will not see any reflected radio waves from the Rafale.

Figure 16.16 Stealth aircraft are designed to be invisible to radar.

Copyright of Ross Harrison Koty/Getty Images, Inc.

Stealth aircraft avoid radar detection because of their body shape and design. The odd shapes and angles of the stealth body are designed to deflect the radio waves around the plane instead of reflecting them to the producer of the radar signal. Some of the stealth material is also designed to absorb the radio waves (Fig. 16.16).

PROBLEMS 16.2

1. Find the period of a wave whose frequency is $50\overline{0}$ Hz.
2. Find the frequency of a wave whose period is 0.550 s.
3. Find the velocity of a wave with wavelength 2.00 m and frequency $40\overline{0}$ Hz.
4. (a) What is the frequency of a light wave with wavelength 5.00×10^{-7} m and velocity 3.00×10^8 m/s? (b) Find the period of the wave.
5. What is the speed of a wave with frequency 3.50 Hz and wavelength 0.550 m?
6. Find the wavelength of water waves with frequency 0.650 Hz and velocity 1.50 m/s.
7. What is the wavelength of longitudinal waves in a coil spring with frequency 7.50 Hz and velocity 6.10 m/s?
8. A wave generator produces 20 pulses in 3.50 s. (a) What is its period? (b) What is its frequency?
9. Find the frequency of a wave produced by a generator that emits 30 pulses in 2.50 s.
10. What is the wavelength of an electromagnetic wave with frequency 50.0 MHz?
11. What is the frequency of an electromagnetic wave with a wavelength of 0.25 m?
12. What is the wavelength in metres of an electromagnetic wave with a frequency of 10^{14}/s?
13. Find the frequency of an electromagnetic wave with wavelength 1.50 m.
14. Find the wavelength of a wave traveling at 2.68×10^6 m/s with a period of 0.0125 s.
15. Find the wavelength of a wave traveling twice the speed of sound (speed of sound = 331 m/s) that is produced by an oscillator emitting 63 pulses every 8.3×10^{-6} min.
16. (a) Find the velocity of X rays emitted with wavelength 3.00×10^{-9} m and frequency 3.00×10^{18} Hz. (b) Find the period of the waves.
17. Find the velocity of microwaves having wavelength 0.750 m and frequency 2.75×10^{10} Hz.

SKETCH

12 cm^2 | w

4.0 cm

DATA

$A = 12 \text{ cm}^2$, $l = 4.0$ cm, $w = ?$

BASIC EQUATION

$A = lw$

WORKING EQUATION

$w = \frac{A}{l}$

SUBSTITUTION

$w = \frac{12 \text{ cm}^2}{4.0 \text{ cm}} = 3.0$ cm

16.3 Sound Waves

Daily we are exposed to many sounds and communicate with other people using the medium of sound. Whether the sound is pleasant music, a voice, or a loud siren, there are three requirements for the detection of sound in a physiological sense. There must be a source of sound, a medium (such as air) for transmitting it, and an ear to receive it. In a physical sense, sound is a vibratory disturbance in an elastic medium which may produce the sensation of sound. The frequency range over which the human ear responds is approximately 20 to 20,000 Hz. Ultrasonic waves have a frequency higher than 20,000 Hz.

Sound refers to those waves transmitted through a medium with frequencies capable of being detected by the human ear and is produced by a vibrating source. A ringing bell, the vibrating head of a drum, and a tuning fork (Fig. 16.17) are common examples of vibrating sources of sound. Other vibrating sources are not as easily recognized. Vibrating vocal cords produce speech. The notes of a clarinet originate with a vibrating reed. An auto horn uses an electrically driven vibrating diaphragm.

Figure 16.17 Sound waves produced by a vibrating tuning fork.

The most common medium for the propagation of sound is air. Sound will also propagate in solids or liquids. A vacuum will not carry sound. A mechanic may listen to the sounds of a running engine by placing one end of a metal rod against the engine and the other end against his or her ear. Sounds transmitted through water are utilized by passive sonar receivers aboard ships or submarines to identify other ships nearby.

Sound waves transmitted through the earth may be detected by an instrument called a *seismograph,* which can detect small motions of the earth's crust. An earthquake produces both longitudinal and transverse waves, which propagate with different velocities through the earth's crust. The distance to the source can be determined by measuring the time interval between the arrival of the two types of waves. Comparison of such data from seismographs at several points on the surface of the earth helps locate the epicenter of the earthquake. Waves can be intentionally set off by buried explosives. Reflections of these sound waves from different rock formations are recorded by seismographs. These recordings allow geologists to determine the underlying structure of the earth and predict the location of possible oil- or gas-producing regions.

Musical instruments utilize vibrating strings and vibrating columns of air to produce regular sounds. Wind instruments use an enclosed or partially enclosed column of air where waves are produced in the confined air by a pressure change (usually, blowing). Standing waves can then be produced in a column of proper length.

You may have watched distant lightning and noticed the time lapse before the sound of thunder reaches you. This is an example of the relatively slow speed of sound compared to the speed of light (3.00×10^8 m/s). The **speed of sound** in dry air at 1 atm pressure and 0°C is 331 m/s. Changes in humidity and temperature cause a variation in the speed of sound. The speed of sound increases with temperature at the rate of 0.61 m/s/°C. The speed of sound in dry air at 1 atm pressure is then given by

$$v = 331 \text{ m/s} + \left(0.61 \, \frac{\text{m/s}}{\text{°C}}\right) T$$

$$v = 1087 \text{ ft/s} + \left(1.1 \, \frac{\text{ft/s}}{\text{°F}}\right) (T - 32\text{°F})$$

where v = speed of sound in air
T = air temperature

EXAMPLE 1

Find the speed of sound in dry air at 1 atm pressure if the temperature is 23°C.

Data:

$$T = 23\text{°C}$$
$$v = ?$$

Basic Equation:

$$v = 331 \text{ m/s} + \left(0.61\frac{\text{m/s}}{°\text{C}}\right)T$$

Working Equation: Same

Substitution:

$$v = 331 \text{ m/s} + \left(0.61\frac{\text{m/s}}{°\cancel{C}}\right)(23°\cancel{C})$$

$$= 345 \text{ m/s}$$

· · · · · · · · · · · · · · · · · ·

What is the time required for the sound from an explosion to reach an observer $190\overline{0}$ m away for the conditions of Example 1?

EXAMPLE 2

Data:

$$v = 345 \text{ m/s}$$
$$s = 190\overline{0} \text{ m}$$
$$t = ?$$

Basic Equation:

$$s = vt \qquad \text{(from Section 4.1)}$$

Working Equation:

$$t = \frac{s}{v}$$

Substitution:

$$t = \frac{190\overline{0} \text{ m}}{345 \text{ m/s}}$$

$$= 5.51 \text{ s}$$

$$\boxed{\frac{\text{m}}{\text{m/s}} = \text{m} \div \frac{\text{m}}{\text{s}} = \cancel{\text{m}} \cdot \frac{\text{s}}{\cancel{\text{m}}} = \text{s}}$$

· · · · · · · · · · · · · · · · · ·

Sound propagates faster in a dense medium such as water than it does in a less dense medium such as air. A list of the speed of sound in various media is given in Table 16.1.

Table 16.1   Speed of Sound in Various Media

Medium	Speed	
	m/s	ft/s
Aluminum	6,420	21,100
Brass	4,7$\overline{0}$0	15,400
Steel	5,960	19,500
Granite	6,0$\overline{0}$0	19,700
Alcohol	1,210	3,970
Water (25°C)	1,5$\overline{0}$0	4,920
Air, dry (0°C)	331	1,090
Vacuum	0	0

All sounds have characteristics that we associate only with that sound. A siren is loud. A whisper is soft. Music can be loud or soft. The physical properties that differ for these sounds are intensity and frequency. The physiological characteristics of these sounds are loudness and pitch. Sound quality is related to the number of frequencies present. Pitch and frequency are closely related terms.

Intensity is the energy transferred by sound per unit time through unit area, thus $\frac{power}{area}$. **Loudness** refers to the strength of the sensation of sound heard by an observer and describes how strong or faint the sensation of sound seems. The ear does not respond equally to all frequencies. Sound must reach a certain intensity before it can be heard. The human ear normally detects sounds ranging in intensity from 10^{-12} W/m^2 (the threshold of hearing) to 10^0 W/m^2 or 1 (the threshold of pain). Levels of intensity are also measured on a logarithmic scale in decibels (dB); the unit "bel" is named after **Alexander Graham Bell**. Table 16.2 shows a range of familiar sounds.

Alexander Graham Bell (1847–1922),

inventor, was born in Scotland. He invented the telephone and the telegraph in 1876 and established the Bell Telephone Company in 1877.

Table 16.2   Range of Sound Levels and Intensities

Situation	Sound Level (dB)	Intensity (W/m^2)	Sound
Threshold of hearing	0	10^{-12}	Scarcely audible
Minimal sounds	10	10^{-11}	
Ticking watch	20	10^{-10}	
Whisper	30	10^{-9}	Faint
Leaves rustling, refrigerator	40	10^{-8}	
Average home, neighborhood street	50	10^{-7}	Moderate
Normal conversation, microwave	60	10^{-6}	
Car, alarm clock, city traffic	70	10^{-5}	Loud
Garbage disposal, vacuum cleaner, outboard motor, noisy restaurant	80	10^{-4}	Very loud
Factory, electric shaver, screaming child	85		
Lawn mower, passing motorcycle, convertible ride on a freeway	90	10^{-3}	
Hair dryer, diesel truck, chain saw, helicopter, subway train	100	10^{-2}	
Car horn, snowblower	110	10^{-1}	Deafening
Rock concert, prop plane	120	10^{0}	Painful
100 ft from jet engine, air raid siren	130	10^{1}	
Shotgun blast	140	10^{2}	

Source: Better Hearing Institute, *Self-Help for Hard of Hearing People.* Each increase of 10 dB is a tenfold increase in sound intensity; that is, 90 dB is ten times noisier than 80 dB. The U.S. government advises wearing earplugs or other hearing protection for anyone exposed to 85 dB for a period of more than a few hours. Some hearing experts have found that hearing damage is done at sound levels as low as 70 dB. Federal regulations require a hearing conservation program in the workplace where employees are exposed to 85 dB or more during an 8-h work period.

The **pitch** of a sound is the effect of the frequency of sound waves on the ear. Higher pitched sounds have a higher frequency than lower ones. The quality of sound can easily be determined by a casual listener. Irregular vibrations tend to produce noise, whereas regular vibrations in multiples of the fundamental vibration rate of an object produce sounds more pleasing to the human ear. These multiples of the fundamental tone are called harmonics. The quality of a sound depends on the number of harmonics produced and their relative intensities.

Sound waves from two sources reaching an observer at the same time will constructively or destructively interfere with each other. Identical waves from two separate sources such as stereo loudspeakers will enhance or reduce each other at different points in space. Sound waves of slightly different frequencies will also enhance or reduce each other at different points in time. The alternating periods of increasing and decreasing volume are called *beats*. The number of times per second the sound reaches a maximum is known as the *beat frequency* and is equal to the difference in frequencies of the two sound waves. This phenomenon is used for tuning musical instruments against a set of standard-frequency tuning forks. When the beat frequency is small, the instrument is well matched with the tuning fork.

Figure 16.18 Sound waves move through all three regions of the ear and are sent to the brain in the form of electrical waves.

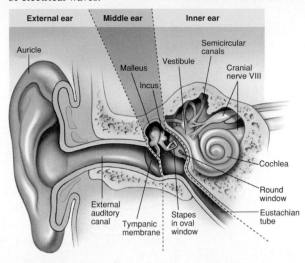

The ear is the human organ used to detect sound. It consists of three sections: the outer ear, the middle ear, and the inner ear (Fig. 16.18). Sound travels through the ear canal, which ends at the eardrum. The sound waves cause the eardrum to vibrate. The sound energy is transferred through the three smallest bones in the body to a fluid in the inner ear, which excites tiny hairs located in the cochlea, called cilia. Cilia resonate in the fluid, sending nerve impulses to the brain in the form of electrical waves. The brain then interprets the sound and we are able to hear.

16.4 The Doppler Effect

The **Doppler effect** refers to the variation of the frequency heard when a source of sound and the ear are moving relative to each other. As an automobile or a train passes by at a high speed sounding its horn or whistle, the frequency or pitch of the sound drops noticeably as it passes the observer. This variation in pitch is called the Doppler effect, named after **Christian Johann Doppler**.

Water waves spread over the flat surface of the water. Sound and light waves, though, travel in three-dimensional space in all directions like an expanding balloon. Just as circular wave crests are closer together in front of a swimming duck, spherical sound or light wave crests ahead of a moving source are closer together than those behind the source and reach a receiver more frequently (Fig. 16.19).

Motion of a source of sound toward an observer increases the rate at which he or she receives the vibrations. The velocity of each vibration is the speed of sound whether the source is moving or not. Each vibration from an approaching source has a shorter distance to travel. The wavelength is shortened when the source is moving toward the observer and is lengthened when the source is moving away from the observer. The vibrations are therefore received at a higher frequency than they are sent. Similarly, sound waves from a receding source are received at a lower frequency than that at which they are sent (Fig. 16.20).

The apparent Doppler-shifted frequency for sound is given by the equation

$$f' = f\left(\frac{v}{v \pm v_s}\right)$$

where f' = Doppler-shifted frequency
f = actual source frequency
v = speed of sound
v_s = speed of the source

The $+$ sign in the denominator is used when the source is moving away from the observer. The $-$ sign is used when the source is moving toward the observer.

Figure 16.19 A source moving to the left creates higher-frequency sound waves ahead and lower-frequency sound waves behind.

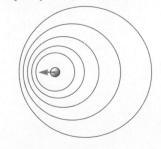

Figure 16.20 As this car moves to the left, Observer A in front of the car hears the car horn at a higher pitch than the driver, whereas Observer B hears a lower pitch than the driver.

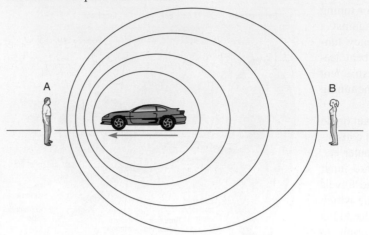

EXAMPLE

An automobile sounds its horn while passing an observer at 25 m/s. The actual horn frequency is $40\bar{0}$ Hz.

(a) What is the frequency heard by the observer while the car is approaching?
(b) What is the frequency heard when the car is leaving?
   Assume that the speed of sound is 345 m/s.

(a) Data:

$$f = 40\bar{0} \text{ Hz}$$
$$v = 345 \text{ m/s}$$
$$v_s = 25 \text{ m/s toward observer}$$
$$f' = ?$$

Basic Equation:

$$f' = f\left(\frac{v}{v - v_s}\right)$$

Working Equation: Same

Substitution:

$$f' = (40\bar{0} \text{ Hz})\left(\frac{345 \text{ m/s}}{345 \text{ m/s} - 25 \text{ m/s}}\right)$$
$$= 431 \text{ Hz}$$

(b) We simply change the sign from $-$ to $+$ in the basic equation of part (a). All other data remain the same. We then find

$$f' = (40\bar{0} \text{ Hz})\left(\frac{345 \text{ m/s}}{345 \text{ m/s} + 25 \text{ m/s}}\right)$$
$$= 373 \text{ Hz}$$

. . . .

The Doppler effect can be easily demonstrated by two students and a toy horn. One is the observer and remains stationary. The other, with the horn, blows it while turning around. The change in pitch heard by the observer is the result of the Doppler effect.

The Doppler effect is usually experienced with sound waves. However, it is common to all waves. For visible light, the Doppler effect is seen as a change in color since the color of light is determined by its frequency. Astronomers use this principle in the study of the universe to determine whether heavenly bodies are approaching or moving away from us.

PHYSICS CONNECTIONS

Ultrasound

Ultrasound is used to examine tissue and liquid-based internal organs and systems without subjecting the patient to invasive procedures. Ultrasound is an extremely high-frequency sound wave ranging from 20 kHz to 5 MHz. A device called a transducer sends sound waves of a particular frequency and measures the frequency as the waves reflect off various media or organs. Since the speed of sound is dependent on the medium through which it travels, the ultrasound processor that takes the information from the device can determine the position and density of the tissue.

Ultrasound is often used for diagnostic purposes in place of X rays because it does not use radiation and is safer for the person and/or the fetus being examined. Along with monitoring the status of a fetus and generating deep heat for therapeutic purposes, ultrasound is also used to observe various human systems such as the nervous, urinary, reproductive, and circulatory systems. Ultrasound is not used to view bone structures because the high-frequency sound waves reflect well off only liquid-based objects. Recent technological improvements in ultrasound have led researchers to advances in three-dimensional imaging and Doppler ultrasound, where the movement of blood can be monitored using sound waves (Fig. 16.21).

Figure 16.21

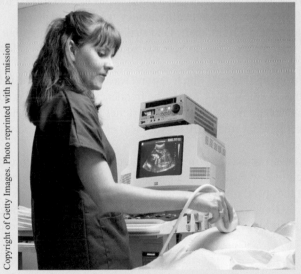

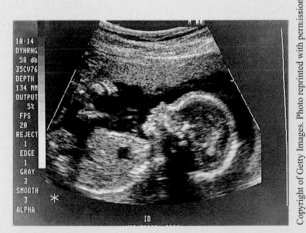

(a) Ultrasound technology is used to visualize tissue inside the human body without subjecting the person to invasive procedures or harmful radiation.

(b) An ultrasound of a human fetus.

PROBLEMS 16.4

1. Find the speed of sound in m/s at $\overline{10}°C$ at 1 atm pressure in dry air.
2. Find the speed of sound in m/s at 35°C at 1 atm pressure in dry air.
3. Find the speed of sound in m/s at −23°C at 1 atm pressure in dry air.
4. How long will it take a sound to travel 21.0 m for the conditions of Problem 1?

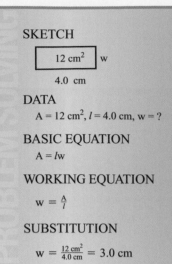

SKETCH

12 cm² | w

4.0 cm

DATA

$A = 12$ cm², $l = 4.0$ cm, $w = ?$

BASIC EQUATION

$A = lw$

WORKING EQUATION

$w = \frac{A}{l}$

SUBSTITUTION

$w = \frac{12 \text{ cm}^2}{4.0 \text{ cm}} = 3.0$ cm

5. How long will it take a sound to travel through $750\overline{0}$ m of water at 25°C?

6. A sound wave is transmitted through water from one submarine, is reflected off another submarine 15 km away, and returns to the sonar receiver on the first submarine. What is the round-trip transit time for the sound wave? Assume that the water temperature is 25°C.

7. A sonar receiver detects a reflected sound wave from another ship 3.52 s after the wave was transmitted. How far away is the other ship? Assume that the water temperature is 25°C.

8. A woman is swimming when she hears the underwater sound wave from an exploding ship across the harbor. She immediately lifts her head out of the water. The sound wave from the explosion propagating through the air reaches her 4.00 s later. How far away is the ship? Assume that the water temperature is 25°C and the air temperature is 23°C.

9. A train traveling at a speed of $4\overline{0}$ m/s approaches an observer at a station and sounds a $55\overline{0}$-Hz whistle. What frequency will be heard by the observer? Assume that the sound velocity in air is 345 m/s.

10. What frequency is heard by an observer who hears the $45\overline{0}$-Hz siren on a police car traveling at 35 m/s away from her? Assume that the velocity of sound in air is 345 m/s.

11. A car is traveling toward you at 40.0 mi/h. The car horn produces a sound at a frequency of $48\overline{0}$ Hz. What frequency do you hear? Assume that the sound velocity in air is 1090 ft/s.

12. A car is traveling away from you at 40.0 mi/h. The car horn produces a sound at a frequency of $48\overline{0}$ Hz. What frequency do you hear? Assume that the sound velocity in air is 1090 ft/s.

13. A jet airplane taxiing on the runway at 13.0 km/h is moving away from you. The engine produces a frequency of $66\overline{0}$ Hz in -6.0°C air. What frequency do you hear?

14. While snorkeling you hear a dolphin's sound as it approaches at 5.00 m/s. If the perceived frequency is $85\overline{0}$ Hz, what is the actual frequency being emitted?

15. Two construction workers stand 112 m apart. One strikes a steel beam with a hammer. How long does it take for the other to hear the sound? Assume the velocity of sound in air is 345 m/s.

16. What is the length of a brass pipe through which a sound wave is transmitted in 0.136 s?

17. A crop duster airplane flies overhead at 44.7 m/s. The frequency of the sound is 605 Hz. (a) What do you perceive as the frequency as it approaches? Assume the velocity of sound in air is 345 m/s. (b) What frequency do you hear as the plane flies away and has accelerated to 55.0 m/s?

18. Two iron workers are on a project 72.0 m apart. To get the other's attention, one worker strikes his lunch pail with his wrench. How long does it take the sound to travel through the air if the speed of sound under these conditions is 345 m/s?

19. A construction worker 30.0 m above the ground drops his hammer. How long does his shouted warning take to reach the ground if the speed of sound under these conditions is 345 m/s?

Figure 16.22 Forced vibration of a guitar sounding board.

16.5 Resonance

The **natural frequency** of an object, such as a tuning fork, is the frequency at which it vibrates when struck by another object, such as a rubber hammer. This frequency depends on the length and thickness of the tuning fork and the material from which it is made. Strings on a guitar also vibrate at a natural frequency. The sounding board of a guitar is forced to vibrate at the same frequency as the strings because of energy transfer from the strings to the sounding board (Fig. 16.22). This is an example of *forced vibration*. The natural frequency of the board is typically different from that of the strings or tuning fork. Because the area of the sounding board is large, energy transfer into sound waves is very efficient.

Figure 16.23 Identical tuning forks with resonant air column. Sound waves of the left fork cause the right fork to vibrate.

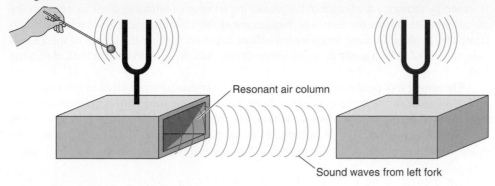

Resonant air column

Sound waves from left fork

Therefore, the vibrating string or tuning fork loses its energy or dies out more rapidly if in contact with a sounding board.

Consider two objects such as tuning forks with the same natural frequency that are set close together (Fig. 16.23). One is set into vibration and then stopped after a few seconds. It is found that the other tuning fork is weakly vibrating. The sound waves of the first fork cause the second to vibrate. This is called *sympathetic vibration* or *resonance*. **Resonance** occurs when the natural vibration rates of two objects are the same. Energy transfer into vibrations of the second fork is found to be much more efficient when both forks have the same frequency than when they have different frequencies. Large vibrations can be set up if the driving force is at the natural frequency of a system. Auto body rattles sometimes occur at certain speeds and disappear for small speed changes. Radio receivers operate on the principle of resonance. The natural frequency of vibration of electrical currents in a circuit may be tuned to that of an incoming radio signal, which is then amplified and converted into sound.

The playing of a musical instrument demonstrates resonance. In wind instruments either the lips or a reed vibrates. Without an air column, however, no music is produced.

TRY THIS ACTIVITY

Singing Wineglasses

Moisten the tip of your finger and gently rub it around the rim of a quality wineglass. At just the right speed, the friction between your finger and the wineglass causes the glass to vibrate at its natural frequency (Fig. 16.24). The glass forms a standing wave, begins to resonate, and produces the pure ringing sound of the wineglass. What can be done to the glass to change the frequency of the resonating sound?

Figure 16.24

Figure 16.25 Resonance of an air column.

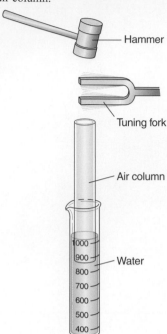

Hammer

Tuning fork

Air column

Water

The tube that makes up the instrument is necessary. Air in the tube vibrates at the same frequency as the lips or reed (resonance) to produce the musical sound. The pitch of the sound is varied by changing the length of the resonating column of vibrating air. The length of the air column determines the resonant frequencies of the vibrating air. This is easily demonstrated by holding a tuning fork above a hollow tube and varying the length of the air column (Fig. 16.25). The sound is louder when the air column is in resonance with the tuning fork.

The tuning fork produces alternating high- and low-pressure variations in the air as it vibrates. Because of the movement of waves up and down the tube with accompanying constructive and destructive interference between the waves, resonance is found at one-fourth wavelengths in a closed tube. Resonance is found at multiples of $\lambda/4$. Open pipes also resonate. However, resonance in open pipes is at half-wavelength multiples ($\lambda/2$, λ, $3\lambda/2$, etc.).

16.6 Simple Harmonic Motion

Periodic motion occurs when an object moves repeatedly over the same path in equal time intervals. Attach a mass m to a spring suspended from a support [Fig. 16.26(a)]. Pull the mass down [Fig. 16.26(b)] and release it [Fig. 16.26(c)]. The mass moves up and down in periodic motion.

Simple harmonic motion is a type of linear motion in which the acceleration of an object is directly proportional to its displacement from its equilibrium position; the motion is always directed toward the equilibrium position. That is, the farther the spring is pulled down, the faster the spring moves when it is released, and the motion is always directed toward the equilibrium (rest) position. The mass on the spring in Fig. 16.26 is an example of an object in simple harmonic motion.

Figure 16.26 Mass suspended by a spring from a support in simple harmonic motion.

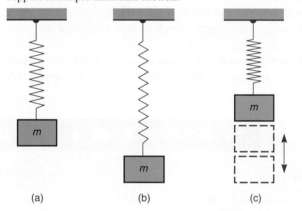

(a)   (b)   (c)

Next, we will compare simple harmonic motion and circular motion and discuss some of the corresponding terms. Assume that a Ferris wheel is rotating uniformly, with you in the only seat and with the sun directly overhead (Fig. 16.27). Now compare your position with the position of your shadow. When you are in position a, your shadow is in position a'; when you are in position b, your shadow is in position b'; and so on. As you complete one revolution on the Ferris wheel, your shadow makes one complete vibration (cycle) on the ground in simple harmonic motion. When your shadow is at b', the *displacement* of your shadow is the distance $b'O'$, which is the distance from your shadow to the midpoint of its vibration, O'. In general, the **displacement** of an object in simple harmonic motion is its distance from its equilibrium, or rest, position. The **amplitude** of the vibration is the maximum displacement $O'P$ or $O'Q$, which is also the radius of the Ferris wheel. The **period** is the time required for one complete vibration—the time required for you to make one complete revolution on the Ferris wheel and the time required for your shadow to make one complete vibration on the ground. The **frequency** is the number of complete vibrations per unit of time or the number of complete revolutions that you make on the Ferris wheel per unit of time. The motion of the rider when graphed over time produces a special curve called a sine wave as shown in Fig. 16.28. The frequency f equals the reciprocal of the period T. That is, $f = \dfrac{1}{T}$. The equilibrium position of the shadow is the midpoint of its path, O'.

Figure 16.27 The simple harmonic motion of the shadow of a person is shown on line *PQ*, where the person is rotating at uniform speed in a circle on a Ferris wheel. When the person is at position *a* on the Ferris wheel, his or her shadow is shown at position *a'* on line *PQ*. Shadow positions *b'*, *c'*, *d'*, and *e'* correspond to positions *b*, *c*, *d*, and *e*, respectively, of the person on the Ferris wheel.

Figure 16.28 Top view of the motion of a Ferris wheel rider graphed over time.

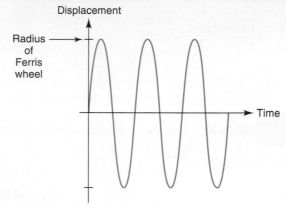

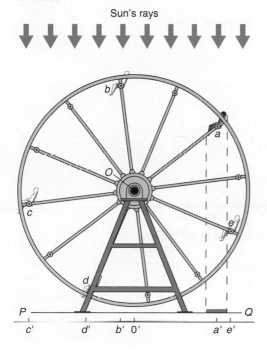

A **pendulum** consists of an object suspended so that it swings freely back and forth about a pivot (Fig. 16.29). Pendulums have been commonly used in clocks for many years. The motion of a pendulum, when the displacement is small, very closely approximates simple harmonic motion. There are three basic properties of a pendulum discovered by Galileo:

Figure 16.29 Free-swinging pendulum of length *l*.

1. Its period is independent of its mass. (Air resistance is more affected by the size and shape of the bob than by its mass.)
2. Its period is independent of the amplitude when the arc is small, that is, when its arc is less than 10°.
3. Subject to these conditions, its period is given by

$$T = 2\pi\sqrt{\frac{l}{g}}$$

where T = period (usually, in seconds)
     l = length of pendulum (m or ft)
    g = 9.80 m/s² or 32.2 ft/s²

The period of any pendulum depends only on its length and the acceleration of gravity. The longer the pendulum, the longer is the time for each complete swing or period. The less the gravitational attraction on a particular planet or moon, the larger is its period.

TRY THIS ACTIVITY

Swing Set Physics

A swing often serves as a good pendulum. Using a stopwatch, measure the period of a swing as it moves forward and backward. Vary the amplitude of the swing, the mass of the person sitting on it, and the length of the swing's chain. Which variable alters the period?

EXAMPLE

Find the length (in cm) of a pendulum with a period of 1.50 s.

Data:

$$T = 1.50 \text{ s}$$
$$g = 9.80 \text{ m/s}^2$$
$$l = ?$$

Basic Equation:

$$T = 2\pi\sqrt{\frac{l}{g}}$$

$$T^2 = 4\pi^2\frac{l}{g} \qquad \text{Square both sides.}$$

$$gT^2 = 4\pi^2 l \qquad \text{Multiply both sides by } g.$$

$$\frac{gT^2}{4\pi^2} = l \qquad \text{Divide both sides by } 4\pi^2.$$

Working Equation:

$$l = \frac{gT^2}{4\pi^2}$$

Substitution:

$$l = \frac{(9.80 \text{ m/s}^2)(1.50 \text{ s})^2}{4\pi^2}$$

$$= 0.559 \text{ m} = 55.9 \text{ cm}$$

· · · · · · · · · · · · · · · · ·

PROBLEMS 16.6

1. Find the length (in cm) of a pendulum with a period of 1.50 s.
2. Find the length (in ft) of a pendulum with a period of 3.00 s.
3. Find the period of a pendulum 1.25 m long.
4. Find the period of a pendulum 2.00 ft long.
5. Find the length (in in.) of a pendulum with a period of 2.25 s.
6. Find the length (in m) of a pendulum with a period of 0.700 s.
7. Find the period of a pendulum 18.0 in. long.
8. Find the period of a pendulum 35.0 cm long.
9. If you double the length of a pendulum, what happens to its period?
10. If you double the period of a pendulum, what happens to its length?
11. A grandfather clock has a 0.750-m pendulum. What is its period?
12. A grandfather clock has a pendulum with period 2.40 s. (a) Find the length of this pendulum. (b) Is the length of this pendulum longer or shorter than the length of the clock in Problem 11?
13. What is the period of a pendulum that is 0.25 m long?
14. What is the length of a pendulum with a period of 0.85 s?

PROBLEM SOLVING

SKETCH

$$\boxed{12 \text{ cm}^2} \text{ w}$$
$$4.0 \text{ cm}$$

DATA

$A = 12 \text{ cm}^2, l = 4.0 \text{ cm}, w = ?$

BASIC EQUATION

$A = lw$

WORKING EQUATION

$w = \frac{A}{l}$

SUBSTITUTION

$w = \frac{12 \text{ cm}^2}{4.0 \text{ cm}} = 3.0 \text{ cm}$

Glossary

Amplitude The maximum displacement of any part of a wave or a vibration from its equilibrium, or rest, position. (pp. 422, 440)

Constructive Interference The superposition of waves to form a larger disturbance (wave) in a medium. Occurs when two crests or troughs of superimposed waves meet. (p. 424)

Destructive Interference The superposition of waves to form a smaller disturbance (wave) in a medium. (p. 424)

Diffraction The property of a wave that describes its ability to bend around obstacles in its path. (p. 427)

Displacement The distance of an object in simple harmonic motion from its equilibrium, or rest, position. (p. 440)

Doppler Effect The variation of the frequency heard when a source of sound and the ear are moving relative to each other. (p. 435)

Electromagnetic Spectrum The entire range of electromagnetic waves classified according to frequency. (p. 430)

Electromagnetic Wave A transverse wave resulting from a periodic disturbance in an electromagnetic field, which has an electric component and a magnetic component, each being perpendicular to the other and both perpendicular to the direction of travel. (p. 429)

Frequency The number of complete waves passing a given point per unit time; the number of complete vibrations per unit time in simple harmonic motion. (pp. 423, 440)

Intensity The energy transferred by sound per unit time through unit area. (p. 434)

Interference The effect of two intersecting waves resulting in a loss of displacement in certain areas and an increase in displacement in others. (p. 426)

Longitudinal Wave A disturbance in a medium in which the motion of the particles is along the direction of the wave travel. (p. 423)

Loudness The strength of the sensation of sound to an observer. (p. 434)

Natural Frequency The frequency at which an object vibrates when struck by another object, such as a rubber hammer. (p. 438)

Pendulum An object suspended so that it swings freely back and forth about a pivot. (p. 441)

Period The time required for a single wave to pass a given point or the time required for one complete vibration of an object in simple harmonic motion. (pp. 423, 440)

Pitch The effect of the frequency of sound waves on the ear. (p. 434)

Propagation Velocity The velocity of energy transfer of a wave, given by the distance traveled by the wave in one period divided by the period. (p. 424)

Pulse Nonrepeated disturbance that carries energy through a medium or through space. (p. 422)

Resonance A sympathetic vibration of an object caused by the transfer of energy from another object vibrating at the natural frequency of vibration of the first object. (p. 439)

Simple Harmonic Motion A type of linear motion of an object in which the acceleration is directly proportional to its displacement from its equilibrium position and the motion is always directed to the equilibrium position. (p. 440)

Sound Those waves transmitted through a medium with frequencies capable of being detected by the human ear. (p. 432)

Speed of Light The speed at which light and other forms of electromagnetic radiation travel: 3.00×10^8 m/s in a vacuum. (pp. 429)

Speed of Sound The speed at which sound waves travel in a medium: 331 m/s in dry air at 1 atm pressure and 0°C. (p. 432)

Standing Waves A special case of superposition of two waves when no energy propagation occurs along the wave. The wave displacements are constant and remain fixed in location. (p. 424)

Superposition of Waves The algebraic sum of the separate displacements of two or more individual waves passing through a medium. (p. 424)

Transverse Wave A disturbance in a medium in which the motion of the particles is perpendicular to the direction of the wave motion. (p. 422)

Wave A disturbance that moves through a medium or through space. (p. 421)

Wavelength The minimum distance between particles in a wave that have the same displacement and are moving in the same direction. (p. 423)

Formulas

16.1 $f = \dfrac{1}{T}$

$v = \lambda f$

16.2 $c = \lambda f$

16.3 $v = 331 \text{ m/s} + \left(0.61\dfrac{\text{m/s}}{\text{°C}}\right)T$

$v = 1087 \text{ ft/s} + \left(1.1\dfrac{\text{ft/s}}{\text{°F}}\right)(T - 32\text{°F})$

16.4 $f' = f\left(\dfrac{v}{v \pm v_s}\right)$

16.5 $T = 2\pi\sqrt{\dfrac{l}{g}}$

Review Questions

1. Which of the following are methods of energy transfer?
   (a) Conduction   (b) Radiation
   (c) Wave motion   (d) None of the above
2. The minimum distance between particles in a wave that have the same displacement and are moving in the same direction is called
   (a) the period.   (b) the frequency.
   (c) the wavelength.   (d) none of the above.
3. Which of the following refers to the time required for a single wave to pass a given point?
   (a) The period   (b) The frequency
   (c) The wavelength   (d) None of the above
4. Which of the following refers to the number of complete waves passing a given point per unit time?
   (a) The period   (b) The frequency
   (c) The wavelength   (d) None of the above
5. An example of a transverse wave is
   (a) a sound wave.   (b) a water wave.
   (c) interference.   (d) none of the above.
6. Which of the following is an example of longitudinal waves?
   (a) Sound waves   (b) Water waves
   (c) Interference   (d) None of the above

7. Which of the following are electromagnetic waves?
    (a) Sound   (b) Water waves
    (c) Radar waves   (d) X rays
    (e) All of the above
8. Explain the difference between interference and diffraction.
9. Explain the difference between constructive and destructive interference.
10. If waves did not exhibit the property of diffraction, under what conditions would your stereo system sound different?
11. Give an example of diffraction of water waves.
12. What happens to the frequency of a vibrating string on a guitar if the length of the string is decreased?
13. Explain the difference between a wave and a pulse.
14. Give an example of a pulse.
15. What happens to the speed of sound when the temperature increases? Explain why this might happen.
16. Explain how a seismograph works.
17. How does the speed of sound differ in water and air? Explain the reason for this difference.
18. In your own words, explain the Doppler effect.
19. Distinguish between sympathetic and forced vibration.
20. In your own words, explain resonance.
21. State a reason that might explain why many stars appear to have their light shifted to the red (longer wavelength) part of the electromagnetic spectrum when viewed from the earth.
22. Distinguish between amplitude and displacement.
23. Distinguish between period and frequency.
24. Does the period of a pendulum depend on its mass, and if so, how?

Review Problems

1. Find the period of a wave with frequency 355 kHz.
2. Find the frequency of a wave with period 0.320 s.
3. (a) What is the frequency of a light wave with wavelength 4.50×10^{-7} m and velocity 3.00×10^{8} m/s? (b) Find the period of the wave.
4. Find the speed of a wave with frequency 8.97 Hz and wavelength 0.654 m.
5. What is the wavelength of longitudinal waves in a coil spring with frequency 4.65 Hz and velocity 5.78 m/s?
6. Find the frequency of a wave produced by a generator that emits 85 pulses in 1.3 s.
7. What is the wavelength of an electromagnetic wave with frequency 65.5 MHz?
8. Find the speed of sound in m/s at 85°C at 1 atm pressure in dry air.
9. Find the speed of sound in m/s at −35°C at 1 atm pressure in dry air.
10. How long will it take a sound wave to travel through 1450 m of water at 25°C?
11. A sound wave is transmitted through water from one ship, is reflected off another ship 22 km away, and returns to the sonar receiver on the first ship. What is the round-trip transit time for the sound wave if the water temperature is 23°C?
12. A train traveling at a speed of 95 mi/h approaches an observer at a station and sounds a 525-Hz whistle. What frequency will be heard by the observer? Assume that the sound velocity in air is 1090 ft/s.
13. A car is traveling toward you at 95 km/h. The car horn produces a sound at frequency 4950 Hz. (a) What frequency do you hear? Assume that the sound velocity in air is 345 m/s. (b) What frequency do you hear if the car is traveling away from you?

14. What is the frequency of the sound waves being emitted from a train whistle while approaching at 45 m/s in air that is 11°C? The perceived frequency is 425 Hz.
15. The taillight on a car produces light with wavelength 5.00×10^{-7} m. What frequency do you observe when the car is departing at 24 m/s at 0°C?
16. A pendulum has a length of 0.450 m. What is its period?
17. A pendulum has a period of 0.700 s. Find the length of the pendulum in inches.

APPLIED CONCEPTS

1. The pendulum on a grandfather clock is calibrated so its period equals 1.00 s. (a) What is the length of the pendulum cable? (b) If the grandfather clock is moved to the moon where the acceleration due to gravity is 1.62 m/s^2, will the clock keep the correct time? (c) If not, what can be done to the pendulum to correct the problem?

Figure 16.30

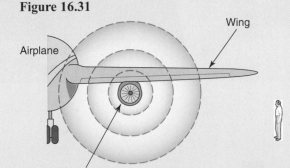

2. The Tacoma Narrows Bridge, built across Puget Sound in Washington, formed a standing wave before it collapsed on November 7, 1940. As seen in the photo in Fig. 16.30, (a) how many full wavelengths were between the two towers, which were spaced 2800 ft apart? (b) If the frequency of the vibrations was 0.20 Hz, find the wave speed for this mechanical wave. (c) Find the wavelength of the bridge's standing wave if the frequency is doubled.

3. Maintaining strong AM radio reception when driving under overpasses or through tunnels is often difficult. FM radio frequencies typically are not as affected as AM frequencies. (a) Using the frequency range for AM radio, 550 kHz to 1650 kHz, and FM radio, 88 MHz to 108 MHz, find the range of wavelengths for both AM and FM bands. (b) In terms of wave diffraction, explain why AM radio waves have a more difficult time passing under overpasses and through tunnels.

4. Dave, a jet engine technician, is exposed to sound intensities of 10.3 W/m^2 (12$\overline{0}$ dB) at 30.4 m from a jet (Fig. 16.31). (a) What is the audible power produced by the jet engine? (b) Since the sound power produced by the engine remains constant and the sound propagates in all directions (spherically), what is the intensity of the sound 60.8 m from the engine? (c) Compare the intensity of the sound at 30.4 m to the intensity of the sound at 60.8 m from the engine.

Figure 16.31

Airplane

Wing

Jet engine

5. The speed of an automobile can be determined using the Doppler effect by sounding the horn. (a) An automobile's horn produces a frequency of 765 Hz. How fast is the car traveling if a stationary microphone measures the horn's frequency as 836 Hz at a temperature of 23.3°C? (b) What is the frequency of the sound wave as the car moves away from the microphone?

Electricity is the best means of transmitting energy for many purposes. We consider the basics of electricity including electric charge, electric fields, static electricity, electric current and circuits, batteries, and electric power.

Objectives

The major goals of this chapter are to enable you to:

1. Describe the nature of electric charges.
2. Distinguish conduction and induction.
3. Use Coulomb's law to find the force between charges.
4. Describe the characteristics of electricity.
5. Use Ohm's law to solve electric flow problems.
6. Use electrical symbols to describe circuits.
7. Find current, voltage, and resistance in simple circuits.
8. Describe the nature of cells and batteries.
9. Analyze circuits with cells in series and parallel.
10. Find electric power.

17.1 Electric Charges

Electrification was first studied 2500 years ago in ancient Greece. It was found that when an amber rod was rubbed with a wool cloth, the rod attracted small objects (Fig. 17.1). When two objects are rubbed together, they become charged.

When you slide rubber-soled shoes on a wool rug on a dry day, you become electrified. That is, you have acquired a static charge. This static charge is usually lost when you touch an object at a different potential as in Fig. 17.2. Part of the static charge may be lost when you touch an object, such as another person, to which charge is transferred.

To understand electricity, we need to know more about the structure of matter. We have seen that all matter is made up of atoms. These atoms are made of electrons, protons, and neutrons. Each **proton** has one unit of positive charge and each **electron** has one unit of negative charge. The **neutron** has no charge. The protons and neutrons are tightly packed into what is called the *nucleus*. Electrons may be thought of as small charged clouds that surround the nucleus of atoms (Fig. 17.3). An atom normally has the same number of electrons as protons and thus is uncharged. If an electron is removed, the atom is left with a *positive charge* (+), that is, an excess of protons. If an extra electron is added, the atom has a *negative charge* (−), that is, an excess of electrons. **Benjamin Franklin** first referred to electric charges as being positive and negative. The study of electric charges and the forces between them is referred to as *electrostatics*. As the name implies, *static electricity* is electricity that does not flow. *Current electricity* is the flow of electric charge through a conductor.

When two materials are rubbed together, the atoms on the two surfaces move across each other and brush off electrons. The electrons are transferred from one surface to the other. One surface is then left with a *positive charge* and the other is *negative*. This process is called *electrification*.

The two types of electric charges can be observed indirectly by using an electroscope. A very simple electroscope is a ball of wood pith on a silk thread [Fig. 17.4(a)]. We can produce a charge on a hard rubber rod by rubbing it with a wool cloth. The rubber rod acquires a negative charge and the wool acquires a positive charge. The universal acceptance

Figure 17.1 Amber rod attracting bits of paper after being rubbed.

Figure 17.2 Stored charge being transferred.

Figure 17.3

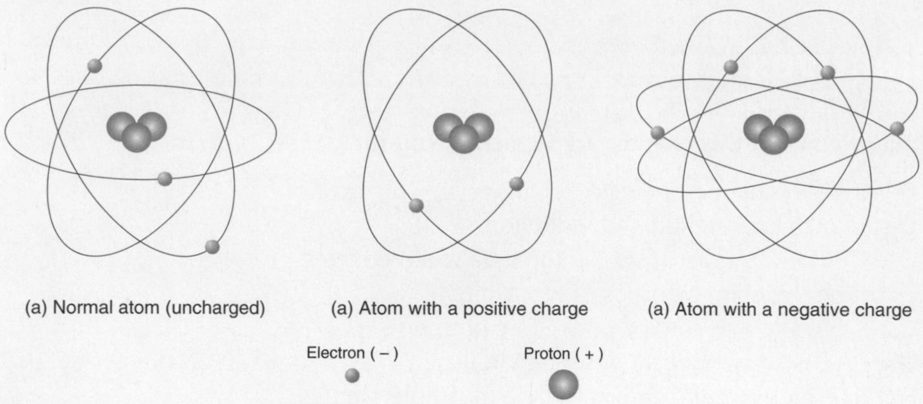

(a) Normal atom (uncharged)   (a) Atom with a positive charge   (a) Atom with a negative charge

Electron (−)   Proton (+)

TRY THIS ACTIVITY

Picking up Dust

Rub a furry or wool fabric against a plastic rod for approximately 15 to 20 s. Bring the rod into a dusty area or near a pile of sawdust. What happens to some of the dust or sawdust? Explain what is taking place in relation to the movement of charge.

Benjamin Franklin (1706–1790),

printer, writer, scientist, and states-man, was born in Boston, Massa-chusetts. He first became a skilled printer and successful business-man. He retired from business in 1748 to devote more time to his scientific interests. He developed the fuel-efficient Franklin stove and conducted a series of experiments in electricity, which brought him international recognition as a sci-entist. Through his famous kite ex-periment, he demonstrated that lightning is an electric discharge. He invented the lightning rod and the bifocal lens. Later in his life, he became a well-known statesman. The two types of electric charge were referred to as positive and negative by Franklin. He arbitrarily called the charge on the rubbed glass rod "positive" and the charge on the rubber rod "negative." We still follow his convention today.

of the description of these charges establishes the convention of positive and negative charge in electric circuits. Now transfer some of this negative charge from the rubber rod to the pith ball [Fig. 17.4(b)]. The pith ball becomes negatively charged by **conduction**, a transfer of charge from one place to another. Another pith ball charged in the same way is repelled by the other pith ball [Fig. 17.4(c)]. This charge is *negative* (−).

Figure 17.4   Charging a simple electroscope

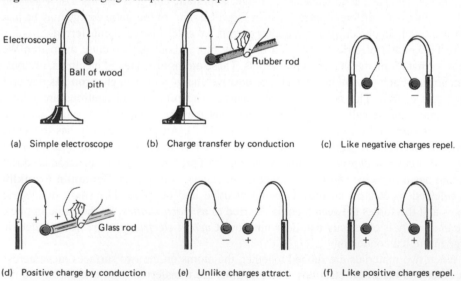

(a) Simple electroscope   (b) Charge transfer by conduction   (c) Like negative charges repel.

(d) Positive charge by conduction   (e) Unlike charges attract.   (f) Like positive charges repel.

Now rub a glass rod with silk. The glass rod acquires a positive charge and the silk acquires a negative charge. Transfer some of the positive charge from the glass rod to a pith ball [Fig. 17.4(d)]. This pith ball is attracted to the negatively charged pith ball [Fig. 17.4(e)]. The charge produced by glass and silk is called a *positive charge* (+). Two pith balls that are positively charged will repel each other [Fig. 17.4(f)].

17.2 Induction

Induction is a method of charging one object by bringing a charged object near to, but not touching, it. The leaf electroscope is more sensitive than the pith ball type and can also be used to show electrification or charging by induction. The leaf electroscope [Fig. 17.5(a)] usually consists of a metal rod with a metal ball on one end and two thin strips of gold foil leaf hanging from the other end. The delicate leaves are enclosed to protect them and the rod is insulated from the enclosure. When charge is placed on the leaves, they diverge because of the force of repulsion of their similar charge.

Electroscopes may be charged by *conduction* by touching a charged object to the metal ball. Electroscopes are charged by *induction* by bringing a charged object near to, but not touching, the metal ball. Like charges on the object are repelled in the electroscope to the leaves, which in turn repel each other and separate, and leave the ball with the unlike charges [Fig. 17.5(b) and (c)].

Figure 17.5 Charging an electroscope by induction

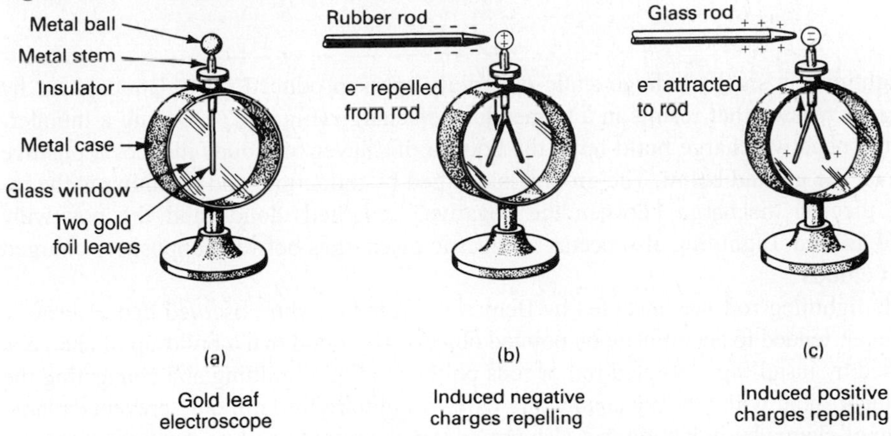

(a)

Gold leaf
electroscope

(b)

Induced negative
charges repelling

(c)

Induced positive
charges repelling

When the charged object is removed, the leaves close as the free electrons redistribute themselves over the ball, rod, and leaves. The electroscope has been only temporarily charged by induction. A residual charge by induction may be obtained by charging the electroscope temporarily as in Fig. 17.6(a) and then touching the electroscope with a neutral object (like your finger) while the charged object is still held close [Fig. 17.6(b)]. The neutral object provides a path for some of the induced charge on the electroscope to escape. When everything is removed, a residual charge by induction remains on the electroscope [Fig. 17.6(d)].

Figure 17.6 Residual charge by induction

Leaves charged
negatively
by induction

(a)

Electrons
repelled
to earth

(b)

Electroscope
with electron
shortage

(c)

Positively
charged
electroscope

(d)

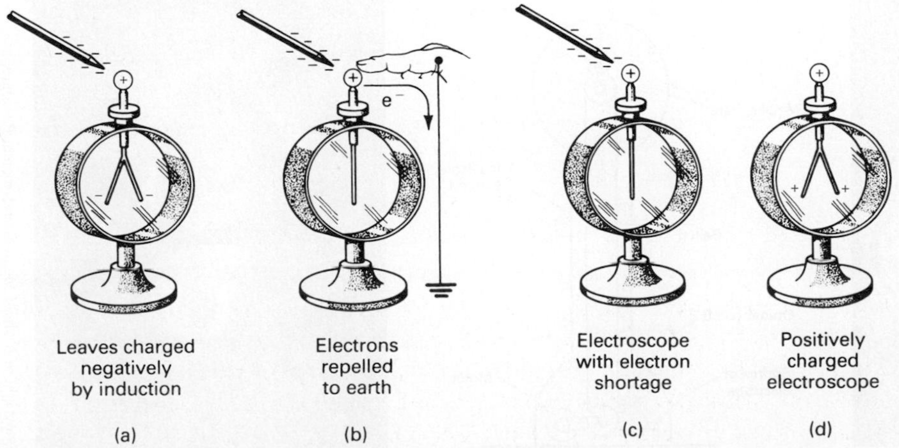

In summary,

Like charges repel each other and unlike charges attract each other.

TRY THIS ACTIVITY

Repelled Balloons

Inflate two latex balloons, tie a piece of light string to each balloon, and suspend them from the ceiling or a horizontal rod so they are 1 ft or so apart. Rub each of the balloons with a furry fabric or your hair. Both balloons are now charged. Slowly reduce the distance between the two balloons. Look for evidence of an electric force. If you see such evidence, explain why the force is either attractive or repulsive. In addition, what are the two main factors that play a role in the strength of the electric force?

Figure 17.7 Lightning: static electric discharge.

Courtesy of Pearson Education-Asia

Lightning is simply a huge static electricity spark produced in the atmosphere by moving air masses that results in a tremendous discharge (Fig. 17.7). During a thunderstorm the negative charge build-up at the bottom of a layer of clouds induces a positive charge on the ground below. The ground is charged by induction, and lightning is the resulting electric discharge between the negatively charged clouds and the positively charged ground. Lightning also occurs as electric discharges between oppositely charged parts of clouds.

The lightning rod was invented by Benjamin Franklin, who observed that electricity from the air tended to accumulate on pointed objects. He found that a build-up of charge is decreased by installing a pointed rod or rods on the roof of a building and connecting the rod to the ground with a heavy conducting wire. A lightning rod not only prevents a large build-up of charge by induction, but also provides a conducting path to the ground for any sudden tremendous discharge of a lightning strike. This often results in protecting the building from damage.

The *van de Graaff generator*, named after **Robert van de Graaff**, is a laboratory machine that is used to produce static electricity and transfer it to a metal sphere by conduction. The van de Graaff consists of an electron source, a rubber belt driven by a motor, and a metal sphere supported by an insulating stand (Fig. 17.8). The electron source charges the rubber belt as it passes by, which carries the electrons up to the sphere and deposits them

Figure 17.8 Van de Graaff generator.

(a) Laboratory model

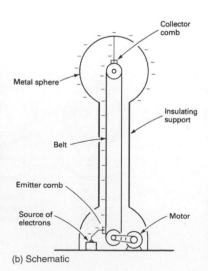

(b) Schematic

(c) Student model

there. This builds up a high potential difference (several hundred thousand volts) between the sphere and the ground. Note that there is no charge on the inside surface of the sphere.

17.3 Coulomb's Law

The charge on a proton is denoted by the symbol e^+ and the charge on an electron by e^-. In the study of electricity, a much larger unit of charge is required and is called the *coulomb*, C, named after **Charles Coulomb**. The measurement of the unit charges gives $e^+ = +1.60 \times 10^{-19}$ C and $e^- = -1.60 \times 10^{-19}$ C. Thus, a collection of 6.25×10^{18} electrons, which is defined by

$$\frac{1}{1.60 \times 10^{-19}} = 6.25 \times 10^{18}$$

has a total charge of 1.00 C. Although this seems like a lot of electrons, it represents only the amount of charge that passes through a 100-W light bulb in about 1 s. We use q to denote amount of electric charge.

In 1789, Coulomb made a scientific study of the forces of attraction and repulsion between charged objects using a very sensitive torsion balance. From his experiments he determined the existence of an *inverse square law* for charged particles that can be used to calculate the forces of attraction or repulsion between charged objects. The inverse-square-law behavior was also found to apply to the much weaker gravitational force studied in Chapter 5. While the gravitational force is always attractive, the electric force can be either attractive or repulsive. These two forces are important because one holds the solar system together (gravity) and the other holds atoms and molecules together (electricity). Electric forces, however, are billions of times greater than the earth's gravitational force.

Robert van de Graaff (1901–1967),

physicist, was born in Tuscaloosa, Alabama. He invented a constant-potential electrostatic generator (later known as the van de Graaff generator) and later developed this generator for use as a particle accelerator. He invented the insulating-core transformer in the late 1950s.

Charles Coulomb (1736–1806),

physicist, was born in France. He pioneered research in electricity and magnetism. The unit of measuring electricity, the coulomb, is named after him.

COULOMB'S LAW OF ELECTROSTATICS

The force between two point charges q_1 and q_2 is directly proportional to the product of their magnitudes and inversely proportional to the square of the distance separating them, r.

Figure 17.9(a) shows the repulsive force between two like (positive in this case) charges separated by a distance r. The attractive force between two unlike charges is shown in Fig. 17.9(b).

We use a *proportionality constant* k in writing Coulomb's law as an equation to take into account the air or other medium between the charges. Written in equation form, **Coulomb's law** becomes

$$F = \frac{kq_1q_2}{r^2}$$

where F = force of attraction or repulsion   (in newtons)
            $k = 9.00 \times 10^9$ N m²/C²   (k was found by experiment)
    q_1, q_2 = electric charges   (in coulombs)
            r = distance between the charges   (in metres)

The force between the charges is a vector quantity that acts on each charge.

Figure 17.9   (a) Two like charges at distance r apart repel each other with force F. (b) Two unlike charges attract.

(a)

(b)

TRY THIS ACTIVITY

Attracting Water

Turn on a faucet so only a very thin stream of water comes out of the tap. Charge a plastic rod, latex balloon, or plastic comb by rubbing it with a piece of wool or fur. Slowly bring the charged object toward the stream of water. Explain what happens to the stream of water as it flows from the tap.

Two charges, each with magnitude $+6.50 \ \mu C$, are separated by a distance of 0.200 cm. Find the force of repulsion between them.

Data:

$$q_1 = q_2 = +6.50 \ \mu C = +6.50 \times 10^{-6} \ C$$
$$r = 0.200 \ cm = 0.00200 \ m = 2.00 \times 10^{-3} \ m$$
$$k = 9.00 \times 10^9 \ N \ m^2/C^2$$
$$F = \ ?$$

Basic Equation:

$$F = \frac{kq_1q_2}{r^2}$$

Working Equation: Same

Substitution:

$$F = \frac{(9.00 \times 10^9 \ N \ m^2/C^2)(6.50 \times 10^{-6} \ C)(6.50 \times 10^{-6} \ C)}{(2.00 \times 10^{-3} \ m)^2}$$
$$= 9.51 \times 10^4 \ N$$

· · · · · · · · · · · · · · · · ·

PROBLEMS 17.3

1. Two identical charges, each -8.00×10^{-5} C, are separated by a distance of 25.0 cm. What is the force of repulsion?
2. The force of repulsion between two identical positive charges is 0.800 N when the charges are 0.100 m apart. Find the value of each charge.
3. A charge of $+3.0 \times 10^{-6}$ C exerts a force of 940 N on a charge of $+6.0 \times 10^{-6}$ C. How far apart are the charges?
4. A charge of -3.0×10^{-8} C exerts a force of 0.045 N on a charge of $+5.0 \times 10^{-7}$ C. How far apart are the charges?
5. When a -9.0-μC charge is placed 0.12 cm from a charge q in a vacuum, the force between the two charges is 850 N. What is the value of q?
6. How far apart are two identical charges of $+6.00 \ \mu C$ if the force between them is 25.0 N?
7. Three charges are located along the x-axis. Charge A ($+3.00 \ \mu C$) is located at the origin. Charge B ($+5.50 \ \mu C$) is located at $x = +0.400$ m. Charge C ($-4.60 \ \mu C$) is located at $x = +0.750$ m. (a) Find the total force (and direction) on charge B. (b) Find the total force (and direction) on charge A. (c) Find the total force (and direction) on charge C.
8. Three charges are located along the x-axis. Charge A ($+5.00 \ \mu C$) is located at the origin. Charge B ($+4.50 \ \mu C$) is located at $x = +0.650$ m. Charge C ($-4.20 \ \mu C$) is located at $x = +0.650$ m. Find the total force (and direction) on charge B.

17.4 Electric Fields

So far, we have discussed electrification due to the brushing of electrons from a surface. The concept of the electric field is also an important part of the study of static electricity. Two magnets may either attract or repel each other even though they may not be touching. This illustrates the idea of a "field": even though they are not touching, there is an invisible region around each magnet that affects the other magnet if that magnet is placed in the region of the first magnet.

In terms of static electricity, an **electric field** exists where an electric force (of attraction or repulsion) acts on a charge brought into the area. A charged balloon put on a wall

SKETCH

| 12 cm² | w |

4.0 cm

DATA

$A = 12 \ cm^2, \ l = 4.0 \ cm, \ w = \ ?$

BASIC EQUATION

$A = lw$

WORKING EQUATION

$w = \frac{A}{l}$

SUBSTITUTION

$w = \frac{12 \ cm^2}{4.0 \ cm} = 3.0 \ cm$

illustrates this principle (Fig. 17.10). Note that the balloon attracts the wall even without physical contact. The balloon has acquired a negative charge through friction, but the wall surface acquires a positive charge produced by the electric field of the charged balloon. Such an invisible electric field is present around every charged object.

Static electricity can be a real hazard in industry as well as a curiosity and sometimes a nuisance in daily life. The electric spark from static electricity, particularly in synthetic fiber textile mills, is extremely dangerous. Also, some workers in cosmetic factories in which aerosol (spray) products are made with hydrocarbon (petroleum type) propellants are required to wear cotton clothes rather than those made with synthetic fibers.

Electric fields can be measured by using test charges. Fields are represented using lines to show the direction and intensity of the field. Field lines do not really exist. They are just a means of providing a model of the field to visualize how the field is stronger where the lines are closer together and also to represent the direction of the field. Electric fields really do exist, though we study them by observing the effects they produce. Keep in mind electric fields exist in three dimensions though our drawings are only two-dimensional models (Fig. 17.11).

We can use the test charge to calculate the strength of a field but we do not yet know why charged bodies exert forces on each other. Still, we can detect and measure fields because fields produce forces on charges placed in the field.

A test charge creates an electric field about it in all directions. If a second charge is placed at some point in the field of the first charge, it interacts with the first charge. Gravitational fields exist around bodies like the earth and exert a force on nearby bodies, like people. We recall that $F_w = mg$, and we can find the gravitational field to be $g = F_w/m$. Similarly an electric field can be described by

$$E = \frac{F}{q'}$$

where E = electric field
F = force on a test charge placed in the field
q' = test charge, measured in coulombs

The result is a vector quantity that is the magnitude of the electric field. By measuring the force on the test charge at different locations in the field, we can map the field and then represent it by using field lines for a model.

A positive test charge of 30.0 μC is placed in an electric field. The force on it is 0.600 N. What is the magnitude of the electric field at the location of the test charge?

Data:

$$q' = 30.0 \ \mu C = 3.00 \times 10^{-5} \ C$$
$$F = 0.600 \ N$$
$$E = ?$$

Basic Equation:

$$E = \frac{F}{q'}$$

Working Equation: Same

Substitution:

$$E = \frac{0.600 \ N}{3.00 \times 10^{-5} \ C}$$
$$= 2.00 \times 10^4 \ N/C$$

Figure 17.10 Common electric field. These two girls are using static electricity to adhere balloons to the wall as shown.

Courtesy of Pearson Learning

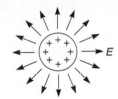

Figure 17.11 An electric field is created in the space around a charged object.

EXAMPLE

Figure 17.12 Electron beam deflection in a television.

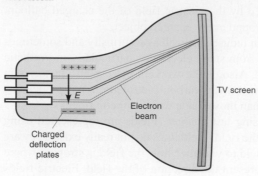

Charged deflection plates

Electron beam

TV screen

Electric fields are used for many applications in electronics and elsewhere. An ink-jet printer uses the deflection caused by an electric field on charged ink droplets to direct ink to the appropriate spot on the paper. In a similar way, the electron beam in a TV picture tube is deflected to the correct spots on the screen to produce a picture (Fig. 17.12).

Energy can be transmitted by means of an electric field through empty space. When charges move in an electric field, they cause a disturbance that produces changes near the moving charges and results in the transfer of energy.

PHYSICS CONNECTIONS

Lightning Safety

The earth experiences about 25 lightning ground strikes every second. Each year in the United States approximately 400 people are struck by lightning, which results in 100 deaths. Many of these strikes take place on ball fields, golf courses, and other open areas.

The safest place to be during a lightning storm is inside a building or a car. The charge from the lightning strike gathers on the outside of the building or car because like electric charges repel and remain as far away from each other as possible. After accumulating on the exterior of the building or car, the charges move quickly to a ground. Although you may be safest inside a building or a car, you should stay away from electric appliances and plumbing fixtures that are connected to the ground and will act as conductors of electricity if the building is struck by lightning.

If you are not able to get inside a building or a car, you should take the following precautions during a lightning storm:

Figure 17.13 The proper position to take if one is not able to get inside during a thunderstorm.

1. Move to the lowest section of ground and crouch down [Fig. 17.13]. By reducing your height, you decrease the likelihood of becoming a lightning target. Do not let your hands or other parts of your unprotected body touch the ground. When lightning strikes the ground, it typically spreads out along the ground and can still reach you. If only your shoes are on the ground, the amount of current passing through your body will be reduced. If you are lying on the ground, the electric current could easily pass through your entire body, including your heart.

2. Remove all metal objects from your body. Metal on your body acts as a conductor of electricity and will attract lightning.

3. Avoid tall objects such as trees and hilltops and open areas like water and fields. Being the tallest object or standing under tall objects makes a person part of a giant lightning rod. Tall objects, like trees and masts, attract the charges from a thundercloud and lightning strikes.

4. If you are on a body of water, get back to shore as quickly as possible. If this is not possible, crouch down and move away from tall metal masts.

New advances in measuring electric fields are being used at ball fields. These devices measure the electric field intensity in the atmosphere and warn of possible lightning strikes. Since many lightning strikes occur at ball fields, such devices are becoming standard equipment at outdoor sporting events.

PROBLEMS 17.4

1. An electric field has a positive test charge of 4.00×10^{-5} C placed on it. The force on it is 0.600 N. What is the magnitude of the electric field at the test charge location?
2. What is the field magnitude of an electric field in which a negative charge of 2.00×10^{-8} C experiences a force of 0.0600 N?
3. An electric field exerts a force of 2.50×10^{-4} N on a positive test charge of 5.00×10^{-4} C. Find the magnitude of the field at the charge location.
4. An electric field exerts a force of 3.00×10^{-4} N on a positive test charge of 7.50×10^{-4} C. Find the magnitude of the field at the charge location.
5. An electric field of magnitude 0.450 N/C exerts a force of 8.00×10^{-4} N on a test charge placed in the field. What is the magnitude of the test charge?
6. An electric field of magnitude 0.370 N/C exerts a force of 6.20×10^{-4} N on a test charge placed in the field. What is the magnitude of the test charge?
7. What force is exerted on a test charge of 3.86×10^{-5} C if it is placed in an electric field of magnitude 1.75×10^4 N/C?
8. What force is exerted on a test charge of 4.00×10^{-5} C if it is placed in an electric field of magnitude 3.00×10^6 N/C?

SKETCH

12 cm²   w

4.0 cm

DATA

A = 12 cm², l = 4.0 cm, w = ?

BASIC EQUATION

A = lw

WORKING EQUATION

w = $\frac{A}{l}$

SUBSTITUTION

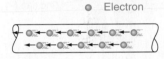

w = $\frac{12 \text{ cm}^2}{4.0 \text{ cm}}$ = 3.0 cm

17.5 Simple Circuits

Electrons moving in a wire produce a current in the wire. When the electron current flows in only one direction (Fig. 17.14), it is called **direct current** (dc). Current that changes direction is called **alternating current** (ac). Alternating current will be considered in Chapter 19.

An electric current is a convenient and cost-effective means of transmitting energy. We all face many daily situations that require energy to do a particular task. To drill a hole in a metal stud (Fig. 17.15), energy must be supplied and transformed into mechanical energy to turn the drill bit. The problem is how to supply energy to the machine being used in a form that the machine can turn into useful work. Electricity is often the most satisfactory means of transmitting energy.

We begin our study of the use of electricity in transferring energy with a circuit of a simple flashlight (Fig. 17.16). An **electric circuit** is a conducting loop in which electrons carrying electric energy may be transferred from a suitable source to do useful work and returned to the source. Energy is stored in the battery. When the switch is closed, energy is transmitted to the light, and the light glows.

Figure 17.14 Current flowing in only one direction is direct current.

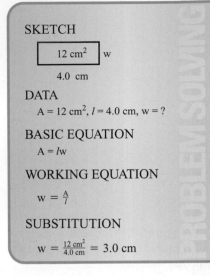

○ Electron

Figure 17.15 Changing stored electric energy to mechanical energy for drilling.

Figure 17.16 Simple electric circuit.

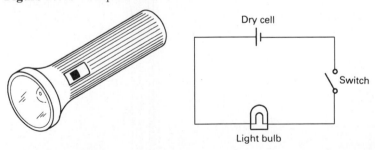

Current electricity is the flow of energized electrons through an electron carrier called a conductor (Fig. 17.17). The electrons move from the energy *source* (the battery, here) to the *load* (where the transmitted energy is turned into useful work). There they lose energy picked up in the source. We now consider each part of the circuit and determine its function.

The Source of Energy

The **source** is the object that supplies electric energy for the flow of electric charge (electrons) in a circuit. The dry cell (Fig. 17.18) is a device that converts chemical energy to

Figure 17.17 Flow of energized electrons through a conductor.

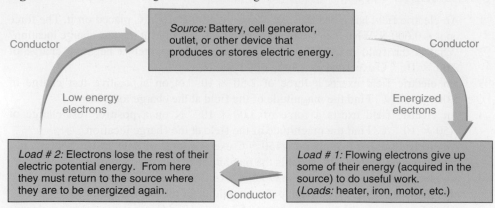

Source: Battery, cell generator, outlet, or other device that produces or stores electric energy.

Conductor

Conductor

Low energy electrons

Energized electrons

Load # 2: Electrons lose the rest of their electric potential energy. From here they must return to the source where they are to be energized again.

Load # 1: Flowing electrons give up some of their energy (acquired in the source) to do useful work. (*Loads:* heater, iron, motor, etc.)

Conductor

Figure 17.18 Chemical energy is changed to electric energy in a dry cell.

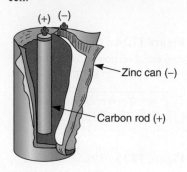

(+) (−)

Zinc can (−)

Carbon rod (+)

electric energy. How the cell does this will be studied later in this chapter. Here, we simply state that, by chemical action, electrons are given energy in the cell. When energy is given to electrons in this manner, their electric potential energy is raised.

What does "electric potential energy" mean? The flow of charge in an electric circuit is often compared to the flow of water in a hydraulic system, as shown in Fig. 17.19(b). Water naturally flows from a position of high potential to a position of low potential and performs work in the process, such as turning a waterwheel or turbine. A pump is needed to return the water from its low-potential position to its high-potential position. There is a *difference in potential* due to its position. Work has been done in lifting it against gravity to the higher position. In a source of electric energy something similar happens. In the source (the battery), work is done on electrons that gives them potential energy. This potential difference between the high–potential energy energized electrons [at the negative (−) pole]

Figure 17.19 The flow of charge in an electric circuit is often compared to the flow of water in a hydraulic system.

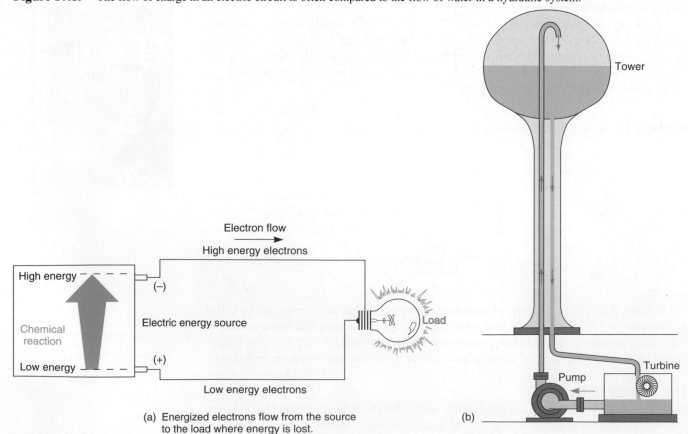

Electron flow

High energy electrons

High energy

(−)

Chemical reaction

Electric energy source

Low energy

(+)

Low energy electrons

Load

Tower

Pump

Turbine

(a) Energized electrons flow from the source to the load where energy is lost.

(b)

and the low–potential energy electrons [at the positive (+) pole] causes the electrons to flow from one point (−) to the other (+) when connected [Fig. 17.19(a)].

Think of this potential difference between the two points as an electric field set up between the poles that drives the electrons in the external circuit from the negative pole to the positive pole. The energized electrons collect at the source's negative pole, repel each other, and flow through the circuit to the positive pole. They lose their potential electric energy to the load.

The Conductor

A conductor carries or transfers the electric charge to the load [Fig. 17.19(a), the light bulb]. A **conductor** is a material (such as copper) through which an electric charge is readily transferred. Such materials have large numbers of free electrons (electrons that are free to move throughout the conductor). As high-energy electrons from the source pass through the conductor, they collide with other electrons in the conductor. These electrons then carry the energy farther along the wire until they collide again and transfer energy on through the wire.

Silver, copper, and aluminum are metals that allow electrons to pass freely through and thus are good conductors. Other metals offer more opposition to the flow of electrons and are poorer conductors (Fig. 17.20). Substances that do not allow electrons to pass readily are called **insulators**. Common insulators are rubber, wool, silk, glass, wood, distilled water, and dry air.

A small number of materials, called **semiconductors**, fall between conductors and insulators in their ability to conduct electric current. Their importance is due to the fact that these materials under certain conditions allow current to flow in one direction only. Silicon is a semiconductor used in transistors and integrated circuits (ICs). Semiconductors are neither good conductors nor good insulators in their pure form. However, they become excellent conductors or insulators when an impurity is added. Transistors are made by layering semiconductor materials together.

Selenium is a semiconductor used in the process of making photocopies. Charge built up on its surface will remain there in the dark. When a certain colored light shines on it, the charge will dissipate very quickly from the areas exposed to the light. A powder sticking only to the charged areas is transferred to a piece of paper, creating a photocopy.

A **superconductor** is a material that continuously conducts electric current without resistance when cooled to typically very low temperatures, usually near absolute zero. H. Kamerlingh-Onnes discovered superconductivity in 1911 shortly after he discovered how to liquefy helium gas. He determined that mercury metal lost its resistance to the flow of electricity at temperatures just below 4.2 K, the boiling point of helium. Scientists are currently finding materials in which superconductivity exists at higher temperatures. The ultimate goal is to find a material in which superconductivity exists at room temperature.

The Load

In the load, electrons lose their energy. The **load** in a circuit converts the electric energy into other forms of energy or work. In a light bulb, electric energy is changed to light and heat (Fig. 17.21). An electric motor changes electric energy to mechanical energy. The load may be a complex motor or only a simple resistor with heat the only new form of energy. The electrons do not collect and remain in the load, but continue back to the low-energy side of the battery (+). There they are energized again for another trip through the circuit.

Current

The flow of electrons through a conductor is called **current**. We could count the electrons passing a point during a certain time to get the rate of flow. This is impractical because the flow of electrons is so large (about 10^{18}/s). To have a workable unit of electric charge, we define a charge of 6.25×10^{18} electrons as 1 *coulomb* (C). The *ampere* (A) is the rate of flow of 1 C of charge passing a point in 1 s. We define a unit for the rate of flow of charge as follows:

$$1 \text{ ampere (A)} = \frac{1 \text{ coulomb (C)}}{1 \text{ second (s)}}$$

Figure 17.20

(a)   Good conductor

(b)   Poor conductor

Figure 17.21 Electric energy is changed to light and heat in the load.

As mentioned earlier, the charge carriers in metals are electrons. In some other conductors, such as electrolytes (conducting liquid solutions), the charge carriers may be positive or negative or both. An agreement must be made to determine which charge carriers should be assumed in our following discussions.

Note that positive charges flow in the opposite direction (toward the negative terminal) from that of negative charges (toward the positive terminal) when a battery is connected to a circuit (Fig. 17.22). A positive current moving in one direction is equivalent for almost all measurements to a current of negative charges flowing in the opposite direction.

In this book we assume that the charge carriers are positive, and we draw our current arrows in the direction that a positive charge would flow. This is the practice of the majority of engineers and technicians. If you encounter the negative-current convention, you should remember that a negative current flows in the opposite direction from that of a positive current. Regardless of the method used, the analysis of a situation by either method will give the correct result. Some of the rules discussed later, such as the right-hand rule for finding the direction of the magnetic field, will be different if the negative-current convention is used.

Figure 17.22 Current flow as positive current or negative current.

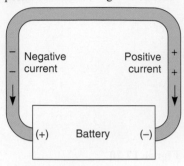

Voltage

We have seen that current flows in a circuit because of the difference in potential of the different points in the circuit. Work is done as a charge moves from one point to another in an electric field. Work is required to move a charge from one point to another when such points differ in electric potential. The *potential difference* between two points in an electric field is the work done per unit of charge as the charge is moved between two points. That is,

$$\text{potential difference} = \frac{\text{work}}{\text{charge}}$$

In *sources*, the raising of the potential energy of electrons that results in a potential difference across a source is called **emf** (E). In *circuits*, the lowering of the potential difference across a load is called **voltage drop**.

The *volt* (V), named after **Allessandro Volta**, is the unit of both emf and voltage drop. We define the volt as the potential difference between two points if 1 J of work is produced or used in moving 1 C of charge from one point to another:

$$1 \text{ volt (V)} = \frac{1 \text{ joule (J)}}{1 \text{ coulomb (C)}}$$

Allessandro Volta (1745–1827),

physicist, was born in Italy. He invented the first electric battery. The unit of electric potential, the volt, is named after him.

Resistance

Not all substances and not even all metals are good conductors of electricity. Those with few free electrons tend to have greater opposition to the flow of charge. This opposition to current flow is called **resistance**. The unit of resistance is the *ohm* (Ω). It is not a fundamental unit and is discussed in Section 17.6.

The resistance of a wire is determined by several factors. Among these are:

1. *Temperature.* An increase in temperature results in an increase in resistance in a wire, for most metals. Other materials, such as semiconductors, show a decrease in resistance with increasing temperature.
2. *Length.* Resistance varies directly with length. If we double the length of a given wire, the resistance is doubled [Fig. 17.23(a)].
3. *Cross-sectional area.* Resistance varies inversely with cross-sectional area. If we double the cross section of a wire, the resistance is *halved*. This is similar to water flowing through two pipes. It flows more easily through the larger pipe. (Note that doubling the radius of a wire [Fig. 17.23(b)] *more than* doubles the cross-sectional area: $A = \pi r^2$.)
4. *Material.* Resistance depends on the nature of the material. For example, copper is a better conductor than steel. The conducting characteristic of various materials is described by resistivity. **Resistivity** (ρ) is the resistance per unit length of a material with uniform cross section.

Figure 17.23

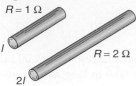
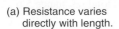

(a) Resistance varies directly with length.

(b) Doubling the radius more than doubles the cross-sectional area.

These factors are related by the equation

$$R = \frac{\rho l}{A}$$

where R = resistance
ρ = resistivity
l = length
A = cross-sectional area

Find the resistance of a copper wire 20.0 m long with cross-sectional area of 6.56×10^{-3} cm^2 at 20°C. The resistivity of copper at 20°C is 1.72×10^{-6} Ω cm.

EXAMPLE

Data:

$$l = 20.0 \text{ m} = 2.00 \times 10^3 \text{ cm}$$
$$A = 6.56 \times 10^{-3} \text{ cm}^2$$
$$\rho - 1.72 \times 10^{-6} \text{ Ω cm}$$
$$R = ?$$

Basic Equation:

$$R = \frac{\rho l}{A}$$

Working Equation: Same

Substitution:

$$R = \frac{(1.72 \times 10^{-6} \text{ Ω cm})(2.00 \times 10^3 \text{ cm})}{6.56 \times 10^{-3} \text{ cm}^2}$$
$$= 0.524 \text{ Ω}$$

· · · · · · · · · · · · · · · ·

PROBLEMS 17.5

1. Find the resistance of 78.0 m of No. 20 aluminum wire at 20°C. ($\rho = 2.83 \times 10^{-6}$ Ω cm, $A = 2.07 \times 10^{-2}$ cm^2.)
2. Find the resistance of 315 ft of No. 24 copper wire with resistance 0.0262 Ω/ft.
3. Find the resistance per foot of No. 22 copper wire if 580 ft has a resistance of 9.57 Ω.
4. At 77°F, 100 ft of No. 18 copper wire has a resistance of 0.651 Ω. Find the resistance of 500 ft of this wire.
5. Find the resistance of 475 m of No. 20 copper wire at 20°C. ($\rho = 1.72 \times 10^{-6}$ Ω cm, $A = 2.07 \times 10^{-2}$ cm^2.)

SKETCH

12 cm^2 w

4.0 cm

DATA
$A = 12$ cm^2, $l = 4.0$ cm, w = ?

BASIC EQUATION
$A = lw$

WORKING EQUATION
$w = \frac{A}{l}$

SUBSTITUTION
$w = \frac{12 \text{ cm}^2}{4.0 \text{ cm}} = 3.0$ cm

PROBLEM SOLVING

6. Find the resistance of $10\overline{0}$ m of No. 20 copper wire at 20°C. ($\rho = 1.72 \times 10^{-6}\ \Omega$ cm, $A = 2.07 \times 10^{-2}$ cm².)

7. Find the resistance of 50.0 m of No. 20 aluminum wire at 20°C. ($\rho = 2.83 \times 10^{-6}\ \Omega$ cm, $A = 2.07 \times 10^{-2}$ cm².)

8. Find the length of copper wire with resistance 0.0262 Ω/ft and total resistance 3.00 Ω.

9. Find the cross-sectional area of copper wire at 20°C that is 60.0 m long and has resistivity $\rho = 1.72 \times 10^{-6}\ \Omega$ cm and resistance 0.788 Ω.

10. Find the length of a copper wire with resistance 0.0262 Ω/ft and total resistance 5.62 Ω.

17.6 Ohm's Law

When a voltage is applied *across* a material that conducts electric current, the relationship between current *through* the material and voltage across it depends upon the type of material as shown in Fig. 17.24. The straight-line relationship shown in Fig. 17.24(a) is typical of many materials, including metal conductors. Other materials, such as semiconductors shown in Fig. 17.24(b), show a nonlinear relationship between I and V. For the materials with a straight-line relationship, the equation relating I and V was determined by **Georg Simon Ohm**. The relationship is called **Ohm's law** (see Fig. 17.25).

Figure 17.24

Figure 17.25   The relationship between the voltage *across* a resistance and the current *through* the resistance is described by Ohm's law.

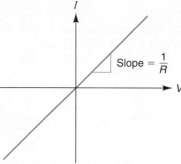

(a) Straight-line (linear) *I–V* characteristic (typical of resistors)

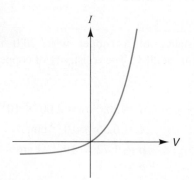

(b) Nonlinear *I–V* characteristic (typical of semiconductor diodes)

Ohm's law	Ohm's law can also be written
$$I = \dfrac{V}{R}$$	$$I = \dfrac{E}{R}$$
where I = current *through* the resistance V = voltage drop *across* the resistance R = resistance	where E = emf of the source of electrical energy

Ohm's law applies to dc circuits containing linear resistors and those ac circuits containing only resistance. It may be applied to the whole circuit or to any part of it.

Ohm's law can aid us in understanding resistance. As we mentioned earlier, the ohm (Ω) is a derived unit. From Ohm's law,

$$I = \frac{V}{R}$$

Solving for R, we obtain

$$R = \frac{V}{I}$$

Substituting units, we obtain

$$\Omega = \frac{V}{A}$$

TRY THIS ACTIVITY

String of Lights

Attach a D-cell battery to a small 2.5-V or 3.5-V light bulb and observe the brightness of the light. Disconnect the circuit and attach another light bulb between the first bulb and the battery. Observe the relative brightness of the bulbs compared to when only one bulb was lit. Repeat the process with two or three additional bulbs. Using Ohm's law, explain what happened to the brightness of each bulb.

A heating element on an electric range operating on 240 V has a resistance of 30.0 Ω. What current does it draw?

EXAMPLE 1

Data:

$$E = 240 \text{ V}$$
$$R = 30.0 \text{ Ω}$$
$$I = ?$$

Basic Equation:

$$I - \frac{E}{R}$$

Working Equation: Same

Substitution:

$$I = \frac{240 \text{ V}}{30.0 \text{ Ω}}$$
$$= 8.0 \text{ V/Ω}$$
$$= 8.0 \text{ A} \qquad \boxed{\frac{V}{\Omega} = A}$$

. . . .

A flashlight bulb is connected to two dry cells with an equivalent voltage of 3.0 V. If it draws 15 mA, what is its resistance?

EXAMPLE 2

Sketch:

R = ?
E = 3.0 V
I = 15 mA

Data:

$$E = 3.0 \text{ V}$$
$$I = 15 \text{ mA} = 15 \times 10^{-3} \text{ A} = 0.015 \text{ A}$$
$$R = ?$$

Basic Equation:

$$I = \frac{E}{R}$$

Working Equation:

$$R = \frac{E}{I}$$

Substitution:

$$R = \frac{3.0 \text{ V}}{0.015 \text{ A}}$$
$$= 2\overline{0}0 \text{ V/A}$$
$$= 2\overline{0}0 \text{ } \Omega$$

· · · · · · · · · · · · · · · · · ·

Electric shock is a very real hazard to electricians. High voltage does not carry the danger that a large current does. The resistance of the human body dry would normally be in the range of 100,000 to 500,000 Ω. If a person is standing in a bathtub, however, his or her resistance is greatly lowered by the water, and any electric appliance falling into the water could deliver sufficient current to cause death. A current as small as 0.070 A can cause serious damage to the nervous system and be fatal. Electricians can protect themselves by using only one hand in some circumstances where a wire must be grasped so that no circuit is completed. Another technique is to touch any questionable wire with the back of one's hand so that if there is any current present, an unexpected shock will not cause a muscular contraction that will keep the hand gripping the wire. In any attempted rescue of a shock victim, the first task is to clear the person from the electric supply with a piece of wood or other nonconductor to avoid the rescuer becoming a second victim.

Appliances today are usually supplied with a third prong on the electric plug to conduct any charge buildup on the appliance to the ground. Sometimes appliances are also made with insulating cases to achieve the same goal—to provide a path not through the human body for electricity to flow.

PROBLEMS 17.6

1. A heating element operates on 115 V. If it has a resistance of 24.0 Ω, what current does it draw?
2. A coffeepot operates on 12.0 V. If it draws 2.50 A, find its resistance.
3. An electric heater draws a maximum of 14.0 A. If its resistance is 15.7 Ω, on what voltage is it operating?
4. A heating coil operates on 22$\overline{0}$ V. If it draws 15.0 A, find its resistance.
5. Find the resistance that draws 0.750 A on 115 V.
6. What current does a 75.0-Ω resistance draw on 115 V?
7. A heater operates on 22$\overline{0}$ V. If it draws 12.5 A, what is its resistance?
8. What current does a 50.0-Ω resistance draw on 115 V?
9. What current does a 175-Ω resistance draw on 22$\overline{0}$ V?
10. A heater draws 3.50 A on 115 V. What is its resistance?
11. (a) What current does a 150-Ω resistance draw on a 1$\overline{0}$-V battery? (b) What voltage battery would produce 3 times the current in (a)? (c) What current would a 75-Ω resistor draw on the 1$\overline{0}$-V battery?
12. A heater draws 4.25 A on 32.0 V. (a) What is the resistance of the heater? (b) What resistance heater would draw 8.50 A on 32.0 V?
13. Electric characteristics of all consumer electric devices must be shown on an attached plate. What is the resistance of an iron that discloses 6.40 A of current used on a 12$\overline{0}$-V line?

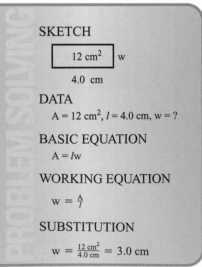

SKETCH

12 cm²   w

4.0 cm

DATA

$A = 12$ cm², $l = 4.0$ cm, $w = ?$

BASIC EQUATION

$A = lw$

WORKING EQUATION

$w = \frac{A}{l}$

SUBSTITUTION

$w = \frac{12 \text{ cm}^2}{4.0 \text{ cm}} = 3.0$ cm

14. What is the effective resistance of a television that draws 2.50 A on a 115-V line?
15. Find the current used by a stereo with resistance 65.0 Ω in a 120-V system.
16. What is the current used by a microwave oven with resistance 20.0 Ω in a 120-V system?

17.7 Series Circuits

In order to communicate about problems in electricity, technicians have developed a "picture language" of their own using symbols and diagrams. The circuit diagram is the most common and useful way to show a circuit. Note how each component (part) of the picture in Fig. 17.26(a) is represented by its symbol in the symbol diagram in its relative position in Fig. 17.26(b). The light bulb can be represented as a resistance. Then the circuit diagram appears as in Fig. 17.26(c). Some of the symbols used most often appear in Appendix C, Table 20.

Figure 17.26   A series circuit.

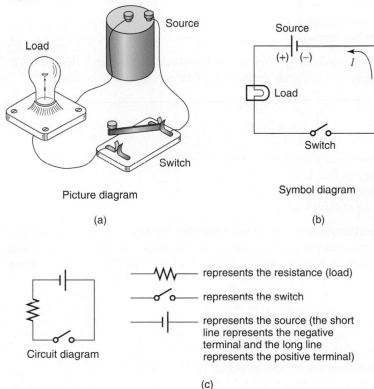

Picture diagram

(a)

Symbol diagram

(b)

Circuit diagram

— represents the resistance (load)

— represents the switch

— represents the source (the short line represents the negative terminal and the long line represents the positive terminal)

(c)

There are two basic types of circuits: series and parallel. A fuse in a house is wired in series with the outlets. The outlets themselves are wired in parallel. A study of series and parallel circuits is basic to a study of electricity.

An electric circuit with only one path for the current to flow (Fig. 17.27) is called a **series circuit**. The current in a series circuit is the same throughout. That is, the current flows out of one resistance and into the next resistance. Therefore, the total current is the same as the current flowing through each resistance in the circuit.

Figure 17.27   Series circuit.

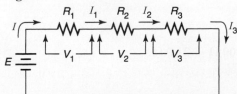

$$\boxed{\begin{array}{c} SERIES \\ I = I_1 = I_2 = I_3 = \cdots \end{array}}$$

where I = total current
I_1 = current through R_1
I_2 = current through R_2
I_3 = current through R_3

In a series circuit, the emf of the source equals the sum of the separate voltage drops in the circuit (Fig. 17.27):

$$\boxed{\begin{array}{c} SERIES \\ E = V_1 + V_2 + V_3 + \cdots \end{array}}$$

where E = emf of the source
V_1 = voltage drop across R_1
V_2 = voltage drop across R_2
V_3 = voltage drop across R_3

The resistance of the conducting wires is very small and will be neglected here. The total resistance of a series circuit equals the sum of all the resistances in the circuit:

$$\boxed{\begin{array}{c} SERIES \\ R = R_1 + R_2 + R_3 + \cdots \end{array}}$$

where R = total or equivalent resistance of the circuit
R_1 = resistance of first load
R_2 = resistance of second load
R_3 = resistance of third load

The **equivalent resistance** is the single resistance that can replace a series and/or parallel combination of resistances in a circuit and provide the same current flow and voltage drop. The equivalent resistance of a series combination is larger than the resistance of any one of the resistances in series.

EXAMPLE 1

Find the total resistance of the circuit shown in Fig. 17.28.

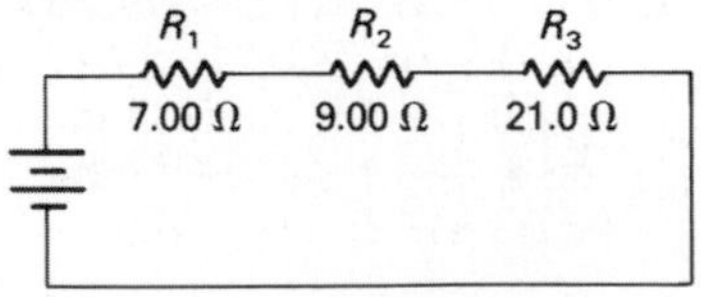

Data:

$$R_1 = 7.00\ \Omega$$
$$R_2 = 9.00\ \Omega$$
$$R_3 = 21.0\ \Omega$$
$$R = ?$$

Figure 17.28

Basic Equation:

$$R = R_1 + R_2 + R_3$$

Working Equation: Same

Substitution:

$$R = 7.00\ \Omega + 9.00\ \Omega + 21.0\ \Omega$$
$$= 37.0\ \Omega$$

Find the current in the circuit shown in Fig. 17.29.

EXAMPLE 2

Data:

$$R_1 = 5.00 \ \Omega$$
$$R_2 = 13.0 \ \Omega$$
$$R_3 = 12.0 \ \Omega$$
$$R_4 = 96.0 \ \Omega$$
$$E = 90.0 \ \text{V}$$
$$I = ?$$

Figure 17.29

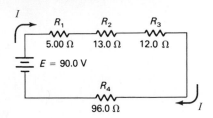

Basic Equations:

$$R = R_1 + R_2 + R_3 + R_4 \quad \text{and} \quad I = \frac{E}{R}$$

Working Equations: Same

Substitutions:

$$R = 5.00 \ \Omega + 13.0 \ \Omega + 12.0 \ \Omega + 96.0 \ \Omega$$
$$= 126.0 \ \Omega$$

$$I = \frac{90.0 \ \text{V}}{126.0 \ \Omega}$$
$$= 0.714 \ \text{A}$$

· · · · · · · · · · · · · · · ·

Find the value of R_3 in the circuit shown in Fig. 17.30.

EXAMPLE 3

Data:

$$I = 3.00 \ \text{A}$$
$$E = 115 \ \text{V}$$
$$R_1 = 23.0 \ \Omega$$
$$R_2 = 14.0 \ \Omega$$
$$R_3 = ?$$

Figure 17.30

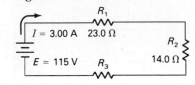

Basic Equations:

$$I = \frac{E}{R} \quad \text{and} \quad R = R_1 + R_2 + R_3$$

Working Equations:

$$R = \frac{E}{I} \quad \text{and} \quad R_3 = R - R_1 - R_2$$

Substitutions:

$$R = \frac{115 \ \text{V}}{3.00 \ \text{A}}$$
$$= 38.3 \ \Omega$$

$$R_3 = 38.3 \ \Omega - 23.0 \ \Omega - 14.0 \ \Omega$$
$$= 1.3 \ \Omega$$

· · · · · · · · · · · · · · · ·

EXAMPLE 4

Find the voltage drop across R_3 in Example 3.

Data:

$$I = I_3 = 3.00 \text{ A}$$
$$R_3 = 1.3 \text{ }\Omega$$
$$V_3 = ?$$

Basic Equation:

$$I_3 = \frac{V_3}{R_3}$$

Working Equation:

$$V_3 = I_3 R_3$$

Substitution:

$$V_3 = (3.00 \text{ A})(1.3 \text{ }\Omega)$$
$$= 3.9 \text{ V}$$

. . . .

Table 17.1 summarizes the characteristics of series circuits.

Table 17.1   Characteristics of Series Circuits

	Series
Current	$I = I_1 = I_2 = I_3 = \cdots$
Equivalent Resistance	$R = R_1 + R_2 + R_3 + \cdots$
Voltage	$E = V_1 + V_2 + V_3 + \cdots$

PROBLEMS 17.7

SKETCH

12 cm² | w

4.0 cm

DATA

A = 12 cm², l = 4.0 cm, w = ?

BASIC EQUATION

A = lw

WORKING EQUATION

w = $\frac{A}{l}$

SUBSTITUTION

w = $\frac{12 \text{ cm}^2}{4.0 \text{ cm}}$ = 3.0 cm

1. Three resistors of 2.00 Ω, 5.00 Ω, and 6.50 Ω are connected in series with a 24.0-V battery. Find the total resistance of the circuit.
2. Find the current in Problem 1.
3. Find the equivalent resistance in the circuit shown in Fig. 17.31.
4. Find the current through R_2 in Problem 3.
5. Find the current in the circuit shown in Fig. 17.32.
6. Find the voltage drop across R_1 in Problem 5.

Figure 17.31

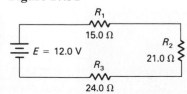

Figure 17.32

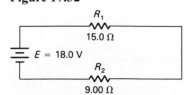

7. What emf is needed for the circuit shown in Fig. 17.33?
8. Find the voltage drop across R_3 in Problem 7.
9. Find the equivalent resistance in the circuit shown in Fig. 17.34.
10. Find R_3 in the circuit in Problem 9.

Figure 17.33

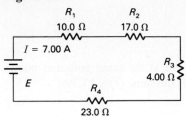

Figure 17.34

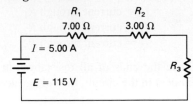

11. Find the values of R_1, R_2, and R_3 in Fig. 17.35.
12. Find the values of V_1, R_2, and V_3 in Fig. 17.36.

Figure 17.35

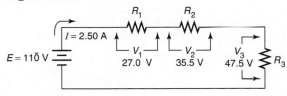

Figure 17.36

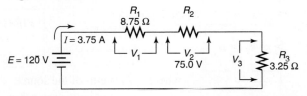

13. Find the values of R_1, V_2, and R_3 in Fig. 17.37.

Figure 17.37

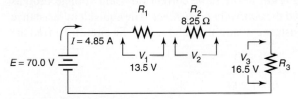

17.8 Parallel Circuits

An electric circuit with more than one path for the current to flow (Fig. 17.38) is called a **parallel circuit**. All resistances connected in parallel have their ends connected to two common points (nodes) in the circuit (points A and B in Fig. 17.38).

Figure 17.38 Different ways to represent a parallel circuit.

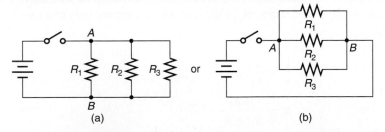

(a)　　　　(b)

The current in a parallel circuit is divided among the branches of the circuit (Fig. 17.39). How it is divided depends on the resistance of each branch. The paths with the least resistance allow the largest currents to flow. Since the current divides, the current from the source equals the sum of the currents through each of the branches:

Figure 17.39 $I = I_1 + I_2 + I_3$.

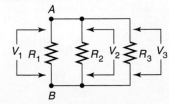

$$\boxed{\begin{array}{c}\textit{PARALLEL}\\ I = I_1 + I_2 + I_3 + \cdots\end{array}}$$

where I = total current in the circuit
I_1 = current through R_1
I_2 = current through R_2
I_3 = current through R_3

Since the ends of all resistances in parallel are connected to the same common points (nodes) in the circuit, the voltage across each resistance is the same (Fig. 17.39):

> **PARALLEL**
>
> $$V_1 = V_2 = V_3 = \cdots$$

Figure 17.40
$E = V_1 = V_2 = V_3$.

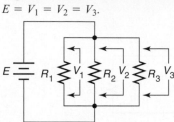

The emf of the source is the same as the voltage drop across each resistance in the circuit if there are no other (series) elements in the circuit (Fig. 17.40):

> **PARALLEL WITH VOLTAGE SOURCE**
>
> $$E = V_1 = V_2 = V_3 = \cdots$$

where E = emf of the source
V_1 = voltage drop across R_1
V_2 = voltage drop across R_2
V_3 = voltage drop across R_3

Therefore, several different loads requiring the same voltage are connected in parallel.

The single resistance that would result in the same current flow and voltage drop as the combination of resistances is called the *equivalent resistance*. The equivalent resistance of a parallel circuit is less than the resistance of any single branch of the circuit. To find the equivalent resistance, use the formula

> **PARALLEL**
>
> $$\frac{1}{R} = \frac{1}{R_1} + \frac{1}{R_2} + \frac{1}{R_3} + \cdots$$

where R = equivalent resistance
R_1 = resistance of R_1
R_2 = resistance of R_2
R_3 = resistance of R_3

If the parallel combination of resistances is replaced by a single resistance with the resistance R, the same current flows in the circuit. In the case where there are only two resistances in parallel, then

Figure 17.41 Resistor R in part (b) is equivalent to the pair of resistances R_1 and R_2 connected in parallel in part (a).

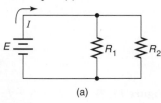

(a)

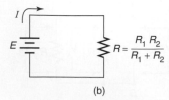

(b)

$$\frac{1}{R} = \frac{1}{R_1} + \frac{1}{R_2}$$

$$R = \frac{R_1 R_2}{R_1 + R_2}$$

(See Fig. 17.41.)

For comparison to parallel circuits, consider the water system shown in Fig. 17.42(a).

1. The total amount of water flowing through $R_1 + R_2 + R_3$ equals the amount flowing through A or B.
2. The water flowing past point A divides into the three branches R_1, R_2, and R_3.
3. The larger pipes have *less* opposition to water flow than do the smaller pipes. Because R_1 has a larger cross-sectional area than R_2 or R_3, it has less opposition to the flow of water and therefore carries more water than R_2 or R_3.

Figure 17.42   A water system may be compared to a parallel electric circuit.

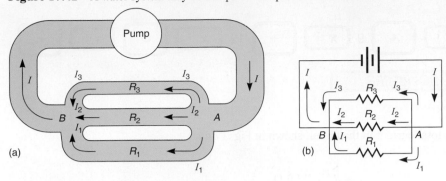

(a)   (b)

Similarly, in a parallel electric circuit as in Fig. 17.42(b):

1. The total amount of current flowing through $R_1 + R_2 + R_3$ equals the amount flowing through A or B.
2. The current flowing past point A divides into the three branches R_1, R_2, and R_3.
3. The smaller resistances have *less* opposition to current flow and therefore carry larger currents.

TRY THIS ACTIVITY

Parallel Bulbs

Attach a D-cell battery to a small 2.5-V or 3.5-V light bulb and observe the brightness of the light. Attach a second light bulb in parallel with the first. After adding a third bulb in parallel with the others, note the brightness of the bulbs. Why, when using the same battery, wires, and bulbs as in the "String of Lights" Try This Activity in Section 17.6 does the brightness of the bulbs differ from the bulbs in the series circuit?

Find the equivalent resistance of the circuit shown in Fig. 17.43.

EXAMPLE 1

Data:

$$R_1 = 7.00 \ \Omega$$
$$R_2 = 9.00 \ \Omega$$
$$R_3 = 12.0 \ \Omega$$
$$R = ?$$

Figure 17.43

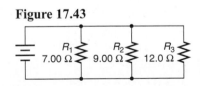

Basic Equation:

$$\frac{1}{R} = \frac{1}{R_1} + \frac{1}{R_2} + \frac{1}{R_3}$$

Working Equation:

When using this formula, you should solve for the reciprocal of the unknown, then substitute.

Substitution:

$$\frac{1}{R} = \frac{1}{7.00 \ \Omega} + \frac{1}{9.00 \ \Omega} + \frac{1}{12.0 \ \Omega}$$
$$R = 2.96 \ \Omega$$

The key entry sequence on a scientific calculator for this calculation is

7 x^{-1} + 9 x^{-1} + 12 x^{-1} = x^{-1} =

2.964705882

EXAMPLE 2

Find the total current in the circuit shown in Fig. 17.44.

Data:

$$R_1 = 23.0 \ \Omega$$
$$R_2 = 14.0 \ \Omega$$
$$R_3 = 5.00 \ \Omega$$
$$E = 90.0 \ \text{V}$$
$$I = ?$$

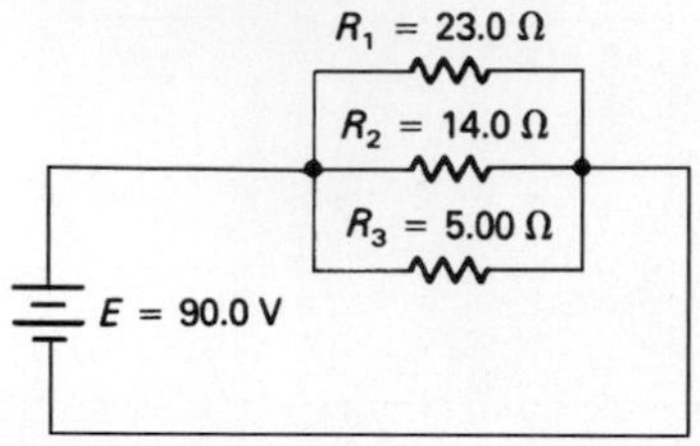

First, find the equivalent resistance, R. Second, find the total current, I. To find R:

Figure 17.44

$R_1 = 23.0 \ \Omega$

$R_2 = 14.0 \ \Omega$

$R_3 = 5.00 \ \Omega$

$E = 90.0 \ \text{V}$

Basic Equation:

$$\frac{1}{R} = \frac{1}{R_1} + \frac{1}{R_2} + \frac{1}{R_3}$$

Working Equation: Same

Substitution:

$$\frac{1}{R} = \frac{1}{23.0 \ \Omega} + \frac{1}{14.0 \ \Omega} + \frac{1}{5.00 \ \Omega}$$

Using a calculator sequence as in Example 1, we find

$$R = 3.18 \ \Omega$$

To find I:

Basic Equation:

$$I = \frac{E}{R}$$

Working Equation: Same

Substitution:

$$I = \frac{90.0 \ \text{V}}{3.18 \ \Omega}$$
$$= 28.3 \ \text{A}$$

EXAMPLE 3

Find the current through R_2 in Fig. 17.44 from Example 2.

Data:

$$R_2 = 14.0 \ \Omega$$
$$E = 90.0 \ \text{V} = V_2$$
$$I_2 = ?$$

Basic Equation:

$$I_2 = \frac{V_2}{R_2}$$

Working Equation: Same

Substitution:

$$I_2 = \frac{90.0 \text{ V}}{14.0 \ \Omega}$$

$$= 6.43 \text{ A}$$

. . . .

Find the equivalent resistance and the value of R_3 in the circuit shown in Fig. 17.45.

EXAMPLE 4

Data:

$$E = 115 \text{ V}$$
$$I = 7.00 \text{ A}$$
$$R_1 = 38.0 \ \Omega$$
$$R_2 = 49.0 \ \Omega$$
$$R_3 = \ ?$$

Figure 17.45

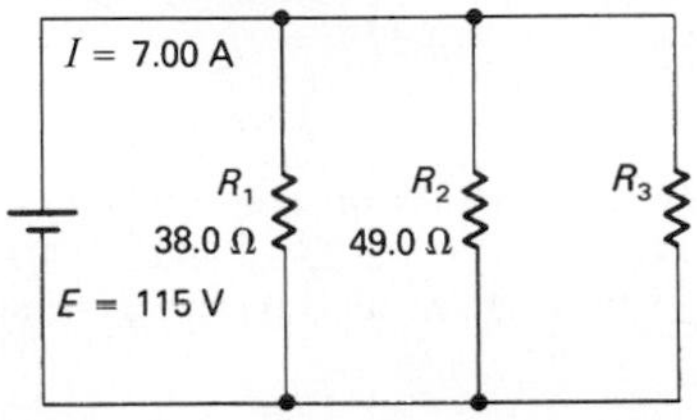

$I = 7.00$ A

R_1 38.0 Ω R_2 49.0 Ω R_3

$E = 115$ V

First find R:

Basic Equation:

$$I = \frac{E}{R}$$

Working Equation:

$$R = \frac{E}{I}$$

Substitution:

$$R = \frac{115 \text{ V}}{7.00 \text{ A}}$$

$$= 16.4 \ \Omega$$

To find R_3:

Basic Equation:

$$\frac{1}{R} = \frac{1}{R_1} + \frac{1}{R_2} + \frac{1}{R_3}$$

Working Equation:

$$\frac{1}{R_3} = \frac{1}{R} - \frac{1}{R_1} - \frac{1}{R_2}$$

Substitution:

$$\frac{1}{R_3} = \frac{1}{16.4 \ \Omega} - \frac{1}{38.0 \ \Omega} - \frac{1}{49.0 \ \Omega}$$

$$R_3 = 70.2 \ \Omega$$

The key entry sequence on a scientific calculator for this calculation is

16.4 $\boxed{x^{-1}}$ $\boxed{-}$ 38 $\boxed{x^{-1}}$ $\boxed{-}$ 49 $\boxed{x^{-1}}$ $\boxed{=}$ $\boxed{x^{-1}}$ $\boxed{=}$

$$\boxed{70.16727941}$$

. . . .

The characteristics of parallel circuits are summarized in Table 17.2.

Table 17.2 Characteristics of Parallel Circuits

	Parallel
Current	$I = I_1 + I_2 + I_3 + \cdots$
Resistance	$\dfrac{1}{R} = \dfrac{1}{R_1} + \dfrac{1}{R_2} + \dfrac{1}{R_3} + \cdots$
Voltage	$E = V_1 = V_2 = V_3 = \cdots$

Overloading is a concern with parallel circuits like those used in household wiring. As more load (bulbs, appliances, etc.) is added to a parallel circuit, the circuit draws more current. Although no change of current occurs in any individual branch, the current in the circuit as a whole increases. As the current increases, more and more heat is produced, which may cause a fire.

Circuit breakers or fuses (Fig. 17.46) are used to prevent overloading in circuits. Circuit breakers have magnets or bimetallic strips that open a switch when the current in the circuit becomes too large. A fuse has a metal strip that melts when the heat from a given amount of current passes through it. All fuses have a rating that describes the amount of current that can pass through the fuse before it melts or "blows" and breaks the circuit.

Figure 17.46 (a) Assorted fuses (b) Circuit breakers.

Photo courtesy of Dorling Kindersley

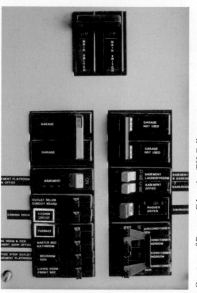

Courtesy of Pearson Education/PH College

PROBLEMS 17.8

1. (a) Find the equivalent resistance in the circuit shown in Fig. 17.47.
   (b) What is the total current in the circuit?
   (c) What is the current through R_1?
   (d) What is the current through R_2?
2. (a) Find I_2 (current through R_2) in the circuit shown in Fig. 17.48.
   (b) Find I_3.
   (c) Find I_1.
   (d) Find the total current in the circuit.
   (e) Find the equivalent resistance in the circuit.

Figure 17.47

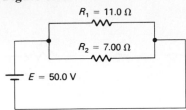

$R_1 = 11.0 \, \Omega$

$R_2 = 7.00 \, \Omega$

$E = 50.0 \, V$

Figure 17.48

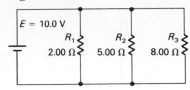

$E = 10.0 \, V$

R_1 R_2 R_3

$2.00 \, \Omega$ $5.00 \, \Omega$ $8.00 \, \Omega$

SKETCH

$12 \, cm^2$ w

$4.0 \, cm$

DATA

$A = 12 \, cm^2, \, l = 4.0 \, cm, \, w = ?$

BASIC EQUATION

$A = lw$

WORKING EQUATION

$w = \frac{A}{l}$

SUBSTITUTION

$w = \frac{12 \, cm^2}{4.0 \, cm} = 3.0 \, cm$

3. (a) Find the resistance of R_3 in the circuit in Fig. 17.49.
   (b) What is the current through R_1?
   (c) What is the current through R_3?
4. (a) What is the equivalent resistance in the circuit shown in Fig. 17.50?
   (b) What emf is required for the circuit?
   (c) What is the voltage drop across each resistance?
   (d) What is the current through each resistance?

Figure 17.49

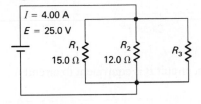

$I = 4.00 \, A$
$E = 25.0 \, V$

R_1 R_2 R_3
$15.0 \, \Omega$ $12.0 \, \Omega$

Figure 17.50

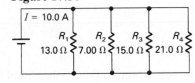

$I = 10.0 \, A$

R_1 R_2 R_3 R_4
$13.0 \, \Omega$ $7.00 \, \Omega$ $15.0 \, \Omega$ $21.0 \, \Omega$

17.9 Compound Circuits

A *compound circuit* contains a combination of resistances in series and parallel arrangements. To simplify solving this kind of circuit, we apply the rules for series and parallel circuits to find an equivalent circuit that reduces to a circuit with one resistance.

Circuit B in Fig. 17.51 is equivalent to circuit A, where $R_4 = R_1 + R_2$. Then, circuit C is equivalent to circuit B, where $\dfrac{1}{R_5} = \dfrac{1}{R_3} + \dfrac{1}{R_4}$.

EXAMPLE 1

Figure 17.51   Circuit A can be replaced by circuit C, where R_5 is the equivalent resistance.

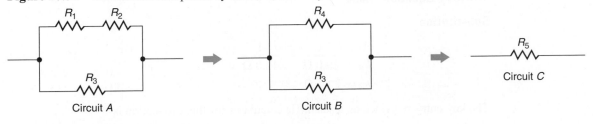

R_1 R_2

R_3

Circuit A

R_4

R_3

Circuit B

R_5

Circuit C

Circuit B in Fig. 17.52 is equivalent to circuit A, where $\dfrac{1}{R_4} = \dfrac{1}{R_2} + \dfrac{1}{R_3}$. Then, circuit C is equivalent to circuit B, where $R_5 = R_1 + R_4$.

EXAMPLE 2

Figure 17.52   Circuit *A* can be replaced by circuit *C*, where R_5 is the equivalent resistance.

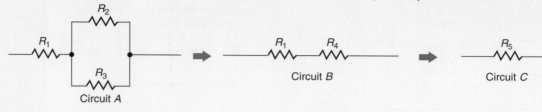

EXAMPLE 3

Find the total current in the circuit shown in Fig. 17.53.

Sketch:

Figure 17.53

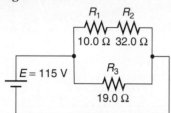

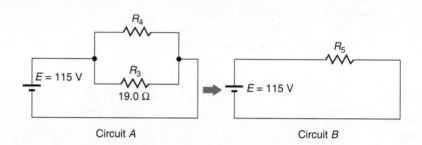

Circuit *A*   Circuit *B*

Circuit *A* is equivalent to the circuit in Fig. 17.53. Then, circuit *B* is equivalent to circuit *A*.

Data:

$$E = 115 \text{ V}$$
$$R_1 = 10.0 \ \Omega$$
$$R_2 = 32.0 \ \Omega$$
$$R_3 = 19.0 \ \Omega$$
$$R_4 = R_1 + R_2 = 10.0 \ \Omega + 32.0 \ \Omega = 42.0 \ \Omega$$
$$I = ?$$

First, find the equivalent resistance, R_5. Second, find the total current, *I*.
    To find R_5:

Basic Equation:

$$\frac{1}{R_5} = \frac{1}{R_3} + \frac{1}{R_4}$$

Working Equation: Same

Substitution:

$$\frac{1}{R_5} = \frac{1}{19.0 \ \Omega} + \frac{1}{42.0 \ \Omega}$$
$$R_5 = 13.1 \ \Omega$$

The key entry sequence on a scientific calculator for this calculation is

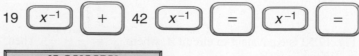

$$19 \quad \boxed{x^{-1}} \quad \boxed{+} \quad 42 \quad \boxed{x^{-1}} \quad \boxed{=} \quad \boxed{x^{-1}} \quad \boxed{=}$$

$$\boxed{13.08196721}$$

To find I:

Basic Equation:

$$I = \frac{E}{R_5}$$

Working Equation: Same

Substitution:

$$I = \frac{115 \text{ V}}{13.1 \text{ } \Omega}$$
$$= 8.78 \text{ A}$$

· · · · · · · · · · · · · · ·

Find the equivalent resistance in the circuit shown in Fig. 17.54.

EXAMPLE 4

Sketch:

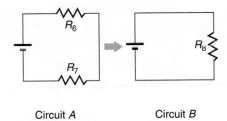

Circuit A   Circuit B

Figure 17.54

Circuit A is equivalent to the circuit in Fig. 17.54, and circuit B is equivalent to circuit A.

Data:

$$R_1 = 7.00 \text{ } \Omega$$
$$R_2 = 35.0 \text{ } \Omega$$
$$R_3 = 25.0 \text{ } \Omega$$
$$R_4 - 20.0 \text{ } \Omega$$
$$R_5 = 62.0 \text{ } \Omega$$
$$E = 12\bar{0} \text{ V}$$
$$R_8 = ?$$

First, find R_6. Second, find R_7. Third, find the equivalent resistance, R_8.
To find R_6:

Basic Equation:

$$\frac{1}{R_6} = \frac{1}{R_1} + \frac{1}{R_2} + \frac{1}{R_3}$$

Working Equation: Same

Substitution:

$$\frac{1}{R_6} = \frac{1}{7.00 \text{ } \Omega} + \frac{1}{35.0 \text{ } \Omega} + \frac{1}{25.0 \text{ } \Omega}$$
$$R_6 = 4.73 \text{ } \Omega$$

The key entry sequence on a scientific calculator for this calculation is

7 [x^{-1}] [+] 35 [x^{-1}] [+] 25 [x^{-1}] [=] [x^{-1}] [=]

| 4.72972973 |

To find R_7:

Basic Equation:

$$\frac{1}{R_7} = \frac{1}{R_4} + \frac{1}{R_5}$$

Working Equation: Same

Substitution:

$$\frac{1}{R_7} = \frac{1}{20.0 \; \Omega} + \frac{1}{62.0 \; \Omega}$$
$$R_7 = 15.1 \; \Omega$$

To find R_8:

Basic Equation:

$$R_8 = R_6 + R_7$$

Working Equation: Same

Substitution:

$$R_8 = 4.73 \; \Omega + 15.1 \; \Omega$$
$$= 19.83 \; \Omega \text{ or } 19.8 \; \Omega$$

· · · · · · · · · · · · · · · · · ·

EXAMPLE 5

Find the total current in Example 4.

Data:

$$E = 12\overline{0} \; \text{V}$$
$$R_8 = 19.8 \; \Omega$$
$$I = ?$$

Basic Equation:

$$I = \frac{E}{R_8}$$

Working Equation: Same

Substitution:

$$I = \frac{12\overline{0} \; \text{V}}{19.8 \; \Omega}$$
$$= 6.06 \; \text{A}$$

· · · · · · · · · · · · · · · · ·

Table 17.3 summarizes the characteristics of series and parallel circuits.

Table 17.3   Characteristics of Series and Parallel Circuits

	Series	**Parallel**
Current	$I = I_1 = I_2 = I_3 = \cdots$	$I = I_1 + I_2 + I_3 + \cdots$
Resistance	$R = R_1 + R_2 + R_3 + \cdots$	$\dfrac{1}{R} = \dfrac{1}{R_1} + \dfrac{1}{R_2} + \dfrac{1}{R_3} + \cdots$
Voltage	$E = V_1 + V_2 + V_3 + \cdots$	$E = V_1 = V_2 = V_3 = \cdots$

PROBLEMS 17.9

Use Fig. 17.55 in Problems 1 through 5.

1. (a) Which resistances are connected in parallel?
    (b) What is the equivalent resistance of the resistances connected in parallel?
2. Find the equivalent resistance of the entire circuit.
3. Find the current in R_1.
4. Find the voltage drop across R_1.
5. (a) Find the current through R_3.
    (b) Find the current through R_2.

Use Fig. 17.56 in Problems 6 through 12.

6. What is the equivalent resistance of the resistances connected in parallel?
7. Find the equivalent resistance of the circuit.
8. Find the current in R_1.
9. What is the voltage drop across the parallel part of the circuit?
10. Find the current through R_3.
11. Find the current through R_5.
12. What is the voltage drop across R_3?

Figure 17.55

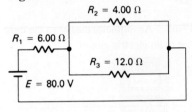

$R_2 = 4.00\ \Omega$
$R_1 = 6.00\ \Omega$
$R_3 = 12.0\ \Omega$
$E = 80.0\ V$

Figure 17.56

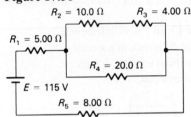

$R_2 = 10.0\ \Omega$ $R_3 = 4.00\ \Omega$
$R_1 = 5.00\ \Omega$
$R_4 = 20.0\ \Omega$
$E = 115\ V$
$R_5 = 8.00\ \Omega$

Use Fig. 17.57 in Problems 13 through 20.

13. Find the equivalent resistance of the parallel arrangement in the upper branch.
14. Find the equivalent resistance of the parallel arrangement in the lower branch.
15. Find the equivalent resistance of the entire circuit.
16. What emf is required for the given current flow in the circuit?
17. Find the voltage drop across the parallel arrangement in the upper branch.
18. Find the voltage drop across R_4.
19. Find the voltage drop across R_6.
20. Find the current through R_6.

Figure 17.57

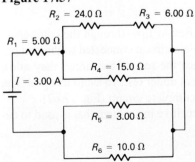

$R_2 = 24.0\ \Omega$ $R_3 = 6.00\ \Omega$
$R_1 = 5.00\ \Omega$
$R_4 = 15.0\ \Omega$
$I = 3.00\ A$
$R_5 = 3.00\ \Omega$
$R_6 = 10.0\ \Omega$

Figure 17.58

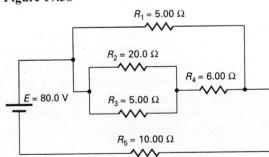

$R_1 = 5.00\ \Omega$
$R_2 = 20.0\ \Omega$
$R_4 = 6.00\ \Omega$
$E = 80.0\ V$ $R_3 = 5.00\ \Omega$
$R_5 = 10.00\ \Omega$

Use Fig. 17.58 in Problems 21 through 25.

21. Find the equivalent resistance in the circuit.
22. Find the current through R_5.
23. Find the voltage drop across R_5.
24. Find the voltage drop across R_4.
25. Find the current through R_2.

SKETCH

12 cm^2 | w

4.0 cm

DATA

$A = 12\ cm^2$, $l = 4.0\ cm$, $w = ?$

BASIC EQUATION

$A = lw$

WORKING EQUATION

$w = \frac{A}{l}$

SUBSTITUTION

$w = \frac{12\ cm^2}{4.0\ cm} = 3.0\ cm$

Figure 17.59 Digital multimeter.

17.10 Electric Instruments

In the laboratory we use several kinds of electric meters for measurements. Great care must be taken to avoid passing a large current through the meters. Meters are fragile instruments, and abuse will ruin them. A large current will burn out the meter. A **multimeter** is an instrument used to measure current flow, voltage drop, and resistance in ac circuits and dc circuits.

Digital instruments have more than one range on which readings are made (Fig. 17.59). Autorange meters adjust ranges automatically. The reading and use of the different modes of operation will be studied in the laboratory.

1. *Voltmeter.* In this mode, the **voltmeter** instrument measures the difference in potential (voltage drop) between two points in a circuit. It should *always* be connected in *parallel* with the part of the circuit across which one wishes to measure the voltage drop (Fig. 17.60). The voltmeter is a high-resistance instrument and draws very little current.
2. *Ammeter.* The **ammeter** mode measures the current flowing in a circuit. Therefore, it is connected in *series* in the circuit (Fig. 17.61). Since all the current flows through the meter, it has very low resistance in this mode so that its effect on the circuit will be as small as possible.
3. *Ohmmeter.* The **ohmmeter** mode is used to measure the resistance of a circuit component. It should only be used when there is *no current* flowing in the circuit. The ohmmeter has a small battery as a built-in source of energy.

Figure 17.60 A voltmeter measures the difference in potential (voltage drop) across two points in an electric circuit.

Figure 17.61 An ammeter measures the current flowing in an electric circuit.

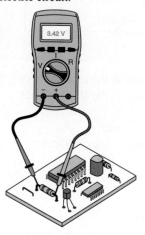

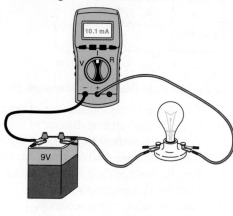

Figure 17.62 An ohmmeter measures the resistance of a component in an electric circuit.

A small current provided by the ohmmeter is caused to flow through the component under test. The presence of another current or a complete circuit connected to the component under test will distort the resistance reading since the rest of the circuit may allow some of this test current to flow "around" the component under test. To avoid this problem, the component should be tested in isolation from any complete circuit (Fig. 17.62).

Another instrument is the *galvanometer*, a very sensitive instrument that is used to detect the presence and direction of *very small* currents.

PROBLEMS 17.10

Using the formulas for series and parallel circuits, fill in the blanks in the tables shown opposite each circuit. In the blanks across from Battery under

V: Write the emf of the battery.
I: Write the total current in the circuit.
R: Write the equivalent or total resistance of the entire circuit.

In the blanks across from R_1 under

    V: Write the voltage drop across R_1.
    I: Write the current flowing through R_1.
    R: Write the resistance of R_1.

In the blanks across from R_2, R_3, ..., fill in the appropriate numbers under V, I, and R. (Begin by looking for key information given in the table and work from there.)

1.

	V	I	R
Battery	12.0 V	A	Ω
R_1	V	A	2.00 Ω
R_2	V	A	4.00 Ω

2.

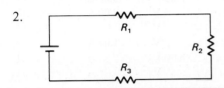

	V	I	R
Battery	V	A	Ω
R_1	V	2.00 A	4.00 Ω
R_2	V	A	6.00 Ω
R_3	V	A	8.00 Ω

3.

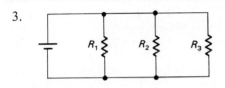

	V	I	R
Battery	V	A	Ω
R_1	V	2.00 A	Ω
R_2	V	3.00 A	12.0 Ω
R_3	V	1.00 A	Ω

4.

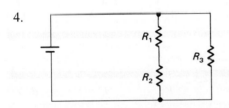

	V	I	R
Battery	12.0 V	2.00 A	Ω
R_1	V	A	6.00 Ω
R_2	V	A	4.00 Ω
R_3	V	A	15.0 Ω

5.

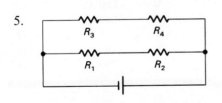

	V	I	R
Battery	50.0 V	5.00 A	Ω
R_1	V	2.00 A	Ω
R_2	25.0 V	A	Ω
R_3	10.0 V	A	Ω
R_4	V	3.00 A	Ω

6.

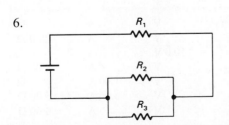

	V	I	R
Battery	24.0 V	A	Ω
R_1	8.00 V	A	Ω
R_2	V	4.00 A	Ω
R_3	V	2.00 A	Ω

7.

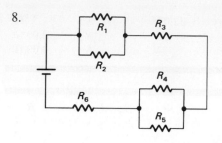

	V	*I*	*R*
Battery	V	A	Ω
R_1	12.0 V	A	2.00 Ω
R_2	V	A	4.00 Ω
R_3	24.0 V	A	4.00 Ω
R_4	V	A	8.00 Ω

8.

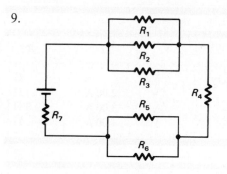

	V	*I*	*R*
Battery	30.0 V	A	Ω
R_1	6.00 V	3.00 A	Ω
R_2	V	2.00 A	Ω
R_3	V	A	3.00 Ω
R_4	V	1.00 A	Ω
R_5	8.00 V	A	Ω
R_6	V	A	Ω

9.

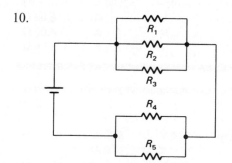

	V	*I*	*R*
Battery	V	12.0 A	Ω
R_1	V	A	Ω
R_2	18.0 V	2.00 A	Ω
R_3	V	A	3.00 Ω
R_4	V	A	4.00 Ω
R_5	V	A	2.00 Ω
R_6	V	8.00 A	Ω
R_7	6.00 V	A	Ω

10.

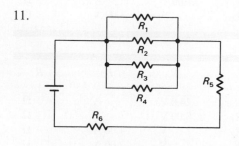

	V	*I*	*R*
Battery	46.0 V	A	Ω
R_1	V	3.00 A	Ω
R_2	V	4.00 A	Ω
R_3	V	A	6.00 Ω
R_4	V	3.00 A	Ω
R_5	V	7.00 A	Ω

11.

	V	*I*	*R*
Battery	V	A	Ω
R_1	V	A	20.0 Ω
R_2	10.0 V	A	Ω
R_3	V	A	4.00 Ω
R_4	V	1.00 A	Ω
R_5	V	5.00 A	5.00 Ω
R_6	V	A	6.00 Ω

17.11 Voltage Sources

Lead Storage Cell

A battery is a group of cells connected together. Each cell consists of a positive plate and a negative plate in a conducting solution. These lead cells are **secondary cells**, which means that they are rechargeable. The passing of an electric current through the cell to restore the original chemicals is called **recharging**. Cells, such as the dry cell, that cannot be efficiently recharged are called **primary cells**. *Note:* The cells are all connected together in series (Fig. 17.63). Six storage cells of 2.0 V each connected in series give 12.0 V for an automobile storage battery.

Lead storage batteries are used to generate voltage in automobiles and in many other types of vehicles and machinery. Lead storage cells are made up of two kinds of lead plates (lead and lead oxide) submerged in a solution of distilled water and sulfuric acid (Fig. 17.64). This acid solution that produces large numbers of free electrons at the negative pole of a cell is called an **electrolyte**. The chemical action between the lead plates and the acid solution produces large numbers of free electrons at the negative (−) pole of the battery. These electrons have a large amount of electric potential energy, which is used in the load in the circuit (for instance, to operate headlights or to turn a starter motor). The work of **Luigi Galvani** led to the production of the electric battery.

As the electric energy is used in the load, the battery must be recharged. This is done by a generator or an alternator. Such devices provide an electric current to reverse the chemical reaction taking place in the battery. The recharging process extends the life of the battery, which would otherwise be very short.

Dry Cell

The dry cell is the most widely used primary cell. This kind of cell is used in flashlights and portable products such as cellular phones, notebook computers, drills, etc. A **dry cell** is a voltage-generating cell that consists of an electrolyte in the form of a chemical paste and two electrodes of unlike materials, one of which reacts chemically with the electrolyte to provide energized electrons. The carbon–zinc dry cell is made of a carbon rod, which is the positive (+) terminal or pole, and a zinc can, which acts as the negative (−) terminal (Fig. 17.65). In between is a paste of chemicals and water that reacts with the terminals to provide energized electrons. These cells are available in a wide range of sizes. Common battery voltages range from 1.5 to 9 V. To achieve 9 V requires a series stack of six cells.

The dry cell, as well as the lead cell, has resistance within the cell itself which opposes the movement of the electrons. This is called the **internal resistance**, r, of the cell. Every cell has internal resistance. Because current flows in the cell, the emf of the cell is reduced by the voltage drop across the internal resistance (Fig. 17.66). The voltage applied to the external circuit is then

$$V = E - Ir$$

where V = voltage applied to the circuit
E = emf of the cell
I = current through the cell
r = internal resistance of the cell

When the current or the voltage available from a single cell is inadequate for a particular job, we usually connect two or more cells in a parallel or series arrangement.

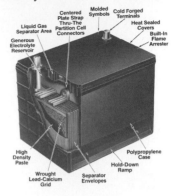

Figure 17.63 Cross-section cutaway of an automobile battery.

Figure 17.64 Simple storage cell

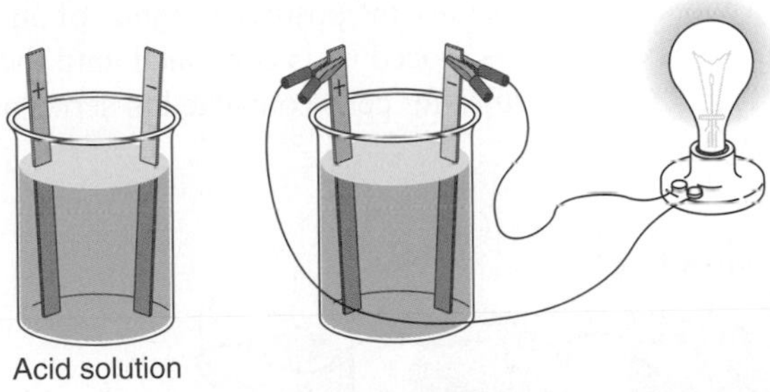

Acid solution

Figure 17.65 Dry cell batteries.

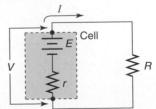

Figure 17.66 Effect of the internal resistance of a dry cell.

Alkaline cells resemble carbon–zinc cells but are five to eight times longer lasting. An alkaline cell has a highly porous zinc anode, which oxidizes more readily than does the carbon–zinc cell's anode. Its electrolyte is a strong alkali solution called *potassium hydroxide*. This compound conducts electricity inside the cell very well and enables the alkaline cell to deliver relatively high currents with greater efficiency than that of carbon–zinc cells.

Nickel–cadmium, metal hydride, and lithium batteries are now also in common use. Their advantage over other types of dry cells is that they are rechargeable. Lithium batteries have greater capacity than nickel–cadmium or hydride batteries.

Another type of dc power source is the solar cell. It is commonly made from a semiconductor material (typically amorphous silicon or gallium arsenide). Many small calculators operate with power supplied by solar cells.

17.12 Cells in Series and Parallel

To connect cells in series, the positive terminal of one is connected to the negative terminal of the next cell. This procedure is continued until the desired number of cells is connected (Fig. 17.67). The rules for cells connected in series and parallel are similar to those for resistances.

Figure 17.67

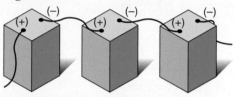
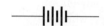

Series-connected dry cells

Circuit diagram for three cells in series

CELLS IN SERIES

1. The current in the circuit equals the current in any single cell:

$$I = I_1 = I_2 = I_3 = \cdots$$

2. The internal resistance of the battery equals the sum of the individual internal resistances of the cells:

$$r = r_1 + r_2 + r_3 + \cdots$$

3. The emf of the battery equals the sum of the emf's of the individual cells:

$$E = E_1 + E_2 + E_3 + \cdots$$

EXAMPLE 1

Two 6.00-V cells with internal resistance of $0.100 \ \Omega$ each are connected in series to form a battery with a current of 0.750 A in each cell (Fig. 17.68).

Figure 17.68

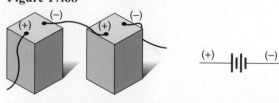

Series

(a)   What is the emf of the battery?
(b)   Find the internal resistance of the battery.
(c)   Find the current in the external circuit.

(a) $E = E_1 + E_2 = 6.00 \text{ V} + 6.00 \text{ V} = 12.00 \text{ V}$ (Rule 3)
(b) $r = r_1 + r_2 = 0.100 \ \Omega + 0.100 \ \Omega = 0.200 \ \Omega$ (Rule 2)
(c) $I = 0.750 \text{ A}$ (Rule 1)

. . . .

To connect cells in parallel, the positive terminals of all the cells are connected together and the negative terminals are all connected together (Fig. 17.69).

Figure 17.69

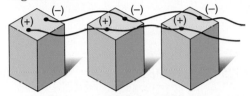  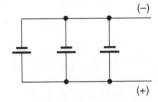

Parallel-connected dry cells   Circuit diagram for
                                                three cells in parallel

A common example of the use of cells in parallel is the practice of jump-starting a car that has a dead battery (Fig. 17.70). It is not common to find cells hooked up in parallel because a mismatch of output voltages could cause problems. The leads from the external circuit may be connected to any positive and negative terminals. (The external circuit is all of the circuit *outside* the battery or cell.)

Figure 17.70   How to properly make connections to jump-start a car that has a dead battery.

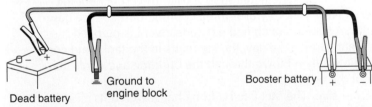

Dead battery   Ground to   Booster battery
                engine block

CELLS IN PARALLEL

1.   The total current equals the sum of the individual currents in each cell:

$$I = I_1 + I_2 + I_3 + \cdots$$

2.   The internal resistance equals the resistance of one cell divided by the number of cells:*

$$r = \frac{r \text{ of one cell}}{\text{number of cells}}$$

3.   The emf of the battery equals the emf of any single cell:

$$E = E_1 = E_2 = E_3 = \cdots$$

---

*This formula works only when all the cells have the same internal resistance. Otherwise, a formula similar to that for resistors in parallel must be used.

EXAMPLE 2

Four cells, each 1.50 V and having an internal resistance of 0.0500 Ω, are connected in parallel to form a battery with a current output of 0.250 A in each cell (Fig. 17.71).

Figure 17.71

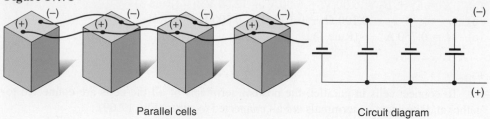

Parallel cells   Circuit diagram

(a) What is the emf of the battery?
(b) Find the internal resistance of the battery.
(c) Find the current in the external circuit.

(a) $E = 1.50$ V (Rule 3)

(b) $r = \dfrac{r \text{ of one cell}}{\text{number of cells}} = \dfrac{0.0500 \ \Omega}{4} = 0.0125 \ \Omega$ (Rule 2)

(c) $I = I_1 + I_2 + I_3 + I_4 = 0.250$ A $+ 0.250$ A $+ 0.250$ A $+ 0.250$ A $= 1.000$ A
(Rule 1)

PHYSICS CONNECTIONS

Ice Block Air Conditioning

In an effort to limit the strain on power grids and lower the cost of cooling, several companies have been using ice blocks to cool office buildings in New York City. This method of cooling also reduces the amount of pollutants in the skies over New York City and the surrounding area.

Instead of using a conventional central air conditioning system that runs constantly throughout the day to keep the building cool, the Met Life Tower in New York City uses three cooling rooms in the basement to freeze 800 gal of water in each of 64 tanks. A solution of chilled ethylene glycol is pumped through pipes in the tanks to freeze the water at night. During the day, the ice melts in the tanks and cools the ethylene glycol solution in the pipes. The solution is then blown through the building and cools the air as it encounters heat exchange coils throughout the building.

The environmental results and savings are impressive. The Met Life building has lowered its electric usage by nearly 2.15 million kilowatt-hours over the course of a year. Although the initial costs are significant, companies are finding that the long-term savings and environmental benefits are tremendous.

PROBLEMS 17.12

1. A cell has an emf of 1.50 V and an internal resistance of 0.0450 Ω. If there is 0.250 A in the cell, what voltage is applied to the external circuit?

2. The voltage applied to a circuit is 11.8 V when the current through the battery is 0.500 A. If the internal resistance of the battery is 0.150 Ω, what is the emf of the battery?

3. The emf of a battery is 12.0 V. If the internal resistance is 0.300 Ω and the voltage applied to the circuit is 11.6 V, what is the current through the battery?

4. Three 1.50-V cells, each with an internal resistance of 0.0500 Ω, are connected in series to form a battery with a current of 0.850 A in each cell.
   (a) Find the current in the external circuit.
   (b) What is the emf of the battery?
   (c) Find the internal resistance of the battery.

5. Five 9.00-V cells, each with internal resistance of 0.100 Ω and current output of 0.750 A, are connected in parallel to form a battery in a certain circuit.
    (a) Find the current in the external circuit.
    (b) What is the emf of the battery?
    (c) Find the internal resistance of the battery.
6. Find the current in the circuit shown in Fig. 17.72.
7. Find the current in the circuit shown in Fig. 17.73.

Figure 17.72

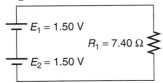

Figure 17.73

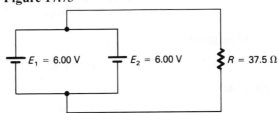

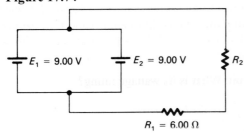
8. If the current in the circuit in Fig. 17.74 is 1.20 A, what is the value of R_2?
9. Find the current in the circuit shown in Fig. 17.75.
10. Find the total resistance in the circuit shown in Fig. 17.75.

Figure 17.74

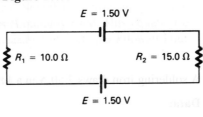

Figure 17.75

E = 1.50 V

R_1 = 10.0 Ω R_2 = 15.0 Ω

E = 1.50 V

17.13 Electric Power

Tremendous quantities of energy are used by industry and sold by power companies. The rate of consuming energy is called **power**. The unit of power is the watt. One *watt* (W) is the power generated by a current of 1 A flowing because of a potential difference of 1 V. A volt is a joule/coulomb (J/C); an ampere is a coulomb/second (C/s). Their product is

$$VA = \frac{J}{C} \cdot \frac{C}{s} = \frac{J}{s}$$

Thus, 1 W = 1 J/s.
Hence, power is

$$P = VI$$

where P = power (watts)
         V = voltage drop
         I = current

This equation applies to components of dc circuits and to whole dc circuits as well as to ac circuits with resistance only.
    Recalling Ohm's law, $I = V/R$, we find two other equations for power:
    Given

$$P = VI$$

Glossary

Alternating Current Current that changes direction. (p. 457)

Ammeter An instrument that measures the current flowing in a circuit. (p. 480)

Conduction A transfer of charge from one place to another. (p. 450)

Conductor A material through which an electron charge is readily transferred. (p. 459)

Coulomb's Law The force between two point charges is directly proportional to the product of their magnitudes and inversely proportional to the square of the distance between them. (p. 453)

Current The flow of charge that passes through a conductor. (p. 459)

Direct Current Current that flows in one direction. (p. 457)

Dry Cell A voltage-generating cell that consists of a chemical paste and two electrodes of unlike materials, one of which reacts chemically with the electrolyte. (p. 483)

Electric Circuit A conducting loop in which electrons carrying electric energy may be transferred from a suitable source to do useful work and returned to the source. (p. 457)

Electric Field An electric field exists where an electric force acts on a charge brought into the area. (p. 454)

Electrolyte An acid solution that produces large numbers of free electrons at the negative pole of a cell. (p. 483)

Electron A negatively charged particle found in every atom. (p. 449)

emf The potential difference across a source. (p. 460)

Energy Work delivered to an electric component or appliance (power $\times$ time). (p. 489)

Equivalent Resistance The single resistance that can replace a series and/or parallel combination of resistances in a circuit and provide the same current flow and voltage drop. (p. 466)

Induction A method of charging one object by bringing a charged object near to, but not touching, it. (p. 451)

Internal Resistance The resistance within a cell that opposes movement of the electrons. (p. 483)

Insulator A substance that does not allow electric current to flow through it readily. (p. 459)

Load The object in a circuit that converts electric energy into other forms of energy or work. (p. 459)

Multimeter An instrument used to measure current flow, voltage drop, and resistance. (p. 480)

Neutron A neutral particle found in the nucleus of most atoms. (p. 449)

Ohm's Law When a voltage is applied across a resistance in an electric circuit, the current equals the voltage drop across the resistance divided by the resistance, $I = V/R$. (p. 462)

Ohmmeter An instrument that measures the resistance of a circuit component. (p. 480)

Parallel Circuit An electric circuit with more than one path for the current to flow. The current is divided among the branches of the circuit. (p. 469)

Power Energy per unit time consumed in a circuit. (p. 487)

Primary Cell A cell that cannot be recharged. (p. 483)

Proton A positively charged particle found in the nucleus of every atom. (p. 449)

Recharging The passing of an electric current through a secondary cell to restore the original chemicals. (p. 483)

Resistance The opposition to current flow. (p. 460)

Resistivity The resistance per unit length of a material with uniform cross section. (p. 460)

Secondary Cell A rechargeable type of cell. (p. 483)

Semiconductors A small number of materials that fall between conductors and insulators in their ability to conduct electric current. (p. 459)

Series Circuit An electric circuit with only one path for the current to flow. The current in a series circuit is the same throughout. (p. 465)

Source The object that supplies electric energy for the flow of electric charge (electrons) in a circuit. (p. 457)

Superconductor A material that continuously conducts electric current without resistance when cooled to typically very low temperatures, often near absolute zero. (p. 459)

Voltage Drop The potential difference across a load in a circuit. (p. 460)

Voltmeter An instrument that measures the difference in potential (voltage drop) between two points in a circuit. (p. 480)

Formulas

17.3 Coulomb's law: $F = \dfrac{kq_1q_2}{r^2}$

17.4 $E = \dfrac{F}{q'}$

17.5 $R = \dfrac{\rho l}{A}$

17.6 Ohm's law: $I = \dfrac{V}{R}$

17.7 *Characteristics of Series Circuits*

Current $I = I_1 = I_2 = I_3 = \cdots$
Resistance $R = R_1 + R_2 + R_3 + \cdots$
Voltage $E = V_1 + V_2 + V_3 + \cdots$

17.8 *Characteristics of Parallel Circuits*

Current $I = I_1 + I_2 + I_3 + \cdots$
Resistance $\dfrac{1}{R} = \dfrac{1}{R_1} + \dfrac{1}{R_2} + \dfrac{1}{R_3} + \cdots$
Voltage $E = V_1 = V_2 = V_3 = \cdots$

17.11 $V = E - Ir$

17.12 *Cells in series:*   *Cells in parallel:*

$I = I_1 = I_2 = I_3 = \cdots$ $I = I_1 + I_2 + I_3 + \cdots$

$r = r_1 + r_2 + r_3 + \cdots$ $r = \dfrac{r \text{ of one cell}}{\text{number of cells}}$

$E = E_1 + E_2 + E_3 + \cdots$ $E = E_1 = E_2 = E_3 = \cdots$

17.13 $P = VI$

$P = I^2R$

$P = \dfrac{V^2}{R}$

energy = power × time

$\text{cost (in cents)} = \text{power (in W)} \times \text{hours} \times \dfrac{1\text{ kW}}{1000\text{ W}} \times \dfrac{\text{cents}}{\text{kWh}}$

Review Questions

1. The atomic particle that carries a positive charge is
   - (a) the neutron.
   - (b) the proton.
   - (c) the electron.
   - (d) none of the above.

2. The atomic particle that carries a negative charge is
   - (a) the neutron.
   - (b) the proton.
   - (c) the electron.
   - (d) none of the above.

3. The process by which an object becomes charged when it comes in contact with a charged object is called
   - (a) induction.
   - (b) electrification.
   - (c) conduction.
   - (d) none of the above.

4. The process by which an object becomes permanently charged when it comes near a charged object requires that the first object be
   - (a) an insulator.
   - (b) touched by another object.
   - (c) a conductor.
   - (d) none of the above.

5. The resistance of a wire is dependent on all of the following except
   - (a) temperature.
   - (b) cross-sectional area.
   - (c) length.
   - (d) material.
   - (e) voltage.

6. Which of the following are good electric conductors?
   - (a) Aluminum
   - (b) Wood
   - (c) Glass
   - (d) Distilled water
   - (e) Silver

7. The total resistance in a circuit containing resistors connected in series is given by
   - (a) the sum of the individual resistances.
   - (b) the sum of the inverse of the individual resistances.
   - (c) the sum of the currents.
   - (d) the sum of the voltages.

8. The current in a parallel circuit is given by
   - (a) the sum of the inverse currents.
   - (b) the sum of the voltages.
   - (c) the sum of the currents in the branches.
   - (d) none of the above.

9. The emf of a battery with cells connected in series equals the sum of the
   - (a) internal resistances of the cells.
   - (b) emf of the individual cells.
   - (c) current in the individual cells.

10. The current in a battery with cells connected in series equals the
    - (a) internal resistances of the cells.
    - (b) emf of the individual cells.
    - (c) current in the individual cells.

11. The current in a battery with cells connected in parallel equals the
    - (a) current in one cell.
    - (b) internal resistance.
    - (c) sum of the currents in each cell.

12. Examples of dry cells include
    - (a) lead–zinc cells.
    - (b) nickel–cadmium cells.
    - (c) carbon–zinc cells.
    - (d) fuel cells.

13. In your own words, describe how materials can become charged by electrification.

14. What particles make up an atom?

15. What particles are located in the nucleus (center) of an atom?

16. Where are electrons located in an atom?

17. What are the two types of charge? What atomic particle carries each type of charge?

18. Describe the process of charging an electroscope by conduction.

19. Describe the process of charging an electroscope by induction.

20. In your own words, describe Coulomb's law of electrostatics.

21. Describe an electric field.

22. Describe lightning.

23. The flow of electrons through a conductor is called _____.

24. (a) The unit of current is the _____. (b) The unit of emf is the _____. (c) The unit of resistance is the _____.
25. What effect does doubling the diameter of a wire have on the wire's resistance?
26. In your own words, explain Ohm's law.
27. Differentiate between a series and a parallel circuit.
28. Differentiate between the equivalent resistance in a series circuit and a parallel circuit.
29. In using an electric instrument, with what range should you start when making a measurement?
30. Explain how a parallel water system compares to a parallel electric circuit.
31. How does the current change in a circuit if the resistance increases by a factor of 2?
32. How does the current change in a circuit if the voltage is increased by a factor of 2?
33. How would the resistance of a wire change if the length were to be increased by a factor of 2?
34. Explain the concept of electric potential.
35. Explain the transfer of energy that occurs in a circuit that includes a dry cell and two lamps in series.
36. Distinguish between a primary and a secondary cell.
37. Explain recharging.
38. Describe the function of an electrolyte.
39. In your own words, describe the manner in which a secondary cell produces electric energy.
40. What is the effect of the internal resistance of a cell?
41. The unit of electric power is the _____.
42. In your own words, explain the relationship between power, voltage, and current.
43. Do we pay the utility company for our power use or our energy use? Explain.
44. Explain the relationship between power, voltage, and resistance.
45. If the current in a circuit is increased by a factor of 2 and the voltage stays constant, how does the power change?
46. If the resistance in a circuit decreases by a factor of 2 and the voltage stays constant, how does the power change?
47. If the voltage and current in a circuit each decrease by a factor of 2, how does the power change?
48. If the current increases in a circuit by a factor of 2 and the voltage stays constant, how would the cost of operating the circuit change?

Review Problems

1. Two charges, each $-4.50\ \mu C$, are 0.150 cm apart. Find their repulsive force.
2. The repulsive force between two identical negative charges is 0.750 N when they are 0.100 m apart. Find the amount of each charge.
3. A charge of 2.50×10^{-8} C exerts a force of 0.0250 N on a second charge of 5.00×10^{-7} C. How far apart are the charges?
4. A positive test charge of $2.50\ \mu C$ is placed in an electric field. The force on it is 0.500 N. What is the magnitude of the electric field at the location of the test charge?
5. Find the magnitude of the electric field in which a negative charge of 1.50×10^{-8} C experiences a force of 0.0500 N.
6. What force is exerted on a test charge of 4.25×10^{-5} C if it is placed in an electric field of magnitude 2.50×10^4 N/C?
7. Find the resistance of 85.5 m of No. 20 aluminum wire ($\rho = 2.83 \times 10^{-6}\ \Omega$ cm, $A = 2.07 \times 10^{-2}$ cm²).
8. At 75°F, $12\overline{0}$ ft of wire has a resistance of 0.743 Ω. Find the resistance of $56\overline{0}$ ft of this wire.
9. Find the resistance of 134 m of No. 20 copper wire at 20°C ($\rho = 1.72 \times 10^{-6}\ \Omega$ cm, $A = 2.07 \times 10^{-2}$ cm²).

PROBLEM SOLVING

SKETCH

12 cm² w

4.0 cm

DATA

A = 12 cm², l = 4.0 cm, w = ?

BASIC EQUATION

A = lw

WORKING EQUATION

$w = \frac{A}{l}$

SUBSTITUTION

$w = \frac{12\ cm^2}{4.0\ cm} = 3.0\ cm$

Figure 17.76

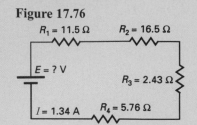

Figure 17.77

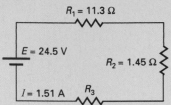

10. Find the length of a copper wire with resistance 0.0273 Ω/ft and total resistance 3.97 Ω.
11. Find the cross-sectional area of copper wire at 20°C that is 55.4 m long and has resistivity $\rho = 1.79 \times 10^{-6}$ Ω cm and resistance 0.943 Ω.
12. A heating element operates on 115 V. If it has a resistance of 15.4 Ω, what current does it draw?
13. A heating coil operates on $22\overline{0}$ V. If it draws 8.75 A, find its resistance.
14. What current does a 234-Ω resistance draw on 115 V?
15. Four resistors of 3.40 Ω, 6.54 Ω, 8.32 Ω, and 1.34 Ω are connected in series with a 12.0-V battery. Find the total resistance of the circuit.
16. Find the current in Problem 15.
17. Find the emf in the circuit shown in Fig. 17.76.
18. Find the equivalent resistance in the circuit shown in Fig. 17.77.
19. Find R_3 in the circuit in Problem 18.
20. Find the equivalent resistance in the circuit shown in Fig. 17.78.
21. Find the current in Fig. 17.78.
22. Find the current through R_1 in Fig. 17.78.
23. Find the current through R_2 in Fig. 17.78.
24. Find the equivalent resistance in the circuit of Fig. 17.79.

Figure 17.78

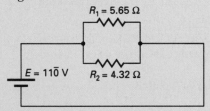

Figure 17.79

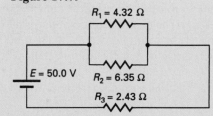

Figure 17.80

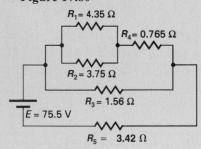

25. Find the current through R_3 in Fig. 17.79.
26. Find the current through R_1 in Fig. 17.79; through R_2.
27. Find the equivalent resistance in Fig. 17.80.
28. Find the current in R_5 in Fig. 17.80.
29. Find the voltage drop across R_5 in Fig. 17.80.
30. Find the current in R_1 in Fig. 17.80.
31. Find the voltage drop across R_1 in Fig. 17.80.
32. Using the formulas for series and parallel circuits, fill in the blanks in the table below for the circuit of Fig. 17.81.

Figure 17.81

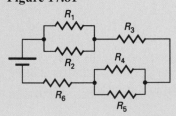

	V	I	R
Battery	35.0 V	A	Ω
R_1	5.00 V	2.75 A	Ω
R_2	V	1.95 A	Ω
R_3	V	A	2.80 Ω
R_4	V	0.97 A	Ω
R_5	7.50 V	A	Ω
R_6	V	A	Ω

33. A cell has an emf of 1.44 V and an internal resistance of 0.0550 Ω. If there is 0.135 A in the cell, what voltage is applied to the external circuit?
34. The voltage applied to a circuit is 12.0 V when the current through the battery is 0.858 A. If the internal resistance of the battery is 0.245 Ω, what is the emf?

35. Six 6.00-V cells, each with an internal resistance of 0.0987 Ω and current output of 0.658 A, are connected in parallel to form a battery in a certain circuit. (a) What is the current in the external circuit? (b) What is the emf of the battery? (c) What is the internal resistance of the battery?
36. Find the current in the circuit shown in Fig. 17.82.
37. Find the total resistance in the circuit shown in Fig. 17.83.

Figure 17.82   **Figure 17.83**

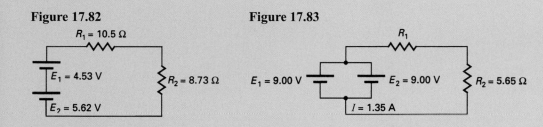

38. What power is needed for a drill that draws 2.45 A and has a resistance of 6.55 Ω on a 11$\overline{0}$-V circuit?
39. How many amperes will a 15$\overline{0}$-W light bulb draw on a 11$\overline{0}$-V circuit?
40. What is the cost to operate the lamp in Problem 39 for 135 h if the cost of energy is $0.05/kWh?
41. If the cost of energy is $0.043/kWh, how long could you operate a motor that draws 0.40 A on a 110-V line for $0.45?
42. How many amperes will a 10$\overline{0}$-W lamp draw on a 110-V line?

APPLIED CONCEPTS

1. A hydrogen atom contains one electron and one proton. (a) Find the electric force between the proton and the electron. (The distance between an electron and a proton in a hydrogen atom is 5.29×10^{-11} m. A proton's positive charge has the same magnitude as the negative charge of an electron.) (b) What is the gravitational force between a proton and an electron? (c) If the electric force is stronger than the gravitational force, why don't people accelerate toward charged combs, balloons, and other electrically charged objects?

2. A rod with charge -4.31×10^{-8} C is held 10.3 cm above a piece of sawdust with mass 3.12×10^{-5} kg. (a) How much does the sawdust weigh? (b) How much electrostatic force must the rod apply to lift the sawdust? (c) If the charged rod lifts the sawdust, what is the electric charge of the sawdust? (d) What happens to the strength of the electric field as the sawdust comes closer to the rod?

3. Hairdryers work by blowing heat that is generated by exposed, high-resistance wires. (a) If a woman using a hairdryer comes in direct contact with the 11$\overline{0}$-V wires, what is the current that travels through her if she is dry and has a resistance of 50$\overline{0}$,000 Ω? (b) If she is wet, the resistance of her body can drop as low as 10$\overline{0}$ Ω. What is the current that passes through her body at this resistance? (Death can result when a current of 0.07 A flows through a person's body for more than 1 s.)

4. A 100$\overline{0}$-W microwave, a 40.0-W fluorescent light bulb, and a 55$\overline{0}$-W computer are plugged into a 12$\overline{0}$-V parallel circuit. (a) What is the current passing through each appliance in the parallel circuit? (b) Find the resistance of each appliance.

5. A 70$\overline{0}$-W toaster is plugged into a 11$\overline{0}$-V outlet. When turned on, the current flowing through the 6.00-m strand of nichrome wire heats the wire and toasts the bread. As the designer of the toaster, what radius do you choose for the nichrome wire? (Assume the temperature of the wire does not affect the resistance. $\rho_{nichrome} = 1.00 \times 10^{-6}$ Ω m.)

CHAPTER 18

MAGNETISM

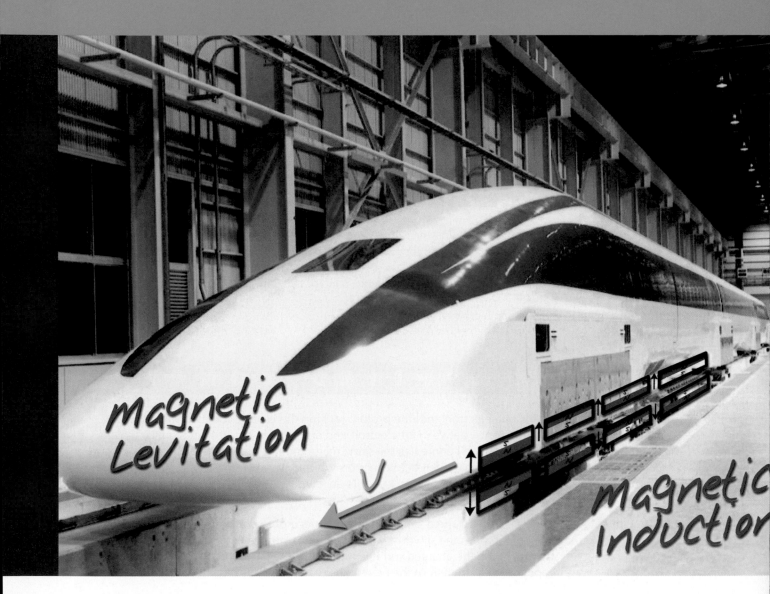

Magnetism and electricity are closely related. Electromagnetism is at the heart of the generation of electric power. We consider how generators and motors use these principles in the production and consumption of electricity.

Objectives

The major goals of this chapter are to enable you to:

1. Describe the nature of magnetism and the magnetic effect of electric current.
2. Describe how induced magnetism and electric current are related.
3. Distinguish between generators and motors, and describe the principles that apply to both.

18.1 Introduction to Magnetism

Many devices that use or produce electric energy depend on the relation of magnetism and electric current. Motors and meters are designed to use the fact that electric currents in wires behave like magnets. Generators produce electric current due to the movement of wires near very large magnets.

In this chapter we investigate the basic properties of magnets and the relation between electric current and magnetism. In later sections on generators, motors, and transformers, we will use the basic principles of magnetism that are developed here.

Metals with the ability to attract pieces of iron or steel are said to be **magnetic** (Fig. 18.1). Iron ore that is naturally magnetic is called *lodestone*.

Artificial magnets can be made from iron, steel, and several special alloys such as Permalloy™ and alnico. We will discuss the process of creating artificial magnets later. Materials that can be made into magnets are called *magnetic materials*. Most materials are nonmagnetic (examples are wood, aluminum, copper, and zinc).

Forces Between Magnets

Suppose that a bar magnet is suspended by a string so that it is free to rotate. It will rotate until one end points north and the other south (Fig. 18.2). The end that points north is called the north-seeking pole, or *north* (N) *pole*. The other end is the south-seeking pole, or *south* (S) *pole*.

If the north pole of another bar magnet is brought near the north pole of this magnet, the two like poles will repel [Fig. 18.3(a)]. The south pole of one magnet will attract the north pole of the other [Fig. 18.3(b)]. In summary:

Figure 18.1 Magnetic materials attract iron and steel.

Figure 18.2 A suspended magnet will rotate to line up north and south.

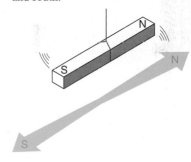

> Like magnetic poles repel each other, and unlike magnetic poles attract each other.

Figure 18.3 (a) Like poles repel each other. (b) Unlike poles attract each other.

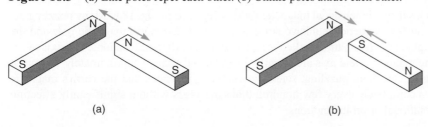

| (a) | (b) |

Figure 18.4 Simple compass.

A **compass** (Fig. 18.4) is simply a small magnetic needle that is free to rotate on a bearing.

Magnetic Fields of Force

A **magnetic field** is a field of force near a magnetic pole or near an electric current that can be detected using another magnet. We can represent this field of force by drawing lines that indicate the direction of the force exerted on a north pole placed in the field. The field of a bar magnet can be mapped by moving a small compass around the magnet as shown in Fig. 18.5(a). These resulting lines are called **flux lines** (lines of force). The flux lines can also be found by sprinkling iron filings near a magnet [Fig. 18.5(b)].

The fields of combinations of magnets can also be found in this way (Fig. 18.6). Although the iron filing patterns are two-dimensional, the magnetic field around a magnet is actually a three-dimensional field, as shown in Fig. 18.7.

Figure 18.5

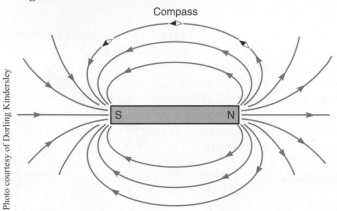

(a) Mapping the field of a bar magnet

(b) Iron filings and plotting compasses reveal the flux lines of magnetic force around a bar magnet

Figure 18.6 Flux lines of (a) unlike poles and (b) like poles near each other shown by iron filings.

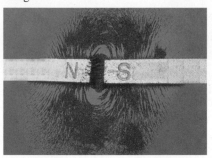

(a) Unlike poles near each other   (b) Like poles near each other

Figure 18.7 A magnetic field is three-dimensional.

Magnetic flux is the total number of magnetic lines of force entering or leaving the pole of a magnet. The symbol for magnetic flux is the Greek letter ϕ. One magnetic line of force is given the unit maxwell (Mx). A more useful unit is 10^8 Mx or one weber (Wb).

The earth's three-dimensional magnetic field is depicted in Fig. 18.8. Many puzzling aspects of the earth's magnetic field have not been resolved. The north magnetic pole and the north geographic pole (sometimes called *true north*) are at different locations. The axis of rotation and the magnetic field axis are slightly different and change approximately 10 min of arc each year. Even more puzzling, scientific evidence indicates that the earth's magnetic field reverses completely every few hundred thousand years without significantly affecting the earth's rotational or orbital motions.

Figure 18.8 The earth's magnetic field is similar to that of other magnets. The geographic and magnetic poles do not coincide; similarly, the geographic and magnetic equators differ. The needle on a compass points to the north magnetic pole.

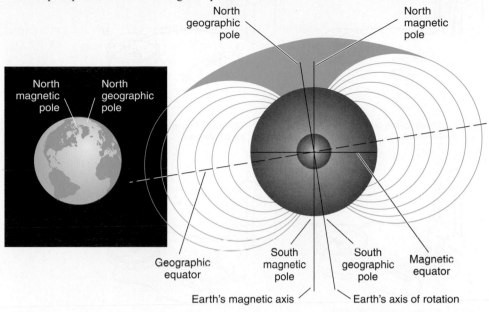

Magnetic poles are similar to the behavior of electric charges in many ways, but there is a very important difference. Electric charges can be isolated, but magnetic poles cannot. A cluster of electrons doesn't need to be accompanied by a cluster of protons. However, a north magnetic pole never exists without the presence of a south magnetic pole. If a magnet is broken in half, each piece still has a north and a south pole.

18.2 Magnetic Effects of Current

When a current passes through a conductor, it forms a magnetic field. A compass placed near the current shows the direction of this magnetic field. **Hans Christian Oersted** was the first to discover the connection between electricity and magnetism. We can show this by connecting a battery to a wire (Fig. 18.9). A compass needle is placed under the wire as in Fig. 18.9(a). When the switch is closed, the needle is deflected as in Fig. 18.9(b). If the terminals of the battery are reversed, the needle is deflected in the opposite direction [Fig. 18.9(c)]. When the compass needle is placed on top of the conductor, the direction of deflection is reversed in each case [Fig. 18.9(d)–(f)]. When the current in a wire flows in a given direction, the flux lines point in one direction below the wire and in the opposite direction above the wire.

The field actually curves around the straight current-carrying wire [Fig. 18.10(a)]. Iron filings on a sheet of paper perpendicular to a current-carrying wire show the shape of the field. The magnetic field is stronger for large currents than for small currents. The direction of the field near a current in a straight wire is shown in Fig. 18.10(b) and given by **Ampère's rule**:

Hans Christian Oersted (1777–1851),

physicist, was born in Denmark. He was the first to discover the connection between electricity and magnetism in 1820.

Andre Ampère (1775–1836),

mathematician and physicist, was born in France. His work became the basis of the science of electrodynamics following Oersted's discovery in 1820 of the magnetic effects of electric currents. The SI unit of electric current, the ampere or amp, is named after him.

AMPÈRE'S RULE
Hold the wire in your right hand with your thumb extended in the direction of the current. Your fingers circle the wire in the direction of the flux lines.

Figure 18.9 Magnetic effects of electric current.

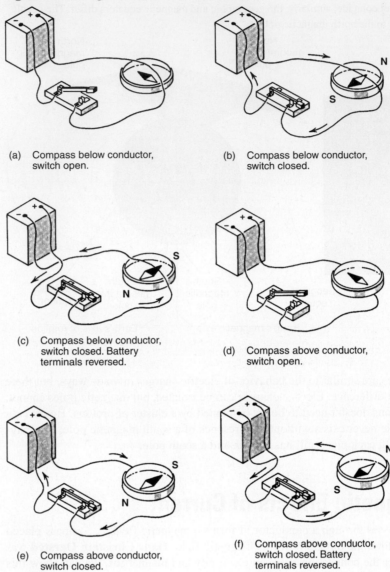

(a) Compass below conductor, switch open.

(b) Compass below conductor, switch closed.

(c) Compass below conductor, switch closed. Battery terminals reversed.

(d) Compass above conductor, switch open.

(e) Compass above conductor, switch closed.

(f) Compass above conductor, switch closed. Battery terminals reversed.

Figure 18.10 (a) Field around a current-carrying wire (b) Direction of the field around a current-carrying wire.

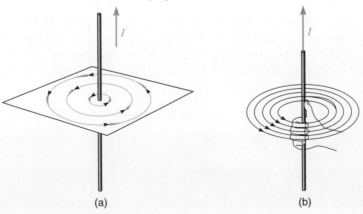

(a)

(b)

The magnetic field near a long current-carrying wire, measured in units of teslas, is circular about the wire and given by Ampère's law:

$$B = \frac{\mu_0 I}{2\pi R}$$

where
B = magnetic field
I = current through the wire
R = perpendicular distance from the center of the wire
$\mu_0 = 4\pi \times 10^{-7}$ T m/A

The magnetic field, B, has the unit *tesla* (T), named after **Nikola Tesla**, and is defined in terms of electric current by the constant μ_0, the permeability constant. The value of μ_0 is not experimentally determined but is an assigned value that explicitly defines magnetic field in terms of electric current.

The magnetic field, B, may also be described in terms of magnetic flux density, or the number of magnetic lines of force passing through a given area:

$$\text{magnetic flux density} = B = \frac{\phi}{A} = \frac{\text{magnetic flux}}{\text{area}}$$

Note that magnetic flux, ϕ, includes the total area, whereas magnetic flux density describes a magnetic field in a specified area. The unit tesla (T) = 1 W/m². If the area is measured in square centimetres, the unit Gauss = 1 Mx/cm² may be used.

Nikola Tesla (1856–1943),

electrical engineer and inventor, was born in Croatia. He invented the electromagnetic motor that became the basis for most alternating-current machinery and sold his patents to Westinghouse in 1888. He then produced important inventions involving high-frequency electricity: the Tesla coil and a resonant air-core transformer. His alternating current system illuminated the Chicago World's Fair in 1893 and led to the construction of the Niagara Falls hydroelectric generating plant in 1896. His later inventions included wireless transmission of electricity and radiocontrolled craft. He also worked on pulsed radar, harnessing solar power, and radio communications to other planets.

EXAMPLE 1

A power line carrying $40\overline{0}$ A is 9.00 m above a transit used by a surveying student [Fig. 18.11(a)].

Figure 18.11

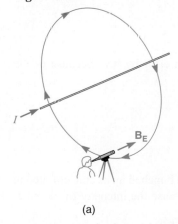

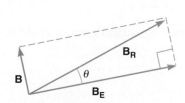

$\mathbf{B_E}$ = magnetic field of earth
$\mathbf{B}$ = magnetic field of power line
$\mathbf{B_R}$ = resultant magnetic field

(a)   (b)

(a)   Find the magnetic field formed by the power-line current above the transit.
(b)   If the earth's horizontal component of magnetic field is 5.20×10^{-5} T at that location, what error could be introduced in the angular measurement? (Assume that the power line runs north–south.)

(a)   Data:

$$I = 40\overline{0} \text{ A}$$
$$R = 9.00 \text{ m}$$
$$\mu_0 = 4\pi \times 10^{-7} \text{ T m/A}$$
$$B = ?$$

Basic Equation:

$$B = \frac{\mu_0 I}{2\pi R}$$

Working Equation: Same

Substitution:

$$B = \frac{(4\pi \times 10^{-7}\ \text{T}\ \cancel{\text{m}}/\cancel{\text{A}})(40\overline{0}\ \cancel{\text{A}})}{2\pi(9.00\ \cancel{\text{m}})}$$

$$= 8.89 \times 10^{-6}\ \text{T}$$

Therefore, the magnetic field from the power line is 8.89×10^{-6} T. With the current from south to north, Ampère's rule shows the direction of B to be east to west.

(b) The angle that the resultant vector [the earth's field plus the wire's field $(B_E + B)$] makes with B_E would be the angular error θ.

Data:

$$B_E = 5.20 \times 10^{-5}\ \text{T} \qquad \text{earth's component}$$
$$B = 8.89 \times 10^{-6}\ \text{T} \qquad \text{[from part (a)]}$$
$$\theta = ?$$

Basic Equation:

$$\tan \theta = \frac{B}{B_E}$$

Working Equation: Same

Substitution:

$$\tan \theta = \frac{8.89 \times 10^{-6}\ \cancel{\text{T}}}{5.20 \times 10^{-5}\ \cancel{\text{T}}} = 0.171$$

$$\theta = 9.7°$$

The bearing on the surveying student's transit could be in error by 9.7° because of the power line.

· · · · · · · · · · · · · · · · ·

Magnetic Field of a Loop

If a wire is bent into a loop, the magnetic field lines become bunched up or concentrated in the center of the loop. A second and subsequent loops increase the intensity of the magnetic field in the loop even more.

To determine the direction of the flux lines of a current in a loop, use Ampère's rule as shown in Fig. 18.12. If several loops are made into a tight spiral as shown in Fig. 18.13, the flux lines add to form the field shown. A coil of tightly wrapped wire is called a **solenoid**. The left side of this solenoid acts like a south magnetic pole. The right side acts like a north magnetic pole. This polarity could be found by using a compass. The rule for finding the polarity of a solenoid is the following:

Hold the solenoid in your right hand so that your fingers circle it in the same direction as the current. Your thumb points to the north pole of the solenoid (Fig. 18.14).

Figure 18.12 Magnetic field around (a) a straight wire and (b) a loop in a current-carrying wire.

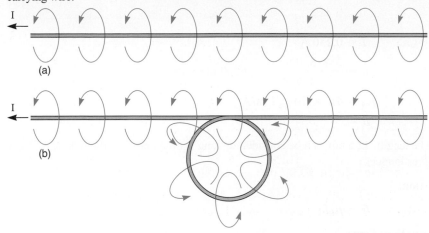

(a)

(b)

Figure 18.13 (a) Magnetic field of a coil (b) Iron filings show the magnetic lines of force from the magnetic field of a coil of red wire charged with an electric current. The coil of wire passes repeatedly through a white surface.

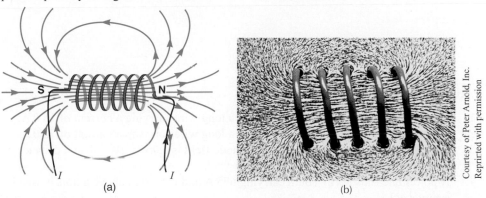

(a)   (b)

Courtesy of Peter Arnold, Inc. Reprirted with permission

Figure 18.14 Polarity of a solenoid.

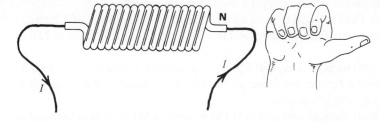

For a long coil that is tightly turned, the field strength at its center is

$$B = \mu_0 I n$$

where B = magnetic field in the region at the center of the solenoid
  μ_0 = permeability constant, $4\pi \times 10^{-7}$ T m/A
  I = current through the solenoid
  n = number of turns per unit length of solenoid

The longer the solenoid in relation to its radius, the more uniform the magnetic field is inside the solenoid; for an infinitely long solenoid, the value of B is uniform throughout.

EXAMPLE 2

Find the magnetic field at the center of a solenoid that is 0.425 m long, 0.0750 m in diameter, and has three layers of $85\overline{0}$ turns each, when 0.250 A flows throughout.

Data:

$$I = 0.250 \text{ A}$$

$$n = \frac{3 \times 85\overline{0} \text{ turns}}{0.425 \text{ m}} = 600\overline{0} \text{ turns/m}$$

$$\mu_0 = 4\pi \times 10^{-7} \text{ T m/A}$$

$$B = ?$$

Note: The length, 42.5 cm, can be considered "long" compared with the radius, 3.75 cm (about 11 times longer).

Basic Equation:

$$B = \mu_0 In$$

Working Equation: Same

Substitution:

$$B = (4\pi \times 10^{-7} \text{ T m/A})(0.250 \text{ A})(600\overline{0}/\text{m})$$

$$= 1.88 \times 10^{-3} \text{ T}$$

SKETCH

12 cm² | w

4.0 cm

DATA

A = 12 cm², *l* = 4.0 cm, w = ?

BASIC EQUATION

A = *l*w

WORKING EQUATION

w = $\frac{A}{l}$

SUBSTITUTION

w = $\frac{12 \text{ cm}^2}{4.0 \text{ cm}}$ = 3.0 cm

PROBLEMS 18.2

1. Find the magnetic field at 0.250 m from a long wire carrying a current of 15.0 A.
2. Find the magnetic field at 0.500 m from a long wire carrying a current of 7.50 A.
3. What is the current in a wire if the magnetic field is 5.75×10^{-6} T at a distance of 2.00 m from the wire?
4. A power line runs north–south carrying 675 A and is 5.00 m above a transit used by a surveyor.
    (a)   What is the magnetic field at the transit because of the power-line current?
    (b)   If the earth's horizontal component of magnetic field is 5.20×10^{-5} T, what error is introduced in the surveyor's angular measurement?
5. Find the magnetic field at 0.350 m from a long wire carrying a current of 3.00 A.
6. Find the current in a wire if the magnetic field is 3.50×10^{-6} T at a distance of 2.50 m from the wire.
7. A solenoid has $100\overline{0}$ turns of wire, is 0.320 m long, and carries a current of 5.00 A. What is the magnetic field at the center of the solenoid? Assume that its length is long in comparison with its diameter.
8. A solenoid has $300\overline{0}$ turns of wire and is 0.350 m long. What current is required to produce a magnetic field of 0.100 T at the center of the solenoid? Assume that its length is long in comparison with its diameter.
9. A small solenoid is 0.150 m in length, 0.0150 m in diameter, and has $60\overline{0}$ turns of wire. What current is required to produce a magnetic field of 1.25×10^{-3} T at the center of the solenoid?
10. A solenoid has $250\overline{0}$ turns of wire and is 0.200 m long. What current is required to produce a magnetic field of 0.100 T at the center of the solenoid? Assume that its length is long in comparison with its diameter.
11. A long solenoid has $100\overline{0}$ turns and is 0.250 m long. If the wire carries a current of 2.50 A, what is the strength of the magnetic field at its center?
12. A small solenoid 0.100 m in length has $100\overline{0}$ turns of wire. What current is required to produce a magnetic field of 1.30×10^{-3} T at the center of the solenoid?

13. An auto mechanic wants to use a solenoid she found on a car starter. If the solenoid is 0.150 m in length and has 750 turns of wire, what current is required to produce a magnetic field of 1.50×10^{-3} T at its center?

14. An earthmover requires a solenoid with 2500 turns and whose length is 0.150 m. What current is required to produce a magnetic field of 0.100 T at the center of the solenoid?

18.3 Induced Magnetism and Electromagnets

Figure 18.15 Simple electromagnet.

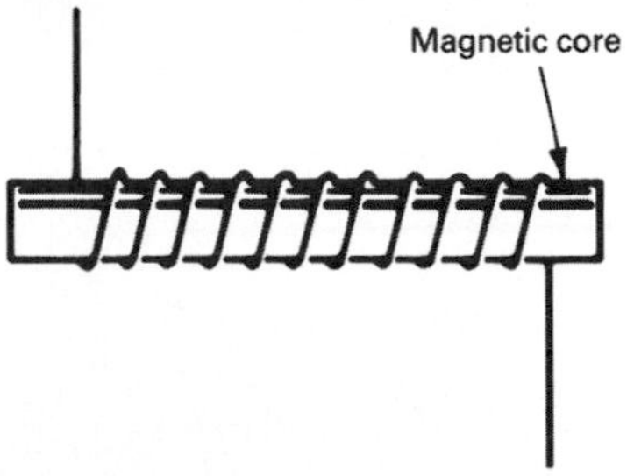

Induced magnetism is produced in a magnetic material such as iron when the magnetic material is placed in a magnetic field, such as that produced in the core of a current-carrying solenoid (Fig. 18.15). An **electromagnet** consists of a solenoid and a magnetic core. When a current is passed through the solenoid, the magnetic fields of the atoms in the magnetic material line up to produce a strong magnetic field. When the current through the coil is turned off, the strength of the induced magnet decreases, but some remains. When the core is removed, a magnetic field remains in the core. In materials such as soft iron, very little magnetic field remains in the core after the current flow stops. In other materials, such as steel, alnico, and Permalloy™, a much stronger field remains. The latter materials are used for permanent magnets. However, they are undesirable for use as a core in an induction motor. Soft-iron cores are often used for this application because less energy is required to reverse the polarity of the induced magnetic field.

A magnet can be thought to consist of many atoms, each behaving like a small magnet. In each atom, the electrons orbit about the nucleus and each electron spins about its own axis, producing small current loops that generate magnetic fields. In most materials, these current loops are arranged so that their magnetic fields point in different directions. The result is that the magnetism of one loop (atom) is canceled out by those of its neighbors.

In magnetic materials, the atomic magnetic fields line up with each other in regions called *magnetic domains* when no field is present [Fig. 18.16(a)]. Each domain has a magnetic field direction to which most atoms in the domain are aligned. When no magnetic field is present, the orientations of the domains are random. However, when an external field is present, the domains tend to line up with the field, causing the domain boundaries to shift [Fig. 18.16(b)]. In high electric fields, nearly all the material is aligned [Fig. 18.16(c)]. When the external magnetic field is removed, some materials such as the alloy alnico retain the aligned domains, creating a permanent magnet.

Figure 18.16 Magnetic fields.

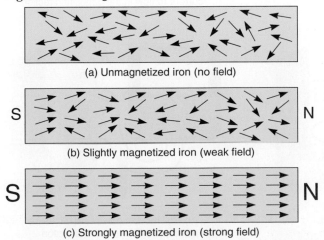

(a) Unmagnetized iron (no field)

(b) Slightly magnetized iron (weak field)

(c) Strongly magnetized iron (strong field)

TRY THIS ACTIVITY

Lifting Paper Clips

Using a large nail, an insulated wire, and a 3.0- to 9.0-V battery, make an electromagnet that will pick up as many paper clips as possible. Without adding more batteries, what can be done to the electromagnet to allow it to pick up more paper clips?

18.4 Induced Current

When a magnet is moved so that its flux lines cut across a wire, an emf is induced in the wire, which is known as **induced current**. The strength of this induced emf depends on the strength of the magnetic field and on the rate at which the flux lines are cut by moving the magnet or wire. Increasing the strength of the field or increasing the rate at which the flux lines are cut also causes the current to increase.

While the magnet shown in Fig. 18.17(a) is moving downward, the galvanometer indicates that a current flows through the wire. If the magnet shown in Fig. 18.17(b) is moving upward, the induced current is in the opposite direction.

A current also flows in the wire if the circuit is closed, the wire is moved, and the magnet is stationary [Fig. 18.17(c)]. The current is produced by the relative motion of the magnet and wire. In commercial generators, magnets are spun inside a set of coils of wire. The induced emf is increased by replacing the single wire with a coil of many turns. For example, tripling the number of turns triples the induced emf [Fig. 18.17(d)].

Electromagnets can be made strong enough to lift very heavy loads, as shown in Fig. 18.18.

Figure 18.17 Induced current in a wire.

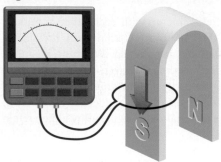

(a) Magnet moving downward

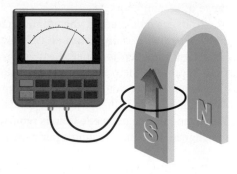

(b) Magnet moving upward causes the current to flow in the opposite direction

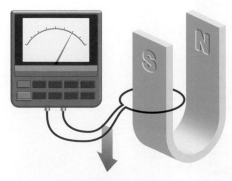

(c) Wire moving downward

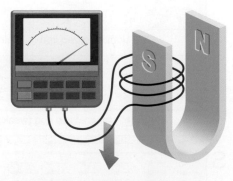

(d) Coil moving downward increases the current flow in same direction

Figure 18.18 A crane with an electromagnet lifts scrap metal from a pile in a salvage yard.

TRY THIS ACTIVITY

Factors Influencing Current

Attach a solenoid to an ammeter to read the current that is flowing through a coil of wire. Then move a bar magnet back and forth inside the coil of wire while observing the current on the ammeter. Instead of increasing the number of coils as discussed in the text, move the magnet in and out of the solenoid at a greater rate of speed. In addition, put one magnet on top of another while moving the magnet back and forth in the solenoid. What impact, if any, do these actions have on the current flowing through the solenoid wire?

PHYSICS CONNECTIONS

Speakers and Electromagnets

A conventional speaker consists of three major components: (a) an amplifier or some other signal source, (b) a cone that vibrates the air to produce a sound wave, and (c) a permanent magnet surrounding a coil of electric wire attached to the back of the cone.

For the speaker to create a sound wave, the amplifier must send an alternating current signal through the wire. As the alternating current travels through the wire, it induces an alternating magnetic field around the coil. The wire's induced magnetic field is repeatedly attracted to and repelled from the magnetic field of the permanent magnet. This repeated motion causes the cone to vibrate at the frequency determined by the amplifier or signal generator (Fig. 18.19).

Figure 18.19 The cross section of a conventional speaker.

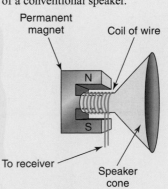

Figure 18.30 Induction motor (Courtesy of Bodine Electric Company, Chicago, IL.).

Figure 18.31

(a) Stator of an 8850-hp synchronous motor driving a steel mill motor

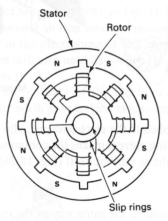

(b) Synchronous motor

PHYSICS CONNECTIONS

Superconducting Electromagnets and Magnetic Resonance Imaging

Superconducting electromagnets are used to accelerate subatomic particles in particle accelerators for nuclear and particle research (Fig. 18.32). The magnetically levitated vehicle, or maglev, now being developed uses electromagnets to levitate the vehicle above a track and virtually eliminates any friction between the vehicle and the track (Fig. 18.33). Air resistance remains a key vehicle design factor.

Magnetic resonance imaging (MRI) is used in medical facilities to provide incredibly clear images of tissues and organs inside the human body without invasive procedures or X-ray exposure. Superconducting electromagnets align hydrogen atoms in the body. After radio waves hit the protons of the hydrogen atoms, the protons return to their previous pattern while emitting electromagnetic signals. Sensors pick up those signals and computers then record and analyze the resulting effects on the different tissues and provide a computer image "slice" of the tissue or organ (Fig. 18.34).

Figure 18.32 Physicists near the Large Electron–Positron Collider (LEP) accelerator used for particle physics experiments at CERN laboratories in Geneva, Switzerland.

Photo copyright of CERN. Reprinted with permission

Figure 18.33 Maglev train.

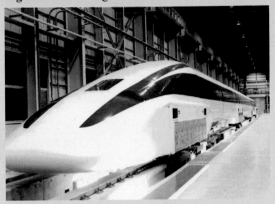

Copyright of Railway Technical Research Institute. Photo reprinted with permission

Figure 18.34 Computer screens display a patient's spinal column and brain images during an MRI exam.

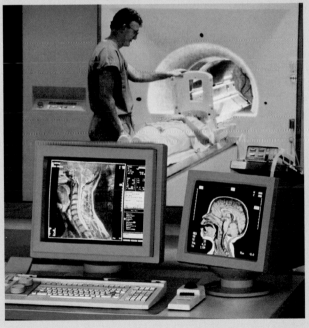

Copyright of Getty Images. Photo reprinted with permission

18.7 Magnetic Forces on Moving Charged Particles

A charged particle that is not moving will produce no interaction with a magnetic field. However, if the charged particle moves through a magnetic field, it is deflected by the magnetic field. The deflection force is greatest when the direction is perpendicular to the magnetic field lines.

TRY THIS ACTIVITY

Moving Electrons

Take a bar magnet and move it close to the front of an old computer monitor, black and white television, or oscilloscope. (Do not risk using a new television set.) What happens to the image on the screen? Why does the magnet influence the electrons' placement on the screen and appear to warp the image?

The force that acts on a moving charged particle does not act in a direction between the sources of interaction but is perpendicular to both the magnetic field and the direction of the charged particle (Fig. 18.35). This behavior is used to spread electrons on a TV tube to produce a TV picture.

Figure 18.35   Deflection of an electron beam by a magnetic field.

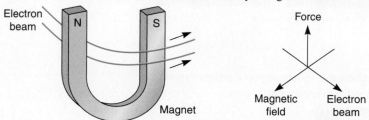

Glossary

Ampère's Rule To find the direction of a magnetic field near a current and a straight wire, hold the wire in your right hand with your thumb extended in the direction of the current. Your fingers circle the wire in the direction of the flux lines. (p. 501)

Armature The rotating coil or electromagnet in a generator. (p. 511)

Commutator A device in an ac generator that produces a direct current. Composed of a split ring that replaces the slip rings in an ac generator and produces a direct current in the circuit connected to the split ring of the generator. (p. 511)

Compass A small magnetic needle that is free to rotate on a bearing. (p. 500)

Electromagnet A combination of a solenoid and a magnetic material, such as iron, in the core of the solenoid. When a current is passed through the solenoid, the magnetic fields of the atoms in the magnetic material line up to produce a strong magnetic field. (p. 507)

Flux Lines Lines indicating the direction of the magnetic field near a magnetic pole. (p. 500)

Generator An apparatus consisting of a coil of wire rotating in a magnetic field. A current is induced in the coil, converting mechanical energy into electric energy. (p. 510)

Induced Current A current produced in a circuit by motion of the circuit through the flux lines of a magnetic field. (p. 508)

Induced Magnetism Magnetism produced in a magnetic material such as iron when the material is placed in a magnetic field, such as that produced in the core of a current-carrying solenoid. (p. 507)

Induction Motor An ac motor with an electromagnetic current induced by the moving magnetic field of the ac current. (p. 513)

Magnetic Property of metals or other materials that can attract iron or steel. (p. 499)

Magnetic Field A field of force near a magnetic pole or a current that can be detected using a magnet. (p. 500)

Motor A device that is composed of an armature and a stator. When a current is passed through the armature, the armature rotates in the magnetic field of the stator and converts electric energy to mechanical energy. (p. 513)

Rotor The rotating coil in a generator. (p. 511)

Solenoid A coil of tightly wrapped wire. Commonly used to create a strong magnetic field by passing current through the wire. (p. 504)

Stator The field magnets in a generator. (p. 511)

Synchronous Motor An ac motor whose speed of rotation is constant and is directly proportional to the frequency of its ac power supply. (p. 513)

Universal Motor A motor that can be run on either ac or dc power. (p. 513)

Formulas

18.2 $\quad B = \dfrac{\mu_0 I}{2\pi R}$

$\quad B = \mu_0 I n$

Review Questions

1. The presence of a magnetic force field may be detected by using
   - (a) a compass.
   - (b) iron filings.
   - (c) a magnet.
   - (d) all of the above.

2. The deflection of a compass needle placed near a current-carrying wire shows
   - (a) the magnetic field of the sun.
   - (b) the magnetic field of the wire.
   - (c) the electric field.

3. Ampère's rule relates
   (a) the strength of a magnetic field to the magnetic pole.
   (b) the direction of a magnetic field surrounding a current-carrying wire.
   (c) the direction of a magnetic field near a bar magnet.
   (d) none of the above.
   (e) all of the above.
4. The unit used to express the strength of a magnetic field is the _____.
5. Describe how a strong magnetic field can be produced in a solenoid.
6. Describe how to determine the direction of a magnetic field in a solenoid.
7. Describe how a magnetic field is induced by a current-carrying coil surrounding a core of magnetic material.
8. Describe how a generator produces current.
9. Describe the function of a commutator.
10. Describe how a motor works.
11. What is a synchronous motor, and how does it work?
12. Distinguish between a universal motor and an induction motor.
13. Distinguish between an armature and a stator.
14. Describe how an electromagnet works.
15. If the current in a solenoid is increased by a factor of 2, how does the magnetic field change?
16. If the radius of a solenoid decreases by a factor of 2, how does the magnetic field change?
17. If the number of turns per inch in a solenoid were increased by a factor of 4, how would the magnetic field change?
18. Describe how to find the flux lines near a bar magnet.
19. How is alternating current produced by a generator?

Review Problems

SKETCH

12 cm² w

4.0 cm

DATA

$A = 12$ cm², $l = 4.0$ cm, $w = ?$

BASIC EQUATION

$A = lw$

WORKING EQUATION

$w = \frac{A}{l}$

SUBSTITUTION

$w = \frac{12 \text{ cm}^2}{4.0 \text{ cm}} = 3.0$ cm

1. Find the magnetic field at 0.255 m from a long wire carrying a current of 1.38 A.
2. Find the magnetic field at 0.365 m from a long wire carrying a current of 8.95 A.
3. What is the current in a wire if the magnetic field is 4.75×10^{-6} T at a distance of 1.75 m from the wire?
4. A solenoid has 2000 turns of wire, is 0.452 m long, and carries a current of 4.55 A. What is the magnetic field at the center of the solenoid?
5. A solenoid has 2750 turns of wire and is 0.182 m long. What current is required to produce a magnetic field of 0.235 T at the center of the solenoid?
6. A power line running north–south carrying 500 A is 7.00 m above a transit used by a surveyor. What error is induced in the compass used by the surveyor? (Assume the earth's horizontal component of magnetic field is 5.20×10^{-5} T.)

APPLIED CONCEPTS

1. A ship's compass is mistakenly placed 8.35 cm away from a wire carrying a current of 8.25 A. (a) What is the strength of the wire's magnetic field on the compass? (b) The strength of the earth's magnetic field is 5.20×10^{-5} T. How far from the wire must the compass be mounted so that it only experiences a magnetic field of 5.20×10^{-7} T ($\frac{1}{100}$ of the magnetic field of the earth) due to the wire?

2. Figure 18.9 shows a compass near a current-carrying wire. (a) What is the strength of the magnetic field when the compass needle is placed 1.03 cm under the wire? Assume the voltage of the battery is 6.00 V and the resistance of the wire is 15.0 Ω. (b) Compare the strength of the earth's magnetic field to the strength of the magnetic field of the wire.

3. A coaxial cable consists of an inner conducting wire encased in an insulating material surrounded by a conducting metal braid and another insulating sheath. The inner wire carries current in one direction, while the outer braid carries current in the opposite direction. Using Ampère's right-hand rule, explain why there is no significant magnetic field outside the coaxial cable. (Coaxial cables are often used to transmit cable TV and other audiovisual signals.)

4. Figure 18.36 shows a picture tube that uses two solenoids to guide electrons as they move through the magnetic field. (a) What is the orientation of the magnetic field for the two solenoids? (b) As a result of the magnetic field, will the electrons travel toward a, b, or c? (Use Fig. 18.35 as a reference.)

Figure 18.36

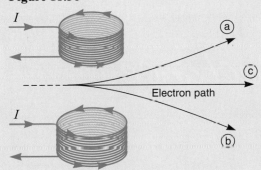

Electron path

5. A copper wire 20.0 m long with radius 4.57×10^{-3} m and $\rho = 1.72 \times 10^{-8}$ Ω cm is used to form a solenoid that will produce a magnetic field. (a) What is the wire's resistance? (b) If the wire is connected to a 4.50-V battery, what is the current that passes through the wire? (c) If the solenoid is designed to have an internal radius of 3.25 cm, how many single-layer coils will be produced? (d) What will be the length of the solenoid if the coils touch one another? (e) Find the strength of the solenoid's magnetic field.

SKETCH

$$12 \text{ cm}^2 \quad w$$

4.0 cm

DATA

$A = 12 \text{ cm}^2, l = 4.0 \text{ cm}, w = ?$

BASIC EQUATION

$A = lw$

WORKING EQUATION

$w = \frac{A}{l}$

SUBSTITUTION

$w = \frac{12 \text{ cm}^2}{4.0 \text{ cm}} = 3.0 \text{ cm}$

PROBLEMS 19.1

1. What is the maximum voltage in an ac circuit in which the instantaneous voltage at $\theta = 35.0°$ is 27.0 V?

2. The instantaneous voltage in an ac circuit at $\theta = 65.0°$ is 82.0 V. What is the maximum voltage?

3. If the maximum ac voltage on a line is 165 V, what is the instantaneous voltage at $\theta = 45.0°$?

4. The maximum current in an ac circuit is 8.00 A. Find the instantaneous current at $\theta = 60.0°$.

5. The instantaneous current in an ac circuit is 6.50 A at $\theta = 45.0°$. Find the maximum current.

6. What is the maximum voltage in an ac circuit where the instantaneous voltage at $\theta = 51.0°$ is 14.5 V?

7. If the maximum ac voltage on a line is 145 V, what is the instantaneous voltage at $\theta = 35.0°$?

8. The maximum current in an ac circuit is 5.75 A. What is the instantaneous current at $\theta = 80.0°$?

9. Find the maximum current in an ac circuit where the instantaneous current at $\theta = 45.0°$ is 4.00 A.

10. The instantaneous voltage in an ac circuit at $\theta = 55.0°$ is $45\overline{0}$ V. Find the maximum voltage.

11. If $I_{max} = 4.59$ A and $I = 4.32$ A, what is θ?

12. If $I = 1.23$ A and $I_{max} = 3.41$ A, what is θ?

13. Find the effective value of an ac voltage whose maximum voltage is 2250 V.

14. Find the maximum current in an ac circuit with an effective value of 6.00 A.

15. Find the effective value of an ac voltage whose maximum voltage is 165 V.

16. Find the maximum current in an ac circuit with an effective value of 4.00 A.

17. Find the effective value of a current in an ac circuit that reaches a maximum of 17.0 A.

18. Find the effective value of an ac voltage whose maximum voltage is 1150 V.

19. What is the maximum current in a circuit in which an ac ammeter reads 8.50 A?

20. Find the maximum current in a circuit in which an ac ammeter reads 7.00 A.

21. A technician uses an oscilloscope to measure an effective voltage in an ac circuit. Find the effective value if the maximum voltage is 135 V.

22. A technician uses a cathode ray oscilloscope to measure current in an ac circuit which reaches a maximum of 125 A. What is the effective value of the current?

23. A maximum voltage of 34.0 V is developed by an ac generator that delivers a maximum current of 0.170 A to a circuit. (a) What is the effective voltage of the generator? (b) Find the effective current delivered to the circuit by the generator. (c) Find the resistance of the circuit.

24. A power plant generator develops a maximum voltage of $17\overline{0}$ V. What is the effective voltage?

25. A maximum current of 0.700 A flows through a 60.0-W light bulb in a circuit. (a) What is the effective current? (b) Find the resistance of the light bulb when it is lit.

26. If the average power dissipated by an electric light is $15\overline{0}$ W, what is the peak power?

27. If the average power dissipation by a hair dryer is $100\overline{0}$ W, what is the peak power?

19.2 ac Power

When the load has only resistance, power in ac circuits is found in the same way as in dc circuits:

$$P = I^2 R$$
$$P = VI \quad \text{(using } V = IR\text{)}$$
$$P = \frac{V^2}{R} \quad \text{(using } I = V/R\text{)}$$

EXAMPLE 1

What power is used in a resistance of 37.0 Ω if it has a current of 0.480 A flowing through it?

Data:

$$R = 37.0 \ \Omega$$
$$I = 0.480 \ A$$
$$P = ?$$

Basic Equation:

$$P = I^2R$$

Working Equation: Same

Substitution:

$$P = (0.480 \ A)^2(37.0 \ \Omega)$$
$$= 8.52 \ W$$

· · · · · · · · · · · · · · · ·

EXAMPLE 2

What power is used in a load of 12.0 Ω resistance if the voltage drop across it is $11\overline{0}$ V?

Data:

$$R = 12.0 \ \Omega$$
$$V = 11\overline{0} \ V$$
$$P = ?$$

Basic Equation:

$$P = \frac{V^2}{R}$$

Working Equation: Same

Substitution:

$$P = \frac{(11\overline{0} \ V)^2}{12.0 \ \Omega}$$
$$= 1010 \ W \quad \text{or} \quad 1.01 \ kW$$

· · · · · · · · · · · · · · · ·

The preceding relationships are true only when e and i are in phase. Phase differences produced by capacitance and inductance in an ac circuit are due to reactance. Capacitance, inductance, and reactance will be studied later.

Note that, in the graphs comparing dc and ac power (Fig. 19.7), ac power varies but is always positive ($+$). The sign indicates only the direction of the current. Even so, p is positive in calculations because the product of $-e$ and $-i$ is positive: $p = (-e)(-i) = ei$.

Transformers and Power

The uses of direct current in industry are somewhat limited. Primary applications are in charging storage batteries, electroplating, generating alternating current, electrolysis, electromagnets, and automobile ignition systems. However, ac can be changed to dc by a simple device called a **rectifier**.

Much more can be done with alternating current. From the kitchen toaster to the largest industrial motors, ac finds wide application. There are very practical reasons for this. The voltage of ac can be easily and efficiently changed in transformers to give almost any desired values. Actually, ac can be used for most purposes just as efficiently as dc. One

Figure 19.7 dc and ac power compared

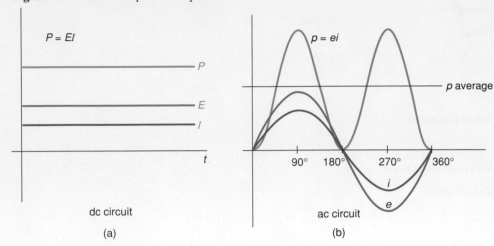

advantage of ac is that it can be transmitted over long distances with very little heat loss. Heat lost in any electric device is found by the formula

$$\text{heat loss (power) } P = I^2 R$$

The energy wasted as heat can be reduced by making the current smaller. Transformers reduce the current by increasing the voltage, since

$$P = \frac{E^2}{R}$$

The major advantage of ac over dc is that ac voltage can easily be changed to meet our needs.

EXAMPLE 3

A plant generates 50.0 kW ($50,\overline{0}00$ W) of power to be sent to a substation on a line with a resistance of 3.00 Ω. We know that some power will be lost as heat during the transmission. The power lost is $P_{\text{lost}} = I^2 R$.

(a)　How much power is lost if the transmission is at 1150 V?
(b)　What percent of the power generated is lost in transmission at 1150 V?
(c)　How much power is lost if the transmission is at 11,500 V?
(d)　What percent of the power generated is lost in transmission at 11,500 V?
(e)　Compare the power losses at the two different transmission voltages.

At 1150 V:

(a)

$$P = VI$$

$$I = \frac{P}{V}$$

$$I = \frac{50,\overline{0}00 \text{ W}}{1150 \text{ V}}$$

$$= 43.5 \text{ A}$$

$$P_{\text{lost}} = I^2 R$$

$$P_{\text{lost}} = (43.5 \text{ A})^2 (3.00 \text{ Ω})$$

$$= 5680 \text{ W}$$

$$= 5.68 \text{ kW}$$

(b)

$$\%_{lost} = \frac{\text{power lost}}{\text{power generated}} \times 100\%$$

$$\%_{lost} = \frac{5.68 \text{ kW}}{50.0 \text{ kW}} \times 100\% = 11.4\%$$

At 11,500 V:

(c)

$$P = VI$$

$$I = \frac{P}{V}$$

$$I = \frac{50,000 \text{ W}}{11,500 \text{ V}}$$

$$= 4.35 \text{ V}$$

$$P_{lost} = I^2 R$$

$$P_{lost} = (4.35 \text{ A})^2 (3.00 \text{ }\Omega)$$

$$= 56.8 \text{ W}$$

$$= 0.0568 \text{ kW}$$

(d)

$$\%_{lost} = \frac{\text{power lost}}{\text{power generated}} \times 100\%$$

$$\%_{lost} = \frac{0.0568 \text{ kW}}{50.0 \text{ kW}} \times 100\% = 0.114\%$$

(e) This example shows that whereas 11.4% of the power is lost during transmission at 1150 V, only 0.114% is lost at 11,500 V; so by increasing the voltage, the current is correspondingly lowered and the power wasted in transmission is greatly reduced.

. . . .

Changing Voltage with Transformers

The transformer is a device that is used to change the voltage to reduce the current and thereby lessen the power loss. A **transformer** consists of two coils of wire wrapped on an iron core (Fig. 19.8) and is used to change the voltage. When an alternating current passes through the **primary coil**, an induced current is produced in the **secondary coil** (Fig. 19.9). The magnitude of the voltage induced in the secondary coil depends on

1. The voltage applied to the primary coil.
2. The number of turns in the primary coil.
3. The number of turns in the secondary coil.
4. The power lost between primary and secondary coils.

Figure 19.8 Basic components of a transformer.

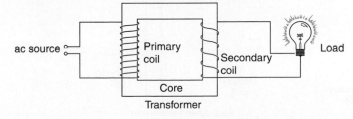

Figure 19.9 An alternating current in the primary coil induces a current in the secondary coil.

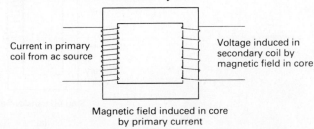

If we assume no power loss between the primary and secondary coils, we have

$$\frac{V_P}{V_S} = \frac{N_P}{N_S}$$

where V_P = primary voltage
 V_S = secondary voltage
 N_P = number of primary turns
 N_S = number of secondary turns

EXAMPLE 4

A transformer on a neon sign has $10\overline{0}$ turns in its primary coil and $15{,}\overline{0}00$ turns in its secondary coil. If the voltage applied to the primary coil is $11\overline{0}$ V, what is the secondary voltage?

Data:

$$V_P = 11\overline{0} \text{ V}$$
$$N_P = 10\overline{0} \text{ turns}$$
$$N_S = 15{,}\overline{0}00 \text{ turns}$$
$$V_S = ?$$

Basic Equation:

$$\frac{V_P}{V_S} = \frac{N_P}{N_S}$$

Working Equation:

$$V_S = \frac{V_P N_S}{N_P}$$

Substitution:

$$V_S = \frac{(11\overline{0} \text{ V})(15{,}\overline{0}00 \text{ turns})}{10\overline{0} \text{ turns}}$$
$$= 16{,}500 \text{ V} \quad \text{or} \quad 16.5 \text{ kV}$$

· · · · · · · · · · · · · · · · · · · ·

Transformers used to raise or lower voltage are called step-up or step-down transformers. *Step-up transformers* are used when a high voltage is needed to operate X-ray tubes or neon signs and to transmit electric power over long distances. A **step-up transformer** increases the voltage by having more turns in the secondary coil than in the primary coil [Fig. 19.10(a)].

A **step-down transformer** lowers the voltage from high-voltage transmission lines to regular 110 V and 220 V for home and industrial use. Voltage is lowered in the step-down transformer because it has more turns in the primary coil than in the secondary coil [Fig. 19.10(b)].

Figure 19.10 Step-up and step-down transformers.

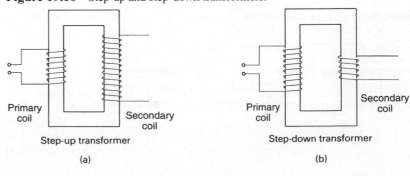

Primary coil

Secondary coil

Step-up transformer

(a)

Primary coil

Secondary coil

Step-down transformer

(b)

Figure 19.11   Some large transformers can be reverse-connected to become either a step-up or a step-down transformer. Since transformers have no moving parts, they have a normal life expectancy of 20–25 years.

(a) A step-down electrical transformer near top of a utility pole

(b) A large transformer at an electrical transmission station

Some common transformers are shown in Fig. 19.11.

Auto transformers are used when a variable output voltage is needed. In this type of transformer, contact can be made across a variable number of the secondary coils using a brush contact. The output voltage is therefore variable from nearly zero to some maximum value. This type of transformer is often used to supply ac power to resistive heater elements to control the heating output.

Transformers do not create energy. In fact, some energy is lost during the change of voltage. Energy losses in transformers are of three types:

1. *Copper losses.* These result from the resistance of the copper wires in the coils and are unavoidable.
2. *Magnetic losses* (called *hysteresis losses*). Some energy is lost (turned into heat) by reversing the magnetism in the core.
3. *Eddy currents.* When a mass of metal (the core) is subjected to a changing magnetic field, *eddy currents* are set up in the metal that do no useful work, waste energy, and produce heat. These losses can be reduced by *laminating* the core. Instead of a solid block of metal for the core [Fig. 19.12(a)], thin sheets of metal with insulated surfaces are used [Fig. 19.12(b)], reducing these induced currents. These ac eddy currents cause the laminations to vibrate, producing the characteristic transformer "hum."

Figure 19.12   Laminating the core reduces the eddy current losses.

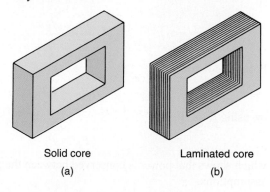

Solid core
(a)

Laminated core
(b)

When a transformer steps up the voltage applied to its primary coil, it reduces the current. Energy is conserved—we cannot get any more electrical energy out of a transformer than we put into it. The relationship between primary and secondary currents is

$$\frac{I_S}{I_P} = \frac{N_P}{N_S}$$

where I_S = current in secondary coil
I_P = current in primary coil
N_P = number of turns in primary
N_S = number of turns in secondary

EXAMPLE 5

The primary current in a transformer is 10.0 A. If the primary coil has $55\overline{0}$ turns and the secondary has $250\overline{0}$ turns, what current flows in the secondary coil?

Data:

$$N_P = 55\overline{0} \text{ turns}$$
$$I_P = 10.0 \text{ A}$$
$$N_S = 250\overline{0} \text{ turns}$$
$$I_S = ?$$

Basic Equation:

$$\frac{I_S}{I_P} = \frac{N_P}{N_S}$$

Working Equation:

$$I_S = \frac{I_P N_P}{N_S}$$

Substitution:

$$I_S = \frac{(10.0 \text{ A})(55\overline{0} \text{ turns})}{250\overline{0} \text{ turns}}$$
$$= 2.20 \text{ A}$$

. . . .

It follows from the last two boxed formulas that

$$\frac{V_P}{V_S} = \frac{N_P}{N_S} = \frac{I_S}{I_P}$$

so

$$\frac{V_P}{V_S} = \frac{I_S}{I_P}$$

From this we obtain,

$$I_P V_P = I_S V_S$$

or, using $P = IV$,

$$P_P = P_S$$

which shows that power is conserved between the primary and secondary coils under these assumptions.

EXAMPLE 6

The power in the primary coil of a transformer is 375 W. If the current in the secondary coil is 11.4 A, what is the voltage in the secondary?

Data:

$$P_P = 375 \text{ W}$$
$$I_S = 11.4 \text{ A}$$
$$V_S = ?$$

Basic Equation:

$$P_P = P_S \quad \text{and} \quad P_S = V_S I_S$$

Working Equations:

$$P_P = V_S I_S \quad (\textit{Note: Substitute for } P_S.)$$

$$V_S = \frac{P_P}{I_S}$$

Substitution:

$$V_S = \frac{375 \text{ W}}{11.4 \text{ A}}$$

$$= 32.9 \text{ V}$$

. . . .

Good transformers are more than 98% efficient. This is very important in power transmission. It is impractical to generate electricity at high voltage, but high voltage is desirable for transmission. Therefore, transformers are used to step up the voltage for transmission. High voltage is unsuitable, though, for consumer use, so transformers are used to reduce the voltage. A simplified diagram of a power distribution system is shown in Fig. 19.13.

Figure 19.13 Power distribution system.

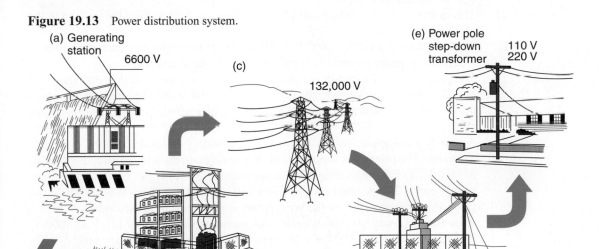

(a) Generating station — 6600 V
(c) 132,000 V
(e) Power pole step-down transformer — 110 V / 220 V
(b) Step-up transformer
(d) Substation step-down transformer — 6000 V

PROBLEMS 19.2

1. A soldering iron is rated at $35\overline{0}$ W. If the current in the iron is 4.00 A, what is the resistance of the iron?
2. What power is developed by a device that draws 6.00 A with resistance 12.0 Ω?
3. Find the output power of a transformer with output voltage $50\overline{0}$ V and current 7.00 A.
4. A heater operates on a $11\overline{0}$-V line and is rated at $45\overline{0}$ W. What is the resistance of the element?
5. A heating element draws 6.00 A on a $22\overline{0}$-V line. What power is used in the element?
6. A 32.0-Ω resistance coil uses 375 W of power. What is the current in the coil?
7. What power is used by a heater that has a resistance of 12.0 Ω and draws a current of 7.00 A?

SKETCH

$$12 \text{ cm}^2 \quad \text{w}$$

4.0 cm

DATA

$A = 12 \text{ cm}^2, l = 4.0 \text{ cm}, w = ?$

BASIC EQUATION

$A = lw$

WORKING EQUATION

$w = \frac{A}{l}$

SUBSTITUTION

$w = \frac{12 \text{ cm}^2}{4.0 \text{ cm}} = 3.0 \text{ cm}$

8. A heater operates on a $11\overline{0}$-V line and is rated at $75\overline{0}$ W. What is the resistance of the heater element?

9. A $11\overline{0}$-Ω resistance coil draws a current of 5.00 A. What power is used?

10. What power is used by a heater with resistance 19.5 Ω and that draws a current of 5.55 A?

11. $V_P = 30.0$ V
$V_S = 45.0$ V
$N_S = 15.0$ turns
Find N_P.

12. $V_P = 25\overline{0}$ V
$N_P = 73\overline{0}$ turns
$N_S = 275$ turns
Find V_S.

13. $I_P = 6.00$ A
$I_S = 4.00$ A
$V_P = 39.0$ V
Find V_S.

14. A step-up transformer on a 115-V line provides a voltage of $230\overline{0}$ V. If the primary coil has 65.0 turns, how many turns does the secondary have?

15. A step-down transformer on a 115-V line provides a voltage of 11.5 V. If the secondary coil has 30.0 turns, how many turns does the primary have?

16. A transformer has 20.0 turns in the primary coil and $220\overline{0}$ turns in the secondary. If the primary voltage is 12.0 V, what is the secondary voltage?

17. If the current is 9.00 A in the primary coil in Problem 13, find the current in the secondary.

18. If the voltage in the secondary coil of a transformer is $11\overline{0}$ V and the current in it is 15.0 A, what power does it supply?

19. A neon sign has a transformer that changes electricity from $11\overline{0}$ V to $15,\overline{0}00$ V. (a) If the primary current is 8.00 A, find the current in the secondary coil. (b) Find the power in the primary coil.

20. A transformer has an output power of 990 W. If the current in the secondary coil is 0.45 A, what is its voltage?

21. The current in the secondary coil of a transformer is 5.00 A. Find the voltage in the secondary if the power is 775 W.

22. A transformer steps down $660\overline{0}$ V to $12\overline{0}$ V. (a) If the secondary current is 14.0 A, what is the primary current? (b) Find the power in the primary coil.

23. The primary coil of a step-down transformer has $750\overline{0}$ turns, and the secondary coil has 125 turns. The voltage across the primary is $720\overline{0}$ V. (a) Find the voltage across the secondary. (b) If the current in the secondary coil is 36.0 A, find the primary current.

24. A step-up transformer has $300\overline{0}$ turns in the secondary coil and $20\overline{0}$ turns in the primary coil. The primary is supplied with alternating current with an effective voltage of 90.0 V. (a) Find the voltage in the secondary coil. (b) If the current in the secondary coil is 2.00 A, what current flows in the primary coil? (c) What power is developed in the primary coil? (d) What power is developed in the secondary coil?

25. A mechanic's tester has a transformer that increases voltage from $11\overline{0}$ V to $300\overline{0}$ V. If the primary coil has $15\overline{0}$ turns, how many turns are in the secondary coil?

26. A construction site generator has a transformer with 175 turns in the primary coil and $75\overline{0}$ turns in the secondary coil. If it has a current of 20.0 A in the primary coil, what is the current in the secondary coil?

19.3 Inductance

Electronic circuitry in televisions, radios, computers, and electronic instruments has many components other than resistors. These components include capacitors, inductors, diodes, and transistors. With these components weak signals can be amplified, noise can be reduced, and signals at certain frequencies can be detected while signals from other frequencies can be rejected (that is, a circuit can be "tuned in" to a frequency). The analysis of the behavior of circuits with these components can become very complex. As a start toward understanding these circuits, we discuss inductors and capacitors. The operation of diodes and transistors will be discussed only briefly. We begin with a discussion of inductors.

Inductance measures the tendency of a coil of wire to resist a change in the current because the magnetism produced by one part of the coil acts to oppose the change of current in other parts of the coil. Inductance is the property of an electric circuit in which a varying current produces a varying magnetic field that induces voltage in the same circuit or in a nearby circuit. An **inductor** is a circuit component, such as a coil of wire, in which an induced emf opposes any change in the current (Fig. 19.14). The emf is induced in the coil itself as the magnetic field of the coil changes.

The unit of inductance, L, is the henry (H), named after **Joseph Henry**. A coil has an inductance of 1 H if an emf of 1 V is induced when the current changes at the rate of 1 A/s. We can express the henry as an ohm second:

$$1 \text{ H} = 1 \, \Omega \text{ s}$$

The henry is a large unit. A more practical unit is the millihenry (mH), which is one one-thousandth of a henry.

Inductance can be illustrated by connecting a coil with a large number of turns and a lamp in series. When connected to a dc source, the lamp burns brightly [Fig. 19.15(a)]. However, when this circuit is connected to an ac power source of the same voltage, the lamp is dimmer because of the inductance of the coil [Fig. 19.15(b)]. The circuit symbol for inductance is shown in Fig. 19.16.

Figure 19.15 A coil in an ac circuit produces inductance.

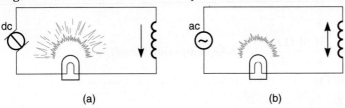

(a)　　　　　(b)

Figure 19.14 Induced emf opposing current change in an ac circuit.

Joseph Henry (1797–1898),

physicist, was born in Albany, New York. He constructed the first electromagnetic motor, discovered electric induction, and demonstrated the oscillatory nature of electric discharges. He also introduced a system of weather forecasting and investigated the propagation of light and sound waves. The unit of induction, the henry, is named after him.

Figure 19.16 Circuit diagram symbol for inductance

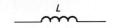

Inductive Reactance

The opposition to ac current flow in an inductor is called **inductive reactance** and is measured in ohms. This is usually represented by X_L. The inductive reactance of a coil is directly proportional to frequency and is found by the following:

$$X_L = 2\pi f L$$

where X_L = inductive reactance
f = frequency of the ac voltage, expressed in hertz (cycles per second), such as $6\overline{0}$ Hz or $6\overline{0}$/s
L = inductance, in henries

If inductance is given in mH, a conversion must be made to H (henries).

The current in a circuit that has only an ac voltage source and an inductor is given by

$$I = \frac{E}{X_L}$$

where I = current
E = voltage
X_L = inductive reactance

A coil with inductance 1.00 mH is connected to a 60.0-kHz ac power source of $11\overline{0}$ V. What is the current in the circuit?

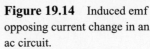

EXAMPLE

Sketch:

$E = 11\overline{0}$ V
$f = 60.0$ kHz
$L = 1.00$ mH

Data:

$$E = 11\overline{0} \text{ V}$$
$$L = 1.00 \text{ mH} = 1.00 \times 10^{-3} \text{ H}$$
$$f = 60.0 \text{ kHz} = 60.0 \times 10^3 \text{ Hz} = 6.00 \times 10^4/\text{s}$$
$$I = ?$$

Basic Equations:

$$X_L = 2\pi fL \quad \text{and} \quad I = \frac{E}{X_L}$$

Working Equations: Same

Substitutions:

$$X_L = 2\pi(6.00 \times 10^4/\text{s})(1.00 \times 10^{-3} \text{ H})$$
$$= 377 \frac{\text{H}}{\text{s}}$$
$$= 377 \frac{\text{H}}{\text{s}} \left(\frac{1 \, \Omega \, \text{s}}{1 \, \text{H}} \right) \quad \text{(note conversion factor)}$$
$$= 377 \, \Omega$$

$$I = \frac{E}{X_L}$$
$$I = \frac{11\overline{0} \text{ V}}{377 \, \Omega}$$
$$= 0.292 \frac{\text{V}}{\Omega} \left(\frac{\text{A} \, \Omega}{\text{V}} \right) \quad \text{(note conversion factor)}$$
$$= 0.292 \text{ A}$$

In an inductor, the current lags behind the voltage by one-fourth of a cycle (Fig. 19.17). For example, in a 60-Hz ac circuit, the frequency is 60 cycles/s; that is, the

Figure 19.17   In an inductive circuit, the current lags behind the voltage by one-fourth of a cycle.

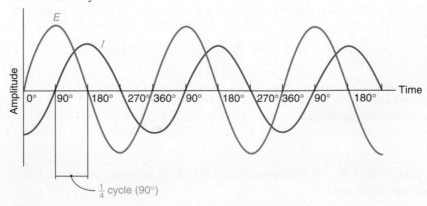

time for one complete cycle is 1/60 s. Thus, the maximum voltage in a 60-Hz circuit occurs

$$\frac{1}{4} \times \frac{1}{60}\, s = \frac{1}{240}\, s$$

before the maximum current. The current lag is usually measured in degrees. One-fourth of a cycle is 90°.

PROBLEMS 19.3

Find the inductive reactance (in ohms) of each inductance at the given frequency.

1. $L = 3.00$ mH, $f = 60.0$ Hz
2. $L = 20.0$ mH, $f = 75.0$ Hz
3. $L = 70.0$ mH, $f = 10.0$ kHz
4. $L = 8.00$ mH, $f = 8.00$ kHz
5. What is the inductive reactance (in ohms) of a 425-μH inductance at a frequency of 15.0 MHz?
6. Find the inductive reactance (in ohms) of a 655-μH inductance at a frequency of 125 MHz.

Find the current (in amperes) in each inductive circuit.

7. $L = 30.0$ mH, $f = 125$ Hz, $E = 14.0$ V
8. $L = 1.00$ mH, $f = 125$ kHz, $E = 145$ V
9. $L = 5.00$ mH, $f = 2.00$ kHz, $E = 50.0$ V
10. $L = 30.0$ mH, $f = 7.00$ MHz, $E = 75.0$ V
11. Find the current (in amperes) in an inductive circuit where $L = 72.0\ \mu$H, $f = 2.00$ MHz, and $E = 105$ V.
12. Find the current (in amperes) in an inductive circuit where $L = 525\ \mu$H, $f = 25.0$ MHz, and $E = 65.0$ V.

SKETCH

12 cm² | w

4.0 cm

DATA

$A = 12$ cm², $l = 4.0$ cm, $w = ?$

BASIC EQUATION

$A = lw$

WORKING EQUATION

$w = \frac{A}{l}$

SUBSTITUTION

$w = \frac{12\ \text{cm}^2}{4.0\ \text{cm}} = 3.0$ cm

PHYSICS CONNECTIONS

Induction and the Ground Fault Interrupter

Ground fault interrupt (GFI) outlets (Fig. 19.18) are required by most building codes for electric outlets around sinks or other devices where water can cause electric shocks. The purpose of a GFI is to detect minor losses of current in a circuit. If a working hairdryer were to fall into a sink or a bathtub, the water would cause the hairdryer to short out and send electric charge into the water and metal pipes. In such a situation, the GFI outlet would automatically shut off the current within milliseconds of the short circuit. If someone touched the hairdryer while in the water, the person would be electrocuted if not for the rapid circuit-breaking ability of the GFI.

All ac electric appliances have current moving back and forth at a frequency of 60 Hz. The GFI is designed to measure the changes in the induced magnetic field of that electric current. In other words, the current going into the appliance establishes a magnetic field in the iron loop, while the current leaving the appliance creates an equal and opposite magnetic field in the loop, reversing direction 60 times per second. These two fields, in a normal operating situation, cancel each other out. However, if even a small amount of current does not return through the GFI outlet, the magnetic fields do not cancel out. In that situation, the magnetic field in the loop creates an electric current in the detection coil. Any current passing through the detection coil causes the breaker to trip, which quickly shuts off the circuit and prevents serious injury.

Figure 19.18 (a) A standard ground fault interrupt outlet. (b) The basic components of a ground fault interrupt outlet.

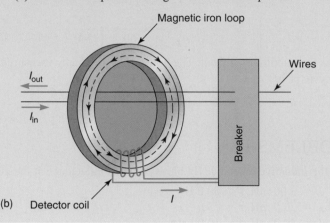

(a)

(b) Detector coil

Figure 19.19 ac circuit with resistance and inductance.

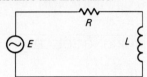

19.4 Inductance and Resistance in Series

In addition to inductance, most ac circuits have resistance in the form of lights or other resistors (Fig. 19.19). The current lags behind the voltage by any amount of time greater than zero and as large as one-fourth of a cycle.

Impedance is a measure of the total opposition to current flow in an ac circuit resulting from the effect of both the resistance and the inductive reactance on the circuit. Ohm's law in an ac circuit can be written as

$$I = \frac{E}{Z}$$

where I = current
E = voltage
Z = impedance

The impedance of a series circuit containing a resistance and an inductance is

$$Z = \sqrt{R^2 + X_L^2}$$
$$Z = \sqrt{R^2 + (2\pi f L)^2}$$

where Z = impedance
R = resistance
X_L = inductive reactance
f = frequency
L = inductance

Figure 19.20 Graphic representation of impedance.

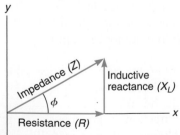

The impedance can be represented as a vector as the hypotenuse of the right triangle shown in Fig. 19.20. The resistance is always drawn as a vector pointing in the positive *x*-direction. The inductive reactance is drawn as a vector pointing in the positive *y*-direction. The angle ϕ shown between the resistance and impedance vectors is the **phase angle** and equals the amount by which the current lags behind the voltage. The phase angle is given by

$$\tan \phi = \frac{X_L}{R}$$

EXAMPLE

A lamp of resistance 40.0 Ω is connected in series with an inductance of 95.0 mH. This circuit is connected to a 115-V, 60.0-Hz power supply. (a) What is the current in the circuit? (b) What is the phase angle?

Sketch:

$$R = 40.0\,\Omega$$

$$E = 115\,\text{V}$$
$$f = 60.0\,\text{Hz}$$

$$L = 95.0\,\text{mH}$$

(a) Data:

$$E = 115\,\text{V}$$
$$f = 60.0\,\text{Hz} = 60.0/\text{s}$$
$$R = 40.0\,\Omega$$
$$L = 95.0\,\text{mH} = 95.0 \times 10^{-3}\,\text{H} = 0.0950\,\text{H}$$
$$Z = ?$$
$$I = ?$$

Basic Equations:

$$Z = \sqrt{R^2 + (2\pi fL)^2} \quad \text{and} \quad I = \frac{E}{Z}$$

Working Equations: Same

Substitutions: First calculate the impedance.

$$
\begin{aligned}
Z &= \sqrt{(40.0\,\Omega)^2 + [2\pi(60.0/\text{s})(0.0950\,\text{H})]^2}\\
&= \sqrt{16\overline{0}0\,\Omega^2 + (35.8\,\text{H/s})^2}\\
&= \sqrt{16\overline{0}0\,\Omega^2 + \left[\left(35.8\,\frac{\text{H}}{\text{s}}\right)\left(\frac{1\,\Omega\,\text{s}}{1\,\text{H}}\right)\right]^2} \quad \text{(note conversion factor)}\\
&= \sqrt{16\overline{0}0\,\Omega^2 + 1280\,\Omega^2}\\
&= \sqrt{2880\,\Omega^2}\\
&= 53.7\,\Omega
\end{aligned}
$$

$$I = \frac{E}{Z}$$

$$I = \frac{115\,\text{V}}{53.7\,\Omega}\left(\frac{1\,\text{A}\,\Omega}{1\,\text{V}}\right)$$

$$= 2.14\,\text{A}$$

(b) To find the phase angle ϕ, first find X_L. Then, construct the vector right triangle as in Fig. 19.21 to find ϕ.

Figure 19.21

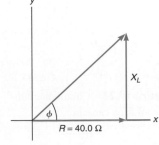

Data:

$$f = 60.0/\text{s}$$
$$L = 0.0950\,\text{H}$$
$$X_L = ?$$
$$\phi = ?$$

Basic Equations:

$$X_L = 2\pi fL \quad \text{and} \quad \tan\phi = \frac{X_L}{R}$$

Working Equations: Same

Substitutions:

$$X_L = 2\pi(60.0/\text{s})(0.0950\ \text{H})$$
$$= 35.8\frac{\text{H}}{\text{s}}$$
$$= 35.8\frac{\cancel{\text{H}}}{\cancel{\text{s}}}\left(\frac{1\Omega\ \cancel{\text{s}}}{1\ \cancel{\text{H}}}\right)$$
$$= 35.8\ \Omega$$

[*Note:* This value can also be taken directly from the first working equation in part (a).]

From Fig. 19.21 we have

$$\tan\phi = \frac{35.8\ \Omega}{40.0\ \Omega} = 0.895$$
$$\phi = 41.8°$$

PROBLEM SOLVING

SKETCH

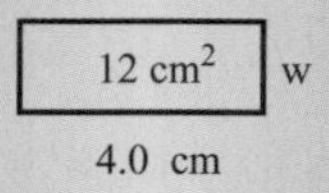

DATA

$A = 12\ \text{cm}^2$, $l = 4.0$ cm, w = ?

BASIC EQUATION

$A = lw$

WORKING EQUATION

$w = \frac{A}{l}$

SUBSTITUTION

$w = \frac{12\ \text{cm}^2}{4.0\ \text{cm}} = 3.0\ \text{cm}$

PROBLEMS 19.4

1. For a circuit with $R = 20\bar{0}\ \Omega$, $L = 10.0$ mH, and $f = 1.25$ kHz:
   (a) find the impedance (in ohms).
   (b) find the phase angle.
   (c) find the current if the voltage is 45.0 V.
2. For a circuit with $R = 12.0\ \Omega$, $L = 1.00$ mH, and $f = 90\bar{0}$ Hz:
   (a) find the impedance (in ohms).
   (b) find the phase angle.
   (c) find the current if the voltage is 10.0 V.
3. For a circuit with $R = 1.00\ \text{k}\Omega$, $L = 50.0$ mH, and $f = 10.0$ kHz:
   (a) find the impedance (in ohms).
   (b) find the phase angle.
   (c) find the current if the voltage is 15.0 V.
4. For a circuit with resistance 2.00 kΩ, inductance 70.0 mH, and frequency 5.00 kHz:
   (a) find the impedance (in ohms).
   (b) find the phase angle.
   (c) find the current if the voltage is 12.0 V.
5. For a circuit with resistance $30\bar{0}\ \Omega$, inductance 2.00 mH, and frequency 3.00 kHz:
   (a) find the impedance (in ohms).
   (b) find the phase angle.
   (c) find the current if the voltage is 6.00 V.

19.5 Capacitance

Figure 19.22 Circuit diagram symbol for a capacitor.

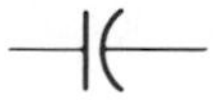

An important component of many ac circuits is the capacitor. A **capacitor** is used to store an electric charge, and consists of two conductors that are usually parallel plates separated by a thin insulator. The plates are often made of a metal foil rolled to a convenient size and inserted in a cylinder. Capacitors are represented in circuit diagrams as shown in Fig. 19.22.

When a capacitor is connected to a battery, electrons flow from the negative terminal to one capacitor plate as shown in Fig. 19.23(b). When the capacitor is removed from the battery, the charges remain on the capacitor. If the capacitor is then connected to a resistor [Fig. 19.23(c)], electrons will flow through the circuit until the capacitor has lost its charge.

Figure 19.23

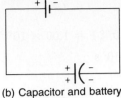

(a) Charged capacitor | (b) Capacitor and battery in a circuit | (c) An electron current flows when the capacitor is connected to a resistance

A capacitor will block a direct current from flowing in a circuit once the capacitor is charged. A low-frequency ac voltage in a capacitive circuit (Fig. 19.24) will cause only a small current to flow because of this blocking nature of a capacitor. A high-frequency ac voltage source will cause a larger current to flow in the capacitive circuit of Fig. 19.24 because a current is required to quickly change the polarity of the capacitor voltage. Capacitors can therefore be used to tune the frequency response of circuits, allowing the blocking of low-frequency electric signals and tuning the resonance frequency of circuits as described in Section 19.8.

Figure 19.24   Capacitor in an ac circuit.

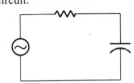

Capacitance is the ratio of the charge on either plate of a capacitor to the potential difference between the plates. That is,

$$C = \frac{Q}{V}$$

where C = capacitance of a capacitor
Q = amount of charge on either plate of the capacitor
V = potential difference between the two plates of the capacitor

The unit of capacitance is the *farad* (F), named after **Michael Faraday**. A more practical unit is the microfarad (μF or 10^{-6} F). **Capacitive reactance** is a measure of the opposition to ac current flow by a capacitor. The effect of a capacitor on a circuit is inversely proportional to frequency and given by

$$X_C = \frac{1}{2\pi f C}$$

where X_C = capacitive reactance (ohms)
f = frequency
C = capacitance (farads)

$$1 \text{ F} = 1 \text{ s}/\Omega$$

In a circuit that contains only capacitors, the current *leads* the voltage by 90° (one-fourth cycle) as shown in Fig. 19.25.

Michael Faraday (1791–1867), *chemist and experimental physicist, was born in England. He introduced the concept of capacitance, discovered electromagnetic induction, and made numerous other contributions to electricity and magnetism. The unit of capacitance, the farad, is named after him.*

Figure 19.25   Current leads the voltage by one-fourth cycle.

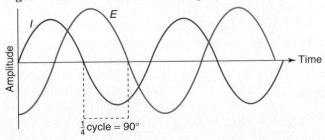

$\frac{1}{4}$ cycle = 90°

EXAMPLE

Find the capacitive reactance of a 10.0-μF capacitor in a circuit of frequency 1.00 kHz.

Data:

$$C = 10.0\ \mu F = 10.0 \times 10^{-6}\ F = 1.00 \times 10^{-5}\ F$$
$$f = 1.00\ \text{kHz} = 1.00 \times 10^{3}/\text{s}$$
$$X_C = ?$$

Basic Equation:

$$X_C = \frac{1}{2\pi f C}$$

Working Equation: Same

Substitution:

$$X_C = \frac{1}{2\pi (1.00 \times 10^{3}/\text{s})(1.00 \times 10^{-5}\ F)}$$

$$= \frac{1}{\left(0.0628\ \dfrac{F}{s}\right)\left(\dfrac{1\ s}{1\ F\ \Omega}\right)} \qquad \text{(note conversion factor)}$$

$$= 15.9\ \Omega$$

·················

PROBLEMS 19.5

Find the capacitive reactance (in ohms) in each ac circuit.

1. $C = 20.0\ \mu F, f = 1.00\ \text{kHz}$
2. $C = 7.00\ \text{mF}, f = 10\overline{0}\ \text{Hz}$
3. $C = 0.600\ \mu F, f = 0.100\ \text{kHz}$
4. $C = 30.0\ \text{mF}, f = 2.50\ \text{MHz}$
5. $C = 0.800\ \mu F, f = 0.250\ \text{MHz}$
6. Find the capacitive reactance of a 15.0-μF capacitor in a circuit of frequency 60.0 Hz.
7. Find the capacitive reactance of a 45.0-μF capacitor in a circuit of frequency 60.0 kHz.
8. Find the capacitive reactance of a 6.00-mF capacitor in a circuit of frequency 10$\overline{0}$ Hz.
9. Find the capacitive reactance of a 33$\overline{0}$-μF capacitor in a circuit of frequency 30$\overline{0}$ Hz.
10. Find the capacitive reactance of a 222-μF capacitor in a circuit of frequency 12$\overline{0}$ Hz.

19.6 Capacitance and Resistance in Series

The combined effect of capacitance and resistance in series is measured by the impedance, Z, of the circuit.

$$Z = \sqrt{R^2 + X_C^2}$$
$$Z = \sqrt{R^2 + \left(\frac{1}{2\pi f C}\right)^2}$$

where
Z = impedance
R = resistance
X_C = capacitive reactance
f = frequency
C = capacitance

SKETCH

12 cm² | w

4.0 cm

DATA

$A = 12\ \text{cm}^2, l = 4.0\ \text{cm}, w = ?$

BASIC EQUATION

$A = lw$

WORKING EQUATION

$w = \frac{A}{l}$

SUBSTITUTION

$w = \frac{12\ \text{cm}^2}{4.0\ \text{cm}} = 3.0\ \text{cm}$

where Z = impedance
R = resistance
X_L = inductive
X_C = capacitive

The vector diagram for th

Figure 19.27 Vector dia
(b) capacitive reactance.

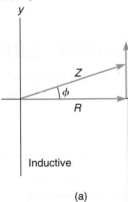

(a)

In a circuit contain
the current lags behind
voltage lags behind the
are *in phase*. If the circ
in phase. The current i

An ac circuit contains
in series with a 25.0-V

Sketch:

Data:

R
C
L
E
f
Z
I

The current is given by Ohm's law:

$$I = \frac{E}{Z}$$

where I = current
E = voltage
Z = impedance

The phase angle can be found by drawing the resistance as a vector in the positive x-direction and the capacitive impedance as a vector in the negative y-direction as shown in Fig. 19.26. The phase angle gives the amount by which the voltage lags behind the current.

$$\tan \phi = \frac{X_C}{R}$$

What current will flow in a 60.0-Hz ac circuit that includes a $11\overline{0}$-V source, a capacitor of 90.0 μF, and a 16.0-Ω resistance in series? Also find the phase angle.

Sketch:

$R = 16.0\ \Omega$

$C = 90.0\ \mu$F

$E = 11\overline{0}$ V
$f = 60.0$ Hz

Data:

$$E = 11\overline{0} \text{ V}$$
$$f = 60.0 \text{ Hz} = 60.0/\text{s}$$
$$R = 16.0\ \Omega$$
$$C = 90.0\ \mu\text{F} = 90.0 \times 10^{-6}\ \text{F} = 9.00 \times 10^{-5}\ \text{F}$$
$$Z = ?$$
$$I = ?$$

Basic Equations:

$$Z = \sqrt{R^2 + \left(\frac{1}{2\pi f C}\right)^2} \quad \text{and} \quad I = \frac{E}{Z}$$

Working Equations: Same

Substitutions:

First, find Z:

$$Z = \sqrt{(16.0\ \Omega)^2 + \left(\frac{1}{2\pi (60.0/\text{s})(9.00 \times 10^{-5}\ \text{F})}\right)^2}$$
$$= \sqrt{256\ \Omega^2 + \left(\frac{1}{0.0339\ \text{F/s}}\right)^2}$$
$$= \sqrt{256\ \Omega^2 + \left(29.5\frac{s}{F} \times \frac{1\ \Omega F}{1\ s}\right)^2}$$
$$= \sqrt{256\ \Omega^2 + 87\overline{0}\ \Omega^2}$$
$$= 33.6\ \Omega$$

Figure 19.26 Determination of the phase angle in an ac circuit.

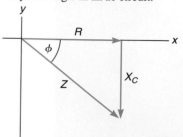

EXAMPLE

EXAMPLE

Find the resonant frequency of a circuit containing a 5.00-nF capacitor in series with a 2.60-μH inductor.

Sketch:

C = 5.00 nF

L = 2.60 μH

Data:

$$C = 5.00 \text{ nF} = 5.00 \times 10^{-9} \text{ F}$$
$$L = 2.60 \ \mu\text{H} = 2.60 \times 10^{-6} \text{ H}$$
$$f = ?$$

Basic Equation:

$$f = \frac{1}{2\pi\sqrt{LC}}$$

Working Equation: Same

Substitution:

$$f = \frac{1}{2\pi\sqrt{(2.60 \times 10^{-6} \text{ H})(5.00 \times 10^{-9} \text{ F})}}$$

$$= \frac{1}{2\pi\sqrt{1.30 \times 10^{-14} \text{ H F}\left(\dfrac{1 \ \Omega \text{ s}}{1 \text{ H}}\right)\left(\dfrac{1 \text{ s}}{1 \text{ F } \Omega}\right)}}$$

$$= \frac{1}{7.16 \times 10^{-7} \text{ s}}$$

$$= 1.40 \times 10^{6} \ \frac{\text{cycles}}{\text{s}} \quad \text{or} \quad 1400 \ \frac{\text{kilocycles}}{\text{s}} \quad \text{or} \quad 1400 \text{ kHz}$$

This frequency is in the AM radio band.

··················

PROBLEMS 19.8

Find the resonant frequency in each ac circuit.

1. $L = 1.00 \ \mu$H and $C = 4.00 \ \mu$F
2. $L = 2.00 \ \mu$H and $C = 35.0 \ \mu$F
3. $L = 2.50 \ \mu$H and $C = 7.00 \ \mu$F
4. $L = 2.65 \ \mu$H and $C = 35.0 \ \mu$F
5. $L = 42.5 \ \mu$H and $C = 40.0 \ \mu$F
6. Find the resonant frequency of a circuit containing a 25.0-μF capacitor in series with a 75.0-μH inductor.
7. Find the resonant frequency of a circuit containing a 33.0-μF capacitor in series with a 43.5-μH inductor.
8. Find the resonant frequency of a circuit containing a 10.0-μF capacitor in series with a 37.5-μH inductor.
9. Find the resonant frequency of a circuit containing an 8.00-μF capacitor in series with a 100-μH inductor.
10. Find the resonant frequency of a circuit containing a 3.75-μF capacitor in series with a 300-μH inductor.

PROBLEM SOLVING

SKETCH

12 cm² | w

4.0 cm

DATA

$A = 12$ cm², $l = 4.0$ cm

BASIC EQUATION

$A = lw$

WORKING EQUAT

$w = \frac{A}{l}$

SUBSTITUTION

$w = \frac{12 \text{ cm}^2}{4.0 \text{ cm}} = 3.0$ cm

PROBLEM SOLVING

SKETCH

12 cm² | w

4.0 cm

DATA

$A = 12$ cm², $l = 4.0$ cm, $w = ?$

BASIC EQUATION

$A = lw$

WORKING EQUATION

$w = \frac{A}{l}$

SUBSTITUTION

$w = \frac{12 \text{ cm}^2}{4.0 \text{ cm}} = 3.0$ cm

19.9 Rectification and Amplification

It is often necessary to change ac into dc to provide dc for charging batteries or to power the integrated circuits (ICs) of computers and other electronic units. This process of changing ac to dc is called **rectification**. A device that accomplishes this is called a *diode*. Early diodes were constructed as vacuum tubes. Modern diodes are made of a semiconductor material and are usually less than 5 mm long (Fig. 19.29).

A **diode** allows current to flow through it in only one direction and not the other. It is similar to a turnstile that revolves in only one direction. People can pass the turnstile in one direction but are blocked when they attempt to pass in the opposite direction (Fig. 19.30).

Thomas Edison found that when a wire filament was heated near a metal plate, an electron charge would begin to flow from the filament across space to the plate. This is called the *Edison effect* and was the beginning of the electronics industry. The electron emitter filament is called the *cathode*. The plate is called the *anode*. The entire device is called a *diode*.

A rectifier changes alternating current to direct current by allowing it to pass in only one direction (Fig. 19.31). Additional circuit devices can be added to the rectifier to smooth out the direct current so that it appears as shown in Fig. 19.32. Rectifiers are used in automobiles to change the alternating current produced by the alternator into direct current.

Figure 19.29

Semiconductor diode (enlarged)

Figure 19.30

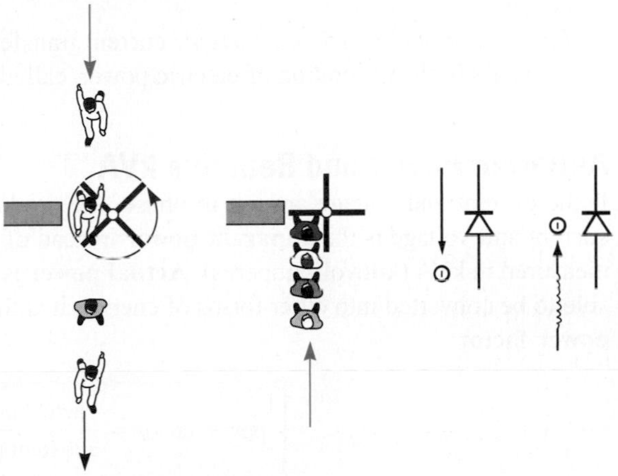

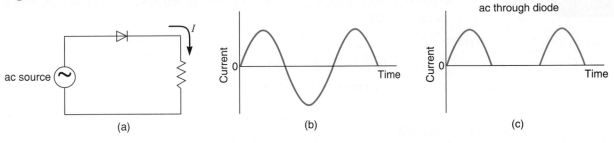

(a) A one-way turnstile allowing people to move through in only one direction is like a diode.

(b) A diode allows electrons to flow in only one direction.

Thomas Edison (1847–1931),

inventor, was born in Milan, Ohio. He held patents to over 1000 inventions, which included the phonograph, the carbon transmitter that made Alexander Graham Bell's telephone practical, an incandescent lamp, a storage battery, and a system for widespread distribution of electric power from central generating stations. His only true scientific discovery in 1883, called "the Edison effect," for which he could see no commercial application, led to the vacuum tube, which became the basis for the radio. He spent his life inventing many of the devices that shaped the modern world.

Figure 19.31 A diode can change ac to dc.

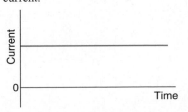

It is often necessary to increase the strength of an electronic signal. This is called **amplification**. Radios, stereos, and many other instruments contain one or more amplifier circuits. Early amplifiers utilized vacuum tubes together with other components. The transistor, developed by **John Bardeen**, has replaced the vacuum tube in most circuitry because of its smaller size and lesser power consumption. A *transistor* amplifier is typically composed of one or more transistors in addition to capacitors, resistors, and possibly inductors to provide the amplifier with the desired gain (amplification), frequency response, and power output. Amplifiers composed of individual transistors, resistors, and capacitors have been replaced for many applications by *integrated circuits* (ICs), which are only slightly larger in size than some transistors. An IC may contain millions of tiny transistors, diodes, resistors, and capacitors on a small chip of silicon less than 1 cm square. In addi-

Figure 19.32 Rectified ac current.

tion to amplifying signals, ICs have been designed to serve as memory or logic units in computers or other applications. These ICs can be programmed to perform arithmetic operations, as in a calculator, or to perform control operations, as in most appliances.

19.10 Commercial Generator Power Output

The power output of the generator is the product of voltage and current. The ac generator converts mechanical energy to electric energy by performing three functions:

1. *Production of voltage:* electric pressure, which pushes the current through the loads
2. *Production of power current:* current converted into heat, light, and mechanical power
3. *Production of magnetizing current:* current transferred back and forth for magnetizing purposes in the generation of electric power, called *reactive kVA* (kilovolt-amperes)

Apparent Power and Reactive kVA

If the current and voltage are not in phase, the product of effective values of alternating current and voltage is the **apparent power** instead of the actual power. Apparent power is measured in kVA (kilovolt-amperes). **Actual power** is a measure of the actual power available to be converted into other forms of energy; it is the product of apparent power and the **power factor**:

$$\text{power factor} = \frac{\text{actual power}}{\text{apparent power}}$$

where the actual power is measured in kW, the apparent power is measured in kVA and is called *reactive kVA*, and the power factor is a unitless ratio less than 1. Note that 1 VA = 1 W.

Mathematically, the power factor is equal to the cosine of the angle by which the current lags behind (or in rare cases leads) the voltage. The power factor is really a correction factor that must be applied to determine actual power produced. The situation is very similar to finding the amount of work done when a force and the motion are not in the same direction.

EXAMPLE

Find the actual power produced by a generating system that produces 13,600 kVA with a power factor of 0.900.

Data:

$$\text{apparent power} = 13,600 \text{ kVA}$$
$$\text{power factor} = 0.900$$
$$\text{actual power} = ?$$

Basic Equation:

$$\text{power factor} = \frac{\text{actual power}}{\text{apparent power}}$$

Working Equation:

$$\text{actual power} = (\text{apparent power})(\text{power factor})$$

Substitution:

$$\text{actual power} = (13,600 \text{ kVA})(0.900)$$
$$= 12,200 \text{ kVA}$$
$$= 12,200 \text{ kW}$$

PROBLEMS 19.10

1. Find the actual power produced by a generating station that produces 12,600 kVA with a power factor of 0.850.
2. A generating station operates with a power factor of 0.910. What actual power is available on the transmission lines if the apparent power is 12,800 kVA?
3. Find the apparent power produced by a generating station whose actual power is $12\overline{0},000$ kW and whose power factor is 0.900.
4. Find the apparent power produced by a generating station whose actual power is $1,9\overline{0}0,000$ kW and whose power factor is 0.800.
5. A generating station operates with a power factor of 0.880. What actual power is available on the transmission lines if the apparent power is 11,500 kVA?
6. Find the apparent power produced by a generating station whose actual power is 2,350,000 kW and whose power factor is 0.850.
7. Find the actual power produced by a generating station that produces 23,800 kVA with a power factor of 0.810.
8. A generating station operates with a power factor of 0.840. What actual power is available on the transmission lines if the apparent power is 13,500 kVA?
9. Find the apparent power produced by a generating station whose actual power is $35\overline{0},000$ kW and whose power factor is 0.860.
10. Find the apparent power produced by a generating station whose actual power is 1,250,000 kW and whose power factor is 0.820.
11. Find the power factor of a generating station whose actual power is 55,800 kW and whose apparent power is 63,400 kVA.
12. Find the power factor of a generating station whose apparent power is 645,000 kVA and whose actual power is 587,000 kW.

SKETCH

12 cm^2	w

4.0 cm

DATA

$A = 12$ cm^2, $l = 4.0$ cm, $w = ?$

BASIC EQUATION

$A = lw$

WORKING EQUATION

$w = \frac{A}{l}$

SUBSTITUTION

$w = \frac{12 \text{ cm}^2}{4.0 \text{ cm}} = 3.0$ cm

Glossary

Actual Power A measure of the actual power available to be converted into other forms of energy. (p. 548)

Alternating Current A current that flows in one direction in a conductor, changes direction, and then flows in the other direction. (p. 521)

Amplification The process of increasing the strength of an electronic signal. (p. 547)

Apparent Power The product of the effective values of alternating current and voltage. (p. 548)

Capacitance The ratio of the charge on either plate of a capacitor to the potential difference between the plates. (p. 539)

Capacitor A circuit component consisting of two parallel plates separated by a thin insulator used to build up and store charge. (p. 538)

Capacitive Reactance A measure of the opposition to ac current flow by a capacitor. (p. 539)

Diode A device that allows current to flow through it in only one direction. (p. 547)

Effective Value The number of amperes of alternating current that produces the same amount of heat in a resistance as an equal number of amperes of a steady direct current. (p. 522)

Impedance A measure of the total opposition to current flow in an ac circuit resulting from the effect of both the resistance and the inductive reactance on the circuit. (p. 536)

Inductance A measure of the tendency of a coil of wire to resist a change in the current because the magnetism produced by one part of the coil acts to oppose the change of current in other parts of the coil. (p. 532)

Inductive Reactance A measure of the opposition to ac current flow in an inductor. (p. 533)

Inductor A circuit component, such as a coil, in which an induced emf opposes any current change in the circuit. (p. 533)

Instantaneous Current The current at any instant of time. (p. 522)

Instantaneous Voltage The voltage at any instant of time. (p. 522)

Phase Angle The angle between the resistance and impedance vectors in a circuit. (p. 536)

Power Factor The ratio of the actual power to the apparent power. (p. 548)

Primary Coil The coil of a transformer that carries an alternating current and induces a current in the secondary coil. (p. 527)

Rectification The process of changing ac to dc. (p. 547)

Rectifier A device that changes ac to dc. (p. 525)

Resonance A condition in a circuit when the inductive reactance equals the capacitive reactance and they nullify each other. The current that flows in the circuit is then at its maximum value. (p. 545)

Secondary Coil The coil of a transformer in which a current is induced by the current in the primary coil. (p. 527)

Step-Down Transformer A transformer used to lower voltage; it has more turns in the primary coil. (p. 528)

Step-Up Transformer A transformer used to increase voltage; it has more turns in the secondary coil. (p. 528)

Transformer A device composed of two coils (primary and secondary) and a magnetic core. Used to step up or step down a voltage. (p. 527)

Formulas

19.1 $i = I_{max} \sin \theta$

$e = E_{max} \sin \theta$

$I = 0.707\, I_{max}$ or $I_{max} = 1.41\, I$

$E = 0.707\, E_{max}$ or $E_{max} = 1.41\, E$

19.2 $P = I^2R = VI = \dfrac{V^2}{R}$

$\dfrac{V_P}{V_S} = \dfrac{N_P}{N_S}$

$\dfrac{I_S}{I_P} = \dfrac{N_P}{N_S}$

$I_P V_P = I_S V_S$

19.3 $X_L = 2\pi f L$

$I = \dfrac{E}{X_L}$

19.4 $I = \dfrac{E}{Z}$

$Z = \sqrt{R^2 + X_L^2}$

$Z = \sqrt{R^2 + (2\pi f L)^2}$

$\tan \phi = \dfrac{X_L}{R}$

19.5 $C = \dfrac{Q}{V}$

$X_C = \dfrac{1}{2\pi f C}$

19.6 $Z = \sqrt{R^2 + X_C^2}$

$Z = \sqrt{R^2 + \left(\dfrac{1}{2\pi f C}\right)^2}$

$I = \dfrac{E}{Z}$

$\tan \phi = \dfrac{X_C}{R}$

19.7 $Z = \sqrt{R^2 + (X_L - X_C)^2}$

$\tan \phi = \dfrac{X_L - X_C}{R}$

$I = \dfrac{E}{Z} = \dfrac{E}{\sqrt{R^2\left(2\pi f L - \dfrac{1}{2\pi f C}\right)^2}}$

19.8 $I = \dfrac{E}{\sqrt{R^2 + (X_L - X_C)^2}}$

$f = \dfrac{1}{2\pi\sqrt{LC}}$

19.10 power factor $= \dfrac{\text{actual power}}{\text{apparent power}}$

Review Questions

1. Which of the following describes alternating current electricity?
    (a) It can be produced by rotating a loop of wire through a magnetic field.
    (b) It flows in one direction for a period of time and then reverses direction.

(c) It goes through one cycle when it flows in one direction and then reverses direction.

(d) All of the above.

2. The voltage, e, and the current, i, in an alternating current circuit are in phase when
   (a) the peak values of both e and i occur at different times.
   (b) the peak values of e and i occur at the same time but their zero values do not.
   (c) the peak values and the zero values of e and i occur simultaneously.
   (d) none of the above.

3. Which of the following affect the voltage induced in the secondary coil of a transformer?
   (a) The current through the primary coil
   (b) The resistance of the primary coil
   (c) The number of turns in the primary coil
   (d) None of the above

4. Which of the following contribute to the energy loss in a transformer?
   (a) Resistance of the copper wires
   (b) Reversing the magnetic field in the core
   (c) Induced currents in the core
   (d) The emf in the outside circuit
   (e) All of the above

5. Explain the difference between maximum current and effective current.

6. Explain the difference between maximum voltage and instantaneous voltage.

7. Explain how power in an ac circuit is related to voltage and current.

8. Explain how power in an ac circuit is related to voltage and resistance.

9. If the number of turns in the secondary coil of a transformer is doubled, how does the output voltage change?

10. The unit of inductance is the _____.

11. Discuss the importance of inductive reactance.

12. How does the inductive reactance depend on frequency?

13. Does the current lead or lag the voltage in an inductive circuit?

14. Describe how energy is stored in a capacitor. How can the stored energy be used?

15. Does the current lead or lag the voltage in a capacitive circuit?

16. How does the reactance of a capacitor depend on frequency?

17. Discuss the condition that leads to resonance.

18. What is the function of a diode in a circuit?

19. Explain the difference between amplification and rectification.

20. Is the phase angle always constant in a circuit containing resistive, capacitive, and inductive elements?

Review Problems

1. What is the maximum voltage in a circuit when the instantaneous value of the voltage is 95.4 V at $\theta = 62°$?

2. If the maximum ac voltage on a line is 185 V, what is the instantaneous voltage at $\theta = 41°$?

3. If the maximum ac voltage on a line is 175 V, what is the instantaneous voltage at $\theta = 23°$?

4. What is the effective value of an ac voltage whose maximum voltage is 135 V?

5. What is the maximum current in a circuit with a current rated at 6.35 A?

6. What power is developed by a device that draws 6.87 A and has a resistance of 15.4 Ω?

7. A heating element draws 4.50 A on a $11\overline{0}$-V line. What power is used in the element?

8. What power is used by a heater with resistance 22.3 Ω and that draws a current of 7.65 A?

9. A step-up transformer on a 115-V line provides a voltage of 2050 V. (a) If the primary coil has 75.0 turns, how many turns does the secondary coil have? (b) If there is a current of 4.55 A in the primary coil, what is the current in the secondary? (c) Find the power in the primary coil.

10. An inductance of 48.0 mH is connected in series with a lamp of resistance 23.0 Ω. This circuit is connected to a 115-V, 60.0-Hz power supply. (a) What is the current in the circuit? (b) What is the phase angle? (c) What is the voltage drop across the inductance?

11. A lamp of resistance 47.5 Ω is connected in series with an inductance of 43.2 mH. This circuit is connected to a 115-V, 60.0-Hz power supply. (a) What is the current in the circuit? (b) What is the phase angle? (c) What is the voltage drop across the resistance?

12. What current will flow in a 60.0-Hz ac series circuit that includes a $11\overline{0}$-V source, a resistor of 19.5 Ω, and a capacitor of 57.4 μF? What is the phase angle? What is the voltage across the resistor? Across the capacitor?

13. A resistor of 21.6 Ω and a capacitor of 38.5 μF are connected in series with a 60.0-Hz ac source with a voltage of $11\overline{0}$ V. What is the current in the circuit? What is the voltage across the capacitor? What is the phase angle?

14. A circuit contains a 175-Ω resistance, a 25.0-μF capacitor, and a 62.0-mH inductance in series with a $11\overline{0}$-V, 60.0-Hz source. Find the impedance and the current.

15. A circuit contains a 115-Ω resistance, a 35.0-μF capacitor, and a 65.0-mH inductance in series with a $11\overline{0}$-V, 60.0-Hz source. Find the impedance and the current.

16. Find the resonant frequency of a circuit containing a 7.50-μF capacitor in series with a 3.70-μH inductance, a 633-Ω resistor, and a $11\overline{0}$-V, 60.0-Hz source. Find the impedance and the current.

17. Find the resonant frequency of a circuit containing a 4.70-μF capacitor in series with a 4.50-μH inductance, a 25.0-Ω resistor, and a $11\overline{0}$-V, 60.0-Hz source. Find the impedance and the current.

18. Find the apparent power produced by a generating station whose actual power is 2,90$\overline{0}$,000 kW and whose power factor is 0.850.

SKETCH

| 12 cm² | w |

4.0 cm

DATA

A = 12 cm², l = 4.0 cm, w = ?

BASIC EQUATION

A = lw

WORKING EQUATION

w = $\frac{A}{l}$

SUBSTITUTION

w = $\frac{12 \text{ cm}^2}{4.0 \text{ cm}}$ = 3.0 cm

APPLIED CONCEPTS

1. A microwave oven is designed to draw 11.8 A of current when connected to a $11\overline{0}$-V ac circuit. (a) What is the average power of the microwave? (b) What is the maximum current drawn by the microwave? (c) Provide a rationale for the current's value 180° after the beginning of a cycle.

2. Before converting alternating current to direct current for an electronic video game, a transformer first must reduce the voltage from $12\overline{0}$ V to 9.00 V. (a) If the primary coil has 307 turns, how many turns are in the secondary coil? (b) What is the power of each coil in the transformer? (c) Explain the relationship between the power for the primary coil and the power for the secondary coil.

3. A neighborhood requires 8.50×10^4 W of power from the local substation. (a) What is the current in the wire if the electricity is delivered at $22\overline{0}$ V? (b) What is the current in the wire if the electricity is delivered at $600\overline{0}$ V? (c) If the resistance of the power line is 0.250 Ω, how much electric power will be converted into heat throughout the $22\overline{0}$-V and $600\overline{0}$-V power lines? (d) Which power line voltage is the more efficient for transmitting electricity?

4. A 65.5-V, 60.0-Hz ac generator is connected to a radio circuit. (a) If the inductor coil has an inductive reactance of 125 Ω, find the inductance of the coil. (b) What is the effective current in the circuit? (c) Find the power of the circuit.

5. An AM radio tuner circuit has an inductance of 275 mH. (a) What is the capacitance of the variable capacitor when tuned to $88\overline{0}$ kHz? (b) What is the capacitance when tuned to 1010 kHz? (c) Are the capacitor plates larger or smaller when the radio is tuned to a higher-frequency station?

CHAPTER 20

LIGHT

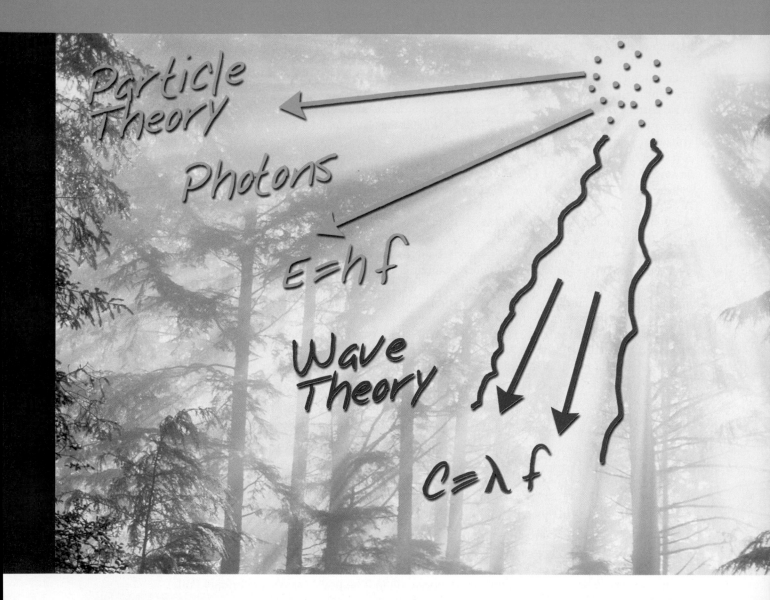

For centuries scientists have sought to explain the nature of light—why it is reflected and why it is refracted. Even the speed of light has been the subject of study since the seventeenth century.

Light seems to exhibit some characteristics of both particles and waves. The study of the measurement of light is photometry. We begin by examining the nature of light.

Objectives

The major goals of this chapter are to enable you to:

1. Describe the nature of light.
2. Solve problems involving the speed of light.
3. Contrast the wave and particle characteristics of light.
4. Apply principles of photometry to technical problems.

20.1 Nature of Light

Light may be defined as radiant energy that can be seen by the human eye. The search for an explanation for the nature of light has been going on for many centuries. A number of famous scientists, including Isaac Newton, Christiaan Huygens, Albert Einstein, and Louis de Broglie, made major contributions to the current theory of light and its interaction with matter.

Many of the foundations for modern scientific theories were developed in the seventeenth century. Two conflicting theories for the nature of light were proposed. The experimental observations that had to be explained were the following:

1. *Straight-line propagation of light.* An application of this property of light is found in survey work, in which sight lines are commonly used (Fig. 20.1).
2. ***Reflection***, *or turning back, of all or a part of a beam of light at the boundary between two different media* [Fig. 20.2]. Examples are the reflection of light by a mirror and by the surface of still water.
3. ***Refraction***, *or bending, of light as it passes through the boundary between two media, such as air and water* [Fig. 20.3]. This bending of light makes objects under water appear to be closer to the surface and farther away from the observer than they really are when viewed from above the surface. It also makes a straight object partially submerged in water appear bent at the surface.

Figure 20.1 Surveyor using the straight-line travel characteristic of light

Photo courtesy of Corbis Corporation. Reprinted with permission

Figure 20.2 Reflection of light by the surface of the water

Figure 20.3 Refraction of light as it passes between media

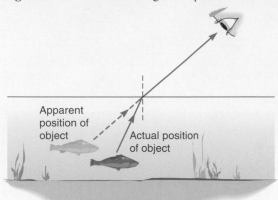

Apparent position of object

Actual position of object

TRY THIS ACTIVITY

Aquarium Tricks

An aquarium is a fun place to observe the refraction of light. The next time you are at an aquarium or have the opportunity to look into a small fish tank, observe an object from above the water line and note its apparent location (Fig. 20.4). Then, attempt to locate the same object by looking from the side of the aquarium. Note the difference in the apparent location of the object. Move your head from side to side and notice how the apparent position changes.

Figure 20.4 Note the apparent positions of the same light-colored stingray when viewed looking down into the aquarium from above the water line and when viewed from the side of the aquarium.

Christiaan Huygens (1629–1695),

physicist and astronomer, was born in The Netherlands. He made the first pendulum clock, proposed the wave theory of light, discovered the polarization of light, and discovered the ring and fourth satellite of Saturn in 1655.

The two conflicting theories referred to above are the wave theory and the particle theory of light. **Christiaan Huygens** proposed the **wave theory** of light, according to which light consists of waves traveling out from light sources like water waves traveling out from the point at which a stone is dropped into still water. The waves continue long after the stone drops to the bottom and thus the succeeding waves cannot be caused by the stone's continuing activity. Huygens developed a geometric model as shown in Fig. 20.5 to explain how a wave front advances and reasoned the following important concept named after him:

Huygens' principle: *Each point on a wave front can be regarded as a new source of small wavelets, which form succeeding waves that spread out uniformly in the forward direction at the same speed.*

Figure 20.5 Huygens' principle is used to determine the future position of both (a) an advancing spherical wave front *CD* and (b) a plane wave front *CD* when *AB* is given.

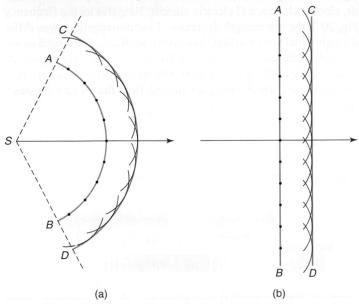

(a)   (b)

The new wave front is the result of all the wavelets. A similar type of wave behavior is sound propagation out from a source of sound. All three types of waves mentioned here—light, water, and sound—travel in straight lines. Waves also reflect and refract when they encounter new media.

Isaac Newton proposed the **particle theory** of light as an alternative explanation of the experimental observations. Newton thought light was made up of streams of particles. These particles of light, which Newton referred to as *corpuscles*, he felt behaved in a manner similar to particles of matter, which travel in a straight line if the net force acting on them is zero. Particles of matter also rebound or reflect off surfaces. An example of this is a rubber ball bouncing off a wall. A change in the velocity of a particle could produce a change in direction. Newton's particle theory persisted until the early nineteenth century, when diffraction and interference of light were observed. Since these properties could only be explained by the wave theory, the particle theory fell out of favor.

A new theory for the nature of light emerged as physicists began to consider light as an oscillating disturbance of an electric field and a corresponding magnetic field. It was discovered that an **electromagnetic wave** consists of two perpendicular transverse waves with one component of the wave being a vibrating electric field and the other being a corresponding vibrating magnetic field; the electromagnetic wave moves in a direction perpendicular to both electric and magnetic field components as shown in Fig. 20.6. All such

Figure 20.6 The electric and magnetic field components of an electromagnetic wave are perpendicular to each other as well as to the direction of travel of the electromagetic wave.

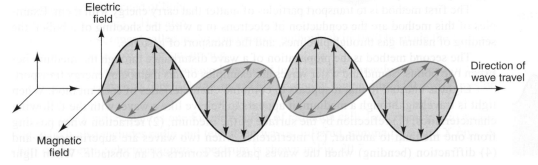

9. Find the energy of a red photon.
10. Find the energy of a blue photon.
11. Find the energy of a yellow photon if the wavelength of its electromagnetic radiation is midway between those of red and blue photons.
12. An AM radio signal has a frequency of 650 kHz. What is the energy of a photon of that electromagnetic radiation?
13. An AM radio station in a nearby town broadcasts a signal with 9.37×10^{-28} J of energy. Find the frequency of this signal in kHz.

PHYSICS CONNECTIONS

Light Amplification by Stimulated Emission of Radiation

Light amplification by stimulated emission of radiation is more commonly known by its acronym, LASER. A **laser** is a light source that produces a narrow beam of light with high intensity and is a good example of how light behaves both as a particle and as a wave (Fig. 20.11). A burst of energy from the laser's flash tube causes an increase in the energy level of some of the electrons in the tube. Immediately following the burst of energy, the energy level of the electrons goes back to its original state. When an electron goes to a lower energy level, it emits a particle of light called a photon. When the emitted photon strikes another atom, it causes a similar event for that atom as well. As this process repeats, more and more photons are generated, causing what is known as "stimulated emission of radiation." The photons located inside the flash tube begin to increase their energy as they travel back and forth and repeatedly reflect off the mirrors on each side of the tube. The increase in the photon energy is called "light amplification." As the energy in the tube increases, a small amount of the coherent laser light escapes out of one of the mirrors that is partially transparent. Coherent light is light in which the crests and troughs of the waves line up and are in synchronization with each other.

The first laser was constructed by T. H. Maiman in 1960. The type of laser described above is the typical ruby laser. Some other common types of lasers are the red helium–neon laser ($\lambda = 633$ nm), the green argon laser ($\lambda = 477$ nm, 488 nm, and 515 nm), and the infrared carbon dioxide laser ($\lambda = 10,600$ nm).

The laser has become an essential tool in virtually all areas of science and technology. In the medical field, lasers are used to destroy cancerous tumors, correct vision defects, and seal capillaries to prevent excessive bleeding during surgery. In manufacturing, lasers are used to measure or align parts for fabrication, drill accurate holes, and weld metals together. The telecommunications and electronics industries use lasers to transmit digital signals in fiber optic cables, create holograms, read bar-codes, and record or read data from CDs, DVDs, and CD-ROMs. Geologists, astronomers, and physicists use lasers to measure great distances or detect minor changes in position.

Figure 20.11   (a) A technician working in a high-tech laser laboratory (b) The inside of a ruby laser.

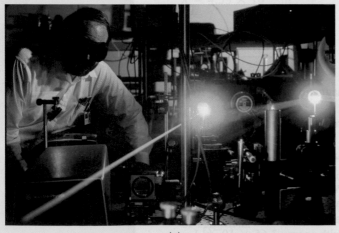

Copyright of Stock Boston. Photo reprinted with permission

(a)

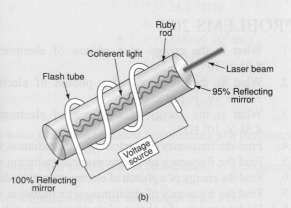

(b)

20.5 Photometry

Recall that light is produced along with other forms of radiation when substances are heated, like our greatest source of natural light, the sun. Light may also be produced by electrons bombarding molecules in a gas such as neon or chemically, as by fireflies. Our most common source of artificial light, however, is the incandescent lamp.

Incandescent lamps (like light bulbs) produce light by the heating of a material (the filament) and the giving off of a wide range of radiation in addition to visible light. Objects that produce light are called *luminous*. On the other hand, objects like the moon, which are not producers of light but only reflect light from another source, are called *illuminated*. When light strikes the surface of most objects, some light is reflected, some is transmitted, and some is absorbed.

The study of the measurement of light is called **photometry**. Two important measurable quantities in photometry are the luminous intensity, I, of a light source, and the illumination, E, of a surface.

Luminous intensity is a measure of the brightness of a light source. The unit for luminous intensity, I, is the candle or *candela* (cd). The early use of certain candles for standards of intensity led to the name of the unit. We now use a platinum source at a certain temperature as the standard for comparison. Another unit, the *lumen*, ℓm, is often used for the measurement of the intensity of a source. One candle produces 4π lumens (ℓm):

$$1 \text{ candle} = 4\pi \ \ell\text{m}$$

To determine the intensity rating in lumens of a 40-W bulb rated at 35 cd, use the following conversion:

$$35 \text{ cd} \times \frac{4\pi \ \ell\text{m}}{1 \text{ cd}} = 440 \ \ell\text{m}$$

Thus, a 40-W light bulb rated at 35 cd has an intensity rating of 440 ℓm.

Illumination is the amount of luminous intensity per unit area. The illumination on a surface may be varied by either changing the luminous intensity of the source (using a brighter bulb) or changing the position of the source (moving it closer to or farther from the surface to be illuminated). Of course, the illumination is also less if the surface illuminated is slanted and not directly facing the light source.

The amount of illumination on a surface varies inversely with the square of the distance from the source. For example, if the distance of the illuminated surface from the source is doubled, the illumination is reduced to one-fourth of its former intensity. This can be illustrated by considering a point source of light at the center of concentric (having the same center) spheres (Fig. 20.12). The solid angle is measured in units called *steradians*.

If the source radiates light uniformly in all directions, the light is uniformly distributed over a spherical surface centered at the source. Since the surface area of a sphere is $4\pi r^2$, the illumination, E, at the surface is given by

$$E = \frac{I}{4\pi r^2}$$

where E = illumination
$\quad\quad I$ = luminous intensity of the source (in lumens)
$\quad\quad r$ = distance between the source and the illuminated surface

The unit of illumination, E, is the *lux*:

$$1 \text{ lux} = 1 \ \ell\text{m/m}^2$$

We assume the surface being illuminated is perpendicular to the source.

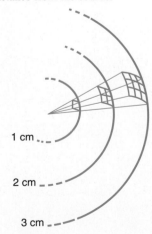

Figure 20.12 The amount of illumination on a surface varies inversely with the square of the distance from the source.

1 cm

2 cm

3 cm

Note that the inverse-square-law behavior of light is similar to the two inverse square laws studied earlier, those of gravity and of the electric force between charges described by Coulomb's law.

A common application of the measure of illumination is the rating given to a camcorder, which measures the sensitivity of the instrument.

EXAMPLE 1

Find the illumination E on a surface located 2.00 m from a source with an intensity of $40\overline{0}$ ℓm.

Data:

$$I = 40\overline{0} \ \ell m$$
$$r = 2.00 \ m$$
$$E = ?$$

Basic Equation:

$$E = \frac{I}{4\pi r^2}$$

Working Equation: Same

Substitution:

$$E = \frac{40\overline{0} \ \ell m}{4\pi(2.00 \ m)^2}$$
$$= 7.96 \ \ell m/m^2 \quad or \quad 7.96 \ lux$$

. . . .

The unit used for illumination is the lux ($\ell m/m^2$) in the metric system as shown above and the foot-candle ($\ell m/ft^2$) in the U.S system:

$$1 \ ft\text{-}candle = 1 \ \ell m/ft^2$$

EXAMPLE 2

Find the illumination 4.00 ft from a source with an intensity of $60\overline{0}$ ℓm.

Data:

$$I = 60\overline{0} \ \ell m$$
$$r = 4.00 \ ft$$
$$E = ?$$

Basic Equation:

$$E = \frac{I}{4\pi r^2}$$

Working Equation: Same

Substitution:

$$E = \frac{60\overline{0} \ \ell m}{4\pi(4.00 \ ft)^2}$$
$$E = 2.98 \frac{\ell m}{ft^2} \times \frac{1 \ ft\text{-}candle}{1 \ \ell m/ft^2}$$
$$= 2.98 \ ft\text{-}candles$$

$$\frac{\ell m}{ft^2} \times \frac{1 \text{ ft-candle}}{1 \ell m/ft^2} = \frac{\ell m}{ft^2} \times 1 \text{ ft-candle} \times \frac{ft^2}{\ell m}$$
$$= \text{ft-candle}$$

....

In photography, photoelectric cells are used in light meters or exposure meters to measure illumination for taking photographs. The electricity produced is proportional to the illumination and is directly calibrated on the instrument scale. The units of measurement of such meters, however, are not standardized, and the scale may be arbitrarily selected.

PROBLEMS 20.5

1. $I = 48.0$ cd
   $I = \_\_\_ \ell m$

2. $I = 342$ cd
   $I = \_\_\_ \ell m$

3. $I = 765 \ell m$
   $I = \_\_\_$ cd

4. $I = 432 \ell m$
   $I = \_\_\_$ cd

5. $I = 75.0$ cd
   $I = \_\_\_ \ell m$

6. $I = 650 \ell m$
   $I = \_\_\_$ cd

7. $I = 900 \ell m$
   $r = 7.00$ ft
   $E = ?$

8. $I = 741 \ell m$
   $r = 6.50$ m
   $E = ?$

9. $I = 893 \ell m$
   $r = 3.25$ ft
   $E = ?$

10. $E = 4.32$ lux
    $r = 9.00$ m
    $I = ?$

11. $E = 10.5$ ft-candles
    $r = 6.00$ ft
    $I = ?$

12. Find the intensity of a light source that produces an illumination of 5.50 ft-candles at 9.85 ft from the source.

13. Find the intensity of a light source that produces an illumination of 2.39 lux at 4.50 m from the source.

14. Find the intensity of a light source that produces an illumination of 5.28 lux at 6.50 m from the source.

15. If an observer triples her distance from a light source:
    (a) Does the illumination at that point increase or decrease?
    (b) In what proportion does the illumination increase or decrease?

16. If the illuminated surface is slanted at an angle of 35.0°, what part of the full-front illumination is lost?

17. Find the illumination on a surface by three light sources, each with intensity 150 ℓm, located at distances of 2.00 m, 2.70 m, and 2.98 m from the surface, respectively.

18. Find the intensity of two identical light sources located 1.40 m and 1.96 m, respectively, from a point where the illumination is 3.54 $\ell m/m^2$.

19. Find the intensity of two identical light sources located 0.880 m and 1.12 m from a point where the illumination is 5.86 $\ell m/m^2$.

20. A desk is 3.35 m below an 1850-ℓm incandescent lamp. What is the illumination on the desktop?

SKETCH

12 cm² | w

4.0 cm

DATA

$A = 12$ cm², $l = 4.0$ cm, $w = ?$

BASIC EQUATION

$A = lw$

WORKING EQUATION

$w = \frac{A}{l}$

SUBSTITUTION

$w = \frac{12 \text{ cm}^2}{4.0 \text{ cm}} = 3.0$ cm

Glossary

Electromagnetic Theory Describes the interrelationship between electric fields, magnetic fields, electric charge, and electric current. (p. 558)

Electromagnetic Wave A wave consisting of two perpendicular transverse waves with one component of the wave being a vibrating electric field and the other component being a corresponding vibrating magnetic field; the electromagnetic wave moves in a direction perpendicular to both electric and magnetic field components. (p. 557)

Frequency The number of complete vibrations or cycles per second of a wave. (p. 561)

Huygen's Principle Each point on a wave front can be regarded as a new source of small wavelets, which form succeeding waves that spread out uniformly in the forward direction at the same speed. (p. 556)

Illumination The luminous intensity per unit area. (p. 565)

Laser A light source that produces a narrow beam with high intensity. An acronym for "light amplification by stimulated emission of radiation." (p. 564)

Light Radiant energy that can be seen by the human eye. (p. 555)

Light-Year The distance that light travels in one earth year: 9.45×10^{15} m or 5.87×10^{12} mi. (p. 560)

Luminous Intensity A measure of the brightness of a light source. (p. 565)

Particle Theory Theory that light consists of streams of particles. (p. 557)

Photoelectric Effect The emission of electrons by a surface when struck by electromagnetic radiation. (p. 558)

Photometry The study of the measurement of light. (p. 565)

Photons Wave packets of energy that carry light and other forms of electromagnetic radiation. (p. 558)

Planck's Constant A fundamental constant of quantum theory (6.626×10^{-34} J s). (p. 563)

Quantum Theory Theory initiated by Planck and Einstein that energy, including electromagnetic radiation, is radiated or absorbed in multiples of certain units of energy. (p. 558)

Reflection The turning back of all or part of a beam of light at the boundary between two different media. (p. 555)

Refraction The bending of light as it passes through the boundary between two media, such as air and water. (p. 555)

Speed of Light The speed at which light and other forms of electromagnetic radiation travel. Equal to 3.00×10^8 m/s in a vacuum. (p. 559)

Wave Theory Theory that light consists of waves traveling out from light sources, like water waves traveling out from the point at which a stone is dropped into water. (p. 556)

Wavelength The distance between two successive corresponding points on a wave. (p. 561)

Formulas

20.2 $s = ct$

20.3 $c = \lambda f$

20.4 $E = hf$

20.5 $E = \dfrac{1}{4\pi r^2}$

Review Questions

1. Which of the following are examples of electromagnetic radiation?
   - (a) Gamma rays
   - (b) Sound waves
   - (c) Radio waves
   - (d) Water waves
   - (e) Visible light

2. The particle theory of light explains
    (a) diffraction of light around a sharp edge.
    (b) refraction of light at a boundary.
    (c) the photoelectric effect.
    (d) none of the above.
3. A light-year equals
    (a) the time it takes light to travel from the sun to the earth.
    (b) the distance from the sun to the earth.
    (c) the distance to the nearest star other than the sun.
    (d) the distance light travels in one earth year.
4. Light behaves
    (a) as a massive particle.
    (b) always as a wave.
    (c) sometimes as a wave, sometimes as a particle.
    (d) as none of the above.
5. Does the wavelength of light depend on its frequency? Explain.
6. How does the energy of a photon of light depend on its frequency?
7. How does the intensity of illumination depend on the distance from a source radiating uniformly in all directions?
8. In your own words, explain how the speed of light has been measured.
9. Does light always travel at the same speed? Explain.
10. What name is given to the entire range of waves that are similar to visible light?
11. Who proposed the particle theory of light?
12. Who developed the wave packet theory of light?
13. Who made the first estimate of the speed of light?
14. How was the first estimate of the speed of light made?
15. What are the units of luminous intensity?
16. In your own words, explain luminous intensity.

Review Problems

1. Find the distance (in metres) traveled by a radio wave in 21.5 h.
2. A radar wave that is bounced off an airplane returns to the radar receiver in 3.78×10^{-5} s. How far (in miles) is the airplane from the radar receiver?
3. How long does it take for a police radar beam to travel to a car and back if the car is 0.245 mi from the radar unit?
4. How long does it take for a pulse of laser light to return to a police speed detector after bouncing off a speeding car 0.274 mi away?
5. How long does it take for a radio signal to travel from the earth to a communications satellite 22,500 mi above the surface of the earth?
6. Find the wavelength of a radio wave from an AM station broadcasting at a frequency of 1230 kHz.
7. Find the frequency of a radio wave if its wavelength is 46.5 m.
8. Find the frequency of a light wave if its wavelength is 5.415×10^{-8} m.
9. What is the energy of a photon with frequency 1.45×10^{11} Hz?
10. What is the frequency of a photon with energy of 4.75×10^{-23} J?
11. What is the energy of a photon with frequency 8.25×10^{-15} Hz?
12. Find the intensity of the light source necessary to produce an illumination of 3.75 ft-candles at 6.75 ft from the source.
13. Find the intensity of the light source necessary to produce an illumination of 4.86 lux at 9.25 m from the source.
14. What is the intensity of the light source required to produce the illumination of Problem 13 if the distance from the light source is doubled?

SKETCH

12 cm² | w

4.0 cm

DATA

$A = 12$ cm², $l = 4.0$ cm, $w = ?$

BASIC EQUATION

$A = lw$

WORKING EQUATION

$w = \frac{A}{l}$

SUBSTITUTION

$w = \frac{12 \text{ cm}^2}{4.0 \text{ cm}} = 3.0$ cm

15. What are the maximum and minimum transit times for light traveling from Jupiter to Mars? The orbital radii are 215 million kilometres for Mars and 725 million kilometres for Jupiter. Assume the planetary orbits are circular. Also make the (nonphysical) assumption that the sun is transparent to the transmission of light between the planets.

16. Find the intensity of two identical light sources located 0.454 m and 0.538 m, respectively, from a point where the illumination is 8.46 $\ell m/m^2$.

17. Find the illumination on a surface by three light sources, each with intensity 125 ℓm, located at 1.85 m, 1.92 m, and 2.43 m from the surface, respectively.

APPLIED CONCEPTS

1. The distance between New York City and London is 3470 mi. (a) If a radio wave from New York City is transmitted directly across the ocean, how long will it take to reach a receiver in London? (b) In fact, due to the curvature of the earth, radio waves cannot be transmitted directly across such large distances on the earth. Instead, a signal is typically transmitted to a communications satellite located 2.20×10^4 mi above the surface of the earth. If the satellite is located midway between the two cities, how long will the radio wave take to reach London? (Ignore the effect that the curvature of the earth has on the calculations.)

2. (a) When the Apollo astronauts landed on the moon, it took the radio signal 1.28 s to reach Mission Control on the earth. How far away were the astronauts from the earth? (b) Since the sun is 1.50×10^{11} m from the earth, how much time does it take for light to travel from the sun to the earth? (c) Light from our next closest star, Alpha Centauri, takes 4.31 years to reach the earth. How far away from the earth is Alpha Centauri?

3. The range of electromagnetic wave frequencies on the FM radio band is 88.0 MHz to 108 MHz. (a) What is the range of wavelengths for the FM radio band? (b) What is the range of photon energy for the FM radio band? (c) Explain the relationship between frequency and photon energy.

4. The individual rods on rooftop antennas are designed to be one-quarter of a wavelength for each television frequency. What is the range of rod lengths needed for television Channels 2 through 6 if their frequencies are between 54.0 MHz and 88.0 MHz?

5. An illumination of 180 lux on student desks and other work areas is the standard that architects use when designing lighting systems for schools. If a ceiling is 2.50 m above a student work area, what intensity light source must be installed? (b) If the ceiling were twice its original height, what light intensity would be needed to meet the standard requirement?

CHAPTER 21

REFLECTION AND REFRACTION

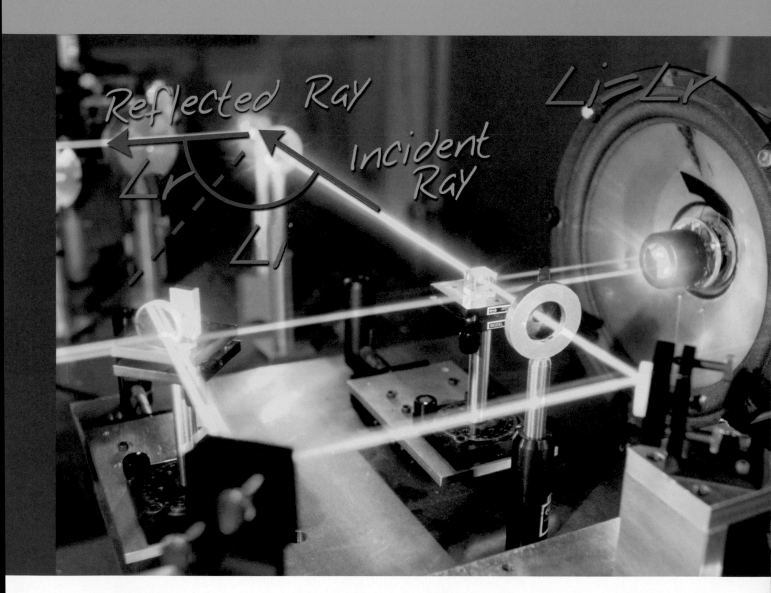

The nature of light may still be somewhat of a mystery. However, its characteristics have been the subject of intensive study for hundreds of years. Light may be transmitted, reflected, or absorbed by a medium.

Anyone wearing glasses can appreciate the refraction of light as it bends upon passing from one medium to another. The index of refraction is a tool that the scientist uses to describe the ability of certain substances to bend light as it passes through them.

Our examination of the behavior of light begins with the study of images and reflection.

Objectives

The major goals of this chapter are to enable you to:

1. Describe the laws of reflection.
2. Locate and describe images formed by plane, convex, and concave mirrors.
3. Apply the mirror formula to image formation.
4. Describe the law of refraction.
5. Describe total internal reflection.
6. Locate and describe images formed by converging and diverging lenses.

21.1 Mirrors and Images

Although the nature of light is complex, we know very well how light behaves. Every day we experience and unconsciously use our knowledge of what light does and depend on the ways it works. **Reflection** is the turning or turning back of all or a part of a beam of light as it strikes a surface. Unlike sound, light does not require a medium (some kind of matter) to travel through and may be transmitted through empty space. When light does strike a medium, the light may be reflected, absorbed, transmitted, or undergo a combination of the three.

Mirrors show how light may be reflected. Any dark cloth shows how light may be absorbed. Window glass illustrates how light may be transmitted through a medium.

A medium can be classified according to how well light is transmitted through it:

1. *Transparent:* almost all light passes through
   Examples: window glass, clear water
2. *Translucent:* some but not all light passes through
   Examples: murky water, light fog, skylight panels for farm buildings, stained glass
3. *Opaque:* almost all light reflected or absorbed
   Examples: wood, metal, plaster

These classifications are relative because for any medium, some light is reflected from the surface and some passes into or through it.

In studying reflection we observe what happens when light is turned back from a surface. The beam of a flashlight directed at a mirror shows several things about reflection. First, upon striking the surface of the glass, some of the light is reflected in all directions. This is called *scattering*. If there were no scattering, no light would reach our eye and we would be unable to observe the beam at all. However, only a very small part of the beam of light is scattered. Rough or uneven surfaces produce more scattering than do smooth ones. This scattering of light by uneven surfaces is called **diffusion**. Diffused lighting has many applications at home and in industry where bright glare is not desirable.

Nearly complete reflection (with very little scattering) is called **regular** (or specular) **reflection**. Regular reflection occurs when parallel or nearly parallel rays of light (such as sunlight and spotlight beams) remain parallel after being reflected from a surface (Fig. 21.1). Note that the incoming rays are referred to as *incident rays*.

Figure 21.1 Regular reflection (Reflected rays are parallel.)

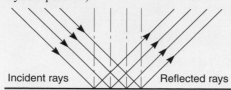

Incident rays   Reflected rays

Figure 21.2 On a regular surface, the reflected rays leave at the same angle as the incident rays.

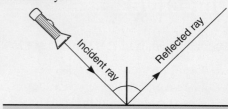

A flashlight beam on a mirror in a darkened room also shows something else about light striking a regular reflecting surface: The reflected rays of light leave the surface at the same angle at which the incident (incoming) rays strike the surface (Fig. 21.2). Expressed another way, the angles measured from the normal (the perpendicular) to the reflecting surface are equal. These angles are shown in Fig. 21.3.

Figure 21.3 The angle of incidence is equal to the angle of reflection.

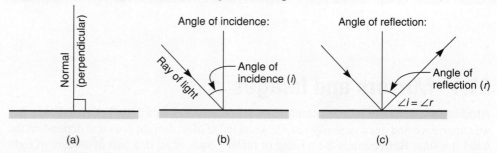

(a)   (b)   (c)

The same principle applies to curved surfaces (Fig. 21.4). This behavior of light rays is defined by the following law:

FIRST LAW OF REFLECTION

The angle of incidence, *i*, is equal to the angle of reflection, *r*; that is,

$$\angle i = \angle r$$

Figure 21.4 Likewise, for curved surfaces, $\angle i = \angle r$.

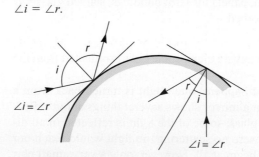

Further observation of the light beam readily shows a second law:

SECOND LAW OF REFLECTION

The incident ray, the reflected ray, and the normal (perpendicular) to the surface all lie in the same plane.

These laws of reflection apply not only to light, but to all kinds of waves.

Figure 21.5 Spherical mirrors.

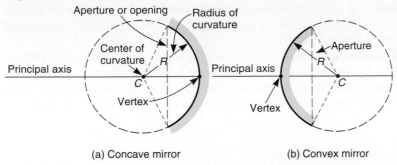

(a) Concave mirror   (b) Convex mirror

We look next at how images are formed by three widely used kinds of mirrors: plane, concave, and convex. **Plane mirrors** are flat. **Concave mirrors** are curved away from the observer [like the inside of a bowl; Fig. 21.5(a)] and **convex mirrors** are curved toward the observer [like ball-shaped Christmas tree ornaments; Fig. 21.5(b)].

Mirrors of glass or any highly reflecting surface have countless practical applications, from rear-view mirrors in automobiles to mirrors to watch for shoplifting in stores (Fig. 21.6).

Figure 21.6 Convex security mirrors are used to watch for shoplifting because they reflect images from over wide areas.

For convenience we will use spherical mirrors (reflecting surfaces that are sections of spheres), although parabolic mirrors (reflecting surfaces in the shape of a parabola; Fig. 21.7) have wider practical use.

We consider next how images are formed by plane, concave, and convex mirrors. Images formed by mirrors may be **real images** (images formed by rays of light) or **virtual images** (images that only appear to the eye to be formed by rays of light).

Real images made by a single mirror are always inverted (upside down) and may be larger than, smaller than, or the same size as the object. They can be shown on a screen. Virtual images are always erect and may be larger than, smaller than, or the same size as the object. They cannot be shown on a screen.

Figure 21.7 Parabolic mirror.

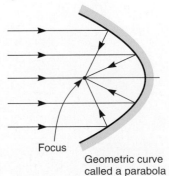

Focus

Geometric curve called a parabola

21.2 Images Formed by Plane Mirrors

Plane mirror images are always erect and virtual and appear as far behind the mirror as the distance the object is in front of the mirror. Note that plane mirrors also reverse right and left, so the right hand held in front of a plane mirror appears in the mirror to be the left hand.

Look at your image in a mirror. Then look at something on the mirror's surface like a speck of dirt. Notice that you have to adjust your eyes and refocus from looking at the image of your face to looking at the mirror surface. It should be clear that the image is not on the surface of the mirror but behind it.

We use light-ray diagrams to illustrate how our eyes see images in the various kinds of mirrors. We do this by representing rays of light and lines of sight with straight lines (Fig. 21.8). This method is used to construct diagrams and locate the images formed. Simply view the object from two or more separate places and construct the light-ray lines.

Figure 21.8

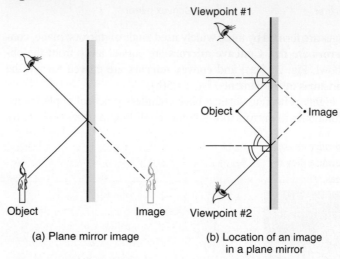

(a) Plane mirror image

(b) Location of an image in a plane mirror

21.3 Images Formed by Concave Mirrors

Find a shiny tablespoon and look at your image in it (Fig. 21.9). Now turn it over and look again. The images are very different; one is erect and the other, inverted (upside down). We use ray diagrams to show why this happens. As we shall see, the kind of image produced depends on the location of the object with respect to the mirror.

Figure 21.9 Reflected images as seen on opposite sides of a large spoon.

Photo courtesy of Visuals Unlimited. Reprinted with permission

Figure 21.10 Spherical mirror.

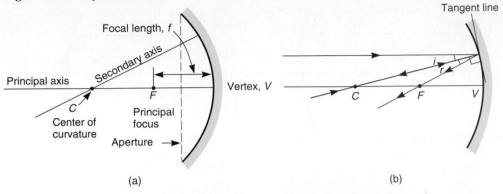

(a)   (b)

We need to define some terms before we discuss how images are formed. Figure 21.10(a) shows a spherical mirror with the key terms identified. The *center of curvature*, *C*, is the center of the sphere that forms a part of the spherical mirror. The *vertex*, *V*, is the center of the mirror (sometimes called its optical center). The *principal axis* is the line *CV* drawn through the center of curvature and the vertex. The *principal focus*, *F*, is the point on the principal axis through which all rays parallel to the principal axis converge in a concave mirror as shown in Fig. 21.10(b) or from which they diverge in a convex mirror. Note that any ray through the center of curvature is along a radius of a spherical mirror and is reflected straight back because any such ray is perpendicular to the surface of the mirror at the point the ray strikes the mirror. The **focal length** is the distance between the principal focus of a mirror (or lens) and its vertex. The *aperture* is that very small portion of the sphere that forms the actual mirror. For mirrors with small apertures, the focal length, *f*, is one-half the radius of curvature, *R*. That is, $f = \dfrac{R}{2}$.

If the object is placed at the focal point, no image will be formed because the rays of light will be reflected parallel to the principal axis (Fig. 21.11). The location of the reflected ray may be found by using the laws of reflection (Fig. 21.12). The angle of incidence is equal to the angle of reflection.

Figure 21.11   No image is formed if the object is located at the focal point.

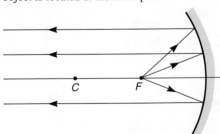

Figure 21.12

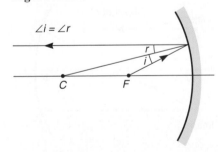

Now consider the more common case, in which the object is beyond the center of curvature [for example, looking into a tablespoon as in Fig. 21.9, whose mirror diagram is shown in Fig. 21.13(a)]. Note that again we use the fact that a ray parallel to the principal axis is reflected through the focal point and a ray through the focal point is reflected parallel to the principal axis. Then, where the two rays intersect, a point on the image is formed. (In this case, it is the flame of the candle. Note that the candle base image lies on the principal axis because the object base is also on that line.) The same method is used for the case where the candle base extends below the principal axis [Fig. 21.13(b)].

Figure 21.13 Images formed in spherical mirrors

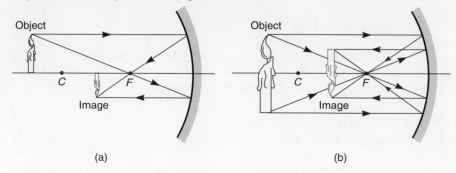

(a)   (b)

Now apply these principles to the diagrams of other images formed by concave mirrors (Fig. 21.14). Decide whether the image is real or virtual; erect or inverted; larger, smaller, or the same size; and where it is located. Note that the only time a virtual image is produced is when the object is between the focal point and the mirror [Fig. 21.14(f)]. The construction of the diagram is the same as the other cases except that the light rays are converging (coming together) and must be extended behind the mirror (where the image appears to be) forming the virtual image.

Figure 21.14 Formation of images in concave mirrors by objects that are located in various positions relative to the center of curvature.

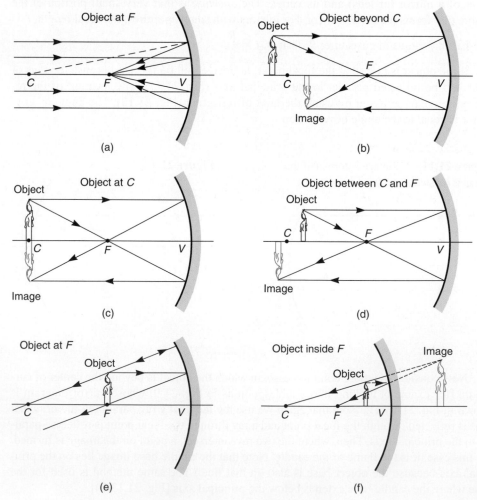

(a)   (b)

(c)   (d)

(e)   (f)

21.4 Images Formed by Convex Mirrors

By looking into the back side of our tablespoon in Fig. 21.9, we see an erect, virtual, smaller image. Use the mirror diagram shown in Fig. 21.15 to see how such an image is formed. Curved surface mirrors are used in some telescopes, spotlights, and automobile headlights. However, because spherical mirrors produce clear images over only a very small portion of their surfaces, the surfaces used commercially are another geometric shape, that of a parabola, or parabolic, as mentioned before. For apertures wider than about 10°, spherical mirrors produce fuzzy images because all parallel rays are not reflected through the focal point. This is called *spherical aberration.*

Figure 21.15 Formation of images in convex mirrors.

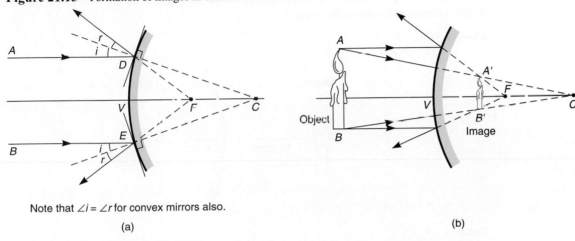

Note that $\angle i = \angle r$ for convex mirrors also.

(a) (b)

Two common applications of convex mirrors are as security devices in convenience stores and on some automobile rearview mirrors because of the wide-angle view they produce.

21.5 The Mirror Formula

As we might expect from the previous cases, the focal length, the distance from the object to the mirror, and the distance from the image to the mirror are all related (Fig. 21.16). This relationship can be expressed as the *mirror formula:*

Figure 21.16 The mirror formula is expressed in terms of f, s_o, and s_i.

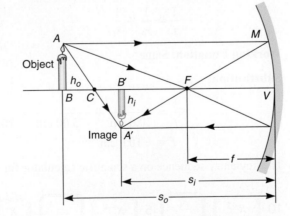

$$\frac{1}{f} = \frac{1}{s_o} + \frac{1}{s_i}$$

where f = focal length of mirror
s_o = distance of object from mirror
s_i = distance of image from mirror

Therefore, if two of the three distances f, s_o, and s_i are known, the third can be found.

A second formula shows the magnification of the mirror and how the height of the object and the height of the image depend on the object distance and the image distance:

$$M = \frac{h_i}{h_o} = \frac{-s_i}{s_o}$$

where M = magnification
h_i = image height
h_o = object height
s_i = image distance
s_o = object distance

In using *both* of the preceding formulas for concave and convex mirrors, remember that the distance to a virtual image is always negative; similarly, the focal length of a convex mirror is also negative. An inverted image has a negative magnification and an erect image has a positive magnification.

EXAMPLE

An object 10.0 cm in front of a convex mirror forms an image 5.00 cm behind the mirror. What is the focal length of the mirror?

Sketch:

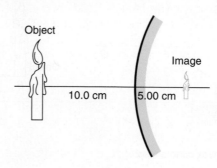

Object

Image

10.0 cm   5.00 cm

Data:

$$s_o = 10.0 \text{ cm}$$
$$s_i = -5.00 \text{ cm}$$

Note: The image is virtual (appears behind the mirror) so s_i is given a (−) sign to show this. [Won't f also be (−)?]

$$f = ?$$

Basic Equation:

$$\frac{1}{f} = \frac{1}{s_o} + \frac{1}{s_i}$$

Working Equation: Same

Substitution:

$$\frac{1}{f} = \frac{1}{10.0 \text{ cm}} + \frac{1}{-5.00 \text{ cm}} = \frac{1}{10.0 \text{ cm}} - \frac{1}{5.00 \text{ cm}}$$
$$f = -10.0 \text{ cm}$$

The key entry sequence on a scientific calculator for this calculation is

10 $\boxed{x^{-1}}$ $\boxed{-}$ 5 $\boxed{x^{-1}}$ $\boxed{=}$ $\boxed{x^{-1}}$ $\boxed{=}$

$$\boxed{-10}$$

Remember that f and s_i may be negative only when forming virtual images and/or using convex mirrors.

PROBLEMS 21.5

Use the formulas $\dfrac{1}{f} = \dfrac{1}{s_o} + \dfrac{1}{s_i}$ and $M = \dfrac{h_i}{h_o} = \dfrac{-s_i}{s_o}$ for Problems 1–8.

1. Given $s_o = 1.65$ cm and $s_i = 6.00$ cm, find f.
2. Given $f = 15.0$ cm and $s_i = 3.00$ cm, find s_o.
3. Given $s_i = 14.5$ cm and $f = 10.0$ cm, find s_o.
4. Given $s_i = -10.0$ cm and $f = -5.00$ cm, find s_o.
5. Given $s_o = 7.35$ cm and $s_i = 17.0$ cm, find f.
6. Given $h_i = 2.75$ cm, $h_o = 4.50$ cm, and $s_i = 6.00$ cm, find s_o.
7. Given $h_o = 12.0$ cm, $s_i = 13.0$ cm, and $s_o = 25.0$ cm, find h_i.
8. Given $h_i = 3.50$ cm, $h_o = 2.50$ cm, and $s_i = 15.5$ cm, find s_o.
9. If an object is 2.50 m tall and 8.60 m from a large mirror with an image formed 3.75 m from the mirror, find the height of the image.
10. An object 30.0 cm tall is located 10.5 cm from a concave mirror with focal length 16.0 cm. (a) Where is the image located? (b) How high is it?
11. An object and its image in a concave mirror are the same height, yet inverted, when the object is 20.0 cm from the mirror. What is the focal length of the mirror?
12. An object 12.6 cm in front of a convex mirror forms an image 6.00 cm behind the mirror. What is the focal length of the mirror?
13. What is the height of an image in a truck mirror when the object is 1.20 m tall and is standing 7.60 m from the mirror and the image is formed 1.50 m from the mirror?
14. A lift truck has a rear view mirror that is 0.76 m from the eye of its driver. How tall is the image of a 1.40-m-tall loaded pallet that is 6.20 m from the mirror?
15. Find the focal length of a convex mirror that forms an image 3.55 cm behind the mirror of an object 24.5 cm in front of the mirror.
16. Find the focal length of a mirror that forms an image 5.66 cm behind a mirror of an object 34.4 cm in front of the mirror.
17. Find the focal length of a mirror that forms an image 2.30 m behind a mirror of an object 6.50 m in front of the mirror.
18. An image of a statue appears to be 11.5 cm behind a convex mirror with focal length 13.5 cm. Find the distance from the statue to the mirror.
19. (a) What is the height of a figurine 7.33 cm in front of a concave mirror that produces an image -2.75 cm high? The image appears to be 5.03 cm in front of the mirror. (b) Find the focal length of the mirror. (c) What distance would an image appear to be from the mirror with double the focal length?

SKETCH

| 12 cm^2 | w |

4.0 cm

DATA

$A = 12$ cm^2, $l = 4.0$ cm, w = ?

BASIC EQUATION

$A = lw$

WORKING EQUATION

$w = \dfrac{A}{l}$

SUBSTITUTION

$w = \dfrac{12 \text{ cm}^2}{4.0 \text{ cm}} = 3.0$ cm

21.6 The Law of Refraction

Does light travel in a straight line? Most people's first answer would be, "Yes." But does it always? See Fig. 21.17. Why does a straw appear to bend at the surface when placed in water as in Fig. 21.17(b)?

TRY THIS ACTIVITY

Bent Pencil

Place a pencil in a glass of water so that it rests against the side of the glass. Observe the glass from a variety of angles. Why does the pencil appear to be broken or bent when viewed at some angles and straight when viewed at other angles?

Figure 21.17 Refraction is the bending of light as it passes from one medium to another.

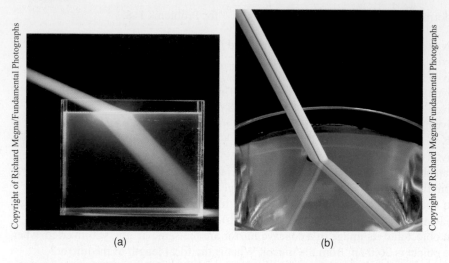

(a)

(b)

The answers to our questions may be found in the study of another property of light—refraction. **Refraction** is the bending of light as it passes at an angle from one medium to another of different optical density. **Optical density** is a property of a transparent material that is a measure of the speed of light through the given material. For example, water is optically denser than air and the speed of light in water is less than the speed of light in air. This change of speed of light when it passes from one medium to another produces refraction. The wave shown in Fig. 21.18 illustrates how this occurs. Note that when passing from one medium to another perpendicular to the surface (called the *interface*) the wave is not bent, although the speed of the wave is slowed. When the light emerges back into the original medium, the wave returns to its original speed.

When the wave passes through at an angle, the entire wave front does not all strike the surface at the same time. The first part of the wave to strike the glass is slowed before the part striking later—thus the bending of the wave (Fig. 21.19). Draw your own diagram to show whether a fish in a pond, as viewed from the bank, is actually nearer the surface or the bottom of the pond than it appears to be. (see Fig. 20.3)

Figure 21.18 The speed of light is different in different media.

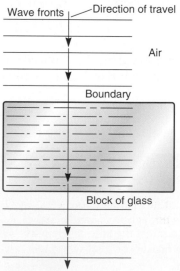

Figure 21.19 The wave bends when all parts of the wave don't strike the glass at the same time. It also bends when leaving the glass.

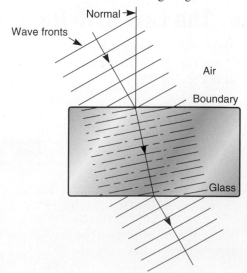

Another way to understand refraction is to compare it to a child's wagon rolling along a sloping sidewalk. The front wheel rolls off the edge and goes into the grass. The interaction of the wheel with the rougher surface, the grass, causes the wagon to change direction and to be pulled toward the grass. The wheel that hits the grass slows down first while the opposite wheel is still rolling on the sidewalk. The wagon pivots, and its path is bent toward the grass.

A mirage is a good example of refraction. It is not a figment of a person's imagination. It is actually formed by light and can be photographed (Fig. 21.20). Light travels faster through the very hot and less dense air near a warm surface such as the ground on a hot day than through the cooler air above it. As a result, the image appears to the observer to shimmer as if it had been reflected from the surface of water. It is not reflected, however, but refracted, forming an image.

As we might expect, since the speed of light increases when a light wave leaves a denser medium to enter the air, bending also occurs. In this case, however, instead of the light bending toward the perpendicular or normal, the light is bent away from the normal when passing from the denser medium to the less dense one (Fig. 21.19). This illustrates the following law:

LAW OF REFRACTION

When a beam of light passes at an angle from a medium of lower optical density to a denser medium, the light is bent *toward* the normal. When a beam of light passes at an angle from a medium of greater optical density to one less dense, the light is bent *away from* the normal.

Figure 21.20 A mirage is actually formed by light and can be photographed.

Copyright of Kent Wood/Photo Researchers, Inc. Photo reprinted with permission

Figure 21.21 The angles of incidence and refraction are measured from the normal.

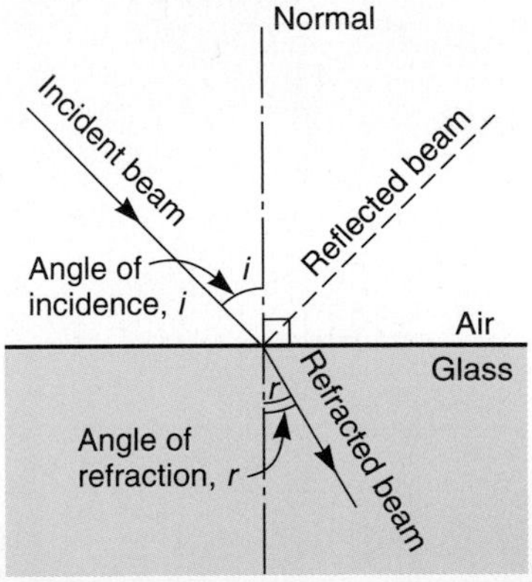

Note: The angles of incidence and refraction are measured from the *normal* as in Fig. 21.21.

Willebord Snell developed a formula to determine the relative optical density of a material, called the **index of refraction**. This index is a constant for a particular material and is independent of the angle at which the light strikes. **Snell's law** may be expressed in two ways:

1. The index of refraction, n, equals the sine of the angle of incidence divided by the sine of the angle of refraction, where the incident ray travels through a vacuum or air:

$$n = \frac{\sin i}{\sin r}$$

Willebord Snell (1580–1626), mathematician, was born in The Netherlands. He discovered the law of refraction known as Snell's law and extensively developed the use of triangulation in surveying.

2. The index of refraction, n, for a given substance is the ratio of the speed of light in a vacuum (nearly the same as in air) to the speed of light in the substance:

$$n = \frac{\text{speed of light in vacuum}}{\text{speed of light in substance}}$$

EXAMPLE 1

The angle of incidence of light passing from air to water is 61.0°. The angle of refraction is 41.0°. What is the index of refraction of the water?

Data:

$$i = 61.0°$$
$$r = 41.0°$$
$$n = ?$$

Basic Equation:

$$n = \frac{\sin i}{\sin r}$$

Working Equation: Same

Substitution:

$$n = \frac{\sin 61.0°}{\sin 41.0°}$$
$$= 1.33$$

EXAMPLE 2

The index of refraction of water is 1.33. What is the speed of light in water?

Data:

$$n = 1.33$$
$$c = 3.00 \times 10^8 \text{ m/s}$$
$$v_{\text{water}} = ?$$

Basic Equation:

$$n = \frac{\text{speed of light in vacuum}}{\text{speed of light in substance}}$$

Working Equation:

$$\text{speed of light in water} = \frac{\text{speed of light in vacuum}}{n}$$

Substitution:

$$\text{speed of light in water} = \frac{3.00 \times 10^8 \text{ m/s}}{1.33}$$
$$= 2.26 \times 10^8 \text{ m/s}$$

The index of refraction for some common substances is given in Table 21.1.

Table 21.1   Indices of Refraction for Various Substances

Substance	Index of Refraction
Air, dry (STP)	1.00029
Alcohol, ethyl	1.360
Benzene	1.501
Carbon dioxide (STP)	1.00045
Carbon disulfide	1.625
Carbon tetrachloride	1.459
Diamond	2.417
Glass, crown flint	1.575
Lucite	1.50
Quartz, fused	1.45845
Water, distilled	1.333
Water vapor (STP)	1.00025
Zircon	1.92

Note: STP indicates standard temperature (0°C) and pressure (101.32 kPa).

TRY THIS ACTIVITY

Twinkling Stars

Look at an object at the bottom of a pool, tank, or shallow area of clear water. If the water is still, the object should appear quite clear; however, if there are ripples in the water, the light from the object becomes irregularly refracted by the shape of the ripples.

This is similar to the explanation of why stars appear to twinkle. Stars themselves really do not twinkle, but they appear to do so when different densities of air get in the way of the light as it comes toward an observer's eye. Instead of a constant beam of light striking the eye, the irregularities in the atmospheric density cause the light to be refracted from time to time, resulting in various intensities of light striking the eye.

TRY THIS ACTIVITY

Apparent Depth

In the previous chapter, a suggested activity involved looking at the apparent location of an object from both above and below the water line (Fig. 21.22). The difference in the apparent location was due to refraction. In this activity, pay particular attention to the location and magnification of the object.

The next time you are at an aquarium, look at something that is protruding out of the water and look at the object from a location where you can see it from above and below the water line. Since water refracts light when the light strikes it at an angle, the submerged portion of the object will appear to be in a different location from each of the two perspectives. Enjoy the wildlife at the aquarium, but also enjoy the physics of optics!

Figure 21.22

(a) A sign and its base as seen from above the water line.

(b) The sign as seen from above the water line and the base as seen from below the water line. Notice the apparent location and size of the sign's base.

21.7 Total Internal Reflection

If the angle of refraction is 90° or greater, a beam of light does not leave the medium but is reflected inside it. This is called **total internal reflection** (Fig. 21.23). Total internal reflection occurs when the angle of incidence is greater than the critical angle. The **critical angle** is the smallest angle of incidence at which all light striking the surface is totally internally reflected. It may be expressed by the formula

$$\sin i_c = \frac{1}{n}$$

Figure 21.23 (a) A point light source S is shown in glass. For all angles of incidence less than the critical angle i_c, the light ray is partially reflected but mostly refracted and leaves the glass as shown at b, c, and d. When the angle of incidence equals the critical angle i_c, the refracted ray points along the glass–air surface as shown at e. For all angles of incidence greater than the critical angle i_c, total internal reflection of the light ray occurs as shown at f, g, and h. (b) Here, the first two beams of light have angles of incidence less than the critical angle [similar to b, c, and d in part (a)] and are mostly refracted upward into the second medium. The third beam of light has an angle of incidence greater than the critical angle [similar to f, g, and h in part (a)] and is totally reflected downward and stays within the first medium.

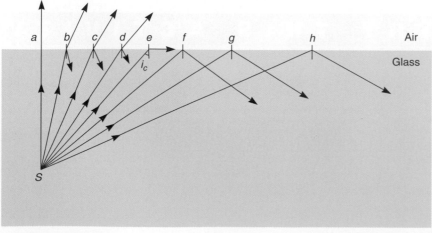

(a)

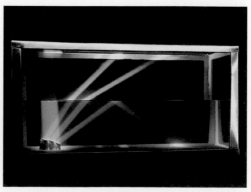

(b)

where i_c = critical angle of incidence
          n = index of refraction of denser medium

Note: The incident ray is traveling through a vacuum or air.

In Fig. 21.24, the angle of incidence from our perspective is greater than the critical angle. Note that the fish in the water are reflected above in the surface boundary.

Figure 21.24

TRY THIS ACTIVITY

Water Fiber Optics

A 2-L plastic soda pop bottle and a pen laser are needed for this activity. Make a small, pencil-sized hole in the lower wall of an empty plastic soda pop bottle. Placing a finger over the hole, fill the bottle with water. Shine the laser pen through the bottle so the laser light emerges from the hole where your finger is located. Remove your finger from the hole. Have someone stand in front of the hole and observe what is happening to the laser light as the water and the light emerge from the hole. Explain, using internal reflection principles, why the laser light behaves as it does when it emerges from the bottle.

What is the critical angle of incidence for water that has an index of refraction of 1.33?   **EXAMPLE**

Data:

$$n = 1.33$$
$$i_c = ?$$

Basic Equation:

$$\sin i_c = \frac{1}{n}$$

Working Equation: Same

Substitution:

$$\sin i_c = \frac{1}{1.33} = 0.752$$
$$i_c = 49°$$

Where there is total internal reflection, no light enters the air; it is totally reflected within the glass [see Figs. 21.25 and 21.28]. The property of having a very small critical angle gives a diamond its brilliance due to multiple internal reflections occurring before the light passes out through the top of the diamond. An example of the practical application of this principle is fiber optics (Fig. 21.25). Light may be transferred inside flexible glass or plastic fibers, which are transparent but keep nearly all the light inside because the light is reflected along the inside surface of the fibers.

Figure 21.25 (a) Light travels inside an optical or glass fiber like a light pipe in which the light is always incident at an angle greater than the critical angle. Thus, no light escapes the optical fiber by refraction. Fiber optics is essential to light-wave communications systems. (b) Total internal reflection within the tiny fibers of this fiber optic cable makes it possible to transmit light along complex paths with minimal loss.

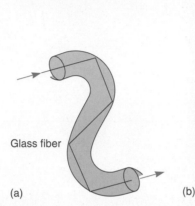

Glass fiber

(a)

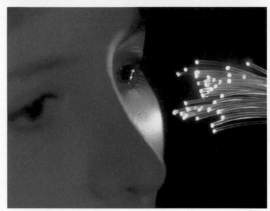

(b)

PHYSICS CONNECTIONS

Fiber Optic Cables

Most transmission of information travels as electric impulses through electric and telephone lines and fiber optic cables. Electric signals travel relatively slowly, cause wires to heat up, and need transformers to boost the voltage of signals traveling over long distances. Electric signals and wires are being replaced with light signals traveling through flexible, low-cost strands of glass. Because light travels through glass optical fibers, there is no electrical resistance to weaken the signal, and the signal travels at the speed of light, which is much faster than the speed of conventional electric signals. Such advances in fiber optics communications are revolutionizing the way we communicate.

Light traveling in the same medium travels in a straight line, whereas fiber optic cables can transmit a signal while twisting and turning, because of total internal reflection. The angle at which the light strikes the cladding of the fiber is always greater than the critical angle of the cladding and the core. The low critical angle allows the light to continually reflect and travel great distances without needing to be reamplified. In order for the cable to maintain a low critical angle, the glass must contain no imperfections or bubbles that would cause the light to be directed out or backward through the cable (see Fig. 21.26).

Fiber optic cables are used in telecommunications, computer networks, and medicine. A few strands of glass fiber can carry thousands of separate digital telephone conversations by slightly altering the frequency of the light for each phone conversation. A digital signal transmitted at one frequency cannot be confused by a signal carried at another frequency. Many computer networks and internal components in computers use fiber optic cables to carry data. By eliminating electric wiring, the fiber optic cable helps to reduce the temperature inside computers and servers. Finally, physicians use fiber optic bundles to perform minimally invasive procedures. A tool called an endoscope, composed of a bundle of fiber optic cables, transmits light into a patient's body while another bundle of fibers on the endoscope functions as a digital camera. The camera picks up the image and sends it back through the fiber optic cable to a monitor in the operating room.

Figure 21.26 (a) The red laser light entering the fiber optic cable is totally internally reflected, which results in the light emerging at the end of the cable. (b) An endoscope is a bundle of fiber optic cables used in many minimally invasive surgeries. Here an endoscope is used in the removal of nose adenoids with a laser therapy procedure.

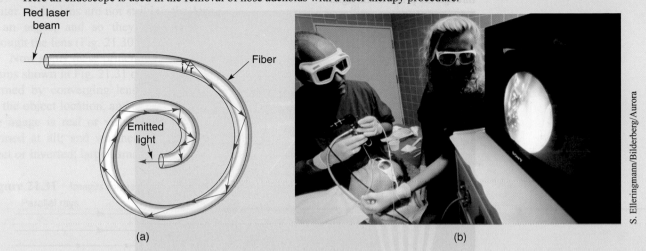

Red laser beam

Fiber

Emitted light

(a)

(b)

S. Elleringmann/Bilderberg/Aurora

21.8 Types of Lenses

Many technical applications, ranging from apparatus to test the nature of liquids to microscopes to eyeglasses, use the principles of refraction. We now consider the use of refraction in applications using lenses. Lenses may be converging or diverging. **Converging lenses** bend the light passing through them to some point beyond the lens. Converging lenses are always thicker in the center than on the edges [Fig. 21.27(a)]. **Diverging lenses** bend the light passing through them so as to spread the light. Diverging lenses are thicker on the edges than at the center [Fig. 21.27(b)].

Figure 21.27

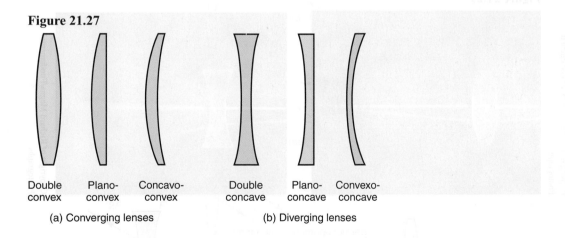

Double convex   Plano-convex   Concavo-convex   Double concave   Plano-concave   Convexo-concave

(a) Converging lenses   (b) Diverging lenses

Understanding how light is bent in lenses may be made easier by observing the bending of light passing through a prism (Fig. 21.28). Recall and apply the law of refraction, noting that light is bent toward the normal when passing into the glass and away from the normal when passing from the glass back to the air.

8. An object 2.50 cm tall is placed 20.0 cm from a converging lens. A real image is formed 9.00 cm from the lens. (a) What is the focal length of the lens? (b) What is the size of the image?

9. The focal length of a lens is 5.00 cm. How far from the lens must the object be to produce an image 1.50 cm from the lens?

10. If the distance from the lens in your eye to the retina is 19.0 mm, what is the focal length of the lens when reading a sign 40.0 cm from the lens?

11. An object 5.00 cm tall is placed 15.0 cm from a converging lens, and a real image is formed 7.50 cm from the lens. (a) What is the focal length of the lens? (b) What is the size of the image?

12. An object 4.50 cm tall is placed 18.0 cm from a converging lens with a focal length of 26.0 cm. (a) What is the location of the image? (b) What is its size?

13. What are the size and location of an image produced by a converging lens with a focal length of 19.5 cm of an object 5.76 cm from the lens and 1.45 cm high?

14. What are the size and location of an image produced by a convex lens with a focal length of 14.5 cm of an object 10.5 cm from the lens and 2.35 cm high?

15. What is the focal length of a convex lens that produces an inverted image twice as large as the object at a distance of 13.3 cm from the lens?

Glossary

Concave Mirror A mirror with a surface that curves away from an observer. (p. 575)

Converging Lens A lens that bends the light passing through it to some point beyond the lens. Converging lenses are thicker in the center. (p. 589)

Convex Mirror A mirror with a surface that curves inward toward an observer. (p. 575)

Critical Angle The smallest angle of incidence at which all light striking a surface is totally internally reflected. (p. 586)

Diffusion Scattering of light by an uneven surface. (p. 573)

Diverging Lens A lens that bends the light passing through it so as to spread the light. Diverging lenses are thicker at the edges than at the center. (p. 589)

First Law of Reflection The angle of incidence equals the angle of reflection. (p. 574)

Focal Length The distance between the principal focus of a mirror or lens and its vertex. (p. 577)

Index of Refraction A measure of the optical density of a material. Equal to the ratio of the speed of light in vacuum to the speed of light in the material. (p. 583)

Law of Refraction When a beam of light passes at an angle from a medium of lower optical density to a denser medium, the light is bent toward the normal. When a beam passes from a medium of greater optical density to one less dense, the light is bent away from the normal. (p. 583)

Opaque Absorbing or reflecting almost all light. (p. 573)

Optical Density A property of a transparent material that is a measure of the speed of light through the given material. (p. 582)

Plane Mirror A mirror with a flat surface. (p. 575)

Real Image An image formed by rays of light. (p. 575)

Reflection The turning or turning back of all or part of a beam of light as it strikes a surface. (p. 573)

Refraction The bending of light as it passes at an angle from one medium to another of different optical density. (p. 582)

Regular Reflection Reflection of light with very little scattering. (p. 573)

Second Law of Reflection The incident ray, the reflected ray, and the normal (perpendicular) to the reflecting surface all lie in the same plane. (p. 574)

Snell's Law The index of refraction equals the sine of the angle of incidence divided by the sine of the angle of refraction. (p. 583)

Total Internal Reflection A condition such that light striking a surface does not pass through the surface but is completely reflected inside it. (p. 586)

Translucent Allowing some but not all light to pass through. (p. 573)

Transparent Allowing almost all light to pass through so that objects or images can be seen clearly. (p. 573)

Virtual Image An image that only appears to the eye to be formed by rays of light. (p. 575)

Formulas

21.5 $\dfrac{1}{f} = \dfrac{1}{s_o} + \dfrac{1}{s_i}$

$M = \dfrac{h_i}{h_o} = \dfrac{-s_i}{s_o}$

21.6 $n = \dfrac{\sin i}{\sin r}$

$n = \dfrac{\text{speed of light in vacuum}}{\text{speed of light in substance}}$

21.7 $\quad \sin i_c = \dfrac{1}{n}$

21.10 $\quad \dfrac{1}{f} = \dfrac{1}{s_o} + \dfrac{1}{s_i}$

$\qquad M = \dfrac{h_i}{h_o} = \dfrac{-s_i}{s_o}$

Review Questions

1. Stained glass is an example of
    (a) a transparent material.
    (b) a translucent material.
    (c) an opaque material.
    (d) none of the above.
2. A virtual image may be
    (a) larger than the object.
    (b) smaller than the object.
    (c) erect.
    (d) all of the above.
    (e) none of the above.
3. A real image may be
    (a) erect.
    (b) shown on a screen.
    (c) formed by a plane mirror.
    (d) none of the above.
4. Explain the difference between diffusion and regular reflection.
5. In your own words, explain the first law of reflection.
6. In your own words, explain the second law of reflection.
7. Describe the type of images formed by plane mirrors.
8. Explain the difference between real and virtual images.
9. Explain the difference between a concave and a convex mirror.
10. Explain the effect of spherical aberration.
11. For a mirror of given focal length, how does the image distance change if the object distance is decreased?
12. For a given object distance from a mirror, how does the image distance change if the focal length is increased?
13. The index of refraction depends on
    (a) the focal length.
    (b) the speed of light.
    (c) the image distance.
    (d) none of the above.
14. Snell's law involves
    (a) the lens equation.
    (b) the index of refraction.
    (c) the focal length.
    (d) none of the above.
15. Explain the difference between converging and diverging lenses.
16. Give several examples of total internal reflection.
17. In your own words, explain the law of refraction.
18. How does the speed of light in a high-index-of-refraction material compare to the speed of light in a vacuum?

19. In your own words, explain why light waves are refracted at a boundary between two materials.
20. What types of images are formed by diverging lenses?
21. What types of images are formed by converging lenses?
22. How do water waves affect the escape of light from below the surface of the water?
23. Explain why a fish under water appears to be at a different depth below the surface than it actually is. Does it appear deeper or shallower?
24. Does light always travel in a straight line? Explain.
25. Explain how total internal reflection allows light in a glass fiber to be guided along the fiber.
26. Under what conditions will a converging lens form a virtual image?
27. Under what conditions will a converging lens form a real image that is the same size as the object?
28. Under what conditions will a diverging lens form a virtual image that is smaller than the object?

Review Problems

1. Using $\dfrac{1}{f} = \dfrac{1}{s_o} + \dfrac{1}{s_i}$, $s_o = 3.50$ cm, and $s_i = 7.25$ cm, find f.

2. Using $\dfrac{1}{f} = \dfrac{1}{s_o} + \dfrac{1}{s_i}$, $s_o = 8.50$ cm, and $f = 25.0$ cm, find s_i.

3. Using $M = \dfrac{h_i}{h_o} = \dfrac{-s_i}{s_o}$, $h_o = 6.50$ cm, $s_i = 7.50$ cm, and $s_o = 14.0$ cm, find h_i.

4. If an object is 3.75 m tall and 7.35 m from a large mirror with an image formed 4.35 m from the mirror, what is the height of the image?

5. An object 43.0 cm tall is located 23.4 cm from a concave mirror with focal length 21.4 cm. (a) Where is the image located? (b) How high is the image?

6. An object and its image in a concave mirror are the same height, but inverted, when the object is 45.3 cm from the mirror. What is the focal length of the mirror?

7. The angle of incidence of light passing from air to a liquid is 41.0°. The angle of refraction is 29.0°. Find the index of refraction of the liquid.

8. If the index of refraction of a liquid is 1.44, find the speed of light in that liquid.

9. If the critical angle of a liquid is 45.6°, find the index of refraction for that liquid.

10. If the index of refraction of a substance is 1.50, find its critical angle of incidence.

11. A converging lens has a focal length of 12.0 cm. If it is placed 36.0 cm from an object, how far from the lens will the image be formed?

12. An object 4.50 cm tall is placed 20.0 cm from a converging lens. A real image is formed 12.0 cm from the lens. (a) What is the focal length of the lens? (b) What is the size of the image?

13. The focal length of a lens is 4.00 cm. How far from the lens must the object be to produce an image 7.20 cm from the lens?

14. What is the focal length of a convex lens that produces an image three times as large as the object at a distance of 25.0 cm from the lens?

15. What is the focal length of a mirror that forms an image 3.44 m behind a convex mirror of an object 5.33 m in front of the mirror?

16. What are the size and location of an image produced by a convex lens with a focal length of 21.0 cm of an object 11.5 cm from the lens and 3.25 cm high?

17. What is the speed of light passing through a diamond? See Table 21.1.

18. Find the critical angle of incidence for Lucite. See Table 21.1.

SKETCH

| 12 cm² | w |

4.0 cm

DATA

$A = 12$ cm², $l = 4.0$ cm, w = ?

BASIC EQUATION

$A = lw$

WORKING EQUATION

$w = \dfrac{A}{l}$

SUBSTITUTION

$w = \dfrac{12 \text{ cm}^2}{4.0 \text{ cm}} = 3.0$ cm

PROBLEM SOLVING

19. Find the focal length of a concave mirror with an object 39.3 cm in front of it that projects an image 17.8 cm in front of the mirror.

20. (a) Find the height and location of an image produced by a concave mirror with focal length 8.70 cm and an object that is 13.2 cm tall and 19.3 cm from the mirror. (b) Find the height of the image produced by the mirror if the object is twice as far from the mirror.

APPLIED CONCEPTS

1. Tamera uses a concave mirror when applying makeup. (a) The mirror has a radius of curvature of 38.0 cm. What is the focal length of the mirror? (b) Tamera's face is located 12.5 cm from the mirror. Where will her image appear? (c) Will the image be upright or inverted? (d) How long will her face appear to be in the mirror if her face is actually 25.0 cm long?

2. A convex security mirror has a radius of curvature of −1.50 m. (a) A shoplifter picks up a 0.255-m-high purse 11.5 m from the mirror. How far behind the mirror is the image of the purse? (b) Is the image real or virtual? (c) How high is the image of the purse? (d) Are convex security mirrors better for viewing detailed images or large fields of view?

3. A fish tank made of crown glass is full of fresh water. (a) What is the angle of refraction when the light travels from the air and enters the crown glass at an incident angle of 30.0°? (b) Determine the angle of refraction when the light leaves the crown glass and enters the water. (c) If the aquarium is full of saltwater, will the angle of refraction be greater or less than the angle of refraction for fresh water?

4. Diamonds are cut to take advantage of internal reflections of light. The angles of a diamond are cut so that all of the light that enters it reflects out the top. The light or brilliance that emerges from the top of a diamond is one of the qualities that determines its monetary value. (a) What is the critical angle for light traveling from a diamond to air? (b) Would diamond jewelers want to create angles smaller or larger than the critical angle? (c) Zircon is an inexpensive substitute for diamond. Zircon's index of refraction is 1.92. What is its critical angle? (d) If a zircon has the same cut as a diamond, which will have a greater brilliance by internally reflecting more light to the top of the stone?

5. A photographer uses a 60.0-mm lens. (a) How far away should the lens be from the film if the object is located 9.20 m from the lens? (b) What will be the height of the image if the object is 2.00 m tall?

Red
Orange
Yellow
Green
Blue
Violet

The Color
Spectrum of
Light

Refraction
&
Dispersion

Understanding why a beam of sunlight forms a spectrum when directed through a glass prism, why the sky is blue, why sunsets are red, why clouds are white, why the ocean is blue, why we see rainbows, how colors of paint are mixed, how Polaroid sunglasses work, and how the LCD screen in a calculator works are basic questions many people commonly ask. Our examination of these basic questions forms the basis of the development and discussion of basic color properties.

Objectives

The major goals of this chapter are to enable you to:

1. Describe how the colors of the visible spectrum are formed through dispersion of light.
2. Describe color as a property of light and how it is related to its frequency or its wavelength.
3. Describe the result of mixing colors.
4. Describe diffraction as a property of light.
5. Describe interference as a property of light.
6. Describe polarization of light and some of its applications.

22.1 The Color of Light

Let us observe a narrow beam of sunlight that is directed into and passes through a glass prism in a dark room as in Fig. 22.1. Note the band of colors with one color shade gradually blending into another. This band of colors is called the **visible spectrum**. This spreading of white light into the full spectrum is called **dispersion**. Dispersion was described by Newton, who observed six colors: red, orange, yellow, green, blue, and violet. Sunlight is an example of white light. The *color* of the light is related to its wavelength or its frequency. Red light has lower frequency and longer wavelength, whereas at the opposite end of the spectrum violet light has higher frequency and shorter wavelength. *Polychromatic light is light consisting of several colors. Monochromatic light consists of* only one color.

As you can see from the dispersion of light, the refraction of the red light is less than that of the others and the refraction of the violet light is the greatest (Fig. 22.2). The shorter wavelengths are refracted more than the longer wavelengths. Thus, **color** is a property of the light that reaches our eyes and is determined by its wavelength or its frequency. Light waves with wavelengths outside the visible spectrum as shown in Fig. 20.7 are not visible to the human eye. Light waves within the visible spectrum have wavelengths in the range of 4.0×10^{-7} m (for violet) to 7.5×10^{-7} m (for red). Some photographic films are sensitive to ultraviolet (UV) light and others, to infrared (IR) light.

Most of the objects we see reflect rather than emit light. A shirt is called red if it reflects only red light and absorbs all other colors [Fig. 22.3(a)]. A shirt is called blue if it reflects only blue light and absorbs all other colors [Fig. 22.3(b)]. The color of an opaque object depends on the color of the light it reflects.

A white object reflects all the colors (light rays of various wavelengths) it receives. A black object absorbs all the colors it receives. A shirt that is blue in sunlight appears black in red light because there is no blue light present for it to reflect, and it absorbs all the other colors. Similarly, this is why a red shirt appears black in blue light.

Figure 22.1 A narrow beam of sunlight directed into a clear glass prism in a dark room is dispersed into the visible spectrum.

Figure 22.2 White light is dispersed by a prism into the visible spectrum on a white screen.

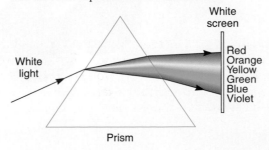

Figure 22.3 (a) A red shirt reflects only red light and absorbs light of all other colors. (b) A blue shirt reflects only blue light and absorbs light of all other colors.

(a)   (b)

Figure 22.4 Red glass transmits only red light and absorbs light of all other colors. Blue glass transmits only blue light and absorbs light of all other colors.

Regular window glass transmits all colors and is often called colorless. Red glass absorbs all colors except red, which it transmits. Blue glass absorbs all colors except blue, which it transmits (Fig. 22.4). The color of a transparent object depends on the color of the light it transmits.

Mixing Colors of Light

Just as polychromatic light, such as sunlight (sometimes called white light), can be dispersed into its separate colors, the separate colors can be combined to form polychromatic light. White light can also be produced by projecting overlapping red, green, and blue light onto a white screen as shown in Fig. 22.5. For this reason, red, green, and blue are called the **primary colors**. This is an additive mixture of the three colors of light.

Note in Fig. 22.5 the colors cyan, magenta, and yellow are produced when only two of the primary colors overlap, that is,

$$\text{blue} + \text{green} = \text{cyan}$$
$$\text{red} + \text{blue} = \text{magenta}$$
$$\text{red} + \text{green} = \text{yellow}$$

Figure 22.5 When the three primary colors overlap (light mixed by addition), white light is produced.

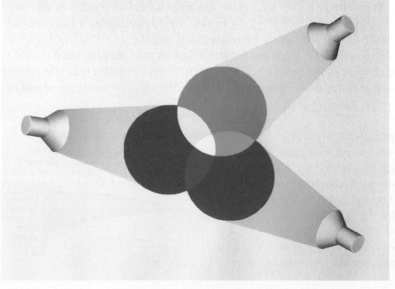

The color cyan is often called the opposite of red, magenta is the opposite of green, and yellow is the opposite of blue. Thus,

$$\text{cyan} + \text{red} = \text{white} = (\text{blue} + \text{green}) + \text{red}$$
$$\text{magenta} + \text{green} = \text{white} = (\text{red} + \text{blue}) + \text{green}$$
$$\text{yellow} + \text{blue} = \text{white} = (\text{red} + \text{green}) + \text{blue}$$

When any two colors are combined to form white, they are called **complementary colors**. Thus, cyan and red, magenta and green, and yellow and blue are complementary colors.

Colors of light are commonly mixed on theatrical stages to produce many different color effects (Fig. 22.6).

Figure 22.6 Bright spotlights at this night club show multiple colors using overlapping colored lights.

Copyright of Gabriel M. Covian/Getty Images, Inc.

Mixing Pigments

Mixing pigments in paints and dyes is completely different from mixing lighting. Artists and painters know that mixing red, green, and blue paint results in a murky dark brown—not white! Pigments are actually small particles that absorb, rather than reflect, given colors. When paint pigments are mixed, each one subtracts, or absorbs, certain colors from white light with the resulting reflected color depending on the light waves *not* absorbed. The **primary pigments** are the complements of the three primary colors, namely, cyan (the complement of red), magenta (the complement of green), and yellow (the complement of blue) (see Fig. 22.7).

Figure 22.7 Primary pigments. The results of the subtractive combination of cyan, magenta, and yellow filters are shown. Note that the combination of any two primary pigments gives the complement of the third by subtraction, namely, red, green, and blue. The subtractive combination of all three primary pigments transmits no light.

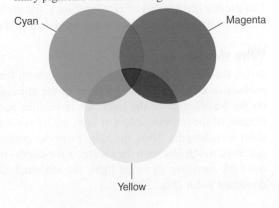

Cyan

Magenta

Yellow

Comparing additive mixtures and the corresponding subtractive mixtures, we have the following:

Additive Mixture	*Corresponding Subtractive Mixture*
cyan + red = white	red = white − cyan
magenta + green = white	green = white − magenta
yellow + blue = white	blue = white − yellow

That is, pigments that produce red absorb its complement, cyan, and reflect red. In other words, taking cyan away from white produces red. Pigments that produce green absorb its complement, magenta, and reflect green. Or, taking magenta away from white produces green. Pigments that produce blue absorb its complement, yellow, and reflect blue. Or, taking yellow away from white produces blue.

When the three primary pigments are mixed in appropriate proportions, a large range of colors can be produced.

Color Printing

All color photographs printed in books, magazines, and newspapers use only four colors of ink: cyan, magenta, and yellow for the colors and black for the shadow areas and definition. Four negatives are made using different filters so that each negative has the given amount of its color corresponding to the original color photograph. A red filter is used for making the cyan-printer negative, a green filter is used for making the magenta-printer negative, and a blue filter is used for making the yellow-printer negative. Each filter makes the color to be printed by its corresponding plate photograph as black and the other two colors photograph as white. The negative for the black plate is made by partial exposures through each of the three color filters so that black and shadows will print where all three colors are present.

A metal printing plate is made from each of these four negatives. Each metal printing plate is treated to hold the printer's ink in areas that have been exposed to light. Each of the four plates is printed in the same position, one after the other, with great precision on white paper. (Recall seeing a color photograph in a newspaper or magazine where either the plates or the paper was not within the proper precision and the colors were out of synchronization.) The lightness and the darkness of each color are controlled by the size of the very small printing dots. Use a magnifying glass to examine the very small printing dots in any color photograph in this text, a magazine, or a newspaper. Note that not only do the dots provide the basic colors, but also the overlapping dots of these basic colors give the appearance of many other colors. Figure 22.8 illustrates how only four colors produce color photographs and illustrations with many vibrant colors.

Inkjet printers deposit simultaneous combinations of cyan, magenta, yellow, and black inks on white paper, which results in a color photograph or illustration using the same four-color principle.

Color digital cameras use filters and record as pixels the three primary colors red, blue, and green of different intensities to create the variety of colors. A pixel is a tiny unit or dot that makes up the image on a television screen, computer monitor, or similar display. The greater the number of pixels per inch, the greater is the resolution, that is, the better the clarity of the detail that can be distinguished in the image on the screen.

Why the Sky Is Blue

When sunlight passes through the atmosphere, the light is *scattered* in all directions by air molecules and other small particles in the atmosphere. The amount of scattering depends on the wavelength of the light. Air molecules and particles much smaller than the wavelengths of the various colors of light are less of an obstruction to long wavelengths than to short wavelengths. Thus, the longer wavelengths of red, orange, yellow, and green light are scattered much less than the shorter wavelengths of blue and violet light. Since our eyes are not very sensitive to violet light, the scattered blue light in the lower atmosphere is the dominate color (Fig. 22.9).

Figure 22.8 Four-color printing process. The colors (a) cyan, (b) magenta, and (c) yellow printed separately. (d) Only these three colors combined. (e) The fourth (black) color. (f) The combined, finished result. Note the importance of black for the background, shadows, and definition.

(a)   (b)   (c)

(d)   (e)   (f)

Figure 22.9 The sky is blue because sunlight is scattered as it passes through the atmosphere. We see the most scattered, shorter wavelengths of blue light, which makes the sky appear blue.

A deep blue sky is most likely on a clear, low-humidity day. When the atmosphere contains water molecules, dust, and other particles larger than the oxygen and nitrogen molecules, the longer wavelength light (red, orange, yellow) is also significantly scattered, which turns the sky into a gray-white haze. This haze is caused by very small particles that scatter and absorb light before it reaches our eyes. As the number of particles increase, more light is absorbed and scattered, which results in less clarity, color, and visual range. Factories and car and truck engines emit pollution in the form of many billions of particles per second. Although these particles are very small, other particles adhere to them and scatter the longer

Figure 22.17 Diffraction patterns resulting from a point source precisely centered behind (a) a dime, (b) a razor blade, and (c) a single slit.

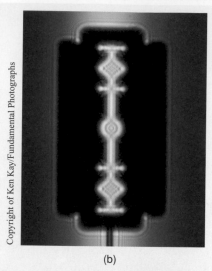

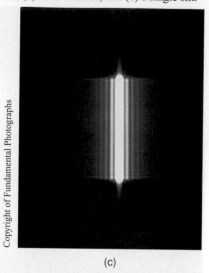

(a)　　　　　　　　(b)　　　　　　　　(c)

Huygens' principle (Section 20.1) also explains what happens when a wave front hits an obstacle and is partially obstructed, such as by a sharp edge, around a fine wire, or through a narrow opening. It shows that waves bend, or diffract, behind the obstacle. Since diffraction only occurs for waves and not for particles, it verifies the wave nature of light. Reminder: Light is an electromagnetic wave that exhibits both wave- and particle-like behavior, as discussed in Chapter 20.

22.3 Interference

We saw the results of the interference of water waves in a ripple tank in Fig. 16.9 in Section 16.1. We saw in Section 16.3 that when sound waves interfere with each other, beats are produced. Now let's study the interference of light waves. In 1801, Thomas Young first performed his famous *double-slit experiment*. Light from a single source (Young used the sun) falls on a screen that contains two closely spaced narrow slits, S_1 and S_2, as shown in a schematic diagram of his experiment in Fig. 22.18. If light consisted of particles, two bright lines would be expected to appear on a viewing screen placed behind the screen with the two slits. However, Young found not two, but a series of bright lines. Young explained his result as a *wave-interference* phenomenon.

Young's double-slit experiment was conducted using sunlight as shown in Fig. 22.19. The narrow slits, S_1 and S_2, are about 1 mm apart on a black piece of paper. A series of dark

Figure 22.18 Basic schematic of Young's double-slit experiment. Young found not two, but a series of bright lines behind the screen with the two slits.

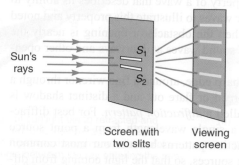

Figure 22.19 The light wave fronts in Young's experiment.

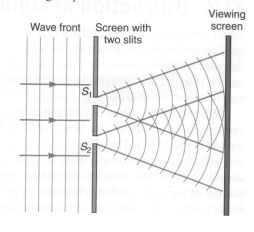

Figure 22.20 Photograph of an interference pattern produced by a double-slit experiment with red laser light.

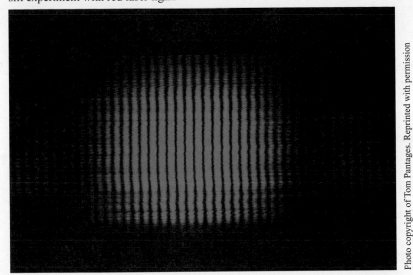

Photo copyright of Tom Pantages. Reprinted with permission

and bright narrow bands is seen on the viewing screen similar to the one in Fig. 22.20. We have used red laser light to produce a dramatic series of red and dark bands to illustrate the constructive and destructive interference patterns.

As we saw in Section 16.1, when two waves of a similar type pass through the same medium, a new wave is created by the *superposition of waves*. This new wave is the algebraic sum of the separate displacements of the individual waves as in Fig. 16.6. *Constructive interference* occurs when the waves add together to form a larger displacement [more intense light (red bands) in Fig. 22.20]. *Destructive interference* occurs when the waves add together to form a smaller displacement [less intense light (dark bands) in Fig. 22.20]. The series of bright and dark bands in Fig. 22.20 results from the different wave paths. Bright areas occur when the waves from both slits arrive in phase, whereas dark areas occur when the overlapping waves arrive out of phase (see Fig. 22.21). If the waves from the two slits travel the same distance, they are in phase and produce the brightest color at the center of the screen. Constructive interference also occurs when one wave travels an extra distance that is a whole number multiple of the wavelength and produces bright lines on the screen. Destructive interference occurs when one wave travels a distance of one-half wavelength (or 3/2, 5/2, 7/2, etc.) more than the other and

Figure 22.21 The bright and dark bands of light on the viewing screen result from the constructive and destructive interference, respectively, of the light waves after they pass through the double slits.

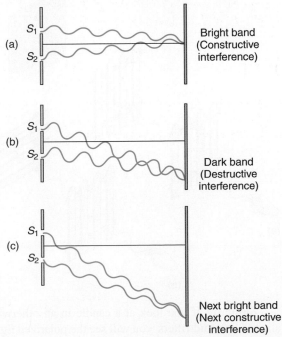

produces dark lines on the screen. In general, constructive interference occurs for a given wavelength of light when

$$d \sin \theta = n\lambda$$

where d = distance between slits
 θ = angle the light rays make with the horizontal
 n = 1, 2, 3, ... (called the *order* of the interference fringes)
 λ = given wavelength of light

Destructive interference occurs for a given wavelength of light when

$$d \sin \theta = (n + \tfrac{1}{2}\lambda)$$

The first-order fringes ($n = 1$) occur on either side of the central bright spot; the second-order fringes occur next; and so forth. The intensity is greatest for $n = 1$ and decreases for higher orders.

22.4 Polarization of Light

Light waves are usually emitted in all directions with many orientations. Some are vertical, some are horizontal, and some are at other angles. An ordinary incandescent light bulb and the sun emit light waves in multiple orientations. **Polarized light** has its waves restricted to a single plane that is perpendicular to the direction of the wave motion. Ordinary light may become polarized using a Polaroid filter (Polaroid materials were invented by Edward Land in 1929). Figure 22.22 shows a simple model to illustrate polarized light. Figure 22.22(a) shows a vibrating rope attached to a wall through two vertically aligned pieces of picket fence. Note that only a vertical transverse wave can pass through the vertical slats of the first fence and then continue *through* the vertical slats of the second. In Fig. 22.22(b) note that the vertical transverse wave that passes through the vertical slats of the first fence is *stopped* by the horizontal slats of the second fence and cannot pass through.

Figure 22.22   A uniform transverse wave generated by a rope (a) passes through uniformly when the filters are aligned but (b) cannot pass through the second filter when they are not aligned.

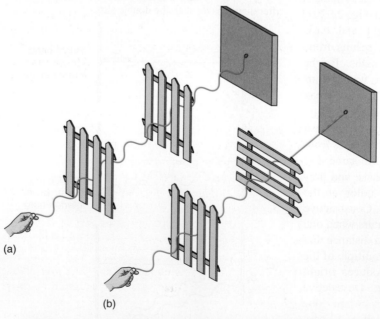

(a)

(b)

Similarly, if you look at a candle in an otherwise dark room through two vertically aligned Polaroid filters, you will see the polarized light from the candle as in Fig. 22.23(a).

Figure 22.23 Light waves generated by a candle in a dark room (a) pass through both filters as polarized light when the filters are aligned but (b) cannot pass through the second filter when they are not aligned.

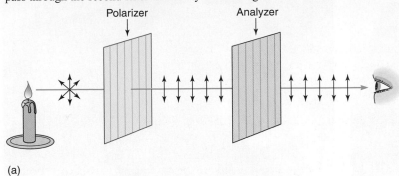

(a)

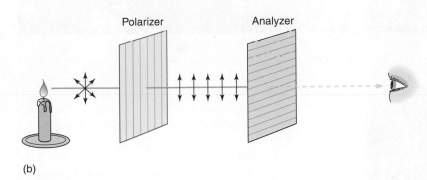

(b)

The two filters will allow only the vertically aligned light waves to pass through from the candle to your eyes. In Fig. 22.23(b), the second filter is horizontally aligned. Note that the vertically aligned light waves now pass through only the first filter and not through the second filter, so that you cannot see the candle. In this arrangement, the first filter is called the *polarizer* and the second is called the *analyzer*. Polaroid sunglasses block out horizontally vibrating light waves and roughly 50% of the light. When two pairs of sunglasses overlap at right angles, no light waves pass through (Fig. 22.24). Whenever unwanted glare occurs, polarized sunglasses are very useful, especially for driving, sailing, or walking on a beach.

Figure 22.25 shows a red flower behind two Polaroid filters. In Fig. 22.25(a) the filters are aligned vertically and one can see the partially shaded red flower through the filters. In Fig. 22.25(b) the second filter is rotated at a 45° angle. Note that one can still see the red flower, but the red flower is more shaded. In Fig. 22.25(c) the second filter is rotated so that the filters are at right angles and no light passes through.

Much of the light that is reflected from nonmetallic surfaces is polarized. Good examples are the glare off smooth water or the glass of a windshield. The reflected light waves from the sun tend to be the horizontal-component waves parallel to the reflecting surface as shown in Fig. 22.26, whereas the vertical-component waves tend to be refracted through the surface into the water or glass. Because most reflected nonmetallic surfaces are horizontal, Polaroid sunglasses are aligned so that the polarized slits are vertical. Figure 22.27(a) shows a photograph of an automobile with glare from its windows. Figure 22.27(b) shows a photograph of the same automobile through a polarized filter; note the absence of glare, so one can more easily see into the automobile.

Electronic devices that display information using a liquid crystal display (LCD) use polarized filters to form the black segments that form the numbers and letters (see Fig. 22.28).

Figure 22.24 When Polaroid sunglasses overlap at right angles, no light gets through.

Copyright of Diane Hirsch/Fundamental Photographs

Figure 22.25 When a red flower is behind two overlapping Polaroid filters that (a) are aligned, one can easily see the red flower; (b) are 45° out of alignment, one can see the partially shaded red flower; and (c) are at right angles, one cannot see the red flower.

(a)

(b)

(c)

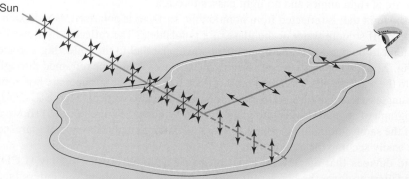

Figure 22.26 Much of the sunlight reflected from water is polarized. The horizontal-component waves reflect parallel to the water's surface, whereas the vertical-component waves tend to be refracted through the surface into the water.

Figure 22.27 (a) An automobile with glare from its windows. (b) The same automobile seen through a polarized filter; note there is no glare, so one can see into the automobile.

Figure 22.28 The segments that form letters and numbers appearing in an LCD are formed using rotating polarized filters.

Courtesy of Research In Motion (RIM)

PHYSICS CONNECTIONS

Polarized Calculators

Polarized filters are often found in sunglasses, but are also used in devices such as stress analyzers, flat-screen computer displays, digital clocks, and calculators. Polarized filters play a part in displaying numbers on calculator screens.

There are three main components in calculators with liquid crystal display screens. A reflective mirror is placed at the rear of the display to reflect the light coming into the display back out to the user. A liquid crystal is located in front of the mirror. Liquid crystals have the ability to gradually twist the orientation of the light 90° from its original orientation. When a small voltage is applied to the liquid crystal, it untwists and maintains the original orientation of the light. The third component of the display is a pair of polarized filters. The polarized filters are oriented perpendicular to one another and are placed on each side of the liquid crystal. The inner polarized filter is aligned horizontally and the outer filter is aligned vertically (Fig. 22.29).

When light enters the display and passes through the first filter, only the vertically polarized light is permitted to pass through the filter and reach the liquid crystal. If there is no voltage placed across the crystal, the light is forced to alter its orientation by 90° as it passes through the liquid crystal. Since the light's orientation is now horizontal, the light passes through the second polarized filter, reflects off the mirror, passes through the second filter again, moves through the liquid crystal, changes its orientation, passes through the initial filter, and is directed to the user. The light emitted is seen as the gray background of the LCD display.

If the calculator is programmed to apply a voltage across the liquid crystal, the crystal will untwist and prevent the light from passing through the second filter and reflecting off the mirror. A black segment would appear to the calculator user because the horizontally polarized filter absorbs the vertically oriented light. Therefore, when a calculator displays a number, voltage must be placed across the liquid crystal to create that black segment.

Figure 22.29 (a) The path that light takes as it enters the liquid crystal display of a calculator. (b) A standard liquid crystal display found on many calculators.

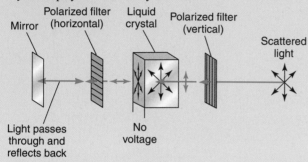

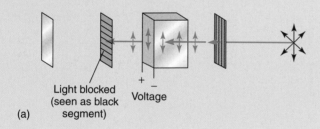

(a)

(b)

Glossary

Color A property of the light that reaches our eyes and is determined by its wavelength or its frequency. (p. 601)

Complementary Colors Two colors that, when combined, form white; for example, cyan and red, magenta and green, and yellow and blue. (p. 603)

Diffraction A property of a wave that describes its ability to bend around obstacles in its path. (p. 609)

Dispersion The spreading of white light into the full spectrum. (p. 601)

Polarized Light Light waves restricted to a single plane that is perpendicular to the direction of the wave motion. (p. 612)

Primary Colors The colors red, green, and blue; an additive mixture of the three colors of light resulting in white. (p. 602)

Primary Pigments The complements of the three primary colors; namely, cyan (the complement of red), magenta (the complement of green), and yellow (the complement of blue). (p. 603)

Rainbow A spectrum of light formed when sunlight strikes raindrops, refracts into them, reflects within them, and then refracts out of them. (p. 608)

Visible Spectrum The colors resulting from the dispersion of white light through a glass prism: red, orange, yellow, green, blue, and violet. (p. 601)

Review Questions

1. Name the colors of the visible spectrum.
2. What property of light determines its color?
3. Name the light waves whose wavelengths are (a) slightly longer than visible light and (b) slightly shorter than visible light.
4. What is the apparent color of a green dress in a closed room with only a red light? Explain.
5. What would the U.S. flag look like through a piece of (a) red glass and (b) blue glass?
6. What are the primary colors?
7. What is a complementary color? Name the complement of each primary color.
8. How are the primary pigments related to the primary colors?
9. What are the four colors used to print color photographs in books, magazines, and newspapers?
10. Why is the sky blue?
11. Why is a sunset red?
12. Why are clouds white?
13. Why is the ocean blue?
14. True or false: For you to see a rainbow, the raindrops must be between the sun and you.
15. Name the property of a wave that describes its ability to bend around obstacles in its path.
16. When a narrow beam of light passes by a sharp edge, around a fine wire, or through a narrow slit or a pinhole, the beam tends to spread or flare out. The distinct shadow formed is called a _____.
17. Name the person who first performed the famous double-slit experiment.
18. When waves add together to form a larger displacement or more intense light, this is called _____.
19. When waves add together to form a smaller displacement or less intense light, this is called _____.
20. How do Polaroid sunglasses reduce the glare of bright sunlight?

CHAPTER 23

SURVEY OF MODERN PHYSICS

To this point we have studied classical or Newtonian physics. Physics is also concerned with the building blocks of all matter, such as atoms, subatomic particles, and the forces between them.

Ancient Greek philosophers thought that all matter could be reduced to four basic elements: air, earth, fire, and water. Modern physics is much more complex and probes deeply into the nature of matter.

Quantum theory, the atom, radioactivity, nuclear fission, and fusion are all topics being studied by physicists today. While an in-depth study of these topics is beyond the scope of this text, some familiarity with the concepts is essential to those in all technical fields.

Objectives

The major goals of this chapter are to enable you to:

1. Describe the basis of quantum theory.
2. Describe the development of the current model of the atom.
3. Describe the structure and properties of the atomic nucleus.
4. Analyze problems of radioactive decay.
5. Describe nuclear fission and fusion.
6. Describe principles of detection and measurement of radioactivity.

23.1 Quantum Theory

In the 1860s, **James Clerk Maxwell** developed a theory relating magnetic fields and electric fields in the transmission of energy across empty space, which became known as the electromagnetic wave theory. Maxwell's theory led to a complete description of electricity and magnetism and was the basis for the development of radio, television, and countless electronic devices. It was also the starting point in the study of the nature of the atom.

Later experiments by Heinrich Hertz confirmed Maxwell's predictions and confirmed that light was a form of wave—almost. Only two problems remained: the wave theory of light could not explain why hot objects change color when heated and why ultraviolet light could discharge electrically charged metal plates. Solving these two problems led to the development of the quantum theory. Its confirmation forever changed our view of the physical world.

Particles Behave Like Waves and Waves Behave Like Particles

Incandescent objects (like a glowing light bulb filament) emit light because of the vibrations of charged particles inside their atoms. The spectrum of color produced by the light depends on the heat of the object and the frequency range of the light. Max Planck explained this with the then-revolutionary idea that particles can have only certain energies, which are whole-number multiples of a constant now known as Planck's constant. Planck proposed that atoms could only emit radiation when their vibration energy changed in multiples of this constant.

In addition, certain metals emit electrons when they are exposed to light. This is called the *photoelectric effect* and could not be explained by simple wave theory. The light produced an effect like a stream of particles. Einstein explained this effect by suggesting that light consists of packets of energy, called *photons*, that act like particles. He thought that although photons have no mass, they travel at the speed of light, have energy related to their frequency, and have momentum. Later experiments confirmed his theory.

Later, **Louis Victor de Broglie** in 1923 suggested that particles have wave properties. This revolutionary idea was largely ignored until Einstein supported it; de Broglie won a Nobel Prize for it in 1929. Two wave properties exhibited by particles are diffraction and interference. These can easily be reproduced in the laboratory.

Most physicists today believe that the particle and wave aspects of light are complementary. That is, they must be viewed together to have a complete picture of the true

James C. Maxwell (1831–1879),

physicist, was born in Scotland. His early work led to the mathematical development of the theory of electricity and magnetism. His greatest work was his theory of electromagnetic radiation and the basic equations of electromagnetism, which established him as the leading theoretical physicist of the nineteenth century.

Louis Victor de Broglie (1892–1987),

physicist, was born in France. In 1923, while a graduate student at the University of Paris, he theorized that since light could be seen to behave under some conditions as particles and other times as waves, atomic particles have both particle and wave properties. This led to his pioneering work on the wave nature of the electron.

nature of light. This complementary nature in exhibiting properties of both waves and particles is the essence of the quantum theory.

23.2 The Atom

In Section 12.1 we saw that an **atom** is the smallest particle of an element that can exist in a stable or independent state. The three most important particles of the atom are (1) the **proton**, a particle with a positive charge, (2) the **electron**, a particle with a negative charge, and (3) the **neutron**, a particle that does not have an electric charge. The **nucleus** is the center part of an atom, made up of protons and neutrons, while the electrons surround the nucleus. Historically, how was this current concept of the atom developed?

By 1900, most scientists agreed atoms exist. In 1890, **J. J. Thomson** discovered small, negatively charged particles, which he called electrons. Between 1909 and 1911, Ernest Rutherford discovered that atoms have a relatively more massive, positively charged center, or nucleus. Another 30 years was required for agreement to be reached on how these parts of the atom were arranged.

The Rutherford Model

Thomson originally thought that electrons were embedded in a spherical volume of positive charge like raisins in a muffin. However, Rutherford, with Hans Geiger and Ernst Marsden, was able to show that the atom was mostly space with most of the mass concentrated in a relatively small region he called the nucleus. He thought electrons orbited the nucleus the way planets orbit the sun, which is the most popular oversimplification of atomic structure still found today. The flaw in Rutherford's model is that if electrons were like planets, they would quickly lose energy and collapse into the nucleus, which does not happen.

Scientists sought a model of the atom that could explain the mystery of atomic spectra. An emission spectrum is a spectrum of electromagnetic radiation emitted by a luminous source. Many substances emit a characteristic color when heated. For example, applying a high voltage to gas atoms, as in a neon light, causes them to emit their characteristic color. The light can be studied with a diffraction grating or prism. A *diffraction grating* is a glass or polished metal surface containing many closely spaced, very fine parallel lines in the surface, which separate the colors of light by interference and produces a light spectrum. In contrast to an emission spectrum, as a gas cools, the opposite process occurs and an absorption spectrum is produced. An absorption spectrum is a spectrum of electromagnetic radiation absorbed by matter when radiation of all frequencies is passed through it. Based on the above concepts, atomic spectra were used to determine the structure of the atom.

Niels Bohr developed the next recognized model of the atom, using Einstein's idea that since emission spectra contain only specific wavelengths, an electron can emit or absorb only specific amounts of energy; that is, the energy is quantized. Bohr knew his model was not complete and continued to search for a better model. The Bohr model did, however, provide an explanation for some of the chemical properties of elements. The foundation of much of our knowledge of chemical reactions and bonding is based on the idea that each atom has a unique electron arrangement.

In 1926, **Erwin Schroedinger** used de Broglie's idea of matter behaving like waves to create a quantum model of the atom based on waves. **Werner Heisenberg** and others developed the theory into a complete description of the atom. The theory does not provide a simple picture like the planetary model. The wave–particle nature of matter means it is impossible to know both the momentum and the position of an electron at the same time. The quantum model of the atom predicts only the probability that an electron is at a specific location. This region in which there is a high probability of finding an electron is called an electron cloud. Though difficult to visualize, quantum mechanics has been very successful in predicting many details of the structure of atoms.

J. J. Thomson (1856–1940),

physicist, was born in England. His experiments in 1897 led to the discovery of the electron, which he found to be 2000 times smaller in mass than the lightest known atomic particle, the hydrogen ion. He received the Nobel Prize in Physics in 1906.

Niels Bohr (1885–1962),

Danish physicist, developed an early model of atomic structure in which electrons travel around the nucleus in given stable orbits determined by quantum conditions.

Erwin Schroedinger (1887–1961),

physicist, was born in Vienna. His study of the wave behavior of matter within quantum mechanics led to the understanding of matter at the subatomic level. He also made significant contributions to molecular biology.

Werner Heisenberg (1901–1976),

theoretical physicist, was born in Germany. He developed a method of expressing quantum mechanics mathematically and formulated his revolutionary uncertainty principle—increasing the accuracy of measurement of one observable quantity increases the uncertainty with which a related quantity may be known.

23.3 Atomic Structure and Atomic Spectra

Niels Bohr suggested the energy of an electron is quantized, that is, restricted only to certain fixed values called *energy levels*. Each of these energy levels was shown by Bohr to be given by the equation

$$E = -\frac{kZ^2}{n^2}$$

where E = energy of electron
  k = a constant (2.179×10^{-18} J or 13.595 eV)
  Z = atomic number (the number of positive charges in the nucleus)
  n = integer that characterizes the energy level, ($n = 1, 2, 3, 4, \ldots$), also called the quantum number.

An *electron volt* (eV) is the energy acquired by an electron as a result of moving through a potential difference of 1 V. One eV is equal to 1.602×10^{-19} J. So, one MeV (one million electron volts) is equal to 1.602×10^{-13} J. The distance from an electron in an atom to the nucleus increases as the integer n increases. (The formula is $r_n = n^2 r_1$ for the hydrogen atom.) In the lowest-energy level or lowest orbit, its energy is minimum. This lowest-energy level ($n = 1$) is called the stable or **ground state** of the electron. The higher-energy levels ($n = 2, 3, 4, \ldots$) are called **excited states** of the electron.

The **Bohr model** was an early model of atomic structure in which electrons travel around the nucleus in a number of discrete stable energy levels determined by quantum conditions. Of the several early proposed models of atomic structure, Bohr's was the first successful model for the atomic structure of the hydrogen atom (Fig. 23.1). The Bohr model is clearly limited in scope, but it was the forerunner to the modern quantum theory.

Figure 23.1   The Bohr model of the hydrogen atom.

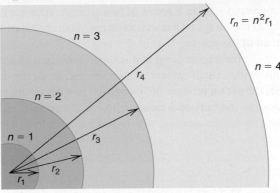

Find the energy of an electron in the $n = 2$ energy level in a hydrogen atom.

EXAMPLE 1

Data:

$$n = 2$$
$$Z = 1$$
$$E = ?$$

Basic Equation:

$$E = -\frac{kZ^2}{n^2}$$

Working Equation: Same

Substitution:

$$E = -\frac{(13.595 \text{ eV})(1)^2}{2^2}$$
$$= -3.3988 \text{ eV}$$

All objects emit radiation with intensity that is proportional to the fourth power of the Kelvin temperature (T_K^4). At normal temperatures, we are not aware of this electromagnetic radiation because of its low intensity. At higher temperatures, sufficient infrared radiation exists so that we feel the heat if we are close to the object. At still higher temperatures (approximately 1000 K), an object, such as molten steel, an electric toaster element, or an electric stove burner, glows. The filament of a light bulb glows at temperatures above 2000 K. This radiation is due to the rapid vibrations of the atoms and molecules.

Rarefied (low density and low pressure) gases can also be excited to emit light by intense heating in a tube. The radiation from excited gases emits light of only certain wavelengths. When the light is analyzed through the slit of a spectroscope, a line spectrum is observed instead of a continuous color spectrum.

Neon lights work on the principle that certain colors of light are given off by different atoms when the electrons in the atoms undergo a transition from the ground state to an excited state due to an electric current passing through the neon tube containing the gas at low pressure. As the electrons return to their ground state, they give off their excess energy in the form of electromagnetic radiation. A neon tube containing hydrogen gives off the color blue. Other atoms give off other colors, leading to the rich colors available through this type of light tube.

Fluorescence is a property of certain substances in which radiation is absorbed at one frequency and then reemitted at a lower frequency. When radiation is absorbed by a particle (an atom or a molecule), the energy of the electrons is increased to a higher energy level. This results in the particle being in an abnormal, excited state. Some absorbed energy is lost by collisions with other particles so that the particle cannot maintain its energy level in its excited state. The electrons give up their remaining extra energy and the particle returns to its ground state in a series of two or more steps and emits radiation in the form of visible light. The image on a television screen is the result of fluorescence due to electron bombardment of the screen.

A *fluorescent lamp* is a source of artificial light that consists of a long, sealed glass tube with a small amount of mercury and an electrode at each end. The inside of the tube is coated with a mixture of fluorescent powders. An electric current through the tube vaporizes the mercury and causes it to emit ultraviolet radiation. That radiation is absorbed by the fluorescent coating, which then emits visible light. See Fig. 23.2.

Figure 23.2   Fluorescent lamp. An electric current through the glass tube vaporizes the mercury and causes it to emit ultraviolet light. That radiation is absorbed by the fluorescent coating, which then emits visible light.

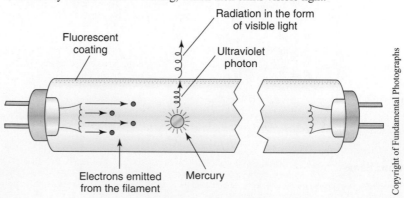

Copyright of Fundamental Photographs

The wavelength of a photon emitted during the transition between energy levels can be determined using concepts developed in earlier chapters: $E = hf$ gives the energy of a photon and $c = \lambda f$ relates the wavelength and frequency of light.

What is the wavelength of a photon given off by a hydrogen atom undergoing a transition from the $n = 2$ to the $n = 1$ energy level?

EXAMPLE 2

Data: For $n = 2$ energy level, $E = -3.3988$ eV (from Example 1)

For $n = 1$ energy level,

$$n = 1$$
$$k = 13.595 \text{ eV}$$
$$Z = 1$$

Basic Equations:

$$E = -\frac{kZ^2}{n^2}$$
$$E = hf$$
$$c = \lambda f$$

Working Equations:

$$E_1 = -\frac{kZ^2}{n^2} \qquad E_{\text{transition}} = E_2 - E_1 \qquad \lambda = \frac{c}{f} = \frac{c}{E/h} = \frac{ch}{E}$$

Substitutions:

$$E_1 = -\frac{(13.595 \text{ eV})(1)^2}{1^2}$$
$$= -13.595 \text{ eV}$$

Therefore,

$$E_{\text{transition}} = -3.3988 \text{ eV} - (-13.595 \text{ eV})$$
$$= 10.196 \text{ eV}$$

So,

$$\lambda = \frac{(3.00 \times 10^8 \text{ m/s})(6.626 \times 10^{-34} \text{ J s})}{10.196 \text{ eV} \times \dfrac{1.602 \times 10^{-19} \text{ J}}{1 \text{ eV}}}$$
$$= 1.22 \times 10^{-7} \text{ m}$$

· · · · · · · · · · · · · · ·

PROBLEMS 23.3

1. Find the energy of an electron in its ground-state energy level in a hydrogen atom.
2. Find the energy of an electron in the $n = 3$ energy level in a hydrogen atom.
3. Find the energy given off by an electron in a hydrogen atom undergoing a transition from the $n = 3$ to the $n = 1$ energy level.
4. Find the wavelength of the photon given off in the transition in Problem 3.
5. Find the wavelength of a photon given off in a hydrogen-filled neon tube undergoing a transition from the $n = 3$ to the $n = 2$ energy level.

23.4 Quantum Mechanics and Atomic Properties

The agreement between the Bohr model and the energy levels of the hydrogen atom does not account for the properties of the more complex atoms, that is, those with more than one electron. The quantum model of the atom better fits the more complex atoms.

SKETCH

12 cm² | w

4.0 cm

DATA

$A = 12$ cm², $l = 4.0$ cm, $w = ?$

BASIC EQUATION

$A = lw$

WORKING EQUATION

$w = \frac{A}{l}$

SUBSTITUTION

$w = \frac{12 \text{ cm}^2}{4.0 \text{ cm}} = 3.0$ cm

PROBLEM SOLVING

Figure 23.3 A shell must contain an integral number of wavelengths of the electron.

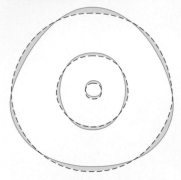

Figure 23.4 The probability of finding the electron at a distance r from the nucleus for the three lowest shells of a hydrogen atom.

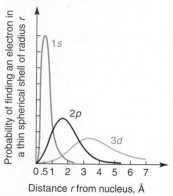

Figure 23.5 Electron density for an electron in the ground state of a hydrogen atom.

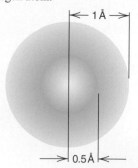

Quantum mechanics is a theory that unifies the wave–particle dual nature of electromagnetic radiation. The quantum model is based on the theory that matter, just like light, behaves sometimes as a particle and sometimes as a wave. The wave theory of the atom can be used in a simple way to understand the Bohr model structure of an atom. In the Bohr model, the electron can have only specific discrete energy levels; other energy levels are not possible. When an electron in an atom behaves as a wave, its distance from the nucleus or its energy level must consist of an integral number of wavelengths; otherwise, destructive interference occurs (Fig. 23.3).

When an electron behaves as a wave, its exact location cannot be determined; instead, the probability can be calculated, using wave equations, of where it is likely to be found. This probability describes for any position in space how likely the electron is to be found at that point at any given time (Fig. 23.4). A thin spherical shell region of space (or electron cloud) surrounding the atomic nucleus in which an electron is likely to be found is its energy level. In the ground state of the hydrogen atom, the electron spends most of its time in a spherical shell region, or energy level, centered about 0.5 Å (angstrom) from the nucleus as shown in Fig. 23.5. *Note:* $1 \text{ Å} = 1 \times 10^{-10}$ m.

The shell regions, or energy levels, are characterized by a number, n, which is called the *principal quantum number*, where $n = 1, 2, 3, 4, \ldots$. Just as in the Bohr model, the shells extend farther from the nucleus as n increases. As the shells extend farther from the nucleus, there are different shell shapes that become possible for each principal quantum number. Each of these shapes is described by a *secondary quantum number* or letter $(s, p, d, f, \ldots)$ as shown in Fig. 23.6.

Each possible shape represents an energy level, or shell, which can contain up to two electrons. In the ground state (lowest energy state) of an atom, the electrons fill up the lowest energy levels. The types of energy levels, or shells, for each principal quantum number are shown in Table 23.1. Table 23.2 summarizes the number of electrons in each shell for the atomic elements with up to 21 protons in the nucleus and therefore up to 21 electrons in energy levels. Note in Table 23.2 that the $n = 4$ shell starts filling up (see potassium) before the $n = 3$ shell is completely filled. This is because the $4s$ shell has lower energy than the $3d$ shells.

Figure 23.6 Shapes for (a) an s subshell, (b) a p subshell, (c) a d subshell, and (d) an f subshell.

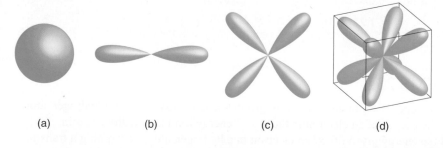

(a) (b) (c) (d)

Table 23.1 Maximum Number of Electrons in Atomic Shells or Energy Levels

n Value	Types of Energy Levels	Maximum Number of Electrons
1	$1s$	2
2	$2s, 2p$	8
3	$3s, 3p, 3d$	18
4	$4s, 4p, 4d, 4f$	32
5	$5s, 5p, 5d, 5f, 5g$	50

Table 23.2   Table of the First 21 Elements

Element	Atomic No.	Number of Protons	Number of Electrons	Electrons in $n = 1$ Shell	Electrons in $n = 2$ Shell	Electrons in $n = 3$ Shell	Electrons in $n = 4$ Shell
Hydrogen	1	1	1	1			
Helium	2	2	2	2			
Lithium	3	3	3	2	1		
Beryllium	4	4	4	2	2		
Boron	5	5	5	2	3		
Carbon	6	6	6	2	4		
Nitrogen	7	7	7	2	5		
Oxygen	8	8	8	2	6		
Fluorine	9	9	9	2	7		
Neon	10	10	10	2	8		
Sodium	11	11	11	2	8	1	
Magnesium	12	12	12	2	8	2	
Aluminum	13	13	13	2	8	3	
Silicon	14	14	14	2	8	4	
Phosphorus	15	15	15	2	8	5	
Sulfur	16	16	16	2	8	6	
Chlorine	17	17	17	2	8	7	
Argon	18	18	18	2	8	8	
Potassium	19	19	19	2	8	8	1
Calcium	20	20	20	2	8	8	2
Scandium	21	21	21	2	8	9	2

The theory of electron shells provides a basis for understanding the properties of different atoms, based on how many electrons are in the outer shell and how much energy it takes to add or remove an electron from that shell during chemical reactions between atoms. Each atom in all matter attempts to have its outer shell complete and does this by attempting to borrow, lend, or share electrons with other atoms surrounding it.

An atom is classified as a metal if it tends to lend electrons. Copper is an example of a metal. If it tends to borrow electrons, it is called a nonmetal. Atoms are classified as inert when they have complete outer shells and therefore tend not to borrow or share electrons. Helium and neon are examples of inert atoms. They both have completely filled outer shells (see Table 23.2). The fewer electrons an atom tends to borrow, lend, or share, the more reactive it tends to be in chemical reactions.

PROBLEMS 23.4

1. Which atom has the $n = 2$ energy level filled?
2. Which atom has the $n = 1$ energy level filled?
3. Which atom is one electron short of having the $n = 2$ energy level filled?
4. Which atom has just one electron in the $n = 3$ energy level?

23.5 The Nucleus—Structure and Properties

Henri Becquerel discovered radioactivity in 1896 when he was working with compounds containing uranium and found that his photographic plates were partially exposed when these uranium compounds were anywhere near his plates. The plates were exposed by penetrating rays that went through the plate coverings. This phenomenon was called radiation. Many scientists then began to study radiation, and one of the first results was an understanding of the composition of the atomic nucleus.

Figure 23.7 The nucleus is composed of neutrons and protons. Electrons surround the nucleus.

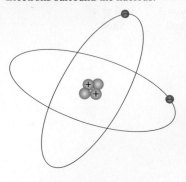

The nucleus of an atom contains over 99.9% of the mass of an atom, yet it occupies less than one-trillionth of 1% of the volume of an atom. The fundamental particles that make up the nucleus of an atom are called *nucleons*. There are two types of nucleons: *protons*, which carry a positive charge, and *neutrons*, which are electrically neutral (Fig. 23.7). Protons have a mass of

$$m_p = 1.6726 \times 10^{-27} \text{ kg}$$

while neutrons have a mass of

$$m_n = 1.6749 \times 10^{-27} \text{ kg}$$

Both are much more massive than the third fundamental particle of an atom, the negatively charged electron, which has a mass of

$$m_e = 9.1094 \times 10^{-31} \text{ kg}$$

Negatively charged electrons are attracted to the nucleus by the positively charged protons (Fig. 23.8). The *size* of the positive charge of a proton is the same as the *size* of the negative charge of an electron. An electrically neutral atom contains an equal number of protons and electrons.

Figure 23.8 The attractive strong force operates at short distances.

Proton   Electron

The simplest atom, hydrogen, contains one proton in its nucleus. Other atoms contain two or more protons together with some number of neutrons in their nuclei. The role of neutrons in the nucleus of these more complex atoms is to bind the positively charged protons together and prevent the nucleus from flying apart under the repulsive electric force between the protons. In addition to the repulsive **Coulomb force** (electric force) between protons, there is a nuclear force, referred to as the **strong force**, which is a very short-range attractive force among all nucleons (neutrons and protons) independent of their charge. The neutrons provide additional strong force to overcome Coulomb repulsion. Two neutrons in a nucleus with two protons provide the additional attractive strong force that holds the nucleus together. For heavier atoms it is necessary to have more neutrons than protons to hold the nucleus together (Fig. 23.9). For example, there are 82 protons and 126 neutrons in an atom of common lead. Atoms heavier than lead cannot be completely stabilized even by the addition of extra neutrons.

Figure 23.9 Number of neutrons versus number of protons for stable nuclei (dots).

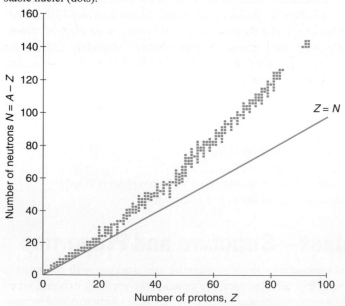

The number of protons in a nucleus is called its **atomic number**, Z. The total number of nucleons (protons and neutrons) in a nucleus is called the **atomic mass number**, A. A

nuclide is a specific type of atom characterized by its nuclear properties, such as its number of neutrons and protons and the energy state of its nucleus. The **neutron number**, N, is the number of neutrons in an atomic nucleus of a particular nuclide. Also, $N = A - Z$, the difference between the atomic mass number and the atomic number. N can be different for nuclei with the same atomic number. Each type of nuclide is specified using a symbol of the form

$$\begin{array}{c}A\\Z\end{array}X$$

where X = chemical symbol for the given element
 A = atomic mass number
 Z = atomic number

For example, $^{14}_{6}C$ refers to a carbon nucleus containing 6 protons with an atomic mass number of 14. This atom thus contains 8 neutrons. Other carbon atoms exist that contain 5, 6, 7, 9, or 10 neutrons. All carbon atoms contain 6 protons. (Otherwise, they wouldn't be carbon!) Nuclei that contain the same number of protons but a different number of neutrons are called **isotopes**. See Fig. 23.10 for an illustration of the isotopes of hydrogen.

Figure 23.10 Three isotopes of hydrogen. Each has a single electron and a single proton, but different numbers of neutrons.

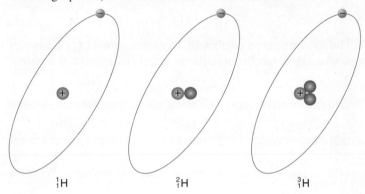

$^{1}_{1}H$ $\qquad\qquad$ $^{2}_{1}H$ $\qquad\qquad$ $^{3}_{1}H$

EXAMPLE 1

For the uranium isotope $^{238}_{92}U$, find (a) its atomic mass number, (b) its atomic number, (c) its neutron number, (d) the number of protons, (e) the number of nucleons, and (f) the number of neutrons.

(a) The atomic mass number is A, 238.
(b) The atomic number is Z, 92.
(c) The neutron number is $N = A - Z = 238 - 92 = 146$.
(d) The number of protons is $Z = 92$.
(e) The number of nucleons is $A = 238$.
(f) The number of neutrons is $N = 146$.

. . . .

The mass of a single atom is called its **atomic mass**. Chemists and physicists have developed a unit of the measure of atomic mass called the **atomic mass unit** (u) to more easily compare the masses of different atoms. This unit is based on the mass of the common carbon atom with its six protons and six neutrons, so that the mass has been given the exact value of 12 u. Thus, 1 u is 1/12 the mass of the common carbon atom and so the approximate mass of a single proton or a single neutron is 1 u. The mass of an atom in atomic mass units is simply the sum of its protons and its neutrons. More precisely, the mass of a proton is 1.007276 u, the mass of a neutron is 1.008665 u, the mass of an electron is 0.00054858 u, and the mass of a neutral hydrogen atom is 1.007825 u.

Periodic Table

In 1889, years before the atomic theory of the atom was developed, **Dmitri Mendeleev** proposed a table containing the elements arranged in a periodic manner that showed various trends in the properties of atoms. This **Periodic Table** (see Table 21 of Appendix C) contains all of the atomic elements arranged according to their atomic numbers, which can be used to predict their chemical properties. The existence of eight columns in his table and the trends in the properties of the atoms were later understood in terms of the atomic structure described in the last section.

The horizontal rows of the Periodic Table are called periods or rows. There are seven periods, each of which starts with an atom having one electron in its outer shell and ends with an atom with a complete outer shell structure (an inert element). The first three rows consist of 2, 8, and 8 elements. Rows 4 and 5 are longer rows consisting of 18 elements, and row 6 has 32 elements. Most row 7 elements are radioactive and do not occur in nature. The number of elements in each row can be understood in terms of the atomic theory described earlier and the number of electrons in each energy level, or shell (see Table 23.2).

Metals are found on the left side of the table, with the most active metals in the lower left corner. Nonmetals are found on the right side. The noble or inert gases are on the far right. The acid-forming properties increase toward the right. Base-forming properties increase toward the left. Other properties, including the atomic weight and the atomic size, are given in a common form in the Periodic Table.

PROBLEMS 23.5

For each given isotope, find (a) its atomic mass number, (b) its atomic number, (c) its neutron number, (d) the number of protons, (e) the number of nucleons, and (f) the number of neutrons.

1. $^{12}_{6}C$
2. $^{13}_{7}N$
3. $^{48}_{22}Ti$
4. $^{141}_{58}Ba$

From the Periodic Table, find (a) the atomic number and (b) the atomic mass for each element.

5. Na (sodium)
6. Fe (iron)
7. Pb (lead)
8. Cl (chlorine)
9. The nucleus of a certain element contains 6 protons and 8 neutrons. Write its chemical symbol.
10. The nucleus of a certain element contains 92 protons and 142 neutrons. Write its chemical symbol.

Use the Periodic Table to determine the following:

11. Which of the elements Br and Ca would be expected to be more acidic?
12. Which of the elements Br and Ca would be expected to be a stronger base?
13. Which of the elements Cl, Ar, I, Ca, and K is inert?
14. Which of the elements C, N, O, and Fe is a metal?

23.6 Nuclear Mass and Binding Energy

When the total mass of a nucleus is compared with the sum of the individual masses of the protons and neutrons when unbound, or broken apart, the nuclear mass is found to be smaller. How can this be? Early in the twentieth century, Einstein stated as a physical law that mass and energy are equivalent forms of "matter." According to this **mass–energy equivalence**, mass can be converted into energy as follows:

$$E = \Delta mc^2$$

where E = energy
  Δm = change in mass
  c = speed of light, 3.00×10^8 m/s

When the total mass in a reaction is decreased by some amount, energy is given off in the form of radiation or kinetic energy. The difference in mass, converted using the equation above, is called the **binding energy of the nucleus**. This energy represents the total energy required to break a nucleus into its separate nucleons. For instance, if the mass of a $^4_2$He nucleus were exactly equal to the mass of two protons and two neutrons, the nucleus would break apart without any additional energy. To be stable, the mass of the nucleus must be less than the mass of its component nucleons. You may think of the binding energy as the energy a nucleus lacks relative to the total mass of its separate nucleon components. The binding energies of nuclei are approximately 10^6 times greater than the binding energies of the electrons in atoms and therefore are much more significant. *Note:* In this chapter we often use more than three significant digits because of the precision of atomic and nuclear measurements.

Find the binding energy of a $^4_2$He nucleus. The mass of the neutral $^4_2$He atom is 6.6463×10^{-27} kg.

EXAMPLE 1

Data: The mass of two individual neutrons, two individual protons, and two individual electrons is

$$m = 2m_p + 2m_n + 2m_e$$
$$= 2(1.6726 \times 10^{-27} \text{ kg}) + 2(1.6749 \times 10^{-27} \text{ kg}) + 2(9.1094 \times 10^{-31} \text{ kg})$$
$$= 6.6968 \times 10^{-27} \text{ kg}$$

Since the mass of neutral $^4_2$He atom $= 6.6463 \times 10^{-27}$ kg,

$$\Delta m = 6.6968 \times 10^{-27} \text{ kg} - 6.6463 \times 10^{-27} \text{ kg}$$
$$= 0.0505 \times 10^{-27} \text{ kg} = 5.05 \times 10^{-29} \text{ kg}$$

Basic Equation:

$$E = \Delta mc^2$$

Working Equation: Same

Substitution:

$$E = (5.05 \times 10^{-29} \text{ kg})(3.00 \times 10^8 \text{ m/s})^2$$
$$= 4.55 \times 10^{-12} \text{ kg m}^2/\text{s}^2 \qquad\qquad (1 \text{ J} = 1 \text{ kg m}^2/\text{s}^2)$$
$$= 4.55 \times 10^{-12} \text{ J}$$

....

The conversion factor relating atomic mass units and metric mass is

$$1 \text{ u} = 1.6605 \times 10^{-27} \text{ kg}$$

The mass of atomic and nuclear particles is also expressed in million electron volts (MeV). Recall that an electron volt is the energy that one electron would gain in passing through an electric potential difference of 1 V. One MeV is given by

$$1 \text{ MeV} = 1.602 \times 10^{-13} \text{ J}$$

Thus, the binding energy of a $^4_2$He nucleus from Example 1 in MeV is

$$4.55 \times 10^{-12} \text{ J} \times \frac{1 \text{ MeV}}{1.602 \times 10^{-13} \text{ J}} = 28.4 \text{ MeV}$$

Express the MeV unit in terms of its equivalent mass in kilograms and atomic mass units using the equation $E = \Delta mc^2$.

EXAMPLE 2

Data:

$$E = 1 \text{ MeV} = 1.602 \times 10^{-13} \text{ J}$$
$$c = 2.998 \times 10^8 \text{ m/s} \qquad\qquad (\textit{Note: } \text{we need to use 4}$$
$$\Delta m = ? \qquad\qquad\qquad\qquad \text{significant digits here.})$$

Basic Equation:

$$E = \Delta mc^2$$

Working Equation:

$$\Delta m = \frac{E}{c^2}$$

Substitution:

$$\Delta m = \frac{1.602 \times 10^{-13} \text{ J}}{(2.998 \times 10^8 \text{ m/s})^2} \times \frac{1 \text{ kg m}^2/\text{s}^2}{1 \text{ J}}$$

$$= 1.782 \times 10^{-30} \text{ kg}$$

Next, convert kilograms to atomic mass units:

$$1.782 \times 10^{-30} \text{ kg} \times \frac{1 \text{ u}}{1.6605 \times 10^{-27} \text{ kg}} = 1.073 \times 10^{-3} \text{ u}$$

Thus, 1 MeV (1.602×10^{-13} J) has an equivalent mass of 1.782×10^{-30} kg or 1.073×10^{-3} u.

················

The masses of some atomic particles are listed in the following table.

	Mass	
Particle	**kg**	**u**
Electron	9.1094×10^{-31}	0.00054858
Proton	1.67262×10^{-27}	1.007276
Neutron	1.67493×10^{-27}	1.008665
$^1_1$H atom	1.67353×10^{-27}	1.007825

The **average binding energy per nucleon** in a nucleus is the total binding energy of the nucleus divided by the total number of nucleons, A. Fig. 23.11 gives the average binding

Figure 23.11 Average binding energy per nucleon for a given mass number A for stable nuclei.

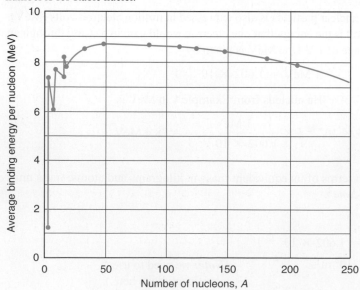

energy per nucleon in MeV for stable nuclei. For $^4_2$He the average binding energy is its total binding energy in MeV, which we found following Example 1, divided by its total number of nucleons; that is, 28.4 MeV/4 = 7.1 MeV. Binding energies are largest for the number of nucleons, A, between about 30 and 60. This allows nuclei below 30 and above 60 to undergo reactions or decay, which produce substantial amounts of energy.

Estimate the average binding energy for $^{238}_{92}$U from Fig. 23.11. ◄

EXAMPLE 3

Note that $^{238}_{92}$U has 238 nucleons as indicated by its atomic mass number, the upper number. Place a ruler or straight edge vertically at 238 on the horizontal axis. Note its point of intersection with the graph. Then, read the number on the vertical axis, 7.6 MeV, corresponding to this point of intersection. Thus, the average binding energy for $^{238}_{92}$U is approximately 7.6 MeV.

·················

PROBLEMS 23.6

1. Find the mass in kilograms of the $^{232}_{92}$U atom if its mass in atomic mass units is 232.037131 u.
2. Find the mass in kilograms of the $^{228}_{90}$Th atom if its mass in atomic mass units is 228.028716 u.
3. Find the binding energy of a $^{14}_7$N nucleus in MeV if the mass of the neutral N atom is 14.00307 u.
4. Find the binding energy of a $^{232}_{92}$U nucleus in MeV if the mass of the neutral U atom is 232.037131 u.
5. Find the binding energy of a $^{228}_{90}$Th nucleus in MeV if the mass of the neutral Th atom is 228.028716 u.

Estimate the average binding energy per nucleon for each of the following nuclei from Fig. 23.11.

6. $^{48}_{22}$Ti
7. $^{141}_{58}$Ba
8. $^{12}_6$C

SKETCH

12 cm² | w

4.0 cm

DATA

$A = 12$ cm², $l = 4.0$ cm, $w = ?$

BASIC EQUATION

$A = lw$

WORKING EQUATION

$w = \frac{A}{l}$

SUBSTITUTION

$w = \frac{12 \text{ cm}^2}{4.0 \text{ cm}} = 3.0$ cm

23.7 Radioactive Decay

The size of an atomic nucleus is limited by the fact that neutrons are unstable. On average, an isolated neutron will decay into a proton and an electron in about 12 min. In the presence of protons, the neutron is more stable (Fig. 23.12). In many atomic nuclei, neutrons are stable for many billions of years. For heavy atoms, the large number of neutrons required to hold the protons together leads to a situation where not enough protons exist to prevent one or more of the neutrons from decaying into a proton and an electron. Such unstable nuclei are called *radioactive*. **Radioactive decay** occurs when an unstable nucleus of an atom is transformed into a new element through the spontaneous disintegration of its atomic nuclei. Examples of radioactive elements are uranium, plutonium, radium, thorium, and their products. Often, one radioactive isotope decays into another isotope that is also radioactive, which decays into yet another radioactive isotope. A *radioactive series* is a series of isotopes of various elements that is successively transformed into lighter elements before the series of elements reaches a stable state, usually lead.

All elements heavier than bismuth exhibit radioactive decay. The radioactive elements exhibit three types of decay in which one or more radioactive rays are emitted from the nucleus: alpha (α), beta (β), and gamma (γ) rays. An $\boldsymbol{\alpha}$ **ray** consists of α particles, each having two protons and two neutrons, and is positively charged; that is, it curves in the direction that known positive charges curve in a magnetic field. The helium nucleus is identical to an α particle. As we discuss later, the α particle (helium nucleus) is the most stable nuclear combination and can therefore be easily formed in a nuclear reaction. A $\boldsymbol{\beta}$ **ray** consists of a stream of β particles (electrons) that are emitted from neutrons in a

Figure 23.12 (a) A neutron by itself is unstable; the lone neutron decays into a proton and an electron. (b) A neutron with a proton is stable.

Unstable   Stable

(a)   (b)

nucleus as the neutrons decay into protons and electrons. A β ray is negatively charged and curves in the opposite direction of an α ray in a magnetic field. A **γ ray** has no mass and is composed of photons of electromagnetic radiation. A γ ray is similar to light but has much higher energy; it is not affected by a magnetic field and is uncharged. (See Fig. 23.13.)

Figure 23.13 Alpha and beta rays curve in opposite directions in a magnetic field; gamma rays are undeflected.

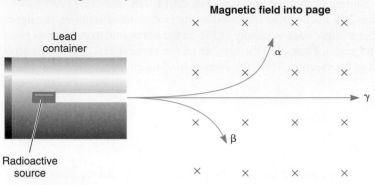

Alpha Decay

Figure 23.14 Uranium 238 is unstable and decays by giving off an α particle.

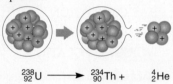

$$^{238}_{92}U \longrightarrow {}^{234}_{90}Th + {}^{4}_{2}He$$

A nucleus that emits an α particle becomes different because it has lost two protons and two neutrons (Fig. 23.14). When radium 226 ($^{226}_{88}Ra$) gives off an α particle, it becomes a nucleus with $A = 226 - 4 = 222$ and $Z = 88 - 2 = 86$. This nucleus is that of the element radon ($^{222}_{86}Rn$). This nuclear reaction is written in the form

$$^{226}_{88}Ra \longrightarrow {}^{222}_{86}Rn + {}^{4}_{2}He$$

Alpha decay occurs because there are not enough neutrons in the nucleus to keep it stable. The total energy released in radioactive decay is called the **disintegration energy**, Q, and is given by

$$Q = (M_p - M_d - m_\alpha)c^2$$

where Q = disintegration energy
   M_p = mass of the parent nucleus
   M_d = mass of the daughter nucleus
   m_α = mass of the α particle
   c = speed of light

The emission of α particles changes the *parent* nucleus into a different nucleus, called a *daughter* nucleus. This disintegration energy is in the form of kinetic energy of (1) the α particle as it moves away from the nucleus and (2) the recoiling nucleus that moves away in the opposite direction.

Beta Decay

A nucleus can also decay by the emission of a β particle (an electron). How can a nucleus made up of only protons and neutrons emit an electron? One of the neutrons in the nucleus can break up into a proton and an electron and give off the electron in a manner similar to the decay of free neutrons. In β decay, a nucleus thus changes its charge, and the atom will have to pick up an additional electron in its charge clouds to remain neutral. An element involved in this process is thereby changed into a different element with one more electron. An example of β decay is the changing of carbon into nitrogen (Fig. 23.15):

$$^{14}_{6}C \longrightarrow {}^{14}_{7}N + e^-$$

Figure 23.15 Carbon 14 can decay into a nitrogen 14 atom by giving off a β particle (an electron).

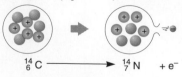

$$^{14}_{6}C \longrightarrow {}^{14}_{7}N + e^-$$

The radioactivity of an isotope is commonly measured by the decay rate or the half-life (Fig. 23.16). The **decay rate** is the probability per unit time that a decay of radioactive isotopes will occur. The decay of radioactive isotopes is a completely random process. That is, it cannot be predicted exactly when a given nucleus will decay. For a large number of nuclei in a sample, the **half-life** is the length of time required for one-half of the original amount of the radioactive atoms in the sample to decay (Fig. 23.16). This means that after the first half-life, one-half of the original amount remains. After the second half-life, one-half of one-half, or one-fourth, of the original amount remains. After the third half-life, one-half of one-fourth, or one-eighth, of the original amount remains. This process continues.

Figure 23.16   An exponential decrease of radioactive nuclei is observed.

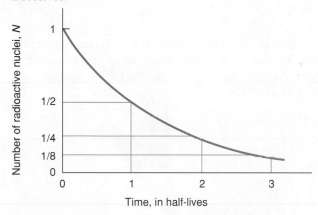

Draw a graph through four half-lives of the radioactive decay of 80.0 g of radium 226, which has a half-life of 1620 years (See Fig. 23.17).

EXAMPLE 1

    Note that the amount of radium is halved as we move down the vertical axis in Fig. 23.17, whereas the number of years doubles as we move across the horizontal axis. That is, one-half of the original amount (80.0 g) of radium 226 decays in the first 1620 years (leaving 40.0 g). During the second half-life, one-half of the remaining 40.0 g decays in the next 1620 years (leaving 20.0 g). During the third half-life, one-half of the remaining 20.0 g decays in the next 1620 years (leaving 10.0 g). During the fourth half-life, one-half of the remaining 10.0 g decays in the next 1620 years (leaving 5.0 g).

Figure 23.17   Radioactive decay of 80.0 g of radium 226 through four half-lives.

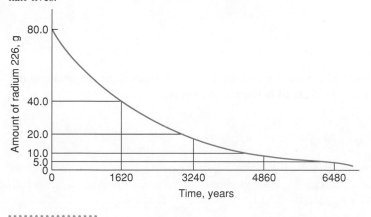

Radioactive decay can be expressed by the equation

$$N = N_0 e^{-\lambda t}$$

where N = number of remaining radioactive atoms
N_0 = original number of radioactive atoms
e = the natural number (e = 2.718 approximately)
λ = decay constant
t = elapsed time

The half-life $T_{1/2}$ is related to the decay constant by the equation

$$T_{1/2} = \frac{0.693}{\lambda}$$

where $T_{1/2}$ = half-life
λ = decay constant

Note: The half-life formula involves an exponential equation with the natural number e, which is beyond the basic mathematics used in this book. However, to gain a basic sense of the results of this formula, exponential calculations involving the natural number e can easily be performed on calculators with an e^x key.

EXAMPLE 2

Find the half-life of $^{13}_{7}N$ if its decay constant is 1.16×10^{-3} decay/s.

Data:

$$\lambda = 1.16 \times 10^{-3}/s$$
$$T_{1/2} = ?$$

Basic Equation:

$$T_{1/2} = \frac{0.693}{\lambda}$$

Working Equation: Same

Substitution:

$$T_{1/2} = \frac{0.693}{1.16 \times 10^{-3}/s}$$
$$= 597 \text{ s}$$

That is, one-half of the original amount will decay in 597 s.

· · · · · · · · · · · · · · · · ·

EXAMPLE 3

Find the remaining amount of radioactive radium after 2450 years if the original amount was 3.54×10^{23} atoms. The half-life of radium is 1620 years.

Data:

$$N_0 = 3.54 \times 10^{23}$$
$$\lambda = \frac{0.693}{T_{1/2}} = \frac{0.693}{1620 \text{ yr}} = 4.28 \times 10^{-4}/yr$$
$$t = 2450 \text{ yr}$$
$$N = ?$$

Basic Equation:

$$N = N_0 e^{-\lambda t}$$

Working Equation: Same

Substitution:

$$N = (3.54 \times 10^{23})e^{-(4.28 \times 10^{-4}/\text{yr})(2450 \text{ yr})}$$
$$= (3.54 \times 10^{23})e^{-1.0486}$$
$$= 1.24 \times 10^{23} \text{ atoms}$$

Note: $\dfrac{1.24 \times 10^{23}}{3.54 \times 10^{23}} = 35.0\%$ of the radioactive atoms remain after 2450 years.

....

Half-lives of atoms range from a small fraction of a second to billions of years. For short times, radioactive half-lives can be measured by waiting until one-half of the atoms decay. For longer times, the rate at which a sample decays can be measured by using one of several types of radiation detectors.

Radiocarbon Dating

Radiocarbon dating is a method used to obtain age estimates of organic materials. The method is used in archaeology, geology, and other branches of science to find the age of wood, charcoal, bone and antler, peat and organic-bearing sediments, and other organic materials. Radioactive carbon 14 is produced when nitrogen 14 is bombarded by cosmic rays in the atmosphere, drifts down to earth, and is absorbed from the air by plants. Animals eat the plants and take the carbon into their bodies. Humans take the carbon into their bodies by eating both plants and animals. When a living organism dies, it stops absorbing the carbon; the carbon already in the organism begins to disintegrate. Scientists then can measure how much carbon 14 has disintegrated and how much remains. Carbon 14 decays at a slow, steady rate (half-life 5730 years) and reverts to nitrogen 14. The method involves counting the number of β radiations per minute per gram of material. Modern carbon 14 emits approximately 15 β radiations per gram of material; carbon material that is 5730 years old will emit only half that amount. Radiocarbon dating of organic materials older than 50,000 years is not reliable because the amount of carbon 14 remaining is so small that the fossil cannot be dated reliably. The accuracy of radiocarbon dating for some materials less than 50,000 years old has been verified through comparison with ancient historical records.

PHYSICS CONNECTIONS

Food Irradiation

The widespread threat of food-borne diseases and contamination of food from bacteria and insects as well as food's relatively short shelf-life have resulted in significant economic losses and health problems. In the 1950s, the U.S. Army began irradiating food to prolong its shelf-life for soldiers. In the 1960s, the U.S. Food and Drug Administration (FDA) approved the irradiation of wheat flour to limit the growth of mold. Since then, the FDA has tested and approved the irradiation of fruits, vegetables, spices, pork, chicken, and red meat to increase shelf-life and reduce or eliminate insects, parasites, and bacteria in food (Fig. 23.18).

Most food is irradiated using γ radiation from cobalt 60 and cesium 137 sources. The γ radiation produced from the radioactive sources breaks up the DNA molecules in the microbes, insects, and bacteria. The exposure to the radiation either kills the organisms or prevents them from reproducing. Insects are quickly killed upon breakup of their DNA, whereas bacteria may take longer to destroy. If fruits and vegetables are exposed to radiation, their DNA is destroyed, which delays unwanted sprouting and/or ripening.

Upon seeing the irradiated food symbol, many people will not purchase the item. When food has been irradiated, it means that the food was exposed to radiation to destroy unwanted bacteria, insects, and microbes, but the food itself does not become radioactive. Irradiating food to increase shelf-life

Figure 23.18 (a) This symbol must be placed on all food that is exposed to food irradiation. (b) Food that has not been exposed and that has been exposed to irradiation, respectively.

reduces the cost of rushed preparation and shipping costs. Such benefits would be realized not only by individuals in the United States, but also by individuals in third-world countries, where a great deal of food is prevented from being imported because it spoils before arriving at the ports. A few studies have indicated that irradiation reduces some essential vitamins and other nutrient levels that were present before the irradiation process. However, other studies have shown that such reductions are also found when food is canned or cooked.

Irradiation, although relatively new for the majority of our population, has been used for years by the military and on NASA space missions. A crop can be preserved for a significant amount of time as a result of irradiation.

PROBLEM SOLVING

SKETCH

12 cm² | w

4.0 cm

DATA

$A = 12 \text{ cm}^2$, $l = 4.0 \text{ cm}$, $w = ?$

BASIC EQUATION

$A = lw$

WORKING EQUATION

$w = \frac{A}{l}$

SUBSTITUTION

$w = \frac{12 \text{ cm}^2}{4.0 \text{ cm}} = 3.0 \text{ cm}$

PROBLEMS 23.7

1. Find the half-life of a radioactive sample if its decay constant is 1.72×10^4 decays/s.
2. Find the half-life of a radioactive sample if its decay constant is 8.25×10^{-6} decay/s.
3. Find the remaining quantity of $^{124}_{55}\text{Cs}$ from an original sample of 50.0 g after 4.00 min. Its half-life is 30.8 s.
4. Find the remaining quantity of radon 222 from an original sample of 75.0 g after 10.0 days. Its half-life is 3.82 days.
5. Find the percent of a sample of $^{124}_{55}\text{Cs}$ that will decay in the next 10.0 s. Its half-life is 30.8 s.
6. Find the percent of a sample of $^{238}_{92}\text{U}$ that will decay in the next 975 years. Its half-life is 4.47×10^9 years.
7. Find the remaining quantity of uranium 238 atoms from an original sample of 5.50×10^{20} atoms after 2.45 billion years. Its half-life is 4.50 billion years.
8. Find the remaining quantity of $^{14}_{6}\text{C}$ from an original sample of 3.75×10^{21} atoms after 1000 years. Its half-life is 5730 years.
9. Find the percent of a $^{14}_{6}\text{C}$ sample that will decay in the next 3000 years. Its half-life is 5730 years.
10. Find the percent of a radioactive sample of half-life 2.35 s that will decay in the next second.

23.8 Nuclear Reactions—Fission and Fusion

Nuclear reactions take place when a nucleus is struck by a neutron, a γ ray, or another nucleus, causing an interaction. Many nuclear reactions have been viewed since Ernest Rutherford's 1919 observation of a reaction in which nitrogen, when bombarded by α particles, emitted a proton and was transformed into an isotope of oxygen. One of the most significant findings occurred during an experiment using neutron bombardment of existing elements in an attempt to produce new elements. Scientists, while attempting to form heavier elements using neutron bombardment of uranium, the heaviest known element at that time, made an amazing discovery, which has had a great impact on the world. This discovery led to the development of nuclear bombs and nuclear reactors used to produce heat energy for electric power, propulsion of nuclear submarines, various industrial processes, and other applications. They found that the bombardment of uranium with neutrons produces lighter nuclei, each nearly half the size of uranium, along with a number of neutrons (Fig. 23.19). The reaction can be written

$$n + {}^{235}_{92}\text{U} \longrightarrow {}^{236}_{92}\text{U} \longrightarrow {}^{141}_{56}\text{Ba} + {}^{92}_{36}\text{Kr} + 3n$$

Figure 23.19   Fission of ${}^{235}_{92}\text{U}$ after neutron capture.

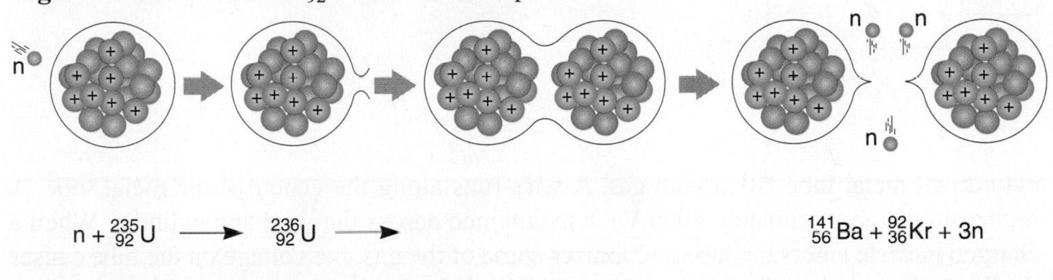

$$n + {}^{235}_{92}\text{U} \longrightarrow {}^{236}_{92}\text{U} \longrightarrow {}^{141}_{56}\text{Ba} + {}^{92}_{36}\text{Kr} + 3n$$

This nuclear reaction, in which an atomic nucleus splits into fragments with the release of energy, is known as **nuclear fission**. A substantial amount of energy is released in this process because the mass of the uranium nucleus plus the mass of the bombarding neutron is larger than the combined mass of the fission fragments produced by the reaction. The total energy release per fission is approximately 200 MeV. On the nuclear scale, this is an extremely large amount of energy per atom!

The neutrons released in each nuclear fission reaction can be used to create further reactions. The process is called a **chain reaction**. **Enrico Fermi** and his associates succeeded in producing the first self-sustaining chain reaction at the University of Chicago in 1942. Chain reactions are used in nuclear reactors to produce electric power, to produce intense neutron sources for research and medical use, and in atomic bombs. **Robert J. Oppenheimer** led the research that resulted in the construction of the first atomic bomb. Many of the end products of fission reactions are radioactive for long periods of time. These radioactive wastes must be disposed of with extreme care.

Another process for releasing energy is **nuclear fusion**, a nuclear reaction in which light nuclei interact to form heavier nuclei with the release of energy. Nuclear fusion can occur because the total mass of the reaction products can be less than that of the initial nuclei and particles. Very high temperatures are required for nuclear fusion to occur. An example of a sequence of fusion reactions is

$$\begin{aligned}
{}^{1}_{1}\text{H} + {}^{1}_{1}\text{H} &\longrightarrow {}^{2}_{1}\text{H} + e^{+} & (0.42\text{ MeV}) \\
{}^{1}_{1}\text{H} + {}^{2}_{1}\text{H} &\longrightarrow {}^{3}_{2}\text{He} + \gamma & (5.49\text{ MeV}) \\
{}^{3}_{2}\text{He} + {}^{3}_{2}\text{He} &\longrightarrow {}^{4}_{2}\text{He} + {}^{1}_{1}\text{H} + {}^{1}_{1}\text{H} & (12.86\text{ MeV})
\end{aligned}$$

This reaction sequence is constantly going on in many stars, including our sun. Fusion is the source of energy in stars. The sun's energy is believed to be produced largely by this sequence of reactions, which starts with protons and produces helium nuclei (α particles) with the release of substantial energy. Nuclear fusion has been used in hydrogen bombs. Fusion is also a subject of considerable research as a "clean" energy source because the end products of the reaction are not radioactive.

23.9 Detection and Measurement of Radiation

Since we cannot see, touch, or feel nuclear radiation, it is necessary to have instruments available to detect its presence. The most common detector of radiation is the Geiger counter (Fig. 23.20), named after **Hans Wilhelm Geiger**. This instrument contains a

Enrico Fermi (1901–1954), *physicist, was born in Italy. He helped explain the theoretical behavior of atomic particles, developed the theory of beta decay, and produced the first controlled nuclear chain reaction at the University of Chicago in 1942.*

Robert J. Oppenheimer (1904–1967), *physicist, was born in New York City. He led the scientific research that resulted in the construction of the first atomic bomb.*

Hans Wilhelm Geiger (1882–1945), *physicist, was born in Germany. He performed experiments on beta-ray radioactivity and devised an instrument (the Geiger counter) to measure it.*

Figure 23.20 Diagram of a Geiger counter.

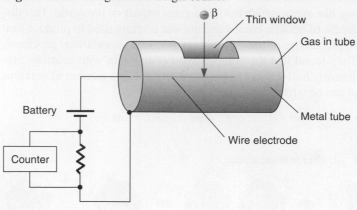

cylindrical metal tube filled with gas. A wire runs along the center of the metal tube. A high voltage (approximately 1000 V) is maintained across the wire and cylinder. When a charged particle enters the tube and ionizes some of the gas, the voltage on the tube causes the gas to "break down" and conduct an electric pulse between the wire and the cylinder. This electric pulse is detected by electronics that count the number of pulses. A loudspeaker is often hooked up to the electronics, allowing a "click" to be heard each time a charged particle enters the tube.

Since radiation can cause severe damage to living organisms and materials, it is necessary to be able to measure the amount, or *dose*, of radiation. There are two common units of *source activity*: the **curie** (Ci), which is defined as

$$1 \text{ Ci} = 3.70 \times 10^{10} \text{ disintegrations/s}$$

and the **becquerel** (Bq), which is the SI unit and defined as

$$1 \text{ Bq} = 1 \text{ disintegration/s}$$

The **source activity** is the strength of a source of radiation that can be specified at a given time according to the equation

$$A = \lambda N = \lambda N_0 e^{-\lambda t} = A_0 e^{-\lambda t}$$

where A = source activity
  N = remaining quantity of radioactive atoms
  N_0 = original quantity
  λ = decay constant
  t = time
  A_0 = original source activity

EXAMPLE

Find the source activity of a $^{222}_{88}\text{Ra}$ (radium) source 6.54 days after it was originally certified to have an activity of 0.356 Ci. Its half-life is 3.82 days.

Data:

$$t = 6.54 \text{ days}$$
$$A_0 = 0.356 \text{ Ci}$$
$$T_{1/2} = 3.82 \text{ days}$$
$$\lambda = \frac{0.693}{T_{1/2}} = \frac{0.693}{3.82 \text{ days}} = 0.181/\text{day}$$
$$A = ?$$

Basic Equation:

$$A = A_0 e^{-\lambda t}$$

Working Equation: Same

Substitution:

$$A = (0.356 \text{ Ci})e^{-(0.181/\text{day})(6.54 \text{ day})}$$
$$= (0.356 \text{ Ci})e^{-1.184}$$
$$= 0.109 \text{ Ci}$$

....

PROBLEMS 23.9

1. Find the source activity of a 1.24-Ci sample of $^{13}_{7}\text{N}$ (nitrogen) 20.0 min after certification. Its half-life is 10.0 min.
2. Find the source activity of a 2.64-Ci sample of $^{14}_{6}\text{C}$ (carbon) 4000 years after certification. Its half-life is 5370 yr.
3. Find the source activity of a 0.476-Ci sample of $^{24}_{11}\text{Na}$ (sodium) 36.5 h after certification. Its half-life is 14.95 h.
4. Find the source activity of a 3.98-Ci sample of $^{11}_{6}\text{C}$ (carbon) 10.3 h after certification. Its half-life is 20.4 min.
5. Find the source activity of a 10.0-Ci sample of $^{226}_{88}\text{Ra}$ (radium) 125 yr after certification. Its half-life is 1590 yr.
6. Find the source activity of a 4.00-Ci sample of $^{131}_{53}\text{I}$ (iodine) 6.00 h after certification. Its half-life is 8.06 days.
7. Find the source activity of a 75.0-μCi sample of $^{214}_{84}\text{Po}$ (polonium) 5.00 μs after certification. Its half-life is 1.50×10^{-4} s.

PROBLEM SOLVING

SKETCH

12 cm^2 | w

4.0 cm

DATA
$A = 12$ cm^2, $l = 4.0$ cm, $w = ?$

BASIC EQUATION
$A = lw$

WORKING EQUATION
$w = \frac{A}{l}$

SUBSTITUTION
$w = \frac{12 \text{ cm}^2}{4.0 \text{ cm}} = 3.0$ cm

23.10 Radiation Penetrating Power

The three types of radiation rays (α, β, and γ) can be stopped by matter. Some are stopped more easily than others (Fig. 23.21). Alpha rays, which are massive and carry a double positive charge, are stopped quite easily during "collisions" with atoms in the material. These collisions occur when the positively charged α particle comes close enough to an atom to feel a Coulomb force from some of the electrons or protons. The α particle gives up its energy to the material through these "collisions." A few pieces of paper or a few centimetres

Figure 23.21 Penetrating power of different forms of nuclear radiation.

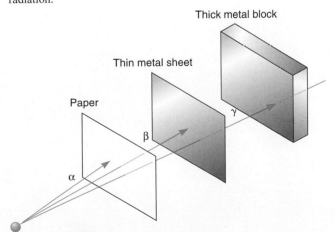

of air are sufficient to stop an α particle. As soon as an α particle slows enough to pick up two electrons while passing through matter, it becomes a harmless helium atom.

Beta rays, which are electrons, can easily be stopped by a thin sheet of metal, such as aluminum. They lose their energy due to many collisions with electrons in atoms in the material they are passing through. They become electrons in the material, indistinguishable from others.

Gamma rays have the greatest penetrating power because they have no charge and therefore cannot be electrically attracted or repelled by electrons or protons. The γ ray, which is an energetic photon, is slowed only through a direct hit of an electron or a nucleus. A dense material, such as a heavy metal, is the best absorber of γ radiation.

As radiation passes through matter and loses its energy, it can do much damage. Many materials become weakened and brittle upon exposure to substantial levels of radiation. This major engineering problem had to be solved in order to build safe nuclear reactors.

Radiation damage to living organisms can be very severe. As radiation is absorbed in material, atoms become ionized as electrons are captured or emitted. If these atoms are basic in bonding molecules together, the molecule may break apart. The functioning of the living cell may be altered substantially if a large number of key molecules are damaged. If the damage occurs to the DNA molecule, which controls the growth and replication of the cell, the cell may be severely damaged and die. Obviously, if a large number of cells die, the organism itself will become sick (radiation sickness) or die. If the damaged cell replicates itself rapidly with faulty DNA, cancer may result. To prevent this type of damage during normal medical X-ray procedures, it is common for patients to wear lead aprons around tissue not being imaged by the X ray. This is also why X-ray technicians step behind a lead shield when X rays are being taken.

Radiation can be useful for medical diagnostics (X rays and radioactive tracers) and radiation therapy. In radiation therapy, cancer cells can be destroyed by the localized application to a tumor.

Glossary

Alpha Ray A ray consisting of alpha particles, each having two protons and two neutrons, and positively charged, that is, it curves in the direction that known positive charges curve in a magnetic field. (p. 631)

Atomic Mass The mass of a single atom, usually expressed in atomic mass units, u. (p. 627)

Atomic Mass Number The total number of nucleons (protons and neutrons) in a nucleus, A. (p. 626)

Atomic Mass Unit (u) A unit of measure of atomic mass based on the mass of the common carbon atom, which has six protons and six neutrons, so the mass of the carbon 12 atom has been given the exact value of 12 u. Thus, the approximate mass of a single proton or a single neutron is 1 u, and the mass of an atom in atomic mass units is simply the sum of the number of its protons and its neutrons. (p. 627)

Atomic Number The number of protons in a nucleus, Z. (p. 626)

Average Binding Energy per Nucleon The total binding energy of the nucleus divided by the total number of nucleons. (p. 630)

Becquerel (Bq) Unit of source activity; 1 Bq = 1 disintegration/s. (p. 638)

Beta Ray A ray consisting of a stream of beta particles (electrons), which are emitted from neutrons in a nucleus as they decay into protons and electrons. This ray is negatively charged and curves in the opposite direction of an alpha ray in a magnetic field. (p. 631)

Binding Energy of the Nucleus The total energy required to break a nucleus apart into separate nucleons. (p. 629)

Bohr Model An early model of atomic structure in which electrons travel around the nucleus in a number of discrete stable energy levels determined by quantum conditions. (p. 621)

Chain Reaction The process of using the neutrons released in each nuclear fission reaction to create further reactions. (p. 637)

Coulomb Force An electric, repulsive force between protons. (p.626)

Curie (Ci) Unit of source activity; 1 Ci = 3.70×10^{10} disintegrations/s. (p. 638)

Decay Rate The probability per unit time that a decay of radioactive isotopes will occur. (p. 632)

Disintegration Energy The total energy released in radioactive decay in the form of kinetic energy. (p. 632)

Electron A fundamental particle of an atom; negatively charged. (p. 620)

Excited State A high-energy level for an electron in an atom. (p. 621)

Fluorescence A property of certain substances in which radiation is absorbed at one frequency and then reemitted at a lower frequency. (p. 622)

Gamma Ray A ray composed of photons of electromagnetic radiation, which have no mass. (p. 632)

Ground State The lowest energy level for an electron in an atom. (p. 621)

Half-Life The length of time required for one-half of the original amount of the radioactive atoms in a sample to decay. (p. 633)

Isotopes Nuclei that contain the same number of protons but a different number of neutrons. (p. 627)

Mass–Energy Equivalence A physical law stating that mass and energy are equivalent forms of matter; $E = \Delta mc^2$; stated by Einstein. (p. 628)

Neutron A fundamental particle in the nucleus of an atom; neutrally charged. (p. 602)

Neutron Number The number of neutrons in an atomic nucleus of a particular nuclide, N. Also, $N = A - Z$, the difference between the atomic mass number and the atomic number. N can be different for nuclei with the same atomic number. (p. 627)

Nuclear Fission A nuclear reaction in which an atomic nucleus splits into fragments with the release of energy. (p. 637)

Nuclear Fusion A nuclear reaction in which light nuclei interact to form heavier nuclei with the release of energy. (p. 637)

Nuclide A specific type of atom characterized by its nuclear properties, such as the number of neutrons and protons and the energy state of its nucleus. (p. 627)

Periodic Table A table that contains all of the atomic elements arranged according to their atomic numbers and which can be used to predict their chemical properties. (p. 628)

Proton A fundamental particle in the nucleus of the atom; positively charged. (p. 620)

Quantum Mechanics A theory that unifies the wave–particle dual nature of electromagnetic radiation. The quantum model is based on the idea that matter, just like light, behaves sometimes as a particle and sometimes as a wave. (p. 624)

Radioactive Decay A type of nuclear decay that occurs when an unstable atom is transformed into a new element through the spontaneous disintegration of its nucleus. (p. 631)

Radiocarbon Dating A method used to obtain age estimates of organic materials using carbon 14 decay. (p. 635)

Source Activity The strength of a source of radiation that can be specified at a given time. (p. 638)

Strong Force An attractive force among all nucleons (neutrons and protons) independent of their charge. (p. 626)

Formulas

23.3 $E = -\dfrac{kZ^2}{n^2}$

23.6 $E = \Delta mc^2$

23.7 $Q = (M_p - M_d - m_\alpha)c^2$

$N = N_0 e^{-\lambda t}$

$T_{1/2} = \dfrac{0.693}{\lambda}$

23.9 $A = \lambda N = \lambda N_0 e^{-\lambda t} = A_0 e^{-\lambda t}$

Review Questions

1. Which of the following are nuclear particles?
   (a) neutrons
   (b) protons
   (c) nucleons
   (d) atoms
   (e) all of the above

2. Einstein's equivalence principle relates to
   (a) weight and time.
   (b) space and gravity.
   (c) mass and energy.
   (d) all of the above.
   (e) none of the above.

3. The amount of radioactive material remaining after a period of time is related to the
   (a) atomic mass.
   (b) pressure.
   (c) half-life.
   (d) volume.
   (e) none of the above.

4. Explain the difference between the ground state and the excited states of electrons in atoms.
5. Describe the Bohr model of the atom.
6. Describe the similarities of protons and neutrons. Describe the differences.
7. Describe the differences between the electric force and the strong force.
8. If the strong force suddenly became much weaker in a nucleus while the electric force remained unchanged, what might happen to the nucleus?
9. Explain the principle of mass–energy equivalence in your own words.
10. What is the difference among the following atoms: $^{11}_{6}C$, $^{12}_{6}C$, and $^{13}_{6}C$? What are the similarities?
11. Explain the term *electron volt* in your own words.
12. Describe the importance of the neutron in atomic nuclei.
13. Describe an α ray in your own words.
14. Describe a γ ray in your own words.
15. Describe a β ray in your own words.
16. What important discovery was made by Enrico Fermi?
17. Explain a self-sustaining chain reaction.
18. Describe nuclear fusion.
19. Describe nuclear fission.
20. What fraction of a radioactive sample has not decayed after four half-lives have elapsed?
21. What damage can be caused to living organisms by radiation?
22. What medical uses does radiation have?

Review Problems

1. Find the energy of the electron in the $n = 5$ energy level of a hydrogen atom.
2. Find the energy of the photon given off when the electron in Problem 1 transitions down to the $n = 3$ energy level.
3. List four inert atoms from the Periodic Table.
4. List four metals from the Periodic Table.
5. Find the mass in kilograms of the $^{15}_{8}O$ atom. Its mass in atomic mass units is 15.003065 u.
6. Find the mass in kilograms of the $^{19}_{9}F$ atom. Its mass in atomic mass units is 18.998404 u.
7. Find the mass in kilograms of the $^{166}_{68}Er$ atom. Its mass in atomic mass units is 165.930292 u.
8. Find the binding energy in MeV of a $^{86}_{38}Sr$ atom. The mass of the neutral Sr atom is 85.909266 u.
9. Estimate the average binding energy for $^{102}_{44}Ru$ from Fig. 23.11
10. Estimate the average binding energy for $^{153}_{63}Eu$ from Fig. 23.11.
11. Estimate the average binding energy for $^{187}_{75}Re$ from Fig. 23.11.
12. Find the remaining quantity of osmium 191 atoms from an original sample of 8.25×10^{13} atoms after 54 days. Its half-life is 15.4 days.
13. Find the remaining quantity of iodine 131 atoms from an original sample of 8.33×10^{18} atoms after 34.4 days. Its half-life is 8.04 days.
14. Find the percent of a strontium 88 sample that will decay in the next 2.40 yr. Its half-life is 29.1 yr.
15. Find the percent of an osmium 191 sample that will decay in the next 2.00 h. Its half-life is 15.4 days.
16. Find the source activity of a 2.43-Ci sample of osmium 191 43.3 days after certification. Its half-life is 15.4 days.

SKETCH

12 cm² | w
4.0 cm

DATA

$A = 12 \text{ cm}^2$, $l = 4.0$ cm, $w = ?$

BASIC EQUATION

$A = lw$

WORKING EQUATION

$w = \frac{A}{l}$

SUBSTITUTION

$w = \frac{12 \text{ cm}^2}{4.0 \text{ cm}} = 3.0$ cm

17. Find the source activity of a 3.79-Ci sample of nitrogen 13 43.0 min after certification. Its half-life is 10.0 min.

18. Find the source activity of a 9.41-Ci sample of carbon 11 95.4 min after certification. Its half-life is 20.4 min.

19. Find the source activity of a 6.75-Ci sample of $^{238}_{92}U$ (uranium) one billion years after certification. Its half-life is 4.5×10^9 yr.

20. Find the source activity of a 50.0-μCi sample of $^{214}_{82}Pb$ (lead) 1.00 h after certification. Its half-life is 26.8 min.

APPLIED CONCEPTS

1. To produce the blue neon light used in many storefront signs, an electron from a hydrogen atom must undergo a transition from one energy level to another. (a) What is the transition energy (eV) if the electron changes its energy level from $n = 4$ to $n = 2$? (b) What is the frequency for the photon that is released? (c) What is the photon's velocity? (d) What is the wavelength for the released photon?

2. A photon is released from a hydrogen atom when an electron moves from a higher energy level to a lower energy level. At what level does an electron begin its journey to the level $n = 1$ if the emitted photon has a wavelength of 1.22×10^{-7} m?

3. The binding energy for a $^4_2$He nucleus is 28.40 MeV. (a) What is the binding energy (in J)? (b) How many protons, neutrons, and electrons are in the $^4_2$He atom? (c) Find the total mass of the various sub-atomic particles in the $^4_2$He atom. (d) What is the mass of the neutral $^4_2$He atom?

4. A source of radon 222 has 2.45×10^{23} atoms remaining after 10.0 days. (a) If its half-life is 3.82 days, what is its decay rate? (b) Find the original number of atoms before the decay.

5. Find the source activity of a 0.875-Ci sample of $^{24}_{11}$Na (sodium) 32.4 h before certification if its half-life is 14.95 h.

CHAPTER 24
SPECIAL AND GENERAL RELATIVITY

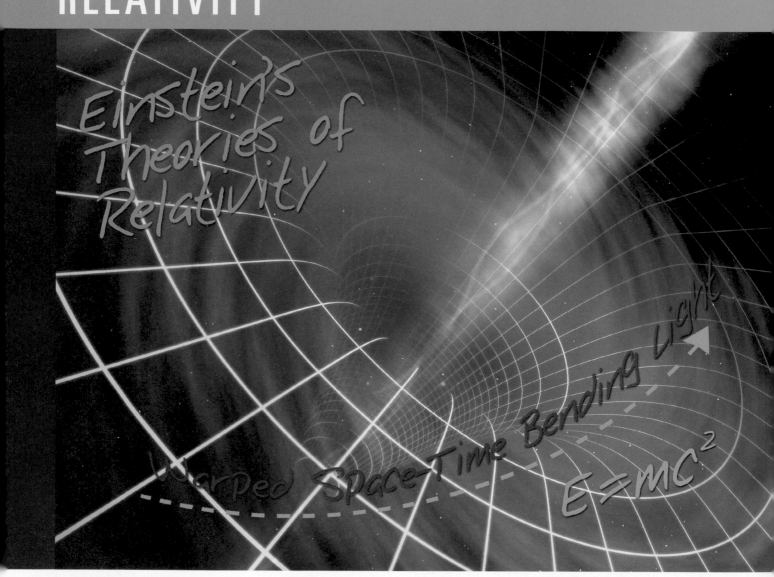

Einstein's Theories of Relativity

Warped Space-Time Bending Light

$E=mc^2$

A t the age of 16, Albert Einstein dreamed of what it would be like to ride on a beam of light. His enthusiasm for science, specifically the topics of light, space, and time, led him to develop several theories that would make him the most famous scientist of all time. In this chapter we will explore the life of Einstein, the warping of space-time, and the formula that changed the world, $E = mc^2$.

The major goals of this chapter are to enable you to:

1. Discuss the theories, successes, and milestones that made Albert Einstein an influential scientist.
2. Explain that the speed of light is constant, but the rate at which time passes can vary according to motion.
3. Find the amount of energy contained in a particular amount of mass.
4. Describe the theory of a four-dimensional space-time universe.
5. Express Einstein's theory that acceleration and gravity are equivalent.

24.1 Albert Einstein

Born in Ulm, Germany, in 1879, Albert Einstein spent most of his younger years (Fig. 24.1) reading and engaging in science experiments, while rebelling against the traditional regimentation found in many German schools at the time. He was given a compass at the age of five by his father and was intrigued by how the compass needle would move even though nothing was physically touching it. The physical mystery surrounding the compass fascinated Einstein and set him on a course that revolutionized science. However, his path was not easy and included being expelled, dropping out of school, and failing academic entrance exams. Einstein was eventually able to enroll in the Zurich Polytechnic Institute, where he was largely left to explore the wonders of science on his own.

After graduation, Einstein was unable to find a full-time job as an assistant in a laboratory. Instead, he obtained a job at the Swiss Patent Office. During this period he did some of his greatest work, including his 1905 paper about the **Special Theory of Relativity**, which is based on the following basic assumptions: (1) *the laws of physics are the same in both moving and nonmoving frames of reference* and (2) *the speed of light is constant no matter what the speed of the observer or the source of light.* The Special Theory boosted Einstein's physics career and caused many universities to take notice of the previously unknown patent clerk.

Ten years later, Einstein produced a paper that extended the special theory to accelerated frames of reference. The **General Theory of Relativity** is based on the assumption that *gravity and acceleration are equivalent and that light has mass and its path can be warped by gravity.* In 1921, Einstein won the Nobel Prize in Physics, although not for relativity, but for his work on the photoelectric effect. He began to focus on a *theory that would link what came to be understood as the five fundamental forces: gravitational, electric, magnetic, strong, and weak forces.* The work that Einstein began is continued today by physicists around the world and is now called the **Grand Unified Theory**.

During the turbulent years leading up to World War II, Germany was not a safe place for the world's most famous Jewish scientist. In the 1930s, Einstein decided to leave Germany and live in the United States. He relocated to Princeton, New Jersey, where he continued to lecture and work on his unified theory. His famous formula $E = mc^2$ was the scientific principle behind the atomic bomb, and he urged the United States to develop such a weapon because he feared that Germany might develop one first. After the war he opposed its use.

Albert Einstein died in Princeton in 1955. Throughout his life, he addressed many major scientific matters and was also a strong moral voice on social and political issues. Einstein showed great compassion for humankind and was known throughout the world as the genius professor who attempted, as Einstein said, "to read the mind of God" (Fig. 24.2).

Figure 24.1 Albert Einstein (right) as a boy with his sister Maja

Photo courtesy of Corbis Corporation/ Betteman. Reprinted with permission

Figure 24.2 Albert Einstein at work in his study at Princeton University in 1943

Photo courtesy of the National Archives and Records Administration

24.2 Special Theory of Relativity

Let's say you are reading this textbook while sitting on the ground. Are you moving? You would probably say no because when you look around, nothing seems to be changing its position. However, you suddenly hear a noise and see a man driving a car down the street. You see that the car is changing position, but how do you know that the car is moving and you are not?

The driver of the car looks out the car window and sees you moving closer and then away from him after passing you. The driver can adopt the perspective that you are moving while he is at rest. In fact, he can see everything outside the car as moving and himself as stationary. This is a simplistic example of **relative motion**, which is the concept that motion can be described differently depending upon the observer's perspective. In order for a motion to be fully described, it must be described in reference to something or someone's perspective. Both the driver and you see the same relative motion in two different ways. You see the car moving toward you, while the driver can argue that you are moving toward him. The difference can be expressed through the notion of *relativity*: the same event can be observed differently depending on the perspective of the observer.

As a result of Einstein's thoughts on motion and light, he concluded that the laws of physics must be the same for all nonaccelerating objects. This is called Einstein's **First Postulate of Special Relativity**: *the laws of physics are the same for both moving and non-moving frames of reference.* If an individual in a moving train drops a ball while the train is moving at a constant velocity, it will appear to that person that the ball accelerates directly down toward the floor of the train. A ball also will fall directly toward the ground if a person standing on the side of the tracks drops it. In other words, Einstein believed that the laws of physics are the same for objects at rest as they are for objects moving at a constant velocity.

Einstein also proposed the **Second Postulate of Special Relativity**: *the speed of light is constant regardless of the speed of the observer or the light source.*

$$c = \text{speed of light} = 3.00 \times 10^8 \text{ m/s} = 1.86 \times 10^5 \text{ mi/s}$$

Einstein created a "thought experiment" to demonstrate the implications of the postulate that the speed of light is constant. In his thought experiment, two telephone poles are struck simultaneously by lightning at the instant a train moving close to the speed of light

PHYSICS CONNECTIONS

Albert Einstein: "The Person of the Century"

"Why is it that nobody understands me, yet everybody likes me?" This quote from Einstein, taken from an interview with *The New York Times* in 1944, is reflective of Einstein's fame and genius. Although Albert Einstein is recognized around the world as a genius, very few people understand what his genius meant for the scientific world. Despite his superior intellect, he was not able to understand why so many people, who did not understand his scientific work, adored him.

Einstein became world famous after two British scientists experimentally proved Einstein's theory that the path of light could be warped due to gravity. This occurred in 1919, just after the end of World War I, when the world needed something to rally around and get excited about. His revolutionary scientific discovery, his "crazy professor" appearance, and his outspoken views on life, human nature, and politics carried his fame to nonscientists throughout the world.

Today, almost a half-century after his death, Albert Einstein's name is still synonymous with "genius." Einstein's unique quirkiness, funny facial expressions, and thoughtful quotes can now be seen on coffee mugs, T-shirts, posters, books, and magazines. In addition, numerous fictional and documentary films have been made about his life and scientific work. *Time Magazine* recognized Albert Einstein's impact on the world by awarding him the title, "Person of the Century."

is passing the midpoint of the two poles. From the perspective of a person standing on the side of the tracks, the lightning strikes both poles at exactly the same time [Fig. 24.3(a)]. A person on the train sees the lightning first strike pole B in front of the train and then strike pole A behind the train an instant later [Fig. 24.3(b)] because, as the train moves closer to pole B, the light from pole B takes less time than the light from pole A to reach the person on the train [Fig. 24.3(c)]. Einstein's thought experiment showed, in theory, that two people can see the same event happen at different points in time. The distance and time may vary, but the speed of light will always be constant. By changing your state of motion, as was the case with the individual on the train relative to the person on the side of the tracks, the temporary relationship of events can be altered.

Figure 24.3 (a) Light travels the same distance from each telephone pole to the observer on the ground, who sees the two lightning strikes at the same time. (b) A person on the moving train experiences the same event differently. (c) As the train moves closer to pole B, the light from pole B reaches the person on the train sooner than the light from pole A because pole B is now a bit closer and the light from it takes less time to reach this observer.

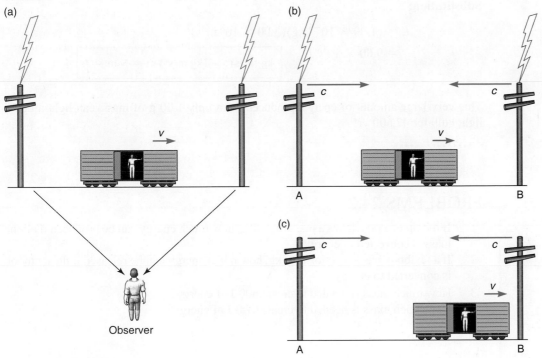

$E = mc^2$

Albert Einstein's development of the equation $E = mc^2$ became one of the most important scientific achievements of the 20th century. This equation defines the equivalence between energy and mass.

$$E = mc^2$$

where E = energy
   m = mass
   $c = 3.00 \times 10^8$ m/s = 1.86×10^5 mi/s (the speed of light)

A very little mass is equivalent to a tremendous amount of energy. However, converting mass to 100% energy is virtually impossible. If you take your pencil eraser and convert it into energy by burning it, some of the eraser's mass would be converted into energy in the form of heat, light, and sound. Yet some quantity of the eraser would remain as burnt ash, which still has mass and therefore energy. Nuclear reactions, as discussed in Chapter 23, are the best methods scientists have for converting mass to pure energy.

EXAMPLE 1

How much energy is contained in 1.00 g of matter?

Data:

$$m = 1.00 \text{ g} = 1.00 \times 10^{-3} \text{ kg}$$
$$c = 3.00 \times 10^8 \text{ m/s}$$
$$E = ?$$

Basic Equation:

$$E = mc^2$$

Working Equation: Same

Substitution:

$$E = (1.00 \times 10^{-3} \text{ kg})(3.00 \times 10^8 \text{ m/s})^2$$
$$= 9.00 \times 10^{13} \text{ J}$$

$$\boxed{\text{kg (m/s)}^2 = (\text{kg m/s}^2) \text{ m} = \text{N m} = \text{J}}$$

This very large amount of energy, produced from only 1.00 g of mass, can light a 60-W light bulb for 47,600 yr!

····················

SKETCH

12 cm^2 w

4.0 cm

DATA

$A = 12 \text{ cm}^2, l = 4.0 \text{ cm, } w = ?$

BASIC EQUATION

$A = lw$

WORKING EQUATION

$w = \frac{A}{l}$

SUBSTITUTION

$w = \frac{12 \text{ cm}^2}{4.0 \text{ cm}} = 3.0 \text{ cm}$

PROBLEMS 24.2

1. If the tip of a pencil has a mass of 2.30 g, how much energy can be produced if all the mass is converted to energy?
2. If a textbook has a mass of 1.30 kg, how much energy can be released if the textbook is converted to energy?
3. How much mass is needed to create $60\overline{0}$ J of energy?
4. How much mass is needed to create 67.0 J of energy?

24.3 Space-Time

Scientists and mathematicians use coordinates to locate objects in our physical world. A pilot defines her location by specifying her longitude, latitude, and altitude. Similarly, mathematicians and scientists use coordinates in relation to the x-, y-, and z-axes to define a position in three dimensions (Fig. 24.4).

Albert Einstein agreed that objects exist in three spatial dimensions, but felt that three spatial dimensions alone could not accurately locate an object at any particular instant. He felt that a time dimension needed to be taken into consideration when defining the dimensions of the physical universe. His argument was that anything could be located in a particular spatial location, but it could also change its position over time. **Space-time** allows an object's position in the universe to be defined using the three spatial dimensions (x, y, z) and one time dimension.

Space-time plays an important role in Einstein's Theory of General Relativity. Einstein's theory describes the universe as a place where massive objects, such as the sun, planets, and stars, warp the four dimensions of space-time. The warping of space-time influences the motion of an object traveling through the universe (Fig. 24.5). Black holes warp space to such an extent that mass and light cannot escape.

Figure 24.4 Spatial location can be determined on a three-axis coordinate system.

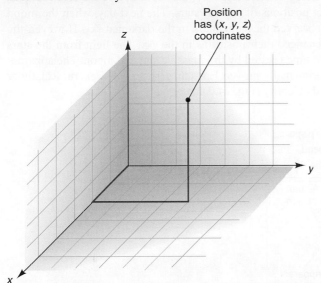

Figure 24.5 Any massive object will warp the space around it, causing objects of mass to deviate from their initial path.

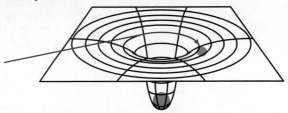

24.4 General Theory of Relativity

Although the Special Theory of Relativity focuses on the relationship between motion and time, Einstein's **General Theory of Relativity** states that gravity is equivalent to acceleration and that the path of light is warped by gravity.

Einstein designed several thought experiments to illustrate General Relativity. In one thought experiment, he pictured a person in a spaceship accelerating at 9.80 m/s². In such a situation, the spaceship applies a force on the astronaut equal to the force of gravity on the earth. As a result, Einstein concluded, gravity and acceleration are equivalent. Einstein continued in his thought experiment. If a beam of light comes through a window as the spaceship accelerates, the light will strike the opposite wall just a bit lower than the height at which it entered. As a result of the accelerating spaceship, the light appears to bend in the accelerating/gravitational field. Finally, since gravity only alters the path of objects with mass, Einstein concluded that light must have mass (Fig. 24.6).

Figure 24.6 (a) If the spaceship accelerates at 9.80 m/s², the force on the person inside will feel the same as the pull of gravity on earth, showing that acceleration is the same as gravity. (b) Although this depiction is exaggerated, the light would strike the wall lower than where it entered, showing that gravity bends light.

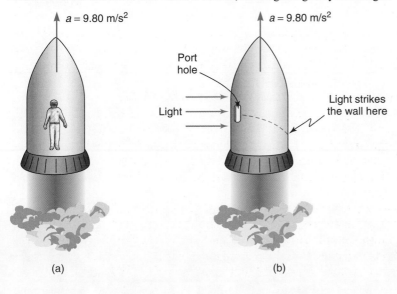

In 1919, British scientists traveled to South America to observe a solar eclipse and experimentally prove Einstein's Theory of Relativity. On the night before the solar eclipse, the scientists observed the exact positions of several stars. The next day, when the moon eclipsed the sun, the scientists observed the same stars in the darkened sky. However, the stars appeared to have slightly changed their positions in the sky. The light from the stars curved around the warp in space-time created by the massive sun and moon. The apparent shift in the stars' positions experimentally proved Einstein's Theory of General Relativity and made him a world-renowned celebrity (Fig. 24.7).

Figure 24.7 Massive objects warp space, causing light from distant stars to bend, altering their apparent position in the sky.

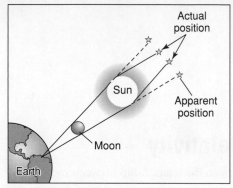

Glossary

$E = mc^2$ The equation that illustrates and defines the equivalence between energy and mass. (p. 649)

First Postulate of Special Relativity The laws of physics are the same for both moving and nonmoving frames of reference. (p. 648)

General Theory of Relativity Extends the special theory of relativity to accelerated frames of reference with the assumption that gravity and acceleration are equivalent and that light has mass and its path can be warped by gravity. (p. 647)

Grand Unified Theory A theory linking the five fundamental forces: gravitational, electric, magnetic, strong, and weak forces. (p. 647)

Relative Motion The concept that motion can be described differently depending upon the observer's perspective. (p. 648)

Second Postulate of Special Relativity The speed of light is constant regardless of the speed of the observer or of the light source. (p. 648)

Space-Time An object's position in the universe can be pinpointed using three spatial dimensions and one time dimension. (p. 650)

Special Theory of Relativity The laws of physics are the same in moving and nonmoving frames of reference and the speed of light is constant no matter what the speed of the observer or the light source. (p. 647)

Formula

24.2 $E = mc^2$

Review Questions

1. What field or fields of physics intrigued Einstein at a very young age?
   - (a) Momentum
   - (b) Thermodynamics
   - (c) Electricity and magnetism
   - (d) Sound and waves
2. Which of the following did Albert Einstein not complete?
   - (a) General Theory of Relativity
   - (b) Special Theory of Relativity
   - (c) Principle of the warping of space-time
   - (d) Grand Unified Theory
3. Describe Einstein's First Postulate of Special Relativity.
4. Explain a situation where you experienced the First Postulate of Special Relativity.
5. If you are riding a bike at 10.0 m/s and throw a ball forward with a velocity of 4.00 m/s, will the ball travel 4.00 m/s relative to the ground, or will it travel at 10.0 m/s plus another 4.00 m/s relative to the ground? Explain your reasoning.
6. Describe Einstein's Second Postulate of Special Relativity.
7. According to Einstein's Second Postulate, if you are moving 10.0 m/s on a bicycle with a headlight on, will the light from the bicycle have a velocity of 3.00×10^8 m/s or will it have a velocity of 3.00×10^8 m/s plus 10.0 m/s? Explain your reasoning.
8. What does traveling close to the speed of light do to the dimension of time?
9. While you are sitting and reading this question, are you traveling through the space dimensions or the time dimension?
10. Explain what $E = mc^2$ represents.

11. What are the four dimensions of our space-time universe? Why is time an important dimension?

12. What does Einstein's General Theory of Relativity state about the relationship between gravity and acceleration?

13. What happens to light and other electromagnetic waves around massive objects such as the sun?

14. Explain how the solar eclipse of 1919 proved Einstein's theory that light has mass.

Review Problems

1. A train is moving at a speed of 65.0 mi/h. The ticket collector is walking 2.00 mi/h toward the front of the train. How fast is the ticket collector moving from the point of view of a person on the train?

2. How fast is the ticket collector in Problem 1 moving relative to an observer on the side of the tracks?

3. The ticket collector in Problem 1 turns around and walks toward the rear of the train at 2.00 mi/h. How fast is he walking relative to an observer on the side of the tracks?

4. Convert the mass of one electron ($m = 9.10 \times 10^{-31}$ kg) to energy.

5. Convert the mass of one proton ($m = 1.67 \times 10^{-27}$ kg) to energy.

6. A particular task requires 9.80 J of energy. Using $E = mc^2$, determine how much mass is needed to accomplish this task.

MATHEMATICS REVIEW

A.1 Signed Numbers

Signed numbers have many applications in the study of physics. The rules for working with signed numbers follow.

Adding Signed Numbers

To add two positive numbers, add their absolute values.* A positive sign may or may not be placed before the result. It is usually omitted.

Add:

EXAMPLE 1

(a) $+4$
       $+7$
       $\overline{+11}$ or 11

(b) $(+3) + (+5) = +8$ or 8

To add two negative numbers, add their absolute values and place a negative sign before the result.

Add:

EXAMPLE 2

(a) -2
       -5
       $\overline{-7}$

(b) $(-6) + (-7) = -13$
(c) $(-8) + (-4) = -12$

To add a negative number and a positive number, find the difference of their absolute values. Place the sign of the number having the larger absolute value before the result.

Add:

EXAMPLE 3

(a) $+4$
       -6
       $\overline{-2}$

(b) -2
       $+8$
       $\overline{+6}$

(c) -8
       $+3$
       $\overline{-5}$

(d) $+9$
       -4
       $\overline{+5}$

(e) $(+7) + (-2) = +5$
(f) $(-9) + (+6) = -3$
(g) $(-3) + (+10) = +7$
(h) $(+4) + (-12) = -8$

---

*The absolute value of a number is its nonnegative value. For example, the absolute value of -6 is 6; the absolute value of $+10$ is 10; and the absolute value of 0 is 0.

To add three or more signed numbers:

1. Add the positive numbers.
2. Add the negative numbers.
3. Add the sums from steps 1 and 2 according to the rules for addition of signed numbers.

EXAMPLE 4

Add: $(-2) + 4 + (-6) + 10 + (-7)$.

Step 1: $\begin{array}{r} +4 \\ +10 \\ +14 \\ \hline +14 \end{array}$ Step 2: $\begin{array}{r} -2 \\ -6 \\ -7 \\ \hline -15 \end{array}$ Step 3: $\begin{array}{r} -15 \\ +14 \\ \hline -1 \end{array}$

Therefore, $(-2) + 4 + (-6) + 10 + (-7) = -1$.

· · · · · · · · · · · · · · · · ·

Subtracting Signed Numbers

To subtract two signed numbers, change the sign of the *number being subtracted* and *add* according to the rules for addition.

EXAMPLE 5

Subtract:

(a) Subtract: $\begin{array}{r} +3 \\ +7 \\ \hline -4 \end{array}$ ↔ Add: $\begin{array}{r} +3 \\ -7 \\ \hline -4 \end{array}$ To subtract, change the sign of the number being subtracted, $+7$, and add.

(b) Subtract: $\begin{array}{r} -9 \\ -6 \\ \hline -3 \end{array}$ ↔ Add: $\begin{array}{r} -9 \\ +6 \\ \hline -3 \end{array}$ To subtract, change the sign of the number being subtracted, -6, and add.

(c) Subtract: $\begin{array}{r} +8 \\ -4 \\ \hline +12 \end{array}$ ↔ Add: $\begin{array}{r} +8 \\ +4 \\ \hline +12 \end{array}$

(d) Subtract: $\begin{array}{r} -6 \\ +8 \\ \hline -14 \end{array}$ ↔ Add: $\begin{array}{r} -6 \\ -8 \\ \hline -14 \end{array}$

(e) $(+6) - (+8) = (+6) + (-8) = -2$ To subtract, change the sign of the number being subtracted, $+8$, and add.

(f) $(-3) - (-5) = (-3) + (+5) = +2$

(g) $(+10) - (-3) = (+10) + (+3) = +13$

(h) $(-5) - (+2) = (-5) + (-2) = -7$

· · · · · · · · · · · · · · · · ·

When more than two signed numbers are involved in subtraction, change the sign of *each* number being subtracted and add the resulting signed numbers.

EXAMPLE 6

Subtract: $(-2) - (+4) - (-1) - (-3) - (+5)$
$= (-2) + (-4) + (+1) + (+3) + (-5)$.

Step 1:   +1   Step 2:   −2   Step 3:   +4
          +3   −4   −11
          ──   −5   ───
          +4   ──   −7
                            −11

Therefore, $(-2) - (+4) - (-1) - (-3) - (+5) = -7$.

· · · · · · · · · · · · · · · ·

     When combinations of addition and subtraction of signed numbers occur in the same problem, change *only* the sign of each number being subtracted. Then add the resulting signed numbers.

◄ **EXAMPLE 7**

Find the result:

$(-2) + (-4) - (-3) - (+6) + (+1) - (+2) + (-7) - (-5)$
$= (-2) + (-4) + (+3) + (-6) + (+1) + (-2) + (-7) + (+5)$

Step 1:   +3   Step 2:   −2   Step 3:   +9
          +1   −4   −21
          +5   −6   ───
          ──   −2   −12
          +9   −7
                            ───
                            −21

Therefore, $(-2) + (-4) - (-3) - (+6) + (+1) - (+2) + (-7) - (-5) = -12$

· · · · · · · · · · · · · · · ·

Multiplying Signed Numbers

To multiply two signed numbers:

1.   If the signs of the numbers are both positive or both negative, find the product of their absolute values. This product is always positive.
2.   If the signs of the numbers are unlike, find the product of their absolute values and place a negative sign before the result.

◄ **EXAMPLE 8**

Multiply:

(a)   +3   (b)   −5   (c)   −6   (d)   +2
      +4   −8   +7   −3
      ───   ───   ───   ──
      +12   +40   −42   −6

(e) $(+3)(+5) = +15$ (f) $(-7)(-8) = +56$
(g) $(-1)(+6) = -6$ (h) $(+4)(-2) = -8$

· · · · · · · · · · · · · · · ·

     To multiply more than two signed numbers, first multiply the absolute values of the numbers. If there is an odd number of negative factors, place a negative sign before the result. If there is an even number of negative factors, the product is positive. *Note:* An *even* number is a number divisible by 2.

◄ **EXAMPLE 9**

Multiply:

(a) $(+5)(-6)(+2)(-1) = +60$
(b) $(-3)(-3)(+4)(-5) = -180$

· · · · · · · · · · · · · · · ·

Dividing Signed Numbers

The rules for dividing signed numbers are similar to those for multiplying signed numbers.
To divide two signed numbers:

1. If the signs of the numbers are both positive or both negative, divide their absolute values. This quotient is always positive.
2. If the two numbers have different signs, divide their absolute values and place a negative sign before the quotient.

Note: Division by 0 is undefined.

EXAMPLE 10

Divide:

(a) $\dfrac{+10}{+2} = +5$ (b) $\dfrac{-18}{-3} = +6$ (c) $\dfrac{+20}{-4} = -5$ (d) $\dfrac{-24}{+2} = -12$

Problems A.1

Perform the indicated operations.

1. $(-5) + (-6)$ 2. $(+1) + (-10)$ 3. $(-3) + (+8)$ 4. $(+5) + (+7)$
5. $(-5) + (+3)$ 6. $0 + (-3)$ 7. $(-7) - (-3)$ 8. $(+2) - (-9)$
9. $(-4) - (+2)$ 10. $(+4) - (+7)$ 11. $0 - (+3)$ 12. $0 - (-2)$
13. $(-9)(-2)$ 14. $(+4)(+6)$ 15. $(-7)(+3)$ 16. $(+5)(-8)$
17. $(+6)(0)$ 18. $(0)(-4)$ 19. $\dfrac{+36}{+12}$ 20. $\dfrac{-9}{-3}$
21. $\dfrac{+16}{-2}$ 22. $\dfrac{-15}{+3}$ 23. $\dfrac{0}{+6}$ 24. $\dfrac{4}{0}$
25. $(+2) + (-1) + (+10)$ 26. $(-7) + (+2) + (+9) + (-8)$
27. $(-9) + (-3) + (+3) + (-8) + (+4)$
28. $(+8) + (-2) + (-6) + (+7) + (-6) + (+9)$
29. $(-4) - (+5) - (-4)$ 30. $(+3) - (-5) - (-6) - (+5)$
31. $(-7) - (-4) - (+6) - (+4) - (-5)$
32. $(-8) - (+7) - (+3) - (-7) - (-8) - (-2)$
33. $(+5) + (-2) - (+7)$ 34. $(-3) - (-8) - (+3) + (-9)$
35. $(-2) - (+1) - (-10) + (+12) + (-9)$
36. $(-1) - (-11) + (+2) - (-10) + (+8)$
37. $(+3)(-5)(+3)$ 38. $(-1)(+2)(+2)(-1)$
39. $(+2)(-4)(-6)(-3)(+2)$ 40. $(-1)(+3)(-2)(-4)(+5)(-1)$

A.2 Powers of 10

The ability to work quickly and accurately with powers of 10 is important in scientific and technical fields.

> When multiplying two powers of 10, add the exponents. That is,
>
> $$10^a \times 10^b = 10^{a+b}$$

EXAMPLE 1

Multiply:
(a) $(10^6)(10^3) = 10^{6+3} = 10^9$ (b) $(10^4)(10^2) = 10^{4+2} = 10^6$
(c) $(10^1)(10^{-3}) = 10^{1+(-3)} = 10^{-2}$ (d) $(10^{-2})(10^{-5}) = 10^{[-2+(-5)]} = 10^{-7}$

When dividing two powers of 10, subtract the exponents as follows:

$$10^a \div 10^b = 10^{a-b}$$

EXAMPLE 2

Divide:

(a) $\dfrac{10^7}{10^4} = 10^{7-4} = 10^3$

(b) $\dfrac{10^3}{10^5} = 10^{3-5} = 10^{-2}$

(c) $\dfrac{10^{-2}}{10^{+3}} = 10^{(-2)-(+3)} = 10^{-5}$

(d) $\dfrac{10^4}{10^{-2}} = 10^{4-(-2)} = 10^6$

· · · · · · · · · · · · · · · ·

To raise a power of 10 to a power, multiply the exponents as follows:

$$(10^a)^b = 10^{ab}$$

EXAMPLE 3

Find each power:

(a) $(10^2)^3 = 10^{(2)(3)} = 10^6$

(b) $(10^{-3})^2 = 10^{(-3)(2)} = 10^{-6}$

(c) $(10^4)^{-5} = 10^{(4)(-5)} = 10^{-20}$

(d) $(10^{-3})^{-4} = 10^{(-3)(-4)} = 10^{12}$

· · · · · · · · · · · · · · · ·

Next, we will show that $10^0 = 1$. To do this, we need to use the substitution principle, which states that

$$\text{if} \quad a = b \quad \text{and} \quad a = c \quad \text{then} \quad b = c$$

First,

$$\dfrac{10^n}{10^n} = 10^{n-n} \qquad \text{To divide powers, subtract the exponents.}$$

$$= 10^0$$

Second,

$$\dfrac{10^n}{10^n} = 1 \qquad \text{Any number other than zero divided by itself equals 1.}$$

That is, since

$$\dfrac{10^n}{10^n} = 10^0 \quad \text{and} \quad \dfrac{10^n}{10^n} = 1$$

then $10^0 = 1$.

We also will use the fact that $\dfrac{1}{10^a} = 10^{-a}$. To show this, we write

$$\dfrac{1}{10^a} = \dfrac{10^0}{10^a} \qquad (1 = 10^0)$$

$$= 10^{0-a} \qquad \text{To divide powers, subtract the exponents.}$$

$$= 10^{-a}$$

We also need to show that $\dfrac{1}{10^{-a}} = 10^a$. We write

$$\frac{1}{10^{-a}} = \frac{10^0}{10^{-a}}$$
$$= 10^{0-(-a)}$$
$$= 10^a$$

In summary,

$$\boxed{\quad 10^0 = 1 \qquad \frac{1}{10^a} = 10^{-a} \qquad \frac{1}{10^{-a}} = 10^a \quad}$$

Actually, any number (except zero) raised to the zero power equals 1.

Problems A.2

Do as indicated. Express the results using positive exponents.

1. $(10^5)(10^3)$
2. $10^6 \div 10^2$
3. $(10^2)^4$

4. $(10^{-2})(10^{-3})$
5. $\dfrac{10^3}{10^6}$
6. $(10^{-3})^3$

7. $10^5 \div 10^{-2}$
8. $(10^{-2})^{-3}$
9. $(10^4)(10^{-1})$

10. $\dfrac{10^0}{10^{-4}}$
11. $(10^0)(10^{-4})$
12. $\dfrac{10^{-4}}{10^{-3}}$

13. $(10^0)^{-2}$
14. 10^{-3}
15. $\dfrac{1}{10^{-5}}$

16. $\dfrac{(10^4)(10^{-2})}{(10^6)(10^3)}$
17. $\dfrac{(10^{-2})(10^{-3})}{(10^3)^2}$
18. $\dfrac{(10^2)^4}{(10^{-3})^2}$

19. $\left(\dfrac{1}{10^3}\right)^2$
20. $\left(\dfrac{10^2}{10^{-3}}\right)^2$
21. $\left(\dfrac{10 \cdot 10^2}{10^{-1}}\right)^2$

22. $\left(\dfrac{1}{10^{-3}}\right)^2$
23. $\dfrac{(10^4)(10^{-2})}{10^{-8}}$
24. $\dfrac{(10^4)(10^6)}{(10^0)(10^{-2})(10^3)}$

A.3 Solving Linear Equations

An equation is a mathematical sentence stating that two quantities are equal. To solve an equation means to find the number or numbers that can replace the variable in the equation to make the equation a true statement. The value we find that makes the equation a true statement is called the *root* or solution of the equation. When the root of an equation is found, we say we have *solved* the equation.

If $a = b$, then $a + c = b + c$ or $a - c = b - c$. (If the same quantity is added to or subtracted from both sides of an equation, the resulting equation is equivalent to the original equation.)

To solve an equation using this rule, think first of undoing what has been done to the variable.

Solve: $x - 5 = -9$. ◄──────────────────── **EXAMPLE 1**

$$x - 5 = -9$$
$$x - 5 + 5 = -9 + 5 \qquad \text{Undo the subtraction by adding 5 to both sides.}$$
$$x = -4$$

·················

Solve: $x + 4 = 29$. ◄──────────────────── **EXAMPLE 2**

$$x + 4 = 29$$
$$x + 4 - 4 = 29 - 4 \qquad \text{Undo the addition by subtracting 4 from both sides.}$$
$$x = 25$$

·················

If $a = b$, then $ac = bc$ or $a/c = b/c$ with $c \neq 0$. (If both sides of an equation are multiplied or divided by the same nonzero quantity, the resulting equation is equivalent to the original equation.)

Solve: $3x = 18$. ◄──────────────────── **EXAMPLE 3**

$$3x = 18$$
$$\frac{3x}{3} = \frac{18}{3} \qquad \text{Undo the multiplication by dividing both sides by 3.}$$
$$x = 6$$

·················

Solve: $x/4 = 9$. ◄──────────────────── **EXAMPLE 4**

$$\frac{x}{4} - 9$$
$$4\left(\frac{x}{4}\right) = 4 \cdot 9 \qquad \text{Undo the division by multiplying both sides by 4.}$$
$$x = 36$$

·················

When more than one operation is indicated on the variable in an equation, undo the additions and subtractions first, then undo the multiplications and divisions.

Solve: $3x + 5 = 17$. ◄──────────────────── **EXAMPLE 5**

$$3x + 5 = 17$$
$$3x + 5 - 5 = 17 - 5 \qquad \text{Subtract 5 from both sides.}$$
$$3x = 12$$
$$\frac{3x}{3} = \frac{12}{3} \qquad \text{Divide both sides by 3.}$$
$$x = 4$$

·················

EXAMPLE 6

Solve: $2x - 7 = 10$.

$$2x - 7 = 10$$

$$2x - 7 + 7 = 10 + 7 \qquad \text{Add 7 to both sides.}$$

$$2x = 17$$

$$\frac{2x}{2} = \frac{17}{2} \qquad \text{Divide both sides by 2.}$$

$$x = \frac{17}{2} = 8.5$$

EXAMPLE 7

Solve: $(x/5) - 10 = 22$.

$$\frac{x}{5} - 10 = 22$$

$$\frac{x}{5} - 10 + 10 = 22 + 10 \qquad \text{Add 10 to both sides.}$$

$$\frac{x}{5} = 32$$

$$5\left(\frac{x}{5}\right) = 5(32) \qquad \text{Multiply both sides by 5.}$$

$$x = 160$$

To solve an equation with variables on both sides:

1. Add or subtract either variable term from both sides of the equation.
2. Add or subtract from both sides of the equation the constant term that now appears on the same side of the equation with the variable. Then solve.

EXAMPLE 8

Solve: $3x + 6 = 7x - 2$.

$$3x + 6 = 7x - 2$$

$$3x + 6 - 3x = 7x - 2 - 3x \qquad \text{Subtract } 3x \text{ from both sides.}$$

$$6 = 4x - 2$$

$$6 + 2 = 4x - 2 + 2 \qquad \text{Add 2 to both sides.}$$

$$8 = 4x$$

$$\frac{8}{4} = \frac{4x}{4} \qquad \text{Divide both sides by 4.}$$

$$2 = x$$

EXAMPLE 9

Solve: $4x - 2 = -5x + 10$.

$$4x - 2 = -5x + 10$$

$$4x - 2 + 5x = -5x + 10 + 5x \qquad \text{Add } 5x \text{ to both sides.}$$

$$9x - 2 = 10$$

$$9x - 2 + 2 = 10 + 2 \qquad \text{Add 2 to both sides.}$$

$$9x = 12$$

$$\frac{9x}{9} = \frac{12}{9}$$ Divide both sides by 9.

$$x = \frac{4}{3}$$

· · · · · · · · · · · · · · · ·

To solve equations containing parentheses, first remove the parentheses and then proceed as before. The rules for removing parentheses follow:

1. If the parentheses are preceded by a plus ($+$) sign, they may be removed without changing any signs.

    Examples:
    $$2 + (3 - 5) = 2 + 3 - 5$$
    $$3 + (x + 4) = 3 + x + 4$$
    $$5x + (-6x + 9) = 5x - 6x + 9$$

2. If the parentheses are preceded by a minus ($-$) sign, the parentheses may be removed if *all* the signs of the numbers (or letters) within the parentheses are changed.

    Examples:
    $$2 - (3 - 5) = 2 - 3 + 5$$
    $$5 - (x - 7) = 5 - x + 7$$
    $$7x - (-4x - 11) = 7x + 4x + 11$$

3. If the parentheses are preceded by a number, the parentheses may be removed if each of the terms inside the parentheses is multiplied by that (signed) number.

    Examples:
    $$2(x + 4) = 2x + 8$$
    $$-3(x - 5) = -3x + 15$$
    $$2 - 4(3x - 5) = 2 - 12x + 20$$

EXAMPLE 10

Solve: $3(x - 4) = 15$.

$$3(x - 4) = 15$$
$$3x - 12 = 15$$ Remove parentheses.
$$3x - 12 + 12 = 15 + 12$$ Add 12 to both sides.
$$3x = 27$$
$$\frac{3x}{3} = \frac{27}{3}$$ Divide both sides by 3.
$$x = 9$$

· · · · · · · · · · · · · · · ·

EXAMPLE 11

Solve: $2x - (3x + 15) = 4x - 1$.

$$2x - (3x + 15) = 4x - 1$$
$$2x - 3x - 15 = 4x - 1$$ Remove parentheses.
$$-x - 15 = 4x - 1$$ Combine like terms.
$$-x - 15 + x = 4x - 1 + x$$ Add x to both sides.
$$-15 = 5x - 1$$

$$-15 + 1 = 5x - 1 + 1 \qquad \text{Add 1 to both sides.}$$
$$-14 = 5x$$
$$\frac{-14}{5} = \frac{5x}{5} \qquad \text{Divide both sides by 5.}$$
$$-2.8 = x$$

Problems A.3
Solve each equation.

1. $3x = 4$
2. $\frac{y}{2} = 10$
3. $x - 5 = 12$
4. $x + 1 = 9$
5. $2x + 10 = 10$
6. $4x = 28$
7. $2x - 2 = 33$
8. $4 = \frac{x}{10}$
9. $172 - 43x = 43$
10. $9x + 7 = 4$
11. $6y - 24 = 0$
12. $3y + 15 = 75$
13. $15 = \frac{105}{y}$
14. $6x = x - 15$
15. $2 = \frac{50}{2y}$
16. $9y = 67.5$
17. $8x - 4 = 36$
18. $10 = \frac{136}{4x}$
19. $2x + 22 = 75$
20. $9x + 10 = x - 26$
21. $4x + 9 = 7x - 18$
22. $2x - 4 = 3x + 7$
23. $-2x + 5 = 3x - 10$
24. $5x + 3 = 2x - 18$
25. $3x + 5 = 5x - 11$
26. $-5x + 12 = 12x - 5$
27. $13x + 2 = 20x - 5$
28. $5x + 3 = -9x - 39$
29. $-4x + 2 = -10x - 20$
30. $9x + 3 = 6x + 8$
31. $3x + (2x - 7) = 8$
32. $11 - (x + 12) = 100$
33. $7x - (13 - 2x) = 5$
34. $20(7x - 2) = 240$
35. $-3x + 5(x - 6) = 12$
36. $3(x + 117) = 201$
37. $5(2x - 1) = 8(x + 3)$
38. $3(x + 4) = 8 - 3(x - 2)$
39. $-2(3x - 2) = 3x - 2(5x + 1)$
40. $\frac{x}{5} - 2\left(\frac{2x}{5} + 1\right) = 28$

A.4 Solving Quadratic Equations
A quadratic equation in one variable is an equation that can be written in the form

$$ax^2 + bx + c = 0 \qquad (\text{where } a \neq 0)$$

EXAMPLE 1

Solve: $x^2 = 16$.

To solve a quadratic equation of this type, take the square root of both sides of the equation.

$$x^2 = 16$$
$$x = \pm 4 \qquad \text{Take the square root of both sides.}$$

In general, solve equations of the form $ax^2 = b$, where $a \neq 0$, as follows:

$$ax^2 = b$$
$$x^2 = \frac{b}{a} \qquad \text{Divide both sides by } a.$$
$$x = \pm\sqrt{\frac{b}{a}} \qquad \text{Take the square root of both sides.}$$

Solve: $2x^2 - 18 = 0$. ◄

EXAMPLE 2

$$2x^2 - 18 = 0$$

$$2x^2 = 18 \qquad \text{Add 18 to both sides.}$$

$$x^2 = 9 \qquad \text{Divide both sides by 2.}$$

$$x = \pm 3 \qquad \text{Take the square root of both sides.}$$

. . . .

Solve: $5y^2 = 100$. ◄

EXAMPLE 3

$$5y^2 = 100$$

$$y^2 = 20 \qquad \text{Divide both sides by 5.}$$

$$y = \pm\sqrt{20} \qquad \text{Take the square root of both sides.}$$

$$y = \pm 4.47$$

. . . .

The solutions of the general quadratic equation

$$ax^2 + bx + c = 0 \qquad \text{(where } a \neq 0\text{)}$$

are given by the formula (called the *quadratic formula*)

$$\boxed{x = \frac{-b \pm \sqrt{b^2 - 4ac}}{2a}}$$

where a = coefficient of the x^2 term
$\quad\quad\; b$ = coefficient of the x term
$\quad\quad\; c$ = constant term

The symbol ($\pm$) is used to combine two expressions or equations into one. For example, $a \pm 2$ means $a + 2$ or $a - 2$. Similarly,

$$x = \frac{-b \pm \sqrt{b^2 - 4ac}}{2a}$$

means

$$x = \frac{-b + \sqrt{b^2 - 4ac}}{2a} \quad \text{or} \quad x = \frac{-b - \sqrt{b^2 - 4ac}}{2a}$$

In the equation, $4x^2 - 3x - 7 = 0$, identify a, b, and c. ◄

EXAMPLE 4

$$a = 4, \quad b = -3, \quad \text{and} \quad c = -7$$

. . . .

Solve $x^2 + 2x - 8 = 0$ using the quadratic formula. ◄

EXAMPLE 5

First, $a = 1$, $b = 2$, and $c = -8$. Then

$$x = \frac{-b \pm \sqrt{b^2 - 4ac}}{2a}$$

$$x = \frac{-2 \pm \sqrt{(2)^2 - 4(1)(-8)}}{2(1)}$$

$$= \frac{-2 \pm \sqrt{4 - (-32)}}{2}$$

$$= \frac{-2 \pm \sqrt{36}}{2}$$

$$= \frac{-2 \pm 6}{2}$$

$$= \frac{-2 + 6}{2} \quad \text{or} \quad \frac{-2 - 6}{2}$$

$$= \frac{4}{2} \quad \text{or} \quad \frac{-8}{2}$$

$$= 2 \quad \text{or} \quad -4 \qquad \text{The solutions are 2 and } -4.$$

If the number under the radical sign is not a perfect square, find the square root of the number by using a calculator and proceed as before.

EXAMPLE 6

Solve $4x^2 - 7x = 32$ using the quadratic formula.

Before identifying a, b, and c, the equation must be set equal to zero. That is,

$$4x^2 - 7x - 32 = 0$$

First, $a = 4$, $b = -7$, and $c = -32$. Then

$$x = \frac{-b \pm \sqrt{b^2 - 4ac}}{2a}$$

$$= \frac{-(-7) \pm \sqrt{(-7)^2 - 4(4)(-32)}}{2(4)}$$

$$= \frac{7 \pm \sqrt{49 - (-512)}}{8}$$

$$= \frac{7 \pm \sqrt{561}}{8}$$

$$= \frac{7 \pm 23.7}{8} \qquad (\sqrt{561} = 23.7)$$

$$= \frac{7 + 23.7}{8} \quad \text{or} \quad \frac{7 - 23.7}{8}$$

$$= 3.84 \qquad \text{or} \quad -2.09$$

The approximate solutions are 3.84 and -2.09.

Problems A.4

Solve each equation.

1. $x^2 = 36$
2. $y^2 = 100$
3. $2x^2 = 98$
4. $5x^2 = 0.05$
5. $3x^2 - 27 = 0$
6. $2y^2 - 15 = 17$
7. $10x^2 + 4.9 = 11.3$
8. $2(32)(48 - 15) = v^2 - 27^2$
9. $2(107) = 9.8t^2$
10. $65 = \pi r^2$
11. $2.50 = \pi r^2$
12. $24^2 = a^2 + 16^2$

Find the values of a, b, and c, in each quadratic equation.

13. $3x^2 + x - 5 = 0$
14. $-2x^2 + 7x + 4 = 0$
15. $6x^2 + 8x + 2 = 0$
16. $5x^2 - 2x - 15 = 0$
17. $9x^2 + 6x = 4$
18. $6x^2 = x + 9$

19. $5x^2 + 6x = 0$ 20. $7x^2 - 45 = 0$ 21. $9x^2 = 64$
22. $16x^2 = 49$

Solve each quadratic equation using the quadratic formula.

23. $x^2 - 10x + 21 = 0$ 24. $2x^2 + 13x + 15 = 0$ 25. $6x^2 + 7x = 20$
26. $15x^2 = 4x + 4$ 27. $6x^2 - 2x = 19$ 28. $4x^2 = 28x - 49$
29. $18x^2 - 15x = 26$ 30. $48x^2 + 9 = 50x$
31. $16.5x^2 + 8.3x - 14.7 = 0$ 32. $125x^2 - 167x + 36 = 0$

A.5 Right-Triangle Trigonometry

A **right triangle** is a triangle with one right angle (90°), two acute angles (less than 90°), two legs, and a hypotenuse (the side opposite the right angle) (Fig. A.1).

When it is necessary to label a triangle, the vertices are often labeled using capital letters and the sides opposite the vertices are labeled using the corresponding lowercase letters (Fig. A.2).

The side opposite angle A is a.
The side adjacent to angle A is b.
The side opposite angle B is b.
The side adjacent to angle B is a.
The side opposite angle C is c and is called the *hypotenuse*.

If we consider a certain acute angle of a right triangle, the two legs can be identified as the side opposite or the side adjacent to the acute angle.

The side opposite angle A is the same as the side adjacent to angle B.
The side adjacent to angle A is the same as the side opposite angle B.
The side opposite angle B is the same as the side adjacent to angle A.
The side adjacent to angle B is the same as the side opposite angle A.

A **ratio** is a comparison of two quantities by division. In a right triangle (Fig. A.3), there are three very important ratios:

$$\frac{\text{side opposite angle } A}{\text{hypotenuse}}$$ is called **sine** A (abbreviated sin A)

$$\frac{\text{side adjacent to angle } A}{\text{hypotenuse}}$$ is called **cosine** A (abbreviated cos A)

$$\frac{\text{side opposite angle } A}{\text{side adjacent to angle } A}$$ is called **tangent** A (abbreviated tan A)

Figure A.1   Parts of a right triangle.

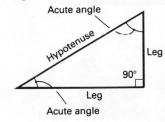

Figure A.2   Labeling a right triangle.

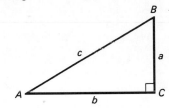

Figure A.3

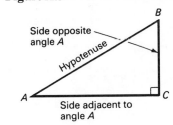

$$\sin A = \frac{\text{side opposite angle } A}{\text{hypotenuse}}$$

$$\cos A = \frac{\text{side adjacent to angle } A}{\text{hypotenuse}}$$

$$\tan A = \frac{\text{side opposite angle } A}{\text{side adjacent to angle } A}$$

The ratios are defined similarly for angle B:

$$\sin B = \frac{\text{side opposite angle } B}{\text{hypotenuse}}$$

$$\cos B = \frac{\text{side adjacent to angle } B}{\text{hypotenuse}}$$

$$\tan B = \frac{\text{side opposite angle } B}{\text{side adjacent to angle } B}$$

EXAMPLE 1

Figure A.4

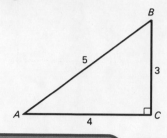

Find the three trigonometric ratios of angle A in Fig. A.4.

$$\sin A = \frac{\text{side opposite angle } A}{\text{hypotenuse}} = \frac{3}{5} = 0.60$$

$$\cos A = \frac{\text{side adjacent to angle } A}{\text{hypotenuse}} = \frac{4}{5} = 0.80$$

$$\tan A = \frac{\text{side opposite angle } A}{\text{side adjacent to angle } A} = \frac{3}{4} = 0.75$$

EXAMPLE 2

Find the three trigonometric ratios of angle B in Fig. A.4.

$$\sin B = \frac{\text{side opposite angle } B}{\text{hypotenuse}} = \frac{4}{5} = 0.80$$

$$\cos B = \frac{\text{side adjacent to angle } B}{\text{hypotenuse}} = \frac{3}{5} = 0.60$$

$$\tan B = \frac{\text{side opposite angle } B}{\text{side adjacent to angle } B} = \frac{4}{3} = 1.33$$

Note: Every acute angle has three trigonometric ratios associated with it.

In this book we assume that you will be using a calculator. When calculations involve a trigonometric ratio, we will use the following generally accepted practice for significant digits:

Angle Expressed to Nearest:	Length of Side Contains:
1°	Two significant digits
0.1°	Three significant digits
0.01°	Four significant digits

A useful and time-saving fact about right triangles is that *the sum of the two acute angles of any right triangle is always* 90°. That is,

$$A + B = 90°$$

Why is this true? We know that the sum of the three interior angles of any triangle is 180°. A right triangle must contain a right angle, whose measure is 90°. This leaves 90° to be divided between the two acute angles. Therefore, if one acute angle is known, the other acute angle may be found by subtracting the known angle from 90°. That is,

$$A = 90° - B$$
$$B = 90° - A$$

EXAMPLE 3

Find angle B and side a in the right triangle in Fig. A.5.

To find angle B, we use

$$B = 90° - A = 90° - 30.0° = 60.0°$$

To find side a, we use a trigonometric ratio. Note that we are looking for the *side opposite* angle A and that the *hypotenuse* is given. The trigonometric ratio having these two quantities is sine.

Figure A.5

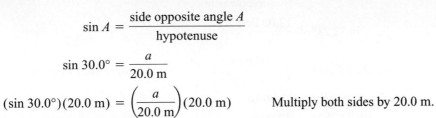

$$\sin A = \frac{\text{side opposite angle } A}{\text{hypotenuse}}$$

$$\sin 30.0° = \frac{a}{20.0 \text{ m}}$$

$$(\sin 30.0°)(20.0 \text{ m}) = \left(\frac{a}{20.0 \text{ m}}\right)(20.0 \text{ m}) \qquad \text{Multiply both sides by 20.0 m.}$$

$$10.0 \text{ m} = a$$

· · · · · · · · · · · · · ·

Find angle A, angle B, and side a in the right triangle in Fig. A.6.

EXAMPLE 4

First, find angle A. The *side adjacent* to angle A and the *hypotenuse* are given. Therefore, we use $\cos A$ to find angle A because $\cos A$ uses these two quantities:

Figure A.6

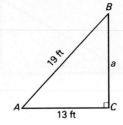

$$\cos A = \frac{\text{side adjacent to } A}{\text{hypotenuse}}$$

$$\cos A = \frac{13 \text{ ft}}{19 \text{ ft}} = 0.684$$

Using a calculator as in Section B.3 of Appendix B, we find that $A = 47°$.

To find angle B, we use

$$B = 90° - A = 90° - 47° = 43°$$

To find side a, we use $\sin A$ because the *hypotenuse* is given and side a is the *side opposite* angle A:

$$\sin A = \frac{\text{side opposite angle } A}{\text{hypotenuse}}$$

$$\sin 47° = \frac{a}{19 \text{ ft}}$$

$$(\sin 47°)(19 \text{ ft}) = \left(\frac{a}{19 \text{ ft}}\right)(19 \text{ ft}) \qquad \text{Multiply both sides by 19 ft.}$$

$$14 \text{ ft} = a$$

· · · · · · · · · · · · · · ·

Find angle A, angle B, and the hypotenuse in the right triangle in Fig. A.7.

EXAMPLE 5

To find angle A, use $\tan A$:

Figure A.7

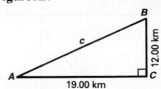

$$\tan A = \frac{\text{side opposite angle } A}{\text{side adjacent to angle } A}$$

$$\tan A = \frac{12.00 \text{ km}}{19.00 \text{ km}} = 0.6316$$

$$A = 32.28°$$

To find angle B,

$$B = 90° - A = 90° - 32.28° = 57.72°$$

To find the hypotenuse, use $\sin A$:

$$\sin A = \frac{\text{side opposite angle } A}{\text{hypotenuse}}$$

$$\sin 32.28° = \frac{12.00 \text{ km}}{c}$$

$$(\sin 32.28°)(c) = \left(\frac{12.00 \text{ km}}{c}\right)(c) \qquad \text{Multiply both sides by } c.$$

$$(\sin 32.28°)(c) = 12.00 \text{ km}$$

$$\frac{c(\sin 32.28°)}{\sin 32.28°} = \frac{12.00 \text{ km}}{\sin 32.28°} \qquad \text{Divide both sides by } \sin 32.28°.$$

$$c = \frac{12.00 \text{ km}}{\sin 32.28°}$$

$$= 22.47 \text{ km}$$

· · · · · · · · · · · · · · · · ·

Figure A.8

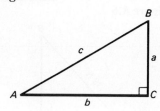

When the two legs of a right triangle are given, the hypotenuse can be found without using trigonometric ratios. From geometry, *the sum of the squares of the legs of a right triangle is equal to the square of the hypotenuse* (**Pythagorean theorem**; see Fig. A.8):

$$a^2 + b^2 = c^2$$

or, taking the square root of each side of the equation,

$$\boxed{c = \sqrt{a^2 + b^2}}$$

Also, if one leg and the hypotenuse are given, the other leg can be found by using

$$\boxed{\begin{array}{l} a = \sqrt{c^2 - b^2} \\ b = \sqrt{c^2 - a^2} \end{array}}$$

EXAMPLE 6 ▶ Find the hypotenuse of the right triangle in Fig. A.9.

Figure A.9

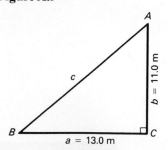

$$c = \sqrt{a^2 + b^2}$$

$$c = \sqrt{(13.0 \text{ m})^2 + (11.0 \text{ m})^2}$$

$$= \sqrt{169 \text{ m}^2 + 121 \text{ m}^2}$$

$$= \sqrt{290 \text{ m}^2}$$

$$= 17.0 \text{ m}$$

· · · · · · · · · · · · · · · · ·

EXAMPLE 7 ▶ Find side b in the right triangle in Fig. A.10.

Figure A.10

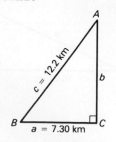

$$b = \sqrt{c^2 - a^2}$$

$$b = \sqrt{(12.2 \text{ km})^2 - (7.30 \text{ km})^2}$$

$$= 9.77 \text{ km}$$

· · · · · · · · · · · · · · · · ·

Problems A.5

Figure A.11

Use right triangle *ABC* in Fig. A.11 to fill in each blank.

1. The side opposite angle *A* is _____.
2. The side opposite angle *B* is _____.
3. The hypotenuse is _____.
4. The side adjacent to angle *A* is _____.
5. The side adjacent to angle *B* is _____.
6. The angle opposite side *a* is _____.
7. The angle opposite side *b* is _____.
8. The angle opposite side *c* is _____.
9. The angle adjacent to side *a* is _____.
10. The angle adjacent to side *b* is _____.

Use a calculator to find each trigonometric ratio rounded to four significant digits.

11.	$\sin 71°$	12.	$\cos 40°$
13.	$\tan 61°$	14.	$\tan 41.2°$
15.	$\cos 11.5°$	16.	$\sin 79.4°$
17.	$\cos 49.63°$	18.	$\tan 53.45°$
19.	$\tan 17.04°$	20.	$\cos 34°$
21.	$\sin 27.5°$	22.	$\cos 58.72°$

Find each angle rounded to the nearest whole degree.

23.	$\sin A = 0.2678$	24.	$\cos B = 0.1046$
25.	$\tan A = 0.9237$	26.	$\sin B = 0.9253$
27.	$\cos B = 0.6742$	28.	$\tan A = 1.351$

Find each angle rounded to the nearest tenth of a degree.

29.	$\sin B = 0.5963$	30.	$\cos A = 0.9406$
31.	$\tan B = 1.053$	32.	$\sin A = 0.9083$
33.	$\cos A = 0.8660$	34.	$\tan B = 0.9433$

Find each angle rounded to the nearest hundredth of a degree.

35.	$\sin A = 0.3792$	36.	$\cos B = 0.06341$
37.	$\tan B = 0.3010$	38.	$\sin A = 0.4540$
39.	$\cos B = 0.8141$	40.	$\tan A = 2.369$

Solve each triangle (find the missing angles and sides) using trigonometric ratios.

41.

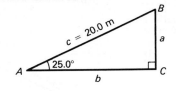

42.

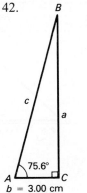

43.

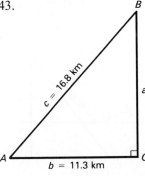

44.

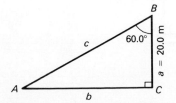

45.

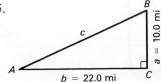

46.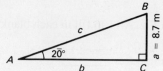

47.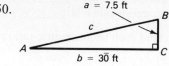

48.

49.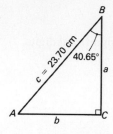

50.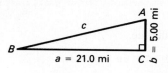

Find the missing side in each right triangle using the Pythagorean theorem.

51.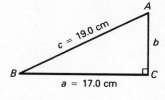

52.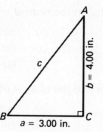

53.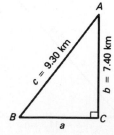

54.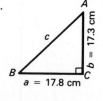

55.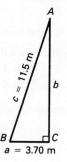

56.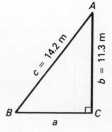

57.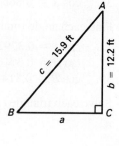

58.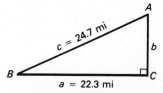

59.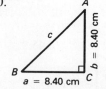

60.

61. A round taper is shown in Fig. A.12.
    (a) Find $\angle BAC$.
    (b) Find the length BC.
    (c) Find the diameter of end x.
62. The distance between two parallel flat surfaces, a, of a hexagonal nut is $\frac{3}{4}$ in. Find the distance between any two farthest corners, b (Fig. A.13).
63. Find distances c and d between the holes of the plate shown in Fig. A.14.

Figure A.12

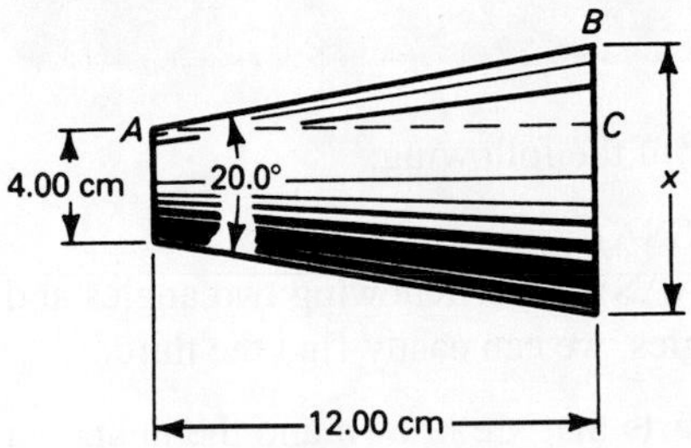

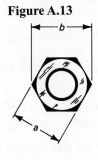

Figure A.13

Figure A.14

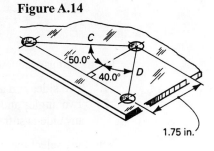

64. A piece of electric conduit cuts across a corner of a room $24\overline{0}$ cm from the corner. It meets the adjoining wall $35\overline{0}$ cm from the corner. Find length AB of the conduit (Fig. A.15).
65. Find the distances between holes on the plate shown in Fig. A.16.
66. Find length x in Fig. A.17. (*Note: AB = BC.*)

Figure A.15

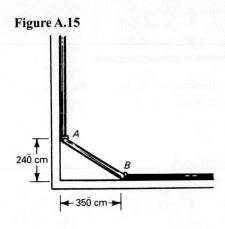

Figure A.16

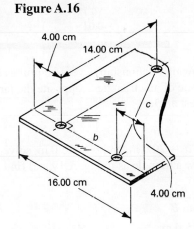

Figure A.17

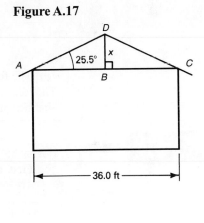

A.6 Law of Sines and Law of Cosines

For those classes and individuals with general triangle trigonometry skills, this section is written as a review. This text is written so that no more than right triangle trigonometry needs to be used, but those with the prerequisite skills may prefer to use the more advanced trigonometry techniques.

An **oblique**, or **general**, triangle is a triangle that contains no right angles. We shall use the standard notation of labeling the vertices of a triangle by the capital letters A, B, and C and using the small letters a, b, and c as the labels for the sides opposite angles A, B, and C, respectively (see Fig. A.18).

Solving a triangle means finding all those sides and angles that are not given or known. To solve a triangle, we need three parts (including at least one side). Solving any oblique triangle falls into one of four cases where the following parts of a triangle are known:

1. Two sides and an angle opposite one of them (SSA).
2. Two angles and a side opposite one of them (AAS).
3. Two sides and the included angle (SAS).
4. Three sides (SSS).

Figure A.18 Oblique, or general, triangle ABC.

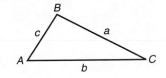

One law that we use to solve triangles is called the **law of sines**. In words, for any triangle the ratio of any side to the sine of the opposite angle is a constant. The formula for the law of sines is as follows:

LAW OF SINES

$$\frac{a}{\sin A} = \frac{b}{\sin B} = \frac{c}{\sin C}$$

In order to use the law of sines, we must know either of the following:

1. Two sides and an angle opposite one of them (SSA).
2. Two angles and a side opposite one of them (AAS). *Note:* Knowing two angles and any side is sufficient because knowing two angles, we can easily find the third.

You must select the proportion that contains three parts that are known and the unknown part.

When calculations with measurements involve a trigonometric function, we shall use the following rule for significant digits:

Angle Expressed to Nearest	Length of Side Contains
1°	Two significant digits
0.1°	Three significant digits
0.01°	Four significant digits

The following relationship is very helpful as a check when solving a general triangle: The longest side of any triangle is opposite the largest angle and the shortest side is opposite the smallest angle.

EXAMPLE 1

If $A = 65.0°$, $a = 20.0$ m, and $b = 15.0$ m, solve the triangle.

First, draw a triangle as in Fig. A.19 and find angle B by using the law of sines:

$$\frac{a}{\sin A} = \frac{b}{\sin B}$$

$$\frac{20.0 \text{ m}}{\sin 65.0°} = \frac{15.0 \text{ m}}{\sin B}$$

$$\sin B = \frac{(15.0 \text{ m})(\sin 65.0°)}{20.0 \text{ m}} = 0.6797$$

$$B = 42.8°$$

Figure A.19

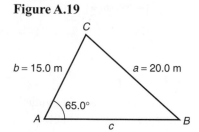

This angle may be found using a calculator as follows:

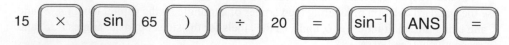

$$42.82261353$$

So $B = 42.8°$ rounded to the nearest tenth of a degree.

To find C, use the fact that the sum of the angles of any triangle is 180°. Therefore,

$$C = 180° - 65.0° - 42.8°$$

$$= 72.2°$$

Finally, find c using the law of sines.

$$\frac{a}{\sin A} = \frac{c}{\sin C}$$

$$\frac{20.0 \text{ m}}{\sin 65.0°} = \frac{c}{\sin 72.2°}$$

$$c = \frac{(20.0 \text{ m})(\sin 72.2°)}{\sin 65.0°} = 21.0 \text{ m} \qquad \text{(Rounded to three significant digits)}$$

This side may be found using a calculator as follows:

$$21.01117096$$

That is, side $c = 21.0$ m rounded to three significant digits.
The solution is $B = 42.8°$, $C = 72.2°$, and $c = 21.0$ m.

....

EXAMPLE 2

If $C = 25°$, $c = 59$ ft, and $B = 108°$, solve the triangle.
First, draw a triangle as in Fig. A.20 and find b:

Figure A.20

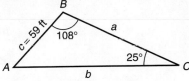

$$\frac{c}{\sin C} = \frac{b}{\sin B}$$

$$\frac{59 \text{ ft}}{\sin 25°} = \frac{b}{\sin 108°}$$

$$b = \frac{(59 \text{ ft})(\sin 108°)}{\sin 25°} = 130 \text{ ft} \qquad \text{(Rounded to two significant digits)}$$

$$132.7730946$$

$$A = 180° - 25° - 108° = 47°$$

Find a:

$$\frac{a}{\sin A} = \frac{c}{\sin C}$$

$$\frac{a}{\sin 47°} = \frac{59 \text{ ft}}{\sin 25°}$$

$$a = \frac{(59 \text{ ft})(\sin 47°)}{\sin 25°} = 1\bar{0}0 \text{ ft}$$

The solution is $A = 47°$, $a = 1\bar{0}0$ ft, and $b = 130$ ft.

....

The Ambiguous Case

The solution of a triangle when two sides and an angle opposite one of the sides (SSA) are given requires special care. There may be one, two, or no triangles formed from the given data. By construction and discussion, let's study the possibilities.

EXAMPLE 3

Construct a triangle given that $A = 35°$, $b = 10$, and $a = 7$.

As you can see from Fig. A.21, two triangles that satisfy the given information can be drawn: triangles ACB and ACB'. Note that in one triangle angle B is acute and in the other triangle angle B is obtuse.

Figure A.21

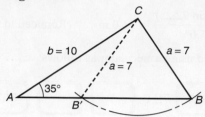

EXAMPLE 4

Construct a triangle given that $A = 45°$, $b = 10$, and $a = 5$.

As you can see from Fig. A.22, no triangle can be drawn that satisfies the given information. Side a is simply not long enough to reach the side opposite angle C.

Figure A.22

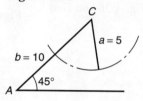

EXAMPLE 5

Construct a triangle given that $A = 60°$, $b = 6$, and $a = 10$.

As you can see from Fig. A.23, only one triangle that satisfies the given information can be drawn. Side a is too long for two solutions.

Figure A.23

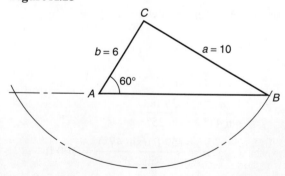

In summary, let's list the possible cases when two sides and an angle opposite one of the sides are given. Assume that *acute* angle A and adjacent side b are given. As a result of $h = b \sin A$, h is also determined. Depending on the length of the opposite side, a, we have the four cases shown in Fig. A.24.

Figure A.24 Possible triangles when two sides and an acute angle opposite one of the sides are given.

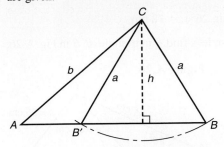

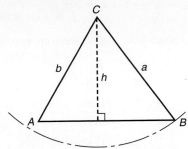

(a) When $h < a < b$, there are two possible triangles. In words, when the side opposite the given *acute* angle is less than the known adjacent side but greater than the altitude, there are two possible triangles.

(b) When $h < b < a$, there is only one possible triangle. In words, when the side opposite the given *acute* angle is greater than the known adjacent side, there is only one possible triangle.

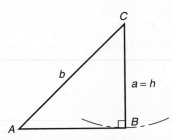

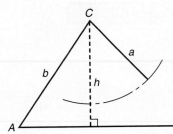

(c) When $a = h$, there is one possible (right) triangle. In words, when the side opposite the given *acute* angle equals the length of the altitude, there is only one possible (right) triangle.

(d) When $a < h$, there is no possible triangle. In words, when the side opposite the given *acute* angle is less than the length of the altitude, there is no possible triangle.

If angle A is *obtuse*, we have two possible cases (Fig. A.25).

Figure A.25 Possible triangles when two sides and an obtuse angle opposite one of the sides are given.

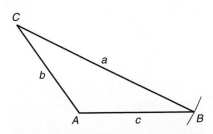

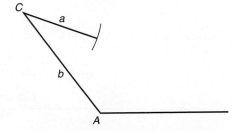

(a) When $a > b$, there is one possible triangle. In words, when the side opposite the given *obtuse* angle is greater than the known adjacent side, there is only one possible triangle.

(b) When $a \le b$, there is no possible triangle. In words, when the side opposite the given *obtuse* angle is less than or equal to the known adjacent side, there is no possible triangle.

Note: If the given parts are not angle A, side opposite a, and side adjacent b as in our preceding discussions, then you must substitute the given angle and sides accordingly. This is why it is so important to understand the general word description corresponding to each case.

EXAMPLE 6

If $A = 26°$, $a = 25$ cm, and $b = 41$ cm, solve the triangle.
First, find h:

$$h = b \sin A = (41 \text{ cm})(\sin 26°) = 18 \text{ cm}$$

Since $h < a < b$, there are two solutions. First, let's find B in triangle ACB in Fig. A.26:

$$\frac{a}{\sin A} = \frac{b}{\sin B}$$

$$\frac{25 \text{ cm}}{\sin 26°} = \frac{41 \text{ cm}}{\sin B}$$

$$\sin B = \frac{(41 \text{ cm})(\sin 26°)}{25 \text{ cm}} = 0.7189$$

$$B = 46°$$

Figure A.26

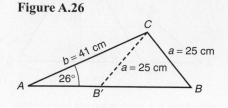

$$C = 180° - 26° - 46° = 108°$$

Find c:

$$\frac{c}{\sin C} = \frac{a}{\sin A}$$

$$\frac{c}{\sin 108°} = \frac{25 \text{ cm}}{\sin 26°}$$

$$c = \frac{(25 \text{ cm})(\sin 108°)}{\sin 26°} = 54 \text{ cm}$$

Therefore, the first solution is $B = 46°$, $C = 108°$, and $c = 54$ cm.

The second solution occurs when B is obtuse, as in triangle ACB'. That is, find the obtuse angle whose sine is 0.7189.

$$B' = 180° - 46° = 134°$$

Then $C = 180° - 26° - 134° = 2\overline{0}°$.
For c,

$$\frac{c}{\sin C} = \frac{a}{\sin A}$$

$$\frac{c}{\sin 2\overline{0}°} = \frac{25 \text{ cm}}{\sin 26°}$$

$$c = \frac{(25 \text{ cm})(\sin 2\overline{0}°)}{\sin 26°} = 2\overline{0} \text{ cm}$$

The second solution is $B' = 134°$, $C = 2\overline{0}°$, and $c = 2\overline{0}$ cm.

EXAMPLE 7

If $A = 62.0°$, $a = 415$ m, and $b = 855$ m, solve the triangle.
First, find h:

$$h = b \sin A$$
$$h = (855 \text{ m})(\sin 62.0°)$$
$$= 755 \text{ m}$$

Since $a < h$, there is no possible solution. What would happen if you applied the law of sines anyway? You would obtain

$$\frac{a}{\sin A} = \frac{b}{\sin B}$$

$$\frac{415 \text{ m}}{\sin 62.0°} = \frac{855 \text{ m}}{\sin B}$$

$$\sin B = \frac{(855 \text{ m})(\sin 62.0°)}{415 \text{ m}} = 1.819 \qquad \text{(Tilt!)}$$

Note: $\sin B = 1.819$ is impossible because $-1 \leq \sin B \leq 1$.

In summary:

1. ***Given two angles and one side (AAS):*** There is only one possible triangle.
2. ***Given two sides and an angle opposite one of them (SSA):*** There are three possibilities. If the side opposite the given angle is:
   (a) greater than the known adjacent side, there is only one possible triangle.
   (b) less than the known adjacent side but greater than the altitude, there are two possible triangles.
   (c) less than the altitude, there is no possible triangle.

Since solving a general triangle requires several operations, errors are often introduced. The following points may be helpful in avoiding some of these errors:

1. Always choose a given value over a calculated value when doing calculations.
2. Always check your results to see that the largest angle is opposite the largest side and the smallest angle is opposite the smallest side.
3. Avoid finding the largest angle by the law of sines whenever possible because it is often not clear whether the resulting angle is acute or obtuse.

Law of Cosines

When the law of sines cannot be used, we use the **law of cosines**. In words, the square of any side of a triangle is equal to the sum of the squares of the other two sides minus twice the product of these two sides and the cosine of their included angle (see Fig. A.27). By formula, the law is stated as follows.

LAW OF COSINES

$$a^2 = b^2 + c^2 - 2bc \cos A$$
$$b^2 = a^2 + c^2 - 2ac \cos B$$
$$c^2 = a^2 + b^2 - 2ab \cos C$$

Figure A.27

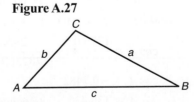

There are two cases when the law of sines does not apply and we use the law of cosines to solve triangles:

1. Two sides and the included angle are known (SAS).
2. All three sides are known (SSS).

Do you see that when the law of cosines is used, there is no possibility of an ambiguous case? If not, draw a few triangles for each of these two cases (SAS and SSS) to convince yourself intuitively.

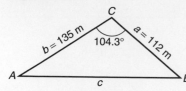

EXAMPLE 8

Figure A.28

If $a = 112$ m, $b = 135$ m, and $C = 104.3°$, solve the triangle.

First, draw a triangle as in Fig. A.28 and find c by using the law of cosines:

$$c^2 = a^2 + b^2 - 2ab \cos C$$
$$c^2 = (112 \text{ m})^2 + (135 \text{ m})^2 - 2(112 \text{ m})(135 \text{ m})(\cos 104.3°)$$
$$c = 196 \text{ m}$$

This side may be found using a calculator as follows:

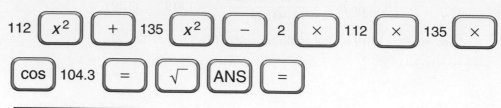

$$\boxed{195.5460308}$$

So $c = 196$ m rounded to three significant digits.

To find A, use the law of sines since it requires less computation:

$$\frac{a}{\sin A} = \frac{c}{\sin C}$$

$$\frac{112 \text{ m}}{\sin A} = \frac{196 \text{ m}}{\sin 104.3°}$$

$$\sin A = \frac{(112 \text{ m})(\sin 104.3°)}{196 \text{ m}} = 0.5537$$

$$A = 33.6°$$

$$B = 180° - 104.3° - 33.6° = 42.1°$$

The solution is $A = 33.6°$, $B = 42.1°$, and $c = 196$ m.

· · · · · · · · · · · · · · · ·

EXAMPLE 9

If $a = 375.0$ ft, $b = 282.0$ ft, and $c = 114.0$ ft, solve the triangle.

First, draw a triangle as in Fig. A.29 and find A by using the law of cosines:

Figure A.29

$$a^2 = b^2 + c^2 - 2bc \cos A$$
$$(375.0 \text{ ft})^2 = (282.0 \text{ ft})^2 + (114.0 \text{ ft})^2 - 2(282.0 \text{ ft})(114.0 \text{ ft}) \cos A$$
$$\cos A = \frac{(375.0 \text{ ft})^2 - (282.0 \text{ ft})^2 - (114.0 \text{ ft})^2}{-2(282.0 \text{ ft})(114.0 \text{ ft})}$$
$$\cos A = -0.7482$$
$$A = 138.43° \quad \text{(Rounded to the nearest hundredth of a degree)}$$

375 $\boxed{x^2}$ $\boxed{-}$ 282 $\boxed{x^2}$ $\boxed{-}$ 114 $\boxed{x^2}$ $\boxed{=}$ $\boxed{\text{ANS}}$ $\boxed{÷}$

(−)2 $\boxed{÷}$ 282 $\boxed{÷}$ 114 $\boxed{=}$ $\boxed{\cos^{-1}}$ $\boxed{\text{ANS}}$ $\boxed{=}$

$$\boxed{138.432994}$$

Next, to find B, use the law of sines:

$$\frac{a}{\sin A} = \frac{b}{\sin B}$$

$$\frac{375.0 \text{ ft}}{\sin 138.43°} = \frac{282.0 \text{ ft}}{\sin B}$$

$$\sin B = \frac{(282.0 \text{ ft})(\sin 138.43°)}{375.0 \text{ ft}} = 0.4990$$

$$B = 29.93°$$

$$C = 180° - 138.43° - 29.93° = 11.64°$$

The solution is $A = 138.43°$, $B = 29.93°$, and $C = 11.64°$.

Problems A.6

Solve each triangle using the labels as shown in Fig. A.30.*

Figure A.30

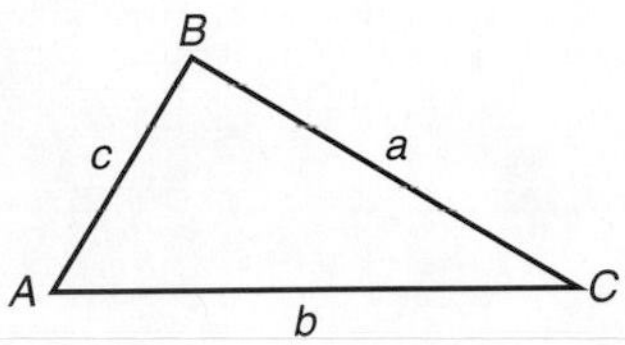

Express the lengths of sides to three significant digits and the angles to the nearest tenth of a degree.

1. $A = 69.0°$, $a = 25.0$ m, $b = 16.5$ m
2. $C = 57.5°$, $c = 166$ mi, $b = 151$ mi
3. $B = 61.4°$, $b = 124$ cm, $c = 112$ cm
4. $A = 19.5°$, $a = 487$ km, $c = 365$ km
5. $B = 75.3°$, $A = 57.1°$, $b = 257$ ft
6. $C = 59.6°$, $B = 43.9°$, $b = 4760$ m
7. $A = 115.0°$, $a = 5870$ m, $b = 4850$ m
8. $A = 16.4°$, $a = 205$ ft, $b = 187$ ft

Express the lengths of sides to four significant digits and the angles to the nearest hundredth of a degree.

9. $C = 72.58°$, $b = 28.63$ cm, $c = 42.19$ cm
10. $A = 58.95°$, $a = 3874$ m, $c = 2644$ m
11. $B = 28.76°$, $C = 19.30°$, $c = 39{,}750$ mi
12. $A = 35.09°$, $B = 48.64°$, $a = 8.362$ km

Express the lengths of sides to two significant digits and the angles to the nearest degree.

13. $A = 25°$, $a = 5\bar{0}$ cm, $b = 4\bar{0}$ cm
14. $B = 42°$, $b = 5.3$ km, $c = 4.6$ km
15. $C = 8°$, $c = 16$ m, $a = 12$ m
16. $A = 105°$, $a = 460$ mi, $c = 380$ mi

For each general triangle, (a) determine the number of solutions and (b) solve the triangle, if possible. Express the lengths of sides to three significant digits and the angles to the nearest tenth of a degree.

17. $A = 37.0°$, $a = 21.5$ cm, $b = 16.4$ cm
18. $B = 55.0°$, $b = 182$ m, $c = 203$ m
19. $C = 26.5°$, $c = 42.7$ km, $a = 47.2$ km
20. $B = 40.4°$, $b = 81.4$ m, $c = 144$ m
21. $A = 71.5°$, $a = 3.45$ m, $c = 3.50$ m
22. $C = 17.2°$, $c = 2.20$ m, $b = 2.00$ m
23. $B = 105.0°$, $b = 16.5$ mi, $a = 12.0$ mi
24. $A = 98.8°$, $a = 707$ ft, $b = 585$ ft

Express the lengths of sides to two significant digits and the angles to the nearest degree.

25. $C = 18°$, $c = 24$ mi, $a = 45$ mi
26. $B = 36°$, $b = 75$ cm, $a = 95$ cm
27. $C = 6\bar{0}°$, $c = 150$ m, $b = 180$ m
28. $A = 3\bar{0}°$, $a = 4800$ ft, $c = 3600$ ft
29. $B = 8°$, $b = 450$ m, $c = 850$ m
30. $B = 45°$, $c = 2.5$ m, $b = 3.2$ m

Express the lengths of sides to four significant digits and the angles to the nearest hundredth of a degree.

31. $B = 41.50°$, $b = 14.25$ km, $a = 18.50$ km
32. $A = 15.75°$, $a = 642.5$ m, $c = 592.7$ m
33. $C = 63.85°$, $c = 29.50$ cm, $b = 38.75$ cm
34. $B = 50.00°$, $b = 41{,}250$ km, $c = 45{,}650$ km

*Because of differences in rounding, your answers may differ slightly from the answers in the text if you choose to solve for the parts of a triangle in an order different from that chosen by the authors.

35. $C = 8.75°, c = 89.30$ m, $a = 61.93$ m
36. $A = 31.50°, a = 375.0$ mm, $b = 405.5$ mm

Express the lengths of sides to three significant digits and the angles to the nearest tenth of a degree.

37. $A = 60.0°, b = 19.5$ m, $c = 25.0$ m   38. $B = 19.5°, a = 21.5$ ft, $c = 12.5$ ft
39. $C = 109.0°, a = 14\overline{0}$ km, $b = 215$ km   40. $A = 94.7°, c = 875$ yd, $b = 185$ yd
41. $a = 19.2$ m, $b = 21.3$ m, $c = 27.2$ m
42. $a = 125$ km, $b = 195$ km, $c = 145$ km
43. $a = 4.25$ ft, $b = 7.75$ ft, $c = 5.50$ ft
44. $a = 3590$ m, $b = 7950$ m, $c = 4650$ m

Express the lengths of sides to two significant digits and the angles to the nearest degree.

45. $A = 45°, b = 51$ m, $c = 39$ m   46. $B = 6\overline{0}°, a = 160$ cm, $c = 230$ cm
47. $a = 7\overline{0}00$ m, $b = 5600$ m, $c = 4800$ m
48. $a = 5.8$ cm, $b = 5.8$ cm, $c = 9.6$ cm
49. $C = 135°, a = 36$ ft, $b = 48$ ft   50. $A = 5°, b = 19$ m, $c = 25$ m

Express the lengths of sides to four significant digits and the angles to the nearest hundredth of a degree.

51. $B = 19.25°, a = 4815$ m, $c = 1925$ m
52. $C = 75.00°, a = 37,550$ mi, $b = 45,250$ mi
53. $C = 108.75°, a = 405.0$ mm, $b = 325.0$ mm
54. $A = 111.05°, b = 1976$ ft, $c = 325\overline{0}$ ft
55. $a = 207.5$ km, $b = 105.6$ km, $c = 141.5$ km
56. $a = 19.45$ m, $b = 36.50$ m, $c = 25.60$ m

SCIENTIFIC CALCULATOR

There are several kinds and brands of calculators. Some are very simple to use; others do more difficult calculations. We demonstrate various operations on common basic scientific calculators that use algebraic logic, which follows the steps commonly used in mathematics. Yours may differ. If so, consult your manual.

To demonstrate how to use a calculator, we show what buttons are pushed and the order in which they are pushed. We assume that you know how to add, subtract, multiply, and divide on the calculator.

B.1 Scientific Notation

Numbers expressed in scientific notation can be entered into many calculators. The results may then also be given in scientific notation.

Multiply $(6.5 \times 10^8)(1.4 \times 10^{-15})$ and write the result in scientific notation. ◄ **EXAMPLE 1**

6.5 ⎡EE⎤ 8 ⎡×⎤ 1.4 ⎡EE⎤ ⎡(−)⎤ 15 ⎡=⎤

$$9.1 \times 10^{-7}$$

The result is 9.1×10^{-7}.

· · · · · · · · · · · · · · · · ·

Divide $\dfrac{3.24 \times 10^{-5}}{7.2 \times 10^{-12}}$ and write the result in scientific notation. ◄ **EXAMPLE 2**

3.24 ⎡EE⎤ ⎡(−)⎤ 5 ⎡÷⎤ 7.2 ⎡EE⎤ ⎡(−)⎤ 12 ⎡=⎤

$$4.5 \times 10^6$$

The result is 4.5×10^6.

· · · · · · · · · · · · · · · · ·

Find the value of $\dfrac{(-6.3 \times 10^4)(-5.07 \times 10^{-9})(8.11 \times 10^{-6})}{(5.63 \times 10^{12})(-1.84 \times 10^7)}$ and write the result in ◄ **EXAMPLE 3**
scientific notation rounded to three significant digits.

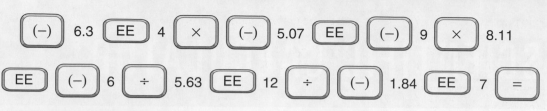

The result rounded to three significant digits is -2.50×10^{-29}.

· · · · · · · · · · · · · · · · ·

B.2 Squares and Square Roots

EXAMPLE 1

Find the value of $(46.8)^2$.

46.8 $\boxed{x^2}$ $\boxed{=}$

> 2190.24

The result is 2190.24.

· · · · · · · · · · · · · · · · ·

EXAMPLE 2

Find the value of $(6.3 \times 10^{-18})^2$.

6.3 $\boxed{\text{EE}}$ $\boxed{(-)}$ 18 $\boxed{x^2}$ $\boxed{=}$

> 3.969 x 10⁻³⁵

The result is 3.969×10^{-35}.

· · · · · · · · · · · · · · · · ·

EXAMPLE 3

Find the value of $\sqrt{158.65}$ and round to four significant digits.

$\boxed{\sqrt{}}$ 158.65 $\boxed{=}$

> 12.59563416

The result rounded to four significant digits is 12.60.

· · · · · · · · · · · · · · · · ·

EXAMPLE 4

Find the value of $\sqrt{6.95 \times 10^{-15}}$ and round to three significant digits.

$\boxed{\sqrt{}}$ 6.95 $\boxed{\text{EE}}$ $\boxed{(-)}$ 15 $\boxed{=}$

> 8.336666 x 10⁻⁸

The result rounded to three significant digits is 8.34×10^{-8}.

· · · · · · · · · · · · · · · · ·

Find the value of $\sqrt{15.7^2 + 27.6^2}$ and round to three significant digits.

$\boxed{\sqrt{}}$ 15.7 $\boxed{x^2}$ $\boxed{+}$ 27.6 $\boxed{x^2}$ $\boxed{=}$

EXAMPLE 5

$\boxed{31.75295262}$

The result is 31.8 rounded to three significant digits.

Note: You may need to use parentheses with some calculators.

· · · · · · · · · · · · · · · · · ·

Find the value of $\dfrac{14}{\sqrt{5}} - \sqrt{\dfrac{15}{8}}$ and round the result to three significant digits.

EXAMPLE 6

14 $\boxed{\div}$ $\boxed{\sqrt{}}$ 5 $\boxed{)}$ * $\boxed{-}$ $\boxed{\sqrt{}}$ 15 $\boxed{\div}$ 8 $\boxed{)}$ * $\boxed{=}$

$\boxed{4.891683943}$

The result is 4.89 rounded to three significant digits.

· · · · · · · · · · · · · · · · · ·

Find the value of $\sqrt{\left(\dfrac{16}{1.3}\right)^2 + \left[\dfrac{1}{2\pi(60)(6 \times 10^{-5})}\right]^2}$ rounded to three significant digits.

EXAMPLE 7

$\boxed{\sqrt{}}$ $\boxed{(}$ 16 $\boxed{\div}$ 1.3 $\boxed{)}$ $\boxed{x^2}$ $\boxed{+}$ $\boxed{(}$ $\boxed{(}$ 2

$\boxed{\times}$ $\boxed{\pi}$ $\boxed{\times}$ 60 $\boxed{\times}$ 6 $\boxed{EE}$ $\boxed{(-)}$ 5 $\boxed{)}$ $\boxed{x^{-1}}$

$\boxed{)}$ $\boxed{x^2}$ $\boxed{)}$ $\boxed{=}$

$\boxed{45.89092973}$

The result is 45.9 rounded to three significant digits.

Note: You will need to insert parentheses to clarify the order of operations. You may need to supply other parentheses.

· · · · · · · · · · · · · · · · · ·

B.3 Trigonometric Operations

Calculators must have sine, cosine, and tangent buttons for use in this book.

Note: Make certain that your calculator is in the degree mode (not in the radian mode) for this section and when working problems in degrees.

Note: You may need to insert a right parenthesis to clarify the order of operations. The square root key sometimes includes the left parenthesis. If not, you may need to supply it.

EXAMPLE 1

Find sin 26° rounded to four significant digits.

| sin | 26 | = |

```
0.438371147
```

That is, sin 26° = 0.4384 rounded to four significant digits.

· · · · · · · · · · · · · · · · ·

EXAMPLE 2

Find cos 36.75° rounded to four significant digits.

| cos | 36.75 | = |

```
0.801253813
```

That is, cos 36.75° = 0.8013 rounded to four significant digits.

· · · · · · · · · · · · · · · · ·

EXAMPLE 3

Find tan 70.6° rounded to four significant digits.

| tan | 70.6 | = |

```
2.839653913
```

That is, tan 70.6° = 2.840 rounded to four significant digits.

· · · · · · · · · · · · · · · · ·

Since we use right triangles almost exclusively, we show how to find the angle of a right triangle when the value of a given trigonometric ratio is known. That is, we will find angle A when $0° \leq A \leq 90°$.

EXAMPLE 4

Given sin A = 0.4321, find angle A to the nearest tenth of a degree.

| sin⁻¹ | .4321 | = |

```
25.60090542
```

Note: Make certain your calculator is in the degree mode.
Thus, A = 25.6° to the nearest tenth of a degree.

· · · · · · · · · · · · · · · · ·

EXAMPLE 5

Given cos B = 0.6046, find angle B to the nearest tenth of a degree.

| cos⁻¹ | .6046 | = |

```
52.79993633
```

Thus, B = 52.8° to the nearest tenth of a degree.

· · · · · · · · · · · · · · · · ·

Given $\tan A = 2.584$, find angle A to the nearest tenth of a degree.

EXAMPLE 6

$\boxed{\tan^{-1}}$ 2.584 $\boxed{=}$

| 68.8437168 |

Thus, $A = 68.8°$ to the nearest tenth of a degree.

· · · · · · · · · · · · · · · · ·

Trigonometric functions often occur in expressions that must be evaluated.

Given $a = (\tan 54°)(25.6 \text{ m})$, find a rounded to three significant digits.

EXAMPLE 7

$\boxed{\tan}$ 54 $\boxed{)}$ $\boxed{\times}$ 25.6 $\boxed{=}$

| 35.23537716 |

Thus, $a = 35.2 \text{ m}$ rounded to three significant digits.

· · · · · · · · · · · · · · · · ·

Given $b = \dfrac{452 \text{ m}}{\cos 37.5°}$, find b rounded to three significant digits.

EXAMPLE 8

452 $\boxed{\div}$ $\boxed{\cos}$ 37.5 $\boxed{=}$

| 569.7335311 |

Thus, $b = 570 \text{ m}$ rounded to three significant digits.

· · · · · · · · · · · · · · · · ·

B.4 Finding a Power

To raise a number to a power, use the $\boxed{\wedge}$ button as follows.

Find the value of 4^5.

EXAMPLE 1

4 $\boxed{\wedge}$ or $\boxed{y^x}$ 5 $\boxed{=}$

| 1024 |

That is, $4^5 = 1024$.

· · · · · · · · · · · · · · · · ·

EXAMPLE 2

Find the value of 1.5^{-4} rounded to three significant digits.

1.5 [∧] or [y^x] [(−)] 4 [=]

```
0.197530864
```

That is, $1.5^{-4} = 0.198$ rounded to three significant digits.

· · · · · · · · · · · · · · · · · ·

Problems B.4

Do as indicated and round each result to three significant digits.

1. $(6.43 \times 10^8)(5.16 \times 10^{10})$ 2. $(4.16 \times 10^{-5})(3.45 \times 10^{-7})$

3. $(1.456 \times 10^{12})(-4.69 \times 10^{-18})$ 4. $(-5.93 \times 10^9)(7.055 \times 10^{-12})$

5. $(7.45 \times 10^8) \div (8.92 \times 10^{18})$ 6. $(1.38 \times 10^{-6}) \div (4.324 \times 10^6)$

7. $\dfrac{-6.19 \times 10^{12}}{7.755 \times 10^{-8}}$ 8. $\dfrac{1.685 \times 10^{10}}{1.42 \times 10^{24}}$

9. $\dfrac{(5.26 \times 10^{-8})(8.45 \times 10^6)}{(-6.142 \times 10^9)(1.056 \times 10^{-12})}$

10. $\dfrac{(-2.35 \times 10^{-9})(1.25 \times 10^{11})(4.65 \times 10^{17})}{(8.75 \times 10^{23})(-5.95 \times 10^{-6})}$

11. $(68.4)^2$ 12. $(3180)^2$

13. $\sqrt{46,500}$ 14. $\sqrt{0.000634}$

15. $(1.45 \times 10^5)^2$ 16. $(1.095 \times 10^{-18})^2$

17. $\sqrt{4.63 \times 10^{18}}$ 18. $\sqrt{9.49 \times 10^{-15}}$

19. $\sqrt{(4.68)^2 + (9.63)^2}$ 20. $\sqrt{(18.4)^2 - (6.5)^2}$

21. $18\sqrt{3} + \left(\dfrac{28.1}{19}\right)^2$ 22. $\dfrac{8}{\sqrt{2}} + \sqrt{\dfrac{58}{14.5}}$

23. $25^2 - \sqrt{\dfrac{29.8}{0.0256}}$ 24. $\dfrac{18.3}{6\sqrt{5}} - \left(\dfrac{225}{147}\right)^2$

25. $(12.6^2 + 21.5^2)^2 + (34.2^2 - 26.4^2)^2$ 26. $\sqrt{21.4^2 + 18.7^2} + \sqrt{31.5^2 - 16.3^2}$

27. $\dfrac{91.4 - 48.6}{91.4 - 15.9}$ 28. $\dfrac{14.7 + 9.6}{45.7 + 68.2}$

29. $\sqrt{\left(\dfrac{80.5}{25.6}\right)^2 + \left[\dfrac{1}{2\pi(60)(1.5 \times 10^{-7})}\right]^2}$

30. $\sqrt{\left(\dfrac{175}{36.5}\right)^2 + \left[\dfrac{1}{2\pi(60)(8.5 \times 10^{-10})}\right]^2}$

31. $\dfrac{(17.2)(11.6) + (8)(17.6) - (6)(16)}{(5)(15) + (8.5)(15) + (10)(26.5)}$

32. $\dfrac{(18.8)(5.5) + (7.75)(16.5) - (9.25)(13.85)}{(6.25)(12.5) + (4.75)(16.5) + (11.5)(14.1)}$

33. $\sin 13°$ 34. $\cos 22°$ 35. $\tan 52.3°$

36. $\tan 31.25°$ 37. $\cos 59.36°$ 38. $\sin 84.55°$

39. $\sin 48°$ 40. $\cos 48°$ 41. $\tan 75°$

42. $\sin 8°$ 43. $\sin 8.7°$ 44. $\cos 35°$

Find each angle rounded to the nearest tenth of a degree.

45. $\sin A = 0.6527$ 46. $\cos B = 0.2577$ 47. $\tan A = 0.4568$

48. $\sin B = 0.4658$ 49. $\cos A = 0.5563$ 50. $\tan B = 1.496$

51. $\sin B = 0.1465$ 52. $\cos A = 0.4968$ 53. $\tan B = 1.987$
54. $\sin A = 0.2965$ 55. $\cos B = 0.3974$ 56. $\tan A = 0.8885$

Find each angle to the nearest tenth of a degree between 0° and 90° and each side to three significant digits.

57. $b = (\sin 58.2°)(296 \text{ m})$

58. $a = (\cos 25.2°)(54.5 \text{ m})$

59. $c = \dfrac{37.5 \text{ m}}{\cos 65.2°}$

60. $b = \dfrac{59.7 \text{ m}}{\tan 41.2°}$

61. $\tan A = \dfrac{512 \text{ km}}{376 \text{ km}}$

62. $\cos B = \dfrac{75.2 \text{ m}}{89.5 \text{ m}}$

63. $a = (\cos 19.5°)(15.7 \text{ cm})$

64. $c = \dfrac{235 \text{ km}}{\sin 65.2°}$

65. $b = \dfrac{36.7 \text{ m}}{\tan 59.2°}$

66. $a = (\tan 5.7°)(135 \text{ m})$

Find the value of each power and round each result to three significant digits.

67. 12^4 68. 1.8^3 69. 0.46^5
70. 9^{-3} 71. 14^{-5} 72. 0.65^{-4}

TABLES

Table 1 U.S. Weights and Measures

Units of Length	Units of Volume	Units of Weight
Standard unit—inch (in. or ″) 12 inches = 1 foot (ft or ′) 3 feet = 1 yard (yd) $5\frac{1}{2}$ yards or $16\frac{1}{2}$ feet = 1 rod (rd) 5280 feet = 1 mile (mi)	*Liquid* 16 ounces (fl oz) = 1 pint (pt) 2 pints = 1 quart (qt) 4 quarts = 1 gallon (gal) *Dry* 2 pints (pt) = 1 quart (qt) 8 quarts = 1 peck (pk) 4 pecks = 1 bushel (bu)	Standard unit—pound (lb) 16 ounces (oz) = 1 pound 2000 pounds = 1 ton (T)

Table 2 Conversion Table for Length

	cm	m	km	in.	ft	mi
1 centimetre =	1	10^{-2}	10^{-5}	0.394	3.28×10^{-2}	6.21×10^{-6}
1 metre =	100	1	10^{-3}	39.4	3.28	6.21×10^{-4}
1 kilometre =	10^5	1000	1	3.94×10^4	3280	0.621
1 inch =	2.54	2.54×10^{-2}	2.54×10^{-5}	1	8.33×10^{-2}	1.58×10^{-5}
1 foot =	30.5	0.305	3.05×10^{-4}	12	1	1.89×10^{-4}
1 mile =	1.61×10^5	1610	1.61	6.34×10^4	5280	1

Table 3 Conversion Table for Area

Metric	U.S.
$1 \text{ m}^2 = 10{,}000 \text{ cm}^2$ $= 1{,}000{,}000 \text{ mm}^2$ $1 \text{ cm}^2 = 100 \text{ mm}^2$ $= 0.0001 \text{ m}^2$ $1 \text{ km}^2 = 1{,}000{,}000 \text{ m}^2$	$1 \text{ ft}^2 = 144 \text{ in}^2$ $1 \text{ yd}^2 = 9 \text{ ft}^2$ $1 \text{ rd}^2 = 30.25 \text{ yd}^2$ $1 \text{ acre} = 160 \text{ rd}^2$ $= 4840 \text{ yd}^2$ $= 43{,}560 \text{ ft}^2$ $1 \text{ mi}^2 = 640 \text{ acres}$

	m^2	cm^2	ft^2	in^2
1 square metre =	1	10^4	10.8	1550
1 square centimetre =	10^{-4}	1	1.08×10^{-3}	0.155
1 square foot =	9.29×10^{-2}	929	1	144
1 square inch =	6.45×10^{-4}	6.45	6.94×10^{-3}	1

1 circular mil = $5.07 \times 10^{-6} \text{ cm}^2 = 7.85 \times 10^{-7} \text{ in}^2$
1 hectare = $10{,}000 \text{ m}^2 = 2.47 \text{ acres}$

Table 4   Conversion Table for Volume

	Metric	U.S.
	$1 \text{ m}^3 = 10^6 \text{ cm}^3$	$1 \text{ ft}^3 = 1728 \text{ in}^3$
	$1 \text{ cm}^3 = 10^{-6} \text{ m}^3$	$1 \text{ yd}^3 = 27 \text{ ft}^3$
	$= 10^3 \text{ mm}^3$	

	m^3	cm^3	L	ft^3	in^3
$1 \text{ m}^3 =$	1	10^6	1000	35.3	6.10×10^4
$1 \text{ cm}^3 =$	10^{-6}	1	1.00×10^{-3}	3.53×10^{-5}	6.10×10^{-2}
$1 \text{ litre} =$	1.00×10^{-3}	1000	1	3.53×10^{-2}	61.0
$1 \text{ ft}^3 =$	2.83×10^{-2}	2.83×10^4	28.3	1	1728
$1 \text{ in}^3 =$	1.64×10^{-5}	16.4	1.64×10^{-2}	5.79×10^{-4}	1

1 U.S. fluid gallon = 4 U.S. fluid quarts = 8 U.S. pints = 128 U.S. fluid ounces = 231 in^3 = 0.134 ft^3
1 L = 1000 cm^3 = 1.06 qt   1 fl oz = 29.5 cm^3 1 ft^3 = 7.47 gal = 28.3 L

Table 5   Conversion Table for Mass

	g	kg	slug	oz	lb	ton
$1 \text{ gram} =$	1	0.001	6.85×10^{-5}	3.53×10^{-2}	2.21×10^{-3}	1.10×10^{-6}
$1 \text{ kilogram} =$	1000	1	6.85×10^{-2}	35.3	2.21	1.10×10^{-3}
$1 \text{ slug} =$	1.46×10^4	14.6	1	515	32.2	1.61×10^{-2}
$1 \text{ ounce} =$	28.4	2.84×10^{-2}	1.94×10^{-3}	1	6.25×10^{-2}	3.13×10^{-5}
$1 \text{ pound} =$	454	0.454	3.11×10^{-2}	16	1	5.00×10^{-4}
$1 \text{ ton} =$	9.07×10^5	907	62.2	3.2×10^4	2000	1

1 metric ton = 1000 kg = 2205 lb

Quantities in the shaded areas are not mass units. When we write, for example, 1 kg "=" 2.21 lb, this means that a kilogram is a mass that weighs 2.21 pounds under standard conditions of gravity ($g = 9.80 \text{ m/s}^2 = 32.2 \text{ ft/s}^2$).

Table 6   Conversion Table for Density

	slug/ft^3	kg/m^3	g/cm^3	lb/ft^3	lb/in^3
$1 \text{ slug per ft}^3 =$	1	515	0.515	32.2	1.86×10^{-2}
$1 \text{ kilogram per m}^3 =$	1.94×10^{-3}	1	0.001	6.24×10^{-2}	3.61×10^{-5}
$1 \text{ gram per cm}^3 =$	1.94	1000	1	62.4	3.61×10^{-2}
$1 \text{ pound per ft}^3 =$	3.11×10^{-2}	16.0	1.60×10^{-2}	1	5.79×10^{-4}
$1 \text{ pound per in}^3 =$	53.7	2.77×10^4	27.7	1728	1

Quantities in the shaded areas are weight densities and, as such, are dimensionally different from mass densities.
Note that

$$D_w = D_m g$$

where D_w = weight density
D_m = mass density
$g = 9.80 \text{ m/s}^2 = 32.2 \text{ ft/s}^2$

Table 7 Conversion Table for Time

	yr	day	h	min	s
1 year =	1	365	8.77×10^3	5.26×10^5	3.16×10^7
1 day =	2.74×10^{-3}	1	24	1440	8.64×10^4
1 hour =	1.14×10^{-4}	4.17×10^{-2}	1	60	3600
1 minute =	1.90×10^{-6}	6.94×10^{-4}	1.67×10^{-2}	1	60
1 second =	3.17×10^{-8}	1.16×10^{-5}	2.78×10^{-4}	1.67×10^{-2}	1

Table 8 Conversion Table for Speed

	ft/s	km/h	m/s	mi/h	cm/s
1 foot per second =	1	1.10	0.305	0.682	30.5
1 kilometre per hour =	0.911	1	0.278	0.621	27.8
1 metre per second =	3.28	3.60	1	2.24	100
1 mile per hour =	1.47	1.61	0.447	1	44.7
1 centimetre per second =	3.28×10^{-2}	3.60×10^{-2}	0.01	2.24×10^{-2}	1
1 mi/min = 88.0 ft/s = 60.0 mi/h					

Table 9 Conversion Table for Force

	N	lb	oz
1 newton =	1	0.225	3.60
1 pound =	4.45	1	16
1 ounce =	0.278	0.0625	1

Table 10 Conversion Table for Power

	Btu/h	ft lb/s	hp	kW	W
1 British thermal unit per hour =	1	0.216	3.93×10^{-4}	2.93×10^{-4}	0.293
1 foot pound per second =	4.63	1	1.82×10^{-3}	1.36×10^{-3}	1.36
1 horsepower =	2550	550	1	0.746	746
1 kilowatt =	3410	738	1.34	1	1000
1 watt =	3.41	0.738	1.34×10^{-3}	0.001	1

Table 11 Conversion Table for Pressure

	atm	Inches of Water	mm Hg	N/m² (Pa)	lb/in²	lb/ft²
1 atmosphere =	1	407	$76\overline{0}$	1.01×10^5	14.7	2120
1 inch of water[a] at 4°C =	2.46×10^{-3}	1	1.87	249	3.61×10^{-2}	5.20
1 millimetre of mercury[a] at 0°C =	1.32×10^{-3}	0.535	1	133	1.93×10^{-2}	2.79
1 newton per metre² (pascal) =	9.87×10^{-6}	4.02×10^{-3}	7.50×10^{-3}	1	1.45×10^{-4}	2.09×10^{-2}
1 pound per in² =	6.81×10^{-2}	27.7	51.7	6.90×10^3	1	144
1 pound per ft² =	4.73×10^{-4}	0.192	0.359	47.9	6.94×10^{-3}	1

[a]Where the acceleration of gravity has the standard value, $g = 9.80$ m/s² = 32.2 ft/s².

Table 12 Mass and Weight Density

Substance	Mass Density (kg/m³)	Weight Density (lb/ft³)
Solids		
Copper	8,890	555
Iron	7,800	490
Lead	11,300	708
Aluminum	2,700	169
Brass	8,700	540
Ice	917	57
Wood, white pine	420	26
Concrete	2,300	140
Cork	240	15
Liquids		
Water	$1,0\overline{0}0$[a]	62.4
Seawater	1,025	64.0
Oil	870	54.2
Mercury	13,600	846
Alcohol	790	49.4
Gasoline	680	42.0
	At 0°C and 1 atm Pressure	At 32°F and 1 atm Pressure
Gases[b]		
Air	1.29	0.081
Carbon dioxide	1.96	0.123
Carbon monoxide	1.25	0.078
Helium	0.178	0.011
Hydrogen	0.0899	0.0056
Oxygen	1.43	0.089
Nitrogen	1.25	0.078
Ammonia	0.760	0.047
Propane	2.02	0.126

[a]Metric weight density of water = $98\overline{0}0$ N/m³.
[b]The density of a gas is found by pumping the gas into a container, measuring its volume and mass or weight, and then using the appropriate density formula.

Table 13   Specific Gravity of Certain Liquids[a]

Liquid	Specific Gravity
Benzene	0.90
Ethyl alcohol	0.79
Gasoline	0.68
Kerosene	0.82
Mercury	13.6
Seawater	1.025
Sulfuric acid	1.84
Turpentine	0.87
Water	1.000

[a]At room temperature (20°C or 68°F)

Table 14   Conversion Table for Energy, Work, and Heat

	Btu	ft lb	J	cal	kWh
1 British thermal unit =	1	778	1060	252	2.93×10^{-4}
1 foot pound =	1.29×10^{-3}	1	1.36	0.324	3.77×10^{-7}
1 joule =	9.48×10^{-4}	0.738	1	0.239	2.78×10^{-7}
1 calorie =	3.97×10^{-3}	3.09	4.19	1	1.16×10^{-6}
1 kilowatt-hour =	3410	2.66×10^6	3.60×10^6	8.60×10^5	1

Table 15   Heat Constants

	Melting Point (°C)	Boiling Point (°C)	Specific Heat cal/g °C or kcal/kg °C or Btu/lb °F	J/kg °C	Heat of Fusion cal/g or kcal/kg	J/kg	Vaporization cal/g or kcal/kg	J/kg
Alcohol, ethyl	−117	78.5	0.58	2400	24.9	1.04×10^5	204	8.54×10^5
Aluminum	660	2057	0.22	920	76.8	3.21×10^5		
Brass	840		0.092	390				
Copper	1083	2330	0.092	390	49.0	2.05×10^5		
Glass			0.21	880				
Ice	0		0.51	2100	$\overline{8}0$	3.35×10^5		
Iron (steel)	1540	3000	0.115	481	7.89	3.30×10^4		
Lead	327	1620	0.031	130	5.86	2.45×10^4		
Mercury	−38.9	357	0.033	140	2.82	1.18×10^4	65.0	2.72×10^5
Silver	961	1950	0.056	230	26.0	1.09×10^5		
Steam			0.48	$200\overline{0}$				
Water (liquid)	0	$10\overline{0}$	1.00	4190	$\overline{8}0$	3.35×10^5	$54\overline{0}$	2.26×10^6
Zinc	419	907	0.092	390	23.0	9.63×10^4		

Table 16   Coefficient of Linear Expansion

Material	α (metric)	α (U.S.)
Aluminum	$2.3 \times 10^{-5}/C°$	$1.3 \times 10^{-5}/F°$
Brass	$1.9 \times 10^{-5}/C°$	$1.0 \times 10^{-5}/F°$
Concrete	$1.1 \times 10^{-5}/C°$	$6.0 \times 10^{-6}/F°$
Copper	$1.7 \times 10^{-5}/C°$	$9.5 \times 10^{-6}/F°$
Glass	$9.0 \times 10^{-6}/C°$	$5.1 \times 10^{-6}/F°$
Pyrex	$3.0 \times 10^{-6}/C°$	$1.7 \times 10^{-6}/F°$
Steel	$1.3 \times 10^{-5}/C°$	$6.5 \times 10^{-6}/F°$
Zinc	$2.6 \times 10^{-5}/C°$	$1.5 \times 10^{-5}/F°$

Table 17   Coefficient of Volume Expansion

Liquid	β (metric)	β (U.S.)
Acetone	$1.49 \times 10^{-3}/C°$	$8.28 \times 10^{-4}/F°$
Alcohol, ethyl	$1.12 \times 10^{-3}/C°$	$6.62 \times 10^{-4}/F°$
Carbon tetrachloride	$1.24 \times 10^{-3}/C°$	$6.89 \times 10^{-4}/F°$
Mercury	$1.8 \times 10^{-4}/C°$	$1.0 \times 10^{-4}/F°$
Petroleum	$9.6 \times 10^{-4}/C°$	$5.33 \times 10^{-4}/F°$
Turpentine	$9.7 \times 10^{-4}/C°$	$5.39 \times 10^{-4}/F°$
Water	$2.1 \times 10^{-4}/C°$	$1.17 \times 10^{-4}/F°$

Table 18   Charge

Charge on one electron $= 1.60 \times 10^{-19}$ coulomb
1 coulomb $= 6.25 \times 10^{18}$ electrons of charge
1 ampere-hour $= 3600$ C

Table 19   Relationships of Metric SI Base and Derived Units

This chart shows graphically how the 17 SI-derived units with special names are derived from the base and supplementary units. It was provided by the National Institute of Standards and Technology.

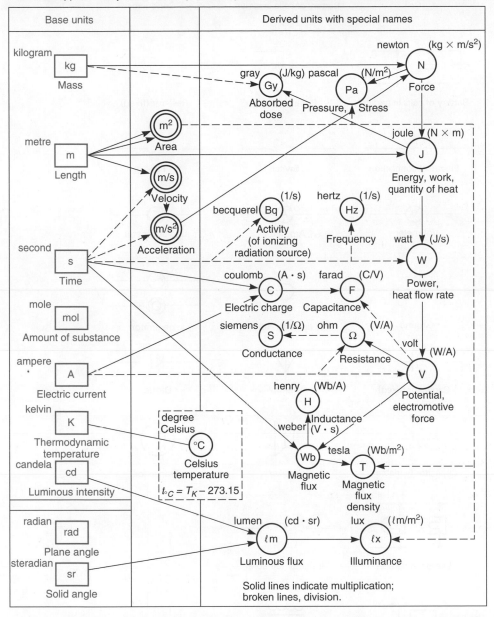

Table 20   Electric Symbols

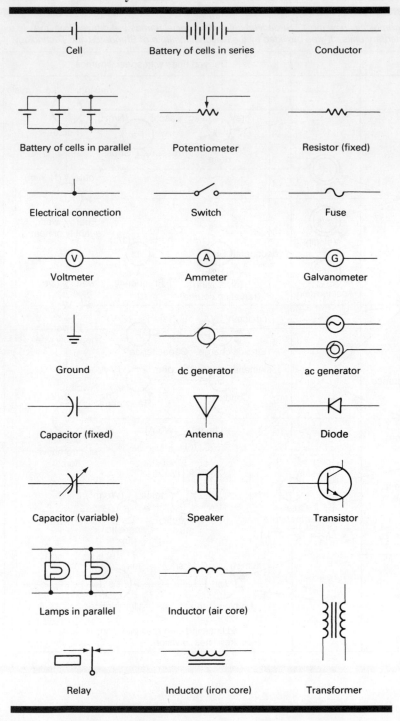

Cell	Battery of cells in series	Conductor
Battery of cells in parallel	Potentiometer	Resistor (fixed)
Electrical connection	Switch	Fuse
Voltmeter	Ammeter	Galvanometer
Ground	dc generator	ac generator
Capacitor (fixed)	Antenna	Diode
Capacitor (variable)	Speaker	Transistor
Lamps in parallel	Inductor (air core)	Transformer
Relay	Inductor (iron core)	Transformer

Table 21 Periodic Table

Key:
Symbol — Cl 17 — Atomic number
Atomic mass* — 35.453
Electron configuration — $3p^5$

Transition Elements

Group I	Group II		Transition Elements									Group III	Group IV	Group V	Group VI	Group VII	Group 0
H 1 1.0079 $1s^1$																	He 2 4.00260 $1s^2$
Li 3 6.94 $2s^1$	Be 4 9.01218 $2s^2$											B 5 10.81 $2p^1$	C 6 12.011 $2p^2$	N 7 14.0067 $2p^3$	O 8 15.9994 $2p^4$	F 9 18.9984 $2p^5$	Ne 10 20.18 $2p^6$
Na 11 22.9898 $3s^1$	Mg 12 24.305 $3s^2$											Al 13 26.9815 $3p^1$	Si 14 28.0855 $3p^2$	P 15 30.974 $3p^3$	S 16 32.06 $3p^4$	Cl 17 35.453 $3p^5$	Ar 18 39.948 $3p^6$
K 19 39.0983 $4s^1$	Ca 20 40.08 $4s^2$	Sc 21 44.9559 $3d^14s^2$	Ti 22 47.9 $3d^24s^2$	V 23 50.9415 $3d^34s^2$	Cr 24 51.996 $3d^54s^1$	Mn 25 54.938 $3d^54s^2$	Fe 26 55.847 $3d^64s^2$	Co 27 58.9332 $3d^74s^2$	Ni 28 58.7 $3d^84s^2$	Cu 29 63.546 $3d^104s^1$	Zn 30 65.39 $3d^104s^2$	Ga 31 69.73 $4p^1$	Ge 32 72.6 $4p^2$	As 33 74.9216 $4p^3$	Se 34 78.96 $4p^4$	Br 35 79.904 $4p^5$	Kr 36 83.80 $4p^6$
Rb 37 85.47 $5s^1$	Sr 38 87.62 $5s^2$	Y 39 88.9059 $4d^15s^2$	Zr 40 91.22 $4d^25s^2$	Nb 41 92.9064 $4d^45s^1$	Mo 42 95.94 $4d^55s^1$	Tc 43 (98) $4d^55s^2$	Ru 44 101.07 $4d^75s^1$	Rh 45 102.906 $4d^85s^1$	Pd 46 106.4 $4d^105s^0$	Ag 47 107.868 $4d^105s^1$	Cd 48 112.41 $4d^105s^2$	In 49 114.82 $5p^1$	Sn 50 118.7 $5p^2$	Sb 51 121.75 $5p^3$	Te 52 127.60 $5p^4$	I 53 126.90 $5p^5$	Xe 54 131.3 $5p^6$
Cs 55 132.905 $6s^1$	Ba 56 137.33 $6s^2$	57–71† La	Hf 72 178.49 $5d^26s^2$	Ta 73 180.95 $5d^36s^2$	W 74 183.85 $5d^46s^2$	Re 75 186.207 $5d^56s^2$	Os 76 190.2 $5d^66s^2$	Ir 77 192.22 $5d^76s^2$	Pt 78 195.08 $5d^96s^1$	Au 79 196.97 $5d^106s^1$	Hg 80 200.59 $5d^106s^2$	Tl 81 204.38 $6p^1$	Pb 82 207.2 $6p^2$	Bi 83 208.980 $6p^3$	Po 84 (209) $6p^4$	At 85 (210) $6p^5$	Rn 86 (222) $6p^6$
Fr 87 (223) $7s^1$	Ra 88 226.025 $7s^2$	89–103‡ Ac	Rf 104 (261) $6d^27s^2$	Ha 105 (262) $6d^37s^2$	106 (263)	107 (262)	108 (265)	109 (266)									

†Lanthanide series

La 57 138.906 $5d^16s^2$	Ce 58 140.12 $5d^14f^16s^2$	Pr 59 140.908 $4f^36s^2$	Nd 60 144.24 $4f^46s^2$	Pm 61 (145) $4f^56s^2$	Sm 62 150.4 $4f^66s^2$	Eu 63 151.96 $4f^76s^2$	Gd 64 157.25 $5d^14f^76s^2$	Tb 65 158.925 $5d^14f^86s^2$	Dy 66 162.50 $4f^106s^2$	Ho 67 164.930 $4f^116s^2$	Er 68 167.26 $4f^126s^2$	Tm 69 168.934 $4f^136s^2$	Yb 70 173.04 $4f^146s^2$	Lu 71 174.967 $5d^14f^146s^2$

‡Actinide series

Ac 89 (227) $6d^17s^2$	Th 90 232.038 $6d^27s^2$	Pa 91 231.036 $5f^26d^17s^2$	U 92 238.029 $5f^36d^17s^2$	Np 93 237.048 $5f^46d^17s^2$	Pu 94 (244) $5f^66d^07s^2$	Am 95 (243) $5f^76d^07s^2$	Cm 96 (247) $5f^76d^17s^2$	Bk 97 (247) $5f^96d^07s^2$	Cf 98 (251) $5f^106d^07s^2$	Es 99 (252) $5f^116d^07s^2$	Fm 100 (257) $5f^126d^07s^2$	Md 101 (258) $5f^136d^07s^2$	No 102 (259) $5f^146d^07s^2$	Lr 103 (260) $6d^17s^2$

*Atomic mass values averaged over isotopes in percentages in which they occur on the earths surface. For many unstable elements, mass number of the most stable known isotope is given in parentheses.

Table 22   The Greek Alphabet

Capital	Lowercase	Name
A	α	alpha
B	β	beta
Γ	γ	gamma
Δ	δ	delta
E	ε	epsilon
Z	ζ	zeta
H	η	eta
Θ	θ	theta
I	ι	iota
K	κ	kappa
Λ	λ	lambda
M	μ	mu
N	ν	nu
Ξ	ξ	xi
O	o	omicron
Π	π	pi
P	ρ	rho
Σ	σ	sigma
T	τ	tau
Υ	υ	upsilon
Φ	ϕ	phi
X	χ	chi
Ψ	ψ	psi
Ω	ω	omega

GLOSSARY

A

Absolute Pressure The actual air pressure given by the gauge reading plus the normal atmospheric pressure.

Absolute Zero The lowest possible temperature.

Acceleration Change in velocity per unit time.

Acceleration Due to Gravity The acceleration of a freely falling object. On the earth's surface the acceleration due to gravity is 9.80 m/s^2 (metric) or 32.2 ft/s^2 (U.S.)

Accuracy The number of digits, called significant digits, in a measurement, which indicates the number of units that we are reasonably sure of having counted. The greater the number of significant digits, the better is the accuracy.

Actual Power A measure of the actual power available to be converted into other forms of energy.

Adhesion The force of attraction between different or unlike molecules.

Alpha Ray A ray consisting of alpha particles, each having two protons and two neutrons, and positively charged, that is, it curves in the direction that known positive charges curve in a magnetic field.

Alternating Current A current that flows in one direction in a conductor, changes direction, and then flows in the other direction.

Ammeter An instrument that measures the current flowing in a circuit.

Ampère's Rule To find the direction of a magnetic field near a current and a straight wire, hold the wire in your right hand with your thumb extended in the direction of the current. Your fingers circle the wire in the direction of the flux lines.

Amplification The process of increasing the strength of an electronic signal.

Amplitude The maximum displacement of any part of a wave or a vibration from its equilibrium, or rest, position.

Angular Acceleration The rate of change of angular velocity (change in angular velocity/time).

Angular Displacement The angle through which any point on a rotating body moves.

Angular Momentum For a rotating body about a fixed axis, the angular momentum is the product of the moment of inertia and the angular velocity of the body.

Angular Velocity The rate of angular displacement (angular displacement/time).

Apparent Power The product of the effective values of alternating current and voltage.

Approximate Number A number that has been determined by some measurement or estimation process.

Archimedes' Principle Any object placed in a fluid apparently loses weight equal to the weight of the displaced fluid.

Area The number of square units contained in a figure.

Armature The rotating coil or electromagnet in a generator.

Astronomy The branch of science that studies everything that takes place outside of the earth's atmosphere.

Atmospheric Pressure The pressure caused by the weight of the air in the atmosphere.

Atom The smallest particle of an element that can exist in a stable or independent state.

Atomic Mass The mass of a single atom, usually expressed in atomic mass units, u.

Atomic Mass Number The total number of nucleons (protons and neutrons) in a nucleus, A.

Atomic Mass Unit (u) A unit of measure of atomic mass based on the mass of the common carbon atom, which has six protons and six neutrons, so the mass of the carbon 12 atom has been given the exact value of 12 u. Thus, the approximate mass of a single proton or a single neutron is 1 u, and the mass of an atom in atomic mass units is simply the sum of the number of its protons and its neutrons.

Atomic Number The number of protons in a nucleus, Z.

Average Binding Energy per Nucleon The total binding energy of the nucleus divided by the total number of nucleons.

B

Becquerel (Bq) Unit of source activity; 1 Bq = 1 disintegration/s.

Bending Consists of both tension and compression stresses. It occurs when a force is placed on a beam causing it to sag.

Bernoulli's Principle For the horizontal flow of a fluid through a tube, the sum of the pressure and energy of motion (kinetic energy) per unit volume of the fluid is constant.

Beta Ray A ray consisting of a stream of beta particles (electrons), which are emitted from neutrons in a nucleus as they decay into protons and electrons. This ray is negatively charged and curves in the opposite direction of an alpha ray in a magnetic field.

Binding Energy of the Nucleus The total energy required to break a nucleus apart into separate nucleons.

Biology The branch of science that studies living organisms.

Bohr Model An early model of atomic structure in which electrons travel around the nucleus in a number of discrete stable energy levels determined by quantum conditions.

Boyle's Law If the temperature of a gas is constant, the volume is inversely proportional to the absolute pressure, $V/V' = P'/P$.

Brinell Method Common industrial method used to measure the hardness of a metal.

Btu (British thermal unit) The amount of heat (energy) necessary to raise the temperature of 1 lb of water 1°F.

Buoyant Force The upward force exerted on a submerged or partially submerged object.

C

Calorie The amount of heat necessary to raise the temperature of 1 g of water 1°C.

Capacitance The ratio of the charge on either plate of a capacitor to the potential difference between the plates.

Capacitive Reactance A measure of the opposition to ac current flow by a capacitor.

Capacitor A circuit component consisting of two parallel plates separated by a thin insulator used to build up and store charge.

Capillary Action The behavior of liquids that causes the liquid level in very small-diameter tubes to be different than in larger-diameter tubes. This behavior is due both to adhesion of the liquid molecules with the tube and to the surface tension of the liquid.

Celsius Scale The metric temperature scale on which ice melts at 0° and water boils at 100°.

Center of Gravity The point of any body at which all of its weight can be considered to be concentrated.

Centripetal Force The force acting on a body in circular motion that causes it to move in a circular path. This force is exerted toward the center of the circle.

Chain Reaction The process of using the neutrons released in each nuclear fission reaction to create further reactions.

Change of Phase (sometimes called *change of state*) A change in a substance from one form of matter (solid, liquid, or gas) to another.

Charles' Law If the pressure on a gas is constant, the volume is directly proportional to its Kelvin or Rankine temperature, $V/T = V'/T'$.

Chemistry The branch of science that studies the composition, structure, properties, and reactions of matter.

Coefficient of Friction The ratio between the frictional force and the normal force of an object. The number represents how rough or smooth two surfaces are when moving across one another.

Coefficient of Linear Expansion A constant that indicates the amount by which a solid expands or contracts when its temperature is changed 1 degree.

Cohesion The force of attraction between like molecules that holds the closely packed molecules of a solid together.

Color A property of the light that reaches our eyes and is determined by its wavelength or its frequency.

Commutator A device in an ac generator that produces a direct current. Composed of a split ring that replaces the slip rings in an ac generator and produces a direct current in the circuit connected to the split ring of the generator.

Compass A small magnetic needle that is free to rotate on a bearing.

Complementary Colors Two colors that, when combined, form white; for example, cyan and red, magenta and green, and yellow and blue.

Component Vector When two or more vectors are added, each of the vectors is called a component of the resultant, or sum, vector.

Compound A substance containing two or more elements.

Compound Machine A combination of simple machines. Its total mechanical advantage is the product of the mechanical advantage of each simple machine.

Compression A stress caused by two forces acting directly toward each other. This stress tends to cause objects to become shorter and thicker.

Concave Mirror A mirror with a surface that curves away from an observer.

Concurrent Forces Two or more forces applied to, or acting at, the same point.

Condensation The change of phase from gas or vapor to a liquid.

Conduction A form of heat transfer from a warmer part of a substance to a cooler part as a result of molecular collisions, which cause the slower-moving molecules to move faster. A transfer of charge from one place to another.

Conductor A material through which an electron charge is readily transferred.

Constructive Interference The superposition of waves to form a larger disturbance (wave) in a medium. Occurs when two crests or troughs of superimposed waves meet.

Convection A form of heat transfer by the movement of warm molecules from one region of a gas or a liquid to another.

Converging Lens A lens that bends the light passing through it to some point beyond the lens. Converging lenses are thicker in the center.

Conversion Factor An expression used to convert from one set of units to another. Often expressed as a fraction whose numerator and denominator are equal to each other although in different units.

Convex Mirror A mirror with a surface that curves inward toward an observer.

Coulomb Force An electric, repulsive force between protons.

Coulomb's Law The force between two point charges is directly proportional to the product of their magnitudes and inversely proportional to the square of the distance between them.

Critical Angle The smallest angle of incidence at which all light striking a surface is totally internally reflected.

Curie (Ci) Unit of source activity; 1 Ci = 3.70×10^{10} disintegrations/s.

Current The flow of charge that passes through a conductor.

Curvilinear Motion Motion along a curved path.

D

Decay Rate The probability per unit time that a decay of radioactive isotopes will occur.

Deceleration An acceleration that indicates an object is slowing down.

Degree An angular unit of measure. Defined as 1/360 of one complete revolution.

Destructive Interference The superposition of waves to form a smaller disturbance (wave) in a medium.

Dew Point The temperature at which air becomes saturated with water vapor and condensation occurs.

Diffraction The property of a wave that describes its ability to bend around obstacles in its path.

Diffusion The process by which molecules of a gas mix with the molecules of a solid, a liquid, or another gas. Scattering of light by an uneven surface.

Diode A device that allows current to flow through it in only one direction.

Direct Current Current that flows in one direction.

Disintegration Energy The total energy released in radioactive decay in the form of kinetic energy.

Dispersion The spreading of white light into the full spectrum.

Displacement The net change in position of an object, or the direct distance and direction it moves; a vector.

Displacement The distance of an object is simple harmonic motion from its equilibrium or rest, position.

Diverging Lens A lens that bends the light passing through it so as to spread the light. Diverging lenses are thicker at the edges than at the center.

Doppler Effect The variation of the frequency heard when a source of sound and the ear are moving relative to each other.

Dry Cell A voltage-generating cell that consists of a chemical paste and two electrodes of unlike materials, one of which reacts chemically with the electrolyte.

Ductility A property of a metal that enables it to be drawn through a die to produce a wire.

E

Effective Value The number of amperes of alternating current that produce the same amount of heat in a resistance as an equal number of amperes of a steady direct current.

Efficiency The ratio of the work output to the work input of a machine.

Effort The force applied to a machine.

Effort Arm The distance from the effort force to the fulcrum of a lever.

Elastic Collision A collision in which two objects return to their original shape without being permanently deformed.

Elastic Limit The point beyond which a deformed object cannot return to its original shape.

Elasticity A measure of a deformed object's ability to return to its original size and shape once the deforming force is removed.

Electric Circuit A conducting loop in which electrons carrying electric energy may be transferred from a suitable source to do useful work and returned to the source.

Electric Field An electric field exists where an electric force acts on a charge brought into the area.

Electrolyte An acid solution that produces large numbers of free electrons at the negative pole of a cell.

Electromagnet A combination of a solenoid and a magnetic material, such as iron, in the core of the solenoid. When a current is passed through the solenoid, the magnetic fields of the atoms in the magnetic material line up to produce a strong magnetic field.

Electromagnetic Spectrum The entire range of electromagnetic waves classified according to frequency.

Electromagnetic Theory Describes the interrelationship between electric fields, magnetic fields, electric charge, and electric current.

Electromagnetic Wave A wave consisting of two perpendicular transverse waves with one component of the wave being a vibrating electric field and the other component being a corresponding vibrating magnetic field; the electromagnetic wave moves in a direction perpendicular to both electric and magnetic field components.

Electron A fundamental particle of an atom; negatively charged.

Element A substance that cannot be separated into simpler substances.

emf The potential difference across a source.

Energy The ability to do work. There are many forms of energy, such as mechanical, electrical, thermal, fluid, chemical, atomic, and sound. Work delivered to an electric component or appliance (power × time).

Equilibrant Force The force that, when applied at the same point as the resultant force, produces equilibrium.

Equilibrium An object is in equilibrium when the net force acting on it is zero. A body that is in equilibrium is either at rest or moving at a constant velocity.

Equivalent Resistance The single resistance that can replace a series and/or parallel combination of resistances in a circuit and provide the same current flow and voltage drop.

Evaporation The process by which high-energy molecules of a liquid continually leave its surface.

Exact Number A number that has been determined as a result of counting or by some definition, such as 1 h = 60 min.

Excited State A high-energy level for an electron in an atom.

Expansion Property of a gas in which the rapid random movement of its molecules causes the gas to completely occupy the volume of its container.

Experimental Physicist A physicist who performs experiments to develop and confirm physical theories.

$E = mc^2$ The equation that illustrates and defines the equivalence between mass and energy forms of matter.

F

Fahrenheit Scale The U.S. temperature scale on which ice melts at 32° and water boils at 212°.

First Condition of Equilibrium The sum of all parallel forces on a body in equilibrium must be zero.

First Law of Reflection The angle of incidence equals the angle of reflection.

First Postulate of Special Relativity The laws of physics are the same for both moving and nonmoving frames of reference.

Fixed Pulley A pulley that is fastened to a fixed object.

Flow Rate The volume of fluid flowing past a given point in a pipe per unit time.

Fluid A substance that takes the shape of its container. Either a liquid or a gas.

Fluorescence A property of certain substances in which radiation is absorbed at one frequency and then reemitted at a lower frequency.

Flux Lines Lines indicating the direction of the magnetic field near a magnetic pole.

Focal Length The distance between the principal focus of a mirror or lens and its vertex.

Force A push or a pull that tends to change the motion of an object or prevent an object from changing motion. Force is a vector quantity with both magnitude and direction.

Formula An equation, usually expressed in letters (called *variables*) and numbers.

Freezing The change of phase from liquid to solid. Also called *solidification*.

Frequency The number of complete vibrations or cycles per second of a wave.

Friction A force that resists the relative motion of two objects in contact caused by the irregularities of two surfaces sliding or rolling across each other.

Fulcrum A pivot about which a lever is free to turn.

Fusion The change of phase from solid to liquid. Also called *melting*.

G

Gamma Ray A ray composed of photons of electromagnetic radiation, which have no mass.

Gas A substance that takes the shape of its container and has the same volume as its container.

Gauge Pressure The amount of air pressure excluding the normal atmospheric pressure.

Gear Train A series of gears that transfers rotational motion from one gear to another.

General Theory of Relativity Extends the special theory of relativity to accelerated frames of reference with the assumption that gravity and acceleration are equivalent and that light has mass and its path can be warped by gravity.

Generator An apparatus consisting of a coil of wire rotating in a magnetic field. A current is induced in the coil, converting mechanical energy into electric energy.

Geology The branch of science that studies the origin, history, and structure of the earth.

Grand Unified Theory A theory linking the five fundamental forces: gravitational, electric, magnetic, strong, and weak forces.

Gravitational Field The area around a massive body in which an object experiences a gravitational force. The more massive and closer an object is to that body, the stronger is the gravitational field.

Gravitational Potential Energy The energy determined by the position of an object relative to a particular reference level.

Ground State The lowest energy level for an electron in an atom.

H

Half-Life The length of time required for one-half of the original amount of the radioactive atoms in a sample to decay.

Hardness A measure of the internal resistance of the molecules of a solid being forced farther apart or closer together.

Heat A form of internal kinetic and potential energy contained in an object associated with the motion of its atoms or molecules and which may be transferred from an object at a higher temperature to one at a lower temperature.

Heat of Fusion The heat required to melt 1 g or 1 kg or 1 lb of a liquid.

Heat of Vaporization The amount of heat required to vaporize 1 g or 1 kg or 1 lb of a liquid.

Heat Pump A device that warms or cools by transferring heat. It contains a vapor (refrigerant) that is easily condensed to a liquid when under pressure. It produces heat during compression and cooling during vaporization.

Hooke's Law A principle of elasticity in solids: The ratio of the force applied to an object to its change in length (resulting in its being stretched or compressed by the applied force) is constant as long as the elastic limit has not been exceeded.

Huygen's Principle Each point on a wave front can be regarded as a new source of small wavelets, which form succeeding waves that spread out uniformly in the forward direction at the same speed.

Hydraulic Principle (Pascal's Principle) The pressure applied to a confined liquid is transmitted without measurable loss throughout the entire liquid to all inner surfaces of the container.

Hydrometer A sealed glass tube weighted at one end so that it floats vertically in a liquid.

Hydrostatic Pressure The pressure a liquid at rest exerts on a submerged object.

Hypothesis A scientifically based prediction that needs testing to verify its validity.

I

Illumination The luminous intensity per unit area.

Impedance A measure of the total opposition to current flow in an ac circuit resulting from the effect of both the resistance and the inductive reactance on the circuit.

Impulse The product of the force exerted and the time interval during which the force acts on the object. Impulse equals the change in momentum of an object in response to the exerted force.

Impulse–Momentum Theorem If the mass of an object is constant, then a change in its velocity results in a change of its momentum. That is, $F\Delta t = \Delta p = m\Delta v = mv_f - mv_i$.

Inclined Plane A plane surface set at an angle from the horizontal used to raise objects that are too heavy to lift vertically.

Index of Refraction A measure of the optical density of a material. Equal to the ratio of the speed of light in vacuum to the speed of light in the material.

Induced Current A current produced in a circuit by motion of the circuit through the flux lines of a magnetic field.

Induced Magnetism Magnetism produced in a magnetic material such as iron when the material is placed in a magnetic field, such as that produced in the core of a current-carrying solenoid.

Inductance A measure of the tendency of a coil of wire to resist a change in the current because the magnetism produced by one part of the coil acts to oppose the change of current in other parts of the coil.

Induction A method of charging one object by bringing a charged object near to, but not touching, it.

Induction Motor An ac motor with an electromagnetic current induced by the moving magnetic field of the ac current.

Inductive Reactance A measure of the opposition to ac current flow in an inductor.

Inductor A circuit component, such as a coil, in which an induced emf opposes any current change in the circuit.

Inelastic Collision A collision in which two objects couple together.

Inertia The property of a body that causes it to remain at rest if it is at rest or to continue moving with a constant velocity unless an unbalanced force acts upon it.

Instantaneous Current The current at any instant of time.

Instantaneous Voltage The voltage at any instant of time.

Insulator A substance that does not allow electric current to flow readily through it.

Intensity The energy transferred by sound per unit time through unit area.

Interference The effect of two intersecting waves resulting in a loss of displacement in certain areas and an increase in displacement in others.

Internal Potential Energy The energy determined by the nature or condition of a substance.

Internal Resistance The resistance within a cell that opposes movement of the electrons.

Isotopes Nuclei that contain the same number of protons but a different number of neutrons.

K

Kelvin Scale The metric absolute temperature scale on which absolute zero is 0 K and the units are the same as on the Celsius scale.

Kilocalorie The amount of heat necessary to raise the temperature of 1 kg of water 1°C.

Kilogram The basic metric unit of mass.

Kinetic Energy The energy due to the mass and the velocity of a moving object.

L

Laser A light source that produces a narrow beam with high intensity. An acronym for "light amplification by stimulated emission of radiation."

Lateral Surface Area The area of all the lateral (side) faces of a geometric solid.

Law The highest level of certainty for an explanation of physical occurrences. A law is often accompanied by a formula.

Law of Acceleration The total force acting on a body is equal to the mass of the body times its acceleration (Newton's second law).

Law of Action and Reaction For every force applied by object A to object B (action), there is an equal but opposite force exerted by object B on object A (reaction) (Newton's third law).

Law of Conservation of Angular Momentum The angular momentum of a system remains unchanged unless an outside torque acts on it.

Law of Conservation of Mechanical Energy The sum of the kinetic energy and the potential energy in a system is constant if no resistant forces do work.

Law of Conservation of Momentum When no outside forces are acting on a system of moving objects, the total momentum of the system remains constant.

Law of Inertia A body that is in motion continues in motion with the same velocity (at constant speed and in a straight line) and a body at rest continues at rest unless an unbalanced (outside) force acts upon it (Newton's first law).

Law of Refraction When a beam of light passes at an angle from a medium of lower optical density to a denser medium, the light is bent toward the normal. When a beam passes from a medium of greater optical density to one less dense, the light is bent away from the normal.

Law of Simple Machines Resistance force × resistance distance = effort force × effort distance.

Lever A rigid bar free to turn on a pivot called a fulcrum.

Light Radiant energy that can be seen by the human eye.

Light-Year The distance that light travels in one earth year: 9.45×10^{15} m or 5.87×10^{12} mi.

Liquid A substance that takes the shape of its container and has a definite volume.

Load The object in a circuit that converts electric energy into other forms of energy or work.

Longitudinal Wave A disturbance in a medium in which the motion of the particles is along the direction of the wave travel.

Loudness The strength of the sensation of sound to an observer.

Luminous Intensity A measure of the brightness of a light source.

M

Machine An object or system that is used to transfer energy from one place to another and allows work to be done that could not otherwise be done or could not be done as easily.

Magnetic Property of metals or other materials that can attract iron or steel.

Magnetic Field A field of force near a magnetic pole or a current that can be detected using a magnet.

Malleability A property of a metal that enables it to be hammered and rolled into a sheet.

Mass A measure of the inertia of a body. A measure of the quantity of material making up an object.

Mass Density The mass per unit volume of a substance.

Mass–Energy Equivalence A physical law stating that mass and energy are equivalent forms of matter; $E = \Delta mc^2$; stated by Einstein.

Matter Anything that occupies space and has mass.

Mechanical Advantage The ratio of the resistance force to the effort force.

Mechanical Equivalent of Heat The relationship between heat and mechanical work.

Melting The change of phase from solid to liquid. Also called *fusion*.

Meniscus The crescent-shaped surface of a liquid column in a tube.

Method of Mixtures When two substances at different temperatures are mixed together, heat flows from the warmer body to the cooler body until they reach the same temperature. Part of the heat lost by the warmer body is transferred to the cooler body and to surrounding objects. If the two substances are well insulated from surrounding objects, the heat lost by the warmer body is equal to the heat gained by the cooler body.

Metre The basic metric unit of length.

Molecule The smallest particle of a substance that exists in a stable and independent state.

Moment of Inertia   Rotational inertia; the property of a rotating body that causes it to continue to turn until a torque causes it to change its rotational motion.

Momentum   A measure of the amount of inertia and motion an object has or the difficulty in bringing a moving object to rest. Momentum equals the mass times the velocity of an object.

Motion   A change of position.

Motor   A device that is composed of an armature and a stator. When a current is passed through the armature, the armature rotates in the magnetic field of the stator and converts electric energy to mechanical energy.

Movable Pulley   A pulley that is fastened to the object to be moved.

Multimeter   An instrument used to measure current flow, voltage drop, and resistance.

N

Natural Frequency   The frequency at which an object vibrates when struck by another object, such as a rubber hammer.

Neutron   A fundamental particle in the nucleus of an atom; neutrally charged

Neutron Number   The number of neutrons in an atomic nucleus of a particular nuclide, N. Also, $N = A - Z$, the difference between the atomic mass number and the atomic number. N can be different for nuclei with the same atomic number.

Newton's Law of Universal Gravitation   All objects that have mass are attracted to one another by a gravitational force.

Normal Force   Force perpendicular to the contact surface.

Nuclear Fission   A nuclear reaction in which an atomic nucleus splits into fragments with the release of energy.

Nuclear Fusion   A nuclear reaction in which light nuclei interact to form heavier nuclei with the release of energy.

Nucleus   The center part of an atom made up of protons and neutrons.

Nuclide   A specific type of atom characterized by its nuclear properties, such as the number of neutrons and protons and the energy state of its nucleus.

Number Plane   A plane determined by the horizontal line called the x-axis and a vertical line called the y-axis intersecting at right angles at a point called the origin. These two lines divide the number plane into four quadrants. The x-axis contains positive numbers to the right of the origin and negative numbers to the left of the origin. The y-axis contains positive numbers above the origin and negative numbers below the origin.

O

Ohm's Law   When a voltage is applied across a resistance in an electric circuit, the current equals the voltage drop across the resistance divided by the resistance, $I = V/R$.

Ohmmeter   An instrument that measures the resistance of a circuit component.

Opaque   Absorbing or reflecting almost all light.

Optical Density   A property of a transparent material that is a measure of the speed of light through the given material.

Orbit   The path taken by an object during its revolution around another object, such as the path of the moon or a satellite about the earth or of a planet about the sun.

P

Parallel Circuit   An electric circuit with more than one path for the current to flow. The current is divided among the branches of the circuit.

Particle Theory   Theory that light consists of streams of particles.

Pendulum   An object suspended so that it swings freely back and forth about a pivot.

Period   The time required for a single wave to pass a given point or the time required for one complete vibration of an object in simple harmonic motion.

Periodic Table   A table that contains all of the atomic elements arranged according to their atomic numbers and which can be used to predict their chemical properties.

Phase Angle   The angle between the resistance and impedance vectors in a circuit.

Photoelectric Effect   The emission of electrons by a surface when struck by electromagnetic radiation.

Photometry   The study of the measurement of light.

Photons   Wave packets of energy that carry light and other forms of electromagnetic radiation.

Physicist   A person who is an expert in or who studies physics.

Physics   The branch of science that describes the motion and energy of all matter throughout the universe.

Pitch   The distance a screw advances in one revolution of the screw. Also the distance between two successive threads. The effect of the frequency of sound waves on the ear.

Planck's Constant   A fundamental constant of quantum theory (6.626×10^{-34} J s).

Plane Mirror   A mirror with a flat surface.

Platform Balance   An instrument consisting of two platforms connected by a horizontal rod that balances on a knife edge. The pull of gravity on objects placed on the two platforms is compared.

Polarized Light   Light waves restricted to a single plane that is perpendicular to the direction of the wave motion.

Potential Energy   The stored energy of a body due to its internal characteristics or its position.

Power   The rate of doing work (work divided by time). Energy per unit time consumed in a circuit.

Power Factor   The ratio of the actual power to the apparent power.

Precision   Refers to the smallest unit with which a measurement is made, that is, the position of the last significant digit.

Pressure   The force applied per unit area.

Primary Cell   A cell that cannot be recharged.

Primary Coil   The coil of a transformer that carries an alternating current and induces a current in the secondary coil.

Primary Colors   The colors red, green, and blue; an additive mixture of the three colors of light resulting in white.

Primary Pigments   The complements of the three primary colors; namely, cyan (the complement of red), magenta (the complement of green), and yellow (the complement of blue).

Principle   A rule or fundamental assumption that has been proven in the laboratory.

Problem-Solving Method   An orderly procedure that aids in understanding and solving problems.

Projectile   A propelled object that travels through the air but has no capacity to propel itself.

Projectile Motion   The motion of a projectile as it travels through the air influenced only by its initial velocity and gravitational acceleration.

Propagation Velocity   The velocity of energy transfer of a wave, given by the distance traveled by the wave in one period divided by the period.

Proton   A fundamental particle in the nucleus of the atom; positively charged.

Pulley   A grooved wheel that turns readily on an axle and is supported in a frame.

Pulse   Nonrepeated disturbance that carries energy through a medium or through space.

Q

Quantum Mechanics   A theory that unifies the wave–particle dual nature of electromagnetic radiation. The quantum model is based on the idea that matter, just like light, behaves sometimes as a particle and sometimes as a wave.

Quantum Theory   Theory initiated by Planck and Einstein that energy, including electromagnetic radiation, is radiated or absorbed in multiples of certain units of energy.

R

Radian   An angular unit of measurement. Defined as that angle with its vertex at the center of a circle whose sides cut off an arc on the circle equal to its radius. Equal to approximately 57.3°.

Radiation   A form of heat transfer through energy being radiated or transmitted in the forms of rays, waves, or particles.

Radioactive Decay   A type of nuclear decay that occurs when an unstable atom is transformed into a new element through the spontaneous disintegration of its nucleus.

Radiocarbon Dating   A method used to obtain age estimates of organic materials using carbon 14 decay.

Rainbow   A spectrum of light formed when sunlight strikes raindrops, refracts into them, reflects within them, and then refracts out of them.

Range   The horizontal distance that a projectile will travel before striking the ground.

Rankine Scale   The U.S. absolute temperature scale on which absolute zero is 0° R and the degree units are the same as on the Fahrenheit scale.

Real Image   An image formed by rays of light.

Recharging   The passing of an electric current through a secondary cell to restore the original chemicals.

Rectification   The process of changing ac to dc.

Rectifier   A device that changes ac to dc.

Rectilinear Motion   Motion in a straight line.

Reflection   The turning back of all or part of a beam of light at the boundary between two different media.

Refraction   The bending of light as it passes at an angle from one medium to another of different optical density.

Regular Reflection   Reflection of light with very little scattering.

Relative Humidity   Ratio of the actual amount of vapor in the atmosphere to the amount of vapor required to reach 100% of saturation at the existing temperature.

Relative Motion   The concept that motion can be described differently depending upon the observer's perspective.

Resistance   The force overcome by a machine. The opposition to current flow.

Resistance Arm   The distance from the resistance force to the fulcrum of a lever.

Resistivity   The resistance per unit length of a material with uniform cross section.

Resonance   A sympathetic vibration of an object caused by the transfer of energy from another object vibrating at the natural frequency of vibration of the first object. A condition in a circuit when the inductive reactance equals the capacitive reactance and they nullify each other. The current that flows in the circuit is then at its maximum value.

Resultant Force   The sum of the forces applied at the same point. The single force that has the same effect as the two or more forces acting together.

Resultant Vector   The sum of two or more vectors.

Revolution   A unit of measurement in rotational motion. One complete rotation of a body.

Rotational Motion   Spinning motion of a body.

Rotor   The rotating coil in a generator.

S

Scalar   A physical quantity that can be completely described by a number (called its magnitude) and a unit.

Science   A system of knowledge that is concerned with establishing accurate conclusions about the behavior of everything in the universe.

Scientific Method   An orderly procedure used by scientists in collecting, organizing, and analyzing new information which refutes or supports a scientific hypothesis.

Scientific Notation   A form in which a number can be written as a product of a number between 1 and 10 and a power of 10. General form is $M \times 10^n$, where M is a number between 1 and 10 and n is the exponent or power of 10.

Screw   An inclined plane wrapped around a cylinder.

Second   The basic unit of time.

Second Condition of Equilibrium   The sum of the clockwise torques on a body in equilibrium must be equal to the sum of the counterclockwise torques about any point.

Second Law of Reflection   The incident ray, the reflected ray, and the normal (perpendicular) to the reflecting surface all lie in the same plane.

Second Postulate of Special Relativity   The speed of light is constant regardless of the speed of the observer or of the light source.

Secondary Cell   A rechargeable type of cell.

Secondary Coil   The coil of a transformer in which a current is induced by the current in the primary coil.

Semiconductors   A small number of materials that fall between conductors and insulators in their ability to conduct electric current.

Series Circuit   An electric circuit with only one path for the current to flow. The current in a series circuit is the same throughout.

Shearing   A stress caused by two forces applied in parallel, opposite directions.

SI (Système International d'Unités) The international modern metric system of units of measurement.

Significant Digits The number of digits in a measurement, which indicates the number of units we are reasonably sure of having counted.

Simple Harmonic Motion A type of linear motion of an object in which the acceleration is directly proportional to its displacement from its equilibrium position and the motion is always directed to the equilibrium position.

Simple Machine Any one of six mechanical devices in which an applied force results in useful work. The six simple machines are the lever, the wheel and axle, the pulley, the inclined plane, the screw, and the wedge.

Snell's Law The index of refraction equals the sine of the angle of incidence divided by the sine of the angle of refraction.

Solenoid A coil of tightly wrapped wire. Commonly used to create a strong magnetic field by passing current through the wire.

Solid A substance that has a definite shape and a definite volume.

Solidification The change of phase from liquid to solid. Also called *freezing.*

Sound Those waves transmitted through a medium with frequencies capable of being detected by the human ear.

Source The object that supplies electric energy for the flow of electric charge (electrons) in a circuit.

Source Activity The strength of a source of radiation that can be specified at a given time.

Space-Time An object's position in the universe can be pinpointed using three spatial dimensions and one time dimension.

Special Theory of Relativity The laws of physics are the same in moving and nonmoving frames of reference and the speed of light is constant no matter what the speed of the observer or the light source.

Specific Gravity The ratio of the density of any material to the density of water.

Specific Heat The amount of heat necessary to change the temperature of 1 kg of a substance 1°C in the metric system or 1 lb of a substance 1°F in the U.S. system.

Speed The distance traveled per unit of time. A scalar described by a number and a unit.

Speed of Light The speed at which light and other forms of electromagnetic radiation travel: 3.00×10^8 m/s in a vacuum.

Speed of Sound The speed at which sound waves travel in a medium: 331 m/s in dry air at 1 atm pressure and 0°C.

Spring Balance An instrument containing a spring, which stretches in proportion to the force applied to it, and a pointer attached to the spring with a calibrated scale read directly in given units.

Standard Position A vector is in standard position when its initial point is at the origin of the number plane. The vector is expressed in terms of its length and its angle, measured counterclockwise from the positive *x*-axis to the vector.

Standard Temperature and Pressure (STP) A commonly used reference in gas laws. Standard temperature is the freezing point of water. Standard pressure is equivalent to atmospheric pressure.

Standards of Measure A set of units of measurement for length, weight, and other quantities defined in such a way as to be useful to a large number of people.

Standing Waves A special case of superposition of two waves when no energy propagation occurs along the wave. The wave displacements are constant and remain fixed in location.

Statics The study of objects that are in equilibrium.

Stator The field magnets in a generator.

Step-Down Transformer A transformer used to lower voltage; it has more turns in the primary coil.

Step-Up Transformer A transformer used to increase voltage; it has more turns in the secondary coil.

Strain The deformation of an object due to an applied force.

Streamline Flow The smooth flow of a fluid through a tube.

Stress The ratio of an outside applied distorting force to the area over which the force acts.

Strong Force An attractive force among all nucleons (neutrons and protons) independent of their charge.

Superconductor A material that continuously conducts electric current without resistance when cooled to typically very low temperatures, often near absolute zero.

Superposition of Waves The algebraic sum of the separate displacements of two or more individual waves passing through a medium.

Surface Tension The ability of the surface of a liquid to act like a thin, flexible film.

Synchronous Motor An ac motor whose speed of rotation is constant and is directly proportional to the frequency of its ac power supply

T

Technology The field that uses scientific knowledge to develop material products or processes that satisfy human needs and desires.

Temperature A measure of the hotness or coldness of an object.

Tensile Strength A measure of a solid's resistance to being pulled apart.

Tension A stress caused by two forces acting directly opposite each other. This stress tends to cause objects to become longer and thinner.

Terminal Speed The speed attained by a freely falling body when the air resistance equals its weight and no further acceleration occurs.

Theoretical Physicist A physicist who predominantly uses previous theories and mathematical models to form new theories in physics.

Theory A scientifically accepted principle that attempts to explain natural occurrences.

Thermal Conductivity The ability of a material to transfer heat by conduction.

Torque A measure of the tendency to produce change in rotational motion. Equal to the applied force times the length of the torque arm.

Torsion A stress related to a twisting motion. This type of stress severely compromises the strength of most materials.

Total Internal Reflection A condition such that light striking a surface does not pass through the surface but is completely reflected inside it.

Total Surface Area The total area of all the surfaces of a geometric solid; that is, the lateral surface area plus the area of the bases.

Transformer A device composed of two coils (primary and secondary) and a magnetic core. Used to step up or step down a voltage.

Translucent Allowing some but not all light to pass through.

Transparent Allowing almost all light to pass through so that objects or images can be seen clearly.

Transverse Wave A disturbance in a medium in which the motion of the particles is perpendicular to the direction of the wave motion.

Turbulent Flow The erratic, unpredictable flow of a fluid resulting from excessive speed of the flow or sudden changes in direction or size of the tube.

U

Universal Motor A motor that can be run on either ac or dc power.

V

Vaporization The change of phase from liquid to a gas or vapor.

Variable A symbol, usually a letter, used to represent some unknown number or quantity.

Vector A physical quantity that requires both magnitude (size) and direction to be completely described.

Velocity The rate of motion in a particular direction. The time rate of change of an object's displacement. Velocity is a vector that gives the direction of travel and the distance traveled per unit of time.

Virtual Image An image that only appears to the eye to be formed by rays of light.

Viscosity The internal friction of a fluid caused by molecular attraction, which makes it resist a tendency to flow.

Visible Spectrum The colors resulting from the dispersion of white light through a glass prism: red, orange, yellow, green, blue, and violet.

Volatility A measure of a liquid's ability to vaporize. The more volatile the liquid, the greater is its rate of evaporation.

Voltage Drop The potential difference across a load in a circuit.

Voltmeter An instrument that measures the difference in potential (voltage drop) between two points in a circuit.

Volume The number of cubic units contained in a figure.

W

Wave A disturbance that moves through a medium or through space.

Wavelength The distance between two successive corresponding points on a wave.

Wave Theory Theory that light consists of waves traveling out from light sources, like water waves traveling out from the point at which a stone is dropped into water.

Wedge An inclined plane in which the plane is moved instead of the resistance.

Weight A measure of the gravitational force or pull exerted on an object by the earth or by another large body.

Weight Density The weight per unit volume of a substance.

Wheel-and-Axle A large wheel attached to an axle so that both turn together.

Work The product of the force in the direction of motion and the displacement.

X

x-component The horizontal component of a vector that lies along the x-axis.

Y

y-component The vertical component of a vector that lies along the y-axis.

ANSWERS TO ODD-NUMBERED PROBLEMS AND TO CHAPTER REVIEW QUESTIONS AND PROBLEMS

Chapter 0 Review Questions Pages 10–11

1. d   **2.** b   **3.** c   **4.** b   **5.** d

6. Archimedes conducted and documented his physical theories.

7. Science is a system of knowledge while technology uses that knowledge to develop material products or processes that satisfy human needs and desires.

8. Behavior of light—fiber optics; electricity—light bulb.

9. The scientific method is used to discover facts about the natural world. The problem-solving method uses the scientific method to create something useful.

10. Knowledgeable of the physical world. It allows one to understand and answer new questions about everyday occurrences. It is also important in many technical fields.

Chapter 1

1.2 Pages 16–17

1. kilo   **3.** hecto   **5.** milli   **7.** mega   **9.** h   **11.** m   **13.** M   **15.** c
17. 135 mm   **19.** 28 kL   **21.** 49 cg   **23.** 75 hm   **25.** 24 metres
27. 59 grams   **29.** 27 millimetres   **31.** 45 dekametres   **33.** 26 megametres
35. metre   **37.** litre and cubic metre   **39.** second

1.3 Page 19

1. 3.26×10^2 **3.** 2.65×10^3 **5.** 8.264×10^2 **7.** 4.13×10^{-3} **9.** 6.43×10^0
11. 6.5×10^{-5} **13.** 5.4×10^5 **15.** 7.5×10^{-6} **17.** 5×10^{-8} **19.** 7.32×10^{17}
21. 86,200   **23.** 0.000631   **25.** 0.768   **27.** 777,000,000   **29.** 69.3
31. 96,100   **33.** 1.4   **35.** 0.0000084   **37.** 700,000,000,000   **39.** 0.00000072
41. 4,500,000,000,000   **43.** 0.000000000055

1.4 Pages 23–24

1. 1 metre   **3.** 1 kilometre   **5.** 1 kilometre   **7.** cm   **9.** m   **11.** mm
13. km   **15.** mm   **17.** m   **19.** km   **21.** cm   **23.** km   **25.** cm   **27.** km
29. mm   **31.** cm   **33.** 1000   **35.** 100   **37.** 0.1   **39.** 1000   **41.** 100
43. 0.001   **45.** 10   **47.** 0.25 km   **49.** 178,000 m   **51.** 8.3 m   **53.** 3750 mm
55. 4,000,000 μm   **57.** (a) 9 ft (b) 3 yd   **59.** (a) 11,000 yd (b) 33,000 ft
61. 412.16 km   **63.** 2.80 in.   **65.** 6.1 m   **67.** 9

1.5 Pages 31–33
1. 40 cm^2 **3.** 39 in^2 **5.** 22 in^2 **7.** 72 in^3 **9.** 40 cm^3
11. 1 litre **13.** 1 cubic centimetre **15.** 1 square kilometre **17.** L
19. m^2 **21.** m^3 **23.** ha **25.** mL **27.** m^3 **29.** L
31. mL **33.** L **35.** m^2 **37.** L **39.** ha **41.** m^3
43. m^2 **45.** L **47.** 1000 **49.** 0.1 **51.** 0.01
53. 10 **55.** 1 **57.** 1,000,000 **59.** 0.001 **61.** 10,000
63. 100 **65.** 0.01 **67.** 100 **69.** 7.5 L **71.** 1600 mL
73. 275,000 mm^3 **75.** 4×10^9 mm^3 **77.** 275 mL **79.** 1000 L
81. 7500 cm^3 **83.** 50 cm^2 **85.** 50,000 cm^2 **87.** 400 km^2
89. 45 ft^2 **91.** 13,935 cm^2 **93.** 0.75 ft^2 **95.** 156,816 in^2
97. 513 ft^3 **99.** 30.1 yd^3 **101.** 13,824 in^3 **103.** 623.2 cm^3
105. 36 cm^2 **107.** 76 cm^2 **109.** 40 mL

1.6 Pages 37–38
1. 1 gram **3.** 1 kilogram **5.** 1 kilogram **7.** kg **9.** kg **11.** metric ton
13. g **15.** mg **17.** kg **19.** g **21.** kg **23.** g **25.** g
27. kg **29.** kg **31.** metric ton **33.** kg **35.** 1000 **37.** 100 **39.** 0.1
41. 1000 **43.** 100 **45.** 0.001 **47.** 1,000,000 **49.** 575,000 mg
51. 0.65 g **53.** 5 g **55.** 30,000,000 mg **57.** 2.5 kg **59.** 0.4 mg
61. 750 g **63.** 15,575 N **65.** 890 N **67.** 450 lb **69.** 7.5 lb **71.** 36 oz
73. 418.3 N **75.** second, s **77.** newton, N **79.** 1 millisecond **81.** 1 ms
83. 45 ns **85.** 3.45×10^{-4} s **87.** 15,915 s **89.** 4×10^9 ns

1.7 Page 40
1. 3 **3.** 4 **5.** 2 **7.** 3 **9.** 5 **11.** 3 **13.** 5 **15.** 4 **17.** 3 **19.** 2
21. 1 **23.** 4 **25.** 2 **27.** 3 **29.** 4

1.8 Page 42
1. 1 ft **3.** 1 m **5.** 0.0001 in. **7.** 10 km **9.** 0.01 m **11.** 0.00001 in.
13. 0.01 m **15.** 1 kg **17.** 0.0001 in. **19.** 1000 N **21.** 1 N **23.** 0.01 m^2
25. 100 kg **27.** 0.000001 kg or 1×10^{-6} kg **29.** 10,000 kg or 1×10^4 kg
31. (a) 15.7 in. (b) 0.018 in. **33.** (a) 16.01 cm (b) 0.734 cm
35. (a) 0.0350 s (b) 0.00040 s **37.** (a) 27,0$\overline{0}$0 L (b) 4.75 L
39. (a) All have one significant digit. (b) 50 N **41.** (a) 0.05 in. (b) 16.4 in.
43. (a) 0.65 m (b) 27.5 m **45.** (a) 0.00005 g (b) 0.75 g
47. (a) 3 N (b) 45,000 N **49.** (a) 20 kg (b) 40$\overline{0}$,000 kg

1.9 Page 46
1. 14,200 ft **3.** 83.3 cm **5.** 7$\overline{0}$,000 N **7.** 802 m or 80,200 cm **9.** 18 s
11. 500 kg **13.** 41.0 g **15.** 3200 km **17.** 900,000 kg **19.** 0.40 m 4$\overline{0}$ cm
21. 4900 m^2 **23.** 1,400,000 km^2 or 1.40×10^6 km^2 **25.** 737.7 m^2
27. 5560 cm^3 **29.** 2.91×10^7 in^3 **31.** 3$\overline{0}$ ft **33.** 3.06 cm **35.** 75 km/h
37. 1$\overline{0}$00 mi/h **39.** 1100 ft lb/s **41.** 370 mi/h **43.** 43.2 m
45. 4530 kg m/s^2 **47.** 10,300 m^3 **49.** 6100 m^2 **51.** 28,800 ft

Chapter 1 Review Questions Pages 47–48
1. c **2.** b **3.** b **4.** c **5.** a
6. (1) Pieces made separately may not fit together.
(2) Workers could not communicate directions to each other.
(3) Workers could not tell each other how much material to buy.

7. It is based on the decimal system.

8. (1) The distance to the moon.

(2) The thickness of aluminum foil.

9. Yes

10. The surface that would be seen by cutting a geometric solid with a thin plate.

11. Yes

12. Hectare

13. Litre

14. (1) Medicines

(2) Perfumes

(3) Wine

15. Mass measures the quantity of matter. Weight measures the gravitational pull on an object.

16. Newton (N)   **17.** Millionth

18. Because nearly all measurements are approximate numbers rather than exact numbers.

19. No   **20.** Yes

Chapter 1 Review Problems Pages 48–49

1. k   **2.** m   **3.** μ   **4.** M   **5.** 45 mg   **6.** 138 cm   **7.** 1 L   **8.** 1 kg

9. 1 m³   **10.** 0.25   **11.** 0.85   **12.** 5400   **13.** 550,000   **14.** 25,000

15. 75,000   **16.** 27,500   **17.** 0.035   **18.** 150,000   **19.** 500   **20.** 68.2

21. 11.0   **22.** 98.4   **23.** 968   **24.** 216   **25.** 212   **26.** 71.2

27. 4 h 20 min   **28.** 3   **29.** 4   **30.** 2   **31.** 3   **32.** 0.1 ft   **33.** 0.0001 s

34. 1000 mi   **35.** 100,000 N

36. (a) 12.00 m   (b) 0.008 m   (c) 0.150 m   (d) 2600 m

37. (a) 18,050 L   (b) 0.75 L   (c) 0.75 L   (d) 18,050 L   **38.** 0.125 s

39. 63,000 N   **40.** 1,800,000 cm³   **41.** 150 m²   **42.** 9.73 kg m/s²

43. 9.90 m²   **44.** 700 cm³

Chapter 2

2.1 Page 54

1. $s = vt$ **3.** $m = \dfrac{w}{g}$ **5.** $R = \dfrac{E}{I}$ **7.** $g = \dfrac{PE}{mh}$ **9.** $h = \dfrac{v^2}{2g}$ **11.** $W = Pt$

13. $t = \dfrac{W}{P}$ **15.** $m = \dfrac{2(KE)}{v^2}$ **17.** $s = \dfrac{W}{F}$ **19.** $I = \dfrac{V - E}{-r}$ or $I = \dfrac{E - V}{r}$

21. $P = \dfrac{\pi}{2R}$ **23.** $C = \dfrac{5F - 160}{9}$ or $C = \dfrac{5}{9}(F - 32)$ **25.** $f = \dfrac{1}{2\pi CX_C}$

27. $R_3 = R_T - R_1 - R_2 - R_4$ **29.** $I_P = \dfrac{I_S N_S}{N_P}$ **31.** $v_i = 2v_{avg} - v_f$

33. $s = \dfrac{v^2 - v_i^2 + 2as_i}{2a}$ **35.** $R = \dfrac{QJ}{I^2 t}$ **37.** $r = \sqrt{\dfrac{A}{\pi}}$ **39.** $d = \sqrt{\dfrac{kL}{R}}$

41. $I = \pm\sqrt{\dfrac{QJ}{Rt}}$

2.2 Pages 56–57

1. (a) $A = bh$ (b) 162 cm²   **3.** (a) $b = \dfrac{A}{h}$ (b) 7.50 cm

5. (a) $c = P - a - b$ (b) 6.0 cm   **7.** (a) $r = \dfrac{C}{2\pi}$ (b) 10.9 yd

9. (a) $b = \dfrac{P - 2a}{2}$ or $b = \dfrac{P}{2} - a$ (b) 33.2 km

11. (a) $h = \dfrac{V}{\pi r^2}$ (b) 6.11 m **13.** (a) $B = \dfrac{V}{h}$ (b) 154 m^2

15. (a) $b = \sqrt{A}$ (b) 21.6 in. **17.** (a) $C = 2\pi r$ (b) 121.6 m

19. (a) $B = \dfrac{3V}{h}$ (b) 122.4 ft^2

2.3 Pages 62–64
1. 25,900 cm^3 **3.** 284 cm^3 **5.** 102.1 cm^2 **7.** 10,100 ft^3 **9.** 864 ft^3
11. 12.0 cm^2 **13.** 1.58 cm^2 **15.** 36.0 m **17.** 52.8 cm^3 **19.** 9.39 m
21. 137 m **23.** 65.5 ft **25.** $6\overline{0}$ panels **27.** 24.1 m^3 **29.** 4.44 yd^3
31. 266 in^3

Chapter 2 Review Questions Page 65
1. c **2.** b **3.** a
4. (1) To find the volume of liquid storage tanks
   (2) To determine the amount of concrete needed for a driveway
5. As a shorthand way to designate different measured quantities of the same type
6. Most mistakes are made in problem solving by missing needed information or misinterpreting the information given.
7. Making a sketch helps to visualize what is happening in the problem.
8. The basic equation
9. The working equation is found by solving the basic equation for the unknown quantity.
10. Carrying the units through a problem shows whether the answer is the kind expected.
11. Making an estimate of the correct answer shows whether the solution is reasonable.

Chapter 2 Review Problems Page 65–66
1. (a) $m = \dfrac{F}{a}$ (b) $a = \dfrac{F}{m}$ **2.** $h = \dfrac{v^2}{2g}$ **3.** $v_f = \dfrac{2s}{t} - v_i$ **4.** $v = \sqrt{\dfrac{2\,KE}{m}}$
5. 18 ft **6.** 12.0 m **7.** 2.19 m **8.** 122 cm^2 **9.** 12.0 cm **10.** 42.3 mm
11. 606 cm^2 **12.** 6.0 cm **13.** 6.27 m **14.** 14.4 m **15.** 430 cm^3 **16.** 4680 m^2

Chapter 2 Applied Concepts Page 67
1. 2.62¢/yd^2 **2.** 2.13 ft^3/s **3.** 2.91 **4.** 16
5. (a) 6.73 m^3 (b) Not safe; the spool's mass is 52,200 kg.

Chapter 3

3.1 Page 72
1. 2.0 **3.** 2.8 **5.** 6.3
7.
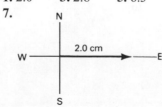
9.
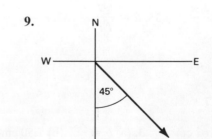
11.
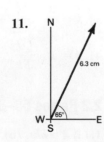

13. $1\frac{1}{4}$ **15.** $2\frac{5}{8}$ **17.** $\frac{15}{16}$

19.
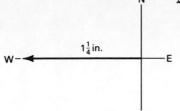

21.
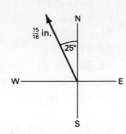

23.
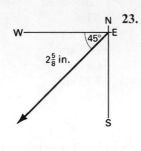

3.2 Pages 78–79

1.
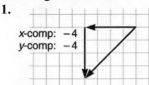

x-comp: −4
y-comp: −4

3.
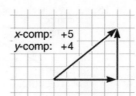

x-comp: +5
y-comp: +4

5.
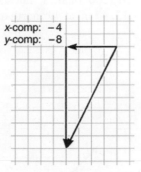

x-comp: −4
y-comp: −8

7.
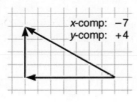

x-comp: −7
y-comp: +4

9.
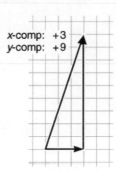

x-comp: +3
y-comp: +9

11.
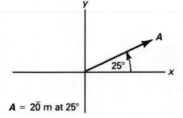

$A = 2\overline{0}$ m at 25°

13.
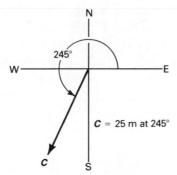

$C = 25$ m at 245°

15.
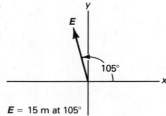

$E = 15$ m at 105°

17.
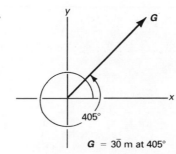

$G = 3\overline{0}$ m at 405°

	x-*component*	y-*component*
19.	9.96 m	8.97 m
21.	−18.2 km	−45.1 km
23.	97.4 km	−14.4 km
25.	38.2 m	7.09 m
27.	6.17 km	−7.35 km
29.	−5.88 m	28.9 m

3.3 Pages 88–91

1. 61 km at 55° north of east   **3.** 1300 mi at 1° west of south

5. 36 km at 5° east of north   **7.** 38 km at 25° north of west

9. 1500 mi at 71° north of east   **11.** 120 km at 72° south of east

13. 47 mi at 49° north of east   **15.**

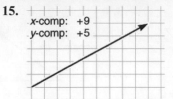

17.

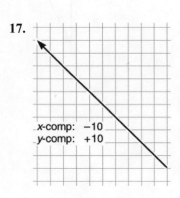

19.

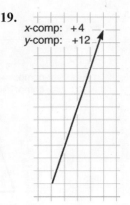

21.

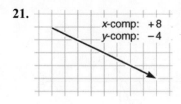

23.

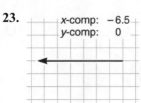

25.

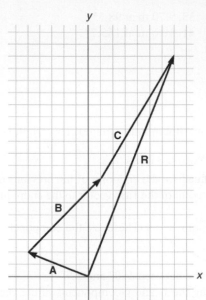

A + B + C = R

Vector	x-component	y-component
A	−5	2
B	6	6
C	6	10
R	7	18

A + B + C = R

27.

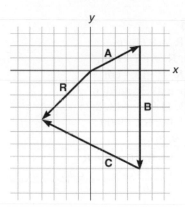

A + B + C = R

Vector	x-component	y-component
A	4	2
B	0	−10
C	−8	4
R	−4	−4

A + B + C = R

29.

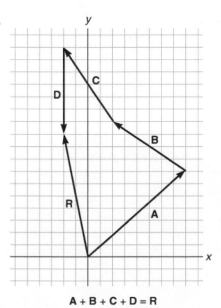

A + B + C + D = R

Vector	x-component	y-component
A	8	7
B	−6	4
C	−4	6
D	0	−7
R	−2	10

A + B + C + D = R

31. 10.0 m at 36.9°   **33.** 26.2 mi at 315.0°   **35.** 9.70 m at 98.3°
37. 53.3 m at 291.5°   **39.** 22.3 mi at 48.8°   **41.** 10.6 m at 155.7°
43. 180 m   **45.** 307 km at 48.4° north of west
47. 155 km at 58.4° south of west

Chapter 3 Review Questions Page 94

1. d   **2.** a   **3.** a
4. Two intersecting lines (axes) determine a plane (the number plane). The origin is the point where the axes intersect.
5. Yes   **6.** Yes
7. Graph each vector placing the initial point of each at the endpoint of the previous one; the resultant is the vector from the beginning point of the first vector to the endpoint of the last vector.
8. Add the x-components and then add the y-components. These are the components of the resultant.
9. No   **10.** Counterclockwise
11. In the third quadrant, the angle measure must be between 180° and 270°.
12. Complete the right triangle with the legs being the x- and y-components of the vector. Then find the lengths of x and y and determine their signs.
13. Complete the right triangle with the legs being the x- and y-axes, respectively.

Chapter 3 Review Problems Pages 94–96

1. $R_x = 11.3$ cm; $R_y = 6.50$ cm   **2.** $R_x = 5.00$ cm; $R_y = 8.66$ cm
3. $R_x = 17.3$ cm; $R_y = 10.0$ cm   **4.** $R_x = -4.50$ cm; $R_y = -7.79$ cm
5. $R_x = 6.89$ cm; $R_y = 5.79$ cm   **6.** $R_x = 10.3$ cm; $R_y = -14.7$ cm
7. 3.00 km north   **8.** 3.00 km south   **9.** 284 km at 39.3° north of west
10. 318 km at 53.4° north of east   **11.** 14.3   **12.** 11.2   **13.** 8.25
14. 5.00   **15.** $R_x = 6.00$; $R_y = -2.00$ **16.** $R_x = 9.00$; $R_y = 9.00$
17. $R_x = -6.00$; $R_y = 2.00$ **18.** $R_x = -9.00$; $R_y = -9.00$

19.

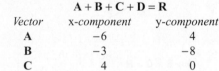

Vector	x-component	y-component
A	−6	4
B	−3	−8
C	4	0
D	0	−5
R	−5	−9

A + B + C + D = R

20.
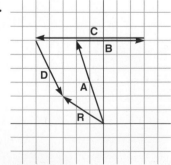

Vector	x-component	y-component
A	−2	6
B	−5	0
C	−8	0
D	2	−4
R	−3	2

A + B + C + D = R

21. 389 km at 63.6° south of west   **22.** 344 km at 25.3° south of east

Chapter 3 Applied Concepts Page 97
1. 417 ft; 184 ft   **2.** 1090 m   **3.** 13.6 m   **4.** 2.80 mi   **5.** 656 km

Chapter 4

4.1 Pages 105–106
1. $5\overline{0}$ mi/h   **3.** 21.6 m/s   **5.** 68.3 ft/s   **7.** 320 km
9. (a) 81 ft/s   (b) 25 m/s   (c) 89 km/h   **11.** 275 km   **13.** 4.42 h
15. $8\overline{0}$ km/h, east   **17.** 125 mi/h, south   **19.** 61.1 km/h at $3\overline{0}°$ south of east
21. (a) 53.7 mi/h   (b) 25.5 mi/h, west   **23.** $37\overline{0}$ km/h at 90.0°
25. 239 km/h at 190.8°   **27.** 128 mi/h at 239.4°   **29.** 227 km/h at 162.4°

4.2 Pages 110–111
1. 15 m/s^2 **3.** $1\overline{0}$ ft/s^2 **5.** 3.2 m/s^2 **7.** 6.25 m/s^2 **9.** 0.206 m/s^2
11. 67.5 mi/h   **13.** $72\overline{0}$ km/h   **15.** 0.534 m/s^2 **17.** 16.7 s   **19.** 0.714 m/s^2
21. 4.00 m/s^2 **23.** −8.33 m/s^2 **25.** −16.7 m/s^2

4.3 Pages 118–120
1. 5.05 m/s   **3.** 127.8 m   **5.** 2.13 ft/s^2 **7.** 9.00 mi/h   **9.** 2.19 m
11. $55\overline{0}$ ft   **13.** 1.1 m/s^2 **15.** −16.4 m/s^2 **17.** −1.6 m/s^2
19. (a) 23.5 m/s   (b) 28.2 m   **21.** (a) 3190 m   (b) 25.5 s   (c) 51.0 s
23. (a) 1.02 s   (b) 5.10 m   (c) 24.3 m/s   (d) 3.50 s
25. (a) 44.0 ft/s   (b) 4.21 s   (c) 135 ft/s   **27.** 2.06 s
29. 116 km/h   **31.** 9.05 s

4.4 Page 124
1. 62.5 m   **3.** 118 m   **5.** 124 m   **7.** 381 m   **9.** 288 ft; the ball will bounce.
11. 0.442 m

Chapter 4 Review Questions Pages 125–126
1. c   **2.** c   **3.** a   **4.** b   **5.** d
6. Gravity accelerates all objects at the same rate.
7. Horizontal motion is independent of the vertical pull of gravity.
8. Velocity has magnitude and direction, whereas speed has only magnitude.
9. No; anything speeding up or slowing down has a changing velocity.
10. Vectors are necessary to determine the direction of velocity, acceleration, and other vector quantities. Examples: navigating airplanes, ships, and spacecraft.
11. (a) An automobile speeding up or slowing down;   (b) anything being dropped to the ground;   (c) a bullet being fired.
12. Acceleration is change in velocity; deceleration is a special case of acceleration where the object is slowing; acceleration need not be uniform and so may be subject to averaging.
13. 9.80 m/s^2 (metric) and 32.2 ft/s^2 (U.S.).

Chapter 4 Review Problems Page 126
1. 25.5 mi   **2.** 5.45 h   **3.** 226 km/h at 107.8°   **4.** 33.4 km/h at 337.1°
5. 4.00 ft/s^2 **6.** 6.9 s   **7.** 15.0 km/h   **8.** 3.90 m/s   **9.** $3\overline{0}$ h
10. 13 m/s   **11.** 4.00 m/s   **12.** 2.6 m/s^2 **13.** 2.52 m/s^2 **14.** 60.4 m/s
15. (a) 205 m/s   (b) 20.9 s   (c) 41.8 s   **16.** (a) 37.0 m/s   (b) 64.6 m
17. 8.55 m   **18.** Since s_x = 52.8 ft, the arrow will miss the bull's eye.

Chapter 4 Applied Concepts Page 127
1. (a) 3.76 m/s   (b) 2.01 m/s   **2.** 16.6 mi/h, 18.8° downriver   **3.** 633 ft
14. 4.71 m/s   **5.** 25.2 m/s

Chapter 5

5.2 Pages 133–134
1. 30.0 N   **3.** 744 lb   **5.** 252 lb   **7.** 23$\overline{0}$ N   **9.** 40.0 m/s^2 **11.** 11.7 m/s^2
13. 0.518 m/s^2 **15.** 1.39 ft/s^2 **17.** 5250 N   **19.** 1320 lb   **21.** 2440 N
23. 1.69 m/s^2 **25.** 5740 N   **27.** 6.00 kg   **29.** 12.8 m/s^2
31. (a) 7.37 kg   (b) It is halved (9.50 m/s^2).   (c) It is doubled (38.0 m/s^2).
33. (a) 9880 N in the opposite direction of the truck.   (b) Double the force would be required to stop the truck in half the distance.

5.3 Pages 137–138
1. 380 N   **3.** 1100 N   **5.** 0.080   **7.** 25,000 lb   **9.** 1600 N   **11.** 26$\overline{0}$0 N
13. 0.350   **15.** (a) −3680 N (the − sign indicates the frictional force acts in opposite direction to that of the truck)   (b) −4.91 m/s^2 (c) 91.6 m

5.4 Page 140
1. 3.0 N; right   **3.** 15.0 N; left   **5.** 4 N; left   **7.** 4.00 ft/s^2 **9.** 0.509 m/s^2
11. 0.504 m/s^2 **13.** 12$\overline{0}$ N

5.5 Pages 142–143
1. 294 N   **3.** 322 lb   **5.** 1.73 kg   **7.** 1220 kg   **9.** 6.8 × 10^{11} kg
11. 11,300 N   **13.** 85.4 slugs   **15.** 765 kg   **17.** (a) 13,200 N   (b) 22$\overline{0}$0 N
19. (a) 65.0 kg   (b) 106 N   **21.** Answers vary.   **23.** Answers vary.
25. (a) 3.57 slugs   (b) 303 lb   **27.** Answers vary.   **29.** 102 kg

Chapter 5 Review Questions Pages 145–146
1. d   **2.** b   **3.** b   **4.** c   **5.** a
6. (a) A bridge being supported over a river   (b) Isometric exercises   (c) A magnet on a refrigerator
7. (a) No   (b) No
8. A car hit from behind is forced into the car ahead of it.
9. A body in motion continues in motion and a body at rest continues at rest unless an outside force acts on it.
10. Acceleration is change of velocity.
11. Only if they have the same mass
12. Yes; 3.00 lb = 13.35 N
13. More difficult; everything would slide.
14. Weight is a measure of gravitational pull. The moon, having less mass than the earth, exerts less gravitational pull.
15. For every force applied by object A to B, there is an equal force applied by B to A in the opposite direction.
16. For every action, there is an equal and opposite reaction.

Chapter 5 Review Problems Page 146
1. 3.00 m/s^2 **2.** 75.0 kg   **3.** 19 ft/s^2 **4.** 19,000 N   **5.** 64.0 N   **6.** 0.17
7. 3800 N   **8.** 11 N   **9.** 127 N   **10.** 12 lb   **11.** 20$\overline{0}$ N   **12.** 896 N
13. 0.273   **14.** 150 N   **15.** 1.20 m/s^2 **16.** 3680 N   **17.** 41.3 kg   **18.** 1.22 kg

Chapter 5 Applied Concepts Page 147
1. (a) 1.55 m/s^2 (b) 39.4 m/s   (c) 68.2 m/s
2. (a) 12.1 m/s^2 (b) 1.36 × 10^5 m or 136 km
3. (a) 12$\overline{0}$ lb   (b) 137 lb   (c) 12$\overline{0}$ lb   (d) 99.6 lb
4. (a) 714 N   (b) −2.94 m/s^2 (c) 13.7 s   **5.** −1.23 × 104 N

Chapter 6

6.1 Pages 157–158

1. 80.0 kg m/s   **3.** 765 slug ft/s   **5.** 9.5×10^8 kg m/s   **7.** 6.89×10^8 kg m/s
9. (a) 12,600 slug ft/s   (b) 158 ft/s   (c) 5800 lb; 2580 lb
11. (a) 55,200 kg m/s   (b) 170 km/h   **13.** 7.67 m/s
15. (a) 14.0 m/s   (b) 1750 kg m/s
17. (a) 0.00287 s   (b) 12,000 N   (c) 34.4 kg m/s   (d) 34.5 kg m/s
19. (a) 8250 N   (b) 52.8 m   **21.** 5830 N

6.2 Pages 163–164

1. 8.58 m/s, right   **3.** 1.75 m/s, right   **5.** 0.964 m/s, right
7. 1.71 m/s, north   **9.** 3.75 m/s, right   **11.** 13.1 m/s
13. (a) 3.95 m/s   (b) 10.4 m/s
15. (a) 2.46×10^4 kg m/s   (b) 1.72×10^4 kg m/s   (c) 24.6 m/s   (d) 17.2 m/s
17. (a) 8730 kg m/s   (b) 7330 kg m/s   (c) 9.19 m/s   (d) 7.72 m/s

Chapter 6 Review Questions Page 165

1. b   **2.** d
3. The slow-moving truck has a large mass and a small velocity, whereas the rifle bullet has a small mass and a large velocity; in both cases, the product of mass and velocity is large.
4. They are the same.
5. The longer the bat (applied force) is on the ball, the greater is the impulse.
6. Total momentum in a system remains constant.
7. Momentum of the escaping gas molecules is equal to the momentum of the rocket.
8. Elastic   **9.** Inelastic   **10.** They are equal.

Chapter 6 Review Problems Page 166

1. 1.23×10^5 slug ft/s   **2.** 0.204 m/s   **3.** 15,000 N s   **4.** 8.5 kg m/s
5. 0.556 m/s   **6.** (a) 12 m/s   (b) 1800 kg m/s
7. (a) 4.62×10^{-4} s   (b) 106,000 N   (c) 49.0 kg m/s   (d) 48.8 kg m/s
8. (a) 70.4 m   (b) 4.74 s   **9.** 0.521 m/s, right   **10.** 1.50 m/s, east
11. 8.12 m/s, right   **12.** 1.42×10^5 kg km/h at 48.0°
13. (a) 0.210 kg m/s   (b) 0.158 kg m/s   (c) 0.600 m/s   (d) 0.451 m/s

Chapter 6 Applied Concepts Page 167

1. (a) 8.46 slug ft/s   (b) The outgoing velocity is less, thereby reducing the change in momentum and the impulse.   **2.** (a) $p_{adult} = -1680$ N s; $p_{child} = 842$ N s
(b) $F_{adult} = 2980$ N; $F_{child} = 3240$ N   **3.** (a) 3690 N   (b) 1010 N   (c) Bungee cords increase the time, thereby decreasing the force of the impulse.   **4.** (a) 6.68 ft/s
(b) It is easier to step out of a heavier canoe. The canoe has a greater inertia and does not move backward with as large a velocity.   **5.** (a) −26.5 m/s = −95.5 km/h (the negative sign represents west)   (b) Yes, the Jeep was speeding.

Chapter 7

7.1 Pages 173–175

1. 30 N (right)   **3.** (a) 400 N   (b) 50 N   **5.** 1640 N at 55.6°   **7.** 1800 N at 33.7°
9. 2730 lb at 140.4°   **11.** 4620 N at 123.2°   **13.** 190 N at 126.3° from $\mathbf{F}_1$

7.2 Pages 182–185

1. $10\overline{0}$ N   **3.** $26\overline{0}$ N   **5.** $69\overline{0}0$ N   **7.** $57\overline{0}$ N   **9.** Yes   **11.** 322 N
13. $F_1 = 70.7$ N; $F_2 = 70.7$ N   **15.** $F_1 = 577$ lb; $F_2 = 289$ lb
17. $F_1 = 433$ lb; $F_2 = 50\overline{0}$ lb   **19.** $T_1 = T_2 = 731$ lb   **21.** $T_1 = 565$ lb; $T_2 = 613$ lb
23. $C = 300\overline{0}$ lb; $T = 260\overline{0}$ lb   **25.** 5540 N   **27.** $T = 2330$ lb; $C = 3690$ lb

7.3 Pages 187–188

1. 96.0 lb ft   **3.** 2.00 m   **5.** 187 N   **7.** 1.60×10^3 N   **9.** 0.357 m
11. 159 lb   **13.** $40\overline{0}$ N   **15.** 217 N   **17.** (a) 25 lb   (b) It is halved.
19. 133 N   **21.** 171 N

7.4 Pages 193–194

1. $10\overline{0}$ lb   **3.** $50\overline{0}$ N   **5.** $45\overline{0}$ N   **7.** $40\overline{0}$ N   **9.** 551 N; 331 N
11. 2.77×10^5 N; 1.07×10^4 N   **13.** 2.67 m   **15.** 872 N; 948 N
17. 34.2 kg; 41.8 kg

7.5 Pages 197–198

1. 22.6   **3.** $F_1 = 99$ lb; $F_2 = 88.0$ lb   **5.** 909 N; 591 N
7. 9.90×10^4 N; 8.82×10^4 N   **9.** 117 lb; 51 lb   **11.** 8.99×10^4 N; 1.01×10^5 N
13. 197 N; 118 N   **15.** 2390 N; 943 N   **17.** 1020 N; 775 N
19. 1525 N (up) 4.54 m from A

Chapter 7 Review Questions Page 200

1. b   **2.** b   **3.** a   **4.** c   **5.** c   **6.** a   **7.** b
8. b   **9.** No (e.g., bridge)
10. They are equal.
11. Equilibrium is the condition of a body where the net force acting on it is zero.
12. Toward the center of the earth.
13. They are in equilibrium.
14. A diagram showing how forces act on a body
15. No, only when the pedals are parallel to the ground.
16. Even if the vector sum of opposing forces is zero, they must also be positioned so there is no rotation in the system.
17. Choose a point through which a force acts to eliminate a variable.
18. (a) Stacking bricks   (b) Riding a bicycle   (c) Lifting any object   (d) Hitting a baseball   (e) Leaning into the wind
19. No; only if the object is of uniform composition and shape
20. The support closer to the bricks

Chapter 7 Review Problems Pages 201–203

1. 569 N (left)   **2.** (a) $50\overline{0}$ lb   (b) $5\overline{0}$ lb   **3.** 3700 N at 156°
4. 8450 lb at 334.5°   **5.** 94,600 N at 80.3°   **6.** 1610 N at 114.2° from F_1
7. 470 N   **8.** 6.0 tons   **9.** −525   **10.** 1080 N at 33.7°
11. $F_1 = 645$ N; $F_2 = 1520$ N   **12.** $F_1 = 5240$ lb; $F_2 = 2210$ lb
13. $T = 1790$ N; $C = 1370$ N   **14.** $T_1 = 348$ lb; $T_2 = 426$ lb; $T_3 = 475$ lb
15. $T_1 = 2900$ N; $T_2 = 1900$ N; $T_3 = 2200$ N
16. $T_1 = 6250$ N; $T_2 = 6250$ N; $C = 6250$ N   **17.** 26.5 N m
18. 27.0 lb   **19.** 880 N   **20.** 160 lb   **21.** 100.0 kg
22. 51.0 cm   **23.** $560\overline{0}$ kg   **24.** End closest to truck, 62,300 N; other end, 47,700 N
25. 343 N; 483 N   **26.** 2580 N

Chapter 7 Applied Concepts Page 203

1. (a) $230\overline{0}$ N m   (b) 1150 N m   (c) Continually push perpendicular to the door
2. (a) 18.6°   (b) 58.3 lb   (c) 79.5 lb   (d) The angle between the ropes in increased so

that Sean and Greg not only pull up but also horizontally.   **3.** (a) 2.28 N m more torque.   (b) 18.0 N less force   **4.** (a) $T = 45.5$ N m, which will not support the torque.   (b) Reduce the angle between the wall and the bracket.   **5.** (a) $F_{fulcrum} = 2210$ N; $F_{bracket} = -1420$ N (negative sign means that the bracket pulls down) (b) Between the bracket and the fulcrum

Chapter 8

8.1 Pages 209–211
1. 34.3 N m   **3.** 917 kJ   **5.** 24.4°   **7.** 16,200 J   **9.** 4410 J   **11.** 34,000 ft lb
13. 24.1 kJ   **15.** 163,000 ft lb   **17.** 6.2 MJ   **19.** 28,900 J
21. (a) 903 J   (b) −903 J   **23.** 10,700 J

8.2 Pages 216–217
1. 18.9 N m/s or 18.9 W   **3.** 0.533 s   **5.** 12.4 W   **7.** (a) 1.68 hp   (b) 219 N
9. 59.4 s   **11.** 1.49 kW   **13.** 6.17 kW   **15.** 1.94 kW
17. (a) 6.9 kW   (b) 9.2 hp   (c) 9.3 kW   **19.** Four times
21. (a) 110 passengers   (b) 22 kW   **23.** 1.15 kW   **25.** 0.350 kW

8.3 Pages 220–221
1. 2460 N m   **3.** 217 J   **5.** (a) 80.7 ft/s   (b) 3.09×10^6 ft lb   **7.** 14.2 ft/s
9. (a) 294 kJ   (b) 392 kJ   **11.** 342 m/s   **13.** 220 kJ
15. (a) 2650 kW   (b) 3550 hp   (c) 1.59×10^8 J or 159 MJ
17. 4   **19.** (a) 1960 J   (b) 1960 J   (c) 2160 N   (d) 1.31×10^5 N
21. (a) 3150 J   (b) 3150 J   (c) The increase in gravitational potential energy is equal to the work done in raising the painter. The work done by the painter is the source of the increase in energy.

8.4 Pages 227–228
1. 7.00 m/s   **3.** 72.7 m/s   **5.** 87.3 ft   **7.** 49.5 m/s   **9.** 8.40 m/s
11. $19\overline{0}$ ft/s
13.

t	s	v	KE	h	PE	Total
0.000	0.00	0.00	0	300.0	11,760	11,760
1.000	4.90	9.80	190	295.1	11,570	11,760
2.000	19.6	19.60	770	280.4	10,990	11,760
3.000	44.1	29.40	1,730	255.9	10,030	11,760
4.000	78.4	39.20	3,070	221.6	8690	11,760
5.000	122.5	49.00	$4,8\overline{0}0$	177.5	6960	11,760
6.000	176.4	58.80	6,910	123.6	4850	11,760
7.000	240.1	68.60	9,410	59.9	2350	11,760
7.800	300.0	76.68	11,760	0.00	0.00	11,760

Chapter 8 Review Questions Pages 229–230
1. c   **2.** a   **3.** a   **4.** d   **5.** c   **6.** No   **7.** No
8. $J = \left(\dfrac{kg\ m}{s^2}\right)(m)$ **9.** No   **10.** No   **11.** Yes

12. By measuring the force applied, the distance traveled, and the time taken
13. It possesses the ability to do work (e.g., turn a wheel) in falling to the lower level.

14. (a) Elevator counterweights   (b) Roller coasters
15. Yes; $KE = \frac{1}{2}mv^2$ **16.** At its lowest point   **17.** At its highest point
18. No   **19.** Yes   **20.** The bolt has accelerated to a higher velocity.

Chapter 8 Review Problems Page 230

1. 3.6×10^6 J   **2.** 0   **3.** 10.2 m   **4.** 9.80×10^5 J
5. (a) 27.2 W   (b) 0.0365 hp   **6.** 0.102 m   **7.** 2.00 ft
8. 1.41 m/s   **9.** 1.40 m/s   **10.** 303,000 ft lb   **11.** 2350 J
12. $30\overline{0}0$ J   **13.** 1670 J   **14.** (a) $30\overline{0}0$ J   (b) 1880 J   **15.** 31.3 m/s

Chapter 8 Applied Concepts Page 231

1. (a) 4.49 W   (b) The mass of the water, the distance the substance is carried, and the time to move the water   **2.** (a) 2.09×10^5 J   (b) A, 34.3 m/s; B, 28.0 m/s; C, 24.3 m/s; D, 19.8 m/s   (c) The higher the elevation, the less is the velocity
3. (a) 4.80×10^7 J (same as the jet's original kinetic energy)   (b) 4.17×10^5 N
(c) The force needed to stop the jet is less because the time is more.   **4.** (a) 5.27×10^9 W (this is enough energy to power all of Paraguay)   (b) 1.05×10^{10} W(c) The higher the dam, the more potential energy changes to kinetic energy.   **5.** (a) 1.56×10^5 J
(b) 1.56×10^5 J   (c) 3.57×10^5 N   (d) When the wrecking ball is at its lowest point, all of its energy has been converted to kinetic energy, and thus it strikes the wall with the greatest velocity.

Chapter 9

9.1 Pages 238–240

1. (a) 40.8 rad   (b) 2340°   **3.** (a) 12.5 rev   (b) 4500°   **5.** 154 rpm
7. 8.38 rad/s   **9.** 354 rev/s   **11.** $645\overline{0}$ rpm   **13.** 14.2 rad/s
15. $66\overline{0}$ rpm   **17.** (a) 6.67 rev/s   (b) $40\overline{0}$ rpm   (c) 41.9 rad/s
19. (a) 0.0571 s   (b) 87.5 rev   **21.** 8.40 rad   **23.** 0.131 m   **25.** 1.33 m/s
27. (a) 68.6 rad/s   (b) 12,300 rad   (c) 157 cm   (d) 3080 m   (e) 17.2 m/s
29. (a) 50.5 rad/s   (b) 1520 rad   (c) 502 m   (d) 502 m   **31.** 188 in./s or 15.7 ft/s
33. (a) 1680 km/h   (b) 1450 km/h   (c) 1190 km/h   (d) 838 km/h
35. 1.33 rad/s^2 **37.** (a) -3.60 rad/s^2 (b) 432 m/s   (c) 18.3 rev

9.3 Pages 244–245

1. 4350 N   **3.** 79.4 slugs   **5.** 5.53 m/s   **7.** 1.92 m   **9.** 3.60×10^3 lb
11. 28.8 m   **13.** 5420 N   **15.** 19 N   **17.** $13\overline{0}0$ N   **19.** $2\overline{0}00$ N
21. 34,400 N

9.4 Pages 248–249

1. 7260   **3.** 18.7   **5.** 1860 N m   **7.** 343 hp   **9.** 4.95/s   **11.** 4600 W
13. (a) 299 hp   (b) 599 hp   **15.** 2.65 N m   **17.** 37 kJ/min   **19.** 251 kW
21. 6.67/s   **23.** 2800 W   **25.** 7.17 N m   **27.** 118 W

9.6 Pages 256–258

1. 52 rpm   **3.** 24 teeth   **5.** 207 rpm   **7.** 42.5 rpm   **9.** 75 teeth   **11.** 63 teeth
13. 144 rpm   **15.** 60 teeth   **17.** 9 teeth   **19.** Counterclockwise   **21.** Clockwise
23. Clockwise   **25.** Clockwise   **27.** Counterclockwise   **29.** 1050 rpm
31. 576 rpm   **33.** 1480 rpm   **35.** 40 teeth   **37.** 20 teeth
39. Gear *B* is reversed in all problems.

9.7 Page 260
1. 2250   **3.** 3600   **5.** 147   **7.** 62.5 rpm   **9.** 22.5 in.   **11.** Clockwise
13. Counterclockwise   **15.** Counterclockwise

Chapter 9 Review Questions Pages 262–263
1. d   **2.** a   **3.** b
4. Curvilinear motion is motion along a curved path; rotational motion occurs when the body itself is spinning.
5. Radians and revolutions
6. A radian is an angle with its vertex at the center of a circle whose sides cut off an arc on the circle equal to its radius.
7. Angular displacement is the angle through which any point on a rotating body moves. It can be measured in radians or revolutions.
8. Linear velocity = angular velocity $\times$ radius
9. They are alike except for the substitution of θ for s, ω for v, and α for a.
10. Law of conservation of angular momentum
11. Yes
12. No; it tends to cause a body to continue in a straight line (tangent to the curve).
13. Number of teeth of driver times number of revolutions of driver = number of teeth of driven gear times number of revolutions of driven gear
14. An idler changes direction of rotation of the driver gear.
15. Opposite
16. Since the gears are connected, they both rotate together.
17. Diameters are used in similar equations rather than teeth.
18. The small pulley
19. The belt is one continuous piece of material.

Chapter 9 Review Problems Pages 263–264
1. (a) 26π or 81.7 rad   (b) 4680°   **2.** 1.7 rad/s   **3.** 0.885 m/s   **4.** 7.50 N
5. 2.75 m/s   **6.** 5.67 ft lb/s   **7.** 6.00 rad/s   **8.** 6.00 rpm   **9.** 20 teeth
10. 57.5 rpm   **11.** Yes   **12.** 105 rpm   **13.** 2.00 cm   **14.** Counterclockwise
15. 720 rpm   **16.** 32 teeth

Chapter 9 Applied Concepts Page 265
1. (a) 1.78 rad/s = 17.0 rpm   (b) 2.51 rad/s = 24.0 rpm   (c) No, the mass of the person is unrelated to the rotational speed.   **2.** (a) 240 Nm   (b) 405 kg m^2 (c) Increase the radius of the waterwheel or make the waterwheel a ring and not a solid disk. More mass farther away from the axis increases the rotational inertia.   **3.** (a) B, 8.00 g's; D, 2.00 g's   (b) B is not safe. The B loop should have a larger radius to reduce the centripetal force. D is safe.   **4.** (a) 86.8 ft/s = 59.2 mi/h   (b) 123 ft/s = 83.9 mi/h (c) As the radius increases, the centripetal force is reduced and the maximum velocity increases.   **5.** (a) 2.26×10^4 lb ft/s = 41.1 hp   (b) 43.2 lb ft   (c) The lower the gear, the lower is the speed and the greater is the torque. More torque is required to move up steep hills.

Chapter 10

10.2 Pages 272–274
1. 14.8   **3.** 36.3   **5.** 52.4   **7.** 2.39   **9.** 48.8   **11.** 1.55   **13.** 2.27
15. (a) 1100 lb   (b) 4.00   **17.** (a) 1.50 ft   (b) 4.00

10.3 Pages 275–276

1. 14.0   **3.** 271   **5.** 524   **7.** 48.8   **9.** 20.4   **11.** 181 N
13. (a) 438 lb   (b) 6.00   **15.** (a) 2.99 cm   (b) 3.34

10.4 Pages 279–280

1. 1   **3.** 3
5. 6   **7.** 2
15. 2
17. (a) 388 N   (b) 82.0 m
19. No

9.   **11.**   **13.**

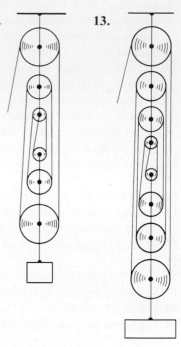

21.   Mechanical Advantage (MA)

Number of								
Pulleys	1	2	3	4	5	6	7	8
Fixed	1	1	2	2	3	3	4	4
Movable	0	1	1	2	2	3	3	4
Fixed		0	1	1	2	2	3	3
Movable		1	1	2	2	3	3	4

10.5 Pages 283–284

1. 12.8   **3.** 21.2   **5.** 36.3   **7.** 4.62   **9.** 5.61
11. (a) 4.00   (b) 182 N   (c) No   **13.** 3.00   **15.** (a) 1.33 m   (b) 4.33
17. 1790 N

10.6 Page 286

1. 2.30   **3.** 14.2   **5.** 2.28   **7.** 29.3   **9.** 35.2
11. (a) 4.84 lb   (b) 754 N   **13.** (a) 37.7   (b) 566 lb   (c) 188 lb

10.8 Pages 288–289

1. 15.0   **3.** 40.0   **5.** 75.0 lb   **7.** 125 N   **9.** 1.44×10^5 N

Chapter 10 Review Questions Pages 291–292

1. d   **2.** a   **3.** b   **4.** d   **5.** b   **6.** b
7. (a) Bicycle   (b) Auto transmission   (c) High-speed drill
8. Resistance force
9. Effort force × effort distance = resistance force × resistance distance
10. Mechanical advantage   **11.** Efficiency   **12.** No

13. The fulcrum   **14.** $MA = \dfrac{\text{Effort arm length}}{\text{Resistance arm length}}$ **15.** First class

16. $F_R \times s_R = F_E \times s_E$

17. The opposite end of the resistance force with the effort force in between

18. Resistance force $\times$ resistance radius = effort force $\times$ effort radius

19. No; it depends on the radii.

20. A fixed pulley does not move. It is suspended by its center axle. A movable pulley is free to move and is suspended by the strand around the groove.

21. No   **22.** $MA = \dfrac{\text{Length of plane}}{\text{Height of plane}}$

23. The distance a screw advances into the wood in one revolution

24. It is greater because the handle of the jackscrew can be longer than the radius of the screwdriver.

Chapter 10 Review Problems Pages 292–293

1. 2.00   **2.** 73.0%   **3.** 3.82   **4.** (a) 540 N   (b) 360 N
5. (a) 6.54 cm   (b) 2.60   **6.** 12   **7.** 675 N   **8.** (a) 292 N   (b) 3.00
9. 2.30 ft   **10.** 16,400 N   **11.** 1.08 mm   **12.** 151
13. (a) 0.112   (b) 0.106   (c) 16.4 N
14. (a) The MA is doubled.   (b) 1.57 cm   **15.** 1.74   **16.** 81.4 cm
17. 5   **18.** 4   **19.** (a) 64.0   (b) 4800 N   **20.** 6.0

Chapter 10 Applied Concepts Pages 294–295

1. (a) 2.62×10^{31} m   (b) $MA_{\text{lever}} = 6.82 \times 10^{22}$ This was clearly a theoretical statement by Archimedes.   **2.** (a) 0.120   (b) 2.54 m   (c) 4.58 m/s   (d) Increases the speed of the resistance end. The tip of the fishing pole needs to be moving at a high speed to cast the lure.   **3.** (a) 7.16   (b) 3.27 lb   (c) The MA would increase.   **4.** (a) 0.118 (b) 6.42 lb   (c) 10.4 in.   (d) An MA less than one is advantageous for speed. The bike must be in a high gear and moving quickly.   **5.** (a) 3.29   (b) 3.53   (c) Increase the length of the lever arm and/or the radius of the wheel.

Chapter 11

11.1 Page 300

1. (a) 826 N   (b) 0.497 N   **3.** 1.23×10^{22} N   **5.** 5.70×10^{16} N
7. 3.37×10^5 N   **9.** 3.61×10^{-51} N

11.3 Page 305

1. 1020 m/s   **3.** 4.79×10^4 m/s   **5.** 9630 m/s   **7.** 7.59×10^6 s or 0.241 yr
9. 9.32×10^8 s or 29.6 yr

Chapter 11 Review Questions Page 306

1. d   **2.** b   **3.** a   **4.** c   **5.** a

6. The earth's mass gives it a greater gravitational force.

7. The increased radius would cause you to weigh one fourth of your original weight.

8. The increased mass would cause you to weigh twice your original weight.

9. The satellite's horizontal velocity enables it to continually miss the earth as it falls.

10. With no horizontal velocity, the apple struck the ground. The moon's horizontal velocity enables it to maintain its orbit.

11. The force at the perigee would be greater than that at the apogee.

12. No, the mass in the period equation is the mass of the object being orbited.

Chapter 11 Review Problems Page 307

1. 9.60×10^{-11} N **2.** 2.82×10^{-7} N **3.** 636 N; 143 lb **4.** 1610 N; 362 lb
5. 49.2 N; 11.1 lb **6.** 8.37×10^5 s or 9.69 days **7.** 6.70×10^6 s or 77.5 days
8. 1.90×10^7 s or $22\overline{0}$ days **9.** 832 N **10.** 745 N

Chapter 11 Applied Concepts Page 307

1. (a) 3.70 m/s^2 (b) 315 N; 833 N (c) The astronaut's muscles would weaken and have
a difficult time supporting the returning astronaut on the earth. **2.** (a) 1.57×10^7 m
(b) 2.46 earth radii **3.** (a) $r = 4.22 \times 10^7$ m from the center of the earth; $r = 3.58 \times 10^7$ m
from the surface of the earth (b) 3.07×10^3 m/s $= 1.11 \times 10^4$ km/h **4.** (a) $1.61 \times$
10^3 m/s $= 5.80 \times 10^3$ km/h (b) 7370 s or 2.05 h (c) Increasing the altitude would decrease the velocity and increase the period. **5.** (a) 565 N (b) The astronauts only feel
weightless because they are in continuous free fall.

Chapter 12

12.2 Pages 321–323

1. 2.50 m $\times$ 0.80 m: 71 kPa; 2.50 m $\times$ 0.45 m: 130 kPa; 0.80 m $\times$ 0.45 m: 390 kPa
3. (a) 46.2 in. (b) 20.3 lb **5.** $20\overline{0}$ N/m. **7.** (a) 0.0250 cm (b) 1.56×10^6 N
9. 3.57×10^5 lb **11.** 5.97×10^7 Pa or 59.7 MPa
13. (a)

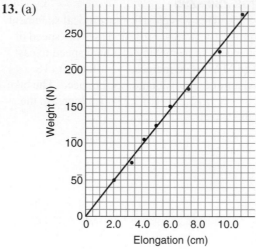

(b) 180 N (c) 9.0 cm (d) 25 N/cm **15.** 16.8 cm; 12.6 cm **17.** 489 N
The breaking point is dependent on the diameter of the wire and not on its length.
19. 0.827 J

12.5 Pages 334–335

1. 2870 kg/m^3 **3.** 1750 lb **5.** 5600 cm^3 **7.** 1210 lb/ft^3 **9.** 2710 kg/m^3
11. 684 kg/m^3 **13.** 3.5 m^3 **15.** 58.8 lb/ft^3 **17.** 1.49 m^3 **19.** 2820 kg/m^3
21. (a) 1000 L (b) 1500 L (c) 73.5 L **23.** 850 kg **25.** 0.917 **27.** 7.8
29. 0.68 **31.** Floats **33.** Floats **35.** 5.8×10^{15} **37.** 2330 kg/m^3

Chapter 12 Review Questions Pages 337–338

1. a, c, and e   **2.** All   **3.** d   **4.** d   **5.** d   **6.** b   **7.** d   **8.** e

9. Mass density refers to the mass per unit volume; weight density refers to weight per unit volume.

10. Yes; no.

11. Capillary action refers to the effect of surface tension of liquids that causes the level of liquid in a small-diameter tube to be higher or lower than that of the liquid in a large-diameter tube due to the adhesive force between the liquid and the tube.

12. Adhesion refers to the attractive force between different molecules; cohesion refers to the attractive force between similar molecules.

13. The surface tension of water allows the base of many insects' legs to be supported by the surface of water, allowing them to "walk" on the surface of a pond.

14. 1800   **15.** Pressure

16. Stress is directly proportional to strain as long as the elastic limit is not exceeded.

17. kPa

18. The specific gravity of an object can be found by dividing the density of the object by the density of water. The density can be found by determining the mass of the object and dividing by its volume.

19. Mass density   **20.** Viscosity   **21.** Elastic limit   **22.** Solid, liquid, gas

23. An atom consists of one nucleus and its surrounding electrons; a molecule consists of two or more atoms.

24. A proton has a positive charge; a neutron has a neutral charge.

25. Tension, compression, shear, bending, and twisting

26. The hydrometer measures the density of the battery electrolyte. The density is related to the amount of sulfuric acid in the electrolyte and therefore the charge on the battery. Temperature does affect the measurement.

Chapter 12 Review Problems Pages 338–339

1. 165 N   **2.** 24.6 cm   **3.** 1.63×10^6 lb/in.; 0.0317 in.
4. 311 MPa   **5.** 2020 lb/ft^3 **6.** 1.3 N   **7.** 8950 kg/m^3
8. 83.4 lb/ft^3 **9.** 55.8 cm^3 **10.** 21.9 m^3 **11.** 2160 kg
12. 59,300 lb   **13.** 18.9 lb/ft^3 **14.** 54.8 lb/ft^3 **15.** 0.117 ft^3
16. 0.694   **17.** 77.8 lb   **18.** 9.26 kg   **19.** 2960 kg/m^3 **20.** Sink

Chapter 12 Applied Concepts Page 339

1. (a) 3.08×10^5 N/m^2 (b) 1.27×10^5 N/m^2 (c) High-speed driving and the high-pressure tire together increase the heat energy in the tire. This could eventually break down the tire, resulting in a blowout.   **2.** (a) A shearing strain causes the warping of the lines.   (b) An automobile exerts a tremendous amount of force on the pavement to begin its acceleration. On the open road, the vehicle exerts much less force on the pavement to continue its motion.   **3.** 3.72 ft^3 **4.** (a) 7.26×10^4 kg   (b) 6.3×10^4 kg (c) 4.9×10^4 kg   **5.** (a) 178 lb/in.   (b) 201 lb   (c) At some point the spring will not be able to compress any farther.

Chapter 13

13.1 Pages 350–351

1. 21.7 lb/in^2 **3.** 16.5 lb/ft^3 **5.** 55.2 lb/in^2 **7.** 245 kPa   **9.** 9.69 m
11. 757 kg/m^3 **13.** 57.8 lb/in^2 **15.** 233 lb/in^2 **17.** 2.22×10^8 N
19. $49\overline{0}$ kPa   **21.** 728 kPa

13.2 Pages 350–351
1. 60.0 lb **3.** 22.0 **5.** 24.0 **7.** 48 N **9.** 23,000 N
11. (a) 146 N (b) 11.6 N/cm^2 (c) 36.0
13. (a) 4800 N (b) 9.55 kPa (c) 9.55 kPa (d) 64.0
(e) The lift will exert twice the force on the large piston; the MA and the pressure will double. (f) The lift will exert four times the force on the large piston; the MA and the pressure will increase by a factor of 4. **15.** Increased by a factor of 4.
17. 2610 N **19.** 1570 cm^2

13.3 Page 354
1. 118 lb/in^2 **3.** 293 kPa **5.** (a) 303.96 kPa (b) 202.64 kPa (c) 88.2 lb/in^2
(d) 506.6 kPa (e) 33.77 kPa (f) 25.33 kPa **7.** 586 kPa **9.** 30.3 lb/in^2
11. 306 kPa **13.** 1174 kPa **15.** 401 kPa

13.4 Page 357
1. 13.0 lb **3.** 3.9 N **5.** 75.5 lb **7.** 3.20 N **9.** 105 lb **11.** 431 N
13. (a) 3.06×10^4 ft^3 (b) 795 tons **15.** 7.65 m^3 **17.** 440 N

13.5 Pages 361–362
1. 358 L/min **3.** (a) 9.6 cm (b) 0.34 m/s **5.** (a) 34.1 ft/min (b) 136 ft/min
7. 16.7 cm **9.** 49,500 L/min

Chapter 13 Review Questions Pages 363–364
1. c **2.** b **3.** b **4.** c **5.** a **6.** kPa
7. Pressure is the force applied per unit area.
8. $F_s = \frac{1}{2} AhD_w$ or one-half the force exerted by the water on the bottom of the tank.
9. External pressure applied to a fluid is transmitted to all inner surfaces of the liquid's container.
10. A ship floats because it is lighter than an equal volume of displaced water.
11. The spinning baseball creates a different air velocity on one side than on the other. This creates a higher pressure on one side, which causes the ball to "curve."
12. The top side of the wing is curved more than the bottom, so the velocity of air rushing past the top must be larger than that going past the bottom. This creates a low-pressure area at the top of the wing and creates lift.
13. In streamline flow, all particles of the fluid passing a given point follow the same path. In turbulent flow, the particles passing a given point may follow different paths from that point.
14. A balloon filled with a gas with lower density than air, such as helium, rises.
15. Absolute pressure is the actual air pressure given by the gauge reading plus normal atmospheric pressure. Gauge pressure is the pressure measured relative to atmospheric pressure. Gauge pressure is used in an automobile tire gauge.
16. The pressures are identical. The forces are different.
17. This force depends on the horizontal surface area and the average pressure on that surface. The average pressure depends on the density and the height of the liquid.
18. The pressure at the bottom is greater than at the top.
19. Smaller. The total force exerted on the brake pads must be larger than the force applied to the brake pedal.
20. Yes, but only if the fluid to be drawn through the straw is in an airtight sealed container.

Chapter 13 Review Problems Pages 364–365
1. 31.8 kPa **2.** 455 ft **3.** 30.3 m **4.** 9300 lb **5.** 196 kPa
6. 9.60×10^7 N **7.** 7.01×10^8 N **8.** 2.80×10^4 N **9.** 2610 kPa
10. 192 lb/in^2 **11.** 9410 kPa **12.** (a) 77.2 lb (b) 8.42 **13.** 35.5 N
14. (a) 103 N (b) 111 N/cm^2 (c) 42.2 **15.** 303 kPa **16.** 554 kPa
17. 299 lb/in^2 **18.** 3.3 N **19.** 6.30 N **20.** 2.4 N **21.** (a) 26,000 ft^3

(b) 654 tons   **22.** 188 L/min   **23.** (a) 3610 L/min   (b) 90,300 L
24. (a) 49.3 N/m^3 (b) Alcohol

Chapter 13 Applied Concepts Page 365

1. (a) 372 m^2 (b) 2.76 × 10^6 N   (c) The bands should be spaced closer together because of the increased pressure at the bottom of the tank.   **2.** (a) 44.6 lb/in^2
(b) 429 N on each piston or brake drum   **3.** (a) 0.703 m^3 (b) 16,000 N   (c) 7060 N
(d) 8900 N   **4.** (a) 50.5 mi/h   (b) The space between the buildings acts as a Venturi tube by limiting the area that the air can pass through.   **5.** (a) 201 in./s = 16.8 ft/s
(b) Limit the number of holes or make the holes smaller. Limiting the area for the water to come out causes it to travel faster.

Chapter 14

14.1 Pages 370–371

1. 25°C   **3.** 125°C   **5.** 293°F   **7.** −12.2°C   **9.** 203°F   **11.** 298 K
13. 133°C   **15.** 223 K   **17.** 5727°C   **19.** −38.9°C   **21.** 29°C
23. −321°F   **25.** 12$\overline{0}$0   **27.** 941   **29.** 1175   **31.** 59$\overline{0}$

14.2 Pages 372–373

1. 23 cal   **3.** 1.21 × 10^6 ft lb   **5.** 3.21 × 10^6 J or 3.21 MJ   **7.** 4450 Btu
9. 2.62 MJ   **11.** 4.82 × 10^7 J   **13.** 42.2 MJ   **15.** (a) 2.77 × 10^5 J/s
(b) 277 kW   **17.** 2.10 × 10^8 J

14.3 Pages 377–378

1. 0.021 ft^2 °F h/Btu   **3.** 0.45 ft^2 °F h/Btu   **5.** 1.7 ft^2 °F h/Btu
7. 8.1 × 10^6 Btu   **9.** 11 J   **11.** 1400 Btu   **13.** 14 MJ   **15.** 6800 J

14.4 Pages 380–381

1. 173 Btu   **3.** 38$\overline{0}$0 Btu   **5.** 2.1 × 10^5 J or 210 kJ   **7.** 280 cal
9. 3.01 × 10^4 J or 30.1 kJ   **11.** 64.0 kcal   **13.** 0.259 kg   **15.** 29,000 kcal
17. 1.89 × 10^8 J or 189 MJ   **19.** 8.07 × 10^7 J or 80.7 MJ   **21.** 252°C
23. 5770 J   **25.** (a) 10.1°C   (b) 22°C   (c) Water is the better coolant because its temperature increase is only about half that of alcohol in absorbing the same amount of heat.

14.5 Pages 383–384

1. 428°F   **3.** 0.051 cal/g °C   **5.** 95°F   **7.** 81°C   **9.** 0.104 cal/g °C
11. 286°C   **13.** 0.259 kg

14.6 Pages 388–389

1. 0.30 ft   **3.** 0.10 m   **5.** 200.10 m   **7.** 0.54 ft   **9.** 0.752 in.
11. 1.8 cm^2 **13.** 88.97 cm^2 **15.** 60.10 cm^3 **17.** 4.9 mm
19. 1.08 × 10^{-5}/°C   **21.** 15 cm^3 **23.** 0.3397 m^3

14.7 Page 391

1. 653 L   **3.** 12.2 m^3 **5.** 3754 ft^3 **7.** 0.58 cm^3 **9.** $416.16   **11.** 23°C
13. 12.5 L

14.8 Page 401
1. 1100 cal   **3.** 10,700 Btu   **5.** 26,000 cal   **7.** 6.70×10^6 J or 6.70 MJ
9. 3.39×10^6 J or 3.39 MJ   **11.** 6070 Btu   **13.** 11,650 Btu
15. 5000 kcal   **17.** 3.12×10^6 J or 3.12 MJ   **19.** 467,000 cal   **21.** 3090 kcal

Chapter 14 Review Questions Pages 403–404
1. a, b, and d   **2.** b and d   **3.** a, b, and d   **4.** a, b, and d   **5.** d   **6.** c
7. The total heat lost by warm objects is gained by cold objects.
8. 778 ft lb of work is equivalent to 1 Btu.
9. The Rankine scale.   **10.** The Kelvin scale.
11. Each Celsius degree is 1.8 times the Fahrenheit degree. The freezing point of water is 0°C and 32°F. The boiling point of water is 100°C and 212°F.
12. Heat is an amount of energy. Temperature is a measure of the average kinetic energy of atoms and molecules in matter.
13. Burning fuel in a furnace. Burning fuel in an engine. Conversion of heat into electricity in a steam-driven generator.
14. Heat generated in a solid by a drill bit. Heat generated by rubbing two objects together. Heat generated in automobile brakes. All are due to friction.
15. Light clothing should be worn. It absorbs less heat.
16. It increases. The metal block increases in size, as does the hole.
17. (a) 4 degrees Celsius.   (b) Most other liquids have their highest density at the temperature at which the change of state to a solid occurs.   (c) Water below 4 degrees Celsius is less dense than water at 4 degrees Celsius and rises. Freezing therefore occurs at the top of a body of water.
18. Ten kilograms of ice at 0°C. The difference is the heat of fusion, which must be added to the ice to turn it to water.
19. The large amount of heat released when steam changes to water can cause severe burns.
20. Because water expands as it solidifies.
21. Conduction is the transfer of kinetic energy from atoms or molecules in a warmer location through nearby atoms or molecules to atoms or molecules in a colder section of the material. Convection is the transfer of kinetic energy from one region to another via the motion of warmer atoms or molecules from one region to another. Radiation is the transfer of energy through the emission and absorption of electromagnetic radiation.
22. Coolants boil at a higher temperature at high pressure. Therefore, the engine coolant can be at a higher temperature and transfer more heat from the engine to the atmosphere because of the greater temperature difference.
23. Heat is extracted from the outside air in winter to vaporize the refrigerant. The heat is given up by the condensing fluid to warm the inside air. In summer, heat is extracted from the inside air to vaporize the refrigerant. The condensing fluid then gives up this heat to the outside air.

Chapter 14 Review Problems Pages 404–405
1. 71°C   **2.** 297 K   **3.** 9230°F   **4.** 335°C   **5.** 10.3 cal   **6.** 1.56 kcal
7. 3.38×10^5 ft lb   **8.** 1.4×10^5 Btu   **9.** 2.4×10^4 J or 24 kJ   **10.** 4.18×10^4 Btu
11. 1.5×10^4 kcal   **12.** 1.22×10^4 kcal   **13.** 1300°F   **14.** 0.353 cal/g °C
15. 85°C   **16.** 12.524 m   **17.** 7.51 mm; 44.3 mm^2 **18.** 0.0295 cm^3
19. 2.57 L   **20.** 123 ft^3 **21.** 11,600 kcal   **22.** 1200 Btu   **23.** 3440 kcal
24. 4.80×10^5 J

Chapter 14 Applied Concepts Page 405
1. 1.25×10^4 J   **2.** 4.52×10^5 kcal   **3.** (a) 8.4×10^6 kcal   (b) 8.9×10^6 kcal
(c) 1.8×10^5 kcal   **4.** (a) Steel will contract less.   (b) 0.0161 ft = 0.193 in.
5. (a) 2.43 gal   (b) $0.14

Chapter 15

15.1 Pages 408–409
1. 288 K **3.** 44°C **5.** 532°R **7.** 9$\overline{0}$°F **9.** 258 cm^3 **11.** −16°F
13. 38$\overline{0}$ m^3 **15.** 1430 L **17.** −59°C **19.** 30.5 L **21.** 107 K
23. −111°C **25.** 154 L **27.** 1920 L **29.** 277 L

15.2 Pages 412–413
1. 265 cm^3 **3.** 75.0 kPa **5.** 2.08 kg/m^3 **7.** 0.180 lb/ft^3 **9.** 90.4 kPa
11. 3.34 kg/m^3 **13.** 801 kPa **15.** 19.0 ft^3 **17.** 162 psi **19.** 2.48 m^3
21. (a) 37.5 in^3 (b) 25.0 in^3 (c) 15$\overline{0}$ in^3 **23.** 4.37 kg/m^3 **25.** 1.16 L
27. 0.352 kg/m^3 **29.** (a) Lower (b) 12.4 kPa

15.3 Pages 415–416
1. 1270 in^3 **3.** 506 m^3 **5.** −39°C **7.** 143°C **9.** 22$\overline{0}$0 psi
11. 399 kPa **13.** (a) 70$\overline{0}$ kPa (b) 891°C (c) 3.59 m^3 (d) 0.563 m^3 **15.** 12.3 L

Chapter 15 Review Questions Page 417
1. c **2.** a **3.** c **4.** c **5.** b **6.** b
7. Standard pressure is 101.32 kPa or 14.7 lb/in^2 and standard temperature is 0°C or 32°F.
8. A temperature increase tends to cause a volume increase. A pressure increase tends to cause a volume decrease. If both temperature and pressure are increased, the volume change is given by the combined Charles' and Boyle's laws.
9. The temperature will increase.
10. Heating a gas increases the kinetic energy of the gas molecules. This causes an increase in pressure and volume.
11. When a gas is compressed, work must be done on it to decrease its volume. This work is transferred into increased kinetic energy of the molecules of the gas, causing a temperature increase.
12. When the volume is increased, the number of gas molecules striking the surface of the container per unit area decreases. Thus the pressure exerted by the gas molecules on the container decreases.

Chapter 15 Review Problems Page 418
1. 14.9 ft^3 **2.** 4.40 m^3 **3.** 131°F **4.** 30.9°C **5.** 34.1°C
6. 138 kPa **7.** 23.9 ft^3 **8.** 3.68 kg/m^3 **9.** 478 in^3
10. 19.9°C **11** (a) 2180 psi (b) 18°F **12.** 304 kPa **13.** 135 kPa
14. 822 L **15.** 3.07 kg/m^3

Chapter 15 Applied Concepts Page 419
1. (a) The temperature of the helium begins to decrease, and the volume of the balloon decreases. (b) 1.66 ft^3 (c) The volume would increase to 1.97 ft^3. (d) Overinflate the balloon in winter; underinflate the balloon in summer. **2.** (a) 39 lb/in^2 (b) 26 lb/in^2
3. (a) 1.88 × 10^{-4} m^3 (b) 138 kPa (c) 148 kPa **4.** (a) 6.68 m^3 (b) 1980 m^3
5. (a) 98$\overline{0}$ L (b) No, the balloon should not be fully inflated. If it were, the balloon would probably burst on the way up.

Chapter 16

16.2 Page 431
1. 2.00 ms   3. 80$\overline{0}$ m/s   5. 1.93 m/s   7. 0.813 m   9. 12.0 Hz
11. 1.2×10^9 Hz   13. 2.00×10^8 Hz   15. 5.21×10^{-3} m   17. 2.06×10^{10} m/s

16.4 Pages 437–438
1. 337 m/s   3. 317 m/s   5. 5.00 s   7. 2640 m   9. 622 Hz
11. 507 Hz   13. 653 Hz   15. 0.325 s   17. (a) 695 Hz (b) 522 Hz
19. 0.0870 s

16.6 Page 442
1. 55.9 cm   3. 2.24 s   5. 4.13 in.   7. 1.36 s   9. It is $\sqrt{2}$ times the original period.   11. 1.74 s   13. 1.0 s

Chapter 16 Review Questions Pages 444–445
1. a, b, and c   2. c   3. a   4. b   5. b   6. a   7. c and d
8. Interference is the result of two or more waves traveling through the same region at the same time. Diffraction is the bending of a single wave passing near an obstacle with an opening nearly the same size as the wavelength.
9. The addition of two or more waves to form a larger wave is constructive interference. The formation of a smaller wave is destructive interference.
10. Sound would not be heard if some obstacle came between you and the stereo speakers and there were no nearby reflecting surfaces.
11. Waves passing through a break in a seawall.
12. It increases.
13. A wave is a periodic disturbance. A pulse is a one-time disturbance.
14. A sharp explosion creates a sound pulse.
15. The speed of sound increases. The higher kinetic energy of air molecules at high temperature leads to a higher velocity of sound through the air.
16. A seismograph detects slight vibrations of the earth's surface by detecting the relative motion of the earth's surface and a massive object.
17. The speed of sound is higher in water than in air. The higher density of water provides a higher speed of sound.
18. Motion toward an oncoming sound wave increases the frequency at which the maximum-pressure regions in the sound wave strike an observer, therefore producing a higher frequency sound. Motion away from an oncoming sound wave decreases the frequency at which the maximum-pressure regions strike an observer, therefore producing a lower frequency sound.
19. Sympathetic vibrations occur when an object vibrates at its natural resonance frequency in response to the vibration of a nearby object at some other frequency. Forced vibration occurs when an object vibrates at the same frequency as another nearby vibrating object.
20. Resonance occurs when an object vibrates at its natural resonance frequency in response to the vibration at the same frequency of a nearby object.
21. The light from these stars is shifted toward the red as a result of their motion away from the earth.
22. Amplitude is maximum displacement.
23. Period is the time required for one full vibration. Frequency is the number of complete vibrations per unit of time.
24. No; the period of a pendulum is independent of its mass.

Chapter 16 Review Problems Pages 445–446

1. 2.82×10^{-6} s or 2.82 μs   **2.** 3.13 Hz   **3.** (a) 6.67×10^{14} Hz   (b) 1.50×10^{-15} s
4. 5.87 m/s   **5.** 1.24 m   **6.** 65 Hz   **7.** 4.58 m   **8.** 383 m/s   **9.** $31\overline{0}$ m/s
10. 0.967 s   **11.** 29 s   **12.** 602 Hz   **13.** (a) 5360 Hz   (b) $46\overline{0}0$ Hz
14. 368 Hz   **15.** 5.59×10^{14} Hz   **16.** 1.35 s   **17.** 4.80 in.

Chapter 16 Applied Concepts Page 447

1. (a) 0.248 m   (b) No   (c) 0.0411 m   **2.** (a) One wavelength   (b) 560 ft/s
(c) 1400 ft   **3.** (a) AM: 182 to 545 m; FM: 2.78 to 3.41 m   (b) FM wavelengths are
closer to the sizes of openings of tunnels and underpasses. As stated in the text,
"Wave diffraction is commonly observed only when the opening is nearly the same
size as the wavelength."   **4.** (a) 1.16×10^5 W   (b) 2.50 W/m^2 (c) Twenty-five
percent of the initial intensity   **5.** (a) 29 m/s   (b) 706 Hz

Chapter 17

17.3 Page 454

1. 922 N   **3.** 1.3 cm   **5.** 1.5×10^{-8} C   **7.** (a) −0.93 N   (b) +0.707 N
(c) −2.08 N

17.4 Page 457

1. 1.50×10^4 N/C   **3.** 0.500 N/C   **5.** 1.78×10^{-3} C   **7.** 0.676 N

17.5 Pages 461–462

1. 1.07 Ω   **3.** 0.0165 Ω/ft   **5.** 3.95 Ω   **7.** 0.684 Ω   **9.** 0.0131 cm^2

17.6 Pages 464–465

1. 4.79 A   **3.** $22\overline{0}$ V   **5.** 153 Ω   **7.** 17.6 Ω   **9.** 1.26 A   **11.** (a) 0.067 A
(b) $3\overline{0}$ V   (c) 0.13 A   **13.** 18.8 Ω   **15.** 1.85 A

17.7 Pages 468–469

1. 13.50 Ω   **3.** 60.0 Ω   **5.** 0.750 A   **7.** 378 V   **9.** 23.0 Ω
11. 10.8 Ω, 14.2 Ω, 19.0 Ω   **13.** 2.78 Ω, 40.0 V, 3.40 Ω

17.8 Pages 474–475

1. (a) 4.28 Ω   (b) 11.7 A   (c) 4.55 A   (d) 7.14 A   **3.** (a) $10\overline{0}$ Ω   (b) 1.67 A
(c) 0.250 A

17.9 Page 479

1. (a) R_2, R_3 (b) 3.00 Ω   **3.** 8.89 A   **5.** (a) 2.23 A   (b) 6.68 A   **7.** 21.24 Ω
9. 45 V   **11.** 5.41 A   **13.** 10.0 Ω   **15.** 17.3 Ω   **17.** 30.0 V   **19.** 6.93 V
21. 13.3 Ω   **23.** 60.2 V   **25.** 0.370 A

17.10 Pages 480–482

1.

	V	I	R
Batt.	12.0 V	9.00 A	1.33 Ω
R_1	12.0 V	6.00 A	2.00 Ω
R_2	12.0 V	3.00 A	4.00 Ω

3.

	V	I	R
Batt.	36.0 V	6.00 A	6.00 Ω
R_1	36.0 V	2.00 A	18.0 Ω
R_2	36.0 V	3.00 A	12.0 Ω
R_3	36.0 V	1.00 A	36.0 Ω

5.

	V	I	R
Batt.	50.0 V	5.00 A	10.0 Ω
R_1	25.0 V	2.00 A	12.5 Ω
R_2	25.0 V	2.00 A	12.5 Ω
R_3	10.0 V	3.00 A	3.33 Ω
R_4	40.0 V	3.00 A	13.3 Ω

7.

	V	I	R
Batt.	36.0 V	9.00 A	4.00 Ω
R_1	12.0 V	6.00 A	2.00 Ω
R_2	12.0 V	3.00 A	4.00 Ω
R_3	24.0 V	6.00 A	4.00 Ω
R_4	24.0 V	3.00 A	8.00 Ω

9.

	V	I	R
Batt.	80.0 V	12.0 A	6.67 Ω
R_1	18.0 V	4.00 A	4.50 Ω
R_2	18.0 V	2.00 A	9.00 Ω
R_3	18.0 V	6.00 A	3.00 Ω
R_4	48.0 V	12.0 A	4.00 Ω
R_5	8.00 V	4.00 A	2.00 Ω
R_6	8.00 V	8.00 A	1.00 Ω
R_7	6.00 V	12.0 A	0.500 Ω

11.

	V	I	R
Batt.	65.0 V	5.00 A	13.0 Ω
R_1	10.0 V	0.500 A	20.0 Ω
R_2	10.0 V	1.00 A	10.0 Ω
R_3	10.0 V	2.50 A	4.00 Ω
R_4	10.0 V	1.00 A	10.0 Ω
R_5	25.0 V	5.00 A	5.00 Ω
R_6	30.0 V	5.00 A	6.00 Ω

17.12 Pages 486–487

1. 1.49 V   **3.** 1.33 A   **5.** (a) 3.75 A   (b) 9.00 V   (c) 0.0200 Ω **7.** 0.160 A
9. 0.120 A

17.13 Pages 490–491

1. 957 W   **3.** 0.682 A   **5.** 6.82 A   **7.** $0.56   **9.** Yes   **11.** $0.026
13. $0.046   **15.** (a) Microwave oven, two fluorescent bulbs, two light bulbs
(b) Projection TV, personal computer, two lightbulbs.   **17.** 2.36 W
19. 60.0 W   **21.** $90\overline{0}$ J   **23.** (a) $24\overline{0}$ W   (b) 50.4 kWh   (c) $5.54
25. (a) 9.58 mA   (b) 1.10 W   (c) $0.07   **27.** 6150 J

Chapter 17 Review Questions Pages 494–495

1. b   **2.** c   **3.** c   **4.** b   **5.** e   **6.** a and e
7. a   **8.** c   **9.** b   **10.** c   **11.** c   **12.** b and c
13. Materials can become charged when a charged object is brought nearby, inducing a polarization (separation) of charge on the material. If one side of the material is touched by another object, the charge at one side of the material can then be "drained" off, leaving the charge at the other side of the material.
14. Protons, electrons, and neutrons
15. Protons and neutrons
16. Electrons are located in charge clouds surrounding the nucleus.
17. Positive and negative. Protons carry a positive charge; electrons carry a negative charge.
18. A charged object is brought into contact with the electroscope, thereby providing some of that charge to the electroscope.

19. A charged object is brought near the conducting ball on an electroscope, causing a polarization (separation) of charge on the electroscope. The conducting ball is touched by a "ground," allowing one type of charge to leave the electroscope and go to ground, leaving the other charge behind.

20. Coulomb's law states that the force between two charges is directly proportional to the product of the magnitude of the charges and inversely proportional to the square of the separation between the two charges.

21. An electric field at a point represents the magnitude and direction of the force that would be exerted on a single unit of charge if placed at that point.

22. Lightning is the discharge of built-up static charge on a portion of a cloud.

23. Current   24. (a) Ampere   (b) Volt   (c) Ohm

25. It decreases the resistance by a factor of 4.

26. The voltage drop across a segment of a circuit equals the product of the current through that segment and the resistance of that segment of the circuit.

27. Current flows sequentially through each portion of a series circuit. The current is divided among different segments of a parallel circuit.

28. The equivalent resistance in a series circuit is the sum of the resistances in the circuit. In a parallel circuit, the equivalent resistance is given by the reciprocal of the sum of the reciprocals of all resistances.

29. The highest range

30. Water flow is split and flows in parallel through different segments of a water distribution system. In a similar manner, current is divided and flows in parallel through resistors connected in parallel.

31. The current decreases by a factor of 2.

32. The current increases by a factor of 2.

33. The resistance increases by a factor of 2.

34. Electrical charges move from regions of higher potential to regions of lower potential. Chemical reactions in batteries raise charges to higher potential energy. These charges can flow through a circuit and do work on circuit elements (create heat, light, or motion).

35. Chemical energy is transferred into electrical potential energy in the dry cell. Charges flow through the two lamps, giving up their energy in the form of light and heat.

36. Primary cells are not rechargeable, whereas secondary cells are.

37. An electric current flows in the reverse direction through a secondary cell, causing the normal chemical reaction to proceed in reverse, thus restoring charge in the battery.

38. The electrolyte causes a chemical reaction at the plates and conducts current between the plates.

39. An electrolyte causes a chemical reaction at the plates that releases energy to force electrical charge to move through the battery and the outside circuit.

40. It decreases the voltage through the outside circuit.

41. Watt

42. Power is given by the product of the voltage and the current.

43. We pay for our energy use. Energy consumed is the total work done. Power is the instantaneous use of electricity; energy is the power multiplied by the duration of time the power is used.

44. Power is the square of the voltage divided by the resistance.

45. The power increases by a factor of 2.

46. The power increases by a factor of 4.

47. The power decreases by a factor of 4.

48. The cost increases by a factor of 2.

Chapter 17 Review Problems Pages 495–497

1. 8.10×10^4 N   2. 9.13×10^{-6} C   3. 0.0671 m
4. 2.00×10^5 N/C   5. 3.33×10^6 C   6. 1.06 N   7. 1.17 Ω
8. 3.47 Ω 9. 1.11 Ω 10. 145 ft   11. 0.0105 cm^2
12. 7.47 A   13. 25.1 Ω 14. 0.491 A   15. 19.60 Ω

16. 0.612 A   **17.** 48.5 V   **18.** 16.2 Ω   **19.** 3.5 Ω
20. 2.45 Ω   **21.** 44.9 A   **22.** 19.5 A   **23.** 25.5 A
24. 5.00 Ω   **25.** 10.0 A   **26.** 5.95 A; 4.05 A   **27.** 4.42 Ω
28. 17.1 A   **29.** 58.5 V   **30.** 2.83 A   **31.** 12.3 V
32.

	V	I	R
Batt.	35.0 V	4.70 A	7.45 Ω
R_1	5.00 V	2.75 A	1.82 Ω
R_2	5.00 V	1.95 A	2.56 Ω
R_3	13.2 V	4.70 A	2.80 Ω
R_4	7.50 V	0.97 A	7.73 Ω
R_5	7.50 V	3.73 A	2.01 Ω
R_6	9.30 V	4.70 A	1.98 Ω

33. 1.43 V   **34.** 12.2 V   **35.** (a) 3.95 A   (b) 6.00 V   (c) 0.0165 Ω
36. 0.528 A   **37.** 6.67 Ω   **38.** 39.3 W   **39.** 1.36 A   **40.** $1.01
41. 240 h   **42.** 0.91 A

Chapter 17 Applied Concepts Page 497

1. (a) -8.23×10^{-8} N   (b) 3.63×10^{-47} N   (c) We would need an extremely large amount of charge on our bodies to feel that attraction. The earth has so much mass that we are noticeably attracted to the earth via gravity.
2. (a) 3.06×10^{-4} N   (b) 3.06×10^{-4} N   (c) 8.38×10^{-9} C   (d) As the distance between the particles decreases, the force increases, which results in a stronger electric field in the sawdust.   **3.** (a) 2.20×10^{-4} A   (b) 1.10 A
4. (a) $I_{\text{microwave}} = 8.33$ A; $I_{\text{bulb}} = 0.333$ A; $I_{\text{computer}} = 4.58$ A   (b) $R_{\text{microwave}} = 14.4$ Ω; $R_{\text{bulb}} = 360$ Ω; $R_{\text{computer}} = 26.2$ Ω   **5.** 3.32×10^{-4} m

Chapter 18

18.2 Pages 506–507

1. 1.20×10^{-5} T   **3.** 57.5 A   **5.** 1.71×10^{-6} T   **7.** 0.0196 T   **9.** 0.249 A
11. 0.0126 T   **13.** 0.239 A

Chapter 18 Review Questions Pages 517–518

1. d   **2.** b   **3.** b   **4.** Tesla
5. A tightly wound solenoid with many turns per unit length carrying a large current produces a strong magnetic field. A magnetic core such as iron significantly increases the magnetic field.
6. Use Ampère's rule to find the direction of the flux line from any single turn in the solenoid. The magnetic field direction of the solenoid is in this direction.
7. The current-carrying coil produces a magnetic field that causes the magnetic domains in the magnetic material to align in the direction of the coil's field. This produces a stronger induced field in the core.
8. A moving magnet induces a current in the generator's coil.
9. The commutator is a split ring that allows the current produced by a generator always to flow in the same direction.
10. An induced magnetic field in the motor's electromagnet is repelled by the permanent magnet, causing the rotor to spin. The commutator in a dc motor allows the current through the electromagnet to change polarity, causing the rotor to continue spinning.

11. A synchronous motor rotates at a frequency that depends on the number of coils and the frequency of the ac power source. The motor has a number of poles along the stator, which cause the rotor to spin at a fixed frequency.

12. A universal motor can operate on either dc or ac. The induction motor can operate only on ac.

13. The stator is a static magnet. The armature is an electromagnet that is free to rotate.

14. An electromagnet produces a strong magnetic field when a current is run through a solenoid. This magnetic field in turn induces a stronger magnetic field in a magnetic core.

15. The magnetic field increases by a factor of 2.

16. The field does not change as long as the length of the solenoid is much greater than the original diameter.

17. The magnetic field increases by a factor of 4.

18. The flux lines can be found by placing a small compass or magnetic filings near the magnet.

19. A spinning armature in the field of a stator crosses the magnetic flux lines of the stator, thereby inducing a current to flow in the armature coil. As the armature rotates and is reversed in the field, the direction of the current in the coil is reversed.

Chapter 18 Review Problems Page 518

1. 1.08×10^{-6} T   2. 4.90×10^{-6} T   3. 41.6 A   4. 0.0253 T
5. 12.4 A   6. 15.4°

Chapter 18 Applied Concepts Page 519

1. (a) 1.98×10^{-5} T   (b) 3.17 m   2. (a) 7.77×10^{-6} T   (b) The wire's magnetic field is 14.9% of the strength of the earth's magnetic field.   3. The magnetic field of the inner cable cancels out the opposite magnetic field orientation of the outer braid.
4. (a) For both solenoids, N is on top and S is on the bottom   (b) "b" The electron is coming out of the page.   5. (a) 0.525 Ω   (b) 8.57 A   (c) 98.0 loops   (d) 0.0896 m
(e) 1.05×10^{-3} T

Chapter 19

19.1 Page 524

1. 47.1 V   3. 117 V   5. 9.19 A   7. 83.2 V   9. 5.66 A   11. 70.3°
13. 1590 V   15. 117 V   17. 12.0 A   19. 12.0 A   21. 95.4 V
23. (a) 24.0 V   (b) 0.120 A   (c) 200 Ω   25. (a) 0.495 A   (b) 242 Ω   27. 2000 W

19.2 Pages 531–532

1. 21.9 Ω   3. 3500 W or 3.50 kW   5. 1320 W or 1.32 kW   7. 588 W
9. 2750 W or 2.75 kW   11. 10.0 turns   13. 58.5 V   15. 300 turns   17. 6.00 A
19. (a) 0.0587 A or 58.7 mA   (b) 880 W
21. 155 V   23. (a) 120 V   (b) 0.600 A   25. 4090 V

19.3 Page 535

1. 1.13 Ω   3. 4400 Ω   5. 40.1 kΩ   7. 0.594 A   9. 0.796 A   11. 0.116 A

19.4 Page 538

1. (a) 215 Ω   (b) 21.4°   (c) 0.209 A   3. (a) 3300 Ω   (b) 72.3°   (c) 4.55 mA
5. (a) 302 Ω   (b) 7.2°   (c) 19.9 mA

19.5 Page 540
1. 7.96 Ω   **3.** 2650 Ω or 2.65 kΩ   **5.** 0.796 Ω   **7.** 58.9 Ω   **9.** 1.61 Ω

19.6 Page 542
1. (a) 1880 Ω   (b) 57.9°   (c) 53.2 mA   **3.** (a) 48$\overline{0}$0 Ω   (b) 0.0241°
(c) 3.13 mA   **5.** (a) 3.18 Ω (b) 89.8°   (c) 4.72 mA

19.7 Pages 544–545
1. 42.4 Ω; 0.118 A   **3.** 1180 Ω; 12.7 mA   **5.** 44$\overline{0}$ Ω; 13.6 mA
7. 206 Ω; 24.3 mA   **9.** 529 Ω; 47.3 mA

19.8 Page 546
1. 79.6 kHz   **3.** 38.0 kHz   **5.** 3.86 kHz   **7.** 4.20 kHz   **9.** 5.63 kHz

19.10 Page 549
1. 10,700 kW   **3.** 133,000 kVA   **5.** 10,100 kW   **7.** 19,300 kW
9. 407,000 kVA   **11.** 0.880

Chapter 19 Review Questions Pages 551–552
1. d   **2.** c   **3.** a and e   **4.** a, b, and c
5. The maximum current is the maximum instantaneous current. The effective value of an alternating current is the number of amperes that produce the same amount of heat in a resistance as an equal number of amperes of a steady direct current.
6. The maximum voltage is the maximum instantaneous voltage. The instantaneous voltage is the voltage at any instant.
7. Power is the product of the effective values of the voltage and the current.
8. Power is the square of the effective value of the voltage divided by the resistance.
9. The output voltage doubles.
10. Henry
11. Inductive reactance allows the analysis of circuits containing inductors.
12. The inductive reactance is directly proportional to the frequency.
13. The current lags the voltage in an inductive circuit.
14. Energy is stored in the form of potential energy associated with a sheet of positive charge on one side of the capacitor and a sheet of negative charge on the other side.
15. The current leads the voltage in a capacitive circuit.
16. The frequency and reactance of a capacitor are inversely proportional.
17. Resonance occurs when the inductive reactance equals the capacitive reactance. The current is then given by its maximum value.
18. A diode allows current to flow in one direction but not in the reverse direction.
19. Amplification produces an increase in the value of a voltage or current in a circuit. Rectification produces a current or voltage in only one direction.
20. No; the phase angle depends on the frequency.

Chapter 19 Review Problems Pages 552–553
1. 110 V   **2.** 120 V   **3.** 68 V   **4.** 95.4 V   **5.** 8.95 A   **6.** 727 W
7. 495 W   **8.** 1310 W   **9.** (a) 1340 turns   (b) 0.255 A   (c) 523 W
10. (a) 3.92 A   (b) 38.2°   (c) 71.0 V   **11.** (a) 2.29 A   (b) 18.9°   (c) 109 V
12. (a) 2.20 A   (b) 67.1°   (c) 42.9 V   (d) 102 V   **13.** (a) 1.52 A   (b) 32.8 V
(c) 72.6°   **14.** 194 Ω; 0.567 A   **15.** 126 Ω; 0.873 A   **16.** (a) 30,200 Hz
(b) 633 Ω, 0.174 A   **17.** (a) 3.46 × 10^5 Hz   (b) 565 Ω, 0.195 A
18. 3.41 × 10^6 kVA

Chapter 19 Applied Concepts Page 553

1. (a) $13\overline{0}0$ W (b) 16.6 A (c) $i = 0$ A because the current is reversing at this instant.
2. (a) 23.0 turns (b) $P_p = 114$ W; $P_s = 114$ W (c) The power is conserved in a transformer.
3. (a) 386 A (b) 14.2 A (c) $P_{L\ 220V} = 37,200$ W; $P_{L\ 6000V} = 50.4$ W (d) The
6000-V power line is better because it only loses 50.4 W of power.
4. (a) 0.332 H (b) 0.524 A (c) 34.3 W **5.** (a) 1.19×10^{-13} F (b) 9.04×10^{-14} F
(c) Tuning to a high frequency lowers the capacitance because the plates are smaller.

Chapter 20

20.2 Pages 560–561

1. 1.50×10^9 m **3.** 0.108 s **5.** $50\overline{0}$ s **7.** 16.1 ms **9.** 7.67×10^{-7} s
11. 3.2×10^{-9} s **13.** (a) $25\overline{0}$ s (b) $120\overline{0}$ s **15.** (a) 1940 s (b) 2890 s
17. 3.84×10^8 m

20.3 Pages 562–563

1. 6.59×10^{12} Hz **3.** 3.09×10^{-4} m **5.** 6.58 m **7.** 4.55×10^{-5} Hz
9. 214 m **11.** 3.51 MHz **13.** 7.5×10^{14} Hz **15.** 5.2×10^{14} Hz

20.4 Pages 563–564

1. 5.93×10^{-23} J **3.** 3.01×10^{-25} J **5.** 83.0 MHz **7.** 5.51×10^{10} Hz
9. 2.7×10^{-19} J **11.** 3.4×10^{-19} J **13.** 1410 kHz

20.5 Page 567

1. 603 ℓm **3.** 60.9 cd **5.** 942 ℓm **7.** 1.46 ft-cd
9. 6.73 ft-cd **11.** 4750 ℓm **13.** 608 ℓm **15.** (a) Decrease
(b) $\frac{1}{9}$ **17.** 5.96 lux **19.** 35.3 ℓm

Chapter 20 Review Questions Pages 568–569

1. a, c, and e **2.** c **3.** d **4.** c
5. Yes; the wavelength varies inversely with the frequency.
6. The energy is directly proportional to the frequency.
7. The intensity falls off as 1 divided by the square of the distance.
8. The speed of light is measured by determining the time light takes to travel a measured
distance and using the relationship $v = s/t$.
9. It always travels at the same speed in a vacuum. It has a slower speed in any medium.
10. Electromagnetic radiation
11. Max Planck
12. Albert Einstein and Max Planck
13. Olaus Roemer
14. By measuring the time difference for the start of the eclipse of the moons of Jupiter as
viewed from different parts of earth's orbit
15. Candela and lumen
16. Luminous intensity is the brightness of a light source.

Chapter 20 Review Problems Pages 569–570

1. 2.32×10^{13} m **2.** 3.52 mi **3.** 2.63×10^{-6} s or 2.63 μs
4. 1.47×10^{-6}s or 1.47 μs **5.** 0.121 s **6.** 244 m
7. 6.45×10^6 Hz or 6.45 MHz **8.** 5.54×10^{15} Hz **9.** 9.61×10^{-23} J
10. 7.17×10^{10} Hz **11.** 5.47×10^{-18} J **12.** 2150 ℓm
13. 5230 ℓm **14.** 20,900 ℓm **15.** 3130 s; $17\overline{0}0$ s **16.** 12.8 ℓm
17. 7.29 lux

Chapter 20 Applied Concepts Page 571
1. (a) 0.0187 s   (b) 0.238 s   **2.** (a) 3.84×10^5 km or 239,000 mi
(b) $50\overline{0}$ s or 8.33 min   (c) 4.08×10^{16} m or 2.53×10^{13} mi
3. (a) 3.41 to 2.78 m   (b) 5.83×10^{-26} to 7.16×10^{-26} J   (c) The higher the frequency, the greater is the energy.   **4.** Channel 2: $\frac{1}{4}\lambda = 1.39$ m to Channel 6: $\frac{1}{4}\lambda = 0.852$ m
5. (a) 14,100 ℓm   (b) 56,500 ℓm (4 times the original)

Chapter 21

21.5 Page 581
1. 1.29 cm   **3.** 32.2 cm   **5.** 5.13 cm   **7.** −6.24 cm   **9.** −1.09 m
11. 10.0 cm   **13.** 0.237 m   **15.** −4.15 cm   **17.** −3.56 m   **19.** (a) 4.01 cm
(b) 2.98 cm   (c) 31.9 cm

21.10 Pages 593–594
1. 1.21   **3.** 2.00×10^8 m/s   **5.** 1.48   **7.** 21.8 cm   **9.** −2.14 cm
11. (a) 5.00 cm   (b) −2.50 cm   **13.** $s_i = -8.17$ cm, $h_i = 2.06$ cm   **15.** 4.43 cm

Chapter 21 Review Questions Pages 596–597
1. a   **2.** d   **3.** a and b
4. Parallel light rays reflected off a rough surface may scatter in many different directions (diffusion), whereas parallel light rays reflect off a smooth surface as parallel rays.
5. At a smooth surface, the incident angle of a light ray is the same as the angle of the reflected ray.
6. At a smooth surface, the normal to the surface and both the incident and reflected rays lie in the same plane.
7. They are virtual, erect, and lie as far behind the mirror as the object is in front of the mirror.
8. A real image can be shown on a screen. A virtual image cannot.
9. Concave mirrors curve away from the observer; convex mirrors curve out toward the observer.
10. For large apertures, not all parallel rays are reflected through the focal point. This produces a fuzzy or aberrant image.
11. The image distance is increased.
12. The image distance is increased.
13. b   **14.** b
15. Converging lenses convert parallel rays into converging rays. Diverging lenses convert parallel rays into diverging rays.
16. Light propagating in an optical fiber, light reflected into a swimming pool from an underwater light
17. Light passing at an angle through an interface from a medium of low optical density to a medium of higher optical density is bent toward the normal to the interface.
18. The speed of light is lower in the high-index material.
19. A wave passing at an angle from one medium to another with a different wave velocity will be bent toward or away from the normal depending on whether the speed is higher or lower in the new material, respectively.
20. Virtual
21. Real or virtual
22. They roughen the surface and scatter the light.
23. It appears shallower because of the refraction of light at the surface.
24. Light travels in a straight line unless it is reflected or refracted.
25. Light traveling at an angle greater than the critical angle is reflected from one side of the fiber to another as it travels down the length of the fiber.

26. When the object is closer to the lens than the focal length.

27. When the object is located a distance from the lens equal to the focal length.

28. When the object is located outside the focal point.

Chapter 21 Review Problems Pages 597–598

1. 2.36 cm   **2.** −12.9 cm   **3.** −3.48 cm   **4.** −2.22 m
5. (a) 250 cm from the mirror   (b) −459 cm   **6.** 22.7 cm   **7.** 1.35
8. 2.08×10^8 m/s   **9.** 1.40   **10.** 41.8°   **11.** 18.0 cm
12. (a) −7.50 cm   (b) 2.70 cm   **13.** 9.00 cm   **14.** 6.25 cm   **15.** −9.70 m
16. −25.4 cm, 7.18 cm high   **17.** 1.24×10^8 m/s   **18.** 41.8°   **19.** 12.3 cm
20. (a) $s_i = 15.8$ cm; $h_i = -10.8$ cm   (b) −3.83 cm

Chapter 21 Applied Concepts Page 599

1. (a) 19.0 cm   (b) $s_i = -36.5$ cm (the minus sign represents the opposite side of the mirror)   (c) Upright   (d) 73.0 cm   **2.** (a) −0.704 m   (b) Virtual   (c) 0.0156 m   (d) Better for increasing the field of view   **3.** (a) 18.5°   (b) 22.1°   (c) The angle will be smaller because light refracts toward the normal line when entering a denser medium.   **4.** (a) 24.44°   (b) Larger than 24.44°   (c) 31.4°   (d) Diamond; a zircon will allow light to escape when angles are less than 31.4°.   **5.** (a) 60.4 mm   (b) −13.1 mm

Chapter 22 Review Questions Page 617

1. Red, orange, yellow, green, blue, and violet

2. The color of light is determined by its wavelength or its frequency.

3. (a) Infrared   (b) Ultraviolet

4. Black. The cloth in the dress absorbs all colors except green and reflects only green. The red light is absorbed and no light is reflected.

5. (a) Red stars on a black field, rest of the flag would be red   (b) Blue field with no stars, rest of the flag would have blue and black stripes

6. Red, green, and blue

7. Two colors that combine to form white are called complementary colors. The complement of red is cyan, the complement of green is magenta, and the complement of blue is yellow.

8. The primary pigments are the complements of the three primary colors.

9. Cyan, magenta, and yellow for the colors and black for the shadow areas and definition

10. The sky is blue because sunlight is scattered as it passes through the atmosphere. The amount of scattering depends on the wavelength of the light. The longer wavelengths of red, orange, and yellow are scattered much less than the shorter wavelengths of blue and violet. Because our eyes are not very sensitive to violet light, the scattered blue light in the atmosphere dominates our vision so that we see a blue sky.

11. At sunset, sunlight travels through significantly more of the atmosphere than at other times of the day. This results in even more scattering of the shorter wavelengths of violet, blue, and green, which are taken out before reaching us, with some orange but primarily red, the light of longest wavelength and least scattered, most readily passing through the atmosphere to reach our eyes for a red sunset.

12. Clouds consist of many different-sized water droplets. The small droplets tend to scatter the short-wavelength (blue) light, the medium droplets tend to scatter the medium-wavelength (green) light, and the large droplets tend to scatter the long-wavelength (red) light. The overall result is that all of the colors are scattered and combined so that we see a general white reflection.

13. The ocean is blue because the water surface acts like a mirror and reflects the blue color of the sky and the water tends to reflect and scatter the short-wavelength yellow, green, blue, and violet light and absorb the long-wavelength orange and red light.

14. False

15. Diffraction

16. Diffraction pattern
17. Thomas Young
18. Constructive interference
19. Destructive interference
20. Polarized sunglasses restrict the light waves passing through to a single plane rather than normal sunlight, which is emitted in all directions with many orientations.

Chapter 23

23.3 Page 623
1. -13.595 eV   **3.** 12.084 eV   **5.** 6.57×10^{-7} m

23.4 Page 625
1. Neon   **3.** Fluorine

23.5 Page 628
1. (a) 12   (b) 6   (c) 6   (d) 6   (e) 12   (f) 6   **3.** (a) 48   (b) 22   (c) 26   (d) 22
(e) 48   (f) 26   **5.** (a) 11   (b) 22.9898 u   **7.** (a) 82   (b) 207.2 u   **9.** ^{14}C
11. Bromine   **13.** Argon

23.6 Page 631
1. 3.8530×10^{-25} kg   **3.** 6680 MeV   **5.** 1750 MeV   **7.** 8.5 MeV

23.7 Page 636
1. 4.03×10^{-5} s   **3.** 0.226 g   **5.** 79.9%   **7.** 3.77×10^{20} atoms   **9.** 30.4%

23.9 Page 639
1. 0.310 Ci   **3.** 0.0875 Ci   **5.** 9.47 Ci   **7.** 73.3 μCi

Chapter 23 Review Questions Pages 642–643
1. a, b, and c   **2.** c   **3.** c
4. The ground state is the lowest energy level for the electron in the atom. The excited states are the higher energy levels, which are unstable. An electron in an excited state will in time decay through lower-energy-level excited states to the ground state. The ground state has the lowest energy level.
5. In the Bohr theory of the atom, the energy levels of the electrons are restricted to certain values; that is, the energy is quantized. The energy levels in a hydrogen atom are given by the equation $E = -kZ^2/n^2$.
6. Protons and neutrons are both nucleons, which exert the attractive strong force on other nucleons. They have similar masses, although the neutron is slightly more massive. The proton is a stable particle as an individual nucleon. The neutron, however, is unstable by itself. The proton has a positive charge, equal in magnitude to that of the electron. The neutron is uncharged.
7. The electric force can be either attractive or repulsive and is exerted between charged particles. The strong force is always attractive and is exerted between nucleons, whether charged or not. The strong force is a very short-range force. The electric force is exerted over larger distances.
8. The nucleus would expand in size due to the repulsive electric force between protons. The nucleus might even break apart.

9. Mass and energy are equivalent forms. Mass can be changed into energy under the proper conditions and vice versa.

10. All three atoms have six protons in the nucleus and six electrons in the orbital shells. They all exhibit the chemical properties of carbon. They each have a different number of neutrons in the nucleus and therefore have a different mass.

11. An electron volt is the energy that is gained or lost by an electron in passing through a potential difference of 1 volt.

12. The neutron is the "glue" that binds the positively charged protons together in the nucleus. Without neutrons, the positively charged protons would be repelled by their similar electric charge.

13. An α ray is composed of particles that have a double positive charge and are composed of two protons and two neutrons. The α particles are identical to the nucleus of a helium atom.

14. A γ ray is composed of very energetic photons of uncharged electromagnetic radiation.

15. A β ray is composed of negatively charged particles (electrons).

16. Enrico Fermi discovered that the neutron bombardment of uranium can result in the formation of lighter nuclei that are approximately one-half the mass of uranium.

17. A self-sustaining chain reaction is a nuclear reaction in which a sufficient number of neutrons are produced to cause subsequent nuclear reactions to continue at a fixed rate.

18. In nuclear fusion reactions, nuclei bombard each other, producing heavier nuclei. The original nuclei "fuse" together.

19. In nuclear fission reactions, nuclei are bombarded by nuclear particles such as neutrons or α particles, causing the original nuclei to split apart.

20. 6.25%

21. Molecules in plant or animal cells can be damaged as a result of changes in the chemicals caused by nuclear reactions produced by the radiation. Some of these damaged molecules may be in the genetic material, which might cause cells produced by the division of this cell to be defective.

22. Radiation is used for diagnostic purposes by the injection or ingestion of radioactive tracer materials that may allow the identification of cancerous or defective cells. It can also be used to treat cancer by bombardment of the cancer cells, therefore destroying the cancerous material.

Chapter 23 Review Problems Pages 643–644

1. −0.5438 eV   2. 0.967 eV   3. He, Ne, Ar, Kr, Xe, Rn
4. Cu, Ag, Au, Fe, Ti, Cr, Mn, etc.   5. 2.4913×10^{-26} kg
6. 3.1547×10^{-26} kg   7. 2.7553×10^{-25} kg   8. 749 MeV
9. 8.7 MeV   10. 8.3 MeV   11. 8.1 MeV   12. 7.26×10^{12} atoms
13. 4.29×10^{17} atoms   14. 5.6%   15. 0.37%   16. 0.346 Ci
17. 0.193 Ci   18. 0.367 Ci   19. 5.8 Ci   20. 10.6 μCi

Chapter 23 Applied Concepts Page 645

1. (a) 2.5491 eV   (b) 6.163×10^{14} Hz   (c) 3.00×10^8 m/s   (d) 4.87×10^{-7} m
2. $n = 2$ 3. (a) 4.550×10^{-12} J   (b) two protons, two neutrons, two electrons
(c) 6.6968×10^{-27} kg   (d) 6.64×10^{-27} kg   4. (a) 0.181/day
(b) 1.50×10^{24} atoms   5. 0.393 Ci

Chapter 24

24.2 Page 650
1. 2.07×10^{14} J   **3.** 6.67×10^{-15} kg

Chapter 24 Review Questions Pages 653–654
1. c   **2.** d

3. Physics is the same for moving and nonmoving objects.

4. If you were in a moving car and flipped a coin, the coin would flip into the air and fall back into your hand the same as would happen if you flipped a coin while standing on the side of the road.

5. It would travel at 14 m/s because the ball was already traveling at 10 m/s while attached to the bike.

6. The speed of light is not relative, but is constant.

7. The light would only travel at 3.00×10^8 m/s.

8. Time passes more slowly than for someone not traveling close to the speed of light.

9. Relative to the ground (considering I am not walking and reading), just the time dimension.

10. That energy and mass are equivalent, just in different forms. Mass has energy and energy has mass.

11. Spatially we exist in three dimensions (x, y, z), yet we also move through the dimension of time. Space and time are needed to locate a particular occurrence in the universe.

12. Gravity and acceleration are the same.

13. Light can be warped around massive objects.

14. The light curved around the warping of space-time that was created by the mass of the sun and the moon.

Chapter 24 Review Problems Page 654
1. 2.00 mi/h   **2.** 67.0 mi/h   **3.** 63.0 mi/h   **4.** 8.19×10^{-14} J   **5.** 1.50×10^{-10} J
6. 1.09×10^{-16} kg

Appendix A

A.1 Page 658
1. -11 **3.** 5   **5.** -2 **7.** -4 **9.** -6 **11.** -3 **13.** 18   **15.** -21
17. 0   **19.** 3   **21.** -8 **23.** 0
25. 11   **27.** -13 **29.** -5 **31.** -8 **33.** -4 **35.** 10
37. -45 **39.** -288

A.2 Page 660
1. 10^8 **3.** 10^8 **5.** $\dfrac{1}{10^3}$ **7.** 10^7 **9.** 10^3 **11.** $\dfrac{1}{10^4}$

13. 1   **15.** 10^5 **17.** $\dfrac{1}{10^{11}}$ **19.** $\dfrac{1}{10^6}$ **21.** 10^8 **23.** 10^{10}

A.3 Page 664
1. $\frac{4}{3}$ **3.** 17   **5.** 0   **7.** 17.5   **9.** 3   **11.** 4
13. 7   **15.** 12.5   **17.** 5   **19.** 26.5   **21.** 9   **23.** 3
25. 8   **27.** 1   **29.** $-\frac{11}{3}$ or $-3\frac{2}{3}$ **31.** 3   **33.** 2
35. 21   **37.** $\frac{29}{2}$ or $14\frac{1}{2}$ **39.** -6

A.4 Pages 666–667

1. ±6   **3.** ±7   **5.** ±3   **7.** ±0.8   **9.** ±4.67
11. ±0.892   **13.** $a = 3$; $b = 1$; $c = -5$ **15.** $a = 6$; $b = 8$; $c = 2$
17. $a = 9$; $b = 6$; $c = -4$ **19.** $a = 5$; $b = 6$; $c = 0$ **21.** $a = 9$; $b = 0$; $c = -64$
23. 3; 7   **25.** $\frac{4}{3}$; $-\frac{5}{2}$ **27.** 1.95; −1.62   **29.** 1.69; −0.855
31. 0.725; −1.23

A.5 Pages 671–673

1. a **3.** c **5.** a **7.** B **9.** B **11.** 0.9455
13. 1.804   **15.** 0.9799   **17.** 0.6477   **19.** 0.3065
21. 0.4617   **23.** 16°   **25.** 43°   **27.** 48°   **29.** 36.6°
31. 46.5°   **33.** 30.0°   **35.** 22.28°   **37.** 16.75°   **39.** 35.50°
41. $B = 65.0°$; $a = 8.45$ m; $b = 18.1$ m   **43.** $A = 47.7°$; $B = 42.3°$; $a = 12.4$ km
45. $A = 24.4°$; $B = 65.6°$; $c = 24.2$ mi   **47.** $B = 70°$; $b = 24$ m; $c = 25$ m
49. $A = 49.35°$; $a = 17.98$ cm; $b = 15.44$ cm   **51.** $b = 8.49$ cm
53. $c = 21.6$ mi   **55.** $a = 10.2$ ft   **57.** $c = 24.8$ cm   **59.** $a = 8.60$ m
61. (a) 10.0° (b) 2.12 cm (c) 8.24 cm   **63.** $C = 2.72$ in.; $D = 2.28$ in.
65. $b = 8.00$ cm; $c = 16.1$ cm

A.6 Pages 681–682

1. $B = 38.0°$, $C = 73.0°$, $c = 25.6$ m   **3.** $C = 52.5°$, $A = 66.1°$, $a = 129$ cm
5. $C = 47.6°$, $a = 223$ ft, $c = 196$ ft   **7.** $B = 48.5°$, $C = 16.5°$, $c = 1840$ m
9. $A = 67.07°$, $B = 40.35°$, $a = 40.72$ cm   **11.** $A = 131.94°$, $a = 89{,}460$ mi,
$b = 57{,}870$ mi   **13.** $B = 20°$, $C = 135°$, $c = 84$ cm   **15.** $A = 6°$, $B = 166°$,
$b = 28$ m   **17.** $B = 27.3°$, $C = 115.7°$, $c = 32.2$ cm   **19.** $A = 29.6°$,
$B = 123.9°$, $b = 79.4$ km; or $A = 150.4°$, $B = 3.1°$, $b = 5.18$ km   **21.** $B = 34.3°$,
$C = 74.2°$, $b = 2.05$ m; or $B = 2.7°$, $C = 105.8°$, $b = 0.171$ m   **23.** $A = 44.6°$,
$C = 30.4°$, $c = 8.64$ mi   **25.** $A = 35°$, $B = 127°$, $b = 62$ mi; or $A = 145°$, $B = 17°$,
$b = 23$ mi   **27.** No triangle   **29.** $A = 157°$, $C = 15°$, $a = 1300$ m; or $A = 7°$, $C = 165°$,
$a = 390$ m   **31.** $A = 59.34°$, $C = 79.16°$, $c = 21.12$ km; or $A = 120.66°$,
$C = 17.84°$, $c = 6.588$ km   **33.** No triangle   **35.** $A = 6.06°$, $B = 165.19°$,
$b = 150.1$ m   **37.** $B = 47.8°$, $C = 72.2°$, $a = 22.8$ m   **39.** $A = 27.0°$,
$B = 44.0°$, $c = 292$ km   **41.** $A = 44.6°$, $B = 51.2°$, $C = 84.2°$ **43.** $A = 32.1°$,
$B = 104.6°$, $C = 43.3°$ **45.** $B = 85°$, $C = 50°$, $a = 36$ m   **47.** $A = 84°$,
$B = 53°$, $C = 43°$ **49.** $A = 19°$, $B = 26°$, $c = 78$ ft   **51.** $A = 148.80°$,
$C = 11.95°$, $b = 3064$ m   **53.** $A = 40.11°$, $B = 31.14°$, $c = 595.2$ mm
55. $A = 113.43°$, $B = 27.84°$, $C = 38.73°$

Appendix B

B.4 Pages 688–689

1. 3.32×10^{19} **3.** -6.83×10^{-6} **5.** 8.35×10^{-11} **7.** -7.98×10^{19}
9. −68.5   **11.** 4680   **13.** 216   **15.** 2.10×10^{10} **17.** 2.15×10^{9}
19. 10.7   **21.** 33.4   **23.** 591   **25.** 609,000   **27.** 0.567
29. 1.77×10^{4} **31.** 0.523   **33.** 0.225   **35.** 1.29   **37.** 0.510
39. 0.743   **41.** 3.73   **43.** 0.151   **45.** 40.7°   **47.** 24.6°   **49.** 56.2°
51. 8.4°   **53.** 63.3°   **55.** 66.6°   **57.** 252 m   **59.** 89.4 m   **61.** 53.7°
63. 14.8 cm   **65.** 21.9 m   **67.** 20,700   **69.** 0.0206   **71.** 1.86×10^{-6}

INDEX

FORMULAS FROM GEOMETRY

Plane Figures

In the following, a, b, c, d, and h are lengths of sides and altitudes, respectively.

		Perimeter	Area
Rectangle		$P = 2(a + b)$	$A = ab$
Square		$P = 4b$	$A = b^2$
Parallelogram		$P = 2(a + b)$	$A = bh$
Rhombus		$P = 4b$	$A = bh$
Trapezoid		$P = a + b + c + d$	$A = \left(\dfrac{a + b}{2}\right)h$
Triangle		$P = a + b + c$	$A = \dfrac{1}{2}bh$
Right triangle		$c^2 = a^2 + b^2$ or $c = \sqrt{a^2 + b^2}$	

		Circumference	Area
Circle		$C = \pi d$	$A = \pi r^2 \quad d = 2r$
		$C = 2\pi r$	$A = \dfrac{\pi d^2}{4}$

Geometric Solids

In the following, B, r, and h are the area of base, length of radius, and height, respectively.

		Volume	Lateral Surface Area
Prism		$V = Bh$	
Cylinder		$V = \pi r^2 h$ $V = \dfrac{\pi d^2 h}{4}$	$A = 2\pi rh$
Pyramid		$V = \dfrac{1}{3}Bh$	
Cone		$V = \dfrac{1}{3}\pi r^2 h$	$A = \pi rs$, s is the slant height.
Sphere		$V = \dfrac{4}{3}\pi r^3$ $V = \dfrac{\pi}{6}d^3$	$A = 4\pi r^2$